Cellular Senescence and Cardiac Aging

Mechanisms, Therapeutic Strategies, and Implications for Cardiovascular Diseases

Cellular Senescence and Cardiac Aging

Mechanisms, Therapeutic Strategies, and Implications for Cardiovascular Diseases

Edited by

Gaurav Gupta
Centre for Research Impact & Outcome, Chitkara College of Pharmacy, Chitkara University, Rajpura, Punjab, India; Centre of Medical and Bio-Allied Health Sciences Research, Ajman University, Ajman, United Arab Emirates

Md Sadique Hussain
Uttaranchal Institute of Pharmaceutical Sciences, Uttaranchal University, Dehradun, Uttarakhand, India; School of Pharmaceutical Sciences, Lovely Professional University, Phagwara, Punjab, India

Moayad Al Shahwan
Centre of Medical and Bio-Allied Health Sciences Research, Ajman University, Ajman, United Arab Emirates

Riya Thapa
Faculty of Pharmacy, AIMST University, Bedong, Kedah, Malaysia

Sanmoy Karmakar
Department of Pharmacology and Toxicology, Jadavpur University, Kolkata, West Bengal, India

ELSEVIER

ACADEMIC PRESS
An imprint of Elsevier

Academic Press is an imprint of Elsevier
125 London Wall, London EC2Y 5AS, United Kingdom
50 Hampshire Street, 5th Floor, Cambridge, MA 02139, United States

ISBN: 978-0-443-45811-8

For Information on all Academic Press publications
visit our website at https://www.elsevier.com/books-and-journals

Publisher: Megan Ball
Acquisitions Editor: Sangeeta Yadav
Editorial Project Manager: Nayanika Saikia
Production Project Manager: Sathyavani Deiva Sigamani
Cover Designer: Mark Rogers

Typeset by MPS Limited, Chennai, India

Contents

List of contributors

S. Afsar Department of Virology, Sri Venkateswara University, Tirupati, Andhra Pradesh, India

Muhammad Afzal Department of Pharmaceutical Sciences, Pharmacy Program, Batterjee Medical College, Jeddah, Saudi Arabia

Anas Ahmad Department of Microbiology, Immunology and Infectious Diseases, Snyder Institute for Chronic Diseases, and Hotchkiss Brain Institute, Cumming School of Medicine, Foothills Medical Centre, University of Calgary, Calgary, AB, Canada

Sivakumar Arumugam School of Biosciences and Technology, Vellore Institute of Technology, Vellore, Tamil Nadu, India

Sumel Ashique Department of Pharmaceutical Technology, School of Health and Medical Sciences, Adamas University, Kolkata, West Bengal, India

Chintan Aundhia Department of Pharmacy, Sumandeep Vidyapeeth Deemed to be University, Vadodara, Gujarat, India

Purushothaman Balakrishnan Department of Biomaterials, Saveetha Dental College and Hospitals, SIMATS, Saveetha University, Chennai, Tamil Nadu, India

Asif Ahmad Bhat Faculty of Pharmacy, AIMST University, Bedong, Kedah, Malaysia

Bindu Madhavi Boddupalli Department of Pharmaceutics, School of Pharmacy, Mount Kenya University, Thika, Kenya

Apporva Chawla School of Pharmacy, Khalsa University, Amritsar, Punjab, India

Rajiv Dahiya School of Pharmacy, Faculty of Medical Sciences, The University of the West Indies, St. Augustine, West Indies, Trinidad and Tobago

Sunita Dahiya Department of Pharmaceutical Sciences, School of Pharmacy, Medical Sciences Campus, University of Puerto Rico, San Juan, PR, United States

Subhajit Dutta Functional Genomics and Metabolism Research Unit, Department of Biochemistry and Molecular Biology, University of Southern Denmark, Odense, Denmark; Center for Functional Genomics and Tissue Plasticity (ATLAS), University of Southern Denmark, Odense, Denmark

Lina Eltaib Department of Pharmaceutics, College of Pharmacy, Northern Border University, Rafha, Saudi Arabia

Rabab Fatima School of Health Science & Technology, UPES, Dehradun, Uttarakhand, India

Shahin Fatima Department of Computer Science and Engineering, Koneru Lakshmaiah Education Foundation, Hyderabad, Telangana, India

Neeraj Kumar Fuloria Faculty of Pharmacy, AIMST University, Bedong, Kedah, Malaysia

Shivkanya Fuloria Faculty of Pharmacy, AIMST University, Bedong, Kedah, Malaysia

Mahak Gayaz Pharmacology Division, Department of Pharmaceutical Sciences, University of Kashmir, Hazratbal, Srinagar, Jammu and Kashmir, India

Gaurav Gupta Centre for Research Impact & Outcome, Chitkara College of Pharmacy, Chitkara University, Rajpura, Punjab, India; Centre of Medical and Bio-Allied Health Sciences Research, Ajman University, Ajman, United Arab Emirates

Saurabh Gupta Department of Pharmacology, Chameli Devi Institute of Pharmacy, Indore, Madhya Pradesh, India

Abdul Hai Department of Medical Laboratory Technology, Faculty of Applied Medical Sciences, Northern Border University, Arar, Saudi Arabia

Md Sadique Hussain School of Pharmaceutical Sciences, Lovely Professional University, Phagwara, Punjab, India; Uttaranchal Institute of Pharmaceutical Sciences, Uttaranchal University, Dehradun, Uttarakhand, India

Mohd Imran Center For Health Research, Northern Border University, Arar, Saudi Arabia

Bharathi Kalidass School of Biosciences and Technology, Vellore Institute of Technology, Vellore, Tamil Nadu, India

Abida Khan Center For Health Research, Northern Border University, Arar, Saudi Arabia

Amna Khan Abbottabad International Medical Institute, Abbottabad, Khyber Pakhtunkhwa, Pakistan

Hamza Khan Department of Medicine, Khyber Medical College, Peshawar, Khyber Pakhtunkhwa, Pakistan

Yumna Khan Saveetha Institute of Basic Medical Sciences (SIBMS), Saveetha Institute of Medical and Technical Sciences (SIMATS), Saveetha University, Thandalam, Chennai, Tamil Nadu, India

Gothandam Kodiveri Muthukaliannan School of Biosciences and Technology, Vellore Institute of Technology, Vellore, Tamil Nadu, India

Mudasir Maqbool Department of Pharmacology, Government Medical College Baramulla, Baramulla, Jammu and Kashmir, India

Janaki Ramaiah Mekala School of Biosciences and Technology, Vellore Institute of Technology, Vellore, Tamil Nadu, India

Reyaz Hassan Mir Pharmaceutical Chemistry Division, Department of Pharmaceutical Sciences, University of Kashmir, Hazratbal, Srinagar, Jammu and Kashmir, India

Riya Mishra Department of Pharmacy, Faculty of Pharmaceutical Science and Nursing, Vivekananda Global University, Jaipur, Rajasthan, India

Roohi Mohi ud din Department of General Medicine, Sher-I-Kashmir Institute of Medical Sciences (SKIMS), Srinagar, Jammu and Kashmir, India

Manisha Nidhar Department of Chemistry, Institute of Science, Banaras Hindu University, Varanasi, Uttar Pradesh, India

Ishtiyaq Nissar Pharmacology Division, Department of Pharmaceutical Sciences, University of Kashmir, Hazratbal, Srinagar, Jammu and Kashmir, India

Surya Nath Pandey Department of Pharmacology, Teerthanker Mahaveer College of Pharmacy, Teerthanker Mahaveer University, Moradabad, Uttar Pradesh, India

Neeraj Patel School of Pharmacy, Suresh Gyan Vihar University, Jaipur, Rajasthan, India

Prasanna Srinivasan Ramalingam School of Biosciences and Technology, Vellore Institute of Technology, Vellore, Tamil Nadu, India

Amita Joshi Rana College of Pharmacy, Graphic Era Hill University, Bhimtal, Uttarakhand, India; Centre for Promotion of Research, Graphic Era (Deemed to be) University, Dehradun, Uttarakhand, India

Doli Rani Das Shree Krishna College of Pharmacy, Sitapur, Uttar Pradesh, India

Mokhtar Rejili Department of Biology, College of Sciences, Imam Mohammad Ibn Saud Islamic University (IMSIU), Riyadh, Saudi Arabia

Pathakota Roja Department of Pharmacology, Sree Dattha Institute of Pharmacy, Sheriguda, Ibrahimpatnam, Hyderabad, Telangana, India

Moayad Al Shahwan Centre of Medical and Bio-Allied Health Sciences Research, Ajman University, Ajman, United Arab Emirates

Rahul Sharma School of Pharmaceutical Sciences, Jaipur National University, Jagatpura, Jaipur, Rajasthan, India

Md Raihan Siddique School of Pharmacy, Suresh Gyan Vihar University, Jaipur, Rajasthan, India

Muhammad Irfan Siddique Department of Pharmaceutics, Faculty of Pharmacy, Northern Border University, Arar, Saudi Arabia

Sachin Kumar Singh School of Pharmaceutical Sciences, Lovely Professional University, Phagwara, Punjab, India

Santosh Kumar Singh ICFAI School of Pharmaceutical Sciences, The ICFAI University, Jaipur, Rajasthan, India

Yogendra Singh Department of Pharmacology, Maharishi Arvind College of Pharmacy, Ambabari Circle, Ambabari, Jaipur, Rajasthan, India

Neelam Singla School of Pharmacy, Suresh Gyan Vihar University, Jaipur, Rajasthan, India

Ayesha Sultana Department of Pharmaceutics, Yenepoya Pharmacy College & Research Centre, Yenepoya (Deemed to be University), Mangalore, Karnataka, India

Nasreen Sulthana Department of Pharmacology, Sree Dattha Institute of Pharmacy, Sheriguda, Ibrahimpatnam, Hyderabad, Telangana, India

Rahamat Unissa Syed Department of Pharmaceutics, College of Pharmacy, University of Ha'il, Hail, Saudi Arabia

Riya Thapa Faculty of Pharmacy, AIMST University, Bedong, Kedah, Malaysia

Muhammad Tufail Department of Oral and Maxillofacial Surgery, Center of Stomatology, Xiangya Hospital, Central South University, Changsha, Hunan, P.R. China

Jasreen Uppal School of Pharmacy, Khalsa University, Amritsar, Punjab, India

Sabina Yasmin Department of Pharmaceutical Chemistry, College of Pharmacy, King Khalid University, Abha, Saudi Arabia

Mohammed Zawiah Department of Clinical Practice, Faculty of Pharmacy, Northern Border University, Arar, Saudi Arabia

Chapter 1

Introduction to cardiac aging and cellular senescence

Chintan Aundhia[1], Muhammad Afzal[2], Surya Nath Pandey[3], Md Raihan Siddique[4], Riya Thapa[5], Asif Ahmad Bhat[5], Md Sadique Hussain[6,7], Yogendra Singh[8], Manisha Nidhar[9], and Gaurav Gupta[10,11]

[1]*Department of Pharmacy, Sumandeep Vidyapeeth Deemed to be University, Vadodara, Gujarat, India,* [2]*Department of Pharmaceutical Sciences, Pharmacy Program, Batterjee Medical College, Jeddah, Saudi Arabia,* [3]*Department of Pharmacology, Teerthanker Mahaveer College of Pharmacy, Teerthanker Mahaveer University, Moradabad, Uttar Pradesh, India,* [4]*School of Pharmacy, Suresh Gyan Vihar University, Jaipur, Rajasthan, India,* [5]*Faculty of Pharmacy, AIMST University, Bedong, Kedah, Malaysia,* [6]*Uttaranchal Institute of Pharmaceutical Sciences, Uttaranchal University, Dehradun, Uttarakhand, India,* [7]*School of Pharmaceutical Sciences, Lovely Professional University, Phagwara, Punjab, India,* [8]*Department of Pharmacology, Maharishi Arvind College of Pharmacy, Ambabari Circle, Ambabari, Jaipur, Rajasthan, India,* [9]*Department of Chemistry, Institute of Science, Banaras Hindu University, Varanasi, Uttar Pradesh, India,* [10]*Centre for Research Impact & Outcome, Chitkara College of Pharmacy, Chitkara University, Rajpura, Punjab, India,* [11]*Centre of Medical and Bio-Allied Health Sciences Research, Ajman University, Ajman, United Arab Emirates*

1.1 Introduction

Aging is an endogenous life measure that is characterized by a progressive loss of physiological uniformity, an increased risk of chronic diseases, and decreased regenerative proficiency. Among all aging-related pathologies, CVDs are the leading cause of morbidity and mortality, particularly in older adults. The importance of this intersection can be supported by epidemiological evidence of high-risk groups of patients aged ≥ 65 years, which is characterized by rapid growth, and age itself has a prevailing independent impact on heart failure, coronary artery disease, and arrhythmias (Yao et al., 2023). Consequently, with the emerging aging demographic structure in the world today, unraveling the molecular pathologies of aging of the heart has become an important public health and research issue. Heart aging can be characterized by systematic changes in the organization and functioning of the myocardium and cellular homeostasis of the heart, even in the absence of overt cardiovascular pathology. These modifications involve wall thickening of the ventricles, diastolic failure, fibrotic remodeling, mitochondrial abnormalities, and decreased heart output (Werbner et al., 2023). At the cellular level, aged cardiac tissue has a reduced regenerative potential, altered calcium metabolism, increased oxidative stress, and the presence of senescent cells. The net result of these cellular changes is performance debilitation and predisposition of the heart to physiological stress, ischemia, and maladaptive remodeling (Izzo et al., 2021).

In the past, the cardiance was assumed to be a postmitotic organ that could not survive. However, modern evidence has disproven this point, showing only limited turnover of cardiomyocytes and the existence of CPCs (Hamledari et al., 2022). However, the regenerative capacity of the aged heart is significantly worse, mostly due to processes of cellular senescence, telomere shortening, mitochondrial dysfunction, and unremitting inflammation. In particular, cellular senescence has recently been identified as a key mediator of cardiac aging. Senescent cells experience cell cycle irreversibility and a hyperproliferative phenotype referred to as the SASP, promoting tissue dysfunction, fibrosis, and paracrine senescence of adjacent cells (Marino et al., 2022). Genomic instability, epigenetic drift, dysregulated proteostasis, impaired nutrient sensing, and stem cell depletion collectively influence the heart and provoke functional impairment, which is typical of the aging process. Therefore cardiac aging is a compilation of the pathological processes associated with endothelial cells, fibroblasts, immune cells, and cardiomyocyte remodeling. These intrinsic mechanisms are altered negatively or favorably by extrinsic factors such as air pollution, sedentary lifestyles, poor diets and their patterns, and psychosocial stress (Dragoumani et al., 2025). Two-way interactions between the corresponding changes at each level, intrinsic molecular modification, and extrinsic environmental exposure. Environmental exposures, such as oxidative stress caused by air pollutants, accelerate their oxidation; however, lifestyle modifications, including aerobic exercise, intermittent fasting, and

Cellular Senescence and Cardiac Aging. DOI: https://doi.org/10.1016/B978-0-443-45811-8.00012-X

following the Mediterranean diet, defend against cardiac senescence by promoting autophagy, inflammation suppression, and maintenance of telomere length. These results explain the pathways of cardiac aging at the molecular level and identify drug opportunities (Bi et al., 2024). Cardiovascular disease is a leading cause of burden among the aged, hence explaining the cellular and molecular mechanisms that drive the aging process in the cardiac system is vital for early detection and formulation of specific interventions. This chapter provides a thorough background on cardiac aging and focuses on how cellular senescence plays a key role in the structural and functional deterioration that accompanies aging.

1.2 Mechanistic foundations of cardiac aging

Cardiac aging is not just a process of the negative effects of time on the cardiovascular aging process. Instead, it is an accumulation of the effects of molecular and cellular injuries, which disrupt cardiac structure and function, as well as natural regenerative powers. These disturbances are driven by the shared underpinnings of metabolism, genetics, and epigenetics, which have been identified in the hallmarks of aging paradigm (Tariq et al., 2022). All hallmarks undermine cellular fitness via collaborative cellular dysfunction and therefore reduce net cardiac output and stress tolerance. These mechanisms include genomic instability, telomere loss, mitochondrial dysfunction, oxidative stress, impaired proteostasis, impaired intercellular communication, and highly critical molecular signaling pathways, NF-kB, and mTOR.

1.2.1 Genomic instability and DNA damage response

Genomic instability is a hallmark of organismal aging and is instigated by the current buildup of DNA damage, which is caused by endogenous and exogenous origins. DNA damage can occur in the heart due to ROS, replication stress, and inadequate DNA repair (Wei et al., 2024). Enhancement of cardiomyocytes and endothelial cells with DNA double-strand breaks occurs in an age-dependent manner, where the two phenomena are defined by two genome-wide assortments, the progression of 2xFocus of 2H indications, and activation of ATM kinase (Wu et al., 2024). DDR is activated by cumulative DNA damage, which eventually leads to the activation of cell cycle inhibitors, the most prominent of which are p53 and p21. This activation leads to the senescence of cardiac cell fibroblasts and progenitor cells. In these senescent cells, irreversible growth arrest is triggered by persistent unrepaired lesions in the DNA, some of which occur at telomeres. With time, the accretion of senescence leads to dysfunctional tissue turnover and a pro-inflammatory microenvironment in the aging heart (Dai et al., 2024).

1.2.2 Telomere attrition and cardiomyocyte senescence

Telomeres are tandem, repetitive nucleotide sequences, often TTAGGG, in vertebrates at chromosome ends, which protect chromosomes as they also prevent fusion of ends together and degradation. As they replicate, the inability to replicate the final parts of these sequences, which is part of what is known as the end-replication problem, leads to a progressive loss of telomeres (Li et al., 2024). Telomere attrition to a critical point occurs in cardiac progenitor cells and some of their dividing counterparts (including fibroblasts), thereby triggering the DNA damage response and cellular senescence by upregulating both the p53-p21 and p16INK4a-Rb axes (Ding et al., 2025). In contrast to the earlier held notion that once cardiomyocytes progress to a state of no longer dividing and hence postmitotic, they remain immune to telomere erosion, recent developments also show that these cells develop telomere dysfunction as they age, although not necessarily associated with any change in length. Instead, the buildup of oxidative stress may trigger the manifestation of telomeric damage, resulting in the occurrence of TAFS without alterations in their significant shortening (Boccardi, 2025). Contractile dysfunction, hypertrophy, and secretion of SASP are characteristics of telomere-based cardiomyocyte senescence (Zhao et al., 2025). LTL in the peripheral blood has been developed as a promising clinical index of systemic aging and cardiovascular risk. Reduced LTL is also associated with high rates of coronary artery disease, heart failure, and adverse cardiovascular outcomes; however, the mechanistic associations are still being examined (Fu et al., 2025) (Fig. 1.1).

1.2.3 Mitochondrial dysfunction and oxidative stress

The deregulation of mitochondria is the major factor of determining cardiovascular aging, which is closely connected to the energy demands of the heart, which is in a large part. The deficiency of mitochondrial dynamics, biogenesis, and quality control systems due to age undermines cellular ATP synthesis, promotes electron leakage, and increases ROS levels in cells (Ding et al., 2024). A prolonged hyperoxidative environment catalyzes mtDNA mutation, respiratory-complex denaturation, and oxidative alteration of nuclear DNA and cytosolic proteins, all of which exacerbate lipid peroxidation and cellular necrosis (Ding et al., 2024). Aging is also characterized by the gradual loss of the potential of the

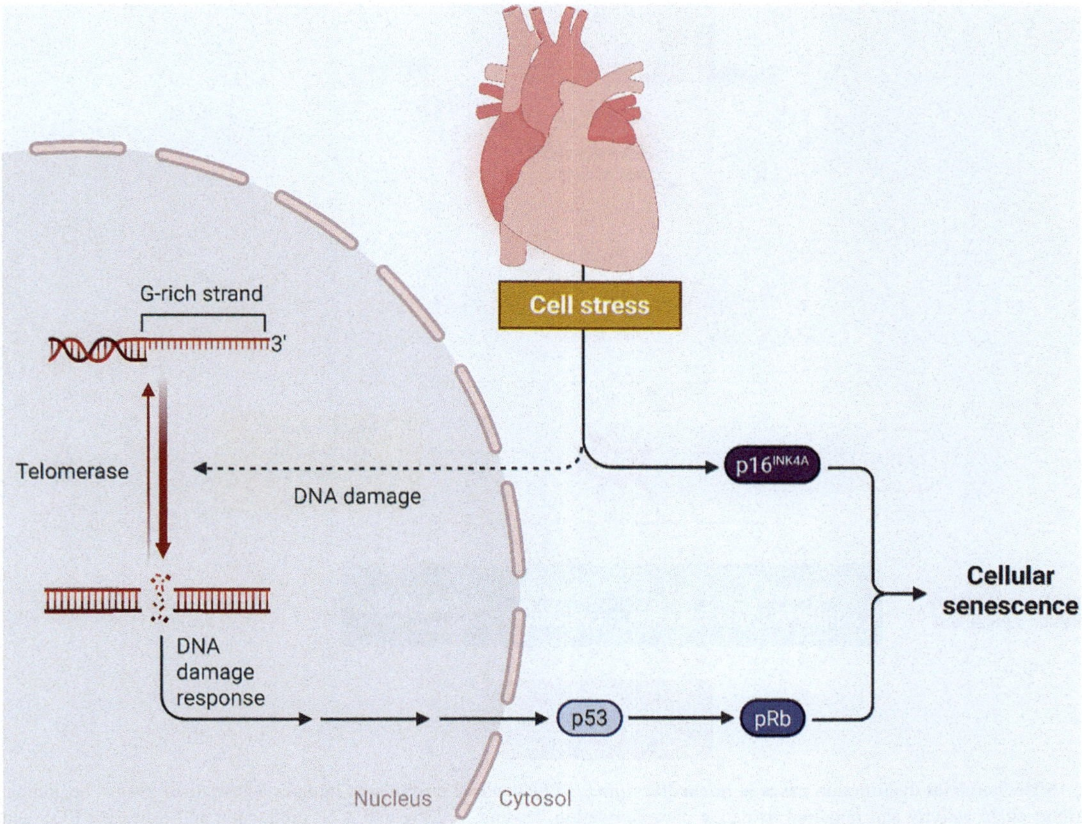

FIGURE 1.1 Telomere shortening and cellular senescence pathway. Under conditions of cardiac cell stress, telomere shortening leads to DNA damage at the G-rich telomeric strand, initiating a DNA damage response in the nucleus. When telomerase fails to maintain telomere integrity, this damage activates p53, which in turn promotes pRb activation, triggering cellular senescence. Independently, cell stress also induces p16INK4A expression, which converges with the p53–pRb axis to enforce irreversible growth arrest. This pathway highlights how telomere dysfunction and stress signaling contribute to senescence in cardiac cells.

mitochondrial membrane, deficiency in the functioning of mitophagy, and the presence of pathologically changed mitochondria within the cell. Consequently, such dysfunctional organelles act as sustained sources of ROS, leading to a positive feedback loop that exacerbates oxidative stress and tissue damage (Shi et al., 2024). ROS overproduction regulates redox-sensitive pathways and promotes inflammation, hypertrophic phenotypes, and extracellular matrix remodeling. Oxidative stress is also involved in the decreased activity of eNOS in the vascular endothelium, leading to endothelial dysfunction and an increased predisposition to atherosclerotic disease. Members of the sirtuin family, SIRT3 and peroxisome PGC-1α, are also key players in mitochondrial biology and redox homeostasis. These molecular downturns, caused by age, impair mitorophagy and dampen mitochondrial biogenesis, thus reducing the cell's ability to produce bioenergy and meet metabolic stress demands (Qian et al., 2024). The mitochondrial dysfunction triggers a pathological cascade in cardiac cells through the inhibition of electron transport chain activity and oxidative phosphorylation, which decreases ATP production and provokes the formation of ROS. High ROS levels also induce the opening of the mPTP, disrupting mitochondrial integrity and function. The result of this state includes damage to mitochondrial DNA, lipid peroxidation, protein damage, and chronic inflammation. In addition, pore formation of mPTP increases NLRP3 inflammasome activation, inhibits mitophagy, and increases DAMP secretion, further increasing cellular damage (Rottenberg and Hoek, 2017) (Fig. 1.2). The effects are even worse owing to dysregulation of mitochondrial fusion and fission dynamics. Therapeutic approaches to induce the restoration of mitophagy or enhance mitochondrial biogenesis are being investigated as possible cardiac aging countermeasures.

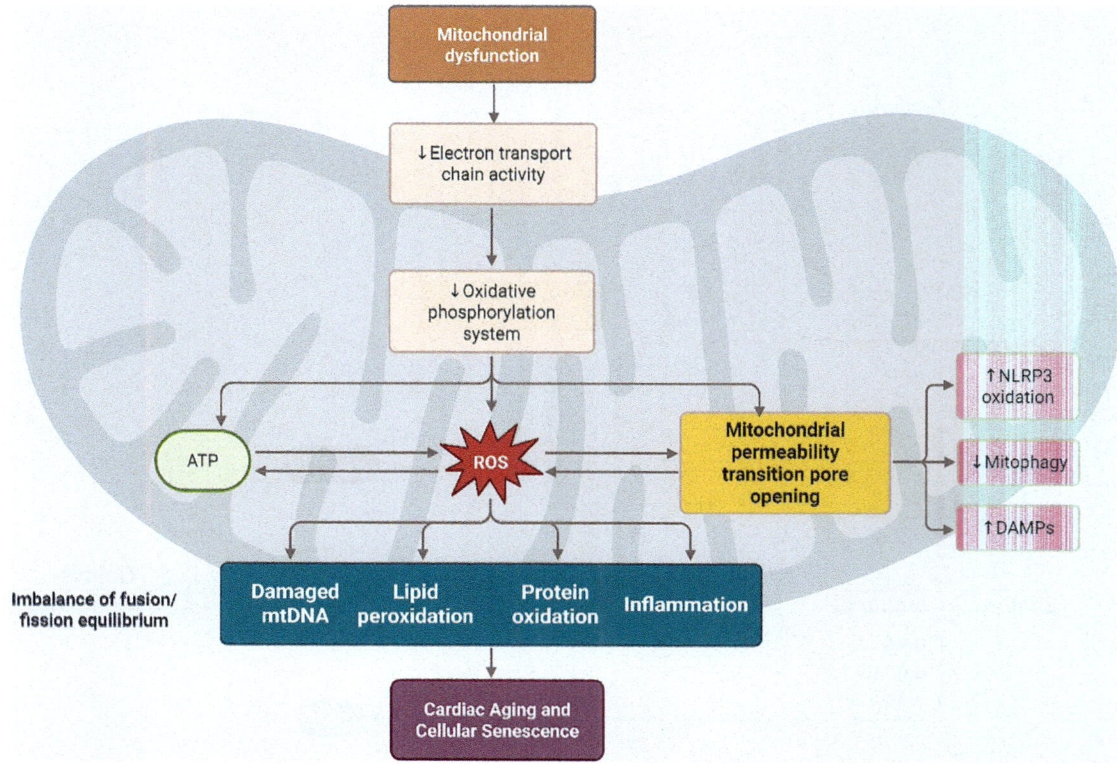

FIGURE 1.2 **Mitochondrial dysfunction cascade in cardiac aging.** Mitochondrial dysfunction initiates a cascade of events, beginning with reduced electron transport chain activity and impaired oxidative phosphorylation, leading to decreased ATP production and excessive ROS generation. ROS accumulation triggers mPTP opening, which further exacerbates mitochondrial damage. This results in mtDNA damage, lipid and protein oxidation, and inflammation. The impaired mitochondrial quality control, including decreased mitophagy and increased DAMPs and NLRP3 oxidation, contributes to disrupted mitochondrial dynamics.

1.2.4 Altered proteostasis and autophagy decline

Proteostasis, or more concretely, protein homeostasis, is a principal cellular process that regulates the general process of protein production, folding, transportation, and degradation. The lack of performance of molecular chaperones, proteasomes, and autophagic degradation pathways in aging cardiomyocytes impairs this regulatory network (Zarzycka et al., 2025). The latter, or lysosome-dependent process, which is vital for digesting unserviceable organelles and protein aggregates, is suppressed in the senescent myocardium, thus permitting the continuing growth of incompetent proteins and cellular clogging. Autophagic flux has been linked to the precipitation of cardiac stiffness, an increased burden of inflammation, and a predisposition to ischemia (Kobak et al., 2025). Impaired proteostasis initiates the UPR and elicits ER stress, which accelerates apoptotic and senescent pathways (Zhang et al., 2024a). The aging myocardium shows an increased level of expression of ubiquitinated proteins and lipofuscin granules, which indicates a lack of proteolytic activity. The focus of proteostasis using calorie restriction and pharmacologic autophagy inducers, including rapamycin or spermidine, and proteasomal activators, represents an effective mechanism of cardiovascular rejuvenation.

1.2.5 Inflammatory signaling and intercellular communication

Aging is linked to a long-term low-grade inflammatory status called inflammaging. The distinguishing features of this condition are increased levels of pro-inflammatory cytokines, such as IL-6, TNF-α, and IL-1β in the circulation. Inflammatory complications associated with inflammaging in the cardiovascular system include exacerbation of endothelial dysfunction by disrupting nitric oxide signaling, oxidative stress-driven myocardial fibrosis, and reduction of cardiac contractility due to extracellular remodeling of the matrix (Zhang et al., 2025). Cellular senescence is a major stimulator of inflammatory mechanisms through the SASP. Senescent cells produce more inflammatory cytokines, chemokines, matrix metalloproteinases, and growth factors, leading to the disruption of tissue homeostasis and the transmission of senescence to adjacent cells. Such a paracrine cascade creates a profibrotic and prohypertrophic environment in the aging myocardium (Kyryk and Parkhomenko, 2025). Simultaneously, aged hearts have impaired intercellular communication through gap junctions and exosomes. Reduced connexin expression and modification of exosomal cargo affect electrical conduction and dampen the actions of cardioprotective signals (Zhou et al., 2024). Senescence in endothelial cells also leads to vascular dysfunction when it causes an increase in the adhesion of leukocytes. Increased levels of intercellular adhesion molecules, such as ICAM-1 and VCAM-1, increase thyroid and vascular tone, leukocyte infiltration, and tissue remodeling (Mazgelytė et al., 2024).

1.2.6 Key signaling pathways in cardiac aging

Cardiac aging is controlled by several intracellular signaling pathways. NF-kB and mTOR are two of these pathways that play especially important roles. The global transcriptional control is NF-kB. This is demonstrated in the aging heart with ongoing activation of NF-kB by cardiomyocytes and endothelial cells, which prompts the expression of SASP, inflammation in the tissues, and heart remodeling. It is also a repressor of autophagy and an activator of mitochondrial abnormalities. According to experimental studies, NF-κB inhibition slows age-related signs and enhances cardiac performance. mTOR is a nutrient-sensing kinase that controls cell growth, metabolism, and autophagy. The process of hyperactivation at this age contributes to protein synthesis, autophagy inhibition, and cardiac hypertrophy and fibrosis potentiation (Grzeczka et al., 2025). mTOR inhibition through pharmacological inhibitors such as rapamycin or metformin has been shown to increase lifespan and improve cardiac output in age-related models. It also modulates additional aging-regulatory elements such as AMPK and sirtuins, further implicating it in the critical pathway of metabolic regulation and environmental stress (Zarzycka et al., 2025).

In addition to NF-KB and mTOR, other signaling pathways play central roles in cardiac aging. The p53/p21 and p16INK4A/Rb pathways are the most crucial regulators of cellular senescence, cell cycle arrest in response to destructive factors and oxidants, and telomere malfunction (Wei et al., 2025). Chronic stimulation of these pathways contributes to the accumulation of senescent cells in the myocardium. The TGF-β/Smad signaling axis contributes to the development of age-related cardiac fibrosis, resulting in myocardial stiffening and reduced diastolic dysfunction by promoting ECM deposition and remodeling (Al-Regaiey, 2024). The PI3K/Akt/FOXO pathway regulates cellular growth and the prevention of apoptosis by activating Akt, which has effects that are opposed by FOXO transcription factors, which also stimulate antioxidant gene expression and longevity. These processes are often inhibited in aging (Jin and Hou, 2024). The Wnt/ 8-catenin available during development is distorted in old age and facilitates maladaptive cardiac progenitor cell differentiation, fibrotic remodeling, and impaired regeneration (Horitani and Shiojima, 2024). The interaction between these pathways constitutes a complex regulatory network that controls the balance between damage and repair in the aging heart.

1.3 Extrinsic factors accelerating cardiac aging

Age-related cardiac changes are not merely the product of intracellular and molecular mechanisms of the heart; they also depend heavily on extrinsic factors, primarily environmental effects, as well as behavioral and psychological pressures. These exogenous factors enhance cardiac degradation of structure and function through pathways that entail oxidative stress, chronic inflammation, and metabolic dysfunction. It is important to note that many of these factors can be modified, and that through behavioral and environmental interventions, cardiovascular risk can be significantly reduced and aging can be slowed down. An understanding of these extrinsic motives is important for developing successful schemes to capitalize on cardiac health and the absence of age-related cardiovascular conditions.

1.3.1 Environmental exposures and pollution

Excessive or chronic heat pollution (PM2.5), industrial poisons, and automobile exhaust have been linked to cardiovascular aging and illness. Of particular significance is the oxidative stress caused by the production of ROS by pollutants in vascular and myocardial cells, which causes DNA damage, telomere shortening, and endothelial dysfunction (Al-Awar and Hussain, 2024). An increase in ambient air pollution stimulates vascular stiffness and promotes the emergence of atherosclerotic plaques, thus increasing the risk of myocardial infarction and cardiac failure (Wong et al., 2024). Epidemiological surveys further show that 61% of residents in polluted cities have shorter LTL, a systemic aging biomarker, in comparison to their counterparts in less polluted areas. At the same time, indoor sources, which include solid-fuel combustion products that are the products of cooking and heating, also increase oxidative stress and events of inflammation (Salih et al., 2024). Daily low-dose exposures during normal activities, such as walking or exercise, may inhibit the cardiovascular effects of physical activity by increasing vascular inflammation and oxidative stress. The best way to curb these adverse consequences is to improve air quality, embrace the use of clean cooking fuels, and engage in protective behaviors, such as avoiding outdoor activities at times when pollution is very high, to protect the cardiovascular system of aging individuals.

In addition to ambient air pollution, cardiac tissue is directly toxic to chemical toxins, including heavy metals, pesticides, and industrial chemicals. Chronic exposure causes mitochondrial dysfunction, failure of ATP production, and induces cell death in the form of apoptosis in cardiomyocytes (Ryoo et al., 2025). Empirical findings have shown that cadmium consumption is associated with arterial calcification, endothelial dysfunction, and hypertension, which increase with age. Furthermore, exposure to solvents or industrial chemicals, especially in poorly regulated environments, increases cardiovascular vulnerability through oxidative stress and systemic inflammation in the long term. All these toxic effects are exacerbated by a downturn in detoxification systems and antioxidant defenses with aging. As a result, workplace regulations regarding safety and regular health checkups are vital preventive measures.

1.3.2 Circadian disruption and lifestyle factors and physical inactivity

The problem of disruption of the circadian rhythm has not received a substantial level of study in the field with regard to its robust extrinsic impact on cardiovascular well-being. Empirical evidence suggests that the cardiovascular system follows regular circadian dynamics that determine metabolic processes, heart rate, and blood pressure (Zhu et al., 2024). Alterations in such rhythms are mainly due to shift work, inconsistent sleep, and persistent sleep deprivation, which have been associated with increased cardiovascular aging and an augmented risk of clinically significant cardiovascular diseases (Young et al., 2025). Chronic discordant circadian rhythms produce systemic metabolic stress, increase cortisol levels in the blood, and initiate oxidative injury in cardiac tissue. The relevance of circadian regulation in the aging of the heart is evidenced by the fact that in rodent models with such disruptions in circadian clock genes as CLOCK and BMAL1, cardiomyopathy has been detected as early as before the age at which cardiomyopathy is typically reported in humans (Li et al., 2025). Moreover, cardiomyopathy is accompanied by fibrosis and mitochondrial dysfunction, which have also been implicated in such cases. In humans, sustained sleep deprivation or disordered sleep is associated with health risks such as high blood pressure, impaired endothelial regeneration, and circulating levels of inflammatory factors, such as CRP, hence the decline in cardiac aging (Xie et al., 2025).

Sedentary behavior is among the most critical modifiable risk factors for cardiac aging. Chronic immobilization is linked to reduced cardiac output, impaired mitochondrial biogenesis, and endothelial dysfunction, all of which enhance age-related deterioration in cardiovascular function. In contrast, natural aerobic training maintains the length of telomeres, increases the efficiency of mitochondria, and makes cardiac cells more effective in the production of autophagy. Exercise also upregulates TERT, which checks the shortening of telomeres and induces Nrf-2, enhancing antioxidant defense systems (Liu et al., 2024). Even mild exercise has been documented to diminish SASP-related inflammation in cardiac tissue, which enhances cardiovascular system resilience. Follow-up investigations on endurance athletes show that those athletes who are on higher levels of physical activity have a so-called younger phenotype of their cardiovascular system with enhanced left ventricular compliance and decreased arterial stiffness when compared to their sedentary counterparts.

1.3.3 Poor nutrition and dietary patterns

Of particular concern are the effects of eating habits on heart aging. Obesity, insulin resistance, and low-grade inflammation, which are repeatedly catalyzed by dietary programs with excessive saturated fat, refined sugar, and processed food content, encourage enhanced cardiovascular aging (Dominguez et al., 2024). These nutrient profiles increase oxidative stress and reduce the bioavailability of nitric oxide, which favors endothelial dysfunction and stiffening of the arteries. Conversely, nutritional habits high in fruits, vegetables, whole grains, nuts, and omega-3 fatty acids, such as the Mediterranean diet, have been associated with a healthier heart and slower aging patterns (Dobroslavska et al., 2024). Evidence has shown that the Mediterranean diet is considered a telomerase indicator, inflammatory marker, and an activator of lipid and metabolic indicators. Mechanistically, caloric restriction and intermittent fasting have beneficial effects on the aging heart by boosting autophagy and mitochondrial function and reducing oxidative stress (Wang et al., 2024). Biological models also indicate that caloric restriction increases lifespan and preserves the youthfulness of the cardiac system by regulating nutrient-sensing routes, such as AMPK and mTOR.

1.3.4 Psychological stress and mental health and sleep deprivation and recovery

Sustained mental stress is a powerful catalyst for aging of the cardiovascular system. Chronic stress via the HPA axis increases stress hormones, particularly cortisol, which stimulates the development of oxidative stress, inflammation, and endothelial dysfunction (Cheon et al., 2025). Coactivation of the sympathetic nervous system increases the heart rate, blood pressure, and oxygen demand of the myocardium, and consequently, unfavorable cardiac remodeling. The connection between psychological stress and reduced telomere length is pertinent to shorter telomere length and greater expression of senescence markers in circulating immune cells (Mbiydzenyuy and Qulu, 2024). Stress-reducing interventions have been shown to positively modify risk factors, including inflammatory markers, vascular health, and telomerase activity. A positive mind frame and social connectedness are also linked to decreased cardiovascular risk and increased survival expectancy, which shows a direct effect of psychosocial factors on cardiac aging.

Sound sleep cannot be dispensed with cardiovascular repair and recovery. Prolonged sleep deprivation or poor-quality sleep increases systemic inflammation, oxidative stress, and blood pressure instability, all of which are negative factors in terms of cardiovascular functional response (Saputro et al., 2025). Individuals with sleep deprivation are characterized by increased frequency of arrhythmias, metabolic syndrome, and faster progression of atherosclerosis. Empirical evidence also shows that a lack of or broken rest disrupts clockwork gene expression in cardiomyocytes, thus interfering with calcium handling and mitochondrial activity (Goette et al., 2024). When these strategies are aimed at improving sleep hygiene, such as sticking to regular sleeping schedules and avoiding screen time before going to bed, their cardiovascular effects are enhanced, and the aging processes are slowed down.

1.4 Intrinsic drivers of cardiac aging

Cardiac tissue aging is driven by exogenous factors and a panel of intrinsic cell-autonomous systems that jointly downregulate cardiomyocyte activity, vascular health, and myocardial plasticity. Endogenous drivers include telomere shortening, epigenetic drift, loss of autophagic capacity, dysregulated proteostasis, and chronic inflammation associated with metabolic imbalance. Unlike external factors, internal factors continue to accumulate over a lifetime, regardless of the environment, making them key players in promoting molecular breakdown in the heart. These inherent modifications can only be understood comprehensively, and vital knowledge of which reveals why the organ in question is more prone to stress, injury, and illnesses during aging.

1.4.1 Telomere dynamics and the shelterin complex

Telomere DNA sequences contain repeated untranscribed nucleotide sequences at the end of chromosomes, which are noncoding by nature (TTAGGG repeats in humans). Their main role is to provide genomic stability by arresting chromosomal end-to-end fusion and inappropriate DNA repair (Spanakis and Tsatsakis, 2025). Telomeres erode during somatic cell division due to the end-replication problem and oxidative stress, reaching a critical length that triggers the DNA damage response and causes cellular senescence (Yılmaz and Bozkurt, 2024). Telomere attrition occurs in both proliferating and postmitotic cardiovascular cell types, including dividing cardiac progenitor cells, fibroblasts, and postmitotic cardiomyocytes in the heart, due to oxidative stress and mitochondrial dysfunction (Abdel-Gabbar and Kordy, 2024). This degeneration of the telomeres, especially that not followed by an obvious decrease in length, leads to the formation of TAFs, which activate the p53/p21 and p16 INK4a/Rb pathways, causing cell cycle arrest and functional dysregulation (Xu et al., 2024).

Telomeres perform structural and regulatory roles, which rely on the shelterin six-protein complex, consisting of lovely TRF1, TRF2, POT1, TIN2, TPP1, and RAP1. It cannot be avoided that this complex prevents recognition of telomeres as the DNA breaks, and it plays a role in allowing telomerase access to the telomeric ends. The loss of shelterin components leads to a loss of shelterin balance, further increasing the vulnerability of telomeres to abuse and shortening the process of senescence in aging (Spanakis and Tsatsakis, 2025). In experimental research on cardiomyocytes, the loss of TRF2 or POT1 in cardiomyocytes is rapidly associated with telomere deprotection, DNA damage signaling activation, and contractile impairment. Although telomerase can increase the length of telomeres, its synthesis in adult somatic tissues, particularly the heart, is mostly inhibited (Spano et al., 2024). Telomere protection by reactivating telomerase in aging models enhances the survival of cardiomyocytes, prevents fibrosis, and maintains cardiac function, highlighting the fact that this is the main approach toward the treatment of cardiac aging.

1.4.2 Histone modifications and DNA methylation

Epigenetic regulation is a heritable change in gene expression that is not accompanied by changes in the DNA sequence. DNA's methylation, histone-dependent changes, and changes in chromatin structure are some of the key mechanisms. In older age, the heart experiences increasing rates of epigenetic drift, where the cardiac epigenome is globally hypomethylated, there is promoter-specific hypermethylation, and patterns of histone modification are interrupted. This drift leads to aberrant gene expression, chromatin instability, and disrupted stress responses (Alrumaihi et al., 2025). Methylation of promoters controlling genes in antioxidant defense, mitochondrial biogenesis, and cell cycle control is often abnormal in old hearts and leads to lower expression of cardioprotective proteins and an increase in pro-aging factors (Desiderio et al., 2024). Chromatin states Chromatin accessibility and transcription depend dynamically on epigenetic histone modifications, including acetylation, methylation, phosphorylation, and ubiquitylation of histones; HDACs and HATs have antagonistic actions (Lv et al., 2024). During aging, the progressive loss of HDAC activity and subsequent hyper-acetylation of inflammatory gene promoters culminate in the SASP and a low-grade state of inflammation found in aging cells in the myocardium (Matacchione et al., 2024). The genetic defects that occur during the loss of H3K27me3 in histone methylation in cardiomyocytes alter the regulation of development-related genes and consequently lead to transcriptional noise and functional impairment. The decrease in sirtuin levels, specifically in cardiac tissues, with age negatively affects mitochondrial functionality, fosters fibrosis, and predisposes to stress-related apoptosis (Zhang et al., 2024c).

DNA methylation age, also known as the epigenetic clock, has a strong correlation with biological age and is an independent cardiovascular risk factor, regardless of chronological age (Lind et al., 2018). Evidence also shows that accelerated epigenetic aging, which was measured by evaluating the methylation patterns of particular CpG sites, is correlated with left ventricular hypertrophy, arterial stiffness, and high mortality. As a result, the idea of countering epigenetic drift of cardiac aging with dietary bioactive components, pharmacological HDAC inhibition, or the use of sirtuin activators as an intervention could form the basis of an attractive approach toward extending healthspan (Joyce et al., 2021).

1.4.3 Autophagy decline and proteostasis imbalance

The functional integrity of cardiomyocytes is essential and dependent on proteostasis, which can be described as the coordinated synthesis, folding, intracellular trafficking, and degradation of proteins. This balance is maintained by the aging process, which causes intracellular leakage of accumulated damaged, misfolded, or aggregated proteins, which is known to tip cellular homeostasis. Autophagy is a key process that restores proteostasis, which is strongly attenuated in aged cardiac tissue. Autophagy is a labeling-dependent catabolism hooked up by a lysosome that delivers dysfunctional organelles and protein aggregates, thereby maintaining cellular quality control. In the hearts of younger individuals, basal autophagy eliminates damaged mitochondria, impairs the generation of reactive oxygen species, and protects contractile activity (Morales et al., 2014). Conversely, elderly cardiomyocytes are associated with reduced autophagic flux due to the inability to trigger autophagic activity, lysosomal unresponsiveness, and mTOR/AMPK signaling disturbance (Dutta et al., 2012).

De-inhibited autophagy allows subsequent toxic protein aggregation in cardiomyocytes, triggering endoplasmic reticulum stress and UPR, making them more prone to apoptosis and necrosis. Moreover, incomplete protein degradation increases the accumulation of lipofuscin, a feature of the aging process, in lysosomes, exacerbating autophagic defects (Tannous et al., 2008). Simultaneously, proteasome impairment, another branch of the proteostatic network, results in increased levels of ubiquitinated proteins and impaired calcium homeostasis in cardiac cells. In animal studies, increasing autophagy through a low-calorie diet, mTOR inhibitors, or autophagy-activating peptide treatment has been able to reverse cardiac dysfunction, cardiac fibrosis, and life extension. Therefore the maintenance of autophagy and proteostasis is a central approach to cardioprotection in old age (Wang and Wang, 2015).

1.4.4 Inflammation and metabolic deregulation

Age is increasingly recognized as a chronic, low-grade inflammatory disease, commonly called inflammaging, and this process modulates progressive myocardial remodeling, fibrosis, and contractile impairment. Fibroblasts and endothelial cells secrete a package of pro-inflammatory factors, known as the SASP, in older cardiac cells, including cytokines, chemokines, and matrix metalloproteinases (Ajoolabady et al., 2023). In this way, they spread systemic and local inflammation, immune infiltration, impair intercellular communication, and activate ECM degradation and fibrosis. The circulating levels of inflammatory mediators, such as IL-6, TNF-α, and MCP-1, are consistently expressed in older hearts and can predict poor cardiac outcomes in elderly individuals (Sorokin, 2010). Metabolic deregulation, which is associated with aging and cardiovascular disease, contributes to an inflammatory milieu. The decline in fatty acid oxidation leads to glycolysis in the aged myocardium, which reduces metabolic efficiency and ATP production. The synergistic effects of mitochondrial dysfunction and insulin resistance increase ROS levels, which activate NF-kB and inflammasome pathways, thereby increasing inflammation (Ferrucci and Fabbri, 2018). Nutrient-sensing pathways, such as AMPK and mTOR, are distorted. The downregulation of AMPK activator is associated with the hyperactivation of mTOR, a facilitator of protein synthesis, cell proliferation, and hypertrophic remodeling. This metabolic disorder is not only dysfunctional in energy production but also enables a setting characterized by cellular senescence and contractile dysfunction (Fernández-Veledo et al., 2013).

Changes in adipokine signaling, such as low adiponectin and leptin resistance, also worsen cardiac performance by causing endothelial dysfunction and cardiomyocyte hypertrophy. These imbalances are further aggravated by obesity and type 2 diabetes, which are more common in aging populations, thus connecting metabolic disease directly to accelerated aging in the heart (Ghantous et al., 2015). Interventions intended to reverse metabolic imbalances, such as caloric restriction, intermittent fasting, and preplanned exercise regimes, have proven helpful in reducing inflammation and reversing the effects of age-related decline in cardiac integrity. Drugs such as metformin and resveratrol have also shown positive results (Mujica et al., 2010).

1.5 Cellular landscape of the aging heart

It is well established that the mature heart is a postmitotic tissue. Nevertheless, it contains a mixed cellular complement, which maintains its functionality, resistance, and regeneration abilities. This component contains cardiomyocytes, endothelial cells, fibroblasts, and cardiac progenitor cells, all of which show different age-related susceptibilities (Foglia and Poss, 2016). These cellular lineages subsequently experience morphological, functional, and molecular changes that impair overall myocardial performance and culminate in cardiac remodeling, fibrosis, and compromised contractility with a reduction in vascular density. The subsequent discussion explains the separate and consequent efforts made by each cellular compartment in the aging heart (Leri et al., 2005).

1.5.1 Cardiomyocytes

Contractile cells of the heart play a central role in age-related structural and functional decline. During the senescent inflammatory state, cardiomyocytes experience cardiomyocyte hypertrophy in response to increased work and reduced regenerative capacity. This swelling is coupled with the disorganization of the sarcomere, swelling of the nuclei, and accumulation of cytoplasmic lipofuscin granules, which together indicate poor proteostasis and lysosomal degradation (Ruiz-Meana et al., 2020). At the functional level, aged cardiomyocytes have impaired calcium management and poor contractility, which are essential for facilitating excitation-contraction coupling. These deficiencies are caused by the alteration of L-type calcium channels and lower expression of SERCA2a, the calcium ATPase that reabsorbs calcium during diastole (Dewenter et al., 2017).

Mitochondrial dysfunction is a key feature of aged cardiomyocytes. In vivo, mitochondria demonstrate reduced membrane potential, defects in oxidative phosphorylation, and structural abnormalities. Such adjustments reduce ATP synthesis and increase the production of ROS, which harm mitochondrial DNA, proteins, and lipids (Lin and Kerkelä, 2020). The consequences of damage resulting in dysfunctional mitochondrial accumulation increase oxidative stress, stimulate pro-apoptotic signatures, and drive senescence. Stress-induced premature senescence is reported to be p53-dependent, with increased expression of p53, p21, and p16 in INK4a, with a rise in senescence-associated galactosidase β- or SA-Bgal, even without significant telomere attrition in both murine pathways and human autopsy samples. These findings imply that postmitotic cardiomyocytes are largely defined by stress-induced senescence, as opposed to replicative senescence (Chistiakov et al., 2018).

In contrast to the long-standing assumption that adult cardiomyocytes are irreversibly nondividing, recent findings indicate that there is a very low rate of turnover in the mature human heart, which is rate-limited. As one grows older, turnover diminishes, but in the adult population, it is estimated that turnover occurs at a rate of less than 1% annually. This poor regenerative ability is further reduced by injury to the mitochondria and pro-inflammatory signaling, which hinders reentry into the cell cycle (Gan et al., 2020). Senescent cardiomyocytes affect contractile performance by building up senescent cardiomyocytes and releasing pro-inflammatory and profibrotic factors that alter adjacent cells, thereby promoting maladaptive hypertrophy and remodeling of the extracellular matrix and recruiting immune cells, which further impairs tissue functionality. In summary, these results present a noncanonical SASP that causes detrimental cardiac remodeling (Fraticelli et al., 1989).

1.5.2 Endothelial cells

One of the key aspects of vascular homeostasis is the endothelial cells that gate vasodilation, nutrient transfer, and an antiinflammatory endothelial tone. However, where vascular aging occurs, there is endothelial dysfunction, which adds to impaired angiogenesis and decreased perfusion of the myocardium (Augustin and Koh, 2024). The evidence collected in both human subjects and animal models has shown that endothelial cells in aged people have greater levels of markers of senescence, for example, p53, p21, and p16^INK4a, with loss of NO bioavailability due to loss of eNOS activity and greater sensitization to oxidative degradation by superoxide. Second, this leads to impaired vasodilation and increased vascular stiffness, predisposing the properties of ischemic injury and impaired stress-adaptive response (Bouïs et al., 2001). Capillary rarefaction commonly denotes a hallmark (thinning) of vascular aging and is associated with the accumulation of microvascular loss in the left ventricular myocardium. This rarefaction reduces the supply of oxygen and nutrients to cardiomyocytes, worsens hypoxia, and causes fibrosis. Left ventricular pressure overload in murine models induces endothelial senescence prior to the development of outward overt heart failure, which is accompanied by high levels of ICAM-1, leading to leukocyte adhesion and a chronically inflamed state. HC senescence also contributes to the pathogenesis of HFpEF, which has an unfair prevalence in the elderly. Such senescence changes to the vasculature in HFpEF models, including cardiac tissue senescence in senescence-accelerated mice fed high-fat/high-salt diets, are accompanied by elevation of senescence markers and the SASP secretome, all of which contribute to myocardial stiffness, diastolic dysfunction, and systemic inflammation (Nair, 2020). EPCs that replenish aging endothelial cells and promote neovascularization are also lost in number and functionality as a normal part of aging. These cells have shorter telomeres and fewer migratory and angiogenic roles; thus they cannot adequately repair and replace the vascular structure. The resultant decreased endothelial regenerative reserve directly affects the decreased flexibility of ageing hearts toward stressors, which include hypertension, ischemia, and oxidative injury (Heiss et al., 2005).

1.5.3 Cardiac fibroblasts

Cardiac fibroblasts, which are the main constructors of the ECM, constitute a significant part of the nonmyocyte cellular environment in the heart. Some of their roles include upholding structural integrity, mechanical compliance, and intercellular communication. During aging, fibroblasts in the heart gain a phenotype characterized by increased collagen synthesis, matrix metalloproteinase activity, and the release of inflammatory mediators. Senescent fibroblasts accumulate in the interstitium and perivascular area, contributing to the myocardial, which is a significant determinant of ventricular stiffening and subsequent diastolic filling impairment (Kohl and Gourdie, 2014). Fibrotic remodeling, which is part of the aging process, is not controlled, and is promoted by cellular senescence and inflammatory crosstalk. Senescent fibroblasts overexpress p53, p21, and ATM; show more active SA-beta-gal; and secrete SASP factors such as IL-6, TGF-beta, and CTGF, which can stimulate neighboring fibroblasts because they can induce collagen deposition (Selman and Pardo, 2021). Myocardial Inf, fibrosis can be reduced and left ventricular function ameliorated in mouse infarct models by experimental reverse senescence of fibroblasts. This observation accentuates the paradox of fibroblast senescence as one event that can rein in runaway increases and fibrosis, leading to collagen accumulation and mechanical failure due to the uncontrolled accumulation of senescent fibroblasts (Liu et al., 2021). In addition to matrix production, cardiac fibroblasts can also play a role in electrical conductivity through interactions with cardiomyocytes and arrhythmogenesis in fibrotic tissue. Aging-related fibrosis directly leads to disturbance of pathways of normal conduction and the formation of reentry circuit substrates, contributing to atrial fibrillation and ventricular arrhythmias. Another area of antifibrotic therapy in elderly patients is the specific targeting of senescence-associated remodeling of fibroblasts and the fibroblast secretory phenotype (Yue et al., 2011).

1.5.4 Cardiac progenitor cells

Collectively, CPCs are a stem-like cellular population that exists within the myocardium and is imperative for cardiac repair and regeneration. These cells have the potential to differentiate into cardiomyocytes, endothelial cells, and smooth muscle lineage under physiological or pathological stress conditions. However, as age progresses, the functional and numerical capacity of CPCs significantly decreases, contributing to the loss of regenerative competence of an older heart (Korski et al., 2019). The changes related to aging are telomere shortening, an increase in the level of expression of senescence biomarkers, p16INK4a and p53, a decrease in clonal and differentiation ability, and sensitivity to DNA damage, oxidative stress, and mitochondrial dysfunction, which completely affect their response to injury or their understanding of the myocardial renewal process (Zglinicki and Martin-Ruiz, 2005). Alongside instrumental shifts, there is a shift in the microenvironment of the heart due to age-related alterations that destabilize progenitor cell maintenance. This niche, known as the hypoxic niche, that acts on the quiescence and potency of CPC is impaired as a person ages, and this impairs signaling pathway sheds like the Notch pathway, Wnt, and IGF-1 that are vital in self-renewal and cardiogenic differentiation (Rota et al., 2015). Experimental data have demonstrated that rejuvenation approaches can partly reverse the loss of CPC functional potential and enhance cardiac performance in animal models of aging and heart failure, including telomerase activation, genetic reprogramming, and epigenetic modification techniques (Ji et al., 2023).

The loss of CPC functionality not only disrupts endogenous regeneration but also limits the effectiveness of autologous cell-based therapy in geriatric patients with cardiovascular disease. CPCs isolated from older adults demonstrate reduced therapeutic potential of transplantation; therefore there is a need for younger phenotype interventions to enhance the therapeutic effect before their use in the treatment of a problem. Ongoing studies are attempting to formulate ex vivo preconditioning schedules, pharmacological agents of senescence-reversal, and combination therapy to maximize the survival, proliferation, and integration of CPCs in aged cardiac tissue (Table 1.1).

1.6 Role of cellular senescence in cardiac aging

Cellular senescence is an environmentally stable and fundamentally irreversible cell cycle arrest triggered by various extrinsic and intrinsic stressors. Originally considered a defensive process to avoid transmitting damage to the surrounding cells, senescence is now positioned as one of the focal markers of the longevity process and age-related disorders, such as cardiovascular diseases. Senescence has a dualistic effect during the aging of the heart; it prevents the division of compromised cells, but this progressive cellular senescence also plays a primary role in the development of myocardial dysfunction, heart fibrosis, inflammation, and reduced regenerative ability (Kumari and Jat, 2021). Unraveling the delicate details of cellular senescence, including its molecular properties, activation factors, and expression down to tissue levels, will thus be vital in cracking the aging and aging pathways in the heart and developing measures to target the senescent phenomena. At the molecular level, senescent cells have several conserved characteristics. Permanent cell cycle arrest is carried out by the activation of the p53/p21 and p16 INK4a/Rb tumor-suppressor pathways, decreased apoptotic susceptibility, changes in chromatin architecture, increased activity of SA-β-gal, and the release of a composite combination of pro-inflammatory cytokines, chemokines, growth factors, and proteins (Kim and Kim, 2024). These hallmarks are additionally characterized by associated TAF, mitochondrial impairment, oxidative stress, and observed aberrations in the autophagic cleansing of cardiomyocytes and fibroblasts. Despite these common senescence pathways being preserved across a wide range of organisms, their expression and functional implications are specific to the cells, stress, and microenvironment.

Two major pathways of cardiac aging have been identified: replicative senescence and SIPS. Strongly related to progressive telomere shortening, replicative senescence exists in dividing lineages, where endothelial cells, fibroblasts, and cardiac progenitor cells are prime examples. As mitoses recur, telomeres shorten until they reach a dangerous level that triggers DDR and cellular growth arrest (Lazzarini et al., 2022). In contrast, SIPS is not dependent on the state of telomeres and is provoked by external or internal stress factors, such as oxidative stress, mitochondrial dysfunction, oncogene activation, and chronic inflammation. This is particularly relevant in the case of postmitotic cardiomyocytes, which develop DNA lesions over time and do not lose telomeres in significant measures. These senescent cells prolong DDR signaling, secrete SASP proteins, and interfere with intercellular signaling, leading to functional deterioration (Nagane et al., 2021). Both replicative senescence and SIPS are concurrent in the aging heart. Endothelial and progenitor cells have replicative dysfunction that negatively affects proliferation and regeneration, leading to capillary rarefaction and stiffening of the vascular system. Mechanical strain, hypoxia, or ROS accumulation in cardiomyocytes results in SIPS and contractile dysfunction, leading to pro-inflammatory signals (Khavinson et al., 2023). Simultaneously, ECM fibrosis, myocardial stiffening, and fibrosis are patronized by senescent fibroblasts. The resultant superposition of replicative and stress-related senescence is the result of the progressive decline in the capacity of the heart, as witnessed in old age (Balaraman et al., 2025).

Senescent cells adopt a unique SASP, a dynamic secretome consisting of inflammatory cytokines, chemokines, growth factors, and MMPs. Despite the fact that the SASP can help maintain growth arrest and attract immune clearance immediately after initial activation, persistent, unregulated SASP secretion is a contributing factor to sterile, or inflammaging, inflammatory processes, and, when it comes to the heart, endothelial dysfunction, leukocyte infiltration, fibroblast activation, and maladaptive hypertrophy (Chen et al., 2025). SASP components exert paracrine and autocrine effects on cardiac morphology and physiology. Secretions from senescent endothelial cells, such as IL-6, affect the

TABLE 1.1 Age-associated cellular and molecular alterations in major cardiac cell populations and their impact on myocardial structure and function.

Cell type	Normal function	Age-related changes	Senescence markers	Impact on cardiac aging	References
Cardiomyocytes	Contractile function and excitation-contraction coupling.	Hypertrophy due to increased workload and reduced regeneration.	Elevated p53, p21, p16INK4a, SA-βGal activity.	Impaired contractility, maladaptive remodeling.	van Meer et al. (2019)
Cardiomyocytes	Sarcomere alignment and calcium cycling.	Sarcomere disorganization, nuclear swelling, lipofuscin accumulation.	No telomere shortening; stress-induced phenotype dominates.	Poor proteostasis and excitation-contraction efficiency.	Rao et al. (2013)
Cardiomyocytes	Mitochondrial ATP production.	Reduced membrane potential, cristae fragmentation, ROS overproduction.	Mitochondrial DNA damage and apoptosis signals.	ATP depletion, oxidative stress, cell death.	Papanicolaou et al. (2012)
Cardiomyocytes	Calcium homeostasis during contraction/relaxation.	Altered L-type Ca^{2+} channels, decreased SERCA2a expression.	Dysregulation of calcium ATPase proteins.	Diastolic dysfunction and arrhythmogenesis.	Periasamy and Huke (2001)
Cardiomyocytes	Limited turnover in adult heart.	Annual turnover <1%, further reduced with age.	Senescence genes restrict cell cycle reentry.	Accumulation of nonfunctional, senescent cells.	Kajstura et al. (2010)
Cardiomyocytes	Maintenance of myocardial structure.	Release of pro-inflammatory, profibrotic SASP.	Elevated cytokines (e.g., IL-6, IL-1β) via SASP.	Fibrosis, ECM disruption, altered adjacent cell behavior.	Dobrev et al. (2023)
Endothelial cells	Regulate vascular tone, nutrient exchange.	Loss of nitric oxide bioavailability, capillary rarefaction.	Loss of eNOS, increased p53, p21, p16INK4a.	Impaired perfusion, hypoxia in myocardium.	Chen et al. (2006)
Endothelial cells	Antiinflammatory barrier function.	Increased vascular stiffness and pro-adhesive molecules (ICAM-1).	SASP-associated inflammation.	Increased risk of ischemia and heart failure (HFpEF).	Pandolfi et al. (2007)
Endothelial cells	Angiogenesis and vessel integrity.	Impaired angiogenesis and neovascularization.	EPC depletion and dysfunction.	Reduced regenerative response after injury.	Herbrig et al. (2006)
Endothelial progenitor cells (EPCs)	Replenish damaged endothelium and promote neovascularization.	Shorter telomeres, loss of migration/angiogenesis capacity.	DNA damage response activation, telomere attrition.	Diminished endothelial regenerative reserve.	Bloom et al. (2022)
Cardiac fibroblasts	Produce ECM and maintain mechanical compliance.	Excessive collagen deposition, matrix remodeling.	Increased p53, p21, ATM, SA-βGal.	Ventricular stiffening, fibrosis.	Adili et al. (2022)
Cardiac fibroblasts	Mediate cell-cell communication.	SASP release promotes inflammation and adjacent fibroblast activation.	IL-6, TGF-β, CTGF secretion.	Exacerbated ECM production and fibrotic signaling.	Ma et al. (2012)

Cardiac fibroblasts	Support electrical insulation and conduction.	Fibrotic tissue disrupts conduction, promotes reentry circuits.	Arrhythmogenic substrate formation.	Increased atrial fibrillation and ventricular arrhythmias.	Vasquez and Morley (2012)
Cardiac progenitor cells	Cardiac regeneration via multilineage differentiation.	Reduced proliferation, telomere shortening, oxidative stress.	High p53, p16INK4a, DNA damage markers.	Reduced myocardial repair capacity postinjury.	Khatiwala et al. (2018)
Cardiac progenitor cells	Maintain cardiac niche and adapt to hypoxia.	Disruption of hypoxic niche, impaired Notch, Wnt, IGF-1 signaling.	Senescence impairs niche communication.	Loss of self-renewal and cardiogenic differentiation.	

calcium regulation and contractility of adjacent cardiomyocytes (Marino et al., 2022). At the same time, the production of TGF-8 by senescent fibroblasts triggers the differentiation and production of myofibroblasts and superfluous collagen, making the walls of the ventricles rigid and causing diastolic dysfunction (Guo et al., 2024). SASP factors upregulate the expression of MMPs, which incidentally break down the ECM and release trapped growth factors, leading to further enhancement of remodeling. The combination of these mechanisms creates an autocrine-paracrine feedback loop that makes the heart less resilient (Balaraman et al., 2025).

Of particular concern is the so-called bystander effect of senescent cells, according to which diffusible SASP molecules cause DNA damage, mitochondrial malfunction, and cell cycle arrest in nearby healthy cells. This effect spreads senescence across cardiac tissues, worsening tissue pathology. Senescent cardiomyocytes and fibroblasts in the aging heart create a microenvironment that encourages the senescence of progenitor cells, endothelial cells, and other surrounding lineages and undermines maintenance and restoration (Redgrave et al., 2023). This bystander senescence helps to explain the mosaic patterns of dysfunction reported in the aged myocardium, where senescent and nonsenescent cells can coexist in domains, but organ dysfunction results.

The growth of old cells in high immune surveillance and clearance is also promoted by age-related immune surveillance and clearance. In most cases, NK cells, macrophages, and T-lymphocytes kill senescent cells using granzyme B, perforin, and death receptor ligands, respectively (Giannoula et al., 2023). However, immune capability is reduced in terms of both quantitative and qualitative nature in older adults, favoring the steady buildup of senescent cells. The avoidance of immune surveillance is also enhanced by immunomodulatory molecules, such as PD-L1 and HLA-E, which protect senescent cells against cytotoxic recognition. The continuous presence of senescent cells leads to hyperinflammation of tissues and the redirection of immune resources toward uncontrolled infection and other pathological processes, contributing to hyperinflammation (Salminen, 2024).

Senescence is a multifactorial accelerator of cardiac aging; therefore increasing attention is being given to senescence-targeted therapeutic interventions, known as senotherapeutics. There are two major subgroups: agents that selectively induce apoptosis in senescent cells and compounds that alter but do not eliminate SASP without inducing apoptosis (Jaba, 2022; Sorrenti et al., 2022). In aged mouse models, the selective elimination of senescent cells ameliorates cardiac parameters, fibrosis, vasculature enhancement, and lifespan. Pharmacological removal of senescent cells in mouse models through senolytic treatment improved left ventricular ejection fraction and decreased myocardial stiffness (Hieda et al., 2021). Likewise, age-related cardiac impairment was significantly reversed when the genetic ablation of p16ink4a senescent cells was achieved in p16ink4a reporter mice. The overall observations point to the harmful effect of the aging process of cardiac senescence and the opportunity of senescent cell-specific therapies in ageless myocardial heart rejuvenation (Oh et al., 2021).

However, the rampant killing of senescent cells must be approached with caution. The population of senescent cells has beneficial effects on wound healing, remodeling of tissues, and tumor suppression. In this regard, widely used clearing approaches can potentially interfere with body homeostasis or hinder repair (Yan et al., 2021). More specific methods may involve temporal and cell-type-specific manipulation of senescence signaling or even the use of combination therapies that only suppress specific parts of the SASP profile without altering the positive aspects of senescent phenotypes. Another limitation is the lack of effective biomarkers to detect senescent cells in human cardiac tissues in vivo. The available markers are useful but not specific and cannot distinguish between transient and persistent senescence with high sensitivity (Marino et al., 2021).

1.7 Biomarkers and diagnostic indicators

Accurate and early diagnosis of cardiac aging is particularly critical in the prevention of progression to blatant cardiovascular disease and in the evaluation of the success of treatment intervention by senescence. Biomarkers are empirical indicators of biological activity and disease progression, aiding in understanding the molecular and cellular defects preceding clinical disease. A comprehensive set of markers has been identified in the aging research field, including a spectrum of canonical cellular senescence markers and telomere-related estimates, epigenetic clocks, transcriptomic and proteomic fingerprints, and more sophisticated imaging and omics-related approaches (Kiseleva et al., 2024). Cellular senescence is a canonical biomarker and the main framework under which the aspect of tissue aging in the myocardium is evaluated. Among them, SA-γ-gal, a lysosomal β-galactosidase whose enzymatic activity is increased under slightly acidic conditions, stands out. The presence of senescence-related markers and their accumulation in cardiac tissue are not unique to senescence, although they are often accompanied by other markers of senescence. The second important hallmark is the initiator of cyclin-dependent kinase inhibitor (p16INK4a), which would contain growth blockage; its promotion has been reported in geriatric cardiomyocytes, fibroblasts, and endothelial cells (Kumar et al.,

2024). Another similar development is that of p21Cip1/Waf1, which is downstream of p53 after DNA damage and is commonly coexpressed with p16 in response to premature senescence due to stress. Despite their intracellular character, these markers may be challenging to use in biopsies of tissues, organ explants, or previously characterized animal models of myocardial aging (Jain et al., 2023).

Long-term weakness of the nucleus gives rise to molecular manifestations. Among them is γH2AX, the phosphorylated variant of histone h2A.X, which is a scar on the location of a double-strand break in DNA and is highly persistent in senescent cells with ongoing damage response signatures. The formation of γH2AX in cardiac tissue is linked to sustained genotoxic stress and mitochondrial dysfunction (Merighi et al., 2021). Another new observation is the third emerging marker, TAF, which indicates chronic signaling of DNA damage at the telomeres (Li et al., 2023). In contrast to transient DNA damage, TAF are permanent and unblocked, and thus constitute a very specific signal to prevent senescence, especially in terminally differentiated cardiomyocytes. Immunohistochemical staining of TAF in heart sections has revealed their enrichment in aged and failing hearts, independent of telomere length, highlighting the importance of telomeric DNA damage, distinct from shortening, in cardiac senescence (Wu et al., 2023).

Telomere length is one of the most cited circulating markers of systemic aging and cardiovascular risk, with the most significant emphasis on the LTL. Shortening of telomeres, the protective capping structures at chromosome ends, during the course of each cell division and under oxidative stress, imparts a practical readout of replicative and oxidative history (Aung et al., 2023). LTL can be determined using quantitative PCR testing of peripheral blood mononuclear cells or FISH. Epidemiological research has proven time and again that shorter LTL is associated with an increased risk of heart failure, coronary artery disease, and hypertension in cohorts of elderly individuals. Although LTL is not directly correlated with cardiomyocyte telomeres, its connection with cardiovascular conditions and inflammatory markers is a good indicator of its application as a minimally invasive biomarker of the biological clock. Notably, LTL is a measure of the overall burden of oxidative and replicative stress, both of which are the main factors of cardiovascular aging (Romaine et al., 2022).

Epigenetic-based surrogates of biological age have been found to be particularly effective in cardiovascular tissues. Epigenetic clocks use the patterns of DNA methylation at specific CpG sites, whose association with chronological age is apparent. The most popular of them, the Horvath clock and GrimAge, are related to cardiovascular morbidity and mortality (Day et al., 2013). Accelerated age from an epigenetic clock has been associated with greater arterial stiffness, hypertrophy of the heart, and vascular malfunction in studies of cardiac aging. Contrary to telomere length, DNA methylation age can be influenced by both lifestyle and treatment measures, making it an encouraging development in detecting the effectiveness of these measures. In addition to DNA methylation, other epigenetic states, such as histone modifications and noncoding RNAs, are being explored for the identification of myocardial aging (Ferrari and Pesce, 2021).

Low- and high-throughput RNA sequencing and proteomics with mass spectrometry widen the molecular landscape of the heart in the elderly. They have shown age-related changes in gene expression and protein levels, especially among pathways related to inflammation, oxidative stress, mitochondrial metabolism, and extracellular matrix remodeling (Song et al., 2024). The upregulation of pro-inflammatory cytokines related to CDKN2A/p16 and fission/fusion proteins in the mitochondria is a constant feature of aged cardiac tissue (Meador et al., 2008). More predictive biomarkers have also been discovered, including the circulating plasma proteins NT-proBNP, galectin-3, and ST2, which increase with age and are indicative of myocardial stress, fibrosis, and remodeling. Transcriptomic and proteomic data integration allows for systems-level insights into the mechanisms of cardiac senescence and makes it possible to discover biomarkers across various biological compartments (Grande et al., 2017). Pluripotent stem cells exhibit the characteristic transcriptional markers that include OCT4, NANOG, and SOX2, at an embryonic stage. Because these cells differentiate in a mesodermal direction, there is the simultaneous upregulation of the Brachyury T as well as MIXL-1, whose focus creates the stage of commitment to the cardiac mesodermal stage and subsequently expresses MESP1/2 and Flk-1. This developmental lineage undergoes further differentiation into representatives of the progenitors of the heart, whose phenotype is laid as NKX2.5, GATA4, TBX5, and MEF2C. The early cardiomyocytes in this population express NKX2.5 concomitant with the production of MYH-proteins and then differentiate into cardiac myocytes with structural and functional cardiac phenotypes that produce cTnT, 8-Actin, MLC2a, MLC2V, SCN5A, CACNA, and IRX4 (Kashyap et al., 2009).

In addition to the analysis of molecular markers, imaging-based diagnostics are essential for imaging the structural and functional correlates of myocardial aging. The most commonly used modality is echocardiography, which can identify age-related disorders such as left ventricular hypertrophy, diastolic filling disorder, and atrial enlargement. Further identification of subclinical dysfunction and strain anomalies of the myocardium, as well as fibrosis, can be performed using advanced echocardiography, especially the speckle-tracking echocardiography method (Del Torto et al., 2022). Complementary information can be derived using MRI of the heart and tissue tagging, as it also identifies pathological changes at the myocardial level. Simultaneously, PET imaging using senescence-specific tracers, such as ^{18}F-labeled galacto-compound imaging with SA-β-gal, has shown promising results as an applicable modality for in vivo measurement of the senescent burden. Such imaging modalities complemented with the information obtained through omics provide longitudinal information on cardiac aging and a broad-based phenotyping approach. The multiomics approach, with its more practical components of metabolomics, lipidomics, and single-cell transcriptomics is yet to transform the way the cardiac aging phenomenon is detected and characterized by offering a high-resolution view of the cell-type-specific pathways of senescence and cell-cell contact networks. The development of machine learning models trained on such multiomics datasets is becoming a potent method to indicate biological age, therapeutic response, and senescence burden with growing precision (Fig. 1.3).

1.8 Therapeutic strategies targeting cardiac aging and senescence

Cardiac aging leads to cell senescence, worldwide malfunction, chronic inflammation, and failure of regeneration, necessitating multifactorial treatment. Modern interventions thus aim not only to alleviate some of the consequences of cardiovascular aging but also to challenge the molecular and cellular roots of aging. The scope of treatments is very wide, starting with mere nonpharmacological behavioral interventions to genetic engineering and cell-based interventions. This discussion evaluates notable treatment strategies that aim to reduce or reverse cardiac aging and senescence.

1.8.1 Lifestyle and behavioral interventions

A wealth of findings supports the theory that regular physical activity, stress management, and sleep optimization are key elements in slowing cardiac aging. Aerobic and resistance exercises have repeatedly been proven to be cardioprotective in both youth and elderly populations. Exercise stimulates mitochondrial biogenesis through PGC-1α, increases autophagy, decreases oxidative stress, and alters inflammatory cytokines. Voluntary wheel running prevents myocardial hypertrophy, lowers p16 INK4a and SA-β-gal levels, and preserves cardiac progenitor cell activity in murine models. Exercise training shows a clinical improvement in left ventricular compliance, a reduction in blood pressure, and improved endothelial function due to elevated availability of nitric oxide in older adults (Yeo et al., 2019). Yoga and meditation are mind-body practices that have shown less research in the case of cardiac senescence, but regulate the HPA axis and lessen the signs of chronic stress. These interventions indirectly suppress SASP by reducing cortisol levels and systemic inflammation (Matacchione et al., 2021). Another area of concern that is underestimated is sleep optimization. Chronic sleep loss precipitates telomere shortening and increased levels of pro-inflammatory cytokines circulating in the blood, factors that

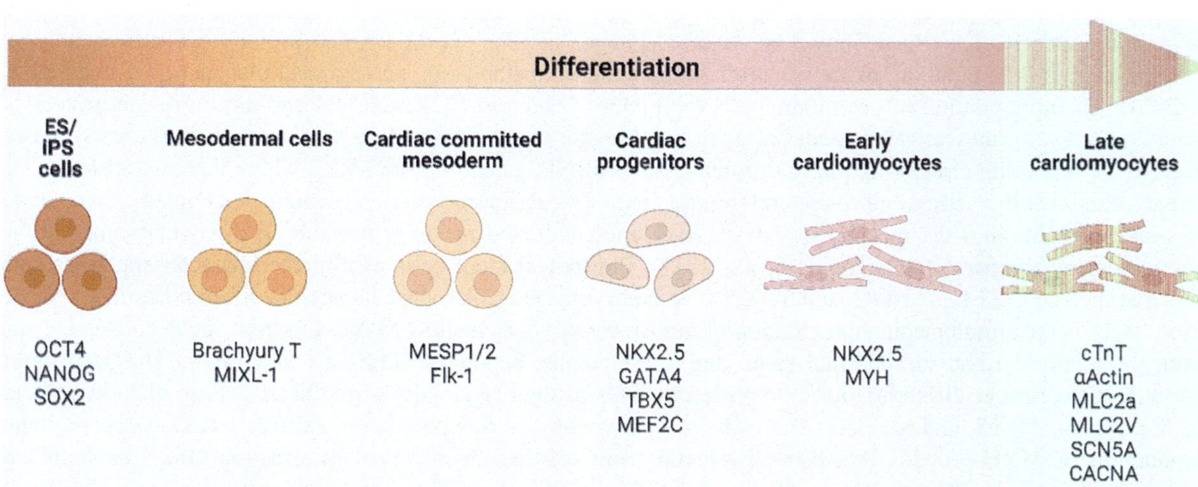

FIGURE 1.3 **Stepwise differentiation from embryonic or induced pluripotent stem cells into mature cardiomyocytes is guided by sequential expression of specific markers.** Pluripotent cells express OCT4, NANOG, and SOX2, which transition into mesodermal cells marked by Brachyury T and MIXL-1. Cardiac-committed mesoderm expresses MESP1/2 and Flk-1, followed by cardiac progenitors expressing NKX2.5, GATA4, TBX5, and MEF2C. Early cardiomyocytes continue expressing NKX2.5 and MYH, while fully differentiated cardiomyocytes exhibit contractile and electrophysiological markers such as cTnT, α-Actin, MLC2a, MLC2V, SCN5A, CACNA, and IRX4.

work synergistically in cardiovascular aging. High-quality sleep promotes glymphatic drainage and mitochondrial recovery, in addition to hormonal regulation, which embraces cardiac homeostasis (Kordinas et al., 2016).

Another factor that affects cardiac aging is healthy nutritional tactics, such as intermittent fasting and following a diet referred to as the Mediterranean diet. These diets recapitulate the effects of CR by the upregulation of sirtuins, activation of AMPK signaling, and inhibition of mTOR signaling, which are causative of longevity and tension mitigation (Testa et al., 2014). Lifestyle interventions are readily available but noninvasive pillars in the process of decelerating cardiac aging and are presently harmonizing with current pharmacological modalities.

1.8.2 Pharmacological interventions

The deployment of pharmacological approaches that target senescent cells has significantly changed the current state of antiaging studies. Senolytic agents that specifically kill senescent cells are especially promising, and their use has been shown to be effective in alleviating cardiac dysfunction in old age. Dasatinib, a tyrosine kinase inhibitor, and quercetin, a well-studied natural flavonoid, are two such compounds. This combination alleviated fibrosis in the myocardium of aged mice and enhanced diastolic activity and normalized the compliance of the left ventricle. Mechanically, the administration targets overexpressed pro-survival networks in senescent cells, inducing apoptosis in senescent cells without damaging healthy tissue (Gu et al., 2023).

Navitoclax (ABT-263) and fisetin are other senolytic drugs that target specific antiapoptotic pathways (BCL-2 and BCL-xL) and kill senescent cells by disabling their antiapoptotic modes of protection. Navitoclax has been shown to deplete the expression of p16 4 a cardiomyocytes in old hearts, but its hematological side effects do not provide a clinical opportunity. Future studies aim to develop more cardiac-selective senolytics or delivery modalities, such as nanoparticle encapsulation or cardiac-targeting ligand delivery, to increase safety and efficacy (Abdelgawad et al., 2022). Simultaneously, CRMs are metabolic strategies that resemble DR without the need for long periods of fasting. Resveratrol, metformin, and rapamycin are examples of agents that alter nutrient-sensing pathways, minimize oxidative harm, and inhibit SASP. The use of the mTOR inhibitor rapamycin has demonstrated the ability to reverse age-related myocardial hypertrophy and improve the state of autophagy in preclinical models; however, chronic administration can reduce immune function, and caution must be taken with the selection of dosing (Sorrenti et al., 2022). The popular type 2 diabetes drug metformin activates AMPK, suppresses complex I in the mitochondria, and diminishes DNA damage signaling proteins, all of which are vascular and cardiac vascular age-protective mechanisms. Clinical trials, such as the TAME study, are currently discussing their effects on senescence and revealing them in the future (Hou et al., 2018). Interventional strategies based on circulating levels of p16INK4a, telomere foci, and epigenetic clocks have been developed as biomarker-driven approaches, in which the efficacy is tracked, and the scheduling of treatment is individualized.

1.8.3 Genetic and epigenetic therapies

New methods for targeting cardiac senescence include epigenetic changes and genetic engineering. Preservation of telomeres is one of the top priorities, and telomerase reactivation, especially by TERT gene therapy, offers hope in terms of therapeutic approaches. Experimental re-expression of TERT in somatic cells is correlated with prolonged cellular senescence. TERT delivery using AAV9 to the heart was sufficient to greatly prolong survival in murine models, increase cardiac output, and reduce DNA damage in aged cardiomyocytes (Wei et al., 2024). Despite lingering safety concerns over the long term, the use of tissue-specific vectors and transient expression systems could render transgenic systems safe by potentially avoiding oncogenicity problems. The FOXO3 transcription factor is another target that regulates oxidative stress resistance, autophagy, and mitochondrial quality control. Activation of FOXO3 extends lifespan and increases cardiac resistance by increasing the expression of genes such as MnSOD, catalase, and BCL2-like proteins (He et al., 2018). Pharmacological and gene-editing interventions, such as CRISPRa-based modulation, have shown promising results in alleviating age-related decline in the myocardium (Huang et al., 2024). At the same time, SIRT1, an epigenetic pivotal regulator of chromatin remodeling and mitochondrial biogenesis, is the subject of study of an NAD+-dependent deacetylase. Activators of SIRT1 are being studied with regard to their suitability for promoting cardiac bioenergetics and inhibiting inflammation (Marin et al., 2017). The role of ncRNAs in the regulation of senescence-related processes and cardiac remodeling is also supported by recent evidence. Improved regulation of gene networks involved in cardiac aging through fine-tuning therapeutic modulation using antagomirs or RNA-based delivery systems is a possibility (Jakubik et al., 2021).

1.8.4 Stem cell-based rejuvenation approaches

In the modern antiaging field of cardiology, regeneration due to stem cell-based treatments is a revolutionary area of research that aims to regenerate myocardial cells whose lives have been destabilized. CPC and iPSC-CMs are at the forefront of these approaches. CPCs are constantly sustained throughout the adult heart, but they progressively decrease and lose the ability to regenerate as they age, in part because of p16INK4a-induced senescence. Isolated, rejuvenated, and re-injected autologous CPCs have already provided promising outcomes, together with accelerated myocardial regeneration and an antifibrotic effect, in preclinical studies (Parrotta et al., 2020). CPCs have a broad platform for use in autologous transplantation. They may be generated from patient-specific somatic cells, reprogrammed into a pluripotent form, and differentiated into functional cardiomyocytes. When found in vivo, iPSC-CMs become integrated into the host myocardium, establish new contractile activity, and improve heart failure phenotypes. To assuage fears of teratoma formation or arrhythmogenicity, recent studies have focused on direct reprogramming, where fibroblasts are reprogrammed in situ into cardiomyocyte-like cells using GMT transcription factors (Joshi et al., 2024).

The next potential intervention is exosome therapy, which is based on the use of young mesenchymal or cardiac stem cells. These extracellular vesicles carry bioactive factors, such as miRNAs, growth factors, and mitochondrial components, that can control the senescence pathway in recipient cells. Exosome therapy in aged hearts, however, has been shown to reduce oxidative stress, decrease the senescence-coupled secretory phenotype, and induce neovascularization without generating the problems of immunogenicity that directly treat transplantation (Giordani et al., 2021). In addition, tissue engineering and organoid strategies are promising. It is possible to insert bioengineered cardiac patches consisting of iPSC-CMs and endothelial cells, which can replace senescent or fibrotic myocardium. Composites with polymers, such as hydrogels or nanofibers, increase retention, vascularization, and coupling of implants to the host myocardium (Sayed et al., 2020).

1.9 Conclusion

Cardiovascular science has experienced rapid development in the past few years, which entails moving the conceptual frameworks into clinically actionable interventions against cardiac aging and senescence. Despite its disease-modifying lifestyle-based basis, which has low risk and systemic effects that can work synergistically with other therapies, new pharmacological and genetic tactics are being developed that are likely to adjust or eliminate senescent cells in cardiac tissue. In particular, both pharmacological senolytics and caloric restriction mimetics have been shown to achieve spectacular functional improvements in preclinical models of cardiac aging. Genetic and epigenetic modulation, in its part, offers the versatile specificity in targeting the senescence pathways, and stem cell-based therapy gives the hope of recovering damaged myocardium and restoring the myocardial youthful mechanical characteristics. These modalities should be planned strategically because of biomarker-based diagnostics and risk factors for patients, which would ultimately prolong cardiac health and delay age-related cardiovascular diseases.

References

Abdel-Gabbar, M., Kordy, M.G.M., 2024. Telomere-based treatment strategy of cardiovascular diseases: imagination comes to reality. Genome Instability & Disease 5 (2), 61–75. Available at https://doi.org/10.1007/s42764-024-00123-x.

Abdelgawad, I.Y., Agostinucci, K., Ismail, S.G., Grant, M.K.O., Zordoky, B.N., 2022. EA.hy926 cells and HUVECs share similar senescence phenotypes but respond differently to the senolytic drug ABT-263. Cells 11 (13). Available at https://doi.org/10.3390/cells11131992.

Adili, A., Zhu, X., Cao, H., Tang, X., Wang, Y., Wang, J., et al., 2022. Atrial Fibrillation underlies cardiomyocyte senescence and contributes to deleterious atrial remodeling during disease progression. Aging and disease 13 (1), 298. Available at https://doi.org/10.14336/ad.2021.0619.

Ajoolabady, A., Pratico, D., Vinciguerra, M., Lip, G.Y.H., Franceschi, C., Ren, J., 2023. Inflammaging: mechanisms and role in the cardiac and vasculature. Trends in Endocrinology & Metabolism 34 (6), 373–387. Available at https://doi.org/10.1016/j.tem.2023.03.005.

Al-Awar, A., Hussain, S., 2024. Interplay of reactive oxygen species (ROS) and epigenetic remodelling in cardiovascular diseases pathogenesis: a contemporary perspective. Frontiers in Bioscience - Landmark 29 (11). Available at https://doi.org/10.31083/j.fbl2911398.

Al-Regaiey, K., 2024. Crosstalk between adipogenesis and aging: role of polyphenols in combating adipogenic-associated aging. Immunity & Ageing 21 (1). Available at https://doi.org/10.1186/s12979-024-00481-w.

Alrumaihi, F., Al-Doaiss, A.A., Ullah, F., Alwanian, W.M., Alharbi, H.O., Alassaf, F.A., et al., 2025. Histone modifications as molecular drivers of cardiac aging: metabolic alterations, epigenetic mechanisms, and emerging therapeutic strategies. Current Problems in Cardiology 50 (7), 103056. Available at https://doi.org/10.1016/j.cpcardiol.2025.103056.

Augustin, H.G., Koh, G.Y., 2024. A systems view of the vascular endothelium in health and disease. Cell 187 (18), 4833–4858. Available at https://doi.org/10.1016/j.cell.2024.07.012.

Aung, N., Wang, Q., van Duijvenboden, S., Burns, R., Stoma, S., Raisi-Estabragh, Z., et al., 2023. Association of longer leukocyte telomere length with cardiac size, function, and heart failure. JAMA Cardiology 8 (9), 808. Available at https://doi.org/10.1001/jamacardio.2023.2167.

Balaraman, A.K., Altamimi, A.S.A., Babu, M.A., Goyal, K., PadmaPriya, G., Bansal, P., et al., 2025. The interplay of senescence and MMPs in myocardial infarction: implications for cardiac aging and therapeutics. Biogerontology 26 (1). Available at https://doi.org/10.1007/s10522-025-10190-6.

Bi, F., Gao, C., Guo, H., 2024. Epigenetic regulation of cardiovascular diseases induced by behavioral and environmental risk factors: mechanistic, diagnostic, and therapeutic insights. FASEB BioAdvances 6 (11), 477–502. Available at https://doi.org/10.1096/fba.2024-00080.

Bloom, S.I., Tucker, J.R., Lim, J., Thomas, T.G., Stoddard, G.J., Lesniewski, L.A., et al., 2022. Aging results in DNA damage and telomere dysfunction that is greater in endothelial versus vascular smooth muscle cells and is exacerbated in atheroprone regions. GeroScience 44 (6), 2741–2755. Available at https://doi.org/10.1007/s11357-022-00681-6.

Boccardi, V., 2025. From telomeres and senescence to integrated longevity medicine: redefining the path to extended healthspan. Biogerontology 26 (3). Available at https://doi.org/10.1007/s10522-025-10246-7.

Bouïs, D., Hospers, G.A.P., Meijer, C., Molema, G., Mulder, N.H., 2001. Endothelium in vitro: a review of human vascular endothelial cell lines for blood vessel-related research. Angiogenesis 4 (2), 91–102. Available at https://doi.org/10.1023/a:1012259529167.

Chen, J., Huang, X., Halicka, D., Brodsky, S., Avram, A., Eskander, J., et al., 2006. Contribution of p16 INK4a and p21 CIP1 pathways to induction of premature senescence of human endothelial cells: permissive role of p53. American Journal of Physiology-Heart and Circulatory Physiology 290 (4), H1575. Available at https://doi.org/10.1152/ajpheart.00364.2005.

Chen, Y., Zhu, Y., Ren, X., Ding, L., Xu, Y., Zhou, M., et al., 2025. Endothelial cell senescence in marfan syndrome: pathogenesis and therapeutic potential of TGF-β pathway inhibition. Journal of the American Heart Association 14 (9). Available at https://doi.org/10.1161/jaha.124.037826.

Cheon, S.Y., Cho, M.R., Kim, S.Y., Koo, B.-N., 2025. The immune-inflammatory responses on the hypothalamic-pituitary-adrenal axis and the neurovascular unit in perioperative neurocognitive disorder. Experimental Neurology 386, 115146. Available at https://doi.org/10.1016/j.expneurol.2025.115146.

Chistiakov, D.A., Shkurat, T.P., Melnichenko, A.A., Grechko, A.V., Orekhov, A.N., 2018. The role of mitochondrial dysfunction in cardiovascular disease: a brief review. Annals of Medicine 50 (2), 121–127. Available at https://doi.org/10.1080/07853890.2017.1417631.

Dai, Z., Ko, T., Fujita, K., Nomura, S., Uemura, Y., Onoue, K., et al., 2024. Myocardial DNA damage predicts heart failure outcome in various underlying diseases. JACC: Heart Failure 12 (4), 648–661. Available at https://doi.org/10.1016/j.jchf.2023.09.027.

Day, K., Waite, L.L., Thalacker-Mercer, A., West, A., Bamman, M.M., Brooks, J.D., et al., 2013. Differential DNA methylation with age displays both common and dynamic features across human tissues that are influenced by CpG landscape. Genome Biology 14 (9). Available at https://doi.org/10.1186/gb-2013-14-9-r102.

Del Torto, A., Guaricci, A.I., Pomarico, F., Guglielmo, M., Fusini, L., Monitillo, F., et al., 2022. Advances in multimodality cardiovascular imaging in the diagnosis of heart failure with preserved ejection fraction. Frontiers in Cardiovascular Medicine 9. Available at https://doi.org/10.3389/fcvm.2022.758975.

Desiderio, A., Pastorino, M., Campitelli, M., Longo, M., Miele, C., Napoli, R., et al., 2024. DNA methylation in cardiovascular disease and heart failure: novel prediction models? Clinical Epigenetics 16 (1). Available at https://doi.org/10.1186/s13148-024-01722-x.

Dewenter, M., von der Lieth, A., Katus, H.A., Backs, J., 2017. Calcium signaling and transcriptional regulation in cardiomyocytes. Circulation Research 121 (8), 1000–1020. Available at https://doi.org/10.1161/circresaha.117.310355.

Ding, P., Xiang, C., Yao, Q., Li, X., Zhang, J., Yin, R., et al., 2024. Aged polystyrene microplastics exposure affects apoptosis via inducing mitochondrial dysfunction and oxidative stress in early life of zebrafish. Journal of Environmental Management 367, 121995. Available at https://doi.org/10.1016/j.jenvman.2024.121995.

Ding, F., Yu, Y., Zhao, J., Wei, S., Zhang, Y., Han, J.H., et al., 2025. The interplay of cellular senescence and reprogramming shapes the biological landscape of aging and cancer revealing novel therapeutic avenues. Frontiers in Cell and Developmental Biology 13. Available at https://doi.org/10.3389/fcell.2025.1593096.

Dobrev, D., Heijman, J., Hiram, R., Li, N., Nattel, S., 2023. Inflammatory signalling in atrial cardiomyocytes: a novel unifying principle in atrial fibrillation pathophysiology. Nature Reviews Cardiology 20 (3), 145–167. Available at https://doi.org/10.1038/s41569-022-00759-w.

Dobroslavska, P., Silva, M.L., Vicente, F., Pereira, P., 2024. Mediterranean dietary pattern for healthy and active aging: a narrative review of an integrative and sustainable approach. Nutrients 16 (11), 1725. Available at https://doi.org/10.3390/nu16111725.

Dominguez, L.J., Veronese, N., Barbagallo, M., 2024. Dietary patterns and healthy or unhealthy aging. Gerontology 70 (1), 15–36. Available at https://doi.org/10.1159/000534679.

Dragoumani, K., Kletsas, D., Chrousos, G.P., Vlachakis, D., Balatsos, N.A.A., 2025. Molecular and environmental modulators of aging: interplay between inflammation, epigenetics, and RNA stability. Genes 16 (7), 796. Available at https://doi.org/10.3390/genes16070796.

Dutta, D., Calvani, R., Bernabei, R., Leeuwenburgh, C., Marzetti, E., 2012. Contribution of impaired mitochondrial autophagy to cardiac aging: mechanisms and therapeutic opportunities. Circulation Research 110 (8), 1125–1138. Available at https://doi.org/10.1161/CIRCRESAHA.111.246108.

Fernández-Veledo, S., Vázquez-Carballo, A., Vila-Bedmar, R., Ceperuelo-Mallafré, V., Vendrell, J., 2013. Role of energy- and nutrient-sensing kinases AMP-activated protein kinase (AMPK) and mammalian target of rapamycin (mTOR) in adipocyte differentiation. IUBMB Life 65 (7), 572–583. Available at https://doi.org/10.1002/iub.1170.

Ferrari, S., Pesce, M., 2021. Stiffness and aging in cardiovascular diseases: the dangerous relationship between force and senescence. International Journal of Molecular Sciences 22 (7), 3404. Available at https://doi.org/10.3390/ijms22073404.

Ferrucci, L., Fabbri, E., 2018. Inflammageing: chronic inflammation in ageing, cardiovascular disease, and frailty. Nature Reviews Cardiology 15 (9), 505–522. Available at https://doi.org/10.1038/s41569-018-0064-2.

Foglia, M.J., Poss, K.D., 2016. Building and re-building the heart by cardiomyocyte proliferation. Development 143 (5), 729–740. Available at https://doi.org/10.1242/dev.132910.

Fraticelli, A., Josephson, R., Danziger, R., Lakatta, E., Spurgeon, H., 1989. Morphological and contractile characteristics of rat cardiac myocytes from maturation to senescence. American Journal of Physiology-Heart and Circulatory Physiology 257 (1), H259. Available at https://doi.org/10.1152/ajpheart.1989.257.1.h259.

Fu, H., Zhu, Y., Lin, L., Jiang, P., Cai, G., Zeng, L., et al., 2025. Shorter leukocyte telomere length is associated with increased major adverse cardiovascular events or mortality in patients with essential hypertension. Journal of Cardiovascular Translational Research 18 (1), 198–208. Available at https://doi.org/10.1007/s12265-024-10558-y.

Gan, P., Patterson, M., Sucov, H.M., 2020. Cardiomyocyte polyploidy and implications for heart regeneration. Annual Review of Physiology 82, 45–61. Available at https://doi.org/10.1146/annurev-physiol-021119-034618.

Ghantous, C.M., Azrak, Z., Hanache, S., Abou-Kheir, W., Zeidan, A., 2015. Differential role of leptin and adiponectin in cardiovascular system. International Journal of Endocrinology 2015, 1–13. Available at https://doi.org/10.1155/2015/534320.

Giannoula, Y., Kroemer, G., Pietrocola, F., 2023. Cellular senescence and the host immune system in aging and age-related disorders. Biomedical Journal 46 (3), 100581. Available at https://doi.org/10.1016/j.bj.2023.02.001.

Giordani, C., Silvestrini, A., Giuliani, A., Olivieri, F., Rippo, M.R., 2021. MicroRNAs as factors in bidirectional crosstalk between mitochondria and the nucleus during cellular senescence. Frontiers in Physiology 12. Available at https://doi.org/10.3389/fphys.2021.734976.

Goette, A., Corradi, D., Dobrev, D., Aguinaga, L., Cabrera, J.-A., Chugh, S.S., et al., 2024. Atrial cardiomyopathy revisited—evolution of a concept: a clinical consensus statement of the European Heart Rhythm Association (EHRA) of the ESC, the Heart Rhythm Society (HRS), the Asian Pacific Heart Rhythm Society (APHRS), and the Latin American Heart Rhythm Society (LAHRS). Europace 26 (9). Available at https://doi.org/10.1093/europace/euae204.

Grande, D., Leone, M., Rizzo, C., Terlizzese, P., Parisi, G., Gioia, M., et al., 2017. A Multiparametric approach based on NT-proBNP, ST2, and Galectin3 for stratifying one year prognosis of chronic heart failure outpatients. Journal of Cardiovascular Development and Disease 4 (3), 9. Available at https://doi.org/10.3390/jcdd4030009.

Grzeczka, A., Graczyk, S., Kordowitzki, P., 2025. Involvement of TGF-β, mTOR, and inflammatory mediators in aging alterations during myxomatous mitral valve disease in a canine model. GeroScience 47 (4), 5401–5433. Available at https://doi.org/10.1007/s11357-025-01520-0.

Gu, Y., Avolio, E., Alvino, V.V., Thomas, A.C., Herman, A., Miller, P.J., et al., 2023. The tyrosine kinase inhibitor Dasatinib reduces cardiac steatosis and fibrosis in obese, type 2 diabetic mice. Cardiovascular Diabetology 22 (1). Available at https://doi.org/10.1186/s12933-023-01955-9.

Guo, D., Zhu, W., Qiu, H., 2024. CC motif chemokine ligand 2 and chemokine receptor 2 in cardiovascular and neural aging and aging-related diseases. International Journal of Molecular Sciences 25 (16), 8794.

Hamledari, H., Asghari, P., Jayousi, F., Aguirre, A., Maaref, Y., Barszczewski, T., et al., 2022. Using human induced pluripotent stem cell-derived cardiomyocytes to understand the mechanisms driving cardiomyocyte maturation. Frontiers in Cardiovascular Medicine 9. Available at https://doi.org/10.3389/fcvm.2022.967659.

He, W., Zhang, A., Qi, L., Na, C., Jiang, R., Fan, Z., et al., 2018. FOXO1, a potential therapeutic target, regulates autophagic flux, oxidative stress, mitochondrial dysfunction, and apoptosis in human cholangiocarcinoma QBC939 cells. Cellular Physiology and Biochemistry 45 (4), 1506–1514. Available at https://doi.org/10.1159/000487576.

Heiss, C., Keymel, S., Niesler, U., Ziemann, J., Kelm, M., Kalka, C., 2005. Impaired progenitor cell activity in age-related endothelial dysfunction. Journal of the American College of Cardiology 45 (9), 1441–1448. Available at https://doi.org/10.1016/j.jacc.2004.12.074.

Herbrig, K., Haensel, S., Oelschlaegel, U., Pistrosch, F., Foerster, S., Passauer, J., 2006. Endothelial dysfunction in patients with rheumatoid arthritis is associated with a reduced number and impaired function of endothelial progenitor cells. Annals of the Rheumatic Diseases 65 (2), 157–163. Available at https://doi.org/10.1136/ard.2005.035378.

Hieda, M., Sarma, S., Hearon, C.M., MacNamara, J.P., Dias, K.A., Samels, M., et al., 2021. One-year committed exercise training reverses abnormal left ventricular myocardial stiffness in patients with stage b heart failure with preserved ejection fraction. Circulation 144 (12), 934–946. Available at https://doi.org/10.1161/circulationaha.121.054117.

Horitani, K., Shiojima, I., 2024. Wnt signaling in cardiac development and heart diseases. In Vitro Cellular and Developmental Biology - Animal 60 (5), 482–488. Available at https://doi.org/10.1007/s11626-024-00917-z.

Hou, W.L., Yin, J., Alimujiang, M., Yu, X.Y., Ai, L.G., Bao, Y.Q., et al., 2018. Inhibition of mitochondrial complex I improves glucose metabolism independently of AMPK activation. Journal of Cellular and Molecular Medicine 22 (2), 1316–1328. Available at https://doi.org/10.1111/jcmm.13432.

Huang, P., Xu, J., Keepers, B., Xie, Y., Near, D., Xu, Y., et al., 2024. Direct cardiac reprogramming via combined CRISPRa-mediated endogenous Gata4 activation and exogenous Mef2c and Tbx5 expression. Molecular Therapy Nucleic Acids 35 (4), 102390. Available at https://doi.org/10.1016/j.omtn.2024.102390.

Izzo, C., Vitillo, P., Di Pietro, P., Visco, V., Strianese, A., Virtuoso, N., et al., 2021. The role of oxidative stress in cardiovascular aging and cardiovascular diseases. Life 11 (1), 60. Available at https://doi.org/10.3390/life11010060.

Jaba, T., 2022. Dasatinib and quercetin: short-term simultaneous administration yields senolytic effect in humans. Issues and Developments in Medicine and Medical Research 2, 22–31.

Jain, A., Casanova, D., Padilla, A.V., Paniagua Bojorges, A., Kotla, S., Ko, K.A., et al., 2023. Premature senescence and cardiovascular disease following cancer treatments: mechanistic insights. Frontiers in Cardiovascular Medicine 10. Available at https://doi.org/10.3389/fcvm.2023.1212174.

Jakubik, D., Fitas, A., Eyileten, C., Jarosz-Popek, J., Nowak, A., Czajka, P., et al., 2021. MicroRNAs and long non-coding RNAs in the pathophysiological processes of diabetic cardiomyopathy: emerging biomarkers and potential therapeutics. Cardiovascular Diabetology 20 (1). Available at https://doi.org/10.1186/s12933-021-01245-2.

Ji, S., Xiong, M., Chen, H., Liu, Y., Zhou, L., Hong, Y., et al., 2023. Cellular rejuvenation: molecular mechanisms and potential therapeutic interventions for diseases. Signal Transduction and Targeted Therapy 8 (1). Available at https://doi.org/10.1038/s41392-023-01343-5.

Jin, L., Hou, P., 2024. Yixin-Fumai granules modulate autophagy through the PI3K/AKT/FOXO pathway and lead to amelioration of aging mice with sick sinus syndrome. Immunity and Ageing 21 (1). Available at https://doi.org/10.1186/s12979-024-00439-y.

Joshi, J., Albers, C., Smole, N., Guo, S., Smith, S.A., 2024. Human induced pluripotent stem cell-derived cardiomyocytes (iPSC-CMs) for modeling cardiac arrhythmias: strengths, challenges and potential solutions. Frontiers in Physiology 15. Available at https://doi.org/10.3389/fphys.2024.1475152.

Joyce, B.T., Gao, T., Zheng, Y., Ma, J., Hwang, S.J., Liu, L., et al., 2021. Epigenetic age acceleration reflects long-term cardiovascular health. Circulation Research 129 (8), 770–781. Available at https://doi.org/10.1161/CIRCRESAHA.121.318965.

Kajstura, J., Gurusamy, N., Ogórek, B., Goichberg, P., Clavo-Rondon, C., Hosoda, T., et al., 2010. Myocyte turnover in the aging human heart. Circulation Research 107 (11), 1374–1386. Available at https://doi.org/10.1161/CIRCRESAHA.110.231498.

Kashyap, V., Rezende, N.C., Scotland, K.B., Shaffer, S.M., Persson, J.L., Gudas, L.J., et al., 2009. Regulation of stem cell pluripotency and differentiation involves a mutual regulatory circuit of the nanog, OCT4, and SOX2 pluripotency transcription factors with polycomb repressive complexes and stem cell microRNAs. Stem Cells and Development 18 (7), 1093–1108. Available at https://doi.org/10.1089/scd.2009.0113.

Khatiwala, R.V., Zhang, S., Li, X., Devejian, N., Bennett, E., Cai, C., 2018. Inhibition of p16INK4A to rejuvenate aging human cardiac progenitor cells via the upregulation of anti-oxidant and NFκB signal pathways. Stem Cell Reviews and Reports 14 (4), 612–625. Available at https://doi.org/10.1007/s12015-018-9815-z.

Khavinson, V., Linkova, N., Dyatlova, A., Kantemirova, R., Kozlov, K., 2023. Senescence-associated secretory phenotype of cardiovascular system cells and inflammaging: perspectives of peptide regulation. Cells 12 (1), 106. Available at https://doi.org/10.3390/cells12010106.

Kim, H.Y., Kim, H.S., 2024. Podoplanin depletion in tonsil-derived mesenchymal stem cells induces cellular senescence via regulation of the p16Ink4a/Rb pathway. Cell Communication and Signaling 22 (1). Available at https://doi.org/10.1186/s12964-024-01705-8.

Kiseleva, O.I., Arzumanian, V.A., Ikhalaynen, Y.A., Kurbatov, I.Y., Kryukova, P.A., Poverennaya, E.V., 2024. Multiomics of aging and aging-related diseases. International Journal of Molecular Sciences 25 (24), 13671. Available at https://doi.org/10.3390/ijms252413671.

Kobak, K.A., Zarzycka, W., King, C.J., Borowik, A.K., Peelor, F.F., Baehr, L.M., et al., 2025. Proteostatic imbalance drives the pathogenesis and age-related exacerbation of heart failure with preserved ejection fraction. JACC: Basic to Translational Science 10 (4), 475–497. Available at https://doi.org/10.1016/j.jacbts.2024.11.006.

Kohl, P., Gourdie, R.G., 2014. Fibroblast-myocyte electrotonic coupling: does it occur in native cardiac tissue? Journal of Molecular and Cellular Cardiology 70, 37–46. Available at https://doi.org/10.1016/j.yjmcc.2013.12.024.

Kordinas, V., Ioannidis, A., Chatzipanagiotou, S., 2016. The telomere/telomerase system in chronic inflammatory diseases. Cause or effect? Genes 7 (9). Available at https://doi.org/10.3390/genes7090060.

Korski, K.I., Kubli, D.A., Wang, B.J., Khalafalla, F.G., Monsanto, M.M., Firouzi, F., et al., 2019. Hypoxia prevents mitochondrial dysfunction and senescence in human c-Kit + cardiac progenitor cells. Stem Cells 37 (4), 555–567. Available at https://doi.org/10.1002/stem.2970.

Kumar, M., Yan, P., Kuchel, G.A., Xu, M., 2024. Cellular senescence as a targetable risk factor for cardiovascular diseases: therapeutic implications: JACC family series. JACC: Basic to Translational Science 9 (4), 522–534. Available at https://doi.org/10.1016/j.jacbts.2023.12.003.

Kumari, R., Jat, P., 2021. Mechanisms of cellular senescence: cell cycle arrest and senescence associated secretory phenotype. Frontiers in Cell and Developmental Biology 9. Available at https://doi.org/10.3389/fcell.2021.645593.

Kyryk, V., Parkhomenko, O., 2025. Potential prognostic markers of cellular senescence in age-associated cardiovascular pathology. Fiziologichnyi Zhurnal 71 (2), 93–105. Available at https://doi.org/10.15407/fz71.02.093.

Lazzarini, E., Lodrini, A.M., Arici, M., Bolis, S., Vagni, S., Panella, S., et al., 2022. Stress-induced premature senescence is associated with a prolonged QT interval and recapitulates features of cardiac aging. Theranostics 12 (11), 5237–5257. Available at https://doi.org/10.7150/thno.70884.

Leri, A., Kajstura, J., Anversa, P., 2005. Cardiac stem cells and mechanisms of myocardial regeneration. Physiological Reviews 85 (4), 1373–1416. Available at https://doi.org/10.1152/physrev.00013.2005.

Li, D., Li, Y., Ding, H., Wang, Y., Xie, Y., Zhang, X., 2023. Cellular senescence in cardiovascular diseases: from pathogenesis to therapeutic challenges. Journal of Cardiovascular Development and Disease 10 (10), 439. Available at https://doi.org/10.3390/jcdd10100439.

Li, B., Xiong, W., Zuo, W., Shi, Y., Wang, T., Chang, L., et al., 2024. Proximal telomeric decompaction due to telomere shortening drives FOXC1-dependent myocardial senescence. Nucleic Acids Research 52 (11), 6269–6284. Available at https://doi.org/10.1093/nar/gkae274.

Li, X., He, Y., Wang, D., Momeni, M.R., 2025. Chronobiological disruptions: unravelling the interplay of shift work, circadian rhythms, and vascular health in the context of stroke risk. Clinical and Experimental Medicine 25 (1). Available at https://doi.org/10.1007/s10238-024-01514-w.

Lin, R., Kerkelä, R., 2020. Regulatory mechanisms of mitochondrial function and cardiac aging. International Journal of Molecular Sciences 21 (4), 1359. Available at https://doi.org/10.3390/ijms21041359.

Lind, L., Ingelsson, E., Sundström, J., Siegbahn, A., Lampa, E., 2018. Methylation-based estimated biological age and cardiovascular disease. European Journal of Clinical Investigation 48 (2). Available at https://doi.org/10.1111/eci.12872.

Liu, C., Lin, K., Xie, Z., Li, D., Fan, J., Chen, Y., et al., 2024. Telomerase reverse transcriptase regulates intracellular Ca2+ homeostasis and mitochondrial function via the p53/PGC-1α pathway in HL-1 Cells. Frontiers in Bioscience-Landmark 29 (7). Available at https://doi.org/10.31083/j.fbl2907263.

Liu, Y., Fan, Y., Li, J., Chen, M., Chen, A., Yang, D., et al., 2021. Combination of LCZ696 and ACEI further improves heart failure and myocardial fibrosis after acute myocardial infarction in mice. Biomedicine & Pharmacotherapy 133, 110824. Available at https://doi.org/10.1016/j.biopha.2020.110824.

Lv, L., Chen, Q.Y., Lu, J., Zhao, Q., Wang, H.Y., Li, J.H., et al., 2024. Potential regulatory role of epigenetic modifications in aging-related heart failure. International Journal of Cardiology 401, 131858. Available at https://doi.org/10.1016/j.ijcard.2024.131858.

Ma, F., Li, Y., Jia, L., Han, Y., Cheng, J., Li, H., et al., 2012. Macrophage-stimulated cardiac fibroblast production of IL-6 is essential for TGF β/Smad activation and cardiac fibrosis induced by angiotensin II. PLoS ONE 7 (5), e35144. Available at https://doi.org/10.1371/journal.pone.0035144.

Marin, T.L., Gongol, B., Zhang, F., Martin, M., Johnson, D.A., Xiao, H., et al., 2017. AMPK promotes mitochondrial biogenesis and function by phosphorylating the epigenetic factors DNMT1, RBBP7, and HAT1. Science Signaling 10 (464). Available at https://doi.org/10.1126/scisignal.aaf7478.

Marino, F., Scalise, M., Cappetta, D., Salerno, L., Molinaro, C., Salerno, N., et al., 2021. Cellular Senescence and Senescence-Associated Secretory Phenotype Drive Multipotent Cardiac Stem/Progenitor Cell Dysfunction in Human Diabetic Cardiomyopathy Independently of Age. Research Square.

Marino, F., Scalise, M., Salerno, N., Salerno, L., Molinaro, C., Cappetta, D., et al., 2022. Diabetes-induced cellular senescence and senescence-associated secretory phenotype impair cardiac regeneration and function independently of age. Diabetes 71 (5), 1081–1098. Available at https://doi.org/10.2337/db21-0536.

Matacchione, G., Gurău, F., Silvestrini, A., Tiboni, M., Mancini, L., Valli, D., et al., 2021. Anti-SASP and anti-inflammatory activity of resveratrol, curcumin and β-caryophyllene association on human endothelial and monocytic cells. Biogerontology 22 (3), 297–313. Available at https://doi.org/10.1007/s10522-021-09915-0.

Matacchione, G., Piacenza, F., Pimpini, L., Rosati, Y., Marcozzi, S., 2024. The role of the gut microbiota in the onset and progression of heart failure: insights into epigenetic mechanisms and aging. Clinical Epigenetics 16 (1). Available at https://doi.org/10.1186/s13148-024-01786-9.

Mazgelytė, E., Černiauskas, L., Burokienė, N., Karčiauskaitė, D., 2024. Association between circulating ICAM-1 and VCAM-1 levels and cardiometabolic risk factors in a group of middle-aged and elderly women: a cross-sectional study. Atherosclerosis 395, 117639. Available at https://doi.org/10.1016/j.atherosclerosis.2024.117639.

Mbiydzenyuy, N.E., Qulu, L.A., 2024. Stress, hypothalamic-pituitary-adrenal axis, hypothalamic-pituitary-gonadal axis, and aggression. Metabolic Brain Disease 39 (8), 1613–1636. Available at https://doi.org/10.1007/s11011-024-01393-w.

Meador, B.M., Krzyszton, C.P., Johnson, R.W., Huey, K.A., 2008. Effects of IL-10 and age on IL-6, IL-1β, and TNF-α responses in mouse skeletal and cardiac muscle to an acute inflammatory insult. Journal of Applied Physiology 104 (4), 991–997. Available at https://doi.org/10.1152/japplphysiol.01079.2007.

Merighi, A., Gionchiglia, N., Granato, A., Lossi, L., 2021. The phosphorylated form of the histone H2AX (γH2AX) in the brain from embryonic life to old age. Molecules 26 (23), 7198. Available at https://doi.org/10.3390/molecules26237198.

Morales, C.R., Pedrozo, Z., Lavandero, S., Hill, J.A., 2014. Oxidative stress and autophagy in cardiovascular homeostasis. Antioxidants and Redox Signaling 20 (3), 507–518. Available at https://doi.org/10.1089/ars.2013.5359.

Mujica, V., Urzúa, A., Leiva, E., Díaz, N., Moore-Carrasco, R., Vásquez, M., et al., 2010. Intervention with education and exercise reverses the metabolic syndrome in adults. Journal of the American Society of Hypertension 4 (3), 148–153. Available at https://doi.org/10.1016/j.jash.2010.02.008.

Nagane, M., Yasui, H., Kuppusamy, P., Yamashita, T., Inanami, O., 2021. DNA damage response in vascular endothelial senescence: implication for radiation-induced cardiovascular diseases. Journal of Radiation Research 62 (4), 564–573. Available at https://doi.org/10.1093/jrr/rrab032.

Nair, N., 2020. Epidemiology and pathogenesis of heart failure with preserved ejection fraction. Reviews in Cardiovascular Medicine 21 (4), 531–540. Available at https://doi.org/10.31083/J.RCM.2020.04.154.

Oh, K.S., Febres-Aldana, C.A., Kuritzky, N., Ujueta, F., Arenas, I.A., Sriganeshan, V., et al., 2021. Cellular senescence evaluated by P16INK4a immunohistochemistry is a prevalent phenomenon in advanced calcific aortic valve disease. Cardiovascular Pathology 52, 107318. Available at https://doi.org/10.1016/j.carpath.2021.107318.

Pandolfi, A., Di Pietro, N., Sirolli, V., Giardinelli, A., Di Silvestre, S., Amoroso, L., et al., 2007. Mechanisms of uremic erythrocyte-induced adhesion of human monocytes to cultured endothelial cells. Journal of Cellular Physiology 213 (3), 699–709. Available at https://doi.org/10.1002/jcp.21138.

Papanicolaou, K.N., Ngoh, G.A., Dabkowski, E.R., O'Connell, K.A., Ribeiro, R.F., Stanley, W.C., et al., 2012. Cardiomyocyte deletion of mitofusin-1 leads to mitochondrial fragmentation and improves tolerance to ROS-induced mitochondrial dysfunction and cell death. American Journal of Physiology - Heart and Circulatory Physiology 302 (1), H167. Available at https://doi.org/10.1152/ajpheart.00833.2011.

Parrotta, E.I., Lucchino, V., Scaramuzzino, L., Scalise, S., Cuda, G., 2020. Modeling cardiac disease mechanisms using induced pluripotent stem cell-derived cardiomyocytes: progress, promises and challenges. International Journal of Molecular Sciences 21 (12), 1–30. Available at https://doi.org/10.3390/ijms21124354.

Periasamy, M., Huke, S., 2001. SERCA pump level is a critical determinant of Ca2+ homeostasis and cardiac contractility. Journal of Molecular and Cellular Cardiology 33 (6), 1053–1063. Available at https://doi.org/10.1006/jmcc.2001.1366.

Qian, L., Zhu, Y., Deng, C., Liang, Z., Chen, J., Chen, Y., et al., 2024. Peroxisome proliferator-activated receptor gamma coactivator-1 (PGC-1) family in physiological and pathophysiological process and diseases. Signal Transduction and Targeted Therapy 9 (1). Available at https://doi.org/10.1038/s41392-024-01756-w.

Rao, C., Prodromakis, T., Kolker, L., Chaudhry, U.A.R., Trantidou, T., Sridhar, A., et al., 2013. The effect of microgrooved culture substrates on calcium cycling of cardiac myocytes derived from human induced pluripotent stem cells. Biomaterials 34 (10), 2399–2411. Available at https://doi.org/10.1016/j.biomaterials.2012.11.055.

Redgrave, R.E., Dookun, E., Booth, L.K., Camacho Encina, M., Folaranmi, O., Tual-Chalot, S., et al., 2023. Senescent cardiomyocytes contribute to cardiac dysfunction following myocardial infarction. npj Aging 9 (1). Available at https://doi.org/10.1038/s41514-023-00113-5.

Romaine, S.P.R., Denniff, M., Codd, V., Nath, M., Koekemoer, A., Anker, S.D., et al., 2022. Telomere length is independently associated with all-cause mortality in chronic heart failure. Heart 108 (2), 124–129. Available at https://doi.org/10.1136/heartjnl-2020-318654.

Rota, M., Goichberg, P., Anversa, P., Leri, A., 2015. Aging effects on cardiac progenitor cell physiology. Comprehensive Physiology 5 (4), 1775–1814. Available at https://doi.org/10.1002/cphy.c140082.

Rottenberg, H., Hoek, J.B., 2017. The path from mitochondrial ROS to aging runs through the mitochondrial permeability transition pore. Aging Cell 16 (5), 943–955. Available at https://doi.org/10.1111/acel.12650.

Ruiz-Meana, M., Bou-Teen, D., Ferdinandy, P., Gyongyosi, M., Pesce, M., Perrino, C., et al., 2020. Cardiomyocyte ageing and cardioprotection: consensus document from the ESC working groups cell biology of the heart and myocardial function. Cardiovascular Research 116 (11), 1835–1849. Available at https://doi.org/10.1093/cvr/cvaa132.

Ryoo, S.-W., Choi, B.-Y., Son, S.-Y., Lee, J.-H., Min, J.-Y., Min, K.-B., 2025. Lead and cadmium exposure was associated with faster epigenetic aging in a representative sample of adults aged 50 and older in the United States. Chemosphere 374, 144194. Available at https://doi.org/10.1016/j.chemosphere.2025.144194.

Salih, A.M., Galazzo, I.B., Menegaz, G., Altmann, A., 2024. Leukocyte telomere length and cardiac structure and function: a mendelian randomization study. Journal of the American Heart Association 13 (3). Available at https://doi.org/10.1161/JAHA.123.032708.

Salminen, A., 2024. The role of the immunosuppressive PD-1/PD-L1 checkpoint pathway in the aging process and age-related diseases. Journal of Molecular Medicine 102 (6), 733–750. Available at https://doi.org/10.1007/s00109-024-02444-6.

Saputro, R.E., Chou, C.-C., Lin, Y.-Y., Tarumi, T., Liao, Y.-H., 2025. Exercise-mediated modulation of autonomic nervous system and inflammatory response in sleep-deprived individuals: a narrative reviews of implications for cardiovascular health. Autonomic Neuroscience 259, 103256. Available at https://doi.org/10.1016/j.autneu.2025.103256.

Sayed, N., Liu, C., Ameen, M., Himmati, F., Zhang, J.Z., Khanamiri, S., et al., 2020. Clinical trial in a dish using iPSCs shows lovastatin improves endothelial dysfunction and cellular cross-talk in LMNA cardiomyopathy. Science Translational Medicine 12 (554). Available at https://doi.org/10.1126/scitranslmed.aax9276.

Selman, M., Pardo, A., 2021. Fibroageing: an ageing pathological feature driven by dysregulated extracellular matrix-cell mechanobiology. Ageing Research Reviews 70, 101393. Available at https://doi.org/10.1016/j.arr.2021.101393.

Shi, L., Tan, Y., Zheng, W., Cao, G., Zhou, H., Li, P., et al., 2024. CTRP3 alleviates mitochondrial dysfunction and oxidative stress injury in pathological cardiac hypertrophy by activating UPRmt via the SIRT1/ATF5 axis. Cell Death Discovery 10 (1). Available at https://doi.org/10.1038/s41420-024-01813-x.

Song, Y., Spurlock, B., Liu, J., Qian, L., 2024. Cardiac aging in the multi-omics era: high-throughput sequencing insights. Cells 13 (20), 1683. Available at https://doi.org/10.3390/cells13201683.

Sorokin, L., 2010. The impact of the extracellular matrix on inflammation. Nature Reviews Immunology 10 (10), 712–723. Available at https://doi.org/10.1038/nri2852.

Sorrenti, V., Benedetti, F., Buriani, A., Fortinguerra, S., Caudullo, G., Davinelli, S., et al., 2022. Immunomodulatory and antiaging mechanisms of resveratrol, rapamycin, and metformin: focus on mTOR and AMPK signaling networks. Pharmaceuticals 15 (8), 912. Available at https://doi.org/10.3390/ph15080912.

Spanakis, M., Tsatsakis, A., 2025. Telomeres, telomerase, and shelterin complex: promising drug targets for longevity, age-related diseases, and cancer. In: Tsatsakis, A. (Ed.), Telomeres: Biomarkers of a Healthy Life and Successful Aging. Jenny Stanford Publishing, New York, pp. 645–667.

Spano, L., Marie-Claire, C., Godin, O., Lebras, A., Courtin, C., Laplanche, J.-L., et al., 2024. Decreased telomere length in a subgroup of young individuals with bipolar disorders: replication in the FACE-BD cohort and association with the shelterin component POT1. Translational Psychiatry 14 (1). Available at https://doi.org/10.1038/s41398-024-02824-z.

Tannous, P., Zhu, H., Nemchenko, A., Berry, J.M., Johnstone, J.L., Shelton, J.M., et al., 2008. Intracellular protein aggregation is a proximal trigger of cardiomyocyte autophagy. Circulation 117 (24), 3070–3078. Available at https://doi.org/10.1161/circulationaha.107.763870.

Tariq, U., Gupta, M., Pathak, S., Patil, R., Dohare, A., Misra, S.K., 2022. Role of biomaterials in cardiac repair and regeneration: therapeutic intervention for myocardial infarction. ACS Biomaterials Science & Engineering 8 (8), 3271–3298. Available at https://doi.org/10.1021/acsbiomaterials.2c00454.

Testa, G., Biasi, F., Poli, G., Chiarpotto, E., 2014. Calorie restriction and dietary restriction mimetics: a strategy for improving healthy aging and longevity. Current Pharmaceutical Design 20 (18), 2950–2977. Available at https://doi.org/10.2174/13816128113196660699.

van Meer, B.J., Krotenberg, A., Sala, L., Davis, R.P., Eschenhagen, T., Denning, C., et al., 2019. Simultaneous measurement of excitation-contraction coupling parameters identifies mechanisms underlying contractile responses of hiPSC-derived cardiomyocytes. Nature Communications 10 (1). Available at https://doi.org/10.1038/s41467-019-12354-8.

Vasquez, C., Morley, G.E., 2012. The origin and arrhythmogenic potential of fibroblasts in cardiac disease. Journal of Cardiovascular Translational Research 5 (6), 760–767. Available at https://doi.org/10.1007/s12265-012-9408-1.

Wang, C., Wang, X., 2015. The interplay between autophagy and the ubiquitin-proteasome system in cardiac proteotoxicity. Biochimica et Biophysica Acta - Molecular Basis of Disease 1852 (2), 188–194. Available at https://doi.org/10.1016/j.bbadis.2014.07.028.

Wang, X., Yan, X., Zhang, J., Pan, S., Li, R., Cheng, L., et al., 2024. Associations of healthy eating patterns with biological aging: national health and nutrition examination survey (NHANES) 1999–2018. Nutrition Journal 23 (1). Available at https://doi.org/10.1186/s12937-024-01017-0.

Wei, X., Zhou, Y., Shao, E., Shi, X., Han, Y., Zhang, Y., et al., 2024. Tert promotes cardiac regenerative repair after MI through alleviating ROS-induced DNA damage response in cardiomyocyte. Cell Death Discovery 10 (1). Available at https://doi.org/10.1038/s41420-024-02135-8.

Wei, Y., Wang, Y., Jiang, X., Sun, Y., Wang, M., Wang, Z., et al., 2025. Xanthones explore the mechanism of p53/p21 signaling pathway to prevent cardiac aging and epigenetic regulation of Nrf2 gene. Archives of Gerontology and Geriatrics 131, 105759. Available at https://doi.org/10.1016/j.archger.2025.105759.

Werbner, B., Tavakoli-Rouzbehani, O.M., Fatahian, A.N., Boudina, S., 2023. The dynamic interplay between cardiac mitochondrial health and myocardial structural remodeling in metabolic heart disease, aging, and heart failure. Journal of Cardiovascular Aging 3 (1). Available at https://doi.org/10.20517/jca.2022.42.

Wong, J.J., Hong, R., Teo, L.L.Y., Tan, R.-S., Koh, A.S., 2024. Atherosclerotic cardiovascular disease in aging and the role of advanced cardiovascular imaging. npj Cardiovascular Health 1 (1). Available at https://doi.org/10.1038/s44325-024-00012-y.

Wu, L., Sowers, J.R., Zhang, Y., Ren, J., 2023. Targeting DNA damage response in cardiovascular diseases: from pathophysiology to therapeutic implications. Cardiovascular Research 119 (3), 691–709. Available at https://doi.org/10.1093/cvr/cvac080.

Wu, Z., Qu, J., Liu, G.H., 2024. Roles of chromatin and genome instability in cellular senescence and their relevance to ageing and related diseases. Nature Reviews Molecular Cell Biology 25 (12), 979–1000. Available at https://doi.org/10.1038/s41580-024-00775-3.

Xie, J., Qian, T., Lin, P., Gao, H., Xu, C., Xu, L., 2025. Accelerometer-measured circadian alignment predicts all-cause and cardiovascular mortality in middle-aged and older adults. European Journal of Preventive Cardiology. Available at https://doi.org/10.1093/eurjpc/zwaf445.

Xu, Y., Zheng, Z., Pan, H., Zhao, M., Zhang, J., Peng, S., et al., 2024. Kielin/chordin-like protein deficiency aggravates pressure overload-induced cardiac dysfunction and remodeling via P53 / P21 / CCNB1 signaling in mice. The FASEB Journal 38 (5). Available at https://doi.org/10.1096/fj.202301841r.

Yan, C., Xu, Z., Huang, W., 2021. Cellular senescence affects cardiac regeneration and repair in ischemic heart disease. Aging and Disease 12 (2), 552. Available at https://doi.org/10.14336/ad.2020.0811.

Yao, X., Zhang, J., Zhang, X., Jiang, T., Zhang, Y., Dai, F., et al., 2023. Age at diagnosis, diabetes duration and the risk of cardiovascular disease in patients with diabetes mellitus: a cross-sectional study. Frontiers in Endocrinology 14. Available at https://doi.org/10.3389/fendo.2023.1131395.

Yeo, D., Kang, C., Gomez-Cabrera, M.C., Vina, J., Ji, L.L., 2019. Intensified mitophagy in skeletal muscle with aging is downregulated by PGC-1alpha overexpression in vivo. Free Radical Biology and Medicine 130, 361–368. Available at https://doi.org/10.1016/j.freeradbiomed.2018.10.456.

Yılmaz, Ş.G., Bozkurt, H., 2024. The expression of shelterin genes and telomere repeat analysis and their effect on Alzheimer's disease. Molecular Biology Reports 51 (1). Available at https://doi.org/10.1007/s11033-024-10063-0.

Young, M.J., Heanue, S., Kanki, M., Moneghetti, K.J., 2025. Circadian disruption and its impact on the cardiovascular system. Trends in Endocrinology and Metabolism 36 (9), 842–854. Available at https://doi.org/10.1016/j.tem.2024.11.010.

Yue, L., Xie, J., Nattel, S., 2011. Molecular determinants of cardiac fibroblast electrical function and therapeutic implications for atrial fibrillation. Cardiovascular Research 89 (4), 744–753. Available at https://doi.org/10.1093/cvr/cvq329.

Zarzycka, W., Kobak, K.A., King, C.J., Peelor, F.F., Miller, B.F., Chiao, Y.A., 2025. Hyperactive mTORC1/4EBP1 signaling dysregulates proteostasis and accelerates cardiac aging. GeroScience 47 (2), 1823–1836. Available at https://doi.org/10.1007/s11357-024-01368-w.

Zglinicki, T., Martin-Ruiz, C., 2005. Telomeres as biomarkers for ageing and age-related diseases. Current Molecular Medicine 5 (2), 197–203. Available at https://doi.org/10.2174/1566524053586545.

Zhang, J., Zhao, Y., Gong, N., 2024a. Endoplasmic reticulum stress signaling modulates ischemia/reperfusion injury in the aged heart by regulating mitochondrial maintenance. Molecular Medicine 30 (1). Available at https://doi.org/10.1186/s10020-024-00869-w.

Zhang, M., et al., 2024c. Inter-and intracellular mitochondrial communication: signaling hubs in aging and age-related diseases. Cellular & Molecular Biology Letters 29 (1), 153.

Zhang, T., Huang, Y., Ji, X., Wu, T., Xiao, P., 2025. CCL11 (Eotaxin) promotes the advancement of aging-related cardiovascular diseases. Reviews in Cardiovascular Medicine 26 (2). Available at https://doi.org/10.31083/RCM26020.

Zhao, S., Zhang, Y., Zhao, Y., Lu, X., 2025. Cellular senescence as a key player in chronic heart failure pathogenesis: unraveling mechanisms and therapeutic opportunities. Progress in Biophysics and Molecular Biology 196, 8–18. Available at https://doi.org/10.1016/j.pbiomolbio.2025.02.002.

Zhou, Y., Yan, J., Qiao, L., Zeng, J., Cao, F., Sheng, X., et al., 2024. Bone marrow mesenchymal stem cell-derived exosomes ameliorate aging-induced BTB impairment in porcine testes by activating autophagy and inhibiting ROS/NLRP3 inflammasomes via the AMPK/mTOR signaling pathway. Antioxidants 13 (2), 183. Available at https://doi.org/10.3390/antiox13020183.

Zhu, X., Ding, L., Zhang, X., Wang, H., Chen, N., 2024. Association between physical frailty, circadian syndrome and cardiovascular disease among middle-aged and older adults: a longitudinal study. BMC Geriatrics 24 (1). Available at https://doi.org/10.1186/s12877-024-04787-8.

Chapter 2

Cardiac aging beyond senescence: structural, functional, and physiological changes

Chintan Aundhia[1], Riya Mishra[2], Surya Nath Pandey[3], Santosh Kumar Singh[4], Neelam Singla[5], Rahul Sharma[6], Gaurav Gupta[7,8], Rajiv Dahiya[9], and Sunita Dahiya[10]

[1]*Department of Pharmacy, Sumandeep Vidyapeeth Deemed to be University, Vadodara, Gujarat, India,* [2]*Department of Pharmacy, Faculty of Pharmaceutical Science and Nursing, Vivekananda Global University, Jaipur, Rajasthan, India,* [3]*Department of Pharmacology, Teerthanker Mahaveer College of Pharmacy, Teerthanker Mahaveer University, Moradabad, Uttar Pradesh, India,* [4]*ICFAI School of Pharmaceutical Sciences, The ICFAI University, Jaipur, Rajasthan, India,* [5]*School of Pharmacy, Suresh Gyan Vihar University, Jaipur, Rajasthan, India,* [6]*School of Pharmaceutical Sciences, Jaipur National University, Jagatpura, Jaipur, Rajasthan, India,* [7]*Centre for Research Impact & Outcome, Chitkara College of Pharmacy, Chitkara University, Rajpura, Punjab, India,* [8]*Centre of Medical and Bio-Allied Health Sciences Research, Ajman University, Ajman, United Arab Emirates,* [9]*School of Pharmacy, Faculty of Medical Sciences, The University of the West Indies, St. Augustine, West Indies, Trinidad and Tobago,* [10]*Department of Pharmaceutical Sciences, School of Pharmacy, Medical Sciences Campus, University of Puerto Rico, San Juan, PR, United States*

2.1 Introduction

The clinical impact of cardiac aging is growing faster worldwide because longevity is increasing, multimorbidity is on the increase, and its clinical presentation is not limited to cellular senescence but to coupled structural, functional, and physiological changes in the myocardium, conduction tissue, microvasculature, and systemic milieu. Cardiovascular disease (CVD) is the most common cause of mortality worldwide, with modern estimates showing approximately 20.5 million deaths annually due to CVD and expected future increases in absolute rates with population aging, although age-standardized rates may become increasingly flat or even decrease in many regions of the world. This is the case in India, where the elderly population is over (Lazzeroni et al., 2022; Subramanian et al., 2025). The population aged ≥ 60 years has surpassed 150 million, with an unabated rate of increase to at least one-fifth of the total Indian population by the mid-century. CVD contributes to approximately one-third of deaths in the country, and the age-adjusted mortality in both sexes is very high. Against this background, the prevalence of heart failure (HF) has steadily increased worldwide (64 million people), accompanied by a progressive transition of HF to preserved ejection fraction (HFpEF) in older adults, a condition characterized by diastolic dysfunction, atrioventricular-vascular stiffening, and systemic inflammation (Jan et al., 2024; Kalra et al., 2023). These demographic changes require a paradigm that transcends senescence to incorporate multi-scale biology, which can be quantified and manipulated by clinicians. Within this integrative lens, cardiomyocyte hypertrophy is coupled with extracellular matrix deposition and microvascular rarefaction in structural remodeling, and yields concentric remodeling and less reserve decades before overt failure; functional changes that impair lusitropy, disrupt calcium handling, and degenerate the conduction system result in age-related arrhythmias and HFpEF; and systems physiology signatures such as autonomic imbalance, metabolic inflexibility, and low-grade inflammaging (inflammaging) (Owens et al., 2021; Puspitasari et al., 2022). Simultaneously, recent HFpEF data show significant mortality and rehospitalization rates in the first year, which underline the necessity of conducting assessments in diastole earlier, managing comorbidities, and identifying multimodal risks. Mechanistically, the converging processes of endothelial microvascular aging and dysfunction of mitochondria and mitophagy, failure of proteostasis through autophagy–lysosome and ubiquitin-proteasome systems, and epigenetic/chromatin remodeling drive the phenotype measured by imaging (strain echocardiography, CMR tissue characterization, PET flow reserve) and biomarkers (NT-proBNP, GDF15, selected microRNAs), and single-cell and spatial omics. In this chapter, we thus describe cardiac aging on three axes structural, functional, and physiological explicitly with links between cellular and

Cellular Senescence and Cardiac Aging. DOI: https://doi.org/10.1016/B978-0-443-45811-8.00006-4

tissue processes and noninvasive readouts on one hand, and noninvasive therapies that go beyond senolytics, such as mitochondrial/metabolic modulators, antifibrotics, lifestyle and autonomic therapies, and regenerative or extracellular vesicle therapies, on the other (Dunlay et al., 2017; Meschiari et al., 2017). This chapter is novel in the sense that it operationalizes beyond senescence as a clinically actionable scaffold, which provides a unification of mechanistic biology with phenotype measures and pragmatic intervention measures in the spectrum between risk and overt disease. To equip clinicians and researchers to address the growing older population, this chapter seeks to equip them with an evidence-based map with clear explanations on how aging alters the structure, function, and systems physiology of the heart. It also incorporates patient stratification, prioritization of therapeutic interventions, and endpoints of trials with biomarkers and imaging (Di Micco et al., 2021; Inci et al., 2022).

2.2 Clinical phenotype and epidemiology of cardiac aging

2.2.1 Age-related HFpEF, conduction disease, and arrhythmia burden

The most common clinical manifestations of cardiac aging are age-related heart failure with preserved ejection fraction (HFpEF), conduction disease, and arrhythmia burden. They all result as structural remodeling, diastolic stiffness, and electrophysiological slowing that are decades in the making. Cohort studies have demonstrated an increased prevalence of HFpEF in the elderly, symptoms induced by deficient lusitropy, left atrial hypertension, abnormalities of pulmonary vascular coupling and exercise intolerability, and conduction tissue fibrosis and ion-channel remodeling, resulting in sinus node dysfunction, atrioventricular delay, and bundle-branch block, which lead to higher pacemaker implantation (Howie et al., 2025). Atrial fibrillation is mainly predominated by atrial dilation, inflammation, and fibrosis, and ventricular ectopy is an indication of diffuse scarring and calcium impaired calcium handling. The multimorbidity of hypertension, diabetes, chronic kidney disease, and obesity increases the risk; each of them increases the microvascular rarefaction and metabolic inflexibility. Natriuretic peptides, echocardiographic diastolic indices and strain, and cardiac MRI tissue characterization are all incorporated in the process of diagnosis, which is supplemented by rhythm monitoring to measure AF burden and brady-tachy syndromes (Karakasis et al., 2025). Investigations include management of diuretics, comorbidity, personalized exercise, and guideline-based cardiovascular prevention, and device therapy of bradyarrhythmia and cautious anticoagulation of AF. All of these phenotypes cooperate to cause hospitalization, frailty, and diminished quality of life, as well as give quantifiable measures of stratified care. Innovations in biomarker-imaging algorithms, rhythm interventions, and beyond-senolytic interventions to alter fibrosis, metabolism, and autonomic tone in aging will be improved in future studies (Camm et al., 2023).

2.2.2 Population trends and multimorbidity patterns

Population aging is changing the demographic environment worldwide and increasing the clinical significance of cardiac aging as an increasing number of individuals live into old age. As populations age, cardiovascular disease has become the leading cause of death and is causing an increasing number of chronic heart diseases that are frequently comorbid with other chronic diseases. These demographic and pathological changes have resulted in rapidly increasing patterns of multimorbidity. Systematic reviews indicate that approximately one-third of the adult population has at least two chronic diseases, and the prevalence increases exponentially with age and tends to be concentrated around cardiovascular, metabolic, respiratory, and neurodegenerative disorders (Wang et al., 2025a; Zhao et al., 2024). The parallel forces of aging, urbanization, and epidemiological transition in India have promoted a significant increase in multimorbidity in middle-aged and older adults, with scoping analyses estimating a prevalence of approximately one-fifth of the population and showing marked sociodemographic gradients in disease clustering. Identified trends highlight the fact that multimorbidity alters the clinical phenotype of cardiac aging by enhancing frailty, complicating diagnosis and guideline implementation, amplifying the risks of polypharmacy, and emphasizing health systems that are built around individual disease care. Therefore prevention and management need to shift toward integrated, age-sensitive cardiovascular care, risk-factor-based interventions that focus on common pathways, and research models that consider dual morbidity, functional progressions, and equitable provision in high-burden contexts. This reorientation of strategies is necessary for the health of the population (Puri et al., 2025; Yogesh et al., 2024).

2.2.3 Distinguishing normative aging from early pathology

The most effective way to draw a line between normative aging and early pathology is to separate those changes that are adaptive and proportionate, and those changes that are predictive of disease. The task is clinically dependent on the integration of the subtle structural, functional, and physiological signals prior to the fixation of the symptoms.

Light concentric remodeling and maintained ejection fraction, slight increases in passive stiffness, and slight decreases in global longitudinal strain without visible scarring or edema are all features of normative aging (Ferrari et al., 2003; Salvioli et al., 2023). Disproportionate left heart expansion, E/e ratio, extracellular volume increase on heart magnetic resonance, or focal late gadolinium enhancement, which suggests fibrosis, all create early pathology. interpretation is useful to interpret biomarkers: When these levels increase with stable weight and blood pressure, then it is likely an increasing diastolic load, but low-to-moderate natriuretic peptides can represent a normal age. Preclinical HFpEF is guided by exercise tests, which demonstrate abnormal blood pressure recovery or rapid changes in filling pressures. Ambulatory rhythm monitoring distinguishes between atrial remodeling due to an event of atrial fibrillation and age-related ectopy (Aquaro et al., 2023; Chan et al., 2024). Microvascular perfusion deficits and low flow reserve are signs of a tendency toward ischemic damage in vascular aging. One of the practical ways to track the change in adaptation to illness is baseline phenotyping, comorbidity auditing, and reevaluation on a routine basis. The combination of all these strategies helps reduce the missed preventive opportunities and adds the degree of follow-up to clinical outcomes and risk perception. Age-adjusted standards should be enhanced, biomarker-imaging methods need to be standardized, and personalized monitoring plans should be established in various mature populations (Hung et al., 2016).

2.3 Structural remodeling of the aging heart

2.3.1 Cardiomyocyte hypertrophy and myofibrillar changes

As cardiac stress increases with age and the number of cardiac contractions accumulates, the compensatory hypertrophy of cardiomyocytes initially normalizes cardiomyocyte contractile output but causes changes in cell architecture over time. Ordered addition of sarcomeres in parallel with hypertrophic growth is accompanied by a shift in myosin heavy chain and troponin isoform expression and an accelerated myofibrillar protein turnover that alters contractile kinetics and force production. Simultaneously, posttranslational changes in proteins, such as oxidation, glycation, and modified phosphorylation, compromise sarcomeric protein activity and facilitate myofibrillar disorganization. Further breakdown of coordinated contraction is caused by the disruption of excitation–contraction coupling by dyadic remodeling and defects in calcium handling, as well as mitochondrial dysfunction and impaired proteostasis, which leads to structural derangement (Haque and Wang, 2017; Weisser-Thomas et al., 2007). Senescent fibroblast intercellular signaling and immune cell intercellular signaling enhance maladaptive remodeling by facilitating extracellular matrix deposition and mechanochemical mismatch between hypertrophic myocytes and their matrix. These changes in structure and molecules augment the stiffness of the myocardium and predispose it to diastolic dysfunction, arrhythmia, and diminished stress-induced reserve. The identification of this progressive cascade informs biomarker development and imaging interventions to detect early myofibrillar injury and contextualizes therapeutics directed at restoring proteostasis and mitochondrial quality, as well as guiding therapeutic delivery to ameliorate the situation between adaptive hypertrophy and heart failure and improve the long-term clinical outcomes of patients (Howes et al., 2024).

2.3.2 Extracellular matrix remodeling and interstitial fibrosis

Aging heart extracellular matrix remodeling and interstitial fibrosis develop on top of the initial changes in matrix production and matrix turnover to an open fibrotic expansion undermining mechanical and electrical activity, which begins with the activation of fibroblasts and differentiation of myofibroblasts, which increases fibrillar collagen (type I and III) and matricellular protein synthesis and promotes enzymatic crosslinking via lysyl oxidase and nonenzymatic crosslinking via advanced glycation end-products, thus raising my At the same time, the lack of balance between matrix metalloproteinases and inhibitors leads to the disruption of physiological matrix degradation, and chronic and low-grade inflammation perpetuated by senescent cell secretomes maintains a profibrotic signaling by TGF-b and related pathways (Horn and Trafford, 2016; Kruszewska et al., 2022). Such cellular and molecular events reorganize the interstitial network, provide nonhomogenous conduction characteristics, and make the heart vulnerable to atrial and ventricular arrhythmias and heart failure with preserved ejection fraction. Modern-day diagnostics use a combination of circulating biomarkers and cardiac magnetic resonance imaging, such as native T1 mapping and extracellular volume quantification, to identify and grade fibrosis. Novel therapies targeting TGF-b signaling, crosslinking enzymes, MMP/TIMP diets, and senescence-targeted therapy require clinical validation in the elderly. Mechanistic findings need to be translated into targeted and safe antifibrotic approaches on a global scale in longitudinal population studies with integrative biomarker frameworks (Mishra and Kass, 2021).

2.3.3 Microvascular rarefaction and endothelial dysfunction

Structural remodeling in the ageing heart focuses on microvascular rarefaction and endothelial dysfunction, starting with the progressive loss of capillaries that decline myocardial perfusion reserve and cause focal hypoxia, triggering fibroblast proliferation and interstitial fibrosis and culminating in endothelial cell damage and endothelial senescence, release of pro-inflammatory/pro-thrombotic cytokines, and loss of nitric oxide bioavailability and vasodilatory capacity, exacerbating micro These interconnected mechanisms impair the provision of oxygen and metabolic support to cardiomyocytes, inhibit angiogenesis and reparative mechanisms, and distort remodeling to concentric hypertrophy and myocardial stiffening, which are clinically manifested by reduced exercise capacity, reduced coronary reserve, increased susceptibility to ischemia, and heart failure with preserved ejection fraction (Mohammed et al., 2015; Fonseka et al., 2025). Systemic factors such as chronic inflammation, oxidative stress, hypertension, and metabolic disease hasten microvascular worsening and increase the amplification of age effects on endothelial status, which explains the role of comorbidity in mediating structural effects. Such priorities in translational research are the maintenance of endothelial health, repair of capillary density by pro-angiogenic or reparative mechanisms, lifestyle requirements, including structured exercise, and control of endothelial senescence, but these are yet to undergo rigorous clinical validation. Early signs of microvascular breakdown identified by sensitive imaging and biomarkers are essential for implementing decisions regarding personal and scaled interventions in vulnerable groups (Wang et al., 2025b).

2.4 Functional and electrophysiological alterations

2.4.1 Diastolic dysfunction and impaired lusitropy

Functional and electrophysiological changes bridge the relationship between cellular changes and mechanical impairment of ventricular relaxation due to age. With progressive cellular senescence, the calcium recycling capability of cardiomyocytes into the sarcoplasmic reticulum via SERCA is suppressed, causing chronic cytosolic calcium and a delayed passive decrease in active tension. Such a calcium-handling deficiency compromises lusitropy, slowing myocardial relaxation and hampering diastolic filling. When relaxation further decays, residual tension continues as diastole, resulting in decreased compliance and increased left ventricular filling pressures (Hamilton and Terentyev, 2019; Wang et al., 2025c). Ultimately, such increased diastolic load initiates structural remodeling of increased fibrosis and stiffening of the extracellular matrix, which further limits ventricular distensibility. Further rigidity worsens diastolic dysfunction by reducing the filling and amplification of filling pressures during hemodynamic load. In addition to mechanical effects, electrophysiological changes, such as slowing of conduction and impaired calcium current modulation, can disrupt relaxation/contraction interactions and further contribute to lusitropic impairment and diastolic heart failure. This series of functional and electrical dysfunctions forms a long chain of cascading molecular decays, leading to clinically apparent diastolic dysfunction (Weil et al., 2025).

2.4.2 Conduction system aging and arrhythmogenic substrate

The progressive impairment of the cardiac conduction system in terms of impulse initiation and propagation through coordinated structural and electrophysiological remodeling associated with aging leads to the formation of arrhythmogenic substrates. The expression of ion channels changes with age, making the pacemaker cells in the sinoatrial node less automatic. This is accompanied by a deceleration of the atrioventricular node and an increase in His-Purkinje conduction. Current discontinuity and overlapping interstitial fibrosis break electrical continuity and introduce spatial heterogeneity. The remodeling of sodium, calcium, and potassium currents increases conduction delay and dispersion in the working myocardium, altering the shape and refractoriness of action potentials (Mesquita et al., 2025; Bhattacharyya and Munshi, 2020). In addition, there is reduced gap-junctional coupling, which further inhibits cell-to-cell conduction. Oxidative stress, mitochondrial dysfunction, and pro-inflammatory signaling associated with cellular senescence add to the problem of anisotropy and heterogeneity and lead to microstructural damage and regional conduction block. The result of the cumulative effect is a substrate with a slower conduction velocity, greater repolarization dispersion, and discrete structural discontinuities that favor ectopic activity and reentrant circuits. Clinically, this predisposes patients to age-related ventricular arrhythmias, atrioventricular conduction disorders, bundle-branch block, sinus node dysfunction, and atrial fibrillation (Camacho-Encina et al., 2024; Miwa et al., 2022).

2.4.3 Mechanoelectric coupling and calcium-handling shifts

Mechanoelectric coupling and calcium-handling shifts explain how the stiffening of the cardiomyocyte structure and a change in the mechanical forces increasingly remodel the electrophysiological response through mechanoelectric coupling, effectively disrupting calcium cycling. Mechanical load transduction through cell stiffening and extracellular remodeling activates more stretch-sensitive ion channels, changing transmembrane currents due to dynamic deformation. The feedback of these disturbed ionic fluxes on intracellular calcium affects sarcolemmal calcium influx and regulates ryanodine receptor gating, changing the balance of calcium release, uptake, and leakage (Peyronnet et al., 2021; Quinn and Kohl, 2021). Consequently, the sarcoplasmic reticulum calcium content decreases, diastolic cytosolic calcium increases, and the fidelity of excitation-contraction coupling decreases. These alterations further destabilize action potentials and conduction due to reduced calcium-dependent currents, where electrophysiological abnormalities and calcium mishandling positively feedback. The overall outcome is a decline in contractile effectiveness, an increased risk of arrhythmia, and the acquisition of the aged phenotype by the cell over time. This cascade of mechanisms shows that mechanical and ionic remodeling are integrally related during cardiac aging, and functional deterioration is coupled with structural degeneration (Kho, Lee and Hajjar, 2012; Koivumäki et al., 2011).

2.5 Systems-physiology adaptations and inflammaging

2.5.1 Autonomic balance and baroreflex sensitivity

Through the aging process, low-grade inflammation and stiffening of the vascular system change autonomic control to sympathetic dominance and suppress parasympathetic tone, and with aging, this reweighting increasingly uncouples baroreflex sensitivity. Endothelial dysfunction and increased arterial pulse wave velocity decrease mechano-transduction at the carotid and aortic baroreceptors, which decreases afferent activity to the nucleus tractus solitarius, resulting in central integration that supports an increase in sympathetic discharge and decreased vagal regulation of the sinoatrial node. Circulating cytokine levels, oxidative stress, and angiotensin II cause additional deterioration of baroreceptor transduction and central autonomic networks, and neuronal resistance is reduced by mitochondrial malfunction in the brainstem and peripheral ganglia (Jiang et al., 2022; Kresnajati et al., 2022). On the cardiac level, cardiovascular acetylcholine signaling is decreased, and norepinephrine spillover increases the slowing of heart rate variability, protracts hemodynamic challenge recovery, and makes the heart more prone to atrial and ventricular arrhythmias. Chronic sympathetic activity stimulates myocardial remodeling by β-adrenergic desensitization, calcium-handling imbalances, and fibrotic signaling, which deteriorate reflex control by modifying mechanosensory signals. These trends can be partially reversed using pharmacologic and lifestyle interventions that reduce arterial stiffness, inhibit the renin-angiotensin-aldosterone cascade, and re-establish the bioavailability of nitric oxide, which increases reflex gain. These mechanisms, in combination, relate inflammaging and arterial aging to dysfunctional autonomic balance and baroreflex, as well as to decreased reflex buffering, increased blood pressure variability, diminished cardiac reserve in older adults, and exercise intolerance (Roy et al., 2013; La Rovere and Pinna, 2014).

2.5.2 Metabolic remodeling: rebalancing fatty acid oxidation and glycolysis

Metabolic remodeling (fatty acid oxidation and glycolytic shift) in which older cardiomyocytes progressively use less fatty acid oxidation to maintain ATP and more glucose to maintain ATP. PPARa/PGC-1a downregulation, decreased carnitine-palmitoyltransferase-1 flux, and accumulation of lipotoxic intermediates inhibit β-oxidation, and electron-transport defects and decreased NAD+/sirtuin activity increase reactive oxygen species and damage mitochondrial DNA. When oxidative phosphorylation loses its elasticity, AMPK signals weaken, and the mTOR tone increases, promoting the expression of glycolytic enzymes and a slight switch towards glycolysis and anaplerosis (Jaswal et al., 2011; Sithara and Drosatos, 2021). This metabolic rebalancing is associated with inflammaging, NF-kB activation, NLRP3 inflammasome priming, and cytokine tumor necrosis factor-alpha (TNF-α) and IL-6 downregulating PPARa targets, disrupting insulin signaling, and enhancing fibrosis. The outcome is impaired cardiac output, increased diastolic stiffness, and increased arrhythmogenicity with ATP-related strain on ion pumps and changes in substrate supply. Systemic adipose and hepatic insulin resistance elevate the levels of circulating free fatty acids, which the aging heart is unable to effectively oxidize, increasing lipotoxic stress, and microvascular endothelial dysfunction restricts oxygen supply and solidifies glycolytic dependence. Exercise training, caloric restriction interventions, and the use of agents to restore AMPK, PPAR, or mitochondrial quality control are potential therapeutic strategies to re-establish substrate usage and reduce sterile inflammation (Shen et al., 2022; Meyers and Zhu, 2020).

2.5.3 Immune-endocrine axes and low-grade chronic inflammation

Remodeling of the crosstalk between the hypothalamic–pituitary–adrenal (HPA) axis, sympathetic outflow, and innate immunity by immune-endocrine axes and low-grade chronic inflammation to maintain inflammaging (Balaraman et al., 2025; Gothwal et al., 2024). As one gets older, the rhythms of cortisol become flatter, and the adrenergic tone increases, which shifts leukocyte trafficking, increases NF-kB signaling, and prepares NLRP3 inflammasome activity.

Insulin resistance and visceral adiposity skew adipokine profiles towards leptin excess and adiponectin depletion, which increases myeloid activity. Senescent cells accumulate in cardiac and vascular niches and secrete a pro-inflammatory SASP containing IL-6, TNF-α, chemokines, and matrix-remodeling enzymes, which feed forward to compromise glucocorticoid responsiveness and disrupt catecholamine responses (Straub et al., 2011; Patel et al., 2025). This leads to overactivation of the renin–angiotensin–aldosterone system, whose signal transduction aggravates oxidative stress, endothelial dysfunction, and profibrotic TGF-b pathways. All these endocrine and immune changes sustain low-grade systemic inflammation to remodel the myocardium, encourage interstitial fibrosis, stiffen the vasculature, impair diastolic reserve, and change the autonomic balance. Interactions between cytokines and ion channels predispose patients to arrhythmias. This results in immune-endocrine dysregulation as a systems-level mediator of senescence biology in cardiac aging and increased cardiovascular risk, and outlines targets for HPA and RAAS modulation, metabolic re-sensitization, and SASP attenuation as preventive/therapeutic interventions (Ekholm and Kahan, 2021; Chen et al., 2023).

2.6 Mechanistic drivers beyond classical senescence

2.6.1 Mitochondrial dysfunction, ROS, and mitophagy

Mitochondrial dysfunction, ROS, and mitophagy explain how sustained mitochondrial loss, oxidative stress, and the inability to eliminate unfit organelles promote cardiac aging (Fig. 2.1). Oxidative alteration of lipids, proteins, and nucleic acids in cardiomyocytes is associated with electron leakage and the growth of reactive oxygen species (ROS) as mitochondrial DNA accumulates mutations and declines in electron transport efficiency with advancing age. Higher ROS levels, in turn, cause mitochondrial permeability transition pore opening and membrane depolarization, which is expected to cause mitophagic pathways facilitated by PINK1/Parkin and BNIP3/NIX; however, aging suppresses this quality-control response. Consequently, dysfunctional mitochondria remain but increase the production of ROS, which can spread

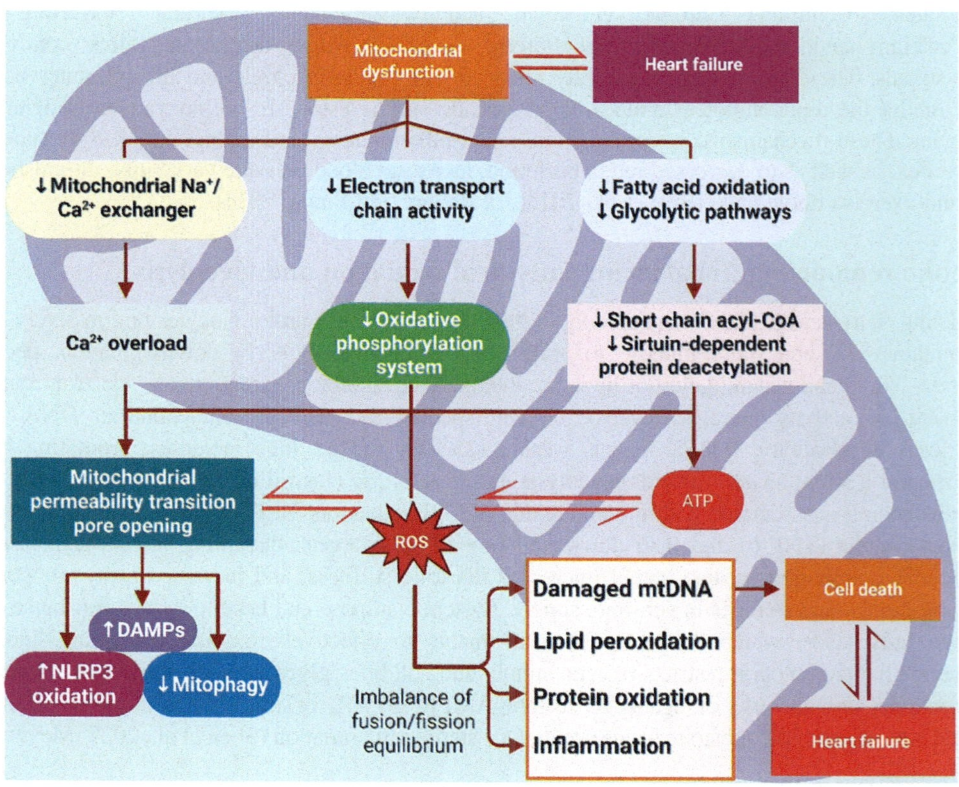

FIGURE 2.1 Mitochondrial dysfunction and inflammation pathways in cardiac aging. Schematic summary of the mitochondrial causes of cardiac aging and failure. Inhibited electron transport and oxidative phosphorylation decrease ATP levels, and low fatty acid oxidation/glycolysis and Na+/Ca2+ exchange lead to metabolic stress and Ca2+ overload. Reduced fusion-fission and disrupted opening of permeability transition pores increase reactive oxygen species, causing damage to the mitochondrial DNA, lipids, and proteins. Inflammation is preserved by suppressed mitophagy and the activation of the NLRP3 inflammasome through the role of DAMP. Mitochondrial quality control is further affected by reduced short-chain acyl-CoA and sirtuin deacetylation. These loops cause cell death and remodeling of the ventricles, resulting in heart failure.

a vicious cycle of damage (Chavda et al., 2022; Zhang et al., 2025). Continued mitochondrial damage also impairs ATP production, buffering of calcium, and redox homeostasis, destabilizes excitation-contraction coupling, and exposes cells to apoptosis or necrosis. Furthermore, ROS serve as signaling messengers that trigger pro-senescence pathways (e.g., p53, p21) and inflammatory responses through NF-kB, thus proving to be a combination of mitochondrial dysfunction with classical senescence phenotypes (Babu et al., 2025; Gupta et al., 2018). Oxidative damage, inefficient mitophagy, and energetic failure converge in this way, crippling the resilience of cardiomyocytes, promoting maladaptive remodeling, and predisposing these cells to heart failure or arrhythmia. This mechanical sequence convincingly connects mitochondrial derangement with the degeneration of cells and organs during aging (Stojanovic et al., 2025).

2.6.2 Proteostasis: autophagy-lysosome and ubiquitin-proteasome

Proteostasis: autophagy-lysosome and ubiquitin-proteasome, describes the role of reduced proteostasis machinery in cardiac senescence with age via progressive disintegration of quality control systems. The first stage is the accumulation of oxidative damage, mitochondrial dysfunction, and misfolded proteins in the heart cells, which overloads autophagic flux and slows lysosomal autophagic degradation. Reduced autophagosome formation, defective lysosome biogenesis, and reduced lysosomal acidification fail to promote protein aggregate and damaged organelle clearance, thereby exerting strain on the ubiquitin-proteasome system (UPS). When the UPS is overloaded with ubiquitinated substrates, the proteasome becomes inactive, ubiquitin recycling is disrupted, and more misfolded proteins accumulate (Zheng and Wang, 2010; Mainali et al., 2023). These aggregates interfere with sarcomeric stability, calcium homeostasis, and mitochondrial dynamics, triggering endoplasmic reticulum (ER) stress and unfolded protein response signaling. Continuous proteotoxic stress subsequently strengthens senescence-associated secretory phenotype (SASP) signaling, inflammatory responses, and oxidative bursts, which enhances cell dysfunction (Altamish et al., 2020; Kakkar et al., 2025). Simultaneously, decreased proteostasis causes compensatory responses of chaperone networks and stress responses, but these mechanisms are insufficient, which accelerates the process of cardiomyocyte dropout, fibrosis, and diastolic dysfunction. In turn, autophagy-lysosome breakdown and UPS deterioration have been linked to the relationship between molecular damage and structural remodeling and functional deterioration of the aging heart, and the possibility of improving proteostatic resilience through autophagy or proteasome activators (Li et al., 2025; McLendon and Robbins, 2015).

2.6.3 Epigenetic/Chromatin aging and clonal haematopoiesis of indeterminate potential

The role of aging in progressive epigenomic drift and chromatin remodeling throughout aging, predisposing hematopoietic stem cells to clonal expansion and cardiovascular risk, in epigenetic/chromatin aging, and clonal hematopoiesis of indeterminate potential (CHIP). First, age-related loss of heterochromatin integrity, global hypomethylation of DNA, and selective hypermethylation of regulatory sites change the transcriptional landscape and impair the maintenance of genomes. These progressive epigenetic alterations disable DNA repair, stimulate somatic mutation accumulation, and eventually favor clones with mutations in epigenetic regulators (such as DNMT3A, TET2, and ASXL1) (Allegra et al., 2023; Yan et al., 2023). With the increasing size of such mutated clones, CHIP is generated and contributes to a pro-inflammatory environment through increased inflammasome activation and augmented release of cytokines, including IL-1b and IL-6, which subsequently feedback to cause further chromatin stress in the vascular and cardiac cells (Alharbi et al., 2022b; Singh et al., 2022). Simultaneously, mutated hematopoietic clones also affect systemic inflammation and endothelial damage, enhancing cardiac remodeling and increasing atherosclerotic and fibrotic responses. In turn, the synergistic effect of epigenetic aging and CHIP promotes heart-blood syncretism between hematopoietic maladjustment and cardiac aging, indicating that age-associated cardiovascular pathology can be alleviated by targeting chromatin modifiers or CHIP clones (Choudhury et al., 2025; Datta et al., 2024). The interconnected mechanisms underlying cardiac senescence are illustrated in Fig. 2.2.

2.7 Biomarkers and imaging readouts for cardiac aging

2.7.1 Circulating and tissue biomarkers (NT-proBNP, GDF15, miRNAs)

Molecular indicators of the structural and functional changes that occur progressively with cardiac aging are circulating and tissue biomarkers (e.g., NT-proBNP, GDF15, miRNAs). An increase in myocardial wall stress results in a greater release of N-terminal pro-B-type natriuretic peptide (NT-proBNP), which is associated with diastolic dysfunction and predicts heart failure with a preserved ejection fraction. Simultaneously, the age-increasing stress-responsive cytokine growth differentiation factor-15 (GDF15), which is elevated due to mitochondrial dysfunction and inflammation, independently predicts cardiovascular morbidity and mortality (Alharbi et al., 2022a; Thapa et al., 2023). Circuit breakers,

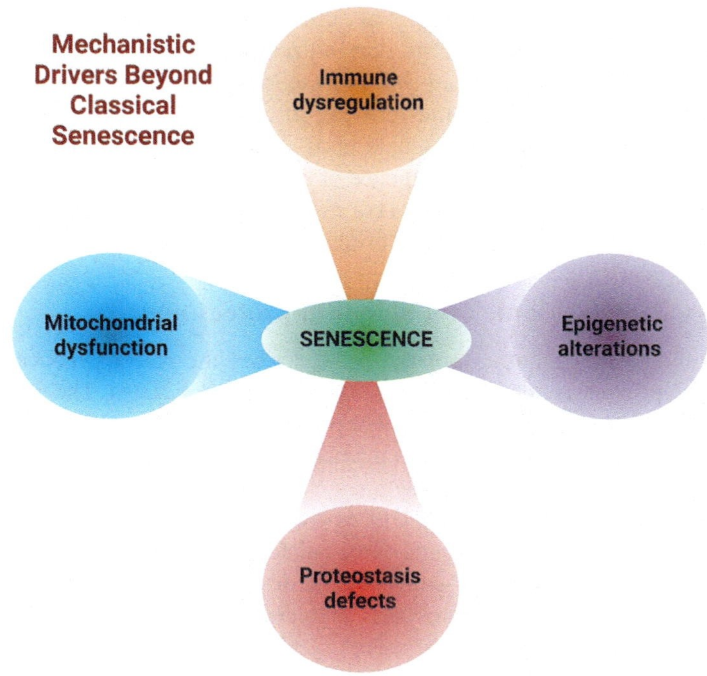

FIGURE 2.2 Integrated cellular drivers and feedback in cardiac senescence and remodeling. Schematic representation of the epicenter of senescence, where four converging inputs are fed out of classical pathways. Impairment of mitochondrial function lowers energy output and increases the generation of ROS. Proteostasis defects indicate a lack of autophagy–lysosome and ubiquitin-proteasome clearance. Epigenetic remodeling restructures gene regulation and chromatin state, and immune dysregulation maintains inflammatory signaling. The senescent phenotype and driver radial color-coded lobes focus on equal weighting and two-way feedback between the individual driver and senescent phenotype. Cumulatively, the axes in combination cast an integrated system perspective between cellular quality control malfunction and senescence and cardiac remodeling with age.

including miR-21, miR-34a, and miR-146a, may be viewed as secondary indicators of fibrosis, apoptosis, and inflammatory signaling, and their expression in plasma, exosomes, and myocardial tissue indicates persistent remodeling before an overt clinical disease manifests (Ibrahim and Januzzi, 2018; Michalska-Kasiczak et al., 2018). These biomarkers, in combination, combine hemodynamic load, metabolic stress, and cellular senescence and provide complementary prognostic information, in addition to traditional risk factors. The study of atrial and ventricular tissue samples also demonstrated concordant changes in the molecular markers of fibrosis and senescence-associated secretory phenotype, which supports the relationship between molecular signatures and histopathology. This integrative circulating-and-tissue methodology aids in the early detection of subclinical cardiac aging, directs risk assessment, and designs therapeutic interventions to delay the process of overt cardiovascular disease (Ţica and Ţica, 2025).

2.7.2 Multimodal imaging (strain echocardiography, CMR T1/T2, PET)

Multimodal imaging (strain echocardiography, CMR T1/T2, and PET) describes the image modalities in terms of their progressive measurement of the structural, functional, and molecular features of cardiac aging. To identify early impaired functional performance, speckle-tracking strain echocardiography first identifies subtle myocardial deformation (global longitudinal strain and circumferential strain) before ejection fraction reduction is detected. Cardiac magnetic resonance imaging (CMR) of native T1 and T2 mapping to measure diffuse interstitial fibrosis and myocardial edema is followed, where an increase in T1 values and extracellular volume fraction indicates collagen deposition, whereas an increase in T2 values indicates low-grade inflammation or increased water content (Jung et al., 2022; Habib et al., 2017). These tissue-remodeling changes are associated with strain perturbation and allow spatial mapping of the remodeling load. Metabolic changes (decreased glucose uptake, altered oxidative metabolism) and inflammation (through radiolabeled ligands to macrophages or fibroblasts) can be detected by positron emission tomography (PET) using tracers such as 18F-FDG or 11C-acetate, which is consistent with changes in the structure. These imaging modalities create an integrated and multi-scale cardiac aging portrait of mechanics from tissue microarchitecture to metabolism and deliver noninvasive readouts of early detection, risk stratification, and tracking of antiaging or cardioprotective therapies (Kundu et al., 2015; Gropler et al., 2010).

2.7.3 Single-cell and spatial multiomics for patient stratification

Single-cell and spatial multiomics to stratify patients, chapter outlines the growing complexity of molecular technologies that are being used to understand cellular heterogeneity in the aging heart and apply it to clinical phenotyping and risk stratification. Introduction Single-cell RNA sequencing detects transcriptomic signatures in specific subpopulations of cardiomyocytes, fibroblasts, endothelial cells, and immune cells, and has shown subclusters with senescence- or stress-related gene expression patterns. Subsequently, this can be followed by integration with single-cell ATAC-seq or methylation profiling to map chromatin accessibility and epigenetic state, and define resilient and vulnerable cellular phenotypes (Wu et al., 2024; Vandereyken et al., 2023). Subsequently, the molecular sites of senescent cell accumulation, fibrotic niches, and inflammatory gradients in myocardial sections were overlaid with spatial transcriptomics and imaging mass cytometry. Spatially resolved omics can be correlated with clinical measures, which enables researchers to stratify patients based on molecular cardiac aging endotypes, including inflammation-dominated, fibrotic, or metabolic dysregulation phenotypes. This joint single-cell and spatial multiomics model will eventually enable personalized prognostication, guide the choice of targeted treatment (e.g., senolytics, epigenetic modulators, and antifibrotics), and enable tracking of intervention responses at the cellular scale (Mahoney et al., 2023; Gurkar et al., 2023). As shown in Fig. 2.3, the multilevel biomarker and imaging cascade links molecular signals to cardiac aging outcomes.

2.8 Modifiers and heterogeneity

2.8.1 Sex differences and hormonal transitions

This section describes how cellular senescence can be modulated by biological sex and endocrine changes that occur with age to determine cardiac aging pathways. Estrogen during the reproductive years helps maintain mitochondrial efficiency, endothelial nitric oxide signaling, and antioxidant defenses, and provides women with relative protection against myocardial hypertrophy and diastolic dysfunction. With the onset of menopause, the precipitous increase in oxidative stress, deterioration of calcium processing, and encouragement of interstitial fibrosis are accelerated, which is consistent with the risk in females being equal to or higher than that in age-matched men (Yusifov et al., 2022; Du et al., 2025). The loss of anabolic signaling and endothelial repair ability in response to a gradual testosterone decline and relative androgen excess in

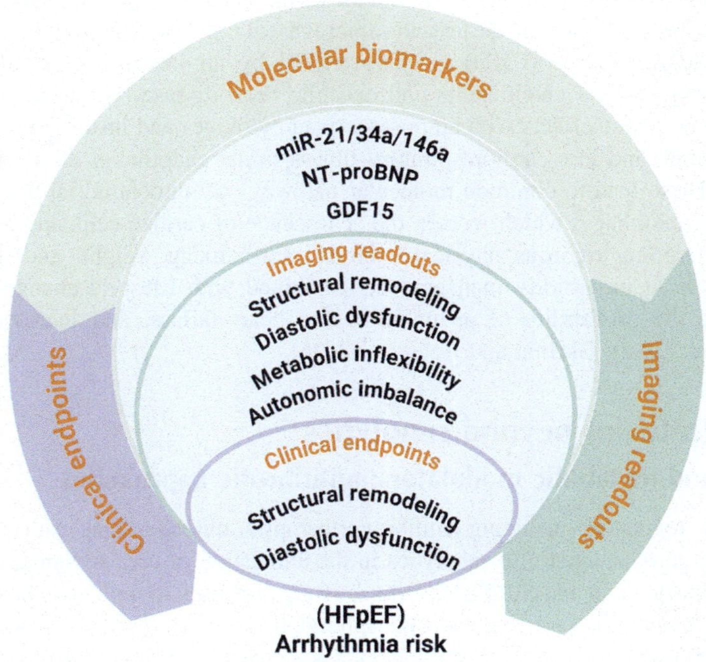

FIGUR 2.3 Multilevel biomarker and imaging cascade linking molecular signals to cardiac aging outcomes. This hierarchical structure associates molecular biomarkers with imaging and clinical biomarkers in cardiac aging. The external ring contained candidate signals of blood/tissue (miR-21/34a/146a, NT-proBNP, GDF15). The intermediate ring projects phenotypes of structural remodeling, diastolic dysfunction, metabolic inflexibility, and autonomic imbalance. The inner ring provides an overview of clinical outcomes, such that structural and diastolic outcomes predict HFpEF and arrhythmic risk. The arrows between rings suggest that there are two-way refinements: biomarkers inform imaging selection and thresholds, and vice versa.

the earlier years of life predisposes adverse remodeling via pro-hypertrophic and pro-inflammatory signaling in men. These divergent timelines are further supported by the genetic regulation of senescence, immune activation, and turnover of the extracellular matrix by sex-specific epigenetic patterns, differences in microRNA expression, and genetic regulation of sex chromosomes. Together, these processes form discrete phenotypes of cardiac aging, including increased heart failure with preserved ejection fraction among postmenopausal women and earlier ischemic illness among men, indicating the importance of sex-differentiated preventive and treatment approaches (Cai et al., 2016; D'Ascenzo et al., 2007).

2.8.2 Genetic background, CHIP, and ancestry considerations

CHIP, ancestry, genetic background, inherited genomic variation, acquired clonal hematopoiesis, and population ancestry are interrelated and influence cardiac aging pathways in individuals. Baseline predisposition to cellular senescence and cardiovascular aging is regulated by genetic polymorphisms in DNA repair, telomere repair, and inflammatory and metabolic pathways. In old age, such predispositions mediate the clonal expansion of hematopoietic stem cells harboring CHIP-associated mutations (e.g., in TET2, DNMT3A, and ASXL1). With CHIP clones, the magnitude of their impact on systemic inflammation and vascular injury depends on the germline predisposition of the host, in the sense that individuals who are carriers of risk alleles in immune response or DNA-damage repair genes can be subjected to exaggerated inflammatory signaling and cardiovascular remodeling (Natarajan, 2023; Jakubek et al., 2023). Simultaneously, ancestral diversity provides different allele frequencies, epigenomic structures, and CHIP mutation penetrance, indicating that individuals of African, Hispanic, or Asian origin might also have various patterns of CHIP incidence, impact of the variants, and cardiac outcomes downstream effects. The combination of inherited genetics, CHIP relationships, and ancestry presents heterogeneity in cardiac aging phenotypes, which accelerates aging, becomes more resilient, and how therapeutic treatment may more specifically require demographic and genomic subgroups (Marnell et al., 2021; Tuncay et al., 2023).

2.8.3 Lifestyle and comorbidities (hypertension, diabetes, CKD, obesity, fitness)

The interplay between cellular senescence and lifestyle and comorbidities (hypertension, diabetes, CKD, obesity, and fitness), as well as environmental exposures and metabolic disorders, influences the process of cardiac aging. Chronic hypertension enhances the rate of arterial stiffening and hypertrophy of the left ventricle, thereby augmenting myocardial oxygen requirements and favoring fibrotic remodeling. Diabetes mellitus increases the accumulation of oxidative stress and advanced glycation end-products, which destroy the microvasculature and disrupt the calcium handling process of cardiomyocytes. Chronic kidney disease (CKD) is characterized by an inflamed uremic load that disrupts mineral metabolism and contributes to vascular calcification and diastolic heart disease (Hu et al., 2022; De la Maza-Bustindui et al., 2025). Obesity reinforces these effects with adipose-derived cytokines and insulin resistance that increase systemic inflammation and dyslipidemia, and low cardiorespiratory fitness limits endothelial bioavailability of nitric oxides and decreases cardiac reserve. These lead to common molecular pathways of mitochondrial dysfunction, chronic low-grade inflammation, and defective autophagy, which worsen the senescence of cardiac cells and their stroma. However, these insults are reduced by undergoing rigorous physical exercise, normalizing weight, and maintaining blood pressure, glucose, and kidney functions. Comorbidity management, combined with lifestyle changes, is essential to decelerate myocardial aging, decrease the probability of arrhythmias and heart failure, and increase the effectiveness of new senotherapies (Zatterale et al., 2020; Gkrinia and Belančić, 2025).

2.9 Therapeutic strategies beyond senolytics

2.9.1 Mitochondrial and metabolic modulators; antifibrotic approaches

Mitochondrial and metabolic regulators, including antifibrotic therapies, mitochondrial resilience, and metabolic flexibility therapies, are also becoming complementary to senolytics in the alleviation of cardiac aging. To begin with, nicotinamide riboside, nicotinamide mononucleotide, and SIRT1/PGC-1a activators are used to replenish the pools of NAD+, increase the efficiency of mitochondrial biogenesis and oxidative phosphorylation, reduce reactive oxygen species, and slow down mitochondrial DNA damage and dysfunction of mitophagy (Wang et al., 2022). The metabolic modulators of AMPK activators, GLP-1 receptor agonists, and PPAR agonists recalibrate substrate use, enhance insulin sensitivity, and switch cardiomyocyte energy use towards either fatty acid oxidation or ketone metabolism in response to stress. Simultaneously, antifibrotic strategies, such as TGF-b signaling inhibitors, connective tissue growth factor, angiotensin receptor neprilysin, and galectin-3 antagonists, target this pathway and suppress extracellular matrix production, collagen crosslinking, and fibroblast-to-myofibroblast differentiation (Kim et al., 2016; Liu et al., 2025). These therapies are synergistic, with healthier mitochondrial energetics, decreased profibrotic signaling, and preserved mitochondrial diffusion and perfusion by attenuated fibrosis. Combined, these combinatorial therapies have the potential to maintain myocardial compliance, decrease stiffness, facilitate contractile reserve, and curb arrhythmogenic substrate formation. Although clinical translation is still in its infancy, mitochondrial and metabolic modulation combined with antifibrotic therapy can be integrated to enhance and synergize senolytic or senostatic therapy to slow cardiac aging and improve cardiovascular outcomes (Li et al., 2020; Jing et al., 2025).

2.9.2 Exercise training, autonomic modulation, and nutrition

Nonpharmacological interventions that reverse the cardiac aging processes that occur in cells and systems include exercise training, autonomic modulation, and nutrition. Popular aerobic and resistance exercises enhance the efficiency of mitochondria, bioavailability of endothelial nitric oxide, and decrease in resting sympathetic tone, which in turn decreases oxidative stress levels and signifies autophagy of cardiomyocytes. These adaptations renew heart rate variability and baroreflex sensitivity, both important signs of balanced autonomic regulation, even as they downregulate inflammatory cytokine synthesis, which catalyzes senescence. Specific autonomic alteration techniques, including vagus nerve stimulation and mindfulness, also move the sympatho-vagal axis towards parasympathetic predominance, normalizing cardiac electrophysiology and eliminating the risk of arrhythmia (Clayton et al., 2022; Wang et al., 2020). Nonpharmacological dietary strategies, such as the Mediterranean or vegetarian diet, adequate omega-3 fatty acids, and regulated caloric intake, help mitigate metabolic stress and enhance insulin sensitivity, as well as supply polyphenols and micronutrients to suppress NF-kB signaling and maintain genomic stability. A combination of such lifestyle interventions can delay the rate of telomere shortening, decrease fibrosis, and maintain diastolic function, thereby reducing the occurrence of heart failure and other age-related cardiovascular issues. Exercise, autonomic modulation, and optimized nutrition are scalable, evidence-based therapeutic strategies that can be used in conjunction with new senescence-targeted drugs to maintain heart resilience throughout the lifespan (Mirabelli et al., 2020; Yuksek et al., 2025).

2.9.3 Regenerative and reprogramming strategies; extracellular vesicle therapies

Regenerative and reprogramming approaches, such as extracellular vesicle transplantation, are novel strategies aimed at reversing or slowing down the aging of the heart by replenishing cell populations and rejuvenating cellular communication. The first attempts focused on transplanting cardiomyocytes or cardiac progenitors generated by induced pluripotent stem cells (iPSCs) as a replacement for the lost or dysfunctional myocardial cells, with the simultaneous use of partial reprogramming (e.g., transient Yamanaka factor expression) to rejuvenate resident cardiac cells without identity loss (Alibhai et al., 2018). Such regenerative responses are augmented by extracellular vesicles (EVs) of stem cells or artificial derivations that transport microRNAs, proteins, and mitochondrial proteins that regulate the survival of may or may not survive of recipient cells, alleviate oxidative stress, inflammation, and senescence cues. EV therapy has been shown to improve angiogenesis, inhibit fibrosis, and augment contractile activity in preclinical models. It is a cell-free modality with reduced immunogenic potential (Lim et al., 2025; Zhang et al., 2024). Hybrid approaches can harness endogenous repair, reset epigenetic topography, and enable paracrine rejuvenation of the aging myocardium by combining reprogramming signals and EV delivery. These two therapeutic innovations are leading the future of senolytic ablation in the direction of regeneration and functional tissue renewal, which holds promise for cardiovascular aging interventions (Saludas et al., 2021).

2.10 Conclusion and future perspectives

Cardiac aging research is moving from mechanistic descriptions to biomarker-guided translation and early human testing. This chapter advances this evolution by integrating pathways, measurable signatures, and trial designs into a coherent, clinically oriented roadmap. Forthcoming work will refine how DNA damage responses, mitochondrial dysfunction, chromatin remodeling, and the senescence-associated secretory phenotype shape myocardial remodeling and electrophysiology, thereby informing the discovery and selection of senescence-linked biomarkers, such as circulating SASP factors, epigenetic clocks, and cell-free DNA features relevant to aging hearts. These candidates must be analytically robust, clinically validated in longitudinal cohorts, and calibrated to quantify biological age and tissue-specific burden, enabling the stratification of older adults for preventive and therapeutic studies. Trial methodology is converging on composite endpoints that couple imaging-based fibrosis and diastolic function with symptoms and clinical events, while embedding safety domains crucial for geriatric populations, including frailty trajectories, renal tolerance, immunosenescence, and drug–drug interactions. We underscore adaptive designs, careful dose escalation, and sentinel monitoring to balance efficacy signals with tolerability in patients with multimorbidity. Translational priorities include selective delivery and timing, whether targeting cardiomyocytes, fibroblasts, or endothelium in preclinical risk states versus established disease, together with optimization of modalities spanning senomorphics, gene and RNA therapies, mitochondrial and metabolic modulators, antifibrotic strategies, and extracellular vesicle platforms. Open questions persist regarding the best way to integrate senescence modulation with standard guideline-directed therapy, how to mitigate established scarring versus prevent progression, and whether intermittent dosing can achieve durable benefits with fewer adverse effects than continuous regimens. In parallel, a rigorous assessment of off-target effects, long-term safety, and equity of access is essential before widespread adoption, particularly as therapies move from tightly controlled trials to real-world practice. Progress will also depend on harmonized data standards, interoperable imaging pipelines, and open biomarker repositories that enable external validation across regions and health systems. Multidisciplinary collaborations linking geroscience, electrophysiology, imaging physics, bioengineering, data science, and regulatory expertise, supported by pragmatic funding mechanisms, will accelerate the path from bench concepts to bedside benefits. Moving forward.

References

Alharbi, K.S., Almalki, W.H., Alzarea, S.I., Kazmi, I., Al-Abbasi, F.A., Afzal, O., et al., 2022. A narrative review on the biology of piezo1 with platelet-rich plasma in cardiac cell regeneration. Chemico-Biological Interactions 363, 110011. Available at https://doi.org/10.1016/j.cbi.2022.110011.

Alharbi, K.S., Singh, Y., Afzal, O., Alfawaz Altamimi, A.S., Kazmi, I., Al-Abbasi, F.A., et al., 2022b. Molecular explanation of Wnt/βcatenin antagonist pyrvinium mediated calcium equilibrium changes in aging cardiovascular disorders. Molecular Biology Reports 49 (11), 11101–11111. Available at https://doi.org/10.1007/s11033-022-07863-7.

Alibhai, F.J., Tobin, S.W., Yeganeh, A., Weisel, R.D., Li, R.K., 2018. Emerging roles of extracellular vesicles in cardiac repair and rejuvenation. American Journal of Physiology - Heart and Circulatory Physiology 315 (4), H733. Available at https://doi.org/10.1152/ajpheart.00100.2018.

Allegra, A., Caserta, S., Mirabile, G., Gangemi, S., 2023. Aging and age-related epigenetic drift in the pathogenesis of leukemia and lymphomas: new therapeutic targets. Cells 12 (19), 2392. Available at https://doi.org/10.3390/cells12192392.

Altamish, M., Samuel, V.P., Dahiya, R., Singh, Y., Deb, P.K., Bakshi, H.A., et al., 2020. Molecular signaling of G-protein-coupled receptor in chronic heart failure and associated complications. Drug Development Research 81 (1), 23–31. Available at https://doi.org/10.1002/ddr.21627.

Aquaro, G.D., De Gori, C., Faggioni, L., Parisella, M.L., Cioni, D., Lencioni, R., et al., 2023. Diagnostic and prognostic role of late gadolinium enhancement in cardiomyopathies. European Heart Journal, Supplement 25, C130. Available at https://doi.org/10.1093/eurheartjsupp/suad015.

Babu, M.A., Jyothi S, R., Kaur, I., Kumar, S., Sharma, N., Kumar, M.R., et al., 2025. The role of GATA4 in mesenchymal stem cell senescence: a new frontier in regenerative medicine. Regenerative Therapy 28, 214–226. Available at https://doi.org/10.1016/j.reth.2024.11.017.

Balaraman, A.K., Altamimi, A.S.A., Babu, M.A., Goyal, K., PadmaPriya, G., Bansal, P., et al., 2025. The interplay of senescence and MMPs in myocardial infarction: implications for cardiac aging and therapeutics. Biogerontology 26 (1). Available at https://doi.org/10.1007/s10522-025-10190-6.

Bhattacharyya, S., Munshi, N.V., 2020. Development of the cardiac conduction system. Cold Spring Harbor Perspectives in Biology 12 (12), 1–21. Available at https://doi.org/10.1101/cshperspect.a037408.

Cai, J.J., Wen, J., Jiang, W.H., Lin, J., Hong, Y., Zhu, Y.S., 2016. Androgen actions on endothelium functions and cardiovascular diseases. Journal of Geriatric Cardiology 13 (2), 183–196. Available at https://doi.org/10.11909/j.issn.1671-5411.2016.02.003.

Camacho-Encina, M., Booth, L.K., Redgrave, R.E., Folaranmi, O., Spyridopoulos, I., Richardson, G.D., 2024. Cellular senescence, mitochondrial dysfunction, and their link to cardiovascular disease. Cells 13 (4), 353. Available at https://doi.org/10.3390/cells13040353.

Camm, A.J., Steffel, J., Virdone, S., Bassand, J.P., Fox, K.A.A., Goldhaber, S.Z., et al., 2023. Guideline-directed medical therapies for comorbidities among patients with atrial fibrillation: results from GARFIELD-AF. European Heart Journal Open 3 (3). Available at https://doi.org/10.1093/ehjopen/oead051.

Chan, R.H., Van Der Wal, L., Liberato, G., Rowin, E., Soslow, J., Maskatia, S., et al., 2024. Myocardial scarring and sudden cardiac death in young patients with hypertrophic cardiomyopathy: a multicenter cohort study. JAMA Cardiology 9 (11). Available at https://doi.org/10.1001/jamacardio.2024.2824.

Chavda, V., Chaurasia, B., Garg, K., Deora, H., Umana, G.E., Palmisciano, P., et al., 2022. Molecular mechanisms of oxidative stress in stroke and cancer. Brain Disorders 5, 100029. Available at https://doi.org/10.1016/j.dscb.2021.100029.

Chen, J., Wei, X., Zhang, Q., Wu, Y., Xia, G., Xia, H., et al., 2023. The traditional Chinese medicines treat chronic heart failure and their main bioactive constituents and mechanisms. Acta Pharmaceutica Sinica B 13 (5), 1919–1955. Available at https://doi.org/10.1016/j.apsb.2023.02.005.

Choudhury, R.P., Hargreaves, R., Chai, J., Fisher, E.A., 2025. Inflammation reprogramming and immunomodulation: next-generation treatments for atherosclerosis. Cell Reports Medicine 6 (9), 102288. Available at https://doi.org/10.1016/j.xcrm.2025.102288.

Clayton, Z.S., Craighead, D.H., Darvish, S., Coppock, M., Ludwig, K.R., Brunt, V.E., et al., 2022. Promoting healthy cardiovascular aging: emerging topics. Journal of Cardiovascular Aging 2 (4). Available at https://doi.org/10.20517/jca.2022.27.

D'Ascenzo, S., Millimaggi, D., Di Massimo, C., Saccani-Jotti, G., Botrè, F., Carta, G., et al., 2007. Detrimental effects of anabolic steroids on human endothelial cells. Toxicology Letters 169 (2), 129–136. Available at https://doi.org/10.1016/j.toxlet.2006.12.008.

Datta, S., Rahman, M.A., Koka, S., Boini, K.M., 2024. High mobility group box 1 (HMGB1): molecular signaling and potential therapeutic strategies. Cells 13 (23), 1946. Available at https://doi.org/10.3390/cells13231946.

De la Maza-Bustindui, N.S., León-Álvarez, M., Ponce-Acosta, C., Zarco-Morales, K.P., Fermín-Martínez, C.A., Antonio-Villa, N.E., et al., 2025. Impact of cardiometabolic risk factors and its management on the reversion and progression of arterial stiffness. npj Cardiovascular Health 2 (1). Available at https://doi.org/10.1038/s44325-025-00074-6.

Di Micco, R., Krizhanovsky, V., Baker, D., d'Adda di Fagagna, F., 2021. Cellular senescence in ageing: from mechanisms to therapeutic opportunities. Nature Reviews Molecular Cell Biology 22 (2), 75–95. Available at https://doi.org/10.1038/s41580-020-00314-w.

Du, J., Liu, J., Wang, X., Wang, X., Ma, Y., Zhang, S., et al., 2025. The role of estrogen in the sex difference for the risk factors of heart failure with preserved ejection fraction. Biology Direct 20 (1). Available at https://doi.org/10.1186/s13062-025-00618-x.

Dunlay, S.M., Roger, V.L., Redfield, M.M., 2017. Epidemiology of heart failure with preserved ejection fraction. Nature Reviews Cardiology 14 (10), 591–602. Available at https://doi.org/10.1038/nrcardio.2017.65.

Ekholm, M., Kahan, T., 2021. The Impact of the renin-angiotensin-aldosterone system on inflammation, coagulation, and atherothrombotic complications, and to aggravated COVID-19. Frontiers in Pharmacology 12. Available at https://doi.org/10.3389/fphar.2021.640185.

Ferrari, A.U., Radaelli, A., Centola, M., 2003. Invited review: aging and the cardiovascular system. Journal of Applied Physiology 95 (6), 2591–2597. Available at https://doi.org/10.1152/japplphysiol.00601.2003.

Fonseka, O., Gare, S.R., Chen, X., Zhang, J., Alatawi, N.H., Ross, C., et al., 2025. Molecular mechanisms underlying heart failure and their therapeutic potential. Cells 14 (5), 324. Available at https://doi.org/10.3390/cells14050324.

Gkrinia, E.M.M., Belančić, A., 2025. The mechanisms of chronic inflammation in obesity and potential therapeutic strategies: a narrative review. Current Issues in Molecular Biology 47 (5). Available at https://doi.org/10.3390/cimb47050357.

Gothwal, S.K., Goyal, K., Garg, A.S., Sahu, B.K., Agrawal, M., Mishra, A., et al., 2024. A rare case of brucellosis with multivalvular endocarditis and complete heart block. Current Cardiology Reviews 20 (6), 34–38. Available at https://doi.org/10.2174/011573403X290326240703100925.

Gropler, R.J., et al., 2010. Imaging myocardial metabolic remodeling. Journal of Nuclear Medicine: Official Publication, Society of Nuclear Medicine 51 (Suppl 1), 88s–101s.

Gupta, G., De Jesus Andreoli Pinto, T., Chellappan, D.K., Mishra, A., Malipeddi, H., Dua, K., 2018. A clinical update on metformin and lung cancer in diabetic patients. Panminerva Medica 60 (2), 70–75. Available at https://doi.org/10.23736/S0031-0808.18.03394-3.

Gurkar, A.U., Gerencser, A.A., Mora, A.L., Nelson, A.C., Zhang, A.R., Lagnado, A.B., et al., 2023. Spatial mapping of cellular senescence: emerging challenges and opportunities. Nature Aging 3 (7), 776–790. Available at https://doi.org/10.1038/s43587-023-00446-6.

Habib, G., Bucciarelli-Ducci, C., Caforio, A.L.P., Cardim, N., Charron, P., Cosyns, B., et al., 2017. Multimodality imaging in restrictive cardiomyopathies: an EACVI expert consensus document In collaboration with the "Working Group on myocardial and pericardial diseases" of the European Society of Cardiology Endorsed by The Indian Academy of Echocardiography. European Heart Journal - Cardiovascular Imaging 18 (10), 1090–1121. Available at https://doi.org/10.1093/ehjci/jex034.

Hamilton, S., Terentyev, D., 2019. Altered intracellular calcium homeostasis and arrhythmogenesis in the aged heart. International Journal of Molecular Sciences 20 (10). Available at https://doi.org/10.3390/ijms20102386.

Haque, Z.K., Wang, D.Z., 2017. How cardiomyocytes sense pathophysiological stresses for cardiac remodeling. Cellular and Molecular Life Sciences 74 (6), 983–1000. Available at https://doi.org/10.1007/s00018-016-2373-0.

Horn, M.A., Trafford, A.W., 2016. Aging and the cardiac collagen matrix: novel mediators of fibrotic remodelling. Journal of Molecular and Cellular Cardiology 93, 175–185. Available at https://doi.org/10.1016/j.yjmcc.2015.11.005.

Howes, A.M., Dea, N.C., Ghosh, D., Krishna, K., Wang, Y., Li, Y., et al., 2024. Fibroblast senescence-associated extracellular matrix promotes heterogeneous lung niche. APL Bioengineering 8 (2). Available at https://doi.org/10.1063/5.0204393.

Howie, J.O., Dziano, J.K., Ariyaratnam, J.P., Abbas, M., Kenny, G.T., Evans, S., et al., 2025. The association of age and left atrial dysfunction in patients with atrial fibrillation. Heart Rhythm: the Official Journal of the Heart Rhythm Society. Available at https://doi.org/10.1016/j.hrthm.2025.07.005.

Hu, C., Zhang, X., Teng, T., Ma, Z.G., Tang, Q.Z., 2022. Cellular senescence in cardiovascular diseases: a systematic review. Aging and Disease 13 (1), 103–128. Available at https://doi.org/10.14336/AD.2021.0927.

Hung, O.Y., Lee, S.K., Eshtehardi, P., Samady, H., 2016. Novel biomarkers of coronary microvascular disease. Future Cardiology 12 (4), 497–509. Available at https://doi.org/10.2217/fca-2016-0012.

Ibrahim, N.E., Januzzi, J.L., 2018. Established and emerging roles of biomarkers in heart failure. Circulation Research 123 (5), 614–629. Available at https://doi.org/10.1161/CIRCRESAHA.118.312706.

Inci, N., Kamali, D., Akyildiz, E.O., Tahir Turanli, E., Bozaykut, P., 2022. Translation of cellular senescence to novel therapeutics: insights from alternative tools and models. Frontiers in Aging 3. Available at https://doi.org/10.3389/fragi.2022.828058.

Jakubek, Y.A., Reiner, A.P., Honigberg, M.C., 2023. Risk factors for clonal hematopoiesis of indeterminate potential and mosaic chromosomal alterations. Translational Research 255, 171–180. Available at https://doi.org/10.1016/j.trsl.2022.11.009.

Jan, B., Imran Dar, M., Choudhary, B., Basist, P., Khan, R., Alhalmi, A., et al., 2024. Cardiovascular diseases among indian older adults: a comprehensive review. Cardiovascular Therapeutics 2024 (1). Available at https://doi.org/10.1155/2024/6894693.

Jaswal, J.S., Keung, W., Wang, W., Ussher, J.R., Lopaschuk, G.D., 2011. Targeting fatty acid and carbohydrate oxidation - a novel therapeutic intervention in the ischemic and failing heart. Biochimica et Biophysica Acta - Molecular Cell Research 1813 (7), 1333–1350. Available at https://doi.org/10.1016/j.bbamcr.2011.01.015.

Jiang, Y., Yabluchanskiy, A., Deng, J., Amil, F.A., Po, S.S., Dasari, T.W., 2022. The role of age-associated autonomic dysfunction in inflammation and endothelial dysfunction. GeroScience 44 (6), 2655–2670. Available at https://doi.org/10.1007/s11357-022-00616-1.

Jing, F., Zhao, M., Xiong, H., Zeng, X., Jiang, J., Li, T., 2025. Mechanisms underlying targeted mitochondrial therapy for programmed cardiac cell death. Frontiers in Physiology 16. Available at https://doi.org/10.3389/fphys.2025.1548194.

Jung, M.H., Chang, S., Han, E.J., Youn, J.C., 2022. Multimodal imaging and biomarkers in cardiac amyloidosis. Diagnostics 12 (3). Available at https://doi.org/10.3390/diagnostics12030627.

Kakkar, C., Sharma, V., Mannan, A., Gupta, G., Singh, S., Kumar, P., et al., 2025. Diabetic cardiomyopathy: an update on emerging pathological mechanisms. Current Cardiology Reviews 21 (2). Available at https://doi.org/10.2174/011573403x331870241025094307.

Kalra, A., Jose, A.P., Prabhakaran, P., Kumar, A., Agrawal, A., Roy, A., et al., 2023. The burgeoning cardiovascular disease epidemic in Indians – perspectives on contextual factors and potential solutions. The Lancet Regional Health - Southeast Asia 12, 100156. Available at https://doi.org/10.1016/j.lansea.2023.100156.

Karakasis, P., Theofilis, P., Vlachakis, P.K., Ktenopoulos, N., Patoulias, D., Antoniadis, A.P., et al., 2025. Atrial cardiomyopathy in atrial fibrillation: mechanistic pathways and emerging treatment concepts. Journal of Clinical Medicine 14 (9). Available at https://doi.org/10.3390/jcm14093250.

Kho, C., Lee, A., Hajjar, R.J., 2012. Altered sarcoplasmic reticulum calcium cycling - targets for heart failure therapy. Nature Reviews Cardiology 9 (12), 717–733. Available at https://doi.org/10.1038/nrcardio.2012.145.

Kim, J., Yang, G., Kim, Y., Kim, J., Ha, J., 2016. AMPK activators: mechanisms of action and physiological activities. Experimental & Molecular Medicine 48 (4), e224. Available at https://doi.org/10.1038/emm.2016.16.

Koivumäki, J.T., Korhonen, T., Tavi, P., McCulloch, A.D., 2011. Impact of sarcoplasmic reticulum calcium release on calcium dynamics and action potential morphology in human atrial myocytes: a computational study. PLoS Computational Biology 7 (1), e1001067. Available at https://doi.org/10.1371/journal.pcbi.1001067.

Kresnajati, S., Lin, Y.-Y., Mündel, T., Bernard, J.R., Lin, H.-F., Liao, Y.-H., 2022. Changes in arterial stiffness in response to various types of exercise modalities: a narrative review on physiological and endothelial senescence perspectives. Cells 11 (22), 3544. Available at https://doi.org/10.3390/cells11223544.

Kruszewska, J., Cudnoch-Jedrzejewska, A., Czarzasta, K., 2022. Remodeling and fibrosis of the cardiac muscle in the course of obesity—pathogenesis and involvement of the extracellular matrix. International Journal of Molecular Sciences 23 (8). Available at https://doi.org/10.3390/ijms23084195.

Kundu, B.K., Zhong, M., Sen, S., Davogustto, G., Keller, S.R., Taegtmeyer, H., 2015. Remodeling of glucose metabolism precedes pressure overload-induced left ventricular hypertrophy: review of a hypothesis. Cardiology 130 (4), 211–220. Available at https://doi.org/10.1159/000369782.

La Rovere, M.T., Pinna, G.D., 2014. Beneficial effects of physical activity on baroreflex control in the elderly. Annals of Noninvasive Electrocardiology 19 (4), 303–310. Available at https://doi.org/10.1111/anec.12170.

Lazzeroni, D., Villatore, A., Souryal, G., Pili, G., Peretto, G., 2022. The aging heart: a molecular and clinical challenge. International Journal of Molecular Sciences 23 (24), 16033. Available at https://doi.org/10.3390/ijms232416033.

Li, X., Zhang, W., Cao, Q., Wang, Z., Zhao, M., Xu, L., et al., 2020. Mitochondrial dysfunction in fibrotic diseases. Cell Death Discovery 6 (1). Available at https://doi.org/10.1038/s41420-020-00316-9.

Li, J.J., Xin, N., Yang, C., Kim, B.G., Tavizon, L.A., Hong, R., et al., 2025. Unveiling the intercompartmental signaling axis: mitochondrial to ER stress response (MERSR) and its impact on proteostasis. PLoS Genetics 21 (5), e1011700. Available at https://doi.org/10.1371/journal.pgen.1011700.

Lim, Y.J., Seo, M.S., Park, W.T., Park, S., Lee, G.W., 2025. Extracellular vesicle-derived MicroRNAs as potential therapies for spinal cord and peripheral nerve injuries. RNA Biology 22 (1), 1–9. Available at https://doi.org/10.1080/15476286.2025.2512618.

Liu, H., et al., 2025. Energy metabolism in health and diseases. Signal Transduction and Targeted Therapy 10 (1), 69.

Mahoney, S.A., Dey, A.K., Basisty, N., Herman, A.B., 2023. Identification and functional analysis of senescent cells in the cardiovascular system using omics approaches. American Journal of Physiology - Heart and Circulatory Physiology 325 (5), H1039. Available at https://doi.org/10.1152/ajpheart.00352.2023.

Mainali, N., Ayyadevara, S., Ganne, A., Reis, R.J.S., Mehta, J.L., 2023. Protein homeostasis in the aged and diseased heart. Journal of Cardiovascular Aging 3 (2). Available at https://doi.org/10.20517/jca.2023.4.

Marnell, C.S., Bick, A., Natarajan, P., 2021. Clonal hematopoiesis of indeterminate potential (CHIP): linking somatic mutations, hematopoiesis, chronic inflammation and cardiovascular disease. Journal of Molecular and Cellular Cardiology 161, 98–105. Available at https://doi.org/10.1016/j.yjmcc.2021.07.004.

McLendon, P.M., Robbins, J., 2015. Proteotoxicity and cardiac dysfunction. Circulation Research 116 (11), 1863–1882. Available at https://doi.org/10.1161/CIRCRESAHA.116.305372.

Meschiari, C.A., Ero, O.K., Pan, H., Finkel, T., Lindsey, M.L., 2017. The impact of aging on cardiac extracellular matrix. GeroScience 39 (1), 7–18. Available at https://doi.org/10.1007/s11357-017-9959-9.

Mesquita, T., Miguel-Dos-Santos, R., Cingolani, E., 2025. Aging and sinus node dysfunction: mechanisms and future directions. Clinical Science 139 (11), 577–593. Available at https://doi.org/10.1042/CS20231025.

Meyers, A.K., Zhu, X., 2020. The NLRP3 inflammasome: metabolic regulation and contribution to inflammaging. Cells 9 (8). Available at https://doi.org/10.3390/cells9081808.

Michalska-Kasiczak, M., Bielecka-Dabrowa, A., von Haehling, S., Anker, S.D., Rysz, J., Banach, M., 2018. Biomarkers, myocardial fibrosis and co-morbidities in heart failure with preserved ejection fraction: an overview. Archives of Medical Science 14 (4), 890–909. Available at https://doi.org/10.5114/aoms.2018.76279.

Mirabelli, M., Chiefari, E., Arcidiacono, B., Corigliano, D.M., Brunetti, F.S., Maggisano, V., et al., 2020. Mediterranean diet nutrients to turn the tide against insulin resistance and related diseases. Nutrients 12 (4), 1066. Available at https://doi.org/10.3390/nu12041066.

Mishra, S., Kass, D.A., 2021. Cellular and molecular pathobiology of heart failure with preserved ejection fraction. Nature Reviews Cardiology 18 (6), 400–423. Available at https://doi.org/10.1038/s41569-020-00480-6.

Miwa, S., Kashyap, S., Chini, E., von Zglinicki, T., 2022. Mitochondrial dysfunction in cell senescence and aging. Journal of Clinical Investigation 132 (13). Available at https://doi.org/10.1172/jci158447.

Mohammed, S.F., Hussain, S., Mirzoyev, S.A., Edwards, W.D., Maleszewski, J.J., Redfield, M.M., 2015. Coronary microvascular rarefaction and myocardial fibrosis in heart failure with preserved ejection fraction. Circulation 131 (6), 550–559. Available at https://doi.org/10.1161/CIRCULATIONAHA.114.009625.

Natarajan, P., 2023. Genomic aging, clonal hematopoiesis, and cardiovascular disease. Arteriosclerosis, Thrombosis, and Vascular Biology 43 (1), 3–14. Available at https://doi.org/10.1161/atvbaha.122.318181.

Owens, W.A., Walaszczyk, A., Spyridopoulos, I., Dookun, E., Richardson, G.D., 2021. Senescence and senolytics in cardiovascular disease: promise and potential pitfalls. Mechanisms of Ageing and Development 198, 111540. Available at https://doi.org/10.1016/j.mad.2021.111540.

Patel, H., et al., 2025. Assessment of hypothalamic–pituitary–adrenal (HPA) axis function in chronic stress: correlation with cortisol rhythms and immune markers. European Journal of Cardiovascular Medicine 15, 823–826.

Peyronnet, R., Solovyova, O., Iribe, G., Katsnelson, L.B., 2021. Editorial: Mechano-calcium, mechano-electric, and mechano-metabolic feedback loops: contribution to the myocardial contraction in health and diseases. Frontiers in Physiology 12. Available at https://doi.org/10.3389/fphys.2021.676826.

Puri, P., Girotra, S., Ghosh, A., 2025. Overview of multimorbidity research in India: a scoping review. Journal of Multimorbidity and Comorbidity 15. Available at https://doi.org/10.1177/26335565251355837.

Puspitasari, Y.M., Ministrini, S., Schwarz, L., Karch, C., Liberale, L., Camici, G.G., 2022. Modern concepts in cardiovascular disease: inflamm-aging. Frontiers in Cell and Developmental Biology 10. Available at https://doi.org/10.3389/fcell.2022.882211.

Quinn, T.A., Kohl, P., 2021. Cardiac mechano-electric coupling: acute effects of mechanical stimulation on heart rate and rhythm. Physiological Reviews 101 (1), 37–92. Available at https://doi.org/10.1152/physrev.00036.2019.

Roy, A., Fields, W.C., Rocha-Resende, C., Resende, R.R., Guatimosim, S., Prado, V.F., et al., 2013. Cardiomyocyte-secreted acetylcholine is required for maintenance of homeostasis in the heart. The FASEB Journal 27 (12), 5072–5082. Available at https://doi.org/10.1096/fj.13-238279.

Saludas, L., Oliveira, C.C., Roncal, C., Ruiz-Villalba, A., Prósper, F., Garbayo, E., et al., 2021. Extracellular vesicle-based therapeutics for heart repair. Nanomaterials 11 (3), 570. Available at https://doi.org/10.3390/nano11030570.

Salvioli, S., Basile, M.S., Bencivenga, L., Carrino, S., Conte, M., Damanti, S., et al., 2023. Biomarkers of aging in frailty and age-associated disorders: state of the art and future perspective. Ageing Research Reviews 91, 102044. Available at https://doi.org/10.1016/j.arr.2023.102044.

Shen, J., Dai, Z., Li, Y., Zhu, H., Zhao, L., 2022. TLR9 regulates NLRP3 inflammasome activation via the NF-kB signaling pathway in diabetic nephropathy. Diabetology & Metabolic Syndrome 14 (1). Available at https://doi.org/10.1186/s13098-021-00780-y.

Singh, Y., Fuloria, N.K., Fuloria, S., Subramaniyan, V., Almalki, W.H., Al-abbasi, F.A., et al., 2022. A European pharmacotherapeutic agent roflumilast exploring integrated preclinical and clinical evidence for SARS CoV-2 mediated inflammation to organ damage. British Journal of Clinical Pharmacology 88 (8), 3562–3565. Available at https://doi.org/10.1111/bcp.15328.

Sithara, T., Drosatos, K., 2021. Metabolic complications in cardiac aging. Frontiers in Physiology 12. Available at https://doi.org/10.3389/fphys.2021.669497.

Stojanovic, B., Milivojcevic Bevc, I., Dimitrijevic Stojanovic, M., Stojanovic, B.S., Lazarevic, T., Spasic, M., et al., 2025. Oxidative stress, inflammation, and cellular senescence in neuropathic pain: mechanistic crosstalk. Antioxidants 14 (10), 1166. Available at https://doi.org/10.3390/antiox14101166.

Straub, R.H., Buttgereit, F., Cutolo, M., 2011. Alterations of the hypothalamic-pituitary-adrenal axis in systemic immune diseases - a role for misguided energy regulation. Clinical and Experimental Rheumatology 29 (5), S23.

Subramanian, V., Tucker, W.J., Peters, A.E., Upadhya, B., Kitzman, D.W., Pandey, A., 2025. Cardiovascular aging and exercise: implications for heart failure prevention and management. Circulation Research 137 (2), 205–230. Available at https://doi.org/10.1161/circresaha.125.325531.

Thapa, R., Afzal, O., Alfawaz Altamimi, A.S., Goyal, A., Almalki, W.H., Alzarea, S.I., et al., 2023. Galangin as an inflammatory response modulator: an updated overview and therapeutic potential. Chemico-Biological Interactions 378, 110482. Available at https://doi.org/10.1016/j.cbi.2023.110482.

Ţica, O., Ţica, O., 2025. Molecular diagnostics in heart failure: from biomarkers to personalized medicine. Diagnostics 15 (14). Available at https://doi.org/10.3390/diagnostics15141807.

Tuncay, I.O., DeVries, D., Gogate, A., Kaur, K., Kumar, A., Xing, C., et al., 2023. The genetics of autism spectrum disorder in an East African familial cohort. Cell Genomics 3 (7), 100322. Available at https://doi.org/10.1016/j.xgen.2023.100322.

Vandereyken, K., Sifrim, A., Thienpont, B., Voet, T., 2023. Methods and applications for single-cell and spatial multi-omics. Nature Reviews Genetics 24 (8), 494–515. Available at https://doi.org/10.1038/s41576-023-00580-2.

Wang, L., Wang, J., Cretoiu, D., Li, G., Xiao, J., 2020. Exercise-mediated regulation of autophagy in the cardiovascular system. Journal of Sport and Health Science 9 (3), 203–210. Available at https://doi.org/10.1016/j.jshs.2019.10.001.

Wang, H., et al., 2022. Nicotinamide mononucleotide supplementation improves mitochondrial dysfunction and rescues cellular senescence by NAD (+)/Sirt3 pathway in mesenchymal stem cells. International Journal of Molecular Sciences 23 (23).

Wang, M.S., Deng, J.W., Geng, W.Y., Zheng, R., Xu, H.L., Dong, Y., et al., 2025a. Temporal trend and attributable risk factors of cardiovascular disease burden for adults 55 years and older in 204 countries/territories from 1990 to 2021: an analysis for the Global Burden of Disease Study 2021. European Journal of Preventive Cardiology 32 (7), 539–552. Available at https://doi.org/10.1093/eurjpc/zwae384.

Wang, W.L., Shih, Y.T., Wei, S.Y., Chiu, J.J., 2025b. Impacts of aging and fluid shear stress on vascular endothelial metabolism and atherosclerosis development. Journal of Biomedical Science 32 (1). Available at https://doi.org/10.1186/s12929-025-01177-z.

Wang, Y., Zhou, T., Zhao, J., Zhu, H., Tan, X., Chen, J., et al., 2025c. Calcium handling remodeling in dilated cardiomyopathy: from molecular mechanisms to targeted therapies. Channels 19 (1). Available at https://doi.org/10.1080/19336950.2025.2519545.

Weil, B.R., Graser, L., Rasam, S., Zimmer, H., Konecny, F., Techiryan, G., et al., 2025. Persistent fibrosis and left ventricular chamber stiffening despite cessation of repetitive pressure overload in swine. JACC: Basic to Translational Science 10 (6), 844–859. Available at https://doi.org/10.1016/j.jacbts.2025.02.009.

Weisser-Thomas, J., Nguyen, Q., Schuettel, M., Thomas, D., Dreiner, U., Grohé, C., et al., 2007. Age and hypertrophy related changes in contractile post-rest behavior and action potential properties in isolated rat myocytes. Age (Melbourne, Vic.) 29 (4), 205–217. Available at https://doi.org/10.1007/s11357-007-9040-1.

Wu, X., Yang, X., Dai, Y., Zhao, Z., Zhu, J., Guo, H., et al., 2024. Single-cell sequencing to multi-omics: technologies and applications. Biomarker Research 12 (1). Available at https://doi.org/10.1186/s40364-024-00643-4.

Yan, B., Yuan, Q., Guryanova, O.A., 2023. Epigenetic mechanisms in hematologic aging and premalignant conditions. Epigenomes 7 (4). Available at https://doi.org/10.3390/epigenomes7040032.

Yogesh, M., Makwana, N., Trivedi, N., Damor, N., 2024. Multimorbidity, health literacy, and quality of life among older adults in an urban slum in India: a community-based cross-sectional study. BMC Public Health 24 (1). Available at https://doi.org/10.1186/s12889-024-19343-7.

Yuksek, E.N., Pereira, A.G., Prieto, M.A., 2025. Dietary supplements derived from food by-products for the management of diabetes mellitus. Antioxidants 14 (10). Available at https://doi.org/10.3390/antiox14101176.

Yusifov, A., Woulfe, K.C., Bruns, D.R., 2022. Mechanisms and implications of sex differences in cardiac aging. Journal of Cardiovascular Aging 2 (2). Available at https://doi.org/10.20517/jca.2022.01.

Zatterale, F., Longo, M., Naderi, J., Raciti, G.A., Desiderio, A., Miele, C., et al., 2020. Chronic adipose tissue inflammation linking obesity to insulin resistance and type 2 diabetes. Frontiers in Physiology 10. Available at https://doi.org/10.3389/fphys.2019.01607.

Zhang, Y., et al., 2024. Engineered extracellular vesicles for tissue repair and regeneration. Burns Trauma 12, tkae062.

Zhang, X., Gao, Y., Zhang, S., Wang, Y., Pei, X., Chen, Y., et al., 2025. Mitochondrial dysfunction in the regulation of aging and aging-related diseases. Cell Communication and Signaling 23 (1). Available at https://doi.org/10.1186/s12964-025-02308-7.

Zhao, D., Wang, Y., Wong, N.D., Wang, J., 2024. Impact of aging on cardiovascular diseases: from chronological observation to biological insights: JACC family series. JACC: Asia 4 (5), 345–358. Available at https://doi.org/10.1016/j.jacasi.2024.02.002.

Zheng, Q., Wang, X., 2010. Autophagy and the ubiquitin-proteasome system in cardiac dysfunction. Panminerva Medica 52 (1), 9–25.

Chapter 3

Molecular mechanisms of senescence in cardiomyocytes

Neeraj Kumar Fuloria
Faculty of Pharmacy, AIMST University, Bedong, Kedah, Malaysia

3.1 Introduction

According to the World Health Organization, CVDs cause approximately 17.9 million deaths annually, representing nearly 31% of all global deaths. Aging is one of the leading risk factors for these conditions, especially heart failure (HF), which significantly increases with age. Approximately 1% of individuals aged >50 years suffer from HF, and this risk doubles with every additional decade (Naeem et al., 2024). In the United States, the population over 65 years has risen from 40 to 51 million in just ten years and is projected to reach 95 million by 2060. This demographic shift poses serious healthcare challenges as age-related heart conditions are on the rise. Moreover, heart failure severely affects the quality of life, causing symptoms such as shortness of breath, fatigue, swelling, insomnia, chest pain, and depression, which are often linked to higher hospitalization and death rates. The heart, which is composed of four chambers, undergoes structural and functional decline with age, even without a history of heart disease. Aging cardiomyocytes develop a senescent phenotype, which leads to myocardial dysfunction. Structural alterations, such as enlarged cardiomyocytes, increased fibrosis, and inflammation, progressively impair the heart's ability to contract and relax properly (Redgrave et al., 2023). In older individuals, exposure to metabolic stress, high blood pressure, or ischemic damage accelerates heart failure. Compounding this problem, the heart has limited regenerative ability, making it more vulnerable in older adults. The purpose of this book chapter is to explore the molecular mechanisms driving cardiac aging and cellular senescence, identify key proteins and signaling pathways involved, and highlight emerging pharmacological targets and therapeutic strategies that may help preserve heart function in the aging population.

3.2 Senescence in the aging heart

Senescence refers to the gradual decline in functional integrity that can be realized at both the cellular and organizational levels. At the cellular level (cellular or replicative senescence), when the cell accumulates damage or there is telomere shortening, the cell is permanently arrested in the cell cycle. In contrast, organismal senescence is characterized by a decline in the ability to balance stress and internal homeostasis, which eventually leads to the impairment of physiological functioning (Ogrodnik, 2021). Cellular senescence is commonly regarded as a central facilitator of aging in organisms and is the chief research subject in studies of aging. Cellular senescence is a form of irreversible cell cycle arrest that occurs when telomeres erode, triggering replicative senescence, or when caused by numerous types of stress, leading to premature senescence (Kumari and Jat, 2021). Senescent cells are also characterized by morphology and hallmarks such as a flattened and enlarged cell phenotype, larger lysosomal content, and active senescence-associated β-galactosidase (SA-gal-act), which is used as a diagnostic marker (Chen et al., 2020). Stress-induced senescence has all the characteristics of replicative senescence and can be triggered experimentally overnight by introducing agents that cause oxidative damage, such as hydrogen peroxide. Cardiac senescence is a physiological aspect of normal aging, but it can be enhanced by pathological factors, including chronic myocardial injury, coronary artery disease, hypertension, myocardial ischemia, hypertrophic cardiomyopathy, and pericardial disorders (Nakou et al., 2016). Such stimuli cause long-term strain on the heart, facilitating premature aging of heart cells.

Cellular Senescence and Cardiac Aging. DOI: https://doi.org/10.1016/B978-0-443-45811-8.00011-8

3.3 Cardiac cell diversity and physiological roles

The heart comprises a diverse array of cell types that work together to maintain its essential functions. Research has shown that noncardiomyocytes make up approximately 70% of the total heart cells, whereas cardiomyocytes account for the remaining 30%, with the latter further categorized into atrial and ventricular myocytes (Varga et al., 2017). Cardiomyocytes are primarily responsible for the contractile activity of the heart; however, their function is closely supported by noncontractile cells. Noncardiomyocytes include fibroblasts, ECs, SMCs, and pericytes, all of which contribute to maintaining the structural and functional integrity of the heart (Zhuang et al., 2020). Fibroblasts, which comprise more than 40% of ventricular cells, are essential for providing structural support and regulating the extracellular matrix to preserve tissue homeostasis. SMCs and pericytes, located within the blood vessel walls, are key to maintaining vascular stability and proper cardiac function (Frangogiannis, 2017).

Endothelial cells line the interior surfaces of blood and lymphatic vessels, where they help control blood flow by regulating vascular permeability and vessel diameter. They also influence the heart's development, contractile performance, and rhythmic function. Mesothelial cells, derived from the mesoderm, share functional characteristics with ECs and contribute to angiogenesis by releasing pro-angiogenic factors (Nagy et al., 2002). Additionally, cardiac adipocytes not only serve as energy reservoirs but also assist in tissue repair, angiogenesis, and immune regulation. Immune cells and neurons are also critical for preserving overall cardiac homeostasis, highlighting the complex interplay between cellular elements within the heart (Panina et al., 2018).

3.4 Cardiac cell senescence

The processes of cellular senescence primarily drive cardiac aging. As cells have a limit to how well they can divide, a halt in cell cycle division, cellular senescence, occurs naturally as an individual grows older. These senescent cells stimulate tissue remodeling, inhibit regeneration, weaken function, and maintain chronic inflammation. This study examined the impact of senescence on cardiomyocytes and nonmyocardial cardiac cell populations.

3.4.1 Senescent phenotypes in cardiac muscle cells

Approximately 80% of the volume of the myocardium volume comprises cardiomyocytes, and an estimation of cardiac cell counts equal to approximately one-third of the total cardiac cell count is composed of cardiomyocytes. The influence of senescence in these cells has a significant impact on organizational performance (Ewoldt et al., 2025). Old cardiac cells have less shortening, spontaneous discharges, irregular contraction style, and secrete senescence synthesis-associated secretory phenotype (SASP) particles, which transform adjacent cells, in particular fibroblasts (Dong et al., 2020). The key mechanisms underlying the deleterious consequences of aging involve the stimulation of the p53-p21 and p16INK4A-Rb tumor-suppressor systems. Furthermore, inflammatory signals and oxidative stress promote cardiomyocyte senescence, and telomere shortening, DNA damage, and mitochondrial dysfunction are among the canonical hallmarks of senescence (Yue et al., 2014). SASP signifies the regression of cytokines, chemokines, growth regulation, and proteins, which could lead to tissue repair or even aid in the progression of aging, depending on the context-responsive reaction. Senescent cardiomyocytes experience contractile impairment, loss of mitochondria, hypertrophy, and erosion of telomeres, which promote inflammation, derailed signaling inflammation, and cell depletion (Alqahtani et al., 2025). Increasing longevity and inflammatory processes in long-living Japanese centenarians are more influential accelerators than telomere loss. As a result, the reduction in senescence in cardiomyocytes plays an important role in maintaining myocardial health (Arai et al., 2015).

An imbalanced metabolism has been revealed as a factor leading to cardiac senescence due to energy deprivation. The imbalance between mitochondrial fission and fusion contributes to increased ROS, and defective mitophagy slows the clearance of damaged organelles. Deletion of three related genes (Drp1, MFN1/2) prolongs mouse survival (Fu et al., 2019). In contrast, FOXO3 and catalase counter oxidative damage. The accumulation of mtDNA mutations increases ROS levels and induces TLR9, resulting in the release of pro-inflammatory cytokines and maintenance of the SASP. Therefore ROS lies at the heart of aging (Dhoke et al., 2018). In glucose processing, senescent destruction triggers signals, such as cytosolic p53, which interfere with Parkin mitophagy. Pharmacological inhibition of p53 is protective in the heart in diabetic models (Hoshino et al., 2014). In addition, cell cycle dynamics are influenced by regulators such as Rb1, Meis2, and epigenetic enzymes, especially KDM4D. Increased miR-34 concentrations in the hearts of elderly individuals are associated with decreased contractile performance. SIRT1 overexpression modestly increases and restricts the development of hypertrophy and apoptosis, and SIRT7 has an anti-aging effect with p53 deacetylation (Kaleem et al., 2025). Senescent cardiomyocytes release pro-inflammatory molecules, including IL-1, IL-6, TNF, Edn3, and GDF15.

In addition to eliciting local inflammation, these factors trigger paracrine-mediated fibrotic transformation of fibroblasts. Therefore the SASP promotes intercellular aging of cardiac tissues (Espitia-Corredor et al., 2022).

3.4.2 Senescent endothelium in cardiovascular aging

ECs represent approximately 60% of the noncardiomyocyte population in the heart and play a vital role in maintaining vascular tone and vasodilation through the release of vasoactive mediators and growth factors. Cellular senescence in ECs is closely linked to cardiovascular pathologies, such as atherosclerosis, heart failure, and atrial fibrillation (Li, 2024). Various stressors, including oxidative stress, vascular inflammation, and metabolic abnormalities such as hyperuricemia or an imbalanced renin–angiotensin system, are known to trigger EC senescence. Epigenetic modifications, including microRNA regulation and protein acetylation, contribute to endothelial aging. For example, miR-217-induced EC dysfunction accelerates atherosclerosis in ApoE-deficient mice (de Yébenes et al., 2020). SIRT1 delays replicative senescence in ECs by deacetylating p53, whereas senescent ECs display reduced nitric oxide production and elevated endothelin-1 levels. These changes enhance crosstalk with neighboring cardiac cells, aggravating vascular inflammation and impairing vasodilation. Therefore targeting EC senescence has emerged as a promising therapeutic approach for mitigating age-associated vascular decline (Zhang et al., 2015). Studies have shown that neuregulin-1 (NRG1) signaling via ErbB tyrosine kinase receptors can effectively reverse EC senescence, whereas telomerase activation helps alleviate vascular dysfunction linked to aging (Lemmens et al., 2007). Endothelial dysfunction plays a central role in the pathology of heart failure by limiting the expression of eNOS and the levels of nitric oxide, thus positively influencing the overproduction of ROS. NF-κB is activated, and ICAM1 and VCAM1 are upregulated as a result of the ROS load. This further causes an increase in endothelial permeability, which boosts myocardial inflammation, hypertrophy, dilation, fibrosis, and systolic and diastolic dysfunction (Liang et al., 2011). Simultaneously, increased ADMA and subsequent inhibition of ATR1/ATR2 receptors reduce antiinflammatory, antifibrotic, and antiapoptotic signaling pathways. ROS in the mitochondria contribute to increased myocardial damage and trigger decreased exercise capacity. An imbalance in the NO sGC-cGMP-PKG cascade was also found, where downstream protective effects were dampened because of ROS, which in turn also limits the production of ECM, further enhancing cardiac pathophysiology (Fig. 3.1) (Yan et al., 2015). Endothelial senescence is also implicated in pulmonary hypertension and in age-related arrhythmias. For instance, SIRT6 depletion or miR-217 upregulation accelerates endothelial dysfunction and atherosclerosis. Senescence-accelerated mouse models (SAMP8/TaHsd) exhibit metabolic derangements, such as diastolic dysfunction, left ventricular hypertrophy, and interstitial fibrosis (D'Onofrio et al., 2018). Furthermore, atrial fibrillation, whose incidence increases with age, is associated with EC and fibroblast senescence, often due to eNOS downregulation. Elevated expression of senescence markers, such as p53 and p16, correlates with atrial fibrillation severity (Fumagalli et al., 2015) (Fig. 3.1).

3.4.3 Cardiac fibroblast senescence and tissue remodeling

In the adult myocardium, fibroblasts represent a population of approximately 20 % of noncardiomyocytes and play crucial roles in ECM homeostasis and cell–cell communication. Despite the fact that it is fibroblast senescence, which can be of use recovery of the tissues following the MI, the same can form part of the myocardial fibrosis which occurs with age (Chu et al., 2023). These cells support matrix activity by producing MMPs and integrins, which cause cellular adhesion and remodeling. Postacute MI stimulates the signaling pathway of p53/p21 to promote premature senescence of fibroblasts. Some of the proteins in SASP, including CCN1, have been demonstrated to reduce fibrosis in what seems to be a paracrine effect that causes senescence of fibroblasts in the period following MI, thus providing a potential solution to fibrosis (Gao et al., 2020a). In contrast, DNA repair NEIL3 has the opposite effect, enhancing post-MI outcomes by increasing MMP2 levels and, therefore exacerbating ECM degradation and increasing the chances of heart rupture (Olsen et al., 2014). Paracrine factors also play a role in fibroblast modulation of cardiomyocyte aging and hypertrophy. After cardiac insult, these cells secrete pro-inflammatory and profibrotic factors, such as IL-1, IL-6, TNF-α, and TGF-β, as well as, at the same time, produce IGF-1, which increases collagen deposition and hypertrophic growth of cardiomyocytes (Turner et al., 2007). As a person ages, senescence of cardiomyocytes and fibroblasts compounds inflammation, fibrosis, vascular rigidity, electrical imbalance, and cardiac deconditioning. Finally, senescent fibroblasts lead to further maladaptive remodeling and worsening of aging heart functionality.

3.4.4 Senescent vascular smooth muscle cells in vascular aging

The process of blood pressure and tonus regulation and general hemodynamics is accomplished with the help of using the action of vascular smooth muscle cells (VSMCs) and ECs. During the aging process, VSMC and other interstitial cells in

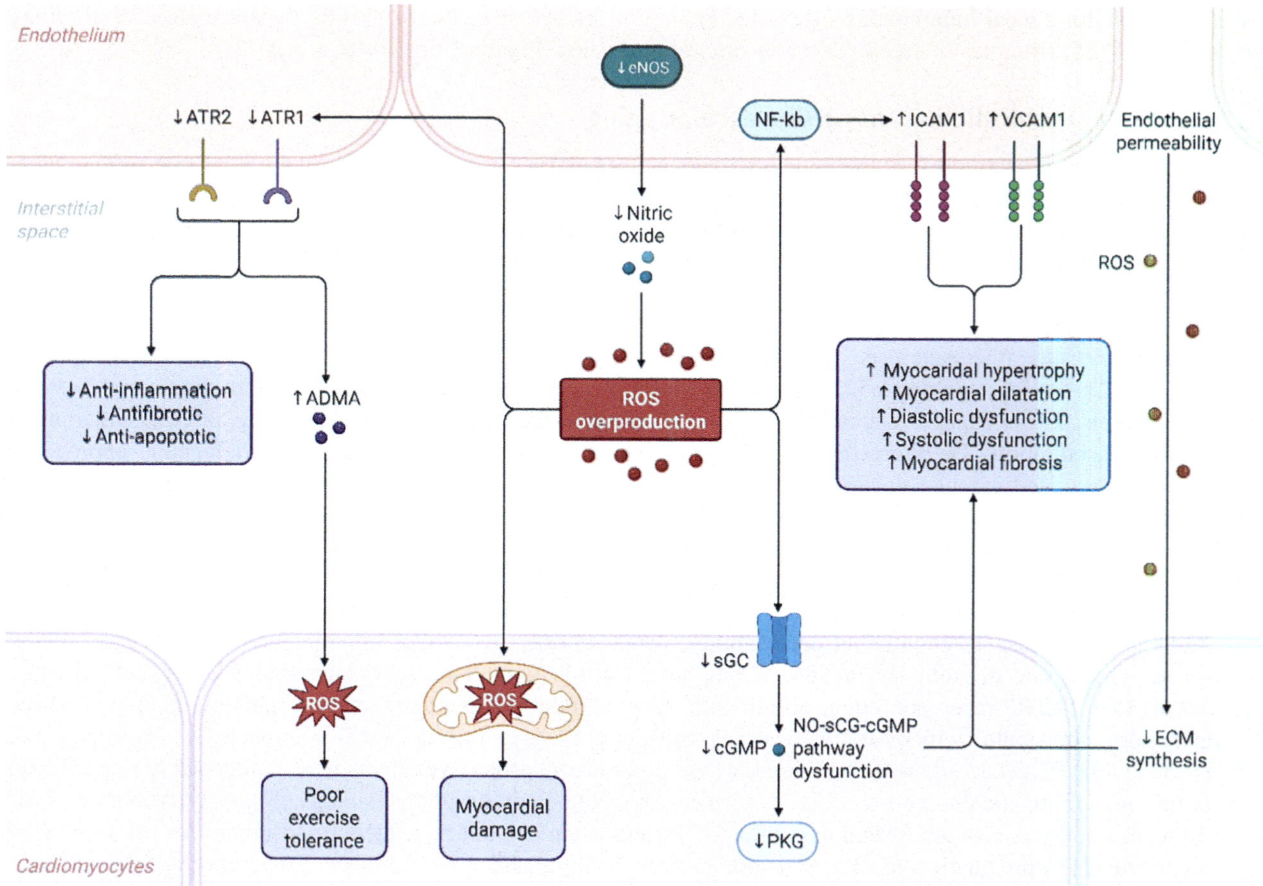

FIGURE 3.1 Endothelial senescence and cardiac vascular pathophysiology.

the vascular and cardiac tissues undergo cell cycle arrest, increase senescence markers, such as p16INK4A, and activate inflammatory pathways. Senescent cells also have a telomere-shrinking pattern and a pro-inflammatory secretory pattern called SASP. The accumulation of senescent VSMCs and further vascular calcification characterize the presence of advanced atherosclerotic lesions (Oh et al., 2021). VSMC senescence is triggered by numerous stressors, such as telomere shortening, oxidation, genomic instability, and defective autophagy. Senescent VSMCs usually have a high percentage of SA-β-Gal activity and an increase in the expression of p16, p21, and Rb. Notably, improper processing of prelamin A impairs the structure of the nucleus and makes cells more vulnerable to DNA damage (Tan et al., 2016). The accumulation of prelamin A caused by mutations in the LMNA gene leads to the initiation of chronic damage responses and release of osteogenic factors as a result of lowering the levels of ZMPSTE, a protease involved in the maturation of prelamin A, leading to chronic DNA damage responses by reducing the levels of ATP and initiating osteogenic factor release, which enhances the process of vascular calcification (Afonso et al., 2016). Reduced SIRT6 in VSMCs is accompanied by hyperacetylation of histones H3K9 and H3K27, worsened telomere maintenance, and enhanced senescence. VSMCs also produce more IL-6, MCP-1, CCL2, CCL3/4, and IL-1/IL-8, and fewer anti-inflammatory molecules during senescence. Upregulation of IL-1a enhances SASP signaling, which effectively increases local paracrine inflammation in the surrounding cells (da Silva Feltran et al., 2024). Similar to ECs, senescent VSMCs are essential in the pathology of the vascular system, including atherosclerotic disease and pulmonary hypertension. Plaque-derived VSMCs have shortened telomeres, increased oxidative stress, and increased senescence markers (Rivard and Andrés, 2000). Low levels of SIRT6 are common, and overexpression of SIRT6 improves senescence-associated phenotypes. Notably, over-accumulation of prelamin A in VSMC occurs in Hutchinson–Gilford progeria, which enhances vascular aging (Takasaka et al., 2014). The rapid evolution of these pathological states in patients with atherosclerosis can cause accelerated aging, and this quality of VSMC senescence highlights this aspect.

3.4.5 Cellular aging in cardiac valve interstitium

The most prevalent cell type in heart valve tissues is VICs, which are essential for the shape maintenance and regular operation of these structures. VICs age gradually and in a process known as senescence, which can interfere with valve integrity and can be involved in most age-related valvular diseases (Liu et al., 2007). Aortic stenosis is an important disease of clinical relevance; it is often caused by the calcification of aortic valve leaflets, and its prevalence escalates significantly after reaching old age (Izquierdo-Gómez et al., 2017). These aortic valves have increased levels of the senescence markers p16 and p53, indicating that cell aging in these aortic valves is directly related to valve dysfunction. Inflammatory macrophages in the valvular tissue can release cytokines (IL-1z, TNF-z, and IL-6) to promote osteogenic differentiation and fibrosis in VICs, which helps initiate valvular calcification. In experimental studies, AMPK activation in porcine aortic VICs causes senescence, characterized by the upregulation of ROS production and formation of calcific nodules (Fang et al., 2009). Various molecular changes that support their pathophysiological phenotype include increased histone H3 acetylation, loss of microRNA, especially miR-17 and miR-30a, and upregulation of cyclin-dependent kinase inhibitor p21 have been found in senescent VICs. These cells also exhibit greater histochemical activity of SA-8-Gal, which also indicates a senescent phenotype (Luan et al., 2024). Overall, these alterations at the cellular level are implicated in the pathological remodeling, stiffening, and calcification of the valves, highlighting the key issue concerning the role of senescent VICs in valvular aging and dysfunction.

3.5 Senescence-associated alterations in cardiac cells

3.5.1 Dysregulation of senescence signaling networks

Cardiac senescence is one of the main characteristics of advanced aging and is related to a series of molecular and cellular disturbances. One of the peculiar features of the aging process is the long-term stimulation of stress-related intracellular pathways, including p38 MAPK, SAPK/JNK, and NF-κB. These pathways trigger and drive cellular senescence and constitute the pathophysiology of cardiovascular disease. Abnormalities in these cascades increase with chronic stress conditions that characterize aging and cause an increase in inflammatory responses and oxidative damage (Han et al., 2007). A prolonged demand for these signaling networks hastens cellular aging through inflammatory pathways and tissue malfunction. Mitochondria produce a large amount of reactive oxygen species (ROS) during aging, and studies have shown that mitochondrial ROS are crucial for activating stress-induced aging pathways. ASK1-signalosome complexes are one such mechanism and act via ROS, their sensitivity leading to the activation of the p38 MAPK pathway and eventually, the senescent phenotypes in cells (Salminen et al., 2012). Research has also shown that oxidative stress occurring in diseases is highly related to the ASK1 p38 MAPK pathway. Malfunctioning mitochondria result in approximately 90% of ROS production in aging tissues, and such ROS induce the p38 MAPK and SAPK/JNK pathways. They are signals that increase the rate of cellular senescence and play a role in the formation of dysfunction in cardiac structures and functions related to age (Hsieh et al., 2003). These pathways constitute major molecular factors leading to cardiac aging and therapeutic targets for protecting cardiac function with age by mitigating oxidative stress.

3.5.2 Chromatin and DNA modifications in senescent heart cells

Cardiac senescence is a major characteristic of cardiovascular aging. In line with this, new evidence has shown that epigenetic alterations play a central role in the pathogenesis of this process. Chronological aging, cellular senescence, and specific epigenetic signatures are significantly interconnected in the myocardium in both murine and human studies. At the core of the molecular dynamics of cardiovascular aging lie variations, such as the control of the expression of genes, and not a change in the actual sequence of the genetic material itself. The epigenetic fingerprints of senescent cardiomyocytes include pronounced distortion of DNA methylation patterns, alterations in histone marks, rearrangement of chromatin structure, and ncRNA expression patterns. Hypomethylation of DNA, which is more marked in late-replicating areas of the genome and in promoter-adjacent sequences, is an obvious trait that might lead to one or more disturbances in gene regulation and increased genomic instability (Alrumaihi et al., 2025). At the chromatin level, there are complementary changes to those at the DNA level: the pattern of methylation and acetylation of histones is significantly altered, and ATP-dependent chromatin remodeling complexes work less efficiently. Simultaneously, the gene expression program controlling aging and senescence is regulated by the dysregulation of ncRNAs (Ikegami et al., 2009). Overall, these adaptations redefine the transcriptional landscape of the aged myocardium, increasing the rate of functional deterioration and predisposing it to cardiovascular diseases. The next section outlines the processes by which epigenetic regulators promote cardiac aging and addresses the possible therapeutic targets in this regard.

3.5.3 Altered protein and metabolite profiles in cellular senescence

Amongst these aspects of cellular senescence are proteomic and metabolomic remodeling, which define changes in the functioning of mitochondria as well as those in metabolic control. Human fibroblasts senesce have been demonstrated to undergo wide-scale changes in their mitochondrial populations, such as significant remodeling of their mitochondrial proteome using proteomic profiling (Karpov et al., 2024). The senescence-induced modification of mitochondrial proteins approximates 40 % of all mitochondrial proteins, of which there is a large upsurge in the membrane-bound proteins compared to that of the matrix-located proteins, and therefore structural and functional mitochondrial malleability in the aged cells (Kamogashira et al., 2017). At the same time, these proteomic changes are coupled with metabolic reprogramming, the senescent cells show an increase of branched-chain amino acids degradation and elevation of lipid metabolism, and processes like 1 C folate metabolism and OXPHOS become suppressed. These alterations are highly associated with a weakened potential to produce proteins and change the homeostasis of energy (Xiu and Field, 2020). Because SASP regulation largely depends on mitochondria, metabolic rewiring has a potential direct impact on SASP expression and the consequences of the senescent cell phenotype in age-related diseases (Salman et al., 2024). In the metabolome, senescent cells exhibit elevated levels of free cytosolic polyunsaturated fatty acids, which are rearranged into pro-inflammatory oxylipins, which are major constituents of SASP. The accumulation of lipid droplets is a feature of senescence. In addition, the deposition of cholesterol in the ER in a p53-dependent manner interferes with the production of SREBP2 and leads to the disruption of lipid homeostasis (Waku et al., 2021). Senescent cells are usually characterized by high amounts of mitochondria or mtDNA, abnormalities in membrane potential, and increased production of ROS, all of which are related to the pro-senescence cellular environment and inflammatory signaling.

3.6 Molecular control mechanisms of senescence

Senescence is a multifactorial biological program that is initiated in response to a heterogeneous set of internal and environmental assaults. Notable mediators of these processes include epigenetic remodeling, metabolic rewiring, shortened telomeres, senescence-associated secretory phenotype (SASP), and other dysregulated basic signaling modules, including mTOR and canonical tumor-suppressor cascades. All such perturbations trigger different intracellular cascades, leading to phenotypic senescence. Notably, there is cell specificity in the molecular structure of senescence (Ossa and Crews, 2006).

3.6.1 Molecular cascades of tumor suppression

Two main tumor-suppressor pathways, p53/p21 and p16/Rb, are located based on which the nucleus of the senescence-inducing and maintaining mechanisms is formed by engineering permanent cell cycle arrest. The p53 protein has an extensive range of activities and regulates cellular metabolic homeostasis, DNA repair, autophagy, apoptosis, and cell cycle progression (Kandhaya-Pillai et al., 2023). Its ability to act is refined through posttranslational modifications, including acetylation, phosphorylation, and ubiquitination. The p53 effector p21 represses cyclin-dependent kinases during the G1/S and G2/M transitions to stabilize cellular senescence. It also interacts with and derails caspase activities, preventing apoptosis and further supporting the cell in establishing its commitment to senescence (Al Bitar and Gali-Muhtasib, 2019). In parallel, the p16/Rb signaling pathway ensures growth arrest by inhibiting CDK4/6 activity, preventing Rb phosphorylation, and maintaining the growth-inhibitory state (Konecny et al., 2011). The combination of these pathways prevents damaged or stressed cells from reentering the mitotic cycle, thereby maintaining tissue integrity in normal physiology. However, chronic activation of these pathways can lead to age-related changes in their function.

3.6.2 mTOR signaling in cellular growth and aging

The mTOR is a serine/threonine kinase that is at the crossroad of cellular growth and survival, metabolism, and responsive to both extracellular and intracellular signals. mTOR is a central regulator of nutrient availability, energy status, and stress messages and regulates diverse cellular mechanisms, such as protein synthesis, autophagy, and mitochondrial activity (Schirrmacher, 2020). The role of mTOR in cellular senescence is more relevant to the expression of the senescence phenotype and the involvement in the maintenance of the senescence phenotype, especially where it is known to exist in inversely correlated relationships with the PI3K/AKT pathway and p53 signaling roles. Simultaneous activation of mTOR and p53 during oncogenic stress leads to cells in a senescent state that manifests itself in permanent cell cycle arrest and metabolic reprogramming (Menendez et al., 2011). Interestingly, mTOR is a molecular switch that determines whether a given cell will senesce or become quiescent. In the case where mTOR and p53 are activated concurrently, a senescence pathway is activated (Li et al., 2016). In contrast, when p53 is activated but when there is not, the cell tends to be in quiescence, a state that is not irreversible and does not have the characteristics of senescence. The significance of mTOR is provided by this distinction in the way it contributes to cell fate determination under stress conditions and provides therapeutic potential against age-related diseases and cancer (Kalinin et al., 2023).

3.6.3 Chromatin remodeling in cellular senescence

Epigenetic regulation, such as DNA methylation, histone modifications, chromatin remodeling, and the function of ncRNAs, is a fundamental regulator of cell senescence, with different patterns in aged cells. Global DNA hypomethylation of late-replicating regions and promoter-proximal locations is an epigenetic signature of senescent cells and a factor in stable cell arrest and disturbed gene expression (Ikegami et al., 2009). The family of NAD+-dependent histone deacetylases, sirtuins, has protective effects against senescence in numerous cell types. SirT1 inhibits the expression of SASP factors in cardiomyocytes and activates eNOS in endothelial cells, which decreases oxidative stress and helps preserve vascular functional capacity (Furukawa et al., 2006). Another type of RNA involved in regulating senescence is miRNA. High miR-22 expression in an aged cardiac state increases senescence in fibroblasts, and miR-29 aggravates cellular aging due to its suppression of histone H4K20 trimethylation (H4K20me3), thus causing a loss of chromatin states. Other miRNAs control the rate of senescence by regulating the expression of major epigenetic and regulatory genes (Jazbutyte et al., 2013). lncRNAs play crucial roles in the regeneration and cardiac aging, where long noncoding RNAs (lncRNAs) are starting to play their crucial roles. They operate by communicating with RNA-binding proteins, serving as miRNA molecular sponges, and forming regulatory lncRNA-miRNA-mRNA axes. For example, lncRNA H19 accelerates the aging of cardiac cells by suppressing miR-19, which unlocks the p53/p21 pathway (Zhuang et al., 2021). Moreover, the loss of lncRNA SNHG12 damages the DNA repair system, increases the rate of endothelial senescence, and promotes the development of atherosclerosis (Haemmig et al., 2020).

3.7 Circadian control of the senescence-associated secretory phenotype

SASP refers to a set of bioactive effectors secreted by senescent cells. Examples of the members of this phenotype are pro-inflammatory cytokines, such as IL-6 and IL-1; chemokines; microRNAs; soluble cytokine receptors, such as TNF receptors; nonprotein factors, such as nitric oxide; growth factors, such as EGF, VEGF, and NGF; and extracellular matrices, such as fibronectin, collagen, and laminin (Radandish et al., 2021). The expression of IL-6 is closely connected to DDR and is independent of p53; it stimulates local cells (paracrine) by the expression of IL-6 receptors (Chen et al., 2019). Another feature of SASP is IL-1, which significantly contributes to the initiation, maintenance, and amplification of inflammatory gene expression and consequently acts as a leading senescent-inflammatory mediator (Prattichizzo et al., 2016). Additionally, SASP factors do not just maintain the senescent state, but also have harmful influences on neighboring cardiomyocytes due to paracrine signaling, supporting the essentiality of noncardiomyocyte cells in cardiac senescence and aging (Khavinson et al., 2022).

3.7.1 Clock genes in the control of senescence

Empirical data reveal a connection between circadian disruption, decreased lifespan, and elevated immune cell senescence in many nonmammalian species. A well-documented example is the finding that mutations in the canonical clock gene Per2 trigger vascular smooth muscle cell (VSMC) senescence as a result of an active response by AKT signaling (Wang et al., 2008). Due to the circadian regulation of cardiac rhythmicity, time positioning can consequently cause a faster rate of cardiovascular aging and predispose a person to cardiovascular disorders. Moreover, senescent VSMCs could also contribute to the disruption of the chronobiotic messages, furthering vascular and cardiac dysfunction in aged tissue (Guo et al., 2022).

3.8 Pathophysiological triggers of heart cell aging

Cellular senescence contributes significantly to cardiovascular pathologies by promoting structural and functional deterioration of the heart. Several factors underlie cardiac aging, including metabolic imbalances, telomere attrition, impaired intercellular communication, genetic susceptibility, and epigenetic modification.

3.8.1 Cardiac energy metabolism and age-related decline

The metabolic state of cardiomyocytes is exceptionally differentiated and undergoes extensive changes during development, acute stress, and as a part of chronological age. Such metabolic alterations have a significant effect on the induction of cellular senescence and a reduction in cardiac function. Significantly, there is a significant downregulation of the membrane-bound enzyme CPT1, which is vital for the use of fatty acids in the mitochondria, in old rat hearts (Virmani et al., 2015). Reduced CPT1 expression is the driving force behind pressure-overload hypertrophy and rate of senescence in the heart. Similarly, lipid metabolism master switches, such as PPARα and its coactivator PGC-1α, are also significantly decreased with age (Corton and Brown-Borg, 2005). The presence of insulin signaling with the IRS-1 pathway enhances the SASP, thereby promoting aging in cardiomyocytes; IL-2, p53, and agonist activation, on the other hand, demonstrate increased glycolytic metabolism that sustains senescence. Conversely, the use of p53 inhibitors reduces aging of the heart, in addition to ameliorating the cardiac dysfunction prevalent in diabetes (Zhang et al., 2022). Heightened uptake of ketone bodies, notably β-hydroxybutyrate and acetoacetate, serves as a substitute form of energy in the old heart, hence saving on ATP levels, such that it is capable of curtailing oxidative and inflammatory stress. Insufficient use of ketone bodies, including those found in SCOT-deficient mice, causes the onset of mitochondrial dysfunction and accelerated aging of the heart (Manolis et al., 2023). The hexosamine biosynthetic pathway (HBP) is also involved in cardiac aging. The activation of HBP and associated protein O-GlcNAcylation by UDP-GlcNAc is initiated by GFAT (Grilo et al., 2024). This signal transduction system is a posttranslational modification that imparts cytoprotective activity based on an alleviating effect on calcium overload, endoplasmic reticulum stress, mitochondrial permeability, and heat shock responses, albeit also playing a role in constituting ischemic injury in some conditions.

3.8.2 Mitochondrial morphology and functional adaptation

Dysfunction is a central phenomenon in cardiac aging, particularly in the senescence process. One of the key mediators of this imbalance is the imbalance between mitochondrial fusion and fission processes. As one ages, this balance shifts such that the mitochondria become elongated and morphologically irregular. The results showed that FIS1 and DRP1 were reduced in expression, and fusion proteins MFN1, MFN2, and OPA1 had an increased expression in aged cardiomyocytes (Chang and Blackstone, 2010). This change favors fusion and encourages the development of a highly networked mitochondrial formation. Despite the fact that fusion plays a role in maintaining mitochondrial integrity and increasing myocardial energy efficiency, fission is critical in the process of eliminating defective organelles via illegitimate mitochondrial digestion. When the balance between fusion and fission is disrupted, dysfunctional mitochondria are left in the cells, which aggregate oxidized proteins and produce oxidative stress (Cho et al., 2013). These dysfunctional mitochondria then potentiate ROS, which cause the destruction of lipids, proteins, and mitochondrial DNA, thereby disrupting metabolism. The progressive acquisition of damaged mitochondria and attendant oxidative damage aggravate the senescent phenotype and directly contribute to the aging of the heart (Victor et al., 2009). Metabolic dysfunction due to altered mitochondrial turnover not only affects energy balance but also serves as a force for the decline of myocardial performance in aging. The preservation of proper mitochondrial dynamics plays an essential role in protecting cardiac cells against senescence-related impairment.

3.8.3 Sirtuin-mediated pathways in cardiac longevity

Sirtuins are nicotinamide adenine dinucleotide (NAD+)-dependent deacetylase enzymes whose presence has been of significant interest and study in association with longevity and metabolic homeostasis, especially that which is achieved through caloric restriction. There are seven family members (SIRT1–7). In this group, SIRT1, SIRT2, SIRT3, SIRT6, and SIRT7 demonstrate cardioprotective effects and participate in processes related to the aging of the heart (Bordone and Guarente, 2005). SIRT1 is the most well-known of these enzymes because of its characteristic influence on prolonging life and protecting the heart. Destruction of SIRT1 in heart-specific models leads to metabolic changes in accordance with what is seen in the aged myocardium, particularly when it is placed under ischemic stress. Moreover, SIRT2 can also contribute to cardioprotection by counteracting angiotensin II–induced cardiac hypertrophy by deacetylating STK11, thereby inducing AMPK signaling, an important regulator of the energy metabolic state (Alcendor et al., 2007). Mitochondrial sirtuins, such as SIRT3, SIRT4, and SIRT5, are mainly involved in the control of mitochondrial metabolism, oxidative stress responses, and mitochondrial integrity, which are some of the key components of the aging of the cardiovascular system (Juszczak et al., 2024). SIRT6 is an enzyme with multiple functions that helps maintain genomic stability and assists with repair, affects glucose metabolism, and helps maintain telomeres in good shape (Khan et al., 2018). Inadequacy in any of SIRT6 or SIRT7 has been linked to premature aging traits, but variants in SIRT5 and

SIRT7 are associated with increased lifespan and caloric restriction responsiveness (Duszka et al., 2020). The issue of longevity modulation has pointed out dietary manipulations, including calorie restriction and famine activities, together with intermittent fasting and pharmacologic interventions, such as OXPHOS inhibitors and other signaling inhibitors that extensively influence the insulin/IGF-1 signaling cascade and the AMP kinase pathway. The resulting blockade of these cascades directly regulates the TOR complex and AMP kinase enzymes, thereby controlling transcription factors such as FOXO, NRF2, HSF1, BMAL1, and CLOCK1. Transcriptional regulators diverge into compositionally and functionally differentiated effector networks, such as autophagy, redox homeostasis, metabolic adaptation, stress responses, and cell cycle regulation (Fig. 3.2) (Trobec et al., 2011). All of these interconnected molecular modules are involved in lifespan

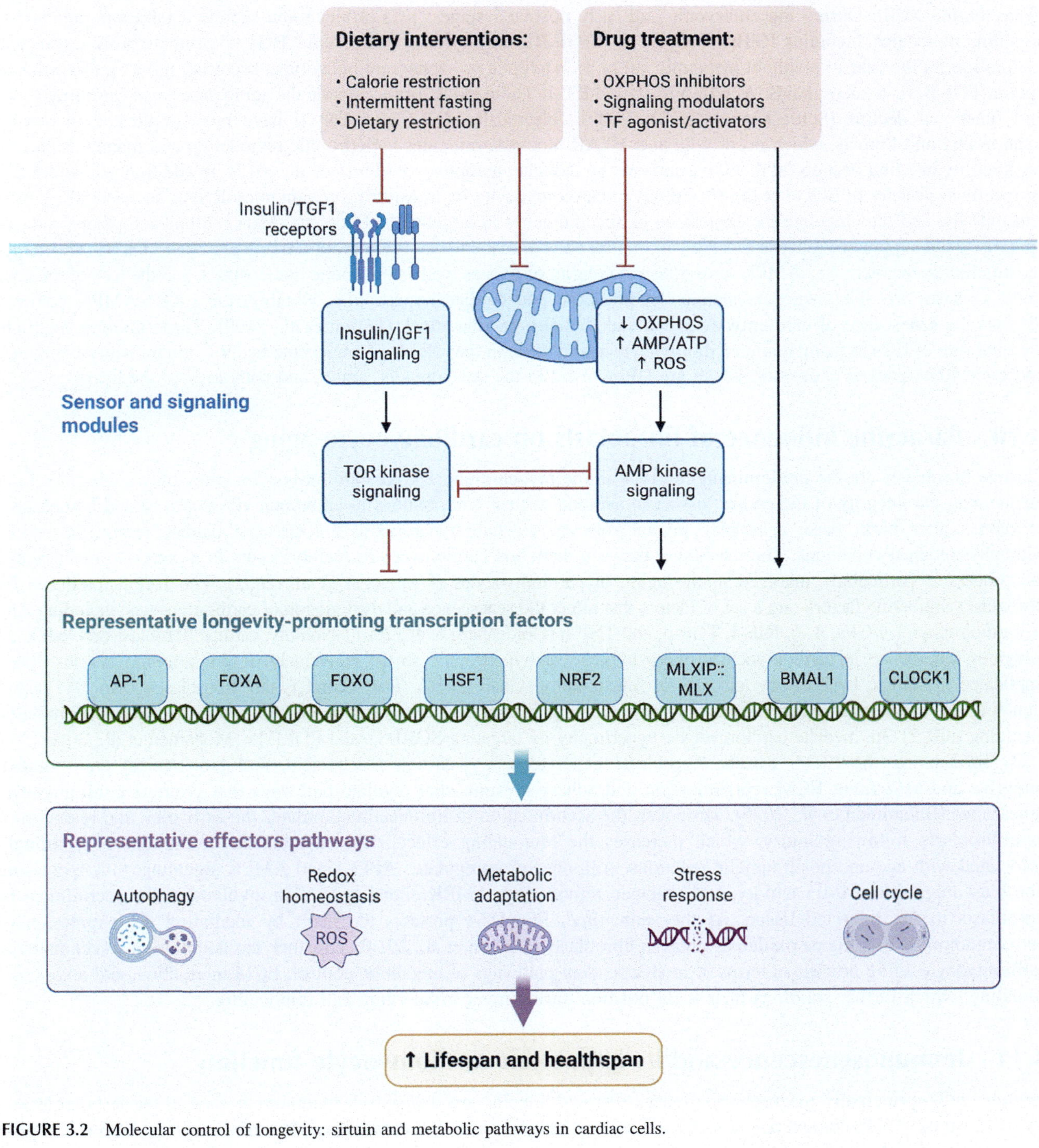

FIGURE 3.2 Molecular control of longevity: sirtuin and metabolic pathways in cardiac cells.

expansion and healthspan preservation in animal models and humans. The results confirm the main idea that sirtuins play a crucial role in cardiac health and metabolic processes. Sirtfoods Dietary compounds that stimulate sirtuin activity through natural means currently attract attention as they are expected to boost metabolic balance and extend healthspan by affecting sirtuin-regulated processes (Fig. 3.2).

3.9 Endothelial influence on cardiomyocyte aging

The powerful aspect of these changes is that metabolic reconfiguration occurs in ECs during the aging process, which is the underlying factor in cardiac senescence. In addition to their classical functions as vascular barriers, ECs also influence the immediate cardiovascular microenvironment via paracrine secretomes, and thus cardiomyocyte performance and aging (Tang et al., 2020). During the embryonic and early postnatal stages, ECs produce and secrete a mixed population of signaling molecules, including PTHrP, NO, ET-1, PDGF-B, Ang II, PGI2, PGE2, and NRG1, to promote proliferation and maturation. In the case of resultant metabolic stress, ECs adopt a pro-senescent phenotype, secreting more factors, such as apelin, TGF-β, IL-6, eicosanoids, Ang II, NRG1, and ET-1. These mediators accelerate the aging process of cardiomyocytes and functional decline (Schreckenberg et al., 2009). Specifically, ET-1 and Ang II have been implicated in cardiac hypertrophy and fibrosis, which occur with age. ETAR blockade mitigates hypertrophic remodeling and maintains cardiac contractility in aging or high-fat diet environments by inducing autophagy (Adiarto et al., 2012). In addition, exosomes and extracellular vesicles of ECs alter the physiology of cardiomyocytes by transferring certain microRNAs, such as miR-126-3p and miR-5p, and thus regulate the expression of specific genes in target cells. Exosomal Mst1 inhibits autophagy, induces apoptosis, and suppresses glucose metabolism in mast cells of diabetic hearts (Improta-Caria et al., 2024). LKB1, an energy-sensing kinase that activates AMPK, preserves the polarity of ECs as well as metabolic homeostasis, and the loss of this axis in ECs causes EC dysfunction, systemic hypertension, and cardiac hypertrophy. Notably, the LKB1-AMPK pathway disrupts the senescence of cardiomyocytes via endothelial paracrine attack (Jansen et al., 2009). These findings highlight the vital role of ECs in maintaining cardiomyocyte health through the release of angiocrine factors, inflammatory cytokines, and microRNA-packed exosomes, which are all involved in the development, aging, and pathology of the heart.

3.10 Paracrine influence of fibroblasts on cardiomyocyte aging

Cardiac fibroblasts are the predominant cell type in the myocardium. Unlike cardiomyocytes, they play a central role in maintaining the integrity of the cardiac tissue. In addition to their contributions to embryonic development and homeostasis of tissues after birth, these cells play pivotal roles in reactive hypertrophy, fibrosis, and cardiac remodeling under pathophysiological conditions. The interaction between them and cardiomyocytes mainly occurs in a paracrine manner, and the process is particularly involved in the aging of cardiomyocytes (Venugopal et al., 2022). The factors produced by fibroblasts-angiocrine factors- are a set of factors that affect the senescence and dysfunction of cardiomyocytes, including Ang II, cardiotrophin-1, FGFs, IL-6, IGF-1, TGF-β, and TNF-α (Goldenberg et al., 2001). Notably, cardiac fibroblast-derived IL-33 mitigates senescence in cardiomyocytes under hypoxic or hypertrophic stress; conversely, IL-11 secretion has also been implicated in cardiac hypertrophy and loss of functionality (Ghali, 2020). The second is that fibroblast-specific ATPase 4 signaling also regulates hypertrophic reactions by increasing the amount of produced sFRP2. Exosomes secreted by fibroblasts, including miR-21-3p, mediate cardiomyocyte hypertrophy by targeting SORBS2 and PDLIM5 (Mohamed et al., 2016).

By modulating the ECM, cardiac fibroblasts generate a myocardial, mechanical, and biochemical environment. Integrins and MM favor ECM reorganization and adhesive events that regulate pathways that promote cardiomyocyte senescence (Balaraman et al., 2025). Moreover, the accumulation of fibronectin stimulates the activation and retention of myofibroblasts following injury, which increases the remodeling effects. The metabolism of fibroblasts is strongly associated with age and posttraumatic reactions. Adiponectin upregulates APPL1 and AMPK signaling, thus increasing fibroblast migration, MMP activity, and collagen remodeling. AMPK signaling is also involved in the recruitment of fibroblasts to the infarcted tissue. At the same time, FOXO3A protects the heart by mediating the expression of peroxiredoxin III to mitigate oxidative stress in fibroblasts (Dadson et al., 2014). Together, cardiac fibroblasts regulate the cardiomyocyte aging process in terms of interdependent pathways of metabolic control, ECM interaction, and angiocrine signaling, which further solidifies their focal position during myocardial aging and remodeling.

3.11 Immunosenescence and its impact on cardiomyocyte function

Immune cells, principally macrophages, mast cells, and T cells, are decisive in maintaining cardiac tissue homeostasis and orchestrating pathological reactions by influencing inflammation and cardiomyocyte senescence regulation. It plays a central role in heart development, regeneration, and remodeling, and macrophages occupy a central position (Varricchi et al., 2020). Deletion of connexin 43 in macrophages leads to enhanced senescence of cardiomyocytes and poor electrical propagation. Activation of the NLRP3 inflammasome in cardiac macrophages aggravates the expression of IL-1b in a diabetic environment, long-term extension of cardiomyocyte action potential, and accelerates senescence (Yang et al., 2022). Adjacent findings in atrial fibrillation can demonstrate that depletion of hepcidin molecules encourages the gathering of CCR2+ inflammatory macrophages, which in turn stimulates the phosphorylation of STAT3 and affects the regeneration of cardiomyocytes. Despite these data strongly implicating macrophages in cardiomyocyte aging and regeneration, their particular subsets and paracrine factors have not yet been fully characterized (Li et al., 2023). Infected or injured cells produce DAMPs during the inflammatory phase, activating PRRs on resident macrophages. Subsequent research on signaling cascades has shown that they enhance the production of pro-inflammatory cytokines, with IL-1-B and IL-18 being the most prominent. CD14 monocytes are activated by cytokines IL-1-B and IL-18, which bind to their cognate receptors and cause the recruitment of monocytes through chemokine receptor CCR2 signaling. During the next proliferative phase, Ly6C monocytes, which secrete an antiinflammatory cytokine, such as IL-10, and transform. Such collagenated environment undergoes the maturation stage, whereby the enzymatic cross-linking process makes collagen fibrils insoluble to stabilize the forming fibrotic tissue (Burmeister and Marriott, 2018). Finally, rapid collagen replacement leads to cardiomyocyte senescence.

Conventional wisdom dictates that T cells are adaptive immune cells that also affect the aging of the heart. T cells, which secrete IL-17A, worsen apoptosis of cardiomyocytes after ischemia/reperfusion injury; however, the blockage of activation in T cells avoids hypertrophic remodeling and cell death of cardiomyocytes in the condition of stress evoked by pressure overload (Liao et al., 2012; Sharma et al., 2025; Subedi et al., 2025). The transfer of regulatory T cells (CD4+CD25+ Tregs) has been demonstrated to tile down Ang II–induced scarring and heart destruction. The secretory products of T-items, such as cystatin F, FGL-2, IL-33, matrilin-2, TNFSF11, and IGF-2, enhance cardiomyocyte proliferation and anti-senescent activity, hence their regenerative capacity (Emmerson et al., 2018). Cardiac aging and tissue remodeling are also associated with mast cells. When they are run down, contractility decreases, and postischemic recuperation is worse than normal. In contrast, mast cell activation results in the discharge of chymase, which enhances hypertrophy and senescence of cardiomyocytes (Wang et al., 2016). The activation of immune cells is also regulated by metabolic pathways. Pro-inflammatory M1 macrophages are glycolytic, and anti-inflammatory M2 macrophages are oxidative phosphorylation-biased in systems with tissue repair. AMPK also plays a regulatory role in these metabolic shifts and remains a moderating factor in inflammation and the correlation between metabolic imbalance and aging cardiac abnormalities (Wang et al., 2018a). Taken together, these results support the role of immune cells and immune cell metabolism in regulating cardiomyocyte senescence and heart health (Fig. 3.3).

3.12 Epigenetic and transcriptional control in cardiac senescence

It has been suggested that genetic factors alone cannot substantively affect overall human longevity or cardiac aging; nonetheless, they have been shown to have an impressive impact, with an estimated proportion of approximately 25% of the interindividual variability in lifespan. The remaining part consists of complex interactions among lifestyle, environmental exposure, and epigenetic changes. Genome-wide association studies have now clarified loci related to late-onset cardiovascular diseases and healthy aging, showing prospective genomic targets for prevention and therapy (Wang et al., 2018a). The most reliably reported variants are APOE and FOXO3, which are associated with a longer life expectancy. Other loci, including ABO, APOE-TOMM40, CDKN2B-ANRIL, and SH2B3-ATXN2, have also been suggested to play a role in longevity (Nygaard et al., 2014). In contrast, Hutchinson–Gilford progeria and Werner syndrome, LMNA, and WRN mutation-mediated syndromes of premature aging, respectively, have devastating impacts on genome stability and DNA repair, which activate aging and predispose individuals to diseases and cardiovascular diseases (Whisenant, 2022). Overlapping variants have been identified across populations in China, Europe, and the United States and are commonly found at 5q33.3, associated with lower midlife blood pressure and diminished cardiovascular risk in midlife. Such results not only emphasize the genetic part of aging but also the realization that environmental and behavioral factors, including smoking, diabetes, education level, and psychological characteristics, all contribute to an equally powerful part (Tang et al., 2015; Pandey et al., 2025a, 2025b).

Epigenetic processes alter gene function without changing the DNA sequence. Among these critical processes are DNA methylation, alterations to histones, chromatin remodeling, and noncoding RNA control. This type of adaptation is dynamic and reversible and often reacts quickly to the environment. In old age, epigenetic changes affect the structure of chromatin, patterns of gene expression, and genomic stability, which have a substantial impact on cardiac aging and pathogenesis (Zingg, 1997). Epigenetics methylation of DNA is a key regulator of gene expression. To carry out

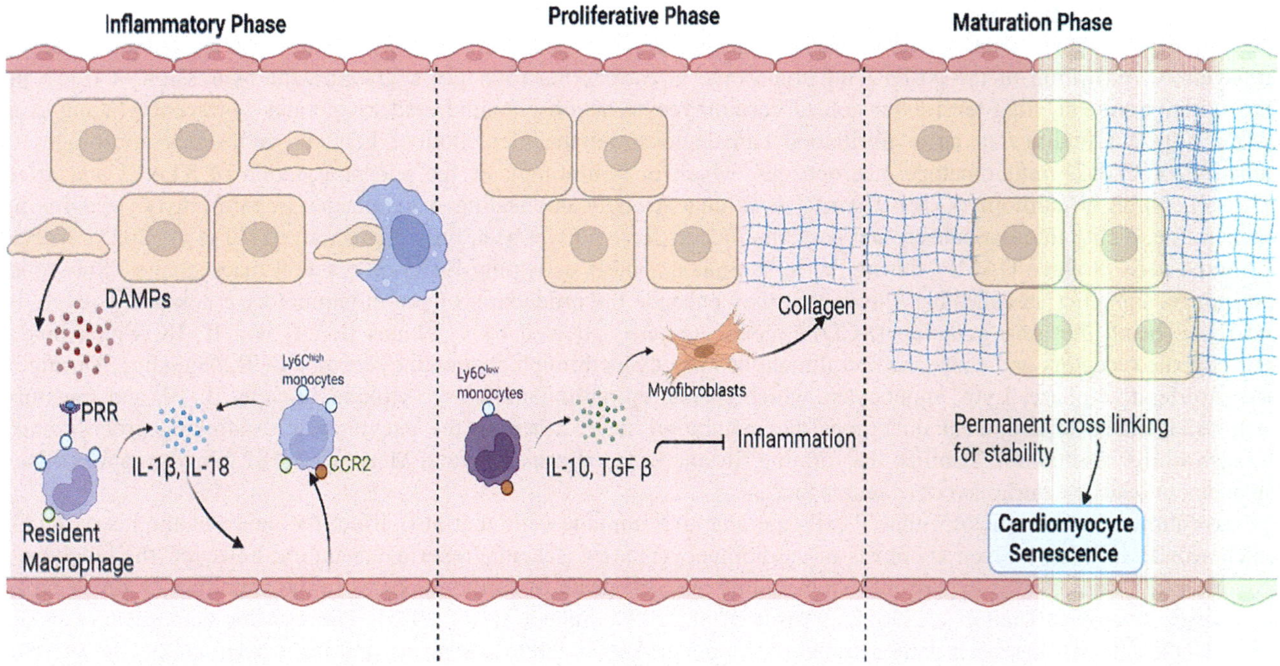

FIGURE 3.3 Immune-mediated cardiac fibrosis and cellular senescence.

transcription, DNA methyltransferases add methyl groups to cytosine residues, especially those in CpG islands neighboring promoters, and thus repress transcription. Aging is also characterized by global hypomethylation and regional hypermethylation, a paradigm associated with replicative senescence (Santos et al., 2005). Dysregulated methylation has also been linked to other cardiovascular pathologies, particularly atherosclerosis. Disordered DNA methylation is coupled with elevated plasma levels of homocysteine, which is associated with heightened levels of coronary artery disease. Abnormal methylation can also be observed in progeroid syndromes (such as HGPS and Werner syndrome), irrespective of the mutated LMNA or WRN genes, indicating that aging diseases involve DNA methylation dysregulation as a primary change (Pietruszyńska-Reszetarska et al., 2024).

Histones are nucleoproteins that form the basic scaffold components of DNA, wrapping to form nucleosomes, the fundamental structural components of chromatin found in eukaryotes. Histone loss in the aging heart has been discovered to be a mark of cell senescence, which is characterized by loss of synthesis, incompetent maturation, or defective deposition of histone proteins (Thoma and Koller, 1981; Hussain et al., 2025; Nagar et al., 2025). DNA damage that accumulates in response to replicative senescence is the primary cause of such depletion and has been described in several species. This leads to chromatin disorganization and tissue instability in the genome. Altered histone dynamics also affect myocardial regeneration. In dividing cardiomyocytes, canonical histone H3 substitution by the variant histone H3.3 was found to inhibit DNA repair and induce a SASP-associated cell phenotype. In addition, another histone variant, referred to as CENPA, which is necessary during the formation of centromeres and segregation of chromosomes, was found to be considerably reduced in old age within human pancreatic islets and murine CPCs (Isermann, 2021). The loss of CENPA in CPCs shortens their lifespan, resulting in increased cell death. In contrast, histone overexpression has also been demonstrated in model organisms, such as yeast, where it increases lifespan, suggesting that histone regulation might be involved in lifespan (Maehara et al., 2010). In addition to the composition of histones, the posttranslational modifications of histone tails also play significant roles in the remodeling of chromatin and gene expression during aging. Histone acetylation enhances transcription by dissolving the chromatin structure, whereas deacetylation facilitates tightening of the chromatin structure, inhibiting gene expression. Histone acetyltransferases and histone deacetylases regulate these alterations, with the latter categorized into four groups based on their dependence on zinc or NAD^+ (Sadakierska-Chudy and Filip, 2015). There are several clear posttranscriptional modifications of RNA, the most significant of which are 5-methylcytosine (m 5 C), N 4 acetylcytidine (ac 4 C), N 6 methyladenosine (m 6 A), pseudouridine (Psi), and A-to-I editing. These modifications are mediated by five different enzymes: DNMT2, NSUN2, NAT10, ADAR1, and METTL3. DNMT2, which modifies chromatin, and NSUN2, which modifies snoRNA-guided, both trigger m5C modification in the 5 UTR and are involved in the stress-induced truncation of tRNAs, cardiac hypertrophy, and fibrosis tumors. A member of

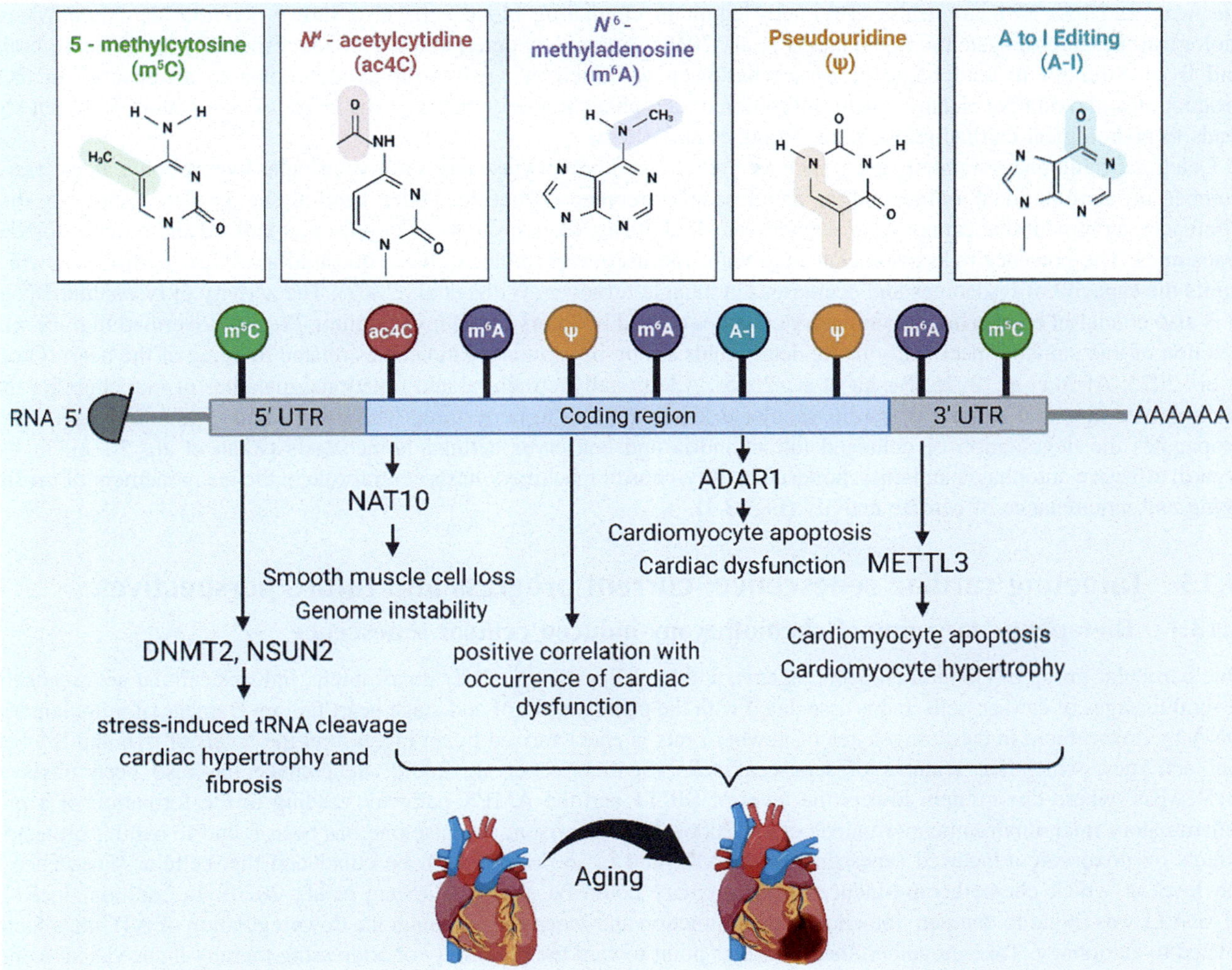

FIGURE 3.4 RNA modifications and epigenetic control in cardiac aging.

the sirtuin family, the enzyme NAT10, mediates ac(R)C formation, and this alteration has been explained with respect to apoptosis in smooth muscle cells and genomic instability. METTL3-dependent m6A and its correlations with cardiac dysfunction and its role in cardiomyocyte apoptosis and hypertrophy appear as modifications in both the coding and 3UTR areas. The editing of A-to-I is carried out by the ADAR1 enzyme, which is also associated with cardiomyocyte apoptosis and dysfunction (Fig. 3.4) (Symeonidis et al., 2025).

Histone methylation also controls transcription and is implicated in the regulation of critical aspects of aging, such as autophagy, DNA damage response, and stress responses. The enzymes that affect these changes in histones are proving to be a potential target area of treatment to prevent or even reverse the deterioration of aging cardiac systems and to maintain the integrity of the genome (Jenwitheesuk et al., 2014). Telomeres are specialized DNA–protein complexes located at the ends of eukaryotic chromosomes. They are composed of repeating nucleotide sequences, arrays of multicopies, the RNA component of telomerase (TERC), and the TERT catalytic subunit, all of which stabilize chromosomes during cell division. One of these is a set of proteins collectively known as shelterins, which are required to maintain telomere structure and enhance genomic stability (Blackburn, 1997; Gupta et al., 2025a, 2025b). As cells continually divide, telomeres become shorter and are further worn off by oxidative stress and constant inflammation. At a certain critical length, there is activation in the cell of DNA-damage reactions, which can lead to cell cycle arrest or induction of senescence. Therefore telomere length has been extensively used as a biomarker of aging (Vaziri and Benchimol, 1996). Oxidative stress, together with inflammatory injuries, predominantly affects genomic biomarkers, including telomere shortening, and the expression of TERT and telomerase activity is notably reduced after birth. Cardiac aging and disease progression; similarly, the length of telomeres in noncardiomyocytes has been suggested as a marker of telomere length in the heart and loss of heart health. Telomere length in circulating blood cells also shows potential as a noninvasive biological measure of systemic age (Ghezzi et al., 2018). Autophagy deteriorates with age, especially in the heart, brain, and liver. Strength in autophagic processes has been identified in aged cardiomyocytes due to an upsurge in ROS production and oxidative endangerment. Inhibition of autophagy not only increases the levels of endogenous ROS but also leads to pathological cardiac remodeling (Wang et al., 2018b).

Deletion of autophagy-related genes, such as Atg5, in cardiomyocytes causes left ventricular hypertrophy, an excessive number of nonfunctional mitochondria, and loss of sarcomere structure. Older heart tissue is also known to store lipofuscin, a pigment that cannot be broken down, which forms when proteins and lipids are oxidized and remain stored in lysosomes. The presence of lysosomes loaded with lipofuscin contributes to the sequestration of new hydrolases, which limits the capacity of lysosomes and decreases autophagic turnover (Willis et al., 2009). The activity of lysosomal DNase II is also crucial in countering inflammatory cascades caused by injury to the myocardium. Pressure overload in mice with deletion of this enzyme specifically in the heart yields strong inflammatory molecules related to aging of the heart (Otoda et al., 2025; Afzal et al., 2025; Bashir et al., 2025). Additionally, structural and functional maladies of mitochondria that are involved in a vast number of cardiovascular diseases are said to be restored with the mediation of autophagy, which propagates the degradation of damaged mitochondria and conserves cellular homeostasis (Zhao et al., 2022). In this regard, effective autophagic and mitochondrial quality-control machinery has a central role in the postponement of cardiac aging and maintenance of cardiac activity (Fig. 3.4).

3.13 Targeting cardiac senescence: current progress and future perspectives

3.13.1 Therapeutic targeting of chemotherapy-induced cellular senescence

One particular group of chemotherapeutic agents, anthracyclines, particularly doxorubicin, induces cellular senescence in several lineages of cardiac cells and is associated with the pathogenesis of end-stage heart failure. Damage of mitochondrial DNA by doxorubicin in cardiomyocytes of newborn rats is characterized by an increase in the levels of p16 and SA-beta Gal activities, which are features of senescent cells (Spallarossa et al., 2009). Senescence has also been observed in VSMCs, where doxorubicin lowers the level of SIRT1 and the AMPK pathway, leading to the formation of a pro-inflammatory microenvironment (Kunieda et al., 2006). In comparison, prednisolone has been found to exhibit protective effects on doxorubicin-incurred senescence in VSMC, and in so doing, it can be concluded that cellular senescence is the level at which chemotherapy-induced cardiotoxicity could be alleviated (Sung et al., 2020). In addition, lncRNA MALAT1 was found to dampen mitochondrial dysfunction and senescence through the downregulation of ATG4a, a factor related to autophagy. Taken together, these findings point toward the possibility of addressing therapy-induced senescence as a method of dealing with cancer treatment-related cardiotoxicity (Xia et al., 2020).

3.13.2 Senolytics and senostatics

An important characteristic of senescence is the occurrence of several physiological changes, such as an increase in glycolysis, stress-induced DNA damage response (DDR), and secretion of factors associated with the senescence-associated secretory phenotype (SASP). These properties make senescent cells popular targets for senolytic (senescence-clear) and senostatic (senescence-modifying) interventions. Dasatinib, a second member of the class of senolytics, and quercetin are the first senolytics discovered, as these molecules only kill senescent cells (Alharbi et al., 2022). Mechanistically, dasatinib inhibits the binding of ephrin receptors and ligands, whereas quercetin destabilizes the PI3K/mTOR axis and simultaneously activates SIRT1, reducing the retention of lipids and thus ameliorating atherosclerosis in mice (Hickson et al., 2019). There was an increase in cardiac performance and a decrease in p16 expression in aged mice exposed to radiation when these agents were combined. This combination also increased vasomotor regulation in hypercholesterolemic mice. Navitoclax is another senolytic agent that induces apoptosis in senescent cells and improves the outcome of MI due to reduced fibrosis, increased systolic function, and increased conduction velocity. The action of cardiac glycosides on senescent cells is due to the inhibition of the Na^+/K^+-ATPase pump with the unusual membrane potential of old cells. In particular, digoxin clears senescent pulmonary fibroblasts in mouse models, and the approaches used are clinically translatable (Deryabin et al., 2021).

3.13.3 Modulating autophagy and metabolism

CR is a thoroughly defined anti-aging intervention, significantly due to its ability to trigger autophagic pathways mediated by IGF-1, sirtuins, AMPK, and mTOR. In the experiments, it was determined that CR leads to the activation of Atg7,

TABLE 3.1 Therapeutic strategies targeting cardiac senescence and their associated mechanisms.

Category	Agent	Mechanism of action	Outcome	Model	References
Chemotherapy-induced senescence	Doxorubicin	Damages mtDNA, increases p16 and SA-β-Gal in cardiomyocytes and VSMCs	Induces senescence and inflammation	Neonatal rat cardiomyocytes, VSMCs	Abdelgawad, (2023)
Chemotherapy protection	Prednisolone	Prevents doxorubicin-induced VSMC senescence	Vascular protection	VSMC culture	Sung et al. (2020)
Senescence suppression	lncRNA MALAT1	Downregulates ATG4a, improves mitochondrial function	Reduces autophagy dysfunction and senescence	Cardiomyocytes	Xia et al. (2020)
Senolytic	Dasatinib	Inhibits ephrin receptor-ligand interaction	Triggers apoptosis in senescent cells	Aged/radiated mice	Mitchell et al. (2024)
Senolytic	Quercetin	Inhibits PI3K/mTOR, activates SIRT1	Improves atherosclerosis, cardiac function	ApoE-deficient mice	Granato et al. (2017)
Senolytic	Navitoclax	Induces apoptosis in senescent fibroblasts	Improves LV function, reduces fibrosis post-MI	MI and Ang II HF mouse models	Takaya et al. (2023)
Senolytic	Digoxin	Inhibits Na^+/K^+-ATPase, targets polarized senescent cells	Removes senescent pulmonary fibroblasts	Pulmonary fibrosis mice	Ren et al. (2021)
Autophagy inducer	Caloric Restriction (CR)	Activates IGF-1, AMPK, sirtuins, mTOR pathways	Elevates Atg7, Atg9, LC3-II in aged heart	Aging rats	Dong et al. (2017)
Senomorphic	Simvastatin	Inhibits HMG-CoA reductase, suppresses SASP	Lowers senescence in endothelial and fibroblast cells	Cell culture, aged models	Rombouts et al. (2003)
Metabolic modulator	Resveratrol	Activates SIRT1, NO signaling, Nrf2 pathway	Prevents apoptosis, reduces inflammation	High-calorie-fed rodents	Tian et al. (2016)
mTOR inhibition	Rapamycin	Inhibits mTOR, enhances autophagy	Reduces progeria-related senescence	Progeria mouse model	Gao et al. (2020b)
Oxidative stress regulator	Celastrol	Reduces ROS, blocks mTOR, enhances autophagy	Prevents Ang II–induced senescence in VSMCs	Vascular smooth muscle cell models	Liu et al. (2019)
Metabolic/Autophagic modulator	Metformin	Restores autophagy, disrupts p62-RIP1-RIP3 complex	Protects against EC senescence, reduces necroptosis	ApoE-/- and ischemic heart models	Li et al. (2020)
SASP suppression	Melatonin	Inhibits PARP-1/CBP, blocks H2BK120 acetylation	Reduces SASP gene expression	Human fibroblast lines	Yu et al. (2017)

Atg9, and LC3-II in aged hearts of rats (Testa et al., 2014). Simultaneously, in combination with their hypocholesterolemic properties, statins decrease the senescence of endothelial cells and SASP inhibition in fibroblasts, independent of lipid control. The metabolic modulator resveratrol has also been shown to mitigate senescence-associated gene expression, extend survival in many high-calorie diet-fed mice, and afford protection against ischemic and inflammatory cardiac injury by SIRT1, Nrf2, and NO signaling (Waldman et al., 2018). Similar findings have been observed with mTOR inhibitors (e.g., rapamycin) through upregulation of autophagy, improvement of Hutchinson–Gilford progeria syndrome-related senescence, and lifespan extension in mice (Wiersma et al., 2016). Celastrol and metformin also decrease oxidative stress and inflammation, which then de-balance autophagy and boost cardiac outcomes. Distinctive discoveries have also shown how melatonin exerts its action by suppressing SASP through PARP-1 and CREB-binding protein pathways, and calcitonin gene-related peptides prevent the senescence of fibroblasts by blocking SASP-related responses (Yu et al., 2017) (Table 3.1).

3.14 Conclusion

Cardiac senescence is a major mediator of age-related cardiovascular dysfunction. Different cardiac cell lineages are senescent with age, causing impairment of functional capacity, exacerbating chronic inflammation, and leading to tissue remodeling. Accurate evaluation of the discrete functions of these cellular populations and the key signaling pathways that drive senescence is therefore vital for devising specific interventions to maintain cardiovascular performance, even as the individual grows old. In addition, current therapeutic proposals impact cardiac senescence using two types of interventions: the first one, senolytics, eliminates senescent cells; and the second one, senostatics, reprograms the secretory phenotype of senescent cells, decreasing the harmful effects of the SASP. In addition, there is an attempt to study the metabolic regulators, autophagy inducers, and sirtuin activators to neutralize the oxidative and metabolic stress that is a trait of the cardiac aging process. The way forward depends on the development of epigenetic, metabolomic, and proteomic approaches to unravel the molecular signatures of cardiac aging so that the practice of precision medicine can be applied with an individualized approach to the treatment of specific patients based on their personal, idiosyncratic aging picture. In the long run, better mechanistic knowledge of cardiac cellular aging will be able to postpone the development of cardiovascular diseases, increase the quality of life, and prolong the period of healthy aging in the older population.

References

Abdelgawad, I.Y., 2023. Molecular Mechanisms and Cardiovascular Ramifications of Doxorubicin-induced Senescence. University of Minnesota.

Adiarto, S., et al., 2012. ET-1 from endothelial cells is required for complete angiotensin II-induced cardiac fibrosis and hypertrophy. Life Sciences 91 (13-14), 651–657.

Afonso, P., et al., 2016. LMNA mutations resulting in lipodystrophy and HIV protease inhibitors trigger vascular smooth muscle cell senescence and calcification: role of ZMPSTE24 downregulation. Atherosclerosis 245, 200–211.

Afzal, M., et al., 2025. Targeting the senescence-associated secretory phenotype to modify osteoarthritis in aging. Inflammopharmacology.

Al Bitar, S., Gali-Muhtasib, H., 2019. The role of the cyclin dependent kinase inhibitor p21cip1/waf1 in targeting cancer: molecular mechanisms and novel therapeutics. Cancers 11 (10), 1475.

Alcendor, R.R., et al., 2007. Sirt1 regulates aging and resistance to oxidative stress in the heart. Circulation Research 100 (10), 1512–1521.

Alharbi, K.S., et al., 2022. A study of the molecular mechanism of quercetin and dasatinib combination as senolytic in alleviating age-related and kidney diseases. Journal of Food Biochemistry 46 (12), e14471.

Alqahtani, S., et al., 2025. SASP modulation for cellular rejuvenation and tissue homeostasis: therapeutic strategies and molecular insights. Cells 14 (8), 608.

Alrumaihi, F., et al., 2025. Histone modifications as molecular drivers of cardiac aging: metabolic alterations, epigenetic mechanisms, and emerging therapeutic strategies. Current Problems in Cardiology 50 (7), 103056.

Arai, Y., et al., 2015. Inflammation, but not telomere length, predicts successful ageing at extreme old age: a longitudinal study of semi-supercentenarians. EBioMedicine 2 (10), 1549–1558.

Balaraman, A.K., et al., 2025. The interplay of senescence and MMPs in myocardial infarction: implications for cardiac aging and therapeutics. Biogerontology 26 (1), 46.

Bashir, B., et al., 2025. Bridging the gap in the management of Alzheimer's disease using fecal microbiota transplantation. Molecular and Cellular Neurosciences 135, 104052.

Blackburn, E., 1997. The telomere and telomerase: nucleic acid–protein complexes acting in a telomere homeostasis system. A review. Biochemistry-New York-English Translation of Biokhimiya 62 (11), 1196–1201.

Bordone, L., Guarente, L., 2005. Calorie restriction, SIRT1 and metabolism: understanding longevity. Nature Reviews Molecular Cell Biology 6 (4), 298–305.

Burmeister, A.R., Marriott, I., 2018. The interleukin-10 family of cytokines and their role in the CNS. Frontiers in Cellular Neuroscience 12, 458.

Chang, C.R., Blackstone, C., 2010. Dynamic regulation of mitochondrial fission through modification of the dynamin-related protein Drp1. Annals of the New York Academy of Sciences 1201 (1), 34–39.

Chen, J.-A., et al., 2020. In vivo imaging of senescent vascular cells in atherosclerotic mice using a β-galactosidase-activatable nanoprobe. Analytical Chemistry 92 (18), 12613–12621.

Chen, X., et al., 2019. IL-6 signaling contributes to radioresistance of prostate cancer through key DNA repair-associated molecules ATM, ATR, and BRCA 1/2. Journal of Cancer Research and Clinical Oncology 145 (6), 1471–1484.

Cho, B., et al., 2013. Physiological and pathological significance of dynamin-related protein 1 (drp1)-dependent mitochondrial fission in the nervous system. Experimental Neurobiology 22 (3), 149.

Chu, L., Xie, D., Xu, D., 2023. Epigenetic regulation of fibroblasts and crosstalk between cardiomyocytes and non-myocyte cells in cardiac fibrosis. Biomolecules 13 (9), 1382.

Corton, J.C., Brown-Borg, H.M., 2005. Peroxisome proliferator-activated receptor γ coactivator 1 in caloric restriction and other models of longevity. The Journals of Gerontology Series A: Biological Sciences and Medical Sciences 60 (12), 1494–1509.

da Silva Feltran, G., et al., 2024. Epigenetic modulation of vascular calcification: Looking for comprehending the role of sirt1 and histone acetylation in VSMC phenotypic transition. Experimental Cell Research 443 (1), 114311.

Dadson, K., et al., 2014. Adiponectin mediated APPL1-AMPK signaling induces cell migration, MMP activation, and collagen remodeling in cardiac fibroblasts. Journal of Cellular Biochemistry 115 (4), 785–793.

Deryabin, P.I., Shatrova, A.N., Borodkina, A.V., 2021. Apoptosis resistance of senescent cells is an intrinsic barrier for senolysis induced by cardiac glycosides. Cellular and Molecular Life Sciences 78 (23), 7757–7776.

Dhoke, N.R., Geesala, R., Das, A., 2018. Low oxidative stress-mediated proliferation via JNK-FOXO3a-catalase signaling in transplanted adult stem cells promotes wound tissue regeneration. Antioxidants & Redox Signaling 28 (11), 1047–1065.

Dong, D., et al., 2017. Alleviation of senescence and epithelial-mesenchymal transition in aging kidney by short-term caloric restriction and caloric restriction mimetics via modulation of AMPK/mTOR signaling. Oncotarget 8 (10), 16109.

Dong, M., et al., 2020. Aging attenuates cardiac contractility and affects therapeutic consequences for myocardial infarction. Aging and Disease 11 (2), 365.

D'Onofrio, N., Servillo, L., Balestrieri, M.L., 2018. SIRT1 and SIRT6 signaling pathways in cardiovascular disease protection. Antioxidants & Redox Signaling 28 (8), 711–732.

Duszka, K., et al., 2020. Peroxisome proliferator-activated receptors and caloric restriction—common pathways affecting metabolism, health, and longevity. Cells 9 (7), 1708.

Emmerson, A., et al., 2018. Nox2 in regulatory T cells promotes angiotensin II–induced cardiovascular remodeling. The Journal of Clinical Investigation 128 (7), 3088–3101.

Espitia-Corredor, J.A., et al., 2022. Resolvin E1 attenuates doxorubicin-induced cardiac fibroblast senescence: a key role for IL-1β Biochimica et Biophysica Acta (BBA)-Molecular Basis of Disease 1868 (11), 166525.

Ewoldt, J.K., et al., 2025. Induced pluripotent stem cell-derived cardiomyocyte in vitro models: benchmarking progress and ongoing challenges. Nature Methods 22 (1), 24–40.

Fang, Y., et al., 2009. Changes of seral TNF-alpha, IL-6 and IL-10 level after implantation of valved bovine jugular vein conduit in complex congenital heart diseases. Chinese Journal of Reparative and Reconstructive Surgery 23 (7), 877–881.

Frangogiannis, N.G., 2017. Fibroblasts and the extracellular matrix in right ventricular disease. Cardiovascular Research 113 (12), 1453–1464.

Fu, W., Liu, Y., Yin, H., 2019. Mitochondrial dynamics: biogenesis, fission, fusion, and mitophagy in the regulation of stem cell behaviors. Stem Cells International 2019 (1), 9757201.

Fumagalli, S., et al., 2015. Age-related differences in presentation, treatment, and outcome of patients with atrial fibrillation in Europe: the EORP-AF General Pilot Registry (EURObservational Research Programme-Atrial Fibrillation). JACC: Clinical Electrophysiology 1 (4), 326–334.

Furukawa, A., et al., 2006. H2O2 accelerates cellular senescence by accumulation of acetylated p53 via decrease in the function of SIRT1 by NAD+ depletion. Cellular Physiology and Biochemistry 20 (1-4), 045–054.

Gao, G., et al., 2020b. Rapamycin regulates the balance between cardiomyocyte apoptosis and autophagy in chronic heart failure by inhibiting mTOR signaling. International Journal of Molecular Medicine 45 (1), 195–209.

Gao, L., et al., 2020a. TNAP inhibition attenuates cardiac fibrosis induced by myocardial infarction through deactivating TGF-β1/Smads and activating P53 signaling pathways. Cell Death & Disease 11 (1), 44.

Ghali, R., 2020. The Role of Interleukin-33 (IL-33) Administration in Cardiac Function Post-myocardial Infarction. Université Paris Cité.

Ghezzi, P., et al., 2018. Oxidative stress and inflammation induced by environmental and psychological stressors: a biomarker perspective. Antioxidants & Redox Signaling 28 (9), 852–872.

Goldenberg, I., et al., 2001. Angiotensin II-induced apoptosis in rat cardiomyocyte culture: a possible role of AT1 and AT2 receptors. Journal of Hypertension 19 (9), 1681–1689.

Granato, M., et al., 2017. Quercetin induces apoptosis and autophagy in primary effusion lymphoma cells by inhibiting PI3K/AKT/mTOR and STAT3 signaling pathways. The Journal of Nutritional Biochemistry 41, 124–136.

Grilo, L.F., et al., 2024. Cardiac molecular analysis reveals aging-associated metabolic alterations promoting glycosaminoglycans accumulation via hexosamine biosynthetic pathway. Advanced Science 11 (38), 2309211.

Guo, Z., et al., 2022. Biomechanical stretch-induced CLOCK upregulation in venous smooth muscle cells promotes phenotypic and functional transformation. Vascular Pharmacology 146, 107097.

Gupta, G., et al., 2025. Exploring the causal relationship between telomere regulation, aging and neurological disorders. Ageing Research Reviews 113, 102930. Available at https://doi.org/10.1016/j.arr.2025.102930.

Gupta, S., et al., 2025b. Harnessing the FOXO-SIRT1 axis: insights into cellular stress, metabolism, and aging. Biogerontology 26 (2), 65.

Haemmig, S., et al., 2020. Long noncoding RNA SNHG12 integrates a DNA-PK–mediated DNA damage response and vascular senescence. Science Translational Medicine 12 (531), eaaw1868.

Han, H.J., et al., 2007. Uric acid inhibits renal proximal tubule cell proliferation via at least two signaling pathways involving PKC, MAPK, cPLA2, and NF-kappaB. American Journal of Physiology Renal Physiology 292 (1), F373–F381.

Hickson, L.J., et al., 2019. Senolytics decrease senescent cells in humans: Preliminary report from a clinical trial of Dasatinib plus Quercetin in individuals with diabetic kidney disease. EBioMedicine 47, 446–456.

Hoshino, A., et al., 2014. Inhibition of p53 preserves Parkin-mediated mitophagy and pancreatic β-cell function in diabetes. Proceedings of the National Academy of Sciences 111 (8), 3116–3121.

Hsieh, C.C., Rosenblatt, J.I., Papaconstantinou, J., 2003. Age-associated changes in SAPK/JNK and p38 MAPK signaling in response to the generation of ROS by 3-nitropropionic acid. Mechanisms of Ageing and Development 124 (6), 733–746.

Hussain, M.S., et al., 2025. Autophagy and cellular senescence in Alzheimer's Disease: key drivers of neurodegeneration. CNS Neuroscience & Therapeutics 31 (7), e70503.

Ikegami, K., et al., 2009. Interplay between DNA methylation, histone modification and chromatin remodeling in stem cells and during development. The International Journal of Developmental Biology 53 (2-3), 203–214.

Improta-Caria, A.C., et al., 2024. Overexpression of miR-126-5p in the serum of patients with systemic arterial hypertension is correlated with creatinine and platelets. Research Square. Available at https://doi.org/10.21203/rs.3.rs-4110557/v1.

Isermann, A.-L., 2021. Elucidating the Role of Histone Variant H2A. J Following Ionizing Radiation. Available at https://publikationen.sulb.uni-saarland.de/bitstream/20.500.11880/31381/1/Thesis_UdS_Anna%20Isermann.pdf.

Izquierdo-Gómez, M.M., et al., 2017. Valve calcification in aortic stenosis: etiology and diagnostic imaging techniques. BioMed Research International 2017 (1), 5178631.

Jansen, M., et al., 2009. LKB1 and AMPK family signaling: the intimate link between cell polarity and energy metabolism. Physiological Reviews 89 (3), 777–798.

Jazbutyte, V., et al., 2013. MicroRNA-22 increases senescence and activates cardiac fibroblasts in the aging heart. Age (Melbourne, Vic.) 35 (3), 747–762.

Jenwitheesuk, A., et al., 2014. Melatonin regulates aging and neurodegeneration through energy metabolism, epigenetics, autophagy and circadian rhythm pathways. International Journal of Molecular Sciences 15 (9), 16848–16884.

Juszczak, F., Arnould, T., Decleves, A.-E., 2024. The role of mitochondrial sirtuins (SIRT3, SIRT4 and SIRT5) in renal cell metabolism: implication for kidney diseases. International Journal of Molecular Sciences 25 (13), 6936.

Kaleem, M., et al., 2025. Epigenetic dynamics and molecular mechanisms in oncogenesis, tumor progression, and therapy resistance. Naunyn-Schmiedeberg's Archives of Pharmacology 398 (10), 13625–13646. Available at https://doi.org/10.1007/s00210-025-04217-5.

Kalinin, A., Zubkova, E., Menshikov, M., 2023. Integrated stress response (ISR) pathway: unraveling its role in cellular senescence. International Journal of Molecular Sciences 24 (24), 17423.

Kamogashira, T., et al., 2017. Functionally and morphologically damaged mitochondria observed in auditory cells under senescence-inducing stress. npj Aging and Mechanisms of Disease 3 (1), 2.

Kandhaya-Pillai, R., et al., 2023. Key elements of cellular senescence involve transcriptional repression of mitotic and DNA repair genes through the p53-p16/RB-E2F-DREAM complex. Aging (Albany NY) 15 (10), 4012.

Karpov, O.A., et al., 2024. Proteomics of the heart. Physiological Reviews 104 (3), 931–982.

Khan, R.I., Nirzhor, S.S.R., Akter, R., 2018. A review of the recent advances made with SIRT6 and its implications on aging related processes, major human diseases, and possible therapeutic targets. Biomolecules 8 (3), 44.

Khavinson, V., et al., 2022. Senescence-associated secretory phenotype of cardiovascular system cells and inflammaging: perspectives of peptide regulation. Cells 12 (1), 106.

Konecny, G.E., et al., 2011. Expression of p16 and retinoblastoma determines response to CDK4/6 inhibition in ovarian cancer. Clinical Cancer Research 17 (6), 1591–1602.

Kumari, R., Jat, P., 2021. Mechanisms of cellular senescence: cell cycle arrest and senescence associated secretory phenotype. Frontiers in Cell and Developmental Biology 9, 645593.

Kunieda, T., et al., 2006. Angiotensin II induces premature senescence of vascular smooth muscle cells and accelerates the development of atherosclerosis via a p21-dependent pathway. Circulation 114 (9), 953–960.

Lemmens, K., Doggen, K., De Keulenaer, G.W., 2007. Role of neuregulin-1/ErbB signaling in cardiovascular physiology and disease: implications for therapy of heart failure. Circulation 116 (8), 954–960.

Li, C., et al., 2020. Metformin mediates cardioprotection against aging-induced ischemic necroptosis. Aging Cell 19 (2), e13096.

Li, L., et al., 2023. M2 macrophage-derived sEV regulate pro-inflammatory CCR2+ macrophage subpopulations to favor post-AMI cardiac repair. Advanced Science 10 (14), 2202964.

Li, M.M., 2024. Development of a Vascularized Heart-on-a-chip for Studying the Impact of Endothelial Cells in Cardiac Homeostasis and Disease. . Available at https://hdl.handle.net/1807/142817.

Li, T., et al., 2016. Loss of p53-mediated cell-cycle arrest, senescence and apoptosis promotes genomic instability and premature aging. Oncotarget 7 (11), 11838.

Liang, C.-J., et al., 2011. Viscolin reduces VCAM-1 expression in TNF-α-treated endothelial cells via the JNK/NF-κB and ROS pathway. Free Radical Biology and Medicine 51 (7), 1337–1346.

Liao, Y.-H., et al., 2012. Interleukin-17A contributes to myocardial ischemia/reperfusion injury by regulating cardiomyocyte apoptosis and neutrophil infiltration. Journal of the American College of Cardiology 59 (4), 420–429.

Liu, A.C., Joag, V.R., Gotlieb, A.I., 2007. The emerging role of valve interstitial cell phenotypes in regulating heart valve pathobiology. The American Journal of Pathology 171 (5), 1407–1418.

Liu, X., et al., 2019. Celastrol mediates autophagy and apoptosis via the ROS/JNK and Akt/mTOR signaling pathways in glioma cells. Journal of Experimental & Clinical Cancer Research 38 (1), 184.

Luan, Y., et al., 2024. Cardiac cell senescence: molecular mechanisms, key proteins and therapeutic targets. Cell Death Discovery 10 (1), 78.

Maehara, K., Takahashi, K., Saitoh, S., 2010. CENP-A reduction induces a p53-dependent cellular senescence response to protect cells from executing defective mitoses. Molecular and Cellular Biology 30 (9), 2090–2104.

Manolis, A.S., Manolis, T.A., Manolis, A.A., 2023. Ketone bodies and cardiovascular disease: an alternate fuel source to the rescue. International Journal of Molecular Sciences 24 (4), 3534.

Menendez, J.A., et al., 2011. mTOR-regulated senescence and autophagy during reprogramming of somatic cells to pluripotency: a roadmap from energy metabolism to stem cell renewal and aging. Cell Cycle (Georgetown, Tex.) 10 (21), 3658–3677.

Mitchell, S.M., et al., 2024. An investigation of binding interactions of tumor-targeted peptide conjugated polyphenols with the kinase domain of ephrin B4 and B2 receptors. Molecular Diversity 28 (2), 817–849.

Mohamed, T.M., et al., 2016. The plasma membrane calcium ATPase 4 signalling in cardiac fibroblasts mediates cardiomyocyte hypertrophy. Nature Communications 7 (1), 11074.

Naeem, A., et al., 2024. Global impact and strategies to reduce the mortality from cardiovascular diseases. Integrated Science for Sustainable Development Goal 3: Empowering Global Wellness Initiatives. Springer, pp. 283–306.

Nagar, L., et al., 2025. Unravelling the role of nanomedicine in attenuating inflammation, oxidative stress and cellular ageing in chronic obstructive pulmonary disease. Archives of Toxicology 99 (10), 4003–4019.

Nagy, J.A., et al., 2002. Vascular permeability factor/vascular endothelial growth factor induces lymphangiogenesis as well as angiogenesis. The Journal of Experimental Medicine 196 (11), 1497–1506.

Nakou, E., et al., 2016. Healthy aging and myocardium: A complicated process with various effects in cardiac structure and physiology. International Journal of Cardiology 209, 167–175.

Nygaard, M., et al., 2014. Birth cohort differences in the prevalence of longevity-associated variants in APOE and FOXO3A in Danish long-lived individuals. Experimental Gerontology 57, 41–46.

Ogrodnik, M., 2021. Cellular aging beyond cellular senescence: markers of senescence prior to cell cycle arrest in vitro and in vivo. Aging Cell 20 (4), e13338.

Oh, K.S., et al., 2021. Cellular senescence evaluated by P16INK4a immunohistochemistry is a prevalent phenomenon in advanced calcific aortic valve disease. Cardiovascular Pathology 52, 107318.

Olsen, M., et al., 2014. P483 The DNA glycosylase Neil3 improves survival in myocardial infarction by regulating fibroblast proliferation and function. Cardiovascular Research 103 (suppl_1), S88.

Ossa, K.L., Crews, D.E., 2006. Biological and genetic theories of the process of senescence throughout life. Human Clocks: The Bio-Cultural Meanings of Age 5, 61.

Otoda, T., Aihara, K.-i, Takayama, T., 2025. Lysosomal stress in cardiovascular diseases: therapeutic potential of cardiovascular drugs and future directions. Biomedicines 13 (5), 1053.

Pandey, S., et al., 2025. The impact of age and gender on cervical vestibular-evoked myogenic potentials using 500 Hz tone bursts stimuli: a cross-sectional study. Cureus 17 (5), e84142.

Pandey, S.N., et al., 2025b. ATM and p53 in aging and cancer: a double-edged sword in genomic integrity. Biogerontology 26 (3), 102.

Panina, Y.A., et al., 2018. Plasticity of adipose tissue-derived stem cells and regulation of angiogenesis. Frontiers in Physiology 9, 1656.

Pietruszyńska-Reszetarska, A., Pietruszyński, R., Irzmański, R., 2024. The significance of genetically determined methylation and folate metabolism disorders in the pathogenesis of coronary artery disease: a target for new therapies? International Journal of Molecular Sciences 25 (13), 6924.

Prattichizzo, F., et al., 2016. Anti-TNF-α treatment modulates SASP and SASP-related microRNAs in endothelial cells and in circulating angiogenic cells. Oncotarget 7 (11), 11945.

Radandish, M., Khalilian, P., Esmaeil, N., 2021. The role of distinct subsets of macrophages in the pathogenesis of MS and the impact of different therapeutic agents on these populations. Frontiers in Immunology 12, 667705.

Redgrave, R.E., et al., 2023. Senescent cardiomyocytes contribute to cardiac dysfunction following myocardial infarction. npj Aging 9 (1), 15.

Ren, Y., et al., 2021. Structural insights into the interactions of digoxin and Na+/K+-ATPase and other targets for the inhibition of cancer cell proliferation. Molecules (Basel, Switzerland) 26 (12), 3672.

Rivard, A., Andrés, V., 2000. Vascular smooth muscle cell proliferation in the pathogenesis of atherosclerotic cardiovascular diseases. Histology and Histopathology 15 (2), 557–571. Available at https://doi.org/10.14670/HH-15.557.

Rombouts, K., et al., 2003. Effect of HMG-CoA reductase inhibitors on proliferation and protein synthesis by rat hepatic stellate cells. Journal of Hepatology 38 (5), 564–572.

Sadakierska-Chudy, A., Filip, M., 2015. A comprehensive view of the epigenetic landscape. Part II: histone post-translational modification, nucleosome level, and chromatin regulation by ncRNAs. Neurotoxicity Research 27 (2), 172–197.

Salman, O., et al., 2024. Prognostic significance and biologic associations of senescence-associated secretory phenotype biomarkers in heart failure. Journal of the American Heart Association 13 (17), e033675.

Salminen, A., et al., 2012. Mitochondrial dysfunction and oxidative stress activate inflammasomes: impact on the aging process and age-related diseases. Cellular and Molecular Life Sciences: CMLS 69 (18), 2999–3013.

Santos, K., Mazzola, T., Carvalho, H., 2005. The prima donna of epigenetics: the regulation of gene expression by DNA methylation. Brazilian Journal of Medical and Biological Research 38, 1531–1541.

Schirrmacher, V., 2020. Mitochondria at work: new insights into regulation and dysregulation of cellular energy supply and metabolism. Biomedicines 8 (11), 526.

Schreckenberg, R., et al., 2009. Cell-specific effects of nitric oxide deficiency on parathyroid hormone-related peptide (PTHrP) responsiveness and PTH1 receptor expression in cardiovascular cells. Endocrinology 150 (8), 3735–3741.

Sharma, S., et al., 2025. Tailoring the biomarkers of Alzheimer's disease using a gut microbiome-centric approach: preclinical, clinical, and regulatory perspectives. Ageing Research Reviews 112, 102888.

Spallarossa, P., et al., 2009. Doxorubicin induces senescence or apoptosis in rat neonatal cardiomyocytes by regulating the expression levels of the telomere binding factors 1 and 2. American Journal of Physiology-Heart and Circulatory Physiology 297 (6), H2169–H2181.

Subedi, S., et al., 2025. Cellular senescence in chronic obstructive pulmonary disease: molecular mechanisms and therapeutic interventions. Ageing Research Reviews 110, 102813.

Sung, J.Y., et al., 2020. Prednisolone suppresses adriamycin-induced vascular smooth muscle cell senescence and inflammatory response via the SIRT1-AMPK signaling pathway. PLoS One 15 (9), e0239976.

Symeonidis, A., Chroni, A., Dereki, I., Chartoumpekis, D., Sgourou, A., 2025. Intracellular mis-localization of modified RNA molecules and non-coding RNAs: Facts from Hematologic Malignancies. Current Issues in Molecular Biology 47 (9), 758. Available at https://doi.org/10.3390/cimb47090758.

Takasaka, N., et al., 2014. Autophagy induction by SIRT6 through attenuation of insulin-like growth factor signaling is involved in the regulation of human bronchial epithelial cell senescence. The Journal of Immunology 192 (3), 958–968.

Takaya, K., et al., 2023. Navitoclax (ABT-263) rejuvenates human skin by eliminating senescent dermal fibroblasts in a mouse/human chimeric model. Rejuvenation Research 26 (1), 9–20.

Tan, P., et al., 2016. The PI3K/Akt/mTOR pathway regulates the replicative senescence of human VSMCs. Molecular and Cellular Biochemistry 422 (1), 1–10.

Tang, X., Li, P.-H., Chen, H.-Z., 2020. Cardiomyocyte senescence and cellular communications within myocardial microenvironments. Frontiers in Endocrinology 11, 280.

Tang, Z.-M., et al., 2015. Association between rs2431697 T allele on 5q33. 3 and systemic lupus erythematosus: case-control study and meta-analysis. Clinical Rheumatology 34 (11), 1893–1902.

Testa, G., et al., 2014. Calorie restriction and dietary restriction mimetics: a strategy for improving healthy aging and longevity. Current Pharmaceutical Design 20 (18), 2950–2977.

Thoma, F., Koller, T., 1981. Unravelled nucleosomes, nucleosome beads and higher order structures of chromatin: influence of non-histone components and histone H1. Journal of Molecular Biology 149 (4), 709–733.

Tian, Y., et al., 2016. Resveratrol supplement inhibited the NF-κB inflammation pathway through activating AMPKα-SIRT1 pathway in mice with fatty liver. Molecular and Cellular Biochemistry 422 (1), 75–84.

Trobec, K., et al., 2011. Growth hormone, insulin-like growth factor 1, and insulin signaling—a pharmacological target in body wasting and cachexia. Journal of Cachexia, Sarcopenia and Muscle 2 (4), 191–200.

Turner, N.A., et al., 2007. Mechanism of TNFα-induced IL-1α, IL-1β and IL-6 expression in human cardiac fibroblasts: effects of statins and thiazolidinediones. Cardiovascular Research 76 (1), 81–90.

Varga, I., Kyselovič, J., Galfiova, P., Danisovic, L., et al., 2017. The non-cardiomyocyte cells of the heart. Their possible roles in exercise-induced cardiac regeneration and remodeling. In: Xiao, J. (Ed.), Exercise for Cardiovascular Disease Prevention and Treatment. Advances in Experimental Medicine and Biology, vol 999. Springer, Singapore, pp. 117–136. Available at https://doi.org/10.1007/978-981-10-4307-9_8.

Varricchi, G., Marone, G., Kovanen, P.T., 2020. Cardiac mast cells: underappreciated immune cells in cardiovascular homeostasis and disease. Trends in Immunology 41 (8), 734–746.

Vaziri, H., Benchimol, S., 1996. From telomere loss to p53 induction and activation of a DNA-damage pathway at senescence: the telomere loss/DNA damage model of cell aging. Experimental Gerontology 31 (1-2), 295–301.

Venugopal, H., et al., 2022. Properties and functions of fibroblasts and myofibroblasts in myocardial infarction. Cells 11 (9), 1386.

Victor, V.M., et al., 2009. Oxidative stress and mitochondrial dysfunction in sepsis: a potential therapy with mitochondria-targeted antioxidants. Infectious Disorders-Drug Targets (Formerly Current Drug Targets-Infectious Disorders) 9 (4), 376–389.

Virmani, A., et al., 2015. The carnitine palmitoyl transferase (CPT) system and possible relevance for neuropsychiatric and neurological conditions. Molecular Neurobiology 52 (2), 826–836.

Waku, T., et al., 2021. NRF3 upregulates gene expression in SREBP2-dependent mevalonate pathway with cholesterol uptake and lipogenesis inhibition. Iscience 24 (10).

Waldman, M., et al., 2018. Regulation of diabetic cardiomyopathy by caloric restriction is mediated by intracellular signaling pathways involving 'SIRT1 and PGC-1α Cardiovascular Diabetology 17 (1), 111.

Wang, C.-Y., et al., 2008. Increased vascular senescence and impaired endothelial progenitor cell function mediated by mutation of circadian gene Per2. Circulation 118 (21), 2166–2173.

Wang, F., et al., 2018a. Interferon gamma induces reversible metabolic reprogramming of M1 macrophages to sustain cell viability and pro-inflammatory activity. EBioMedicine 30, 303–316.

Wang, H., et al., 2016. Mast cell inhibition attenuates cardiac remodeling and diastolic dysfunction in middle-aged, ovariectomized Fischer 344× Brown Norway rats. Journal of Cardiovascular Pharmacology 68 (1), 49–57.

Wang, J.-P., et al., 2018b. The role of endogenous reactive oxygen species in cardiac myocyte autophagy. Physiological Research 67 ((1), 31–40.

Whisenant, D., 2022. From Gene Therapy to Somatic Mutagenesis: Insights into the Common Hutchinson-Gilford Progeria Syndrome Mutation. Karolinska Institutet (Sweden).

Wiersma, M., et al., 2016. The orphan drug 4-phenyl butyrate protects the heart against Atrial Fibrillation by blocking ER stress-induced autophagy. Implications of the Cardiomyocyte Stress Response on Protein Homeostasis in Atrial Fibrillation 116.

Willis, M.S., et al., 2009. Build it up–tear it down: protein quality control in the cardiac sarcomere. Cardiovascular Research 81 (3), 439–448.

Xia, W., et al., 2020. Long-noncoding RNA MALAT1 sponges microRNA-92a-3p to inhibit doxorubicin-induced cardiac senescence by targeting ATG4a. Aging (Albany NY) 12 (9), 8241.

Xiu, Y., Field, M.S., 2020. The roles of mitochondrial folate metabolism in supporting mitochondrial DNA synthesis, oxidative phosphorylation, and cellular function. Current Developments in Nutrition 4 (10), nzaa153.

Yan, S., et al., 2015. Clematichinenoside inhibits VCAM-1 and ICAM-1 expression in TNF-α-treated endothelial cells via NADPH oxidase-dependent IκB kinase/NF-κB pathway. Free Radical Biology and Medicine 78, 190–201.

Yang, H.-T., et al., 2022. MicroRNA-155 inhibition attenuates myocardial infarction-induced connexin 43 degradation in cardiomyocytes by reducing pro-inflammatory macrophage activation. Cardiovascular Diagnosis and Therapy 12 (3), 325.

de Yébenes, V.G., et al., 2020. Aging-associated miR-217 aggravates atherosclerosis and promotes cardiovascular dysfunction. Arteriosclerosis, Thrombosis, and Vascular Biology 40 (10), 2408–2424.

Yu, S., et al., 2017. Melatonin regulates PARP 1 to control the senescence-associated secretory phenotype (SASP) in human fetal lung fibroblast cells. Journal of Pineal Research 63 (1), e12405.

Yue, Z., et al., 2014. Gene expression of the p16 (INK4a)-Rb and p19 (Arf)-p53-p21 (Cip/Waf1) signaling pathways in the regulation of hematopoietic stem cell aging by ginsenoside Rg1. Genetics and Molecular Research: GMR 13 (4), 10086–10096.

Zhang, E., et al., 2015. Metformin and resveratrol inhibited high glucose-induced metabolic memory of endothelial senescence through SIRT1/p300/p53/p21 pathway. PLoS One 10 (12), e0143814.

Zhang, L., et al., 2022. Insulin-like growth factor-binding protein-7 (IGFBP7) links senescence to heart failure. Nature Cardiovascular Research 1 (12), 1195–1214.

Zhao, M., et al., 2022. The regulatory mechanism between lysosomes and mitochondria in the aetiology of cardiovascular diseases. Acta Physiologica 234 (2), e13757.

Zhuang, L., et al., 2020. Comprehensive integration of single-cell transcriptional profiling reveals the heterogeneities of non-cardiomyocytes in healthy and ischemic hearts. Frontiers in Cardiovascular Medicine 7, 615161.

Zhuang, Y., et al., 2021. LncRNA-H19 drives cardiomyocyte senescence by targeting miR-19a/socs1/p53 axis. Frontiers in Pharmacology 12, 631835.

Zingg, J.-M., Jones, P.A., 1997. Genetic and epigenetic aspects of DNA methylation on genome expression, evolution, mutation and carcinogenesis. Carcinogenesis 18 (5), 869–882.

Chapter 4

Cellular senescence in atherosclerosis and heart failure: pathophysiology and clinical implications

Shivkanya Fuloria
Faculty of Pharmacy, AIMST University, Bedong, Kedah, Malaysia

4.1 Introduction

Atherosclerosis and chronic heart failure (CHF) are the leading causes of cardiovascular mortality worldwide, presenting significant clinical and socioeconomic challenges. Atherosclerosis is the major etiology of ischemic sequelae, such as myocardial infarction or cerebrovascular accident, and CHF refers to the manifestation of the end stage of different heart issues (Abdelsayed et al., 2022). Although conventional risk factors, including high blood pressure, diabetes mellitus, hypercholesterolemia, and tobacco use, remain crucial in the pathogenesis of these ailments, chronological ageing has been rated as a distinct and strong predictor of atherosclerosis and CHF (Hajar, 2017). Notably, the influence of traditional risk factors is moderated by biological aging processes, which accelerate the onset of vascular dysfunction, atherosclerotic plaque formation, and maladaptive cardiac remodeling. At the heart of this biological mechanism is cellular senescence, an irreversible condition of growth arrest caused by genomic instability, telomere loss, oxidative stress, and metabolic malfunctions (Xu et al., 2025). The deposition of senescent cells in vascular smooth muscle, endothelial cells, cardiomyocytes, and fibroblasts results in the release of a senescence-associated secretory phenotype (SASP) rich in cytokines, chemokines, and proteases. Acute senescence aids in tissue repair and tumor suppression; however, unresolved senescence in the long term produces an inflammatory and fibrotic microenvironment (Khavinson et al., 2022). This microenvironment favors the onset of plaque, instability, and vascular stiffening in atherosclerosis. In congestive heart failure, SASP signaling, mitochondrial dysfunction, disrupted autophagy, and dysregulated epigenetics converge to trigger contraction degradation, fibrosis, and adverse remodeling. Therefore, emerging evidence places senescence as a common mechanistic nexus between vascular and myocardial aging, connecting the pathophysiology of atherosclerosis and congestive heart failure (Pandit and Yurdagul, 2025). The discovery of these similarities, besides improving our understanding of cardiovascular aging, highlights new areas of research to address therapy. The objective of this review is to summarize the current understanding of the pathways of senescence-related mechanisms, including mitochondrial dysfunction, defective autophagy, the dynamics of the SASP, and epigenetic changes, and to evaluate their clinical implications, with a special emphasis on senescence-targeted interventions that have the potential to transform atherosclerosis and heart failure prevention and treatment.

4.2 Senescence-driven pathology in atherosclerosis and chronic heart failure

Atherosclerosis and CHF pathology are good examples of how cellular senescence coordinates vascular and myocardial aging. Senescent VSMCs, endothelial cells, and infiltrating immune cells in advanced atherosclerotic plaques break tissue homeostasis by reducing proliferation, suppressing the production of extracellular matrix, and maintaining a continuous pro-inflammatory secretory state (Machado-Oliveira et al., 2020). These changes weaken the fibrous cap, stimulate plaque instability, lead to repeated rupture–repair cycles of the fibrous cap, and culminate in luminal narrowing, ischemia, and acute thrombotic events. In CHF, a similar pattern emerges within the myocardium, where cardiomyocytes, endothelial cells, VSMCs, and fibroblasts adopt distinct senescence phenotypes under sustained stressors, such as oxidative damage,

Cellular Senescence and Cardiac Aging. DOI: https://doi.org/10.1016/B978-0-443-45811-8.00019-2

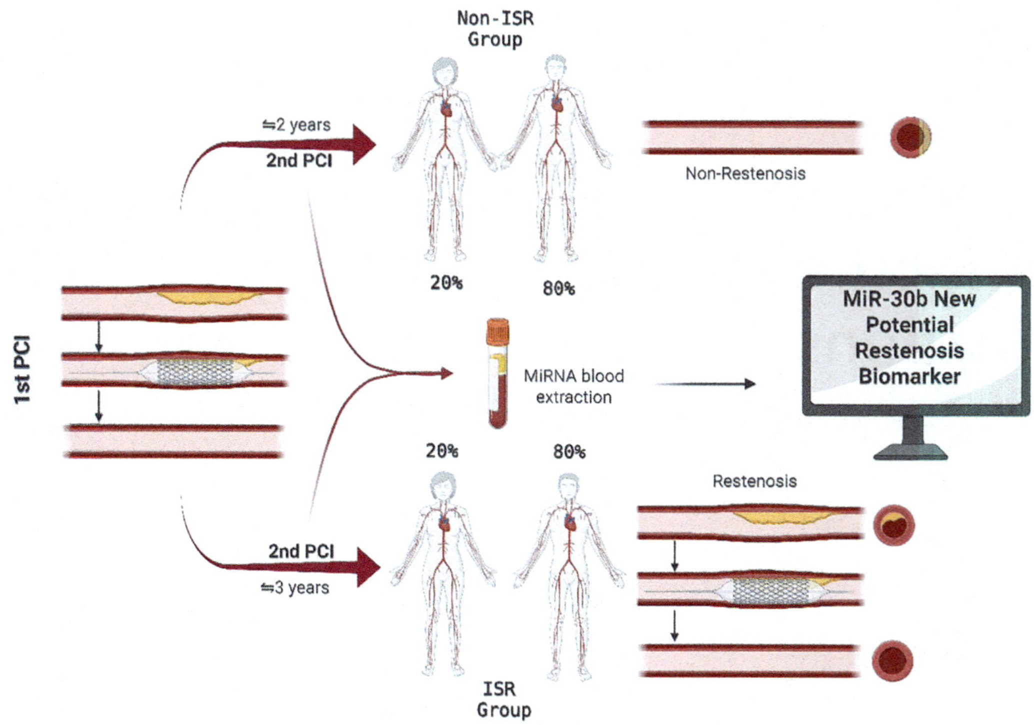

FIGURE 4.1 miR-30b as a circulating biomarker for restenosis.

telomere attrition, and mitochondrial dysfunction (Xie et al., 2023). Senescent cardiomyocytes lose contractile competence and regenerative potential, endothelial senescence disrupts nitric oxide signaling and microvascular function, and VSMC senescence contributes to maladaptive hypertrophy in the heart. In parallel, senescent fibroblasts release cytokines and matrix-degrading enzymes that exacerbate fibrosis and ventricular stiffening (Redgrave et al., 2023).

Despite their disease-specific features, both vascular plaques and failing hearts share a common pathogenic thread: Senescent cells create a chronic inflammatory and degenerative microenvironment that accelerates disease progression and undermines tissue repair. This mutual communication between senescent cellular populations supports oxidative stress, fibrosis, and structural remodeling, creating a vicious cycle that is a major process in the pathophysiology of both atherosclerosis and CHF (Kumar et al., 2024). These general processes of senescence have been identified, providing a conceptual framework for developing specific pharmacotherapeutic interventions to stabilize atherosclerotic plaques, sustain myocardial activity, and ultimately relieve the burden of cardiovascular aging (Fig. 4.1).

4.3 Senescence in vascular and myocardial aging

Normal aging of the vasculature is a complication, even in the absence of clinically apparent atherosclerosis, and is marked by structural and functional deterioration. Intimal and medial thickening, loss of elasticity, elastin fragmentation, and collagen accumulation gradually impair vascular compliance and promote arterial stiffness. Synergistic vascular inflammation and remodeling through endothelial dysfunction, elevated matrix metalloproteinase activity, and enhanced angiotensin 2 signaling contribute to a pro-atherogenic environment (Stary et al., 1995). These physiological changes reacidify early atherogenic processes and explain the mechanism by which senescence alters vessels to become pathologically plaque-causing. This association is supported by premature vascular senescence in Hutchinson–Gilford progeria syndrome (HGPS) and Werner syndrome, as inherited genomic abnormalities increase telomere erosion, collagen deposition, and lesion formation (Sarkar and Shinton, 2001). Collectively, these processes highlight a bidirectional relationship in which vascular aging and atherosclerosis reinforce each other, converging on cellular senescence as a unifying mechanism. Parallel biology operates in the myocardium, where cardiac progenitor cells (CPCs) and cardiomyocytes undergo senescence under sustained stress. Senescent CPCs driven by telomere attrition, mTOR dysregulation, and impaired CXCL12/CXCR4 signaling lose their regenerative capacity and secrete profibrotic and pro-inflammatory mediators (IL-1β, IL-6, IL-8, PAI-1, and MMP-3) (Lewis-McDougall et al., 2019). This SASP amplifies fibrosis, impairs neovascularization, and accelerates pathological remodeling in CHF mice. Similarly, senescent cardiomyocytes, characterized by the activation of ^{p16}INK4a and ^{p21}CIP pathways, accumulate DNA damage and mitochondrial dysfunction. Their atypical SASP promotes fibroblast activation, maladaptive hypertrophy, and interstitial fibrosis, reinforcing the cycle of ventricular stiffening and loss of contractile reserve (Zhai and Sadoshima, 2024). Despite affecting different tissues, vascular and myocardial senescence are mechanistically linked. In both settings, the accumulation of senescent cells undermines the repair capacity, drives fibrosis and inflammation, and creates a feed-forward loop of oxidative stress and structural decline. This convergence explains why vascular aging predisposes individuals to atherosclerosis and why cardiomyocyte/CPC senescence accelerates CHF. Importantly, these insights position senescence not only as a biomarker of cardiovascular aging but also as a potential therapeutic target. Strategies to delay or modulate vascular and myocardial senescence may stabilize plaques, preserve cardiac function, and ultimately improve outcomes in patients with atherosclerosis and heart failure.

4.4 Mechanisms of cellular senescence in atherosclerosis

The development of cellular senescence in atherosclerosis arises from a complex interplay of multiple mechanisms that accumulate in vascular cells over time. One major driver of vascular aging is telomere shortening, which progressively limits the proliferative capacity of endothelial and smooth muscle cells, ultimately triggering replicative senescence. Simultaneously with the above findings, chronic nuclear and mitochondrial genomic damage, often facilitated by ROS, accelerates the process of senescence, especially in settings of defective DNA repair machinery. ROS-guided genomic instability not only causes cellular growth arrest but also facilitates pro-inflammatory transduction chains, thus exacerbating the formation of plaques (Ali et al., 2024). Moreover, epigenetic changes, including aberrant DNA methylation and histone remodeling, can reprogram transcriptional programs that regulate cell cycle progression, apoptosis, and SASP. These processes do not often occur singly, but rather telomere dysfunction, oxidative stress, impaired DNA repair, and epigenetic reprogramming combine in the vascular microenvironment to drive cellular senescence (Kane and Sinclair, 2019). Together, they provide a feed-forward mechanism of inflammation, impaired repair, and vascular pathophysiology, thus acting not only as correlates but also as unlikely causal factors of atherosclerotic pathogenesis.

4.4.1 Telomere maintenance

The relationship between aging and atherosclerosis is largely dependent on telomere dynamics. It has been demonstrated that progressive telomere shortening is not confined to VSMCs and ECs of atherosclerotic plaques but also in the systemic circulation of leukocytes, indicating systemic aging and exhaustion of progenitor cells. Telomere shortening can impair cell regeneration capacity, augment DNA damage repair, and favor senescence, accelerating the development of vascular dysfunction. Experimental models have emphasized the subtle role of telomerase. On the one hand, the restoration of proliferative capacity and slowing of tissue degeneration in TERT-deficient conditions by ectopic telomerase expression has been discovered. On the other hand, it would seem that reduced telomerase activity in TERC knockout mice is atheroprotective, in part because of the restraint of macrophage and inflammatory cell overgrowth (Satyanarayana et al., 2003). These seemingly conflicting results would indicate that telomerase can regulate the progression of diseases in a context- and cell-type-specific manner, in which early proliferative stimulation is advantageous, but excessive proliferation of immune cells causes plaque instability. Moreover, the immune subset responses to perturbation of telomere-related proteins, such as the Shelterin complex, show different responses, which underscores the complexity of the role of telomeres in the regulation of the vascular and hematopoietic lineages (Mir et al., 2020). Importantly, telomerase overexpression does not universally prevent senescence in VSMCs and ECs, suggesting that telomere attrition is only one of the components of vascular aging. Additional telomere-independent pathways, including genomic instability, mitochondrial dysfunction, and persistent DNA damage, converge to drive the senescent phenotype and ultimately contribute to atherosclerotic progression.

4.4.2 DNA damage recognition and repair

DNA damage is an inevitable biological process arising from both intrinsic factors, such as spontaneous mutations and replication errors, and extrinsic insults, including ultraviolet radiation, chemical exposure, and free radicals. While transient DNA lesions are usually resolved by robust DDR pathways, persistent or defective repair has profound implications for vascular pathology (Al-Khayyat et al., 2025). In the atherosclerosis paradigm, it is becoming increasingly apparent that DNA damage is not an associative feature but a direct catalyst of plaque development and medial erosion.

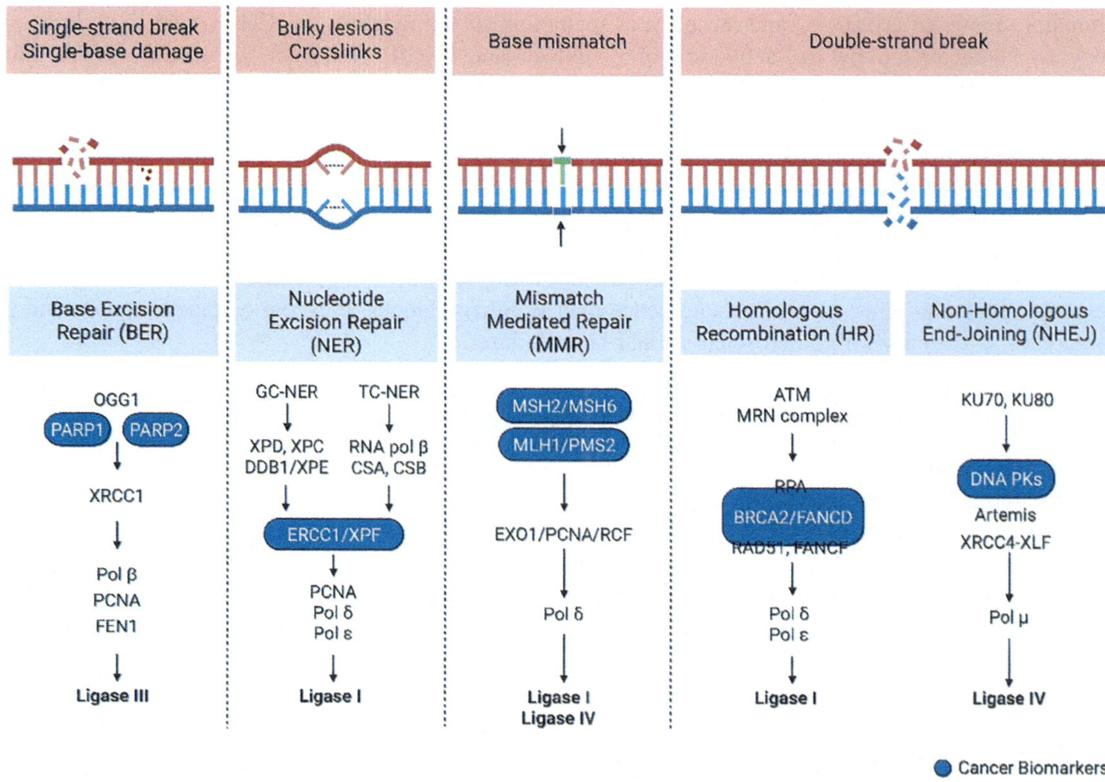

FIGURE 4.2 DNA damage and repair pathways in vascular aging.

Experimental models have indicated that genetic impairments in DDR elements correlate with disease progression. For example, heterozygous deletion of ATM accelerates atherosclerotic disease in murine models, resulting in increased nuclear and mitochondrial DNA damage and metabolic defects, which are at least partially corrected by transplanting wild-type ATM into the bone marrow (Davies and Menzies, 2025). Nuclear abnormalities with age also lead to nuclear abnormalities that induce vascular senescence, as the buildup of prelamin-A prevents the compartmentalization of nuclear signaling, leading to the impairment of DNA repair mechanisms and the accumulation of genomic damage (Hao et al., 2025). This mechanistic pathway is reminiscent of the pathology based on human progeroid syndromes, especially HGPS and Werner Syndrome (WS), where impairment of lamin-A or lamin helicase activity initiates genomic instability, accelerated cellular senescence, and accelerated vascular aging, including early onset atherosclerosis (Ikeda et al., 2025). Together, these findings emphasize that DNA damage and poor repair mechanisms serve as driving factors of vascular aging and that genomic instability is causally related to atherogenesis. Therefore, DNA damage response pathways may be effective therapeutic interventions (Fig. 4.2).

4.4.3 Reactive oxygen species and oxidative stress

ROS and oxidative stress play pivotal roles in linking vascular aging with the pathogenesis of atherosclerosis. Evidence from in vitro studies demonstrates that VSMCs derived from aged individuals accumulate greater oxidative damage than those from younger individuals, primarily due to enhanced ROS generation coupled with weakened antioxidant defense systems (Liu et al., 2025a). Mitochondria are the major intracellular source of ROS, and mitochondrial DNA damage can amplify ROS output, establishing a vicious cycle of oxidative stress and mitochondrial dysfunction. In atherosclerosis, oxidized lipoproteins, particularly oxidized LDL, further drive ROS production, with hydroxyl radicals (•OH) being the most damaging species (Menon and Gueble, 2025). Importantly, the vascular microenvironment influences the ROS balance: endothelial cells exposed to laminar shear stress upregulate protective antioxidant enzymes, whereas disturbed or turbulent flow at sites of plaque initiation impairs this defense, enhancing the local oxidative burden (Mladenov et al., 2025; Alharbi et al., 2022; Hussain et al., 2023).

At the molecular level, ROS inflict DNA damage in both nuclear and mitochondrial genomes, targeting telomeric and nontelomeric regions, with elevated 8-oxo-deoxyguanosine levels serving as a biomarker of such injury in

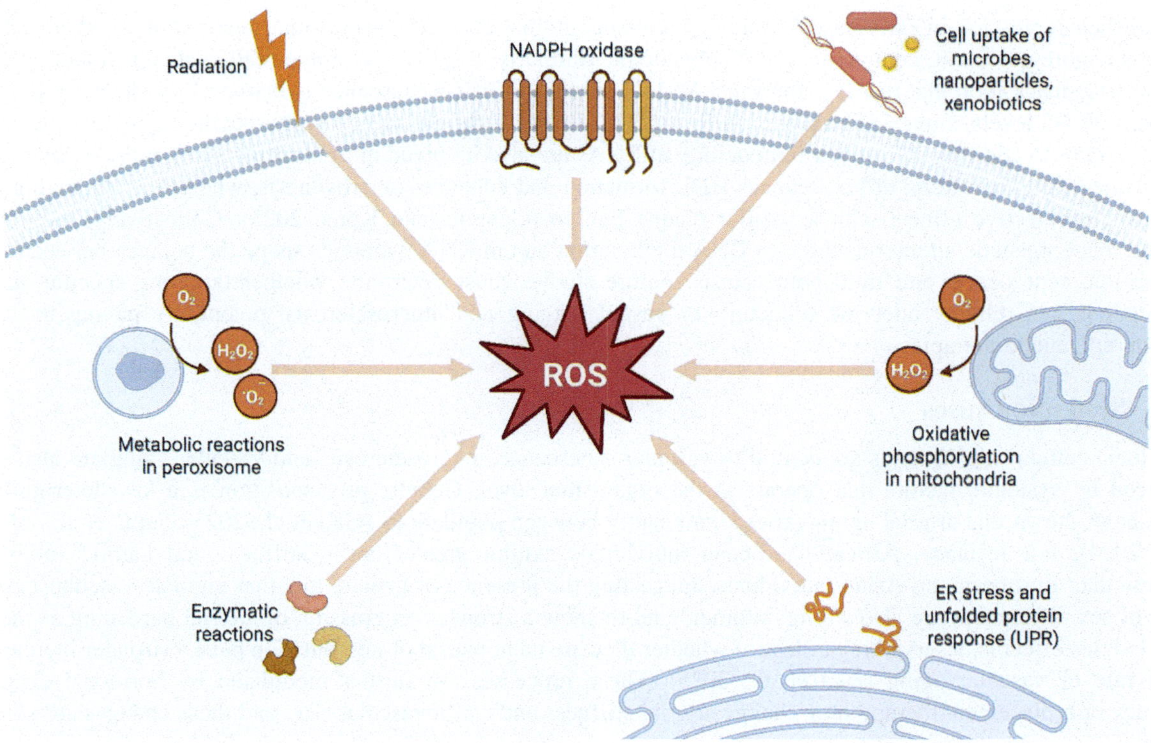

FIGURE 4.3 Cellular sources of reactive oxygen species in atherosclerosis.

plaque-associated VSMCs and macrophages (Negoi, 2025). This oxidative DNA damage triggers DDR, leading to cellular senescence and apoptosis. High ROS concentrations can induce stress-induced premature senescence (SIPS) in vascular cells. In addition to the direct localized DNA damage, ROS affect atherosclerosis development through several signaling pathways (Panigrahi et al., 2025). The PI3K/Akt pathway is suppressed, which suppresses telomerase activity, suppressing downstream survival signaling and activating apoptosis. Activation of the tumor suppressor p53 through ROS leads to further downregulation of IGF-1 receptor expression, thus increasing the susceptibility of vascular cells (Raghunathan and Krishnan, 2025). Moreover, the active form of latent TGF-B conversion into latent TGF-B by ROS stimulates fibrotic and pro-atherogenic effects. Finally, ROS increase the activity of scavenger receptors that facilitate the uptake of oxidized low-density lipoprotein (oxLDL), thus enhancing the level of lipid deposition in atheromatous plaques (Shan et al., 2025; Pandey et al., 2025). Overall, ROS are not only harmful by-products of aging and mitochondrial malfunction, but can also mediate vascular senescence and atherosclerosis development. Such a dual role makes oxidative stress one of the key mechanistic targets, and the restoration of redox balance can slow down vascular aging and avert the progression of atherosclerotic disease (Fig. 4.3).

4.4.4 Epigenetic modifications

Epigenetic changes are key regulators of premature vascular senescence and atherosclerosis. Some of the main processes include alterations in DNA methylation, histone acetylation, and noncoding RNA activity, which regulate the behavior of VSMCs and ECs. For example, hypermethylation of the MCT3 gene results in reduced gene expression in atherosclerotic lesions, which induces the proliferation of VSMC (Chen et al., 2025a; Bhat et al., 2024; Goyal et al., 2024). Similarly, hypermethylation of ER subunits with age and disease correlates with changes in the abundance of the receptors, where ERB hypermethylation correlates with early VSMC proliferation and ERalpha hypermethylation with later vascular senescence in advanced plaques (Cheraghi et al., 2025).

Another epigenetic control involved in vascular pathology is miRNA. The miR-21 and miR-221/222 families are always upregulated after vascular injury, and in the situation of coronary artery disease, they contribute to the survival and proliferation of smooth muscle cells by suppressing the tumor suppressor pathways, especially PTEN and p27 kip (Ding et al., 2025a). In contrast, miR-143/145, typically induced by TGF-β signaling, restrain VSMC proliferation by inhibiting

KLF4/5; their repression in early vascular injury favors neointimal growth, whereas their restoration in advanced lesions promotes a more quiescent, contractile VSMC phenotype. Similarly, miR-133 reinforces cell cycle arrest and senescence, forming a feedback loop that restricts abnormal VSMC growth. In ECs, senescence is marked by reduced miR-21, miR-214, and miR-92 levels, alongside increased miR-221/222, further highlighting cell-type–specific regulatory patterns (Fan and Shi, 2025). In addition to vascular remodeling, miRNAs are also involved in lipid metabolism. miR-33, by repressing ABCA1-mediated cholesterol efflux, reduces HDL formation and enhances atherogenesis, whereas antagonizing miR-33 yields atheroprotective effects with a smaller plaque burden (Gangula and Kona, 2025). Collectively, these findings illustrate that epigenetic alterations through DNA methylation and miRNA dynamics shape the balance between vascular proliferation, senescence, and lipid homeostasis. Future studies must determine whether targeting specific epigenetic modifications can reliably alter the trajectory of vascular aging and atherosclerosis, potentially paving the way for precision epigenetic therapies.

4.4.5 Systemic causes

While local cellular mechanisms are central to vascular senescence, atherosclerosis, and vascular aging are also strongly influenced by systemic factors that operate at the organismal level. Genetic predisposition is a key determinant, and studies have shown that arterial aging varies significantly between populations (Gao et al., 2025; Gupta et al., 2025a; Jha et al., 2024). For instance, African-Caribbean individuals exhibit greater aortic stiffness and carotid intima-media thickness than European and Asian individuals, suggesting the presence of genetic loci that regulate vascular aging. The effects of sex differences are also strong: Women tend to show a stronger progression of arterial hardening as they grow older, and this effect is observed regardless of whether they are menopausal or not, but men have a stronger increase in the overall rate of vascular aging (Hua et al., 2025). These processes are further modulated by hormonal changes, as deficiency or replacement therapy may change arterial stiffness and cardiovascular risk, and these changes are sometimes reversible.

In addition to inherited factors, acquired risk factors also contribute significantly to premature vascular senescence. Smoking enhances the rate of arterial stiffness, pulse pressure, and senescence-related processes of endothelial cells and fibroblasts, and long-term heavy alcohol use raises the incidence of arterial stiffness, regardless of the protective effect of low-to-moderate alcohol consumption (Ismaeel et al., 2025). High glucose levels, which are either age-, diet-, or diabetes-dependent, also expose people to vascular senescence and stiffness, especially in younger groups. Hyperglycemia triggers the generation of AGEs, which are produced as people age and as their metabolic diseases progress (Jakobsche-Policht et al., 2025). AGEs promote vascular stiffening via collagen cross-linking, impair nitric oxide bioavailability, enhance endothelial permeability, and stimulate vascular smooth muscle migration. They also increase oxidative stress, inflammation, and lipoprotein oxidation, amplifying atherogenic processes (Lahariya et al., 2025). Importantly, reducing dietary AGE intake or targeting their formation has shown protective effects in experimental models, highlighting the modifiable systemic drivers of vascular aging.

4.5 Functional consequences of cell aging

Cellular aging induces profound alterations in the structure and function of diverse vascular and immune cell types, changes that are conserved from rodent models to humans. Beyond the generalized decline in proliferative capacity and repair efficiency, senescent cells acquire a pro-inflammatory and pro-thrombotic phenotype, releasing cytokines, chemokines, and matrix-degrading enzymes collectively termed the SASP (Agrawal et al., 2025). In endothelial cells, aging leads to impaired nitric oxide production, endothelial dysfunction, and increased permeability, all of which accelerate lipid infiltration and plaque formation. Vascular smooth muscle cells undergo reduced contractility, abnormal proliferation, and calcification, weakening vessel wall stability and predisposing it to rupture (Amanda et al., 2024). Senescent macrophages exhibit defective efferocytosis, resulting in impaired clearance of apoptotic cells and expansion of the necrotic cores within plaques. Collectively, these functional consequences directly promote atherosclerotic progression by amplifying vascular inflammation, stiffness, and plaque vulnerability. External stressors, such as smoking, metabolic disorders, and oxidative stress, further exacerbate these age-associated cellular defects, accelerating premature vascular aging and deteriorating cardiovascular outcomes.

4.5.1 Endothelial cells

ECs undergo profound structural and functional alterations with aging, which strongly contribute to vascular dysfunction and atherosclerosis. Morphologically, old ECs are enlarged and flattened, and in most cases, they have a polypoid nucleus, which is a characteristic of cell senescence. These changes are closely associated with impaired cytoskeletal organization, reduced proliferative capacity, defective angiogenesis, and diminished migratory potential (Albinsson et al., 2025). Functionally, senescent ECs exhibit a marked decline in endothelial NO production, undermining vasodilation and vascular homeostasis, while simultaneously increasing endothelin-1 release, which promotes vasoconstriction and endothelial activation. With continued aging or repeated passage in culture, ECs demonstrate reduced expression of key adhesion molecules, such as VCAM-1 and ICAM-1, but paradoxically show heightened NF-κB activation and greater susceptibility to apoptosis, collectively driving a pro-inflammatory, proapoptotic phenotype (Chang et al., 2025). Importantly, age-related changes in ICAM-1 function further enhance leukocyte adhesion and transendothelial migration, creating a permissive environment for monocyte infiltration into the vessel wall. Endothelial cell senescence is a critical driver of vascular inflammation, impaired barrier integrity, and atherosclerotic plaque development.

4.5.2 Vascular smooth muscle cells

VSMCs play a central role in maintaining vascular integrity; however, their functional decline with aging contributes significantly to atherosclerosis progression. VSMCs isolated from human atherosclerotic plaques display diminished proliferative capacity, premature senescence, and heightened susceptibility to apoptosis, impairing their ability to stabilize and repair vulnerable plaques that are prone to rupture (Chen et al., 2025b). Their pathogenic phenotype is further increased during aging, with aged rodent aortas and human vascular smooth muscle cells secreting high amounts of IL -6, a marker of the SASP. In addition to IL, senescent vascular smooth muscle cells also activate chemokines such as CCL2, adhesion molecules such as ICAM-1, and innate immune receptors such as TLR-4, thereby creating a pro-inflammatory environment that brings immune cells into plaques (Hou et al., 2025). Enhanced inflammatory signaling accelerates atherogenesis and, at the same time, disrupts vascular homeostasis. Together, vascular smooth muscle cell senescence is one of the most important factors predisposing plaque, as it links impaired healing scope and perpetuated inflammation, highlighting the cells as plausible treatment objectives in the context of vascular aging and atherosclerosis.

4.5.3 Inflammatory cells

The involvement of inflammatory cells plays an important role in the onset and development of atherosclerosis, and aging increases their pathogenicity. Patient-isolated monocytes exhibit increased oxidative bursts and augmented release of pro-inflammatory cytokines, such as MCP-1, IL-6, IL-1, and TNF-2, thereby creating a microenvironment that favors endothelial dysfunction and plaque instability (Ihan, 2025). Strikingly, these pro-inflammatory traits are also characteristic of aged monocytes, highlighting the direct link between immune senescence and vascular pathology. Experimental evidence shows that telomere disruption alone can recapitulate age-associated inflammatory phenotypes, underscoring cellular aging as a driver of immune dysfunction (Kologrivova et al., 2025; Thapa et al., 2024a, 2025). In parallel, age-related alterations in endothelial adhesion molecules facilitate greater monocyte recruitment and retention within the vascular walls, thereby accelerating macrophage accumulation and activation in plaques (Lim et al., 2025). Collectively, these findings suggest that vascular inflammation in atherosclerosis is not merely a byproduct of risk factors but is potentiated by aging-induced immune remodeling, positioning senescent monocyte/macrophage activity as a critical target for therapeutic interventions.

4.5.4 Endothelial cell senescence and chronic heart failure

Endothelial cells form the fundamental barrier of the cardiovascular system and are essential for vascular tone, permeability, and tissue perfusion regulation. Under physiological conditions, endothelial cells maintain vascular homeostasis; however, during aging or chronic stress, they undergo structural and functional decline, characterized by a flattened, enlarged morphology with polyploid nuclei and reduced proliferative capacity (Lu et al., 2025). Endothelial senescence is a major pathological trigger for CHF, a condition in which the heart cannot meet systemic circulatory demands. Experimental evidence indicates that senescent endothelial cells accelerate CHF via pro-inflammatory and pro-fibrotic pathways. IL-17A, secreted by activated T-helper lymphocytes, induces endothelial senescence via activation of the NF-κB/p53/Rb axis (Młynarska et al., 2025). This signaling cascade not only promotes cell cycle arrest but also downregulates calcium-handling proteins, such as SERCA2α and Cav1.2, leading to impaired excitation–contraction coupling and subsequent deterioration of cardiac function (Nàger et al., 2025). In preclinical models, these alterations are correlated with increased myocardial fibrosis, cardiomyocyte apoptosis, larger infarct size, and worsening ventricular performance, thereby accelerating CHF progression. Dietary stress further exacerbates these changes. In murine models exposed to high-fat and high-salt diets, endothelial cell inflammation and senescence are accompanied by atrial dilatation, ventricular hypertrophy, and myocardial fibrosis, which are hallmark features of CHF (Sutanto et al., 2025; Thapa et al., 2024b). These observations highlight the strong link between vascular aging, endothelial dysfunction, and the transition to overt heart failure. In summary, endothelial cell senescence drives maladaptive vascular remodeling by amplifying inflammation, thrombosis, and atherosclerotic burdens. Through these mechanisms, senescent endothelial cells contribute to vascular stiffening, impaired perfusion, and the initiation and progression of CHF.

4.5.5 Vascular smooth muscle cell senescence and chronic heart failure

VSMCs play a pivotal role in maintaining vascular tone, producing extracellular matrix, and orchestrating arterial repair following injury. When VSMCs undergo senescence, they acquire distinct morphological and functional changes, including a flattened and hypertrophic appearance, diminished proliferative capacity, increased expression of cell cycle inhibitors such as p16 and p21, and elevated senescence-associated β-galactosidase activity (Vance et al., 2025). These alterations compromise their reparative potential and contribute to pathological vascular remodeling. Senescent VSMCs are also a major source of SASP, releasing pro-inflammatory mediators, including IL-6, IL-8, and MCP-1, while reducing antiinflammatory signals, such as RANTES and IL-1 receptor 2 (Bayraktutan, 2025). This secretory profile promotes the recruitment of monocytes and macrophages, amplifies cytokine release, and induces adhesion molecule expression in neighboring endothelial and nonsenescent smooth muscle cells, thereby perpetuating vascular inflammation (Chang et al., 2025).

Beyond inflammatory signaling, senescent VSMCs directly affect the stability of plaques. VSMCs are central to collagen synthesis within the fibrous cap of atherosclerotic plaques. During senescence, collagen synthesis decreases, leading to fibrous cap weakening, necrotic core formation, and ectopic calcification (Chen et al., 2025c). Simultaneously, the release of matrix-degrading proteins and signaling intermediates, such as extracellular signal-regulated kinase (ERK), increases the destabilization of the atheromatous environment. As a result, the risk of plaque rupture and thus precipitating acute ischemic cardiovascular events is increased even in diseased arteries with moderate luminal constriction, hence increasing the risk of adverse sequelae such as malignant heart failure (Choi et al., 2025).

Vascular smooth muscle cell senescence is an accepted cause of atherosclerotic plaque susceptibility and cardiovascular disease; however, its mechanistic role in the development of CHF has not been well established (Deng et al., 2025). Further research is warranted to clarify the interplay between senescent VSMCs, cardiomyocytes, endothelial cells, and systemic inflammatory networks, and thus to fully describe their contribution to the pathogenesis of heart failure and the possible therapeutic tools.

4.5.6 Fibroblast senescence and chronic heart failure

Cardiac fibroblasts play a central role in regulating myocardial architecture and function to maintain ECM homeostasis in a normal physiological state and coordinate pathological remodeling in disease conditions. Under normal physiological conditions, the synthesis and catabolism of ECM components in a physiologically normal myocardium are dynamic and controlled by fibroblasts to ensure tissue integrity (Bak et al., 2025). However, with advancing age or sustained stress, fibroblasts undergo senescence, which is characterized by impaired proliferative capacity and heightened secretion of SASP factors. These senescent fibroblasts progressively deposit ECM components, promoting interstitial fibrosis, stiffening of the myocardium, and deterioration of diastolic and systolic functions. Persistent fibrotic remodeling not only compromises contractility but also establishes a feed-forward cycle of maladaptation that culminates in the development and progression of CHF (Chang et al., 2025).

Recent mechanistic studies have highlighted the importance of fibroblast-derived regulatory proteins in modulating fibrotic pathways. High-temperature requirement A3 (HTRA3), a serine protease, has been identified as a critical regulator of fibrosis and cardiac dysfunction. Experimental models of pressure overload have demonstrated that reduced HTRA3 expression in cardiac fibroblasts activates TGF-β signaling, driving excessive ECM deposition and maladaptive remodeling (Floryan et al., 2025). Conversely, restoration of HTRA3 expression attenuates TGF-β activity, limits fibrosis, and preserves ventricular function following hemodynamic stress (Gupta et al., 2025b). These findings suggest that targeting fibroblast senescence and restoring protective mediators, such as HTRA3, could represent promising therapeutic avenues to delay or mitigate CHF progression by interrupting the cascade of fibrosis-driven remodeling.

4.5.7 Immune cell senescence and chronic heart failure

The immune system plays a central role in the progression of CHF, as age-related senescence of both innate and adaptive immune cells amplifies inflammation, disrupts homeostasis, and promotes maladaptive remodeling of cardiac tissue.

Within the innate immune system, neutrophils, monocytes, macrophages, and NK cells undergo profound senescence-associated changes that influence the pathogenesis of CHF (Altab et al., 2025). Senescent neutrophils exhibit heightened oxidative stress, telomere dysfunction, and aberrant release of neutrophil extracellular traps (NETs). In parallel, lipid accumulation, such as 7-ketocholesterol, accelerates vascular damage, predisposing to hyperlipidemia, atherosclerosis, and compromised myocardial perfusion (Cao et al., 2025). Macrophages, which normally preserve myocardial integrity through the phagocytosis of damaged cardiomyocytes and mitochondria, lose their efficiency with aging. Instead, they acquire pro-inflammatory characteristics, secrete matrix metalloproteinases and chemokines, and contribute to cardiomyocyte hypertrophy and ventricular dysfunction (El Kassar et al., 2025). The decline in macrophage-mediated mitochondrial quality control and aberrant inflammasome activation further aggravate cardiomyocyte injury and energy imbalances. Senescent NK cells, which ordinarily eliminate dysfunctional or senescent cells, display reduced cytotoxicity and impaired cytokine release (Hu et al., 2025a). This immune decline not only heightens the risk of infection but also weakens the clearance of senescent cardiac and vascular cells, thereby promoting inflammation and structural deterioration.

The adaptive immune system, composed of T and B lymphocytes, also contributes to CHF through senescence-related dysfunction. Senescent $CD4^+$ T cells secrete excessive interferon-gamma, infiltrate myocardial tissue, activate macrophages, and increase metalloproteinase production, which collectively exacerbate inflammation and tissue stress (Jung et al., 2025). These cells also destabilize arterial plaques and accelerate vascular damage, predisposing patients to coronary ischemia and adverse ventricular remodeling. Cytotoxic T cell activity, mediated by perforin and granzymes, damages vascular smooth muscle cells and endothelial cells, promoting atherosclerosis. Immunometabolic reprogramming of senescent T cells marked by cholesterol accumulation and impaired efflux pathways induces a senescence-like phenotype, fueling systemic inflammation and worsening myocardial dysfunction (Kraus et al., 2025). Clinical studies have shown an expansion of hyperdifferentiated $CD4^+$ T cells in advanced CHF, which correlates with poor ventricular performance. In addition, the reduced expression of regulatory microRNAs, such as miR-181c, and a decline in circulating B cells point toward impaired adaptive immunity in CHF, although the precise role of B cell senescence remains incompletely defined (Linkner et al., 2025). Taken together, these findings underscore that senescent immune cells act through multiple interlinked pathways to drive chronic inflammation, maladaptive remodeling, and progressive cardiac failure. While existing evidence strongly implicates immune cell senescence in CHF, further mechanistic and translational studies are required to clarify the causality and identify therapeutic strategies to mitigate immune-driven deterioration in heart failure.

4.5.8 Mitochondrial dysfunction in senescent cells and chronic heart failure

Mitochondria are central regulators of cellular energy metabolism, maintaining their structural and functional integrity through dynamic cycles of fission and fusion, collectively termed mitochondrial dynamics. These processes ensure energy homeostasis, coordinate stress responses, and regulate survival signals. Fusion of the outer and inner mitochondrial membranes preserves cristae integrity, prevents the release of apoptotic factors, and enables content exchange, whereas fission facilitates the removal of damaged mitochondria and adaptation to energy demands. However, disruption of this balance drives mitochondrial dysfunction and contributes to cellular senescence (Chang et al., 2025).

During senescence, excessive fission mediated by aberrant interactions of regulatory proteins promotes mitochondrial fragmentation, elevates ROS accumulation, and induces progressive oxidative stress. In parallel, suppression of mitochondrial fusion through impaired inner membrane remodeling and reduced expression of fusion-related proteins leads to structural abnormalities, functional decline, and impaired mitochondrial bioenergetics (Denk et al., 2025). The downstream consequences include the aberrant release of mitochondrial nucleic acids, activation of innate immune pathways such as cGAS-STING, and amplification of the SASP, creating a self-perpetuating cycle of inflammation and cellular dysfunction (Ding et al., 2025b).

Senescent cells display reduced mitochondrial respiratory capacity, loss of membrane potential, and compromised repair mechanisms. Impaired mitophagy further exacerbates the accumulation of dysfunctional organelles, aggravating oxidative stress and disrupting metabolic networks (He et al., 2025). External insults, such as cigarette smoke, accelerate this process by blocking mitophagy, increasing the damaged mitochondrial burden, and sustaining ROS generation. Antioxidant interventions targeting mitochondria have been shown to delay senescence phenotypes, underscoring the central role of mitochondrial health in cellular longevity (Kraus et al., 2025).

In the myocardium, mitochondrial dysfunction directly translates into pathological remodeling. Cardiomyocytes with damaged mitochondrial DNA and increased ROS levels show lipid and protein oxidation, telomere shortening, and progressive contractile decline. These events accelerate fibrosis, vascular stiffening, and cardiomyocyte loss, ultimately contributing to CHF. The imbalance between mitochondrial biogenesis, turnover, and energy conversion drives cellular senescence and fuels systemic maladaptation in the aging heart (Liang et al., 2025). Thus mitochondrial dysfunction serves as a critical nexus between senescence and CHF pathogenesis. By restoring mitochondrial dynamics and preserving bioenergetic function, it may be possible to interrupt the feedback loop of oxidative stress, SASP signaling, and inflammation, which underlies disease progression. Targeted modulation of mitochondrial fission and fusion, enhancement of mitophagy, and development of mitochondria-specific antioxidants or gene therapies are promising avenues for therapeutic innovation in CHF.

4.6 Senescence-associated secretory phenotype and chronic heart failure

Cellular senescence is increasingly recognized as a pathological driver of CHF through the secretion of SASP. The SASP encompasses a spectrum of pro-inflammatory cytokines, chemokines, proteases, and growth factors that disrupt tissue homeostasis and create a chronic, low-grade inflammatory environment. In the cardiovascular system, SASP secretion amplifies local and systemic inflammation via pathways such as NF-κB, IL-1α/IL-6, and TGF-β, which act in both autocrine and paracrine manners to promote atherosclerosis, impair vascular function, and accelerate myocardial remodeling (Bie and Tao, 2025). These maladaptive responses ultimately increase the susceptibility to CHF. Importantly, SASP expression has been identified as an independent risk factor for CHF progression in aging populations, underscoring its role as both a biomarker and therapeutic target. A growing body of evidence highlights the central role of metabolic reprogramming in the regulation of SASP expression. Shifts between oxidative phosphorylation and glycolysis alter chromatin dynamics, redox balance, and mitochondrial function, thereby sustaining SASP activity in cardiac and vascular cells (Chen et al., 2025c). For example, modifications in lactate-dependent chromatin remodeling and pyruvate dehydrogenase kinase-driven glucose metabolism have been shown to activate pro-senescent signaling cascades and accelerate vascular dysfunction (Fan et al., 2025a). Energy-sensing regulators, such as AMPK and sirtuins, further integrate oxidative stress responses with metabolic rewiring, linking mitochondrial dysfunction to sustained SASP activation (Gagliardi et al., 2025). Collectively, these mechanisms demonstrate how senescent cardiomyocytes and vascular cells aggravate the cardiovascular microenvironment by continuously propagating the SASP through altered energy metabolism.

Chemotherapy-induced cardiotoxicity exemplifies the contribution of SASP to CHF. Doxorubicin, a widely used anticancer agent, triggers ROS-mediated p53 activation and enhances IL-6 secretion, thereby amplifying SASP expression and promoting late-onset cardiomyopathy. Modulation of upstream regulators, such as the suppression of GATA4 to reduce SASP production, has been shown to attenuate senescence-driven cardiac dysfunction (Gulej et al., 2025). Moreover, emerging technologies, such as artificial intelligence-assisted electrocardiography, are beginning to reveal correlations between SASP signatures, telomerase activity, inflammatory burden, and CHF severity, suggesting their potential utility for risk stratification (Liu et al., 2025b). Despite these advances, several challenges remain to be addressed. Key gaps include the limited in vivo validation of SASP-targeted interventions, incomplete understanding of secretion dynamics, and difficulties in identifying precise molecular targets. However, targeting the SASP using senomorphic agents, metabolic modulators, or pathway-specific inhibitors offers promising avenues for CHF prevention and treatment, with the potential to move beyond symptomatic relief toward disease modification.

4.7 Epigenetic regulation and chronic heart failure

Cardiac aging is a multifactorial process driven by intricate interactions between genetic regulatory pathways and epigenetic control mechanisms. Epigenetic modifications, including DNA methylation, histone modifications, chromatin remodeling, and the regulation of noncoding RNAs, play decisive roles in shaping chromatin structure and transcriptional activity (D'Incal et al., 2025). These changes contribute to pathological outcomes, such as cardiomyocyte functional decline, mitochondrial metabolic impairment, and cardiac fibrosis, all of which promote the progression of CHF. DNA methylation undergoes profound alterations with age, marked by global hypomethylation and site-specific hypermethylation (Drekolia et al., 2025). Such shifts can silence essential housekeeping genes, activate retrotransposable elements, and accelerate the degeneration of the cardiovascular system. Histone posttranslational modifications add another critical regulatory layer by modulating nucleosome accessibility and chromatin compaction (Ferraro et al., 2025). The loss of repressive histone marks, such as H3K9me3, increases genomic instability in senescent cardiomyocytes, whereas the abnormal enrichment of active enhancer marks, such as

H3K27ac, at pro-inflammatory loci amplifies chronic inflammation (Gao et al., 2025). The balance between the histone acetyltransferases and the deacetylases is a major factor in the functioning of the cells; the loss of the deacetylase functions impairs the process of mitochondrial biogenesis, and mitigated SIRT3 activity increases the vulnerability to the influence of oxidative stress (Hickenlooper et al., 2025).

lncRNAs, miRNAs, and circRNAs form an important category of RNAs that regulate the expression and cellular fate of cells via their interaction with the epigenetic machinery. Their dysregulation leads to telomere shortening, DNA damage, fibroblast activation, maladaptive ventricular remodeling, and increased cardiac dysfunction (Jiang et al., 2025). In addition to structural remodeling, epigenetic mechanisms refine inflammatory responses, which affect the severity and duration of myocardial inflammation with the progression of congestive heart failure (Li et al., 2025). In brief, the epigenomic state of cardiac aging and congestive heart failure is an extremely dynamic and integrative system. It is always critical to delineate the mechanisms through which DNA methylation, histone posttranslational modifications, and noncoding RNA species interact to trigger cardiac dysfunction in order to be able to design targeted therapies rationally (Łoboda et al., 2025). Future studies should chart these regulatory interactions with greater analytical accuracy and identify targets for explicit epigenetic intervention that may delay cardiac senescence and improve long-term cardiovascular outcomes.

4.8 Current and emerging senescence-targeted therapies in atherosclerosis and heart failure

4.8.1 Senescence-associated β-galactosidase to multiomic profiling

Cellular senescence can now be elucidated and measured in a much more comprehensive way than by conventional senescence-associated 2-galactosidase (SA 2gal) staining. Although once the hallmark of senescence detection, SA-β-gal suffers from limited specificity, pH sensitivity, and the inability to capture dynamic cellular changes. Multi-omic and imaging platforms have now been incorporated into modern technologies to describe senescent subpopulations in an exceptionally precise manner (Chang et al., 2025). Circulating biomarkers, such as GDF15, IGFBP7, CCL17, and IL-23R, have emerged as robust indicators of systemic aging and cardiac dysfunction. At the tissue level, FOXC1 in telomere-shortened cardiomyocytes is associated with increased p16 4A and p21CIP1, mitochondrial dysfunction, and reduced contractility, which places FOXC1 at the heart of myocardial senescence (Zhang et al., 2025). Simultaneously, methodological advances, such as fluorescent probes for lysosomal activity or iron accumulation, molecular sensors coupled with CRISPR-based gene editing, and spatial transcriptomics, have allowed real-time visualization and lineage tracing of senescent cells (Yatsu et al., 2025). Single-cell sequencing has distinguished endothelial, smooth muscle, and stromal senescence signatures, linking them to fibrotic and inflammatory remodeling (Xue et al., 2025). Collectively, these innovations mark a decisive shift from static single-marker assays to dynamic multidimensional profiling, which enables more accurate disease monitoring and individualized therapeutic strategies.

4.8.2 Senolytic and senomorphic therapies

Translating the biology of senescence into clinical therapy has given rise to senolytic agents that selectively eliminate senescent cells and senomorphics,which suppress their deleterious secretory activity. Among senolytics, BCL-2 family inhibitors, such as navitoclax (ABT-263), and tyrosine kinase inhibitors, such as dasatinib, particularly when paired with the natural flavonoid quercetin, have demonstrated efficacy in reducing senescent cell burden, alleviating fibrosis, and improving ventricular compliance (Fernández-Moreno et al., 2025). Early phase trials in heart failure with preserved ejection fraction (HFpEF) reported functional gains and decreased NT-proBNP levels following intermittent dasatinib-quercetin therapy, although hematologic toxicity, especially thrombocytopenia, limits its widespread use (Flores-Torres et al., 2025).

Natural senolytics, such as quercetin and fisetin, have attracted attention because of their favorable safety profiles. In randomized pilot studies, fisetin modestly improved left ventricular ejection fraction, suggesting cardioprotective benefits with minimal adverse effects. Despite these encouraging results, definitive conclusions await large-scale randomized trials capable of evaluating long-term safety and efficacy (Gold et al., 2025). Complementary to senolysis, senomorphics aim to modulate the SASP rather than eradicating senescent cells (Hu et al., 2025b). Targeting inflammatory mediators, such as IL-6, matrix metalloproteinases, and NF-κB signaling, can attenuate chronic inflammation while preserving transient reparative senescence, which is beneficial for tissue healing.

4.8.3 Metabolic and autophagic modulation

Metabolic reprogramming and impaired autophagy are central features of vascular and myocardial senescence. Overactivation of the mTOR pathway promotes hypertrophy, mitochondrial stress, and maladaptive remodeling, whereas its inhibition through rapalogs, such as everolimus, restores autophagic balance and improves contractile performance. The EVERHEART trial reported that 6 months of low-dose everolimus increased ejection fraction by approximately five percent in patients with heart failure with reduced ejection fraction (HFrEF), although metabolic side effects such as stomatitis and hyperglycemia limited tolerability (Fan et al., 2025b). Parallel attention has been paid to nicotinamide adenine dinucleotide (NAD^+) metabolism, which is a crucial determinant of mitochondrial function and redox homeostasis. Supplementation with NAD + precursors, such as nicotinamide mononucleotide (NMN) and nicotinamide riboside (NR), enhances mitochondrial respiration, decreases oxidative stress, and supports cardiomyocyte survival (Ferri et al., 2026). Early clinical studies have shown improvements in exercise tolerance and peak oxygen uptake, although variability in outcomes underscores the need for standardized dosing and mechanistic clarification of NAD^+-senescence interactions.

4.8.4 Repurposing established cardiometabolic drugs as antisenescence Agents

Several mainstay cardiovascular drugs exert secondary antiaging effects that extend their therapeutic scope beyond risk factor management. Statins, traditionally prescribed for dyslipidemia, delay vascular smooth muscle and endothelial cell senescence through telomere stabilization, DNA repair enhancement, and oxidative stress attenuation (Keshri et al., 2025). ACEIs and ARBs similarly counteract angiotensin II-induced reactive oxygen species production and DNA damage while promoting nitric oxide-mediated endothelial protection via bradykinin signaling (Mu et al., 2025).

Nonpharmacological interventions, such as caloric restriction (CR) and its mimetics, also contribute to vascular rejuvenation. CR enhances nitric oxide bioavailability, limits the oxidative burden, and maintains extracellular matrix integrity. These benefits are largely mediated through sirtuin activation, particularly SIRT1, an NAD^+-dependent deacetylase that improves endothelial function and suppresses atherogenesis (Qiu et al., 2025). Resveratrol, a polyphenolic SIRT1 activator, recapitulates many of these effects, reinforcing the link between metabolic adaptation, redox control, and vascular longevity (Malicevic et al., 2025).

4.8.5 Targeting vascular aging and atherosclerosis Through senescence modulation

In atherosclerosis, targeting aging as a modifiable biological process rather than a byproduct of plaque accumulation has gained momentum. Agents such as pioglitazone and chloroquine can influence telomerase and DDR signaling, thereby limiting plaque formation in experimental models (Qiu et al., 2025). Nevertheless, global telomerase activation carries a carcinogenic risk, prompting a shift toward safer strategies that integrate oxidative stress reduction, DNA repair enhancement, and SASP modulation. A practical near-term approach emphasizes "senescence-informed cardiometabolic care," wherein conventional agents statins, ACEIs, ARBs, and antioxidants are layered on top of standard lipid and blood pressure control to attenuate vascular aging (Mao et al., 2025). Precision therapeutic interventions based on intermittent senolytic agents or localized DNA damage response modulators can provide the following benefits, especially under the direction of biomarker panels that comprise telomere stability, SASP markers, and DNA damage marks (Młynarska et al., 2025). Although extensive trials on exogenous antioxidant supplementation have produced conflicting cardiovascular evidence, evidence of modification of intrinsic redox pathways, by means of Nrf2 activation or glutathione replacement, provides a more physiologically based approach to therapeutic intervention.

4.8.6 Gender-specific aspects of cardiac senescence

The effect of sexual dimorphism on cardiovascular senescence and treatment intervention efficacy is significant. The protective effect of estrogen on the vascular endothelium is through mechanisms that augment endothelial nitric oxide synthase activity and reduce oxidative stress, a situation that is lost after menopause and, therefore contributes to the rapid vascular narrowing and increased prevalence of heart failure in women (Ohayon-Steckel et al., 2025). In contrast, men tend to exhibit earlier hypertrophy and metabolic dysregulation driven by androgen-induced oxidative stress (Tanley et al., 2025). Despite these well-recognized differences, most clinical regimens remain gender-neutral, underscoring the need for sex-specific dosing, biomarker validation, and hormone-responsive therapeutic designs to optimize senescence-targeted interventions.

4.9 Conclusion and future perspective

Cellular senescence is a fundamental biological process that bridges vascular and myocardial aging, linking the development of atherosclerosis and CHF through shared molecular mechanisms. In both diseases, chronic oxidative stress, genomic instability, and metabolic dysregulation lead to the accumulation of senescent endothelial cells, vascular smooth muscle cells, fibroblasts, and cardiomyocytes. These cells adopt a SASP characterized by pro-inflammatory cytokines, growth factors, and matrix-degrading enzymes that propagate inflammation, fibrosis, and tissue remodeling. This results in a feed-forward loop in which senescence contributes to and is reinforced by the evolving disease microenvironment. In atherosclerosis, vascular senescence undermines endothelial barrier integrity, reduces smooth muscle cell reparative capacity, and weakens fibrous caps, thereby accelerating plaque instability and ischemic complications. In CHF, persistent senescence amplifies mitochondrial dysfunction, impairs autophagy, and disrupts cellular homeostasis, culminating in contractile failure and maladaptive remodeling of the heart. Together, these processes exemplify how senescence transforms localized injuries into systemic cardiovascular decline.

However, the dual nature of senescence, protective when transient and detrimental when chronic, poses major therapeutic challenges. Although senolytics, senomorphics, autophagy inducers, and SASP modulators show promise, the broad ablation of senescent cells risks impairing essential repair mechanisms. Therefore, precision strategies are needed, which selectively target harmful senescent cell subsets, modulate SASP components such as IL-6 or MMPs, and restore metabolic resilience without disrupting physiological regeneration. Future research must move beyond descriptive associations toward mechanistic integration. Advanced single-cell multi-omics, spatiotemporal imaging, and organoid or in vivo lineage-tracing models can delineate the dynamic interplay between senescent and nonsenescent cells in the vascular and cardiac compartments. Equally important is the development of reliable circulating and imaging biomarkers to stratify patients based on senescence burden and therapeutic responsiveness. Ultimately, uniting vascular and myocardial research within the framework of cellular senescence offers a paradigm shift in cardiovascular medicine from treating the symptoms of ischemia and failure to targeting the molecular aging processes that underlie them. Precision modulation of senescence has the potential to preserve vascular integrity, sustain myocardial performance, and extend the cardiovascular health span across aging populations.

References

Abdelsayed, M., et al., 2022. Repurposing drugs to treat cardiovascular disease in the era of precision medicine. Nature Reviews Cardiology 19 (11), 751–764.

Agrawal, N., et al., 2025. The role of VEGF in vascular dementia: impact of aging and cellular senescence. Biogerontology 26 (2), 77.

Albinsson, S., et al., 2025. Hippo in smooth muscle - a therapeutic target in vascular diseases driven by aging and hypertension. Frontiers in Physiology 16, 1674714.

Alharbi, K.S., et al., 2022. Molecular explanation of Wnt/βcatenin antagonist pyrvinium mediated calcium equilibrium changes in aging cardiovascular disorders. Molecular Biology Reports 49 (11), 11101–11111.

Ali, I., et al., 2024. Understanding the intricacies of cellular senescence in atherosclerosis: mechanisms and therapeutic implications. Ageing Research Reviews 96, 102273.

Al-Khayyat, W., et al., 2025. FRA1 (FOSL1) suppresses neoplastic transformation and modulates radiation responses via transcriptional control of mitogenic and stress-responsive networks. Frontiers in Cell and Developmental Biology 13, 1659986.

Altab, G., et al., 2025. miRNA changes with ageing and caloric restriction in male rat skeletal muscle: potential roles in muscle cell function. Biogerontology 26 (6), 202.

Amanda, B., et al., 2024. The hippo signaling pathway, reactive oxygen species production, and oxidative stress: a two-way traffic regulation. Cells 13 (22), 1868. Available at https://doi.org/10.3390/cells13221868.

Bak, D.H., et al., 2025. Methyl jasmonate elicitation enhances photoprotective and anti-inflammatory properties in Lavandula angustifolia cell suspension cultures. Scientific Reports 15 (1), 39177.

Bayraktutan, U., 2025. Angiotensin II and cardiovascular disease: balancing pathogenic and protective pathways. Current Issues in Molecular Biology 47 (7), 501. Available at https://doi.org/10.3390/cimb47070501.

Bhat, A.A., et al., 2024. Therapeutic approaches targeting aging and cellular senescence in Huntington's disease. CNS Neuroscience & Therapeutics 30 (10), e70053.

Bie, H., Tao, Z., 2025. The pathophysiological mechanisms of immunosenescence in coronary artery disease. Frontiers in Cell and Developmental Biology 13, 1686947.

Cao, L., et al., 2025. The controversial role of senescence-associated secretory phenotype (SASP) in cancer therapy. Molecular Cancer 24 (1), 283.

Chang, D., et al., 2025. Cellular senescence and cell therapy in cardiovascular diseases. Stem Cell Research & Therapy 16 (1), 613. Available at https://doi.org/10.1186/s13287-025-04731-6.

Chen, J., et al., 2025. METTL3-mediated m6A modification of TNFAIP3 promotes endothelial to mesenchymal transition and atherosclerosis via the NF-κB/Snail signaling pathway. Journal of Lipid Research 66 (12), 100931. Available at https://doi.org/10.1016/j.jlr.2025.100931.

Chen, J., et al., 2025c. Macrophage-derived KIF13B interacts with USP9X to attenuate abdominal aortic aneurysm development by potentiating TFEB stability. Theranostics 15 (17), 9114–9130.

Chen, M., et al., 2025b. Interleukin-17F induces ischemic injury in aged mice by causing endothelial cells' senescence and dysfunction. The FASEB Journal 39 (21), e71197.

Cheraghi, M., et al., 2025. The interplay between diabetes and atherosclerosis: a review of pathophysiological mechanisms. ARYA Atheroscler 21 (5), 62–80.

Choi, H., et al., 2025. Smooth muscle LRRC8A knockout preserves vascular function in Ang II hypertension. Hypertension 83 (1), 40–53. Available at https://doi.org/10.1161/HYPERTENSIONAHA.125.25222.

Davies, J., Menzies, G.E., 2025. Local sequence context at KRAS codons modulates DNA repair efficiency: insights from molecular dynamics simulations. Frontiers in Molecular Biosciences 12, 1654434. Available at https://doi.org/10.3389/fmolb.2025.1654434.

Deng, S., Yin, X., Zhu, R., 2025. Integration of vascular smooth muscle cell phenotypic switching and senescence. Journal of Cardiovascular Pharmacology 86 (5), 435–447.

Denk, D., et al., 2025. Effect of the mitophagy inducer urolithin A on age-related immune decline: a randomized, placebo-controlled trial. Nature Aging 5 (11), 2309–2322. Available at https://doi.org/10.1038/s43587-025-00996-x.

D'Incal, C.P., et al., 2025. A frameshift variant in activity-dependent neuroprotective protein (ADNP) causes nucleocytoskeletal alterations in a dizygotic male twin: a case study. Clinical Epigenetics 17 (1), 185. Available at https://doi.org/10.1186/s13148-025-01998-7.

Ding, J., et al., 2025a. Protein acylation in inflammatory diseases: from mechanisms to therapeutic strategies. Cell Communication and Signaling: CCS 23 (1), 488.

Ding, X.Q., et al., 2025b. Galangin alleviates vascular aging and endothelial cell senescence by targeting SMAD3 to initiate mitophagy. Phytomedicine: International Journal of Phytotherapy and Phytopharmacology 148, 157466.

Drekolia, M.K., et al., 2025. Cystine import and oxidative catabolism fuel vascular growth and repair via nutrient-responsive histone acetylation. Cell Metabolism 38 (1), 115–134.e20. Available at https://doi.org/10.1016/j.cmet.2025.10.003.

El Kassar, N., et al., 2025. A proposed path to explaining the unexplained anemia of aging. The Journals of Gerontology. Series A, Biological Sciences and Medical Sciences 81 (1), glaf251. Available at https://doi.org/10.1093/gerona/glaf251.

Fan, C., et al., 2025a. Rosmarinic acid alleviates doxorubicin-induced cellular senescence and cardiotoxicity by targeting the 14-3-3/Foxo1 signaling axis. Phytomedicine: International Journal of Phytotherapy and Phytopharmacology 148, 157482.

Fan, S., et al., 2025b. SIK2 activates the autophagy–apoptosis pathway through SP1 regulation to inhibit the progression of hepatocellular carcinoma. Frontiers in Pharmacology 16, 1635953.

Fan, S., Shi, J., 2025. Research progress on DNA hydroxymethylation in atrial fibrillation: a review. Frontiers in Pharmacology 16, 1591675. Available at https://doi.org/10.3389/fphar.2025.1591675.

Fernández-Moreno, M., et al., 2025. The senotherapeutic effects of APPA (Apocynin [AP] and Paeonol [PA]) on senescent human chondrocytes. Pharmaceuticals (Basel) 18 (9), 1386. Available at https://doi.org/10.3390/ph18091386.

Ferraro, E., et al., 2025. Inhibition of bromodomain and extra-terminal motif (BET) proteins in pediatric sarcoma: a systematic review of in vitro and in vivo studies. Drug Discovery Today 30 (12), 104516. Available at https://doi.org/10.1016/j.drudis.2025.104516.

Ferri, G., et al., 2026. Insights into the procedures and interpretation of autophagy in Trypanosoma cruzi. Methods in Molecular Biology 2982, 409–425.

Flores-Torres, R.P., et al., 2025. Senotherapy as a multitarget intervention in chronic obesity: Modulation of senescence, neuroinflammation, dysbiosis, and synaptic integrity in middle-aged female Wistar rats. Experimental Neurology 392, 115331.

Floryan, M., et al., 2025. Long-term physiological flow rescues regressed microvascular networks and increases their longevity. npj Biological Physics and Mechanics 2, 24. Available at https://doi.org/10.1038/s44341-025-00028-0.

Gagliardi, A., et al., 2025. Emerging radioligands as tools to track multi-organ senescence. Diagnostics (Basel) 15 (19), 2518. Available at https://doi.org/10.3390/diagnostics15192518.

Gangula, A.B., Kona, L., 2025. Role of homeobox (HOX) genes in adult cardiovascular disease: from developmental regulators to therapeutic targets. Cureus 17 (10), e95730.

Gao, X., et al., 2025. The METTL Family as Regulators of Methylation and Therapeutic Targets in Cardiovascular Diseases. European Journal of Pharmacology 1008, 178344. Available at https://doi.org/10.1016/j.ejphar.2025.178344.

Gold, N.M., et al., 2025. Therapeutic potential of nicotinamide and ABT263 in alcohol-associated liver disease through targeting cellular senescence. MedComm 6 (2), e70086. 2020.

Goyal, K., et al., 2024. Chronic kidney disease and aging: dissecting the p53/p21 pathway as a therapeutic target. Biogerontology 26 (1), 32.

Gulej, R., et al., 2025. Irradiation-induced brain senescence accelerates cardiac aging via systemic mechanisms: insights from transcriptomic profiling. Geroscience 48 (1), 951–971. Available at https://doi.org/10.1007/s11357-025-01953-7.

Gupta, S., et al., 2025a. Harnessing the FOXO-SIRT1 axis: insights into cellular stress, metabolism, and aging. Biogerontology 26 (2), 65.

Gupta, P., et al., 2025b. The aging microenvironment is a determinant of immune exclusion and metastatic fate in pancreatic cancer. Cancer Research 86 (4), 909–924. Available at https://doi.org/10.1158/0008-5472.CAN-25-1904.

Hajar, R., 2017. Risk factors for coronary artery disease: historical perspectives. Heart Views 18 (3), 109–114.

Hao, S., et al., 2025. Multi-target pharmacological mechanisms of Iris tectorum Maxim. (CSG) against tonsillitis and dyspepsia: a network pharmacology approach. Medicine (Baltimore) 104 (39), e44516.

He, Z., et al., 2025. The remodeling of mitochondrial-endoplasmic reticulum contacts by omega-3 fatty acids mitigates dietary advanced glycation end product-driven sertoli cell senescence and oligoasthenozoospermia. International Journal of Biological Sciences 21 (14), 6042–6061.

Hickenlooper, S., et al., 2025. Expression profiles of histone H4K20 methylation and its associated enzymes in mouse cardiac disease and human heart failure. Epigenetics: Official Journal of the DNA Methylation Society 20 (1), 2578553.

Hou, Y., et al., 2025. T cells in cardiovascular disease: solved and unsolved mysteries. Science China. Life Sciences 69 (1), 17–36. Available at https://doi.org/10.1007/s11427-024-2976-8.

Hu, Y., et al., 2025a. Downregulation of PDIA4 inhibits proliferation and migration in human oral squamous cell carcinoma. Hereditas 162 (1), 222.

Hu, Y., et al., 2025b. An effective system for senescence modulating drug development using quantitative high-content analysis and high-throughput screening. Communications Biology 8 (1), 1316.

Hua, J., et al., 2025. ALYREF inhibits ferroptosis in vascular endothelial cells and improves atherosclerosis by epigenetic modification of CISD1. Clinical Epigenetics 17 (1), 151.

Hussain, M.S., et al., 2023. Probing the links: Long non-coding RNAs and NF-κB signalling in atherosclerosis. Pathology, Research and Practice 249, 154773.

Ihan, A., 2025. Senescent polarization of macrophages and inflammatory biomarkers in cardiovascular disease. Cells 14 (17).

Ikeda, H., et al., 2025. Continuous fractionated irradiation with irradiation-free intervals enhances survival and clonogenicity in osteosarcoma MG-63 cells via adaptive DNA damage response. Advances in Biological Regulation 98, 101114.

Ismaeel, A., et al., 2025. Multiomics analysis of skeletal muscle identifies dysregulation of hypoxia-induced genes in peripheral artery disease. Journal of the American Heart Association 14 (19), e042350.

Jakobsche-Policht, U., et al., 2025. The role of protease-activated receptors (PARs) in the functioning of platelets and platelet-derived microparticles (PMPs). Frontiers in Molecular Biosciences 12, 1636893.

Jha, S.K., et al., 2024. Cellular senescence in lung cancer: molecular mechanisms and therapeutic interventions. Ageing Research Reviews 97, 102315.

Jiang, Y., et al., 2025. Advances in BRD proteins in cardiovascular pathophysiology and therapy: a review. International Journal of Biological Macromolecules 333, 148798.

Jung, J.H., Yang, Y., Kim, Y., 2025. cPLA2α-driven Cox-2/PGE2 axis promotes hypoxia-induced senescence in Jurkat T cells. Biochemical and Biophysical Research Communications 790, 152894.

Kane, A.E., Sinclair, D.A., 2019. Epigenetic changes during aging and their reprogramming potential. Critical Reviews in Biochemistry and Molecular Biology 54 (1), 61–83.

Keshri, P.K., et al., 2025. 18β-glycyrrhetinic acid regulates endoplasmic reticulum stress and autophagy dysregulation in the MPTP/p-induced model of parkinson disease. Molecular Neurobiology 62 (12), 16659–16678.

Khavinson, V., et al., 2022. Senescence-associated secretory phenotype of cardiovascular system cells and inflammaging: perspectives of peptide regulation. Cells 12 (1), 106. Available at https://doi.org/10.3390/cells12010106.

Kologrivova, I., et al., 2025. Associations between regulatory immune cells, thymus cellular remodeling, and vascular aging in advanced coronary atherosclerosis: a pilot study. Diagnostics (Basel) 15 (19), 2494. Available at https://doi.org/10.3390/diagnostics15192494.

Kraus, R.F., et al., 2025. Age affects the immune system more than a moderate surgical trauma and anesthesia. Scientific Reports 15 (1), 38993.

Kumar, M., et al., 2024. Cellular senescence as a targetable risk factor for cardiovascular diseases: therapeutic implications: JACC family series. JACC. Basic to Translational Science 9 (4), 522–534. Available at https://doi.org/10.1016/j.jacbts.2023.12.003.

Lahariya, R., Anand, G., Kumari, B., 2025. Trained Immunity in sepsis: exploring the molecular link to long-term cardiometabolic disorders. Immunologic Research 73 (1), 139.

Lewis-McDougall, F.C., et al., 2019. Aged-senescent cells contribute to impaired heart regeneration. Aging Cell 18 (3), e12931.

Li, B., et al., 2025. RNA 5-methylcytosine modification in myocardial fibrosis. Reviews in Cardiovascular Medicine 26 (10), 44079.

Liang, M., et al., 2025. A mitochondria-targeted fluoropolymer nanoparticle with inherent mitophagy-inducing and red fluorescence properties for treatment of atherosclerosis. Nature Communications 16 (1), 9845.

Lim, J., et al., 2025. Late-life aerobic exercise attenuates DNA damage and telomere dysfunction in non-atheroprone but not in atheroprone aortic regions. Aging Cell 24 (10), e70196.

Linkner, T.R., et al., 2025. Circulating tumor cells: indicators of cancer progression, plasticity and utility for therapies. Pathology Oncology Research: POR 31, 1612181.

Liu, X., et al., 2025a. Control of the archaeal DNA damage-responsive pathway by phosphorylation of Orc1-2, the global regulator in Saccharolobus islandicus. Nucleic Acids Research 53 (17).

Liu, J., Yu, H., Xu, Y., 2025b. Targeting cellular senescence: pathophysiology in multisystem age-related diseases. Biomedicines 13 (7), 1727. Available at https://doi.org/10.3390/biomedicines13071727.

Łoboda, A., Zieliński, T., Dulak, J., 2025. Cell-free strategies for cardiomyocyte proliferation and heart repair. Pharmacological Reports: PR 78 (1), 65–89. Available at https://doi.org/10.1007/s43440-025-00805-7.

Lu, J., et al., 2025. Pgk1 activation restores endothelial metabolic homeostasis to alleviate vascular aging and atherosclerosis. Cardiovascular Diabetology 24 (1), 427.

Machado-Oliveira, G., et al., 2020. Cell senescence, multiple organelle dysfunction and atherosclerosis. Cells 9 (10), 2146. Available at https://doi.org/10.3390/cells9102146.

Malicevic, U., et al., 2025. Intricate interplay between ORMDL3, ER stress, and autophagy in the diabetic intestine. Molecular and Cellular Biochemistry 481 (2), 791–807. Available at https://doi.org/10.1007/s11010-025-05424-7.

Mao, J., et al., 2025. Neutrophil extracellular traps as therapeutics target in vascular aging. Frontiers in Immunology 16, 1657938.

Menon, V., Gueble, S.E., 2025. DNA repair helicases: from mechanistic understanding to therapeutic implications. NAR Cancer 7 (4), zcaf034.

Mir, S.M., et al., 2020. Shelterin complex at telomeres: implications in ageing. Clinical Interventions in Aging 15, 827–839.

Mladenov, E., et al., 2025. CRISPR/Cas9 generated DSB clusters mimic complex lesions induced by high-LET radiation and shift repair from c-NHEJ to mutagenic repair pathways. Scientific Reports 15 (1), 36480.

Młynarska, E., et al., 2025. Inflammaging and senescence-driven extracellular matrix remodeling in age-associated cardiovascular disease. Biomolecules 15 (10), 1452. Available at https://doi.org/10.3390/biom15101452.

Mu, J., et al., 2025. Nrf2 promotes BHBA-induced pyroptosis in bovine endometrial epithelial cells by modulating autophagy mediated through oxidative stress. Journal of Agricultural and Food Chemistry 73 (44), 28126–28142.

Nàger, M., et al., 2025. The multifaceted role of autophagy and mitophagy in cardiovascular health and disease. Autophagy Reports 4 (1), 2572511.

Negoi, I., 2025. From infection to invasion: the role of deleted in malignant brain tumors 1 in Helicobacter pylori-driven gastric cancer. World Journal of Clinical Oncology 16 (10), 108377.

Ohayon-Steckel, L., et al., 2025. miR-499 in platelet-derived extracellular vesicles augments inflammatory cell generation and cardiac remodeling after myocardial infarction. Circulation 153 (4), 243–259. Available at https://doi.org/10.1161/CIRCULATIONAHA.124.073527.

Pandey, S.N., et al., 2025. Targeting NLRP3 inflammasome with curcumin: mechanisms and therapeutic promise in chronic inflammation. Inflammopharmacology 33 (10), 5667–5687.

Pandit, R., Yurdagul, A., 2025. The atherosclerotic plaque microenvironment as a therapeutic target. Current Atherosclerosis Reports 27 (1), 47.

Panigrahi, R., et al., 2025. Structural insights into γH2Ax containing nucleosomes. Nucleic Acids Research 53 (19).

Qiu, Y., et al., 2025. Calcium silicate induces mitophagy-mediated metabolic shifts toward oxidative phosphorylation in BMSCs to facilitate osteogenesis and bone regeneration. Regenerative Biomaterials 12, rbaf101.

Raghunathan, A., Krishnan, M., 2025. Minimum energy pathway for lesion recognition and DNA binding by RAD4/XPC. Journal of Chemical Information and Modeling 65 (19), 10375–10389.

Redgrave, R.E., et al., 2023. Senescent cardiomyocytes contribute to cardiac dysfunction following myocardial infarction. NPJ Aging 9 (1), 15.

Sarkar, P., Shinton, R., 2001. Hutchinson-Guilford progeria syndrome. Postgraduate Medical Journal 77 (907), 312–317.

Satyanarayana, A., et al., 2003. Telomere shortening impairs organ regeneration by inhibiting cell cycle re-entry of a subpopulation of cells. The EMBO Journal 22 (15), 4003–4013.

Shan, X., et al., 2025. DNA - histone cross-link locks the nucleosome structure and disrupts its recognition and processing. Protein & Cell 17 (5), 476–482. Available at https://doi.org/10.1093/procel/pwaf094.

Stary, H.C., et al., 1995. A definition of advanced types of atherosclerotic lesions and a histological classification of atherosclerosis: a report from the Committee on Vascular Lesions of the Council on Arteriosclerosis, American Heart Association. Circulation 92 (5), 1355–1374.

Sutanto, H., et al., 2025. The role of the senescence-associated secretory phenotype in cardiovascular disease among the elderly. European Journal of Internal Medicine 141, 106488.

Tanley, J., et al., 2025. Subclinical vascular burden in relation to cognitive decline and impairment: the multi-ethnic study of atherosclerosis (MESA). Alzheimer's & Dementia: the Journal of the Alzheimer's Association 21 (10), e70866.

Thapa, R., et al., 2024a. Epithelial-mesenchymal transition to mitigate age-related progression in lung cancer. Ageing Research Reviews 102, 102576.

Thapa, R., et al., 2024b. The role of sirtuin 1 in ageing and neurodegenerative disease: a molecular perspective. Ageing Research Reviews 102, 102545.

Thapa, R., et al., 2025. Hypoxia-inducible factor and cellular senescence in pulmonary aging and disease. Biogerontology 26 (2), 64.

Vance, M.L., et al., 2025. Endothelial-to-mesenchymal transition in the central nervous system: a potential therapeutic target to combat age-related vascular fragility. The Journal of Pharmacology and Experimental Therapeutics 392 (11), 103747.

Xie, S., et al., 2023. Metabolic landscape in cardiac aging: insights into molecular biology and therapeutic implications. Signal Transduction and Targeted Therapy 8 (1), 114.

Xu, C., et al., 2025. The role of cellular senescence in cardiovascular disease. Cell Death Discovery 11 (1), 431.

Xue, Y., et al., 2025. Comparative evaluation of extraction methods for Scorias spongiosa polysaccharides: yield, bioactivity, and anti-aging potential. Ultrasonics Sonochemistry 123, 107655.

Yatsu, T., Koyama, M., Miyata, Y., 2025. VDR is a novel regulator of cellular senescence in lens epithelial cells. Journal of Nutritional Science and Vitaminology 71 (5), 395–401.

Zhai, P., Sadoshima, J., 2024. Cardiomyocyte senescence and the potential therapeutic role of senolytics in the heart. Journal of Cardiovascular Aging 4 (2).

Zhang, L., Luo, D., Bai, L.F.Z., 2025. Apelin-12 attenuates LPS-induced cellular senescence in human dental pulp cells via SIRT6-mediated pathways: implications for gingivitis management. Molecular Oral Microbiology e70012.

Chapter 5

Senescence-associated secretory phenotype in cardiac tissue

Riya Thapa[1], Asif Ahmad Bhat[1], Yogendra Singh[2], Surya Nath Pandey[3], Santosh Kumar Singh[4], Md Raihan Siddique[5], Neeraj Patel[5], Doli Rani Das[6], and Muhammad Afzal[7]

[1]*Faculty of Pharmacy, AIMST University, Bedong, Kedah, Malaysia,* [2]*Department of Pharmacology, Maharishi Arvind College of Pharmacy, Ambabari Circle, Ambabari, Jaipur, Rajasthan, India,* [3]*Department of Pharmacology, Teerthanker Mahaveer College of Pharmacy, Teerthanker Mahaveer University, Moradabad, Uttar Pradesh, India,* [4]*ICFAI School of Pharmaceutical Sciences, The ICFAI University, Jaipur, Rajasthan, India,* [5]*School of Pharmacy, Suresh Gyan Vihar University, Jaipur, Rajasthan, India,* [6]*Shree Krishna College of Pharmacy, Sitapur, Uttar Pradesh, India,* [7]*Department of Pharmaceutical Sciences, Pharmacy Program, Batterjee Medical College, Jeddah, Saudi Arabia*

5.1 Introduction

Cellular senescence is a piece of protective biological process that ceases the increase of broken cells and, therefore beats their unregulated growth. It is induced by a variety of noxious stimuli, which include oxidative stress, ionizing radiation, and chemotherapeutic drugs in terminally differentiated cells (Besse et al., 1994). The accumulation of senescent cells occurs during tissue aging, and they develop as a natural part of the aging process. They subsequently prevail in age-related diseases, including cardiovascular diseases (CVD), type 2 diabetes mellitus (T2DM), neurodegenerative diseases, malignancies, and musculoskeletal diseases (Biggs and Booth, 1990).

Having empirical data that show that when these senescent cells are selectively destroyed, the life-span of animals may be increased, thereby involving the involvement of their involvement in age-related pathologies. Senescence-associated secretory phenotype (SASP) is one of the potential mechanisms through which senescent cells secrete signaling molecules, such as pro-inflammatory cytokines and extracellular matrix-degrading enzymes (Brand, 2019). Such secretory activity contributes to a low-grade, chronic, systemically based inflammation known as inflammaging, which is a dominant disease risk factor in CVD and other such illnesses of age-related disease (Cannon and Bodmer, 2016).

Another mostly central area of inquiry in modern gerontology and molecular medicine is the formulation of agents in the form of peptides that will inevitably facilitate longevity and prolong the maintenance of physiological integrity. These strategies have found special interest in the short peptides, in particular, dipeptides, tripeptides, and tetrapeptides, as potential ways in counteracting premature aging. Dried tissue extracts of these molecules normally exhibit organ selectivity in both in vitro and in vivo studies of young and old animal models (Carbonin et al., 1990; Carrier et al., 1992).

Compared to the larger polypeptides, short regulatory peptides in a number of ways are not immunogenic, are not species-specific, and, in other ways, act as peptide hormones do. Their efficacy has been validated and their safety confirmed by empirical studies on cells, tissues, and intact organisms, as well. The clinical application of peptide formulations is traditionally accompanied by the enhancement of physiological functioning and decreased mortality at different ages in the context of long-term follow-ups of patients (Chang et al., 2025; Chen et al., 2022). The identifications make the regulatory peptides promising in a study in relation to geroprotective ability, which can be a leap in finding a way to delay the occurrence of age-related diseases and prolong the healthy life of elderly people (Cianflone et al., 2020).

Cardiovascular diseases, and especially coronary artery disease and myocardial infarction, are among the age-related diseases that cause the greatest number of deaths worldwide. The diseases have atherosclerosis as a hallmark of aging, which is characterized by chronic inflammation of the endothelial lining tissue and the subendothelial formation of plaques in the artery walls. It has been believed that senescence-associated secretory phenotype (SASP) factors may contribute considerably to such chronic inflammatory environment and are therefore central to the process of atherosclerosis development and its resulting cardiovascular complications (Conroy, 2024; Cowling, 2015).

Cellular Senescence and Cardiac Aging. DOI: https://doi.org/10.1016/B978-0-443-45811-8.00015-5

The review discusses the signaling molecules that have been involved in SASP and inflammaging in the cardiovascular tissues and how specific peptides can regulate such processes. It focuses on the molecular pathway leading to SASP in the most important cardiovascular cells—endothelial cells, vascular smooth muscle cells (VSMCs), and cardiomyocytes—and evaluates the vasoprotective peptides as therapy in cardiovascular aging.

5.2 Senescence-associated secretory phenotype-driven mechanisms in cardiovascular senescence

Consistent observation studies show that senescence cells modify their environments by bidirectional secretion of diverse catalog of bioactive items. This amalgamated production is known as the senescence-associated secretory phenotype (SASP), an invariable set of productions that regulates physiological and pathological properties like tissue restoration, restoration, pro-inflammatory conditions, developmental morphogenesis controls, and ability development. The SASP consists of cytokines, chemokines, proteases, growth factors, and structural matrix proteins, and each one of them functions as an agent of the cross-talk between cells (Czibik et al., 2021; Douin-Echinard et al., 2020).

A classification system used in academia discriminates SASP components based on the composition of the molecules and effects on function. Signaling molecules involved in the modulation of recipient-cell behavior are soluble molecules whose effect is achieved after interacting with receptors at the cell surface; these are the cytokines, chemokines, and growth factors. Proteolytic enzymes, which are directly acting molecules (metalloproteinases and serine proteases in particular) and reactive oxygen species (ROS), along with nitric oxide, which also acts as direct-acting molecule, were shown to have mechanistic effect on the integrity of the matrix itself and on the cellular integrity (Ellison-Hughes, 2023; Fatmi et al., 2024). Regulatory factors, in particular, tissue inhibitor of metalloproteinases (TIMPs) and plasminogen activator inhibitors, attenuate or streamline proteolysis of the SASP origin. Most recently, microRNA-laden extracellular vesicles (miRNA-EVs) have been identified as a new form of effectors of SASP activities that have both pro- and antisenescence activities depending on the miRNA content (Gabillard-Lefort et al., 2025; Geokas et al., 1990).

The secretion phenotype is called Senescent Associated Secretory Phenotype (SASP), and it will occur in sequential phases: initial, early, and mature. The first process, which begins directly after the DNA lesion and takes about 36 h, is induction. The initial stages take a period of a few days, during which there is a gradual production of cytokines. The mature stage happens 4–10 days after the initiating senescence-inducing stimulus, whereby autocrine feedback enhances the release of the components of SASP. SASP is regulated both transcriptionally and post-transcriptionally, and the key regulatory system of SASP gene expression is nuclear factor kappa B (NF-κB) (Groban, 2005; Gude et al., 2018).

Repetitive deposition of senescent cells and constant expression of SASP undermine tissue performance, accelerate biological aging, and increase the likelihood of growing a variety of age-related pathologies, especially cardiovascular disease. In the cardiovascular system, senescent cells excessively synthesize the products of matrix metalloproteinases that weaken the structures of the circulation system and worsen coronary heart disease. Further, the senescent vascular smooth muscle cells release pro-inflammatory cytokines that promote growth and advancement of atherosclerosis (Hastings et al., 2025; He et al., 2024). Each of the cellular components of the cardiovascular tissues has specific signatures on SASP. These phenomena suggest the importance of cell-specific SASP in the development of chronic inflammation and structural remodeling in the age-related cardiovascular pathology, highlighting the different molecular profiles proposed by Senescent endothelial cells, vascular smooth muscle cells, and cardiomyocytes in the general inflammatory and degenerative environment (Hedge et al., 2025; Heist et al., 2003) (Fig. 5.1).

5.2.1 Senescence-associated secretory phenotype expression in aging endothelium

Senescence in endothelial cells (ECs) involves morphological and functional changes, which are distinct in aged ECs, as compared to young ones. Due to these changes, there is a shift towards pro-inflammatory and prothrombotic phenotypes, and these are the factors contributing to the development of age-related disorders. In terms of the metabolism, elderly ECs are characterized by impaired glycolysis, and production media consumption intensifies (Ho et al., 1986; Holmstrom and Chugh, 2023).

Several aspects could enhance premature senescence and the SASP emergence. These are chronic strand breakage in DNA, mild mitochondrial dysfunction, hemodynamic disruption, oxidative damage, shortening of telomeres, and radiation. The secretion of the SASP components by ECs also depends on the type and severity of the external stressor and the microenvironment. As another example, human umbilical vein endothelial cells (HUVECs) in replicative aging secrete pro-inflammatory factors including IL-6, IL-8, PAI-1, and MCP-1. However, in comparison to replicative senescence, stress-induced aging of HUVECs, which occurs in response to chemokines or mechanical stress, does increase CXCL11

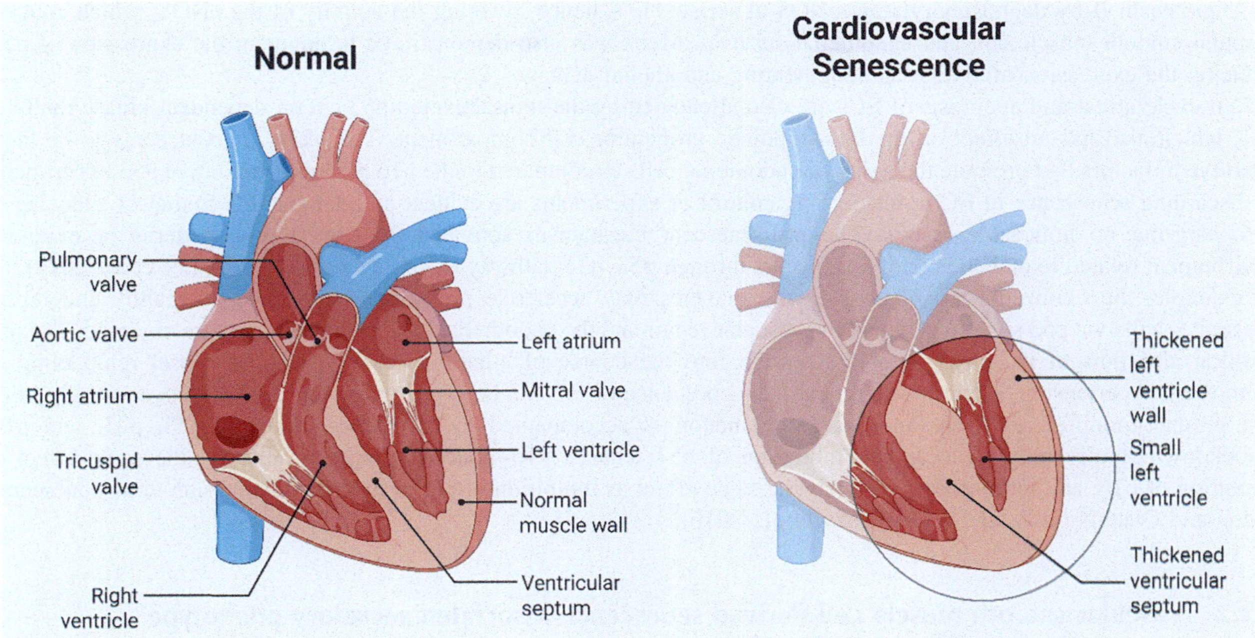

FIGURE 5.1 **Structural changes in the senescent heart.** This figure contrasts a normal heart with a heart affected by cardiovascular senescence. The senescent heart shows left ventricular wall thickening, reduced chamber size, and septal thickening that can impair diastolic function and overall performance.
No permission required.

levels and reactive oxygen species (ROS), which is not usually the case in replicative senescence. Interestingly, there are molecules that are stably expressed under both aging settings, such as IL-1 alpha, so there are shared signaling points (Hong et al., 2024; Horn, 2015).

As compared to EC having not been exposed to TGF-alpha, growth factor-stimulated EC exhibit different senescence-associated secretory phenotype patterns which include increased secretion of adhesion molecules like E-Selectin and ICAM-1, pro-inflammatory cytokines like IL-6 and IL-8, as well as several regulators like PAI-1 and IGFBP-5. The chemotherapeutic agent doxorubicin has the potential to trigger the SASP, as a result of which platelets can be activated, which reflects that there may be a connection between age-associated cardiovascular disease and endothelial senescence as a risk factor in thrombosis. Moreover, there seems to be growing evidence pointing to MMP-9 being involved in chronic inflammation and becoming part of an inflammaging marker to the vascular endothelium. Another chemokine, MCP-1, which activates through CCR-2 receptors, also works along with EC senescence (Huang et al., 2021; Hudlicka and Brown, 1996).

Recently, other data have also indicated the gendered differences in the SASP expression. It has been found that these sex differences in endothelial aging profiles are related to divergent hormone levels and receptor expression according to the animal studies. Determination of consistent biomarkers has discovered unique transcriptional patterns; Transcriptomic replaced dozens of variously expressed RNA-markers between ECs, and fibroblasts, in senescence. There are two major signaling axes, which regulate endothelial senescence: the p53–p21 axis and the p16-retinoblastoma (pRb) axis (Isoyama, 1996; Jahangir et al., 2007). The p53-p21 axis is mostly held as the first driver of senescence; the p16-pRb path further maintains the senescence phenotype. Most importantly, p16-pRb cascade induction marks a switch to an irreversible p phenotype, where senescence was originally initiated and maintained by p53. p53 The tumor suppressor protein encoded by the TP53 gene is activated in response to DNA damage and responds to the level of stress by inducing cell cycle arrest or apoptosis. Without the damage, C-terminal controllers MDM2 and MDM4 are described to manage the levels of p53 through degradation facilitated by ubiquitination. Whereas under an oxidative and genotoxic stress, p53 is phosphorylated and acetylated, forming stabilization, thus overcoming MDM2 inhibition. Histone deacetylase- SIRT1 also forms repression of p 53 functions by removal of acetyl moieties of activating lysines. Notably, p53 also leads to vascular impairment by affecting the ability of endothelium-dependent vasodilation that plays a significant role in circulatory homeostasis (Kane and Howlett, 2016; Kitzman and Edwards, 1990). In addition, p53 supervises endothelial dysfunction of Angiotensin II by dephosphorylating eNOS at serine-1177, hence lowering the activity of the eNOS, which controls vascular smooth muscle cell and endothelial survival. Meox was also demonstrated to augment the expression of p53, reducing the expression of eNOS, and augmenting endothelial damage.

Cell cycle arrest and apoptosis of ECs are also attenuated by the transactivated p53 cyclin-dependent kinase inhibitor p21, which also has an effect on the hematopoietic progenitor cell compartment. The p53-p21 axis seems to be more sensitive to factors that promote the aging of endothelial cells as compared to the p16 pathway. Inhibitive p53 experiments in discarding senescence in EC in undergoing control of experiments are evident and comparable to that of silencing of p16, carrying no noticeable results. The pro-senescent mechanism activated by altered flow patterns or oxidative environment related to diabetic conditions occurs through p53–p21 pathway (Kurokawa et al., 1998; Lee et al., 2024). As an example, the oxidized LDL of people with diabetes may accelerate premature senescence of healthy endothelial progenitor cells via p53 signaling so that the vascular repair and the endothelial barrier maintenance are hindered. The p53 position also puts in its intrusion of the protein into the course of atherosclerosis as well as vessel remodeling of pulmonary hypertension. Experiments reveal that senescence of EC can be speed up by contact with advanced glycation end product-modified collagen, and this phenomenon is accompanied by the upregulation of p53, p21, and p16. Knockdown studies make it clear that inhibition of p53 reduces p16-induced aging centrally has the same effect as inhibition of p16, and inhibition of pRb has the same effect as inhibition of p16 or p21 overexpression upon senescence (Leon and Gustafsson Å, 2016; Lesnefsky et al., 2016).

5.2.2 Vascular smooth muscle cell-derived senescence-associated secretory phenotype

Arteries: An arterial architecture (and there is no arterial architecture without) forms the basis of how the vascular smooth muscle cells (VSMCs) organize the tone and strength of the vessel. Underlying the phenomena of senescence, VSMCs experience a complex of biological and molecular changes due to oxidative stress, chronic inflammation, and angiotensin II (Ang II), DNA damage, and contact with low-molecular-weight compounds. The evidence suggests that these stimuli alter VSMC behavior via the genetic and epigenetic routes, eventually leading to arterial dysfunction, prolonged vascular inflammation, and age-related cardiovascular disease.

The senescent VSMCs have impaired proliferation and cellular senescence with changes in crucial signaling events (Lewis-McDougall et al., 2019; Li et al., 2020). There is an increase in senescence-associated beta-galactosidase (SA-beta-gal) activity, increase in the expression of antiproliferative proteins (p16, p21, p38, and p53), and an increasing phosphorylation of histone H2A.X. Senescent phenotype is also characterized by reduced sensitivity to vasoactive stimuli, alteration in phenotype towards synthetic- rather than contractile, disorder of protein kinase G-1 (PKG-1) receptor signaling, and inactivation of the large-conductance calcium-activated potassium (BKCa) channels. At the same time, the binding between VSMC and the extracellular matrix decreases (Li et al., 2021; Libertini, 2014).

In aging VSMCs, the secretomes are enriched with an inflammatory pattern with elevated secretion of interleukin-1alpha (IL-1 alpha), IL-1 beta, IL-6, IL-8, IL-18, and TNF-alpha. It is pertinent to note that IL-1alpha secretion by the senescent aortic VSMCs stimulates the inflammatory and adhesive changes in surrounding cells. At the same time, high concentrations of reactive oxygen species (ROS) and low antioxidant abilities exacerbate chronic inflammation, the accelerating process of atherosclerosis, as well as the aging of vessels. Angiotensin II is one of the key constituents of the renin-angiotensin-aldosterone system (RAAS), which becomes an active agent of VSMC senescence and SASP. Ang II also promotes inflammatory processes and structural degradation by enhancing chemotactic factors like MCP-1, activating calpain-1, and activating matrix metalloproteinases (MMPs) (Lin and Kerkelä, 2020; Lodi et al., 2021).

There are three major pathways that mediate the effects of renal angiotensin II on vascular smooth muscle cell (VSMC) aging, namely, matrix metalloproteases (MMPs), monocyte chemoattractant protein-1 (MCP-1), and transforming growth factor-beta 1 (TGF-beta 1). MMPs alter the functions of VSMCs, such as growth (via MMP9), migration and cellular adhesion remodeling (via MMP1, 2, and 9), and vasodilation (via MMP2 and 9). Angiotensin II predisposes activation of MMP2 that increases vascular fibrosis and calcification, which is characteristic of vascular aging (Ma and Li, 2015; Maggiorani et al., 2024). VSMC senescence is also stimulated by TGF-b1 that is released during inflammation or after activation of MMP. Angiotensin 2 stimulates TGF-beta 1 directly, and TGF-beta 1 enhances the production of collagen-degrading agents (collagenase and stromelysin) and reduces the activities of TIMP-1, a natural inhibitor of MMPs. The resultant imbalance in the matrix causes senescence and forms part of vascular stiffness and dysfunction. Additionally, TGF-beta-1 released by senescent cells has also been shown to cause premature aging in adjacent cells through a mechanism of paracrine transmission, and by increasing the senescence-associated secretory phenotype (SASP) in the vascular tissue (Maharajan et al., 2021; Mapuskar et al., 2025).

Angiotensin II enhances the process of VSMC senescence by activating the p53-p21 pathway to induce vascular aging and atherosclerosis. With age, VSMCs produce less collagen and secrete more proteases, destabilize components of the extracellular matrix and may be the reason behind the rupture of atherosclerotic plaques. SMC calcification is another critical aging sign in cells as aged cells overexpress osterogenic markers like RUNX-2, alkaline phosphatase(ALP), bone morphogenetic protein-2(BMP-2), and type I collagen and thus partially transform into osteoblast-like cells. This calcific transformation worsens the elasticity of the vessels and is a cause of cardiovascular dysfunction (Marino et al., 2022; Mendelsohn and Larrick, 2013).

NF-kB is the master transcription factor involved with the changes during the process of inflammation-related gene expression in aging VSMCs. Its activation increases the transcription of pro-vascular-pathology and oxidative-stress-based genes, and it contributes to vascular pathology. According to the previous studies, there have been stable findings of elevated NF-C activity in cardiovascular tissues of the older rodents and in vitro aging processes in VSMCs. Along with C/EBPbeta, NF-kB coordinates the expression of major SASP proteins like IL-6 and IL-8. These cytokines then activate NF-kB and C/EBP beta in positive feedback that enhances the signal of SASP. EBP does not only regulate CCL2 but also other SASP mediators such as IL-1 beta, GRO-alpha, and NAP-2, and thus upregulates chronic inflammatory pathways (Mialet-Perez et al., 2015; Morley and Reese, 1989).

Recent data point ever more to the involvement of extracellular vesicles (EVs) in the senescence of vascular smooth muscle cells (VSMCs). VSMC premature aging mediated by miRNA or long noncoded RNA sequences (lncRNAs) encapsulated into microvesicles has been demonstrated. The overexpression of miR-34A in human aortic VSMCs also results in cell-cycle arrest, high levels of p21 expression, and induction of senescence-associated secretory phenotype (SASP), as well as the inhibition of SIRT1, which plays a pivotal role in controlling longevity and inflammation. Reexpression of SIRT1 inhibits miR-34a-dependent senescence and SASP activation, which means that the miR-34a SIRT1 axis forms a possible target in manipulating vascular aging. Senescence of VSMC is accompanied by substantial nuclear architecture interference (Nguyen and Sussman, 2015; Nishimura et al., 2011). Accumulation of a mutant form of lamin A, named prelamin A, which is present in the Hutchinson–Gilford progeria syndrome (HGPS), is one such example. It is a unique genetic condition that leads to improper processing of nuclear lamina and presents itself as premature atherosclerosis, calcification, fibrosis, as well as VSMC ablation. HGPS patients often succumb to cardiovascular events in the form of myocardial infarction or stroke during adolescence. Clinical and experimental models have indicated that excessive production of prelamin A lowers the rate at which DNA damage, mitotic aberrations, and premature senescence occur in VSMCs, thus facilitating vascular degeneration (Obas and Vasan, 2018; Osiewacz, 2015).

Finally, SASP of VSMCs is defined by up-regulated senescence-related proteins, i.e., p16, p21, p38, and p53; pro-inflammatory factors, i.e., IL-1α, IL-1β, IL-6, IL-8, IL-18, TNF-a, TGF-b1 and NF-kB; and matrix-degrading enzymes including MMP1, MMP2, and MMP9. The increased expression of miR-34a and simultaneous reduction, in turn, of SIRT1 additionally intensifies cellular aging and inflammation. All these structural and functional alterations of the molecular and cellular nature of blood vessels lead to the loss of their structure and functionality in old age and age-related cardiovascular diseases (Osorio et al., 2023; Papp et al., 2012).

5.2.3 Senescence-associated secretory phenotype in cardiomyocytes

The earlier senescence of cardiomyocytes is one of the major factors that determine the age of cardiac tissue and significantly contributes to the rate and the development of heart failure. Demonstrations of this aging process are age-related cardiomyocyte hypertrophy, electrical instability, and structural abnormalities that reflect various disease states with the term cardiomyopathy being used to describe many of them. The senescent cardiomyocytes have unique morphological and molecular senescence features, which can serve as both aging biomarkers and therapeutic targets. At the morphological level, senescent cardiomyocytes exhibit flattened and enlarged morphological appearance, whereas, at the molecular level, senescence is marked by the increased activity of the senescence-associated 8-galactosidase (SA-8-gal) and a corresponding shortening of the telomeres (Pavanello et al., 2020; Pesce et al., 2017).

The senescence of cardiomyocytes is also characterized by damage to the DNA, endoplasmic reticulum stress, mitochondrial dysfunction, a decrease in contractility, and the change of cellular growth patterns. At this aging profile, cardiomyocytes release the senescence-associated secretory phenotype (SASP) compounds, and hence, further compromise cardiac functions. The senescence process of cardiomyocytes is highly reliant on the microenvironment around them, interactions with endothelial cells, fibroblasts, and immune cells, etc. More notably, senescent cardiomyocytes mutually condition the behavior of other neighboring cells leading to the development of negative cardiac remodeling and pathological fibrosis of the myocardium (Peters et al., 2020; Picca et al., 2018).

Endothelial cell death leads to faster aging of cardiomyocytes, which release the pro-inflammatory substances TGF-beta, IL-6, IL-33, and angiotensin 2 (Ang II). Furthermore, extracellular vesicles released by endothelial cells, especially exosomes that are enriched with miRNAs and kinases such as Mst1 are also able to alter the activity of cardiomyocytes. An essential regulatory role is therefore also played in paracrine signaling and extracellular matrix remodeling by fibroblasts. They also produce MMPs, integrins, fibronectin, and some proinflammatory cytokines, IL-11 and IL-33, besides the secretion of extracellular vesicles carrying miR-21-3p, EGFR, and osteopontin (Pillai et al., 2021; Pun et al., 2025). Immune cells also play a role in enhancing cardiomyocyte senescence through the discharge of cytokines that spotlight applicable stress mechanisms. On the other hand, the cells become aged, and senescent cardiomyocytes release SASP factors and other chemokines that activate the immune cells and make neighboring endothelial cells and fibroblasts age in a pathological loop of cellular dysfunction. Some chemotherapeutic agents like doxorubicin have also been found to induce premature senescence in cardiomyocytes, which have shown an increase in the levels of SA-beta-gal, elevated levels of cyclin-dependent kinase inhibitors, and a drop in levels of telomerase activity and phosphorylation of cardiac troponin I. Doxorubicin also increases the mitochondrial DNA damage and glycine-independent histamine elevations in animal models, which proves the mast cell activation during chemotherapy-induced cardiac stress; this aging phenotype shows strong correlation with p53 acetylation and DNA damage response activation (Quarles et al., 2015; Renlund and Gerstenblith, 1987).

Another crucial pathway that leads to aging of cardiomyocytes is metabolic dysregulation. The major enzymes involved in the metabolism of carbohydrates and lipids, namely, carnitine palmitoyl transferase-1 (CPT1) and pyruvate dehydrogenase (PDH), display extreme modifications during cardiac aging. The reduction of CPT1, especially in the aged cardiac tissue, is very steep, thus negatively affecting the fatty acids usage, and making one more prone to cardiac complications related to lipotoxity. In animal models of hypertension, the reduced CPT1 exacerbates cardiomyocyte senescence, supports hypertrophic reversion, and is the cause of cardiac impairment (Ribera Casado, 1997; Roberts, 1988).

The process of cardiomyocyte senescence is not cell autonomic, but it is closely controlled by vascular smooth muscle cells, endothelial cells, fibroblasts, and even immune cells. These adjacent cells enhance the attrition of telomeres, endoplasmic reticulum stress, hypertrophy, and the formation of SASP in cardiomyocytes by secreting the signaling molecules, including TGF-beta, IL-6, IL-11, IL-33, Ang II, and MMPs. All of these changes result in cumulating toward a pathological myocardial condition (Rochette and Vergely, 2016; Rodriguez Morales et al., 2025). External injuries, such as oxidative stress, DNA damage, telomere losses, mitochondrial problems, chemotherapy, and ionizing radiation, are common deliverers of SASP in endothelial cells, VSMCs, and in cardiomyocytes. It results in overall alterations in the presentation of genes, including high concentration of pro-inflammatory cytokines, growth factors, matrix metalloproteinases, RAAS, antiproliferative protein, and extracellular matrixs molecules (Roffe, 1998; Rolle et al., 2020).

Through paracrine communication, the secreted factors of SASP spread to the neighboring nonsenescent cells of the organism, causing their premature ageing and resulting in chronic systematic inflammation. Further, this inflammaging environment increases SASP secretion, creating more support for the senescence cascade. Exposure of neighboring cells to the components of SASP for a long time ages them more quickly and makes up the risk of cardiovascular disease, namely atherosclerosis, coronary artery disease, hypertension, and heart attack.

5.3 Chronic inflammation and cardiovascular aging: the inflammaging link

The term inflammaging referenced by Franceschi and others in 2000 is used to describe a systematically escalating low-level, chronic inflammatory status as one ages. The senescence-associated secretory phenotype (SASP) has now been identified as a main favorite amid a controversial use of etiology. It is marked by the continued release of pro-inflammatory mediators, increased NLRP3 inflammasome activation, the expression of acetylcholine receptor sensitivity disruption, and a change in NAD+ metabolic process. NLRP3 inflammasome is a main intracellular sensor of damage-associated molecular patterns (DAMPs), intracellular damage products, e.g., oxidized proteins and defective mitochondria that escape compromised lysosomes. When activated, NLRP3 triggers procaspase-1, leading to the proteolytic activation of IL-1 beta, IL-1 alpha, and IL-18 through the activation of caspase-1. Activation of this cascade is rigidly regulated with the NF-adaptor signaling and requires an earlier priming and oligomerization step (Rupp and Schranz, 2015; Sanz-Ros et al., 2025).

There is evidence that the mutual impacts of endothelial cell and vascular smooth muscle cell (VSMC) senescence, senescence-associated secretory phenotype (SASP), and senescence-induced inflammaging are vital to the pathogenesis of a continuum of cardiovascular diseases, encompassing atherosclerosis, aortic aneurysms, and fibrous neointrace develop. Atherosclerosis has its roots in damage to the endothelial wall, which opens the way to the deposition and oxidation of cholesterol-bearing particles in the arterial wall, which stimulates a chronic immune and inflammatory response (Schafer et al., 2017; Schaffer et al., 2015). Innate, as well as adaptive components of the immune system, play a pathogenic role, transforming the original endothelial dysfunction into a complicated form of the endothelial dysfunction, predisposed to thrombosis. The infiltration of the intima with monocytes differentiates into macrophages and transforms into lipid-laden foam cells comprising the necrotic core in the plaque. In such an environment, NLRP3 inflammasomes are triggered in macrophages by cholesterol crystals and DAMPs to produce secretions of IL-18, IL-18, and other chemotactic cytokines to recruit T and B lymphocytes. Increased lesions are characterized by Aggregation of apoptotic and senescent cells,

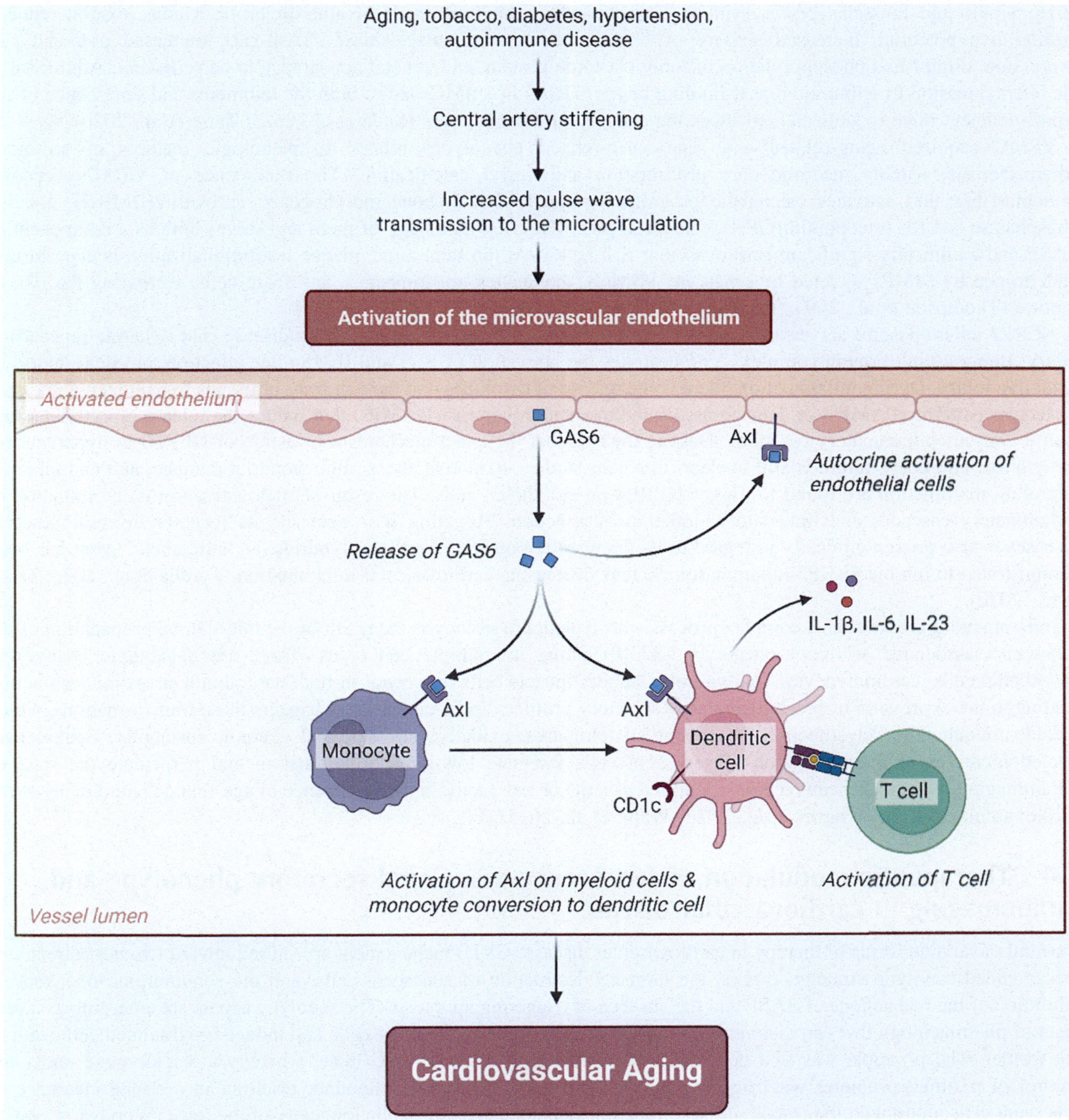

FIGURE 5.2 **Endothelial activation and cardiovascular aging pathways.** This figure summarizes how aging and risk factors drive central arterial stiffness, increased pulse wave transmission, and microvascular endothelial activation. GAS6–Axl signaling and downstream dendritic cell and T-cell activation amplify vascular inflammation, promoting chronic cardiovascular aging.
No permission required.

increased inflammation, and destabilization of plaques; therefore increasing the danger of rupture and the hurt of vessels (Sen et al., 2025; Shah et al., 2023).

In atherosclerotic arteries, the inflammatory environment is determined to have high levels of inflammatory cytokines, as well as matrix metalloproteinases (MMPs), in comparison with noninvolved arterial tissue. A major cellular component of plaques, senescent VSMCs produce interleukin-6 (IL-6), interleukin-8 (IL-8), and growth factors, vascular endothelial growth factor (VEGF), and platelet-derived growth factor (PDGF), hence perpetuating local inflammation and extracellular matrix remodeling. Also, VSMCs cultured out of older individuals express more PDGFR-alpha receptor and respond less to a given stimulus to trigger apoptosis, for example, nitric oxide or nutrient deprivation (Shinmura, 2016; Sipido and Janssens, 2007). Primary vascular cells, which are made of atherosclerotic lesions, exhibit reduced proliferative potential, increased activity of senescence-related β-galactosidase (SA-β-gal), increased p16 and p21 expression, diminished phosphorylation of retinoblastoma protein, and general appearance to be senescent. Additionally, the overexpression of telomeric repeat-binding factor (TRF2) in VSMCs slows both the telomeres and senescence of the ApoE-deficient mice models and enhances the stability of these plaques (Suda et al., 2023; Tang et al., 2020).

VSMC acquired aging-related and senescence-related phenotypes related to pathologic features of advanced atherosclerosis, namely, necrotic core proliferation and vessel calcification. The senescence of VSMCs becomes osteoblast-like and activates osteogenic programs by the RUNX-2, bone morphogenetic protein-2(BMP-2), alkaline phosphatase (ALP), osteopontin (OPN), osteoprotegerin (OPG). OPG is one of them and serves both as a component of SASP and a clinically significant cardiovascular risk factor. At the same time, plaque institutionalization is even further undermined by MMPs secreted by senescent VSMCs, monocytes, macrophages, and foam cells, increasing the risk of rupture (Thompson et al., 2003; Thurmon, 1998).

NLRP3 inflammasome serves as a central activator of atherosclerosis, coronary heart disease, and ischemia-reperfusion injury. Being a multiprotein complex, it orchestrates the activation of IL-1 and IL-18 after infection, physical injury, or oxidative injury. During inflammation, there is the release of cytokines and the exposure of tissues to pathogen-associated molecular patterns (PAMPs) or damage-associated molecular patterns (DAMPs) that induce the release of cytokines and lead to various inflammatory reactions. Despite the fact that the exact mechanism of action of NLRP3 in hypertension, arrhythmia, and heart failure is still unclear, in recent studies, oxidative stress, mitochondrial damage, and endoplasmic reticulum malfunction are found to trigger NLRP3 in endothelial cells. The result of such activation is the initiation of inflammatory cascades and linked to endothelial dysfunction. Targeting IL-1 beta and its receptor in cardiovascular diseases is also proven clinically in regard to its therapeutic potential. Further, glyburide, an antidiabetic agent has been demonstrated to inhibit NLRP3 inflammasomes, thus decreasing cardiovascular inflammation (Torella et al., 2023; Tracy et al., 2020).

Inflammaging is a systemic, complex process whose pathophysiology is the result of the multilateral propagation of the senescence-associated secretory phenotype (SASP) acting in multiple cell types. There are phenotypic changes in endothelial cells, cardiomyocytes, and vascular smooth muscle cells that occur in both natural and stress-induced aging, leading to the expression of pro-inflammatory secretory profiles. Endogenous stress triggers these transformations, which include mitochondrial dysfunction, shortening of telomeres, oxidative stresses, and genomic instability. Senescence-related increases in the population of senescent cells facilitate low-grade inflammation and perpetuate the state of inflammaging, which has been recently identified as a major risk factor in the occurrence of age-related pathologies of the cardiovascular system (Wagner et al., 2023; Wang et al., 2015) (Fig. 5.2).

5.4 Therapeutic modulation of senescence-associated secretory phenotype and inflammaging in cardiovascular disease

Currently available forms of therapy in cardiovascular disease (CVD) management are shifted toward two new directions, the so-called senolytic strategy, that is, the targeted destruction of senescent cells, and the senomorphic one, that is, inhibition of the bad actions of SASP and the absence of triggering apoptosis. The senolytic agents are emerging as a new facet of pharmacology that can distinguish between the aged cells and healthy cells and induce programmed cell death in the former. The principle was first demonstrated in transgenic INK-ATTAC mice, whereby a suicide gene under the control of p16Ink2 promoter was triggered by the use of the AP20187 compound, resulting in a strong clearance of senescent cells along with the rescue of tissue functions and an extension of the longevity of the mice (Wood et al., 2025; Wu et al., 2025).

Since then a group of senolytic candidates has been identified, specifically antiapoptotic proteins of the BCL-2 family. Some memorable ones are ABT263, A1331852, A1155463, and UX1325. Dasatinib is a tyrosine kinase inhibitor that has been found to be senolytic, as it targets the ephrin B2 protein, which is involved in senescence cell survival mechanisms.

Senolytic activity in specific cell types may be obtained by feeding compounds naturally present in many fruits, such as quercetin and fisetin, which downregulate PI3K and PAI-1. As senescent cells are heterogeneous in the underlying apoptotic machinery required to kill them, the combination of multi-agent senolytic may be necessitated in specific tissues. Interestingly, the cardiac glycoside digoxin, which has been used traditionally in the treatment of heart failure, is known to have senolytic effects since it alters intracellular pH that is a sole weakness in senescent cells (Xie et al., 2023; Yang et al., 2005).

Tissue damages that are caused by a significant percentage in aging are mediated through senescence-associated secretory phenotype (SASP). In accordance with that, therapeutic approaches are oriented to de-emphasize rather than eliminate senescent cell phenotype. These senomorphic agents will have analgesic effects similar to senolytic but will work without as much undesirable cytotoxicity. It is possible that clinical translation may be hindered by compromised efficacy in intermittent administration, as opposed to continuous one. Such medicines are still in experimental stages, but a number of drug candidates have advanced into clinical trials, both in malignant and nonmalignant diseases. Types of diseases in which the BCL-2 family inhibitor UX1325 is being evaluated include diabetic macular edema; those in which fisetin, quercetin, and dasatinib are being tested include skeletal muscle dysfunction; and those in which the effectiveness of fisetin, quercetin, and dasatinib is being investigated include Alzheimer's disease, as well as osteoarthritis. On the other hand, senolytic treatment in CVD has not been studied in any clinical trials so far in humans. The lack here postulates the minimal safety margin of the candidate drugs, another limit that hinders the usefulness of the same in chronic atherosclerosis. Recent preclinical evidence nonetheless indicates the therapeutic value of senolytics in heart repair and implies that conjugated senolytics like INK-ATTAC and P16 -3MR are less toxic and effective in atherosclerosis models than conventional senolytic agents (Yoshida et al., 2022; Yousef et al., 2024; Zhai et al., 2024).

Recent researchers have shown that senescent endothelial cells produce higher levels of reactive oxygen species (ROS) and senescence-associated secretory phenotype (SASP), which are hemostatic pro-inflammatory agents. Nrf2 agonists, which are the pharmacological activators of Nrf2, such as resveratrol, have been shown to alleviate DNA damage and senescence triggered by ROS in cardiovascular cells. Similarly, the administration of phosphate binders like lanthanum carbonate and calcium carbonate can decrease vascular smooth muscle cell (VSMC) aging secondary to phosphate, and the overexpression of SIRT1 can mediate the negative effect of aging on vascular calcification in response to high phosphate levels. Taken together, these data are consistent with the idea of systemic inflammation (inflammaging) being partially resolved by ablation of senescent cells. Besides specific senolytic therapy, there is a variety of classic agents with senomorphic effects. Both antioxidants and statins, ACE inhibitors, and angiotensin receptor blockers decrease the production of ROS and DNA damage, thereby postponing senescence in the cardiovascular tissue. Senomorphic agents have multiple targets into which they modulate down SASP such as mTOR, p38-MAPK, nuclear factor kappa-light-chain-enhancer of activated B cells (NF-x B), janus kinase/signal transducer and activator of transcription (JAK/STAT), Rho-associated coiled-coil-containing protein kinase (ROCK), and glucocorticoid receptor, among others. Since the composition of the SASP varies on the basis of phenotypic cell type, senescence-stimulating factor, and hormonal environment, it should be intervened in different ways. Rapamycin, metformin, glucocorticoids, and JAK inhibitors, among others, have shown potential to mediate SASP expression (Thompson et al., 2003; Zhai et al., 2025).

Rapamycin is a powerful compound that can inhibit mTOR dependent transfer of interleukin-1 alpha (IL-1 alpha) as well as the JAK/STAT3 pathway, which decreases SASP in senescent fibroblasts. The compound also has more general antiinflammatory and geroprotective effects. Similarly, inhibitors of NF-kB like BMS-06 and vinpocetine have proved to slow down the process of atherosclerosis in experimental animals. Similarly, inhibition of p38-mitogen-activated protein kinase (MAPK) signaling can decrease the SASP production and its paracrine activity. Statins, commonly prescribed to control cholesterol levels in an aged population, slow the ageing process of endothelial cells and T-cells by inhibiting the p38-mediated SASP factors as well as by altering cellular cycle patterns (Sen et al., 2025; Wagner et al., 2023).

Their effects are mediated by inhibition of the enzyme HMG-CoA reductase in the mevalonate pathway and thus limit the production of farnesyl pyrophosphate and geranylgeranyl pyrophosphate that are crucial in the activation of Ras and Rac proteins. Statins reduce the development of SASP in senescent fibroblasts. They also have an effect on endothelial ageing since they affect cholesterol biosynthesis, which decreases the expression of markers of senescence in endothelial progenitor cells. However, statins may increase radiosensitivity, and this may increase the vulnerability of tumor cells to radiation and also have inconsistent impact on signaling pathways that contribute to cardiovascular diseases and cellular senescence. Activation of P53 by ageing of endothelial cells suppresses synthesis of mevalonate, reducing the cholesterol production, consequently, halting SASP (Ribera Casado, 1997; Suda et al., 2023).

Recent research associates the expression of SASP with the formation of chromatin in the cytoplasm caused by damage of DNA. The clearance of chromatin-containing cytoplasmic material or recruitment of autophagy activators like metformin and rapamycin bears the capacity to eliminate chromatin-containing cytoplasmic material and prevents

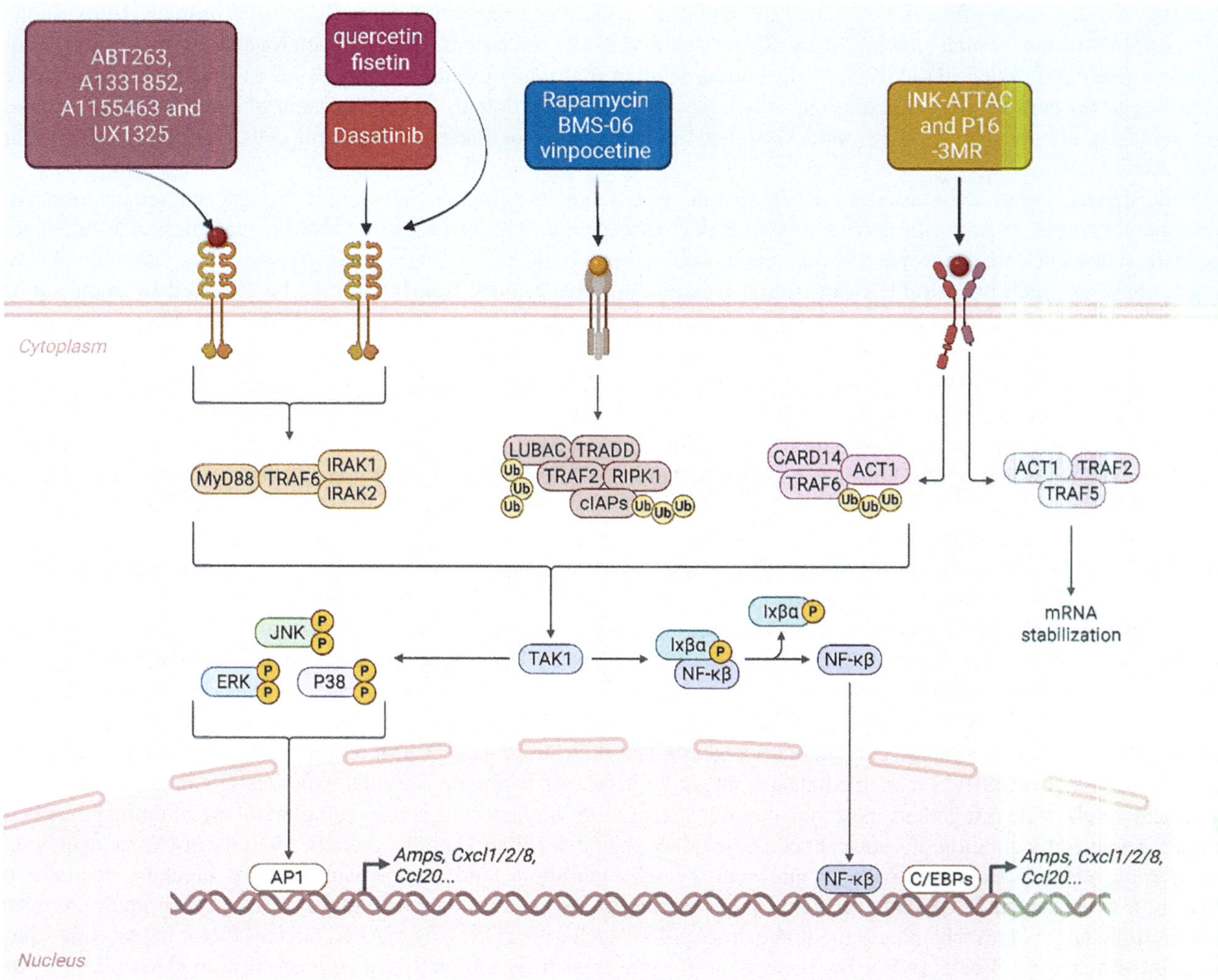

FIGURE 5.3 **SASP signaling cascades and targeted pharmacologic interventions.** This figure depicts key receptor-mediated pathways that activate MAPKs and NF-κB/AP-1 to drive the inflammatory secretome of senescent cells. It also highlights senolytic and senomorphic agents that modulate these nodes to suppress SASP and inflammation.
No Permission Required.

activation of the cGAS-STING-NF-kB pathway - an essential mechanistic driver of cellular senescence. These findings are counterbalanced by inhibitors of lysosomal autophagy such as bafilomycin A1, which validates the evidence of the critical role of lysosomal autophagy in SASP regulation. Metformin can also improve the work of mitochondria and activate autophagy in T-cells, which improves the level of pro-inflammatory cytokine (Wang et al., 2015; Zhai et al., 2024) (Fig. 5.3).

5.5 Modulating cardiovascular aging with therapeutic peptides

Due to physiological mechanism of action and relatively widespread low rates of side effects, peptides have been identified as potentially therapeutically useful compounds and geroprotective agents. In the framework of this type of peptide, there are peptides that can be biologically active and can cross cytoplasm, nucleus, and even nucleoli, which interact with histones, nucleosomes, and DNA. This characteristic renders such agents as senomorphic substances in which they can regulate gene expression and cell activity on normal and aging tissues (Singh et al., 2022).

One of the main epigenetic contributions of the peptides is DNA methylation. Peptides can become immunomodulatory, neuroprotective agents with antimicrobial, antiviral, and cardioprotective functions because they can have an effect on the activity of genes. Recent works also underline therapeutic potential of short peptides in aging and tissue pathophysiology mediated by SASP. It can be well illustrated by glucagon-like peptide-1 (GLP-1). GLP-1 is a 30-amino-acid peptide hormone produced following post-translational processing of proglucagon and released by L-cells of the intestine and certain neurons in response to the intake of food. It consists of two biologically active forms, namely GLP-1 (7–36) and GLP-1 (7–37), which assume a dual helix conformation critical in the process of binding receptors (Alharbi et al., 2022b; Singh et al., 2020). The GLP-1 receptor agonists, especially liraglutide, a stable derivative with increased affinity of receptors, have proven to be effective in the prevention of cardiovascular risk in patients with type 2 diabetes. Liraglutide reduces blood pressure, alleviates cardiac hypertrophy and fibrosis, enhances endothelial dysfunction, and reduces oxidative stress and inflammation. It normalizes serum glucose, decreases the production of ROS, inhibits the accumulation of collagen, and dampens the expression of AT1R in CAT tissues in models of Ang II-induced hypertension. In vitro, liraglutide blocks Ang II-driven ROS accumulation and collagen production in fibroblasts, thus disclosing an antifibrostatic mechanism of action related to the redox balance (Altamish et al., 2020; Thapa et al., 2023).

Laboratory experiments of liraglutide in obese diabetic patients with type 2 diabetes mellitus proved that it has antiinflammatory potential, namely, by its role in upregulating SIRT1 levels and reducing NF-kB signaling pathway activation. At the end of a 6-week regimen, significant reductions in pro-inflammatory cytokine expressions (TNF-a, TLR2, TLR4, and ceruloplasmin) were witnessed, and there was upregulation of SIRT1. Amazingly, the decline in mRNA expression of various markers of inflammation remained 6 weeks after the therapy was discontinued, and the results were interpreted as an indication of the continued regulatory action of SIRT1 at an epigenetic level. Extensive research on the topic of antiinflammatory properties has focused on the atrial natriuretic peptide (ANP), a 28-amino-acid hormone secreted and produced by the cardiomyocytes as a result of atrial stretch. An in vitro experiment establishes the fact that ANP exhibits inhibitory effects on iNOS, COX-2, and TNF-alpha expression by macrophages and prevents the attachment of leukocytes to the endothelial lining (Singh et al., 2019). The recombinant analog of ANP is the carperitide, which is authorized in a number of countries to treat acute heart failure. Although certain studies affirm its usefulness in the improvement of postoperative hypertension and renal dysfunction, large, observational studies slur its long-run appropriateness and efficacy when compared to conventional vasodilators.

Besides, urocortin 1 (Ucn1), a 40 amino-acid member of the corticotropin-releasing factor, was found to induce vasodilation and extensive hypotension. Ucn1 inhibited the development of plaques and the accumulation of foam cells and reduced expression of lipid-uptake proteins (CD36 and cholesterol acyltransferase 1) in atherosclerosis-prone ApoE-deficient mice. Ucn1 also inhibited VSMC migration and proliferation and enhanced MMP2 and MMP9 activities, which also indicated that ucn1 may act in two ways as a controller of vascular remodeling as well as the inflammatory swing (Gupta et al., 2018).

Vasoprotective and antifibrotic peptides Relaxin A. In addition to the well-known vasoprotective and antifibrotic effects, relaxin peptides increase nitric oxide-dependent endothelial relaxation and prevent vasoconstriction. Despite their effectiveness in preclinical models of hypertension, relaxin-based therapeutics are hampered by poor oral bioavailability and high manufacturing costs. Serelaxin is a recombinant relaxin-2, which has proven successful in acute heart failure but its clinical value is limited by the necessity of intravenous route of administration due to peripheral short half-life. To overcome such limitations, new single-chain lipidized peptide mimetics have been constructed, resulting in greater stability, receptor selectivity, and subcutaneous bioavailability of the drug (Gothwal et al., 2024).

Adropin is a 76-amino-acid polypeptide present in the circulation that has been described as a modulator of endothelial physiology in cardiovascular disease, with this protein being highly conserved. In endothelial cells, adropin via the mitogen-activated protein kinase (MAPK) pathway has a significant effect on vascular endothelial growth factor receptor 2 (VEGFR2). Adropin binds to G protein-coded receptor 19 (GPR19) in cardiomyocytes, thus regulating the effect of pyruvate dehydrogenase (PDH), which increases the expression of other important metabolism enzymes (Balaraman et al., 2025). Despite the lack of clinically utilized adropin due to pharmacokinetic limitations, it does suppress inflammatory and atherogenic parameters with high levels in circulation being inversely proportional to homocysteine and high-sensitive C-reactive protein (hs-CRP), in addition to pro-inflammatory cytokines. Experimental studies also report that adropin suppresses the expression of TNF-alpha and IL-6, and suppresses the inflammatory signals downstream of the CD36 receptors that have been implicated in both atherosclerotic inflammation and neuroinflammatory diseases (Babu et al., 2025; Singh et al., 2020).

The other peptide is KED (Lys-Glu-Asp), which is a tripeptide shown to possess vasoprotective properties. KED regulates the genes related to the process of endothelial regulation and antioxidant protection in a cell culture epigenetically. There is normalization of endothelin-1 mRNA expression, increase in connexin, and up-regulation of

TABLE 5.1 Summary of therapeutic actions of peptide-based therapies in cardiovascular diseases.

Peptide	Structure/Origin	Mechanism of action	Cardiovascular effects	References
GLP-1 (Lsiraglutide)	30-aa from proglucagon	GLP-1 receptor agonist, ↑SIRT1, ↓NF-κB	↓BP, ↓hypertrophy, ↑endothelial function	Davari et al. (2020)
ANP (Carperitide)	28-aa hormone from cardiomyocytes	Blocks iNOS, COX-2, TNF-α	Improves acute heart failure	Kerkelä et al. (2015)
Urocortin 1 (Ucn1)	40-aa CRF family	↓Foam cells, ↓CD36, ↑MMP2/9	↓Plaque formation, ↑vascular remodeling	Hasegawa et al. (2014)
Relaxin/Serelaxin	Relaxin-2 analog	NO-mediated vasodilation	Antifibrotic, ↓vasoconstriction	Teerlink et al. (2013)
KED	Tripeptide (Lys-Glu-Asp)	Epigenetic gene regulation, ↑SIRT1	Promotes endothelial renewal	Heo et al. (2016)
AEDR	Tetrapeptide from heart	↑Cytoskeletal/mitochondrial proteins	↓Infarct size, ↑cardiac recovery	Khavinson et al. (2022)
BNP	32-aa peptide	Activates natriuretic peptide receptors	↓BP, ↑natriuresis	Hawkridge et al. (2005)
Angiotensin-(1–7)	Heptapeptide	Binds Mas receptor	Vasodilation, ↓fibrosis	Schiavone et al. (1988)
Apelin	77-aa prepropeptide	Binds APJ receptor	Vasodilatory, ↑contractility	Le Gonidec et al. (2017)
Oxytocin	Nonapeptide	Binds oxytocin receptor	Cardioprotective, vasodilation	Giełdoń et al. (2001)

Source: No permission required.

Ki67, the proliferation marker, indicating increased endothelial renewal and prevention of early-stage atherosclerosis. Besides, the peptide increases the rate of SIRT1 and hence enhancement of DNA repair in the process of reducing p53-induced cell-cycle arrest in senescing endothelia. All of this evidence is to the effect that KED has a part in endothelial repair (Altamish et al., 2020; Singh et al., 2019).

Isolated cardioprotective tetrapeptide (Ala-Glu-Asp-Arg) found in polypeptide complexes of the heart (AEDR) also shows that it has cardioprotective activity during experimental myocardial infarction. AEDR treatment with rat models decreases the size of the infarct, maintains myocardial glycogen, improves mitochondrial activity, and increases the synthesis of actin, tubulin, vimentin, lamin A, and lamin C, which are cytoskeleton and nuclear proteins, indicating involvement in the repair and regeneration processes after damage occurs. Collectively, these data show that a variety of compounds based on peptides could modulate molecular targets that relate to SASP and inflammaging. A GLP-1 analog (Liraglutide) inhibits the NF- kappa B signal and increases the activity of SIRT1. ANP also blocks the production of inflammatory mediators and is currently used in clinical treatment of heart failures (Alharbi et al., 2022b; Solanki et al., 2024). Adropin and urocortin 1 have lipid-modifying capacities and have the ability to inhibit endothelial dysfunction and atherosclerosis. The Relaxin peptides remodel blood circulation by reversing the vasodilation using NO-dependent processes. KED stimulates the repair of cells of the vascular endothelium by enhancing cell division and inhibiting apoptosis of these cells, and AEDR enhances the recovery of the myocardium by increasing cytoskeletal and mitochondrial integrity. Within this class, these peptides provide a new pharmacological platform against the aging-related vascular and inflammation-induced cardiovascular diseases (Alharbi et al., 2022a; Thapa et al., 2023) (Table 5.1).

5.6 Conclusions

The SASP is a typical global of subsisting cytokine, chemokine, proteolytic enzyme, adhesion protein, and extracellular vesicle issuance in replicative and stress-induced senescent components of the cardiovascular system (CVS), specifically vascular smooth muscular cells (VSMCs), endothelial cells, and cardiomyocytes. These molecules interact in a paracrine manner with the related fibroblasts and immune cells, driving the secretion of each other's inflammatory mediators and creating a self-reinforcing inflammatory circuit that is characteristic of the inflamaging state. The result of progressive inflammammaging is deterioration of vascular and cardiac structure; over the years, progressive inflammammaging underlies the incidence of development of hypertension, atherosclerosis, heart disease, and heart attacks. The present evidence suggests that targeted elimination of senescent VSMCs with SASP phenotypes can slow down the process of biological aging in vivo; therefore it is of the utmost importance that future work is done to organizing/developing therapy approaches to block SASP production and interruption of the inflammaging process. The development of one of the new research directions involves small biologically active molecules that could regulate the synthesis and secretion of the components of SASP. Peptides that are biologically active play a leading role in this respect. Along with the endogenous peptide hormones like urocortin, adropin, and GLP-1, short peptides derived out of the vascular and cardiac tissue, including KED and AEDR, are under experimentation due to immunomodulatory and cardioprotective properties. By regulating chromatin remodeling and gene expression, these peptides can increase the synthesis of proteins in the context of structural and functional integrity of the cardiovascular cells, thus reducing any pathological damage caused by the process of aging. The peptides have also been found to significantly improve the outcomes of animals used as models of myocardial infarction, atherosclerosis, and hypertension. The peptide size and molecular weight reduce adverse allergic reactions, and the cellular entry is proposed to be through transporters like POTs and LATs. Significant headway has already been achieved in clinical trials with similar peptides, and this gives a factual basis that they will be explored to treat cardiovascular aging and inflammaging in the days ahead.

References

Alharbi, K.S., Almalki, W.H., Alzarea, S.I., Kazmi, I., Al-Abbasi, F.A., Afzal, O., et al., 2022. A narrative review on the biology of piezo1 with platelet-rich plasma in cardiac cell regeneration. Chemico-biological Interactions 363, 110011.

Alharbi, K.S., Singh, Y., Afzal, O., Alfawaz Altamimi, A.S., Kazmi, I., Al-Abbasi, F.A., et al., 2022b. Molecular explanation of Wnt/βcatenin antagonist pyrvinium mediated calcium equilibrium changes in aging cardiovascular disorders. Molecular Biology Reports 49, 11101–11111.

Altamish, M., Samuel, V.P., Dahiya, R., Singh, Y., Deb, P.K., Bakshi, H.A., et al., 2020. Molecular signaling of G-protein-coupled receptor in chronic heart failure and associated complications. Drug Development Research 81, 23–31.

Babu, M.A., Jyothi, S.R., Kaur, I., Kumar, S., Sharma, N., Kumar, M.R., et al., 2025. The role of GATA4 in mesenchymal stem cell senescence: a new frontier in regenerative medicine. Regenerative Therapy 28, 214–226.

Balaraman, A.K., Altamimi, A.S.A., Babu, M.A., Goyal, K., PadmaPriya, G., Bansal, P., et al., 2025. The interplay of senescence and MMPs in myocardial infarction: implications for cardiac aging and therapeutics. Biogerontology 26, 46.

Besse, S., Delcayre, C., Chevalier, B., Hardouin, S., Heymes, C., Bourgeois, F., et al., 1994. Is the senescent heart overloaded and already failing? Cardiovascular Drugs and Therapy 8, 581–587.

Biggs, R.B., Booth, F.W., 1990. Protein synthesis rate is not suppressed in rat heart during senescence. The American Journal of Physiology 258, H207–H211.

Brand, T., 2019. Length doesn't matter-telomere damage triggers cellular senescence in the ageing heart. The EMBO Journal 38.

Cannon, L., Bodmer, R., 2016. Genetic manipulation of cardiac ageing. The Journal of Physiology 594, 2075–2083.

Carbonin, P., Cocchi, A., Zuccalà, G., Menichelli, P., 1990. Heart aging and its clinical implications. Recenti Progressi in Medicina 81, 215–220.

Carrier, L., Boheler, K.R., Chassagne, C., de la Bastie, D., Wisnewsky, C., Lakatta, E.G., et al., 1992. Expression of the sarcomeric actin isogenes in the rat heart with development and senescence. Circulation Research 70, 999–1005.

Chang, D., Wang, J., Zhu, S., Shentu, Y., Wang, Y., Zhu, P., et al., 2025. Cellular senescence and cell therapy in cardiovascular diseases. Stem Cell Research & Therapy 16, 613.

Chen, M.S., Lee, R.T., Garbern, J.C., 2022. Senescence mechanisms and targets in the heart. Cardiovascular Research 118, 1173–1187.

Cianflone, E., Torella, M., Biamonte, F., De Angelis, A., Urbanek, K., Costanzo, F.S., et al., 2020. Targeting cardiac stem cell senescence to treat cardiac aging and disease. Cells 9.

Conroy, G., 2024. The human heart shows signs of ageing after just a month in space. Nature 634, 17–18.

Cowling, R.T., 2015. The aging heart, endothelin-1 and the senescent cardiac fibroblast. Journal of Molecular and Cellular Cardiology 81, 12–14.

Czibik, G., Mezdari, Z., Murat Altintas, D., Bréhat, J., Pini, M., d'Humières, T., et al., 2021. Dysregulated phenylalanine catabolism plays a key role in the trajectory of cardiac aging. Circulation 144, 559–574.

Davari, M., Hashemi, R., Mirmiran, P., Hedayati, M., Sahranavard, S., Bahreini, S., et al., 2020. Effects of cinnamon supplementation on expression of systemic inflammation factors, NF-kB and Sirtuin-1 (SIRT1) in type 2 diabetes: a randomized, double blind, and controlled clinical trial. Nutrition Journal 19 (1).

Douin-Echinard, V., Lefevre, L., Parini, A., 2020. In the heart of cardiac stromal senescence. Aging (Albany NY) 12, 1039–1041.

Ellison-Hughes, G.M., 2023. Targeting cell senescence to improve cardiac regeneration. Texas Heart Institute Journal 50 (5), e238262.

Fatmi, M.K., Rouhi, N., Lozonschi, L., Li, J., 2024. Cardiac metabolism in the elderly: effects and consequences. Aging (Albany NY) 16, 11773–11775.

Gabillard-Lefort, C., Thibault, T., Lenaers, G., Wiesner, R.J., Mialet-Perez, J., Baris, O.R., 2025. Heart of the matter: mitochondrial dynamics and genome alterations in cardiac aging. Mechanisms of Ageing and Development 224, 112044.

Geokas, M.C., Lakatta, E.G., Makinodan, T., Timiras, P.S., 1990. The aging process. Annals of Internal Medicine 113, 455–466.

Giełdoń, A., Kaźmierkiewicz, R., Slusarz, R., Ciarkowski, J., 2001. Molecular modeling of interactions of the non-peptide antagonist YM087 with the human vasopressin V1a, V2 receptors and with oxytocin receptors. Journal of Computer-Aided Molecular Design 15, 1085–1104.

Gothwal, S.K., Goyal, K., Garg, A.S., Sahu, B.K., Agrawal, M., Mishra, A., et al., 2024. A rare case of brucellosis with multivalvular endocarditis and complete heart block. Current Cardiology Reviews 20 (6), e030724231559. Available at https://doi.org/10.2174/011573403X290326240703100925.

Groban, L., 2005. Diastolic dysfunction in the older heart. Journal of Cardiothoracic and Vascular Anesthesia 19, 228–236.

Gude, N.A., Broughton, K.M., Firouzi, F., Sussman, M.A., 2018. Cardiac ageing: extrinsic and intrinsic factors in cellular renewal and senescence. Nature Reviews Cardiology 15, 523–542.

Gupta, G., de Jesus Andreoli Pinto, T., Chellappan, D.K., Mishra, A., Malipeddi, H., Dua, K., 2018. A clinical update on metformin and lung cancer in diabetic patients. Panminerva Medica 60, 70–75.

Hasegawa, A., Sato, K., Shirai, R., Watanabe, R., Yamamoto, K., Watanabe, K., et al., 2014. Vasoprotective effects of urocortin 1 against atherosclerosis in vitro and in vivo. PLoS ONE 9, e110866.

Hastings, M.H., Zhou, Q., Wu, C., Shabani, P., Huang, S., Yu, X., et al., 2025. Cardiac ageing: from hallmarks to therapeutic opportunities. Cardiovascular Research 121, 1474–1488.

Hawkridge, A.M., Heublein, D.M., Bergen 3rd, H.R., Cataliotti, A., Burnett Jr, J.C., Muddiman, D.C., 2005. Quantitative mass spectral evidence for the absence of circulating brain natriuretic peptide (BNP-32) in severe human heart failure. Proceedings of the National Academy of Sciences of the United States of America 102, 17442–17447.

Hedge, E.T., Brazile, T.L., Hughson, R.L., Levine, B.D., 2025. Plasticity of the heart in response to changes in physical activity. The Journal of Physiology 603, 3665–3677.

Heist, E.K., Huq, F., Hajjar, R., 2003. Telomerase and the aging heart. Science of Aging Knowledge Environment: SAGE KE 2003 (19), PE11. Available at https://doi.org/10.1126/sageke.2003.19.pe11.

Heo, K.S., Berk, B.C., Abe, J., 2016. Disturbed flow-induced endothelial proatherogenic signaling via regulating post-translational modifications and epigenetic events. Antioxidants & Redox Signaling 25, 435–450.

He, S., Yan, L., Yuan, C., Li, W., Wu, T., Chen, S., et al., 2024. The role of cardiomyocyte senescence in cardiovascular diseases: a molecular biology update. European Journal of Pharmacology 983, 176961.

Holmstrom, L., Chugh, S.S., 2023. Can artificial intelligence identify physiologically "old" hearts? Mayo Clinic Proceedings. Mayo Clinic 98, 360–362.

Hong, Y.X., Wu, C., Li, J.Z., Song, F., Hu, Y., Han, Y., et al., 2024. SUMOylation of TP53INP1 is involved in miR-30a-5p-regulated heart senescence. Experimental & Molecular Medicine 56, 2519–2534.

Horn, M.A., 2015. Cardiac physiology of aging: extracellular considerations. Comprehensive Physiology 5, 1069–1121.

Ho, A.K., Shang, K., Duffield, R., 1986. Calmodulin regulation of the cholinergic receptor in the rat heart during ontogeny and senescence. Mechanisms of Ageing and Development 36, 143–154.

Huang, P., Bai, L., Liu, L., Fu, J., Wu, K., Liu, H., et al., 2021. Redd1 knockdown prevents doxorubicin-induced cardiac senescence. Aging (Albany NY) 13, 13788–13806.

Hudlicka, O., Brown, M.D., 1996. Postnatal growth of the heart and its blood vessels. Journal of Vascular Research 33, 266–287.

Isoyama, S., 1996. Age-related changes before and after imposition of hemodynamic stress in the mammalian heart. Life Sciences 58, 1601–1614.

Jahangir, A., Sagar, S., Terzic, A., 2007. Aging and cardioprotection. Journal of Applied Physiology: Respiratory, Environmental and Exercise Physiology 103 (1985), 2120–2128.

Kane, A.E., Howlett, S.E., 2016. Novel cardioprotection strategies for the aged heart: evidence from pre-clinical studies. Clinical and Experimental Pharmacology and Physiology 43, 1251–1260.

Kerkelä, R., Ulvila, J., Magga, J., 2015. Natriuretic peptides in the regulation of cardiovascular physiology and metabolic events. Journal of the American Heart Association 4, e002423.

Khavinson, V., Linkova, N., Dyatlova, A., Kantemirova, R., Kozlov, K., 2022. Senescence-associated secretory phenotype of cardiovascular system cells and inflammaging: perspectives of peptide regulation. Cells 12.

Kitzman, D.W., Edwards, W.D., 1990. Age-related changes in the anatomy of the normal human heart. Journal of Gerontology 45, M33–M39.

Kurokawa, T., Ozaki, N., Ishibashi, S., 1998. Difference between senescence-accelerated prone and resistant mice in response to insulin in the heart. Mechanisms of Ageing and Development 102, 25–32.

Le Gonidec, S., Chaves-Almagro, C., Bai, Y., Kang, H.J., Smith, A., Wanecq, E., et al., 2017. Protamine is an antagonist of apelin receptor, and its activity is reversed by heparin. FASEB Journal 31, 2507–2519.

Lee, W.S., Abel, E.D., Kim, J., 2024. New insights into IGF-1 signaling in the heart. Physiology (Bethesda, Md.) 39 (5). Available at https://doi.org/10.1152/physiol.00003.2024.

Leon, L.J., Gustafsson, Å., B., 2016. Staying young at heart: autophagy and adaptation to cardiac aging. Journal of Molecular and Cellular Cardiology 95, 78–85.

Lesnefsky, E.J., Chen, Q., Hoppel, C.L., 2016. Mitochondrial metabolism in aging heart. Circulation Research 118, 1593–1611.

Lewis-McDougall, F.C., Ruchaya, P.J., Domenjo-Vila, E., Shin Teoh, T., Prata, L., Cottle, B.J., et al., 2019. Aged-senescent cells contribute to impaired heart regeneration. Aging Cell 18, e12931.

Libertini, G., 2014. The programmed aging paradigm: how we get old. Biochemistry (Mosc 79, 1004–1016.

Lin, R., Kerkelä, R., 2020. Regulatory mechanisms of mitochondrial function and cardiac aging. International Journal of Molecular Sciences 21 (4), 1359. Available at https://doi.org/10.3390/ijms21041359.

Li, H., Hastings, M.H., Rhee, J., Trager, L.E., Roh, J.D., Rosenzweig, A., 2020. Targeting age-related pathways in heart failure. Circulation Research 126, 533–551.

Li, W.W., Wang, H.J., Tan, Y.Z., Wang, Y.L., Yu, S.N., Li, Z.H., 2021. Reducing lipofuscin accumulation and cardiomyocytic senescence of aging heart by enhancing autophagy. Experimental Cell Research 403, 112585.

Lodi, R.S., Xia, L., Zhang, Y., Liu, F., 2021. Evolving roles of cardiac fibroblasts in cardiogenesis and immunology, electrophysiology, and aging. Reviews in Cardiovascular Medicine 22, 1173–1183.

Maggiorani, D., Santin, Y., Formoso, K., Drapé, E., Martini, H., Brun, S., et al., 2024. Identification of Prominin-2 as a new player of cardiomyocyte senescence in the aging heart. Aging cell 23, e14204.

Maharajan, N., Ganesan, C.D., Moon, C., Jang, C.H., Oh, W.K., Cho, G.W., 2021. Licochalcone D ameliorates oxidative stress-induced senescence via AMPK activation. International Journal of Molecular Sciences 22 (14), 7324. Available at https://doi.org/10.3390/ijms22147324.

Mapuskar, K.A., London, B., Zacharias, Z.R., Houtman, J.C.D., Allen, B.G., 2025. Immunometabolism in the aging heart. Journal of the American Heart Association: Cardiovascular and Cerebrovascular Disease 14, e039216.

Marino, F., Scalise, M., Salerno, N., Salerno, L., Molinaro, C., Cappetta, D., et al., 2022. Diabetes-induced cellular senescence and senescence-associated secretory phenotype impair cardiac regeneration and function independently of age. Diabetes 71, 1081–1098.

Ma, Y., Li, J., 2015. Metabolic shifts during aging and pathology. Comprehensive Physiology 5, 667–686.

Mendelsohn, A.R., Larrick, J.W., 2013. Rejuvenation of aging hearts. Rejuvenation Research 16, 330–332.

Mialet-Perez, J., Douin-Echinard, V., Cussac, D., Bril, A., Parini, A., 2015. Ageing: a matter of heart? Medecine Sciences: M/S 31, 1006–1013.

Morley, J.E., Reese, S.S., 1989. Clinical implications of the aging heart. The American Journal of Medicine 86, 77–86.

Nguyen, N., Sussman, M.A., 2015. Rejuvenating the senescent heart. Current Opinion in Cardiology 30, 235–239.

Nishimura, M., Ocorr, K., Bodmer, R., Cartry, J., 2011. Drosophila as a model to study cardiac aging. Experimental Gerontology 46, 326–330.

Obas, V., Vasan, R.S., 2018. The aging heart. Clinical Science (London) 132, 1367–1382.

Osiewacz, H.D., 2015. Quality control systems in aging. Ageing Research Reviews 23, 1–2.

Osorio, J.M., Espinoza-Pérez, C., Rimassa-Taré, C., Machuca, V., Bustos, J.O., Vallejos, M., et al., 2023. Senescent cardiac fibroblasts: a key role in cardiac fibrosis. Biochimica et Biophysica Acta - Molecular Basis of Disease 1869, 166642.

Papp, Z., Czuriga, D., Balogh, L., Balogh, Á., Borbély, A., 2012. How cardiomyocytes make the heart old. Current Pharmaceutical Biotechnology 13, 2515–2521.

Pavanello, S., Campisi, M., Fabozzo, A., Cibin, G., Tarzia, V., Toscano, G., et al., 2020. The biological age of the heart is consistently younger than chronological age. Scientific Reports 10, 10752.

Pesce, M., Messina, E., Chimenti, I., Beltrami, A.P., 2017. Cardiac mechanoperception: a life-long story from early beats to aging and failure. Stem Cells and Development 26, 77–90.

Peters, C.H., Sharpe, E.J., Proenza, C., 2020. Cardiac pacemaker activity and aging. Annual Review of Physiology 82, 21–43.

Picca, A., Mankowski, R.T., Burman, J.L., Donisi, L., Kim, J.S., Marzetti, E., et al., 2018. Mitochondrial quality control mechanisms as molecular targets in cardiac ageing. Nature Reviews Cardiology 15, 543–554.

Pillai, V.B., Samant, S., Hund, S., Gupta, M., Gupta, M.P., 2021. The nuclear sirtuin SIRT6 protects the heart from developing aging-associated myocyte senescence and cardiac hypertrophy. Aging (Albany NY) 13, 12334–12358.

Pun, R., Haas, A.L., Thapa, A., Takafuji, S.R., Suzuki, R.M., Kay, G.F., et al., 2025. BubR1 insufficiency drives transcriptomic alterations and pathology associated with cardiac aging and heart failure. Aging Cell 24, e70160.

Quarles, E.K., Dai, D.F., Tocchi, A., Basisty, N., Gitari, L., Rabinovitch, P.S., 2015. Quality control systems in cardiac aging. Ageing Research Reviews 23, 101–115.

Renlund, D.G., Gerstenblith, G., 1987. Exercise and the aging heart. Cardiology Clinics 5, 331–336.

Ribera Casado, J.M., 1997. The heart as a model of aging. The clinical consequences. Anales de Medicina Interna 14, 491–494.

Roberts, W.C., 1988. The aging heart. Mayo Clinic Proceedings. Mayo Clinic 63, 205–206.

Rochette, L., Vergely, C., 2016. "Pro-youthful" factors in the "labyrinth" of cardiac rejuvenation. Experimental Gerontology 83, 1–5.

Rodriguez Morales, D., Larcher, V., Ruz Jurado, M., Arifaj, D., Tombor, L.S., Zanders, L., et al., 2025. Vascular niches are the primary hotspots in cardiac aging. Circulation Research 137, 1353–1367.

Roffe, C., 1998. Ageing of the heart. British Journal of Biomedical Science 55, 136–148.

Rolle, I.G., Crivellari, I., Caragnano, A., Cervellin, C., Aleksova, A., Cesselli, D., et al., 2020. Cell senescence in cardiac repair and failure. Current Stem Cell Research & Therapy 15, 685–695.

Rupp, S., Schranz, D., 2015. Cardiac regeneration in children. Pediatric Cardiology 36, 713–718.

Sanz-Ros, J., Huete-Acevedo, J., Mas-Bargues, C., Romero-García, N., Dromant, M., van Weeghel, M., et al., 2025. Small extracellular vesicles from young adipose-derived stem cells ameliorate age-related changes in the heart of old mice. Stem Cell Research & Therapy 16, 138.

Schafer, M.J., Miller, J.D., LeBrasseur, N.K., 2017. Cellular senescence: implications for metabolic disease. Molecular and Cellular Endocrinology 455, 93–102.

Schaffer, S.W., Ramila, K.C., Jong, C.J., Shetewy, A., Shimada, K., Ito, T., et al., 2015. Does taurine prolong lifespan by improving heart function? Advances in Experimental Medicine and Biology 803, 555–570.

Schiavone, M.T., Santos, R.A., Brosnihan, K.B., Khosla, M.C., Ferrario, C.M., 1988. Release of vasopressin from the rat hypothalamo-neurohypophysial system by angiotensin-(1-7) heptapeptide. Proceedings of the National Academy of Sciences of the United States of America 85, 4095–4098.

Sen, I., Trzaskalski, N.A., Hsiao, Y.T., Liu, P.P., Shimizu, I., Derumeaux, G.A., 2025. Aging at the crossroads of organ interactions: implications for the heart. Circulation Research 136, 1286–1305.

Shah, M., de, A.I.M.H., Lu, C., Schiratti, P.R., Zheng, S.L., Clement, A., et al., 2023. Environmental and genetic predictors of human cardiovascular ageing. Nature Communications 14, 4941.

Shinmura, K., 2016. Cardiac senescence, heart failure, and frailty: a triangle in elderly people. The Keio Journal of Medicine 65, 25–32.

Singh, Y., Fuloria, N.K., Fuloria, S., Subramaniyan, V., Almalki, W.H., Al-Abbasi, F.A., et al., 2022. A European pharmacotherapeutic agent roflumilast exploring integrated preclinical and clinical evidence for SARS CoV-2 mediated inflammation to organ damage. British Journal of Clinical Pharmacology 88, 3562–3565.

Singh, Y., Gupta, G., Mishra, A., Chellappan, D.K., Dua, K., 2020. Gender and age differences reveal risk patterns in COVID-19 outbreak. Alternative Therapies in Health and Medicine 26, 54–55.

Singh, Y., Samuel, V.P., Dahiya, S., Gupta, G., Gillhotra, R., Mishra, A., et al., 2019. Combinational effect of angiotensin receptor blocker and folic acid therapy on uric acid and creatinine level in hyperhomocysteinemia-associated hypertension. Biotechnology and Applied Biochemistry 66, 715–719.

Sipido, K.R., Janssens, S.P., 2007. How old is your heart? Circulation Research 101, 323–325.

Solanki, N., Gupta, G., Chellappan, D.K., Singh, S.K., Gulati, M., Paudel, K.R., et al., 2024. Boswellic acids: a critical appraisal of their therapeutic and nutritional benefits in chronic inflammatory diseases. Endocrine, Metabolic & Immune Disorders Drug Targets 24, 116–129.

Suda, M., Paul, K.H., Minamino, T., Miller, J.D., Lerman, A., Ellison-Hughes, G.M., et al., 2023. Senescent cells: a therapeutic target in cardiovascular diseases. Cells 12 (9), 1296. Available at https://doi.org/10.3390/cells12091296.

Tang, X., Li, P.H., Chen, H.Z., 2020. Cardiomyocyte senescence and cellular communications within myocardial microenvironments. Frontiers in Endocrinology 11, 280.

Teerlink, J.R., Cotter, G., Davison, B.A., Felker, G.M., Filippatos, G., Greenberg, B.H., et al., 2013. Serelaxin, recombinant human relaxin-2, for treatment of acute heart failure (RELAX-AHF): a randomised, placebo-controlled trial. Lancet (London, England) 381, 29–39.

Thapa, R., Afzal, O., Alfawaz Altamimi, A.S., Goyal, A., Almalki, W.H., Alzarea, S.I., et al., 2023. Galangin as an inflammatory response modulator: an updated overview and therapeutic potential. Chemico-Biological Interactions 378, 110482.

Thompson, R.P., Reckova, M., deAlmeida, A., Bigelow, M.R., Stanley, C.P., Spruill, J.B., et al., 2003. The oldest, toughest cells in the heart. Novartis Foundation Symposium 250, 157–174 discussion 174-156, 276-159.

Thurmon, T.F., 1998. Genetics, aging, and the heart. The Journal of the Louisiana State Medical Society 150, 356–366.

Torella, D., Salerno, N., Cianflone, E., 2023. Senescent cells enhance ischemic aging in the female heart. Aging (Albany NY) 15, 2364–2366.

Tracy, E., Rowe, G., LeBlanc, A.J., 2020. Cardiac tissue remodeling in healthy aging: the road to pathology. American Journal of Physiology. Cell Physiology 319, C166–C182.

Wagner, J.U.G., Tombor, L.S., Malacarne, P.F., Kettenhausen, L.M., Panthel, J., Kujundzic, H., et al., 2023. Aging impairs the neurovascular interface in the heart. Science 381, 897–906.

Wang, H., Bei, Y., Shi, J., Xiao, J., Kong, X., 2015. Non-coding RNAs in cardiac aging. Cellular Physiology and Biochemistry 36, 1679–1687.

Wood, C., Salter, W.Z., Garcia, I., Nguyen, M., Rios, A., Oropeza, J., et al., 2025. Age-associated changes in the heart: implications for COVID-19 therapies. Aging (Albany NY) 17, 1340–1367.

Wu, W., Lei, R., Xu, H., Lu, L., Song, C., Wu, D., et al., 2025. Malic enzyme 1 reverses cardiomyocyte senescence in a nitric oxide-dependent manner. Free Radical Biology & Medicine 239, 202–212.

Xie, S., Xu, S.C., Deng, W., Tang, Q., 2023. Metabolic landscape in cardiac aging: insights into molecular biology and therapeutic implications. Signal Transduction and Targeted Therapy 8, 114.

Yang, X., Sreejayan, N., Ren, J., 2005. Views from within and beyond: narratives of cardiac contractile dysfunction under senescence. Endocrine 26, 127–137.

Yoshida, Y., Shimizu, I., Minamino, T., 2022. Capillaries as a therapeutic target for heart failure. Journal of Atherosclerosis and Thrombosis 29, 971–988.

Yousef, A., Sosnowski, D.K., Fang, L., Legaspi, R.J., Korodimas, J., Lee, A., et al., 2024. Cardioprotective response and senescence in aged sEH null female mice exposed to LPS. American Journal of Physiology. Heart and Circulatory Physiology 326, H1366–H1385.

Zhai, P., Sung, E.A., Shiheido-Watanabe, Y., Takayama, K., Tian, Y., Sadoshima, J., 2024. Suppression of autophagy induces senescence in the heart. Journal of Molecular and Cellular Cardiology 195, 83–96.

Zhai, P., Sung, E.A., Takayama, K., Ikeda, S., Matsushita, M., Nakada, Y., et al., 2025. S913 phosphorylation of Ulk1 protects the heart from aging through inhibition of cardiac senescence. Cardiovascular Research 121, 1609–1622.

Chapter 6

Mitochondrial dysfunction, oxidative stress, and cardiac aging

Riya Thapa[1], Md Sadique Hussain[2], Md Raihan Siddique[3], Neelam Singla[3], Sachin Kumar Singh[4], Saurabh Gupta[5], Muhammad Afzal[6], Gaurav Gupta[7,8], and Moayad Al Shahwan[8]

[1]*Faculty of Pharmacy, AIMST University, Bedong, Kedah, Malaysia,* [2]*Uttaranchal Institute of Pharmaceutical Sciences, Uttaranchal University, Dehradun, Uttarakhand, India,* [3]*School of Pharmacy, Suresh Gyan Vihar University, Jaipur, Rajasthan, India,* [4]*School of Pharmaceutical Sciences, Lovely Professional University, Phagwara, Punjab, India,* [5]*Department of Pharmacology, Chameli Devi Institute of Pharmacy, Indore, Madhya Pradesh, India,* [6]*Department of Pharmaceutical Sciences, Pharmacy Program, Batterjee Medical College, Jeddah, Saudi Arabia,* [7]*Centre for Research Impact & Outcome, Chitkara College of Pharmacy, Chitkara University, Rajpura, Punjab, India,* [8]*Centre of Medical and Bio-Allied Health Sciences Research, Ajman University, Ajman, United Arab Emirates*

6.1 Introduction

The cardiovascular diseases (CVDs) are now the leading cause of death in the West, and this fact is recognized by bodies like the American Heart Association and the World Health Organization (WHO). Theirs is higher than those of neurodegenerative ones, and it is projected that there would be an increment in trends. Even as of 2012, ischemic heart disease already constituted the greatest cause of deaths at an earlier age in the high-income nations, and current predictions expect an over 23 million death rates based on ischemic heart disease to be experienced in various parts of the world by the year 2030 (Jan et al., 2024). It is anticipated that the number of people who die of CVDs will reach a volume of 40% in the same year, making it the leading cause of mortalities in people over the age of 70 years as per WHO global health report of 2014 (Roth et al., 2017). Although the exposure to modifiable lifestyle-related risk factors, such as an unhealthy diet and sedentary behavior develop cumulatively over time, this contribution of cumulative exposure to cardiovascular pathology is trumped by chronological age itself, which tends to be the greatest determinant of risk (Stamatakis et al., 2023). Cardiovascular damage related to aging is further increased by age-related comorbidities, which include dyslipidemia and type 2 diabetes. Over the last 30 years, there has been an improvement in life expectancy at birth across the world by 6 years, and this has created more prevalence of age-related diseases, particularly cardiovascular diseases (Triposkiadis et al., 2022). This trend is likely to happen in future with estimates projected that by the end of next 20 years, about 20% of the world population shall be aged 65 years or older (Lan et al., 2023).

The human heart is largely made up of terminally differentiated nonreproducing cells and has a high propensity toward aerobic energy generation. It could be hypothesized that such cellular feature makes the organ very susceptible to either cellular or mitochondrial efficiency (Lan et al., 2023). Since it is in need of constant energy, the mitochondrial functionality has been a major consideration of cardiac aging studies to the extent that mitochondria is the body organ that facilitates essential myocardial economy and cell survival. In normal heart tissue, mitochondria make up 2030 % of the volume of the cardiomyocytes and grow larger when the metabolic demands are increased (González-Juanatey et al., 2023).

The most important energying route of cardiomyocytes is the synthesis of adenosine triphosphate (ATP) through mitochondrial oxidative phosphorylation with the help of lipid and carbohydrate metabolism. Each day, the heart of a human being uses around 6 kg of ATP, which is used to perform such processes as contraction, homeostasis of ions, and biosynthesis. Myocardial energy pro-reservoir (Persad and Lopaschuk, 2022). PCr also acts as a temporal store equivalent and a spatial ATP shuttle, which means it can shuttle energy to energy-demanding regions quickly. Inside the mitochondria, the phosphate group donated by ATP is transferred to a creatine molecule by creatine kinase, and the PCr produced is carried by the higher diffusion coefficient into the mitochondria than the ATP itself. The PCr has

Cellular Senescence and Cardiac Aging. DOI: https://doi.org/10.1016/B978-0-443-45811-8.00004-0

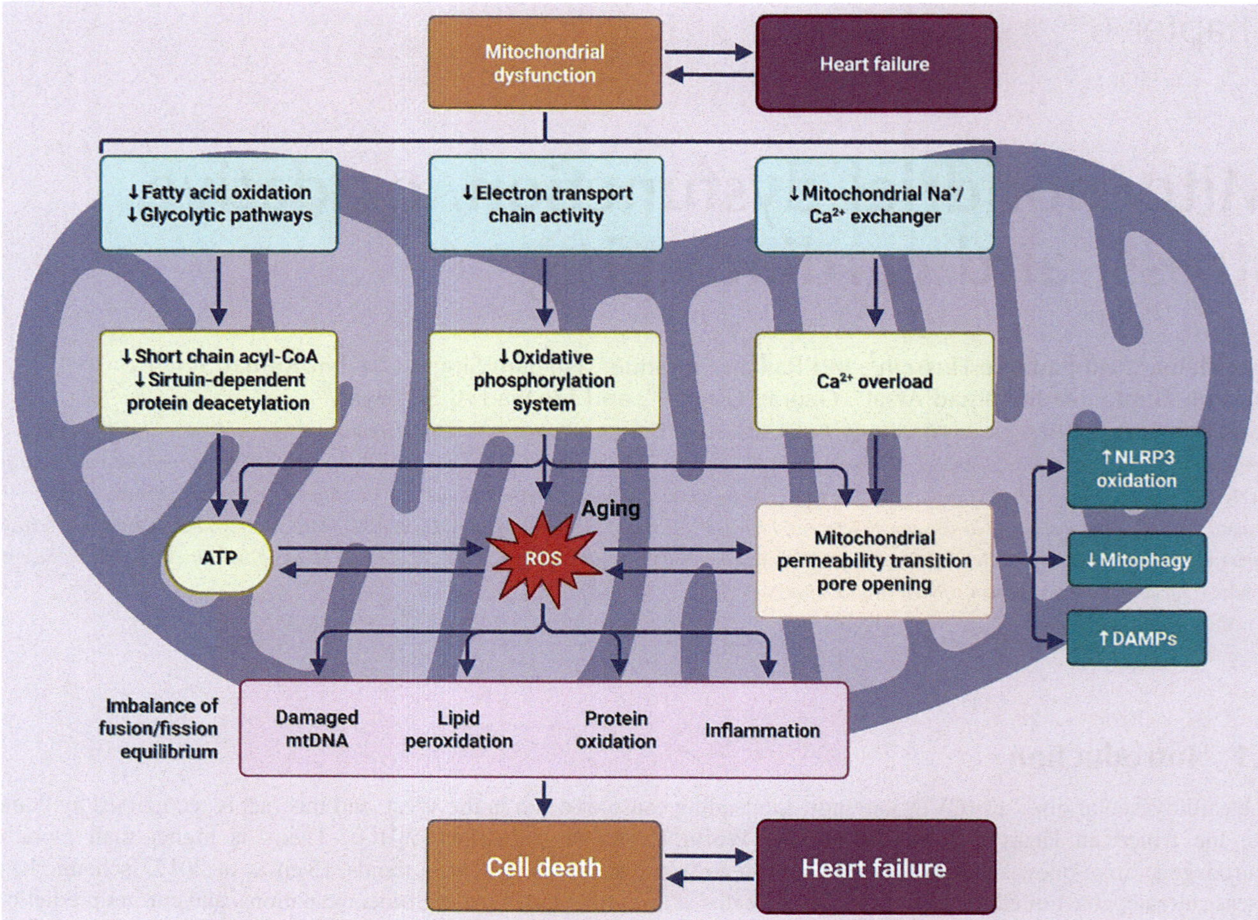

FIGURE 6.1 Mitochondrial dysfunction contributes to heart failure through impaired fatty acid oxidation, glycolytic pathways, and electron transport chain activity, leading to reduced oxidative phosphorylation, ATP depletion, and decreased sirtuin-dependent protein deacetylation. A decline in mitochondrial Na^+/Ca^{2+} exchanger activity causes Ca^{2+} overload, which, together with mitochondrial permeability transition pore opening, enhances ROS generation. Excess ROS promotes aging, damaged mtDNA, lipid peroxidation, protein oxidation, inflammation, and imbalance of mitochondrial fusion/fission equilibrium. Such alterations impair mitophagy, enhance NLRP3-related oxidative changes, and promote the release of damage-associated molecular patterns (DAMPs). Together, these processes accelerate cell death and contribute to the progression of heart failure in a vicious cycle.

the potential of converting back to ATP once within the cytosol mediated by cytosolic creatine kinase, hence maintaining an energy balance in a contractile cell (Correia et al., 2022). Appropriate functioning of mitochondria is highly based on orchestrated nuclear and mitochondrial genome balance in order to attain proper assembly and functioning of the oxidative phosphorylation complexes (Nesci et al., 2021). During high energy demand, e.g., during exercise, the ratio of ATP to ADP and the activation of AMP-activated protein kinase (AMPK) ensure the induction of the process. Such signaling cascade transcriptionally activates the regulatory proteins such as peroxisome proliferator-activated receptor gamma coactivator 1-alpha (PGC-1) and nuclear respiratory factor 1 (NRF1) (Moretti-Horten et al., 2024). These trancribing factors increase the expression of genes of nuclear and mitochondrial DNA (mtDNA), and higher mitochondrial activated oxidative phosphorylation capacity to cope with the energetic demands of the heart. Fig. 6.1 shows mitochondrial dysfunction driving ROS, ATP loss, Ca^{2+} overload, inflammation, cell death, and heart failure.

6.2 Age-related mitochondrial alterations

Because cardiovascular pathology is closely tied to the biology of aging, it is important to clarify the underlying processes in order to design effective interventions. Senescence arises from several interacting factors, most notably mitochondrial dysfunction, genomic instability, and epigenetic alterations, which need to be considered together rather than in isolation (Tenchov et al., 2023). Growing experimental evidence indicates that damage to both nuclear and mitochondrial genomes, triggered by chronic metabolic stress or external insults, is a hallmark of the aging process (Hajam et al., 2022). Mitochondria are particularly susceptible to oxidative stress, as they continuously generate reactive oxygen species (ROS) during electron transport (Chenna et al., 2022). Instability of mitochondrial DNA (mtDNA) is strongly linked to impaired mitochondrial performance and is recognized as a key contributor to both inherited mutations and acquired abnormalities. Such defects are frequently accompanied by energy deficiencies and muscle-related symptoms, underscoring the critical importance of maintaining mtDNA integrity (Tirichen et al., 2021). There is an overlapping set of accelerated aging syndromes that begins with premature cardiovascular decline, attributed to increased genomic instability. The mutant mouse model of mitochondrial DNA polymerase deficiency has also been used to demonstrate that somatic mitochondrial genome mutations can be causative factors in the development of systemic aging (Kallai et al., 2025). In addition to ATP generation via oxidative phosphorylation, mitochondria play pivotal roles in cellular homeostasis; they act as hubs of apoptosis regulation in coordinating the processes of apoptosis, Calcium homeostasis, and redox signaling. The maintenance of mitochondrial functions is therefore essential to the maintenance of cardiac integrity among other cellular integrity during the aging process (Li et al., 2023).

As a person ages, their mitochondrial respiratory chain deteriorates progressively, implying its reduced oxygen consumption in various tissues, which are associated with decreased ATP generation. Although mutations of mitochondrial DNA (mtDNA) form a leading etiology, there are other mechanisms that play auxiliary roles, including oxidative damage of proteins, complex instability of electron transport chain complexes, changes in membrane composition of lipids, and malfunction of mitophagy, which is the process that selectively removes impaired mitochondria (Dabravolski et al., 2022). Notably, mitochondrial malfunction is not an exclusive result of instability of the mtDNA. Because the majority of mitochondrial proteins are coded by euchromatic (nuclear) DNA (nDNA), there is also an immense influence of age-induced damage of the nuclear encoded genes that correlate with the mitochondrial functionality (Bris et al., 2021). In addition to the metabolic abnormalities, mitochondrial impairments regulate programmed cell death pathways. Abnormalities in apoptosis are one of the marked characteristics of the ageing tissue, and their extent and pathological consequences vary depending on tissue (Nguyen et al., 2023). Increased apoptosis causes functional loss of cells of organs, for example, skeletal muscle, the heart, and the brain. This wasting comes along with various age-related disorders such as sarcopenia, heart and nerve disorders (Cheng et al., 2021). Fig. 6.1 shows age-related mitochondrial dysfunction causing oxidative stress, mtDNA mutations, impaired mitophagy, and apoptosis.

6.3 Altered cardiac metabolism during aging

The process known as long-chain fatty acid β-oxidation is the main cardiac metabolism pathway in the healthy adult heart, being responsible for providing between 7070 and 90 of the ATP needed in order to sustain the contractile and cellular processes (Alrumaihi et al., 2025). The rest of ATP is the result of glucose and lactose oxidation, and there are minor contributions of the ketone bodies and some amino acids. A staffing balance between fatty acid carbohydrate metabolism in this pathway is generally known asangesquaranc themselves referred to as the Randle cycle, where the usage of one of the substances suppresses the use of the other (Xie et al., 2023). This metabolic balance between these energy substrates takes on special importance in cardiac disease since the utilization of glucose is more efficient in terms of oxygen extraction than that of the fatty acids (Chung, 2021). A significant change occurs in the metabolism of the mitochondria during the aging of the heart. The expression of respiratory chain components and enzymes that mediate fatty acid metabolism decreases with age, just as in heart failure in younger people, and proteins thought to occur in glucose metabolism and extracellular remodeling are upregulated (Amorim et al., 2022).

Due to the sustained high energy consumption of the heart, there is a close relationship between normal age-related changes in cardiac bioenergetics and age-related impairment. One such mechanism is based upon the idea that a stress, which leads to increased sympathetic activity, can raise the levels of circulating free fatty acids. This metabolic condition interacts with the use of glucose and facilitates insulin resistance, an additional problem in the aging myocardium that operates with poor effectiveness energetically (Poznyak et al., 2022). The oxidized proteins and advanced glycation end products that are formed at an accelerated pace by metabolic disorders, which occur with aging, such as diabetes, make the cardiac functional capacity to be reduced. These are additive metabolic alterations that are of great importance in the structural and functional cardiological deterioration of the aging heart (Xu and Van Remmen, 2021). Fig. 6.2 shows fatty acid uptake, activation, oxidation, storage, lipid droplet metabolism, and ceramide-mediated signaling.

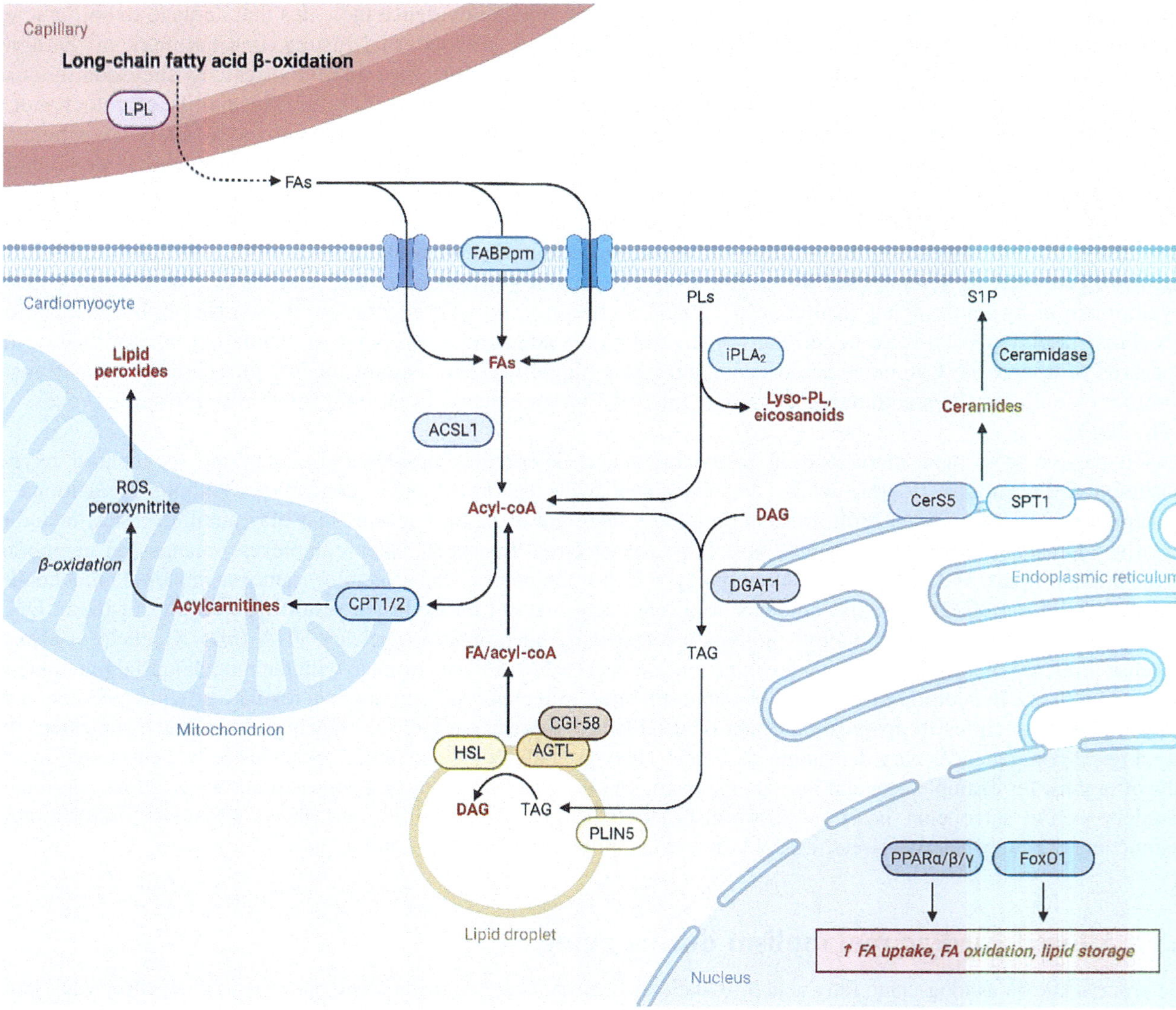

FIGURE 6.2 Lipoprotein lipase (LPL) hydrolyzes triglycerides in the capillary to release fatty acids (FAs), which are transported into cardiomyocytes through plasma membrane proteins such as FABPpm. Inside the cell, FAs are activated by ACSL1 to form acyl-CoA, which can either enter mitochondria via CPT1/2 for β-oxidation—generating energy along with reactive oxygen species (ROS) and lipid peroxides—or be directed toward lipid droplet storage. Acyl-CoA contributes to triacylglycerol (TAG) synthesis through DGAT1, while enzymes including HSL, ATGL, CGI-58, and PLIN5 regulate dynamic lipid droplet metabolism, generating diacylglycerols (DAG) as intermediates. Phospholipid remodeling via iPLA2 produces lysophospholipids and eicosanoids, while ceramide metabolism involves SPT1 and CerS5 in synthesis and ceramidase in breakdown, producing signaling molecules such as S1P. Nuclear regulators, including PPARα/β/γ and FoxO1, coordinate transcriptional responses to increase FA uptake, oxidation, and lipid storage, maintaining energy balance but also contributing to lipotoxic stress under pathological conditions.

6.4 Cardiac tissue reorganization with advancing age

With advancing age, the heart undergoes structural remodeling, most notably within the extracellular matrix (ECM), where both organization and composition are altered. In the heart, collagen maintains tissue integrity; however, too much accumulation around vessels and cardiomyocytes is frequently observed with aging (Alrumaihi et al., 2025). In older hearts, fibroblasts multiply, and collagen accumulates more readily. This fibrotic remodeling is promoted by higher collagen levels, reduced solubility, and greater cross-linking. Alterations in collagen and elastin also contribute to vascular calcification, a process that becomes more pronounced with age (Xie et al., 2023). Such matrix changes are often linked to valve sclerosis, a condition affecting many older individuals. As a result, the myocardium becomes stiffer, ventricular compliance decreases, and diastolic function is reduced (Islam et al., 2023).

Another feature of age-related remodeling is left ventricular hypertrophy (LVH), marked by a thickening of the ventricular walls. Clinical evidence shows that aging is associated with greater LVH, even in individuals without hypertension. Some of the factors take the form of increased mechanical load, decreased cardiomyocytes, and interstitial fibrosis (Guo et al., 2022). During senescence, cardiomyocytes enlarge and show structural damage. A key factor in this decline is impaired autophagy, driven in part by lipofuscin buildup that lysosomes cannot degrade. This accumulation weakens lysosomal function and limits the removal of damaged organelles and proteins (Gao et al., 2024). Similarly, mitochondrial dysfunction worsens with age, as older mitochondria generate more ROS and less ATP. These alterations increase oxidative stress and lead to a buildup of protein aggregates that the cell cannot clear effectively (Yamamoto and Sano, 2022). This sequence of events adds to lipofuscin buildup, raises oxidative stress, and makes cardiomyocytes more vulnerable to apoptosis. As a result, energy production falls, cell damage intensifies, and overall survival of the cells is reduced (Evangelou et al., 2023). Calcium homeostasis in cardiomyocytes is also impaired. While the release of cytosolic calcium is necessary for contraction, its reuptake into the sarcoplasmic reticulum via SERCA2a is essential for relaxation (Izzo et al., 2021). In aging hearts, calcium cycling becomes impaired, and reduced SERCA2a activity leads to delayed ventricular relaxation. Oxidative stress further damages the SERCA pump, lowering its efficiency in calcium reuptake and extending the relaxation phase (Chen et al., 2024).

In aged animals therefore the research indicated that the long-term contraction of cardiomyocytes in aging hearts was associated with the suppressed decay of calcium transient and upregulation of slow contracting isosomes of myosin. The changes affect contractile dynamics as well as the electrical phenomena of the myocardium (Alim, 2021). In addition, decreasing myofilament calcium sensitivity including decreased phosphorylation of troponin I is also known to be associated with poor relaxation and poor contractility. The change in the morphology of calcium transients and relaxation kinetics associated with age therefore interferes with mechanical and electrical synchronization of the heart, exacerbating the functional performance of aging myocardium (Chung et al., 2016).

6.5 Age-associated reactive oxygen species accumulation and mitochondrial DNA damage in the heart

Mitochondria were known to be major age determinants because they produce ROS (ROS) during normal energy production. Despite the reality that free radicals called are produced in many different compartments of a cell, mitochondrion is the major source of free radicals in healthy flesh, considering that the electron transport chain is located in the inner mitochondrial membrane (Jackson and McArdle, 2011). Generation of mitochondrial ROS is always unavoidable during oxidative phosphorylation. These species were initially described based on their harmful actions, but later on, they were realized to promote cellular homeostasis by acting as signaling molecules (Hervouet et al., 2008). In recent studies, it has been hypothesized that low-level ROS can have longevity-promoting qualities due to the stressor-adaptive response it triggers when used in specific situations. These conclusions are based in great part on studies of short-lived model organisms such as nematodes and fruit flies (Ristow and Schmeisser, 2014). In comparison, minimal consequences of antioxidant over- or underexpression in mammalians leave little doubt about the degree to which oxidative stress in isolation defines the aging trajectory. Debate notwithstanding, mitochondria continue to dominate the aging field because of their wider functions in bioenergetics, apoptosis, and redox state (West, 2004).

Even though the exact role of ROS in the aging process is unclear, there is outstanding evidence from human experiments and animal models in support of the idea that oxidative stress and mitochondrial dysfunction are both characteristic of aging. DNA located in the mitochondria (intended as mtDNA) is one of the most susceptible targets of ROS among all cellular targets (Alexeyev, 2009). This is because ROS triggers a repertoire of lesions in mtDNA such as oxidized bases, abasic sites, and single-strand breaks. Such structural changes are mutagenic or cytotoxic; they can delay DNA replication and transcription and thus be conducive to genomic instability (Huang et al., 2020).

The cell organisms use several DNA-repair mechanisms to protect the mitochondrial genome. Base excision repair (BER) is the main defense mechanism in response to relatively simple oxidative processes like single-base changes. There are also complementary repair mechanisms of mitochondria (mismatch repair and single-strand break repair) (Bennett, 2017). When lesions are more extensive, in particular in the form of the double-stranded breaks, mitochondria call into play recombination-based repair pathways, although these are associated with an increased risk of error and may structure deletions. Mammalian, especially in mammalian cells, the recombination mechanism of the mitochondria has not yet been fully understood (Chen, 2013). The buildup of damage in the mtDNA can have serious repercussions, where most cardiomyocytes are no longer able to divide and thus have a low regenerative capacity, which is the case in the heart. In cardiomyocyte mitochondria, the electron transport chain, particularly complexes I and III, acts as a major producer of

ROS (Elorza and Soffia, 2021). Electrons that infiltrate to react with oxygen would produce superoxide that would be further transformed to hydrogen peroxide and become highly reactive hydroxide radicals when mixed with iron. The amount of iron in mitochondria has been found to rise toward old age, which causes malice oxidative attack (Asghar et al., 2015).

The persistence of the production of ROS in the aged cardiac tissue worsens oxidative damage through protein carbonylation, lipid peroxidation, and mutations of mtDNA. Exploration in aged murine groups has shown that the frequency of mtDNA point mutations and deletions in the heart was very high compared to its younger counterpart (Rizvi et al., 2021). As it has been found in empirical examinations, the mitochondrial genomic instability may compromise the functional integrity and trigger cardiac dysfunction. As examples, mice genetically depleted of expression of mitochondrial transcription factor NRF1 show respiratory chain dysfunction, buildup of abnormal mitochondria, and apoptosis in excess, which are clinical demonstrations of mitochondrial myopathies (De Filippis et al., 2015). One model that is very educational is the PolG mouse, which has a mutated mitochondrial DNA polymerase. The animals develop a huge amount of mutations and deletions in their mtDNA, and they have premature phenotypic manifestations. They have defective mitochondrial electron transport chain activity in their cardiac mitochondria, excessive production of ROS, and augmented apoptosis, all of which result in cardiac hypertrophy and loss of cardiac functionality (Anagnostou et al., 2016). Some of these deficiencies are mitigated secondarily after interventions that include mitochondria-targeted antioxidant expression or endurance exercise. The strategies activate mitochondrial biogenesis, reduce oxidative damage, enhance the efficiency of the respiratory chain, as well as reduce age-related degeneration of both skeletal and cardiac muscle (Williamson and Davison, 2020). In more recent years, there are new animal models produced that are designed to create faster deletion of mtDNA in the heart only. Such mice pile up accessory cardiomyocytes with insufficiently performing mitochondria and therefore induce premature arrhythmias and organization-based heart disease (Mohamed et al., 2006). Overall, even though the exact causal mechanism of ROS in aging is debated, it is highly clear that in the aging heart, there is a direct correlation with the mitochondrial genome instability, oxidative stress, and functional decline.

6.6 Pathological role of mitochondrial dysfunction in aged cardiac tissue

As one grows old, physical efficiency becomes gradually weaker, thus disturbing homeostatic stability and predisposing the person to developing chronic illness situations. This loss is more distinct in tissues that have postmitotic cells, for example, heart and brain due to their limited regeneration ability (Livneh, 2023). Aging of the heart can be defined by a set of structural and functional changes in the cardiac muscle and the vasculature associated with it. Its morphological features include thickening of the left ventricle, thickening of the ventricular walls and of the interventricular septum, as well as the stiffening of valve leaflets due to fibrosis and calcification (Lakatta and Yin, 1982). At the same time, the electrical activity of the cardiac system is impaired in part due to the presence of the calcific deposits and due to the presence in the sinoatrial and atrioventricular node of the pacemaker cells, which have been lost with age and are replaced by collagen and fat and amyloid deposits (Khayungarnnawee, 2017).

6.6.1 Cardiac hypertrophy

The development of injury to cardiomyocytes is susceptible to the action of ischemia, toxicants, infectious agents, or increased mechanical loads. The myocardium reacts by making up for the stimuli by initiating compensatory remodeling called cardiac hypertrophy. Faster protein and increased sarcomeres, as well as the enlargement of the cell, give this adaptation increased compounded walls of the ventricles and enhanced contractile power. Early in the process, hypertrophy will alleviate the hemodynamic load (Canty, 2022). Chronic hypertrophic remodeling is associated with poor outcomes such as diastolic dysfunction, heart failure, and arrhythmias. In this state, energy balance is disturbed: ATP levels remain stable at first, but phosphocreatine (PCr) reserves decline quickly. As remodeling advances, the falling ATP/PCr ratio signals a reduced ability to maintain sufficient energy production (Melenovsky et al., 2007). Research has shown that antioxidant treatments can lessen hypertrophic changes, suggesting that oxidative stress plays a central part in driving pathological remodeling (Martins et al., 2022). Hypertrophy can also be triggered in cultured heart cells by mechanical overload or by exposure to bioactive compounds such as angiotensin II, norepinephrine, endothelin-1, or pro-inflammatory cytokines. The resulting rise in ROS activates signaling pathways, including mitogen-activated protein kinases and transcription factors like NF-κB, which drive cell growth and fibrosis, ultimately weakening both the structure and function of the heart (de Souza et al., 2019).

6.6.2 Heart failure

The events of cardiac hypertrophy were traditionally considered in the sense of the adaptation to an increased cardiac workload, but current facts prove that left ventricular hypertrophy (LVH) significantly increases the risks of developing heart failure (HF) and the occurrence of life-threatening arrhythmias (Tsuda, 2021). At the same time, the increasing life expectancy in the world has triggered HF as an alarming issue on the public-health agenda, especially in older generations. Moreover, in low-resource areas, about 2% of the adult population is affected by HF, and the prevalence increases to 6%–10% in people aged 65 or more (Bekele et al., 2024). HF pathogenesis is multifactorial, which does not exclude a range of physiological failures. The most significant phenomena are the slowed-down velocity of relaxation in myocardium and reduced ventricular compliance. Mitochondrial dysfunction is a central mechanism of HF pathogenesis, which creates more deregulated oxidative stress and makes mitochondria a desirable therapeutic target (Kitzman and Upadhya, 2014). The morphopathological defects in the mitochondria establish LVH and facilitate the shift between deceased and overt pathology. Reduced mitochondrial respiration limits ATP synthesis, exacerbates oxidative injury, and dysorganizes calcium-handling proteins, and this causes a reduction in cardiac performance (Budde et al., 2022).

Maladaptive influx of calcium and distortions in myofilament activities in turn exacerbate myocardial dysfunction. Mitochondria are also oxygen sensors and nitric oxide (NO) production regulators within vascular endothelium situations, and so affect vascular tone and continue coronary pathology (Niort, 2022). Oxidative stress destabilizes cardiomyocytes through a mechanism that causes maladaptive cardiac remodeling and worsens myocardial failure. Taken together, these results suggest the mechanistic association between ROS production, mitochondrial impairment, and HF (Shah et al., 2021). In subjects with old age, ejection fraction is normally found to be in the reference range even in the presence of diastolic dysfunction, a condition known as heart failure with preserved ejection fraction (HFpEF). Such clinical morphology is linked to elevated vascular pressures, lung congestion, dyspnea, and reduced exercise capability even without ventricular dilation (Obokata et al., 2020). Prevalence of HFpEF is increasing along with the increase of the aging population, and the illness is expected to increase at a larger scale due to the contribution of women to new cases (Van Ommen et al., 2022).

Once believed to be due to the diastolic impairment issue alone, HFpEF today is understood to be a whole-body disease that includes both cardiac and extracardiac abnormalities. Its pathogenesis is complex and multifactor by coalescence of aging (intrinsic) mechanisms, various comorbidities, and low reserves across a variety of organ systems. Even though there is a similar prognosis to HFrEF, the patients with HFpEF have higher hospitalizations.

The advancements in treatment approaches to HFpEF have led to the formation of clinically applicable animal models. Aortic banding and systemic hypertension models are widely used, since an increase in arterial blood pressure is one of the key factors that predispose HFpEF (Tadic et al., 2018). Dahl salt-sensitive rat, which has exhibited development of hypertension and left ventricular hypertrophy after being exposed to high-salt diets and has then gone on to develop HFpEF, is a well-established experimental model (Nakashima et al., 2025). Likewise, relieving moderate transverse aortic constriction causes concentric hypertrophy and early diastolic failure that offer a powerful platform to study mechanisms and possible interventions in HFpEF (Lesko, 2014).

6.6.3 Diabetic cardiomyopathy

Diabetic cardiomyopathy is a term used when defining the autonomous series of cardiac dysfunction because of obesity or diabetes mellitus without the presence of overt vascularism dysfunction, hypertension, or other organizational defects of the heart. The disease is marked by the poor performance of the left ventricle of diabetics (Spallone, 2019). As one grows older, type 2 diabetes prevalence grows drastically. It strikes almost 1 in 5 adults aged 60 years and above and well over 1/3 of the total number of adult diabetes cases in the world; the figures are expected to be even higher (Standl et al., 2019). In comparison to the putative controls that are nondiabetic, older adults with diabetes exhibit increased instances of cardiovascular disease, stroke, heart failure, neuropathy, visual loss, and joint disease (Zimmet et al., 2003). The most important role of insulin resistance in the pathogenesis of diabetic cardiomyopathy should be mentioned, and it is possible to distinguish between the systemic and cardiac insulin resistance. The effects of cardiac insulin resistance are shown as a diminished response to insulin (insulin signaling) or bad insulin-stimulated glucose uptake in cardiac tissue, regardless of the general metabolic derangement (obesity, hyperglycemia) (Karuna et al., 2025). Inhibition of insulin receptor or signaling proteins ablation in cardiomyocytes, which are experimental models of this effect, is invariably found to reduce glucose utilization and leads to the development of cardiac dysfunction (Witteles and Fowler, 2008). Similarly, the deficiency in mice of the major cardiac glucose transporter has shown the characteristics of heart failure, which emphasizes the role of insulin resistance in diabetic heart disease (Gao et al., 2025). Such a strong correlation exists between insulin resistance and the onset and progression of heart failure, more so in diabetics. It is interesting to note that a significant number of patients with heart failure who are admitted to hospitals have insulin resistance, and with the aging of the population, its rate is expected to rise (He et al., 2025). Additionally, insulin resistance has been connected with the development of type 2 diabetes in old age, as well as the development of the disease in cardiac pathophysiology of diabetes mellitus, and the pathogenesis of coronary artery disease (Chew et al., 2025). There are several mechanisms offered to explain the connection between insulin resistance and cardiac dysfunction in the aging situation; they embrace endothelial dysfunction due to the inefficient signaling of insulin and shortening of leukocyte telomeres that are connected to the chronological high blood pressure and vascular aging (Strazhesko et al., 2017).

Mitochondrial anomaly is largely considered to be instigator of diabetic cardiomyopathy. Since the heart is also very dependent on mitochondrial metabolism to generate ATP, any impairment in the efficiency of mitochondria has a trivial effect on cardiac performance (Wang et al., 2025). Under normal physiological conditions, the heart will decrease free fatty acid and carbohydrate metabolism and increase energy demand in response to the available substrates and energetic demands. However, during insulin-resistant states, this metabolic flexibility is reduced, thus resulting in decreased hepatic glucose uptake, therefore requiring a higher rate of fatty acid oxidation to take place (Lopaschuk and Ussher, 2016). This metabolic shift leads to the accumulation of toxic intermediates within the cell, such as long-chain acyl-CoA and acylcarnitines. Such byproducts interfere with the ATP/ADP ratio in the mitochondria and thus suppress energy production by so doing, undermining contractile activity. Mitochondrial inefficiency also contributes to energy shortages, oxidative stress, structural changes, and the development of diabetic cardiomyopathy (Gupta and Houston, 2017).

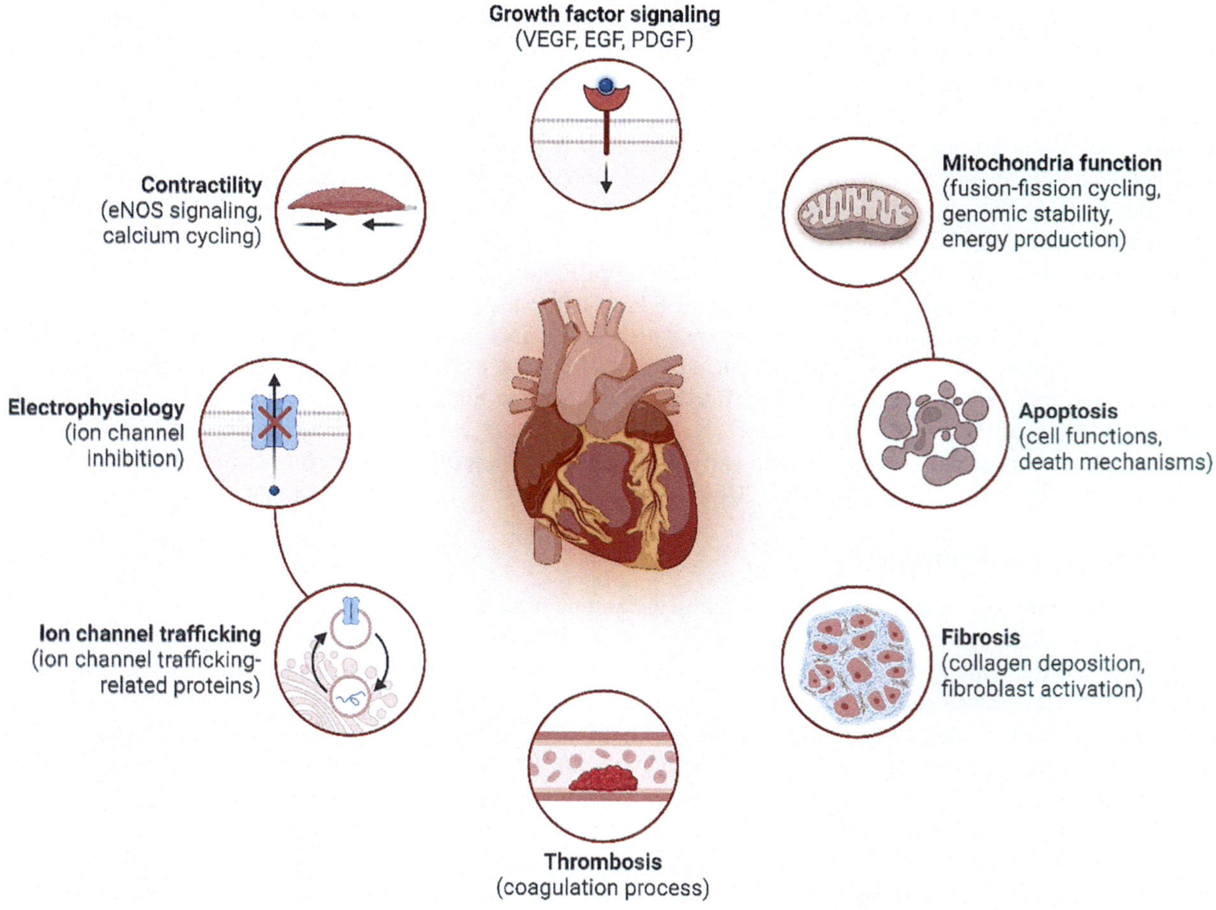

FIGURE 6.3 Mitochondrial dysfunction associated with aging contributes significantly to the development of cardiac disorders by reducing energy output, disturbing calcium handling, and impairing contractile function. Alterations in mitochondrial fusion–fission balance and genomic stability disrupt cellular homeostasis, leading to diminished ATP production and excess reactive oxygen species, which in turn worsen myocardial injury. Age-related changes in ion channel regulation, including both inhibition and defective trafficking, further compromise electrical stability and increase the risk of arrhythmias. At the same time, activation of apoptotic pathways promotes cardiomyocyte loss through programmed cell death. These processes act in concert with fibroblast activation and collagen accumulation, producing fibrosis that stiffens the myocardium and limits compliance. Endothelial dysfunction is also aggravated by disturbed growth factor signaling, including VEGF, EGF, and PDGF pathways, while enhanced thrombosis adds an additional burden on vascular integrity and cardiac perfusion.

TABLE 6.1 Pathological role of mitochondrial dysfunction in aged cardiac tissue.

Condition	Structural and functional features	Mechanisms	Mitochondrial dysfunction role	References
Aging and tissue vulnerability	Decline in physical efficiency; postmitotic tissues (heart, brain) affected	Limited regeneration capacity	Energy imbalance increases stress susceptibility	Livneh (2023)
Cardiac aging morphology	LV thickening, ventricular wall thickening, valve stiffening	Fibrosis, calcification	Alters tissue compliance	Lakatta and Yin (1982)
Electrical conduction defects	Pacemaker cell loss, replaced by collagen/fat/amyloid	Node degeneration	Energy deficit worsens conduction	Cheung (1981)
Early cardiac hypertrophy	Enlarged cardiomyocytes, more sarcomeres	Adaptive remodeling to overload	Energy demand ↑, initially compensated	Canty (2022)
Chronic hypertrophy	Diastolic dysfunction, arrhythmia, HF	ATP maintained but PCr depleted	↓ATP/PCr ratio → energy failure	Melenovsky et al. (2007)
Antioxidant therapy	Reduced hypertrophy with ROS scavengers	Oxidative stress central in remodeling	Limits mitochondrial ROS	Martins et al. (2022)
Stress/hormone-induced hypertrophy	AngII, NE, endothelin-1, cytokines trigger hypertrophy	ROS activate MAPKs & NF-κB	Mito ROS drive growth/fibrosis	de Souza et al. (2019)
LV hypertrophy & HF risk	LVH increases HF and arrhythmia risk	Maladaptive remodeling	ROS-mediated mitochondrial failure	Tsuda (2021)
HF epidemiology	2% adults; 6%–10% ≥65 years	Aging population driver	Bioenergetic decline	Bekele et al. (2024)
HF pathogenesis	Relaxation slowed, ↓compliance	ROS deregulation, Ca^{2+} mishandling	Mito dysfunction = central driver	Kitzman and Upadhya (2014)
Mito defects in HF	LVH progression	↓respiration, Ca^{2+} dysregulation	↓ATP synthesis, ↑oxidative injury	Budde et al. (2022)
Oxygen/NO regulation	Mito = oxygen sensors	NO imbalance alters vascular tone	ROS destabilize signaling	Clarke et al. (2015)
ROS–HF link	ROS destabilize cardiomyocytes	ROS-mediated remodeling	Mito impairment worsens HF	Shah et al. (2021)
HFpEF phenotype	Preserved EF, diastolic dysfunction, lung congestion	Vascular stiffening, fibrosis	Mito inflexibility contributes	Obokata et al. (2020)
HFpEF epidemiology	Higher in women, elderly	Aging + comorbidities	ROS exacerbate dysfunction	Van Ommen et al. (2022)
Aortic constriction model	Concentric hypertrophy, early diastolic failure	Mechanical overload	ROS + energy failure	Stevens et al. (2015)

(*Continued*)

TABLE 6.1 (Continued)

Condition	Structural and functional features	Mechanisms	Mitochondrial dysfunction role	References
Diabetic cardiomyopathy	LV dysfunction in diabetics	Independent of HTN or CAD	Mito inefficiency central	Spallone (2019)
Diabetes prevalence	1 in 5 ≥60 years; 1/3 global burden	Aging ↑ T2DM prevalence	More mito dysfunction cases	Standl et al. (2019)
Diabetes & comorbidities	Elderly diabetics → ↑CVD, stroke, HF, neuropathy	Multisystem disease	Energy imbalance worsens complications	Karuna et al. (2025)
Insulin resistance	Systemic & cardiac insulin resistance	↓Glucose uptake in heart	Mito energy failure	Karuna et al. (2025)
Experimental insulin resistance	Insulin receptor KO in cardiomyocytes	↓Insulin signaling	Impaired glucose → energy deficit	Gao et al. (2025)
Glucose transporter deficiency	Knockout mice show HF features	↓Cardiac glucose uptake	Bioenergetic failure	He et al. (2025)
HF & insulin resistance	High IR in HF patients	Aging + IR	Mito dysregulation accelerates HF	Chew et al. (2025)
IR & CAD link	IR linked with CAD in elderly	Vascular aging	ROS & bioenergetic imbalance	Strazhesko et al. (2017)
IR mechanisms in aging	Endothelial dysfunction, telomere shortening	Insulin signaling inefficiency	Oxidative stress + mito decline	Wang et al. (2025)
Mito role in diabetic heart	ATP-dependent cardiac metabolism	Inefficient mito impairs energy	Bioenergetic deficiency	Lopaschuk and Ussher (2016)
Insulin resistance & metabolism	↓Glucose uptake, ↑FA oxidation	Accumulation of intermediates	Alters ATP/ADP ratio	Gupta and Houston (2017)
Mito inefficiency in DCM	Structural remodeling, oxidative stress	Accumulation of toxic intermediates	Energy suppression + ROS	Livneh (2023)

Fig. 6.3 shows mitochondrial dysfunction driving hypertrophy, heart failure, and diabetic cardiomyopathy through energy impairment. Fig. 6.3 shows mitochondrial dysfunction in aged cardiac tissue driving impaired energy production, apoptosis, fibrosis, electrophysiological instability, thrombosis, and contractile decline (Table 6.1).

6.7 Conclusion

Cardiovascular diseases are strongly linked to profound metabolic disturbances within myocardial cells. In heart failure, metabolic remodeling is typically marked by reduced energy generation, often arising from impaired substrate utilization, diminished mitochondrial biogenesis, and loss of mitochondrial efficiency. Many cardiac disorders begin and worsen with changes in both the quantity and function of mitochondria in cardiomyocytes. Preserving healthy mitochondria and clearing out damaged ones are essential for maintaining cell health and overall heart function. Although the mechanisms of cardiac aging are not yet fully understood and its effects can vary widely between individuals, several interconnected processes are known to drive this decline. Many questions remain unanswered, making cardiac aging a complex phenomenon that still requires deeper investigation. Understanding why protective signals are lost in the aging heart remains a major challenge. Contributing factors include abnormal gene and protein expression, disturbances in signaling pathways, impaired mitochondrial activity, excess production of ROS, and a decline in respiratory capacity. A major unknown is which mitochondrial pathways or substrates are most vulnerable to aging. Research points to oxidative stress, mitochondrial DNA mutations, and cardiomyocyte decline as key contributors to cardiac aging. Gaining clearer insight into these processes could lead to new strategies for preventing and treating age-related heart disease.

References

Alexeyev, M.F., 2009. Is there more to aging than mitochondrial DNA and reactive oxygen species? The FEBS Journal 276 (20), 5768–5787.

Alim, C., 2021. CaMKII Nitrosylation in Cardiac Myocytes at Cys273 Suppresses Activation, but at Cys 290 Mediates Intrinsic Mechanical Afterload-Induced Enhancement of Calcium Transients. University of California, Davis.

Alrumaihi, F., et al., 2025. Histone modifications as molecular drivers of cardiac aging: metabolic alterations, epigenetic mechanisms, and emerging therapeutic strategies. Current Problems in Cardiology 50 (7), 103056.

Amorim, J.A., et al., 2022. Mitochondrial and metabolic dysfunction in ageing and age-related diseases. Nature Reviews Endocrinology 18 (4), 243–258.

Anagnostou, M.E., et al., 2016. Epilepsy due to mutations in the mitochondrial polymerase gamma (POLG) gene: a clinical and molecular genetic review. Epilepsia 57 (10), 1531–1545.

Asghar, A., Raman, A.A.A., Daud, W.M.A.W., 2015. Advanced oxidation processes for in-situ production of hydrogen peroxide/hydroxyl radical for textile wastewater treatment: a review. Journal of Cleaner Production 87, 826–838.

Bekele, F., et al., 2024. Prevalence and risk factors of mortality among heart failure patients in low resource setting hospitals: a multicenter prospective observational study. Frontiers in Cardiovascular Medicine 11, 1429513.

Bennett, L.G., 2017. Epigenetic Mechanisms Involved in the Cellular Response to DNA Damage Processed by Base Excision Repair. The University of Liverpool, United Kingdom.

Bris, C., et al., 2021. Improved detection of mitochondrial DNA instability in mitochondrial genome maintenance disorders. Genetics in Medicine 23 (9), 1769–1778.

Budde, H., et al., 2022. Current understanding of molecular pathophysiology of heart failure with preserved ejection fraction. Frontiers in Physiology 13, 928232.

Canty Jr, J.M., 2022. Myocardial injury, troponin release, and cardiomyocyte death in brief ischemia, failure, and ventricular remodeling. American Journal of Physiology-Heart and Circulatory Physiology 323 (1), H1–H15.

Chen, X., et al., 2024. Small extracellular vesicles from young plasma reverse age-related functional declines by improving mitochondrial energy metabolism. Nature Aging 4 (6), 814–838.

Chen, X.J., 2013. Mechanism of homologous recombination and implications for aging-related deletions in mitochondrial DNA. Microbiology and Molecular Biology Reviews 77 (3), 476–496.

Cheng, M., et al., 2021. PGAM5: a crucial role in mitochondrial dynamics and programmed cell death. European Journal of Cell Biology 100 (1), 151144.

Chenna, S., et al., 2022. Mechanisms and mathematical modeling of ROS production by the mitochondrial electron transport chain. American Journal of Physiology-Cell Physiology 323 (1), C69–C83.

Cheung, D.W., 1981. Electrical activity of the pulmonary vein and its interaction with the right atrium in the guinea-pig. The Journal of Physiology 314, 445–456.

Chew, N.W.S., et al., 2025. Cardiovascular-liver-metabolic health: recommendations in screening, diagnosis, and management of metabolic dysfunction-associated steatotic liver disease in cardiovascular disease via modified delphi approach. Circulation 151 (1), 98–119.

Chung, J.-H., et al., 2016. Myofilament calcium sensitivity: role in regulation of in vivo cardiac contraction and relaxation. Frontiers in Physiology 7, 562.

Chung, K.W., 2021. Advances in understanding of the role of lipid metabolism in aging. Cells 10 (4), 880.

Clarke, J.D., et al., 2015. Perturbed atrial calcium handling in an ovine model of heart failure: potential roles for reductions in the L-type calcium current. Journal of Molecular and Cellular Cardiology 79, 169–179.

Correia, M., et al., 2022. Metabolic determinants in cardiomyocyte function and heart regenerative strategies. Metabolites 12 (6), 500.

Dabravolski, S.A., et al., 2022. The role of mitochondrial DNA mutations in cardiovascular diseases. International Journal of Molecular Sciences 23 (2), 952.

De Filippis, B., et al., 2015. Mitochondrial free radical overproduction due to respiratory chain impairment in the brain of a mouse model of Rett syndrome: protective effect of CNF1. Free Radical Biology and Medicine 83, 167–177.

de Souza, P., et al., 2019. Recent advances in the knowledge of naturally-derived bioactive compounds as modulating agents of the renin-angiotensin-aldosterone system: therapeutic benefits in cardiovascular diseases. Current Pharmaceutical Design 25 (6), 670–684.

Elorza, A.A., Soffia, J.P., 2021. mtDNA heteroplasmy at the core of aging-associated heart failure. An integrative view of OXPHOS and mitochondrial life cycle in cardiac mitochondrial physiology. Frontiers in Cell and Developmental Biology 9, 625020.

Evangelou, K., et al., 2023. Cellular senescence and cardiovascular diseases: moving to the "heart" of the problem. Physiological Reviews 103 (1), 609–647.

Gao, C., et al., 2025. The mitochondrial hub gene UCHL1 May serve as a potential biomarker for diagnosing diabetic cardiomyopathy: a comprehensive integration of biological pathways. BMC Medical Genomics 18 (1), 129.

Gao, S., et al., 2024. Animal models of heart failure with preserved ejection fraction (HFpEF): from metabolic pathobiology to drug discovery. Acta Pharmacologica Sinica 45 (1), 23–35.

González-Juanatey, C., et al., 2023. Impact of advanced age on the incidence of major adverse cardiovascular events in patients with type 2 diabetes mellitus and stable coronary artery disease in a real-world setting in Spain. Journal of Clinical Medicine 12 (16), 5218.

Guo, J., et al., 2022. Aging and aging-related diseases: from molecular mechanisms to interventions and treatments. Signal Transduction and Targeted Therapy 7 (1), 391.

Gupta, A., Houston, B., 2017. A comprehensive review of the bioenergetics of fatty acid and glucose metabolism in the healthy and failing heart in nondiabetic condition. Heart Failure Reviews 22 (6), 825–842.

Hajam, Y.A., et al., 2022. Oxidative stress in human pathology and aging: molecular mechanisms and perspectives. Cells 11 (3), 552.

He, Y., et al., 2025. Mechanisms and therapeutics of insulin signaling transduction genes in diabetic cardiomyopathy: a comprehensive updated review. Frontiers in Endocrinology 16, 1589695.

Hervouet, E., et al., 2008. HIF and reactive oxygen species regulate oxidative phosphorylation in cancer. Carcinogenesis 29 (8), 1528–1537.

Huang, Z., Chen, Y., Zhang, Y., 2020. Mitochondrial reactive oxygen species cause major oxidative mitochondrial DNA damages and repair pathways. Journal of Biosciences 45 (1).

Islam, M.T., et al., 2023. Senolytic drugs, dasatinib and quercetin, attenuate adipose tissue inflammation, and ameliorate metabolic function in old age. Aging Cell 22 (2), e13767.

Izzo, C., et al., 2021. The role of oxidative stress in cardiovascular aging and cardiovascular diseases. Life (Chicago, Ill.: 1978) 11 (1), 60.

Jackson, M.J., McArdle, A., 2011. Age-related changes in skeletal muscle reactive oxygen species generation and adaptive responses to reactive oxygen species. The Journal of Physiology 589 (9), 2139–2145.

Jan, B., et al., 2024. Cardiovascular diseases among Indian older adults: a comprehensive review. Cardiovascular Therapeutics 2024 (1), 6894693.

Kallai, A., et al., 2025. Genomic instability and genetic heterogeneity in aging: insights from clonal hematopoiesis (CHIP), monoclonal gammopathy (MGUS), and monoclonal B-cell lymphocytosis (MBL). Geroscience 47 (1), 703–720.

Karuna, N., et al., 2025. Deep phenotyping of a modified diabetic cardiomyopathy mouse model which reflects clinical disease progression. Diabetology and Metabolic Syndrome 17 (1), 334.

Khayungarnnawee, A., 2017. The Electrical Activity of the Pulmonary Vein and its Potential Role in Cardiac Arrhythmia.

Kitzman, D.W., Upadhya, B., 2014. Heart failure with preserved ejection fraction: a heterogenous disorder with multifactorial pathophysiology. Journal of the American College of Cardiology 63 (5), 457–459.

Lakatta, E.G., Yin, F., 1982. Myocardial aging: functional alterations and related cellular mechanisms. American Journal of Physiology-Heart and Circulatory Physiology 242 (6), H927–H941.

Lan, Y., et al., 2023. Supra-additive effect of chronic inflammation and atherogenic dyslipidemia on developing type 2 diabetes among young adults: a prospective cohort study. Cardiovascular Diabetology 22 (1), 181.

Lesko, P., 2014. Optimizing Exercise Training in Prevention of Heart Failure with Preserved Ejection Fraction: Cardio-renal Function and Survival in Dahl salt-sensitive Animal Model. Norges teknisk-naturvitenskapelige universitet, Det medisinske fakultet.

Li, J., et al., 2023. The crosstalk between ferroptosis and mitochondrial dynamic regulatory networks. International Journal of Biological Sciences 19 (9), 2756.

Livneh, H., 2023. Homeostasis versus growth: a view of their roles in adaptation to adversity, trauma, chronic illness, and physical disability. Part I: history, structure, and processes. Psychological Trauma: Theory, Research, Practice, and Policy 15 (4), 584.

Lopaschuk, G.D., Ussher, J.R., 2016. Evolving concepts of myocardial energy metabolism: more than just fats and carbohydrates. Circulation Research 119 (11), 1173–1176.

Martins, D., et al., 2022. Oxidative stress as a therapeutic target of cardiac remodeling. Antioxidants 11 (12), 2371.

Melenovsky, V., et al., 2007. Cardiovascular features of heart failure with preserved ejection fraction versus nonfailing hypertensive left ventricular hypertrophy in the urban Baltimore community: the role of atrial remodeling/dysfunction. Journal of the American College of Cardiology 49 (2), 198–207.

Mohamed, S.A., et al., 2006. Mitochondrial DNA deletions and the aging heart. Experimental Gerontology 41 (5), 508–517.

Moretti-Horten, D.N., et al., 2024. Synchronized assembly of the oxidative phosphorylation system controls mitochondrial respiration in yeast. Developmental Cell 59 (8), 1043–1057. e8.

Nakashima, H., et al., 2025. Establishment of a HFpEF model using female Dahl salt-sensitive rats: a valuable tool for elucidating the pathophysiology of HFpEF in women. Hypertension Research 48 (2), 672–680.

Nesci, S., et al., 2021. Molecular and supramolecular structure of the mitochondrial oxidative phosphorylation system: implications for pathology. Life (Chicago, Ill.: 1978) 11 (3), 242.

Nguyen, T.T., et al., 2023. Mitochondria-associated programmed cell death as a therapeutic target for age-related disease. Experimental and Molecular Medicine 55 (8), 1595–1619.

Niort, B.C., 2022. Perturbed Calcium Handling in an Ovine Model of Cardiac Failure Following Myocardial Infarction: Role of Phosphodiesterases. The University of Manchester, United Kingdom.

Obokata, M., Reddy, Y.N., Borlaug, B.A., 2020. Diastolic dysfunction and heart failure with preserved ejection fraction: understanding mechanisms by using noninvasive methods. JACC: Cardiovascular Imaging 13 (1 Part 2), 245–257.

Persad, K.L., Lopaschuk, G.D., 2022. Energy metabolism on mitochondrial maturation and its effects on cardiomyocyte cell fate. Frontiers in Cell and Developmental Biology 10, 886393.

Poznyak, A.V., et al., 2022. Mitochondrial implications in cardiovascular aging and diseases: the specific role of mitochondrial dynamics and shifts. International Journal of Molecular Sciences 23 (6), 2951.

Ristow, M., Schmeisser, K., 2014. Mitohormesis: promoting health and lifespan by increased levels of reactive oxygen species (ROS). Dose-Response 12 (2). p. dose-response. 13-035. Ristow.

Rizvi, F., et al., 2021. Effects of aging on cardiac oxidative stress and transcriptional changes in pathways of reactive oxygen species generation and clearance. Journal of the American Heart Association 10 (16), e019948.

Roth, G.A., et al., 2017. Global, regional, and national burden of cardiovascular diseases for 10 causes, 1990 to 2015. Journal of the American College of Cardiology 70 (1), 1–25.

Shah, A.K., et al., 2021. Oxidative stress as a mechanism for functional alterations in cardiac hypertrophy and heart failure. Antioxidants 10 (6), 931.

Spallone, V., 2019. Update on the impact, diagnosis and management of cardiovascular autonomic neuropathy in diabetes: what is defined, what is new, and what is unmet. Diabetes and Metabolism Journal 43 (1), 3–30.

Stamatakis, E., Rezende, L.F., Rey-López, J.P., 2023. Sedentary behaviour and cardiovascular disease. Sedentary Behaviour Epidemiology. Springer, pp. 213–250.

Standl, E., et al., 2019. The global epidemics of diabetes in the 21st century: current situation and perspectives. European Journal of Preventive Cardiology 26 (2_suppl), 7–14.

Stevens, A.L., et al., 2015. Exercise improves cardiac function and attenuates insulin resistance in Dahl salt-sensitive rats. International Journal of Cardiology 186, 154–160.

Strazhesko, I.D., et al., 2017. Growth hormone, insulin-like growth factor-1, insulin resistance, and leukocyte telomere length as determinants of arterial aging in subjects free of cardiovascular diseases. Frontiers in Genetics 8, 198.

Tadic, M., et al., 2018. The role of arterial hypertension in development heart failure with preserved ejection fraction: just a risk factor or something more? Heart Failure Reviews 23 (5), 631–639.

Tenchov, R., et al., 2023. Aging hallmarks and progression and age-related diseases: a landscape view of research advancement. ACS Chemical Neuroscience 15 (1), 1–30.

Tirichen, H., et al., 2021. Mitochondrial reactive oxygen species and their contribution in chronic kidney disease progression through oxidative stress. Frontiers in Physiology 12, 627837.

Triposkiadis, F., et al., 2022. Pathogenesis of chronic heart failure: cardiovascular aging, risk factors, comorbidities, and disease modifiers. Heart Failure Reviews 27 (1), 337–344.

Tsuda, T., 2021. Clinical assessment of ventricular wall stress in understanding compensatory hypertrophic response and maladaptive ventricular remodeling. Journal of Cardiovascular Development and Disease 8 (10), 122.

Van Ommen, A., et al., 2022. Diastolic dysfunction and sex-specific progression to HFpEF: current gaps in knowledge and future directions. BMC Medicine 20 (1), 496.

Wang, B., et al., 2025. Mitochondrial ultrastructural pathology in diabetic cardiomyopathy: integrated analysis via scanning electron microscopy and 3D visualization imaging. Cardiovascular Diabetology 24 (1), 331.

West, M.J., 2004. The Effect of Oxidative Stress on Ageing in Drosophila melanogaster. University of London, University College, London (United Kingdom).

Williamson, J., Davison, G., 2020. Targeted antioxidants in exercise-induced mitochondrial oxidative stress: emphasis on DNA damage. Antioxidants 9 (11), 1142.

Witteles, R.M., Fowler, M.B., 2008. Insulin-resistant cardiomyopathy clinical evidence, mechanisms, and treatment options. Journal of the American College of Cardiology 51 (2), 93–102.

Xie, S., et al., 2023. Metabolic landscape in cardiac aging: insights into molecular biology and therapeutic implications. Signal Transduction and Targeted Therapy 8 (1), 114.

Xu, H., Van Remmen, H., 2021. The SarcoEndoplasmic Reticulum Calcium ATPase (SERCA) pump: a potential target for intervention in aging and skeletal muscle pathologies. Skeletal Muscle 11 (1), 25.

Yamamoto, T., Sano, M., 2022. Deranged myocardial fatty acid metabolism in heart failure. International Journal of Molecular Sciences 23 (2), 996.

Zimmet, P., Shaw, J., Alberti, K.G., 2003. Preventing type 2 diabetes and the dysmetabolic syndrome in the real world: a realistic view. Diabetic Medicine: A Journal of the British Diabetic Association 20 (9), 693–702.

Chapter 7

Telomere attrition and cardiovascular senescence

Abida Khan[1], Abdul Hai[2], Mohammed Zawiah[3], Muhammad Irfan Siddique[4], and Mohd Imran[1]
[1]*Center For Health Research, Northern Border University, Arar, Saudi Arabia,* [2]*Department of Medical Laboratory Technology, Faculty of Applied Medical Sciences, Northern Border University, Arar, Saudi Arabia,* [3]*Department of Clinical Practice, Faculty of Pharmacy, Northern Border University, Arar, Saudi Arabia,* [4]*Department of Pharmaceutics, Faculty of Pharmacy, Northern Border University, Arar, Saudi Arabia*

7.1 Introduction

Telomere biology provides a unifying framework for understanding how aging and stress remodel the cardiovascular system, with direct implications for population health. Globally, cardiovascular diseases (CVDs) remain the leading cause of death, accounting for an estimated 19.8 million deaths in 2022 (~32% of all deaths), with the majority occurring in low- and middle-income countries (LMICs). In India, CVDs are the leading cause of mortality amid suboptimal control of hypertension and diabetes, and the aging population is accelerating. Approximately 11% of Indians will be aged ≥60 years by 2025, amplifying their vulnerability to telomere-linked senescence phenotypes (Bhattacharyya et al., 2017; Yeh and Wang, 2016). At the chromosome terminus, telomeres are protected by the six-protein shelterin complex TRF1, TRF2, TIN2, TPP1, POT1, and RAP1, which organizes T-loops, shields the end from DNA damage recognition, and coordinates the replication of G-rich repeats. Recent structural and functional studies have delineated distinct TRF1, TIN2, TPP1, POT1, and TRF2-RAP1 behaviors that tune end protection in response to stressors. In cardiovascular cells, partial loss of shelterin function or replication stress exposes chromosome ends, recruits DDR mediators, and seeds persistent telomere dysfunction-induced foci that stabilize the growth arrest (Bandaria et al., 2016; Xin et al., 2008). Telomere length (TL) is primarily maintained by telomerase, a ribonucleoprotein comprising TERT and the RNA template TERC. In most adult somatic tissues, including the myocardium and vascular wall, telomerase activity is low, which favors gradual attrition during cell division and oxidative injury. A minority of human cells, notably in certain cancers, preserve telomeres through recombination-based alternative lengthening of telomeres (ALT), a mechanism increasingly resolved by multi-omics and imaging, and is relevant as a biological contrast to telomerase-dependent maintenance. Telomere dynamics are tissue- and cell-type specific throughout the lifespan (Tan et al., 2023; Cao et al., 2025). Cardiomyocytes rapidly exit the cell cycle postnatally with limited telomerase activity, making their telomeres susceptible to oxidative damage and establishing long-lived DNA damage foci. Endothelial and vascular smooth muscle cells retain their proliferative capacity but are subjected to shear, metabolic, and inflammatory stresses that accelerate cellular shortening. Progressive erosion triggers ATM/ATR signaling, stabilizes p53, and induces p21/p16-RB programs that fix senescence, while mitochondrial dysfunction elevates reactive oxygen species, reinforcing telomere injury and propagating a paracrine SASP that promotes fibrosis, calcification, and microvascular rarefaction features that map onto heart failure, atherosclerosis, and arterial stiffness trajectories highlighted by contemporary global and Indian burden estimates (Aix et al., 2016; Wei et al., 2024). The novelty and aim of this chapter are to integrate telomere architecture, maintenance mechanisms (TERT/TERC versus ALT), and lifespan dynamics with cell-specific cardiovascular contexts to produce a mechanistic arc from molecular end-protection to tissue-level remodeling and population-level risks. We synthesized emerging structural insights into shelterin with translational biomarkers and imaging readouts to identify actionable nodes where telomere stabilization, improved mitochondrial quality control, and redox correction may attenuate senescence programs in the heart and vasculature (Hoffmann et al., 2021). The aim is to provide a rigorous, clinically oriented scaffold that links basic telomere biology to trial-ready endpoints relevant to India and the global CVD landscape. The evidence presented here encourages the development of targeted combination strategies aimed at slowing the attrition of the shortest telomeres, minimizing

Cellular Senescence and Cardiac Aging. DOI: https://doi.org/10.1016/B978-0-443-45811-8.00013-1

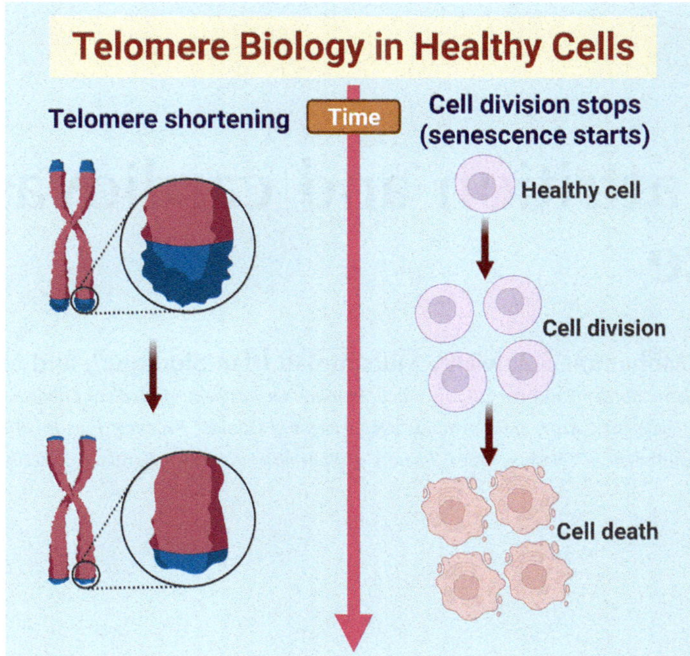

FIGURE 7.1 **Telomere shortening and cellular senescence.** The figure shows age-related shortening of telomeres and cellular implications of telomere shortening. Progressive degradation of terminal repeats at the end of chromosomes over time is depicted on the left. As telomeres wear off and protective capping becomes weak, cells show signs of damage, preventing them from proliferating. Healthy cycling cells on the right gradually stop dividing and enter senescence before finally moving on to cell death in the case of sustained stress. The graph is a time-wise summary of the intact telomeres to growth arrest, as the integrity of telomeres limits the replicative capacity in the somatic tissues.

telomere-related DNA damage signaling, and enhancing endothelial and myocardial function. The subsequent sections detail the necessary biomarker standards, therapeutic approaches, and Fig. 7.1 study designs required to effectively translate these mechanisms into long-lasting cardioprotection (Abdel-Gabbar and Kordy, 2024; Doroschuk et al., 2021; Alharbi et al., 2022a; Wadhwa et al., 2020). The conceptual relationship between telomere attrition and the transition from proliferative capacity to growth arrest is summarized in Fig. 7.1.

7.2 Mechanistic links between telomere attrition and cellular senescence

7.2.1 DNA damage responses (ATM/ATR-p53-p21/p16) and replicative arrest

DNA damage responses regulated by ATM/ATR-p53-p21/p16 constitute the central checkpoint response axis between telomere instability and long-term cell-cycle arrest. Telomere shortening or uncapping in cardiovascular tissues triggers foci generation of telomere dysfunction in cells, leading to ATM/ATR phosphorylation of H2AX and 53BP1 recruitment to stabilize p53, transactivate p21, inhibit cyclin CDK activity, and prevent G1/S or G2/M transition. With the continued presence of replication and oxidative stress, p16INK4a increases, retains RB in a hypophosphorylated form, and transforms temporary arrest into senescence (Wang et al., 2013; Guan et al., 2025). Malfunctioning mitophagy increases the production of reactive oxygen species (ROS) by mitochondria, which further oxidizes telomeric repeats. These checkpoints in endothelial and vascular smooth muscle cells suppress eNOS signaling, favorably change leukocyte adhesion, and induce a synthetic matrix remodeling phenotype. Mitochondrial dysfunction and calcium-handling defects, which occur in cardiomyocytes, coordinate irreversible cell cycle exit and the impairment of contractility. The continued presence of DDR cues the sustained SASP of IL-6, IL-8, MCP-1, and matrix proteases, enhancing the damage, microvascular rarefaction, and fibrosis of the vessels and myocardium, and hardening the vessels and myocardium (Liu et al., 2022; Sutanto et al., 2025; Alharbi et al., 2022b; Thapa et al., 2023). In models and cohorts, the shortening of leukocyte or tissue telomeres co-purifies with gH2AX/53BP1 foci and p21/p16 expression, which is associated with the activation of pathways through vascular dysfunction, adverse remodeling, and clinical risks. These studies led to the conclusion that the ATM/ATR-p53-p21/p16 pathway is the primary pathway that connects telomere shortening to the termination of cell replication and degradation of tissue functionality. Emphasizing the stabilization of telomeres and modulation of redox processes, alongside bolstering mitophagy and minimizing SASP to combat aging, can be seen as a strategic path forward to transform pathway repair into enduring cardioprotection (Xu et al., 2025a; Herbig et al., 2004).

7.2.2 Mitochondria–ROS–telomere feedback loops

The mitochondria–ROS–telomere feedback loop is a positive feedback loop in which mitochondrial impairment increases ROS levels, damaging guanine-rich telomeric DNA, and telomere erosion increases signal transduction, leading to additional mitochondrial quality control impairment. Exposure of the telomere caps and uncapping of the telomere in cardiovascular cells leads to activation of p53, subsequent inhibition of the PGC-1a/PGC-1b programs, and decreased biogenesis and respiratory efficiency, which elevates ROS generation. Oxidative damage and replication stress enhance telomere loss, maintain gH2AX/53BP1 foci, and stabilize p21/p16-RB arrest, but stalled mitophagy (such as PINK1/Parkin constraints) permits the accumulation of dysfunctional mitochondria and sustains redox imbalance (Moslehi et al., 2012; Lin and Epel, 2022). There is a decrease in endothelial nitric oxide signaling, vascular smooth muscle becomes synthetic and matrix-remodeling phenotype, and cardiomyocytes undergo calcium-handling defects with decreasing ATP supply, moving toward rarefaction, stiffness, and reduced contractile reserve. SASP mediators that are discharged in the case of a persistent damage response increase the oxidative burden and paracrine telomere stress throughout the myocardial-vascular niche (Chellappan et al., 2023; Singh et al., 2023). Altogether, these studies indicate the presence of a vicious circle involving p53-PGC signaling, augmented ROS load, and instability of telomeres, which becomes a typical route to vascular dysfunction and myocardial structural alterations. Future studies should focus on telomere stabilization, increased mitochondrial turnover with improved mitophagy, and alleviation of oxidative stress using specific antioxidants and NAD+ supplementation. They will be assessed using single-cell and spatial methods, as well as trial-ready endpoints, to determine whether interfering with this cycle is a possibility for achieving long-term cardioprotection (Sutanto et al., 2025; Sawasaki et al., 2024).

7.2.3 Epigenetic drift and telomere position effects

Epigenetic drift and telomere position effects overlap with age-related changes in DNA methylation and chromatin organization, with telomere status reprogramming the gene expression. Telomere shortening or uncapping in cardiovascular

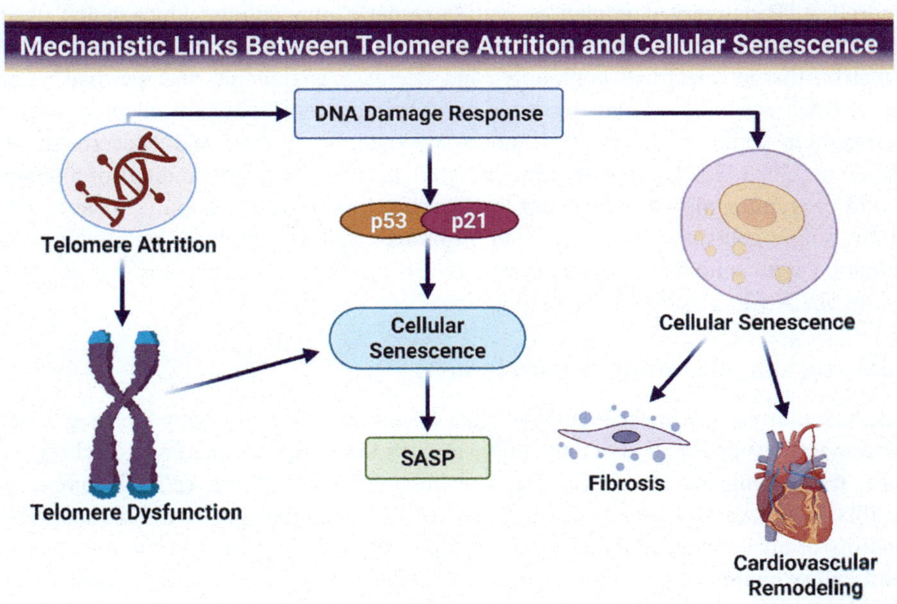

FIGURE 7.2 **Telomere attrition and cardiovascular remodeling.** This diagram shows how telomere attrition is connected to cardiovascular remodeling in a mechanistic cascade. Telomere dysfunction is caused by telomere shortening or uncapping, which stimulates the DNA damage response (ATM/ATR), p53 stabilization, and p21 induction. These checkpoints create cellular senescence that is maintained after the initial injury. The SASP of cytokines and proteases released by senescent cells enhances fibroblast activation and extracellular matrix remodeling, which causes fibrosis. Senescent propagation of tissues via paracrine signaling results in structural and functional remodeling of the heart and vessels.

cells deregulates shelterin, subtelomeric heterochromatin, and allows the telomere position effect over long distances (TPE-OLD), which regulates genes many base pairs away from the chromosome ends. Repressive histone losses (such as H3K9me3 and H4K20me3) and lamina dislocation enhance TERRA transcription, whereas CpG-sub-telomeric methylation changes and CTCF/cohesin loop rearrangement reverse enhancer-promoter interactions (Robin et al., 2014; Chevalier et al., 2025). These modifications collaborate with the long-term DDR response to stabilize p53, decrease SIRT1/SIRT6 action, and dysregulate the DNMT/TET ratio, which expands epigenetic drift in stress reaction, metabolic, and matrix pathways. Mitochondrial ROS also increase the further oxidation of telomeric guanine, telomere erosion, and histone acetylation patterns, permitting SASP transcription, thereby enhancing endothelial pathophysiology, fibroblast activation, and proinflammatory remodeling of the extracellular matrix (Gadhave et al., 2024; Samuel et al., 2019). Chromatin relaxation at telomeres initiates distal transcriptional noise, which condenses senescence and vascular-myocardial resilience (Alves-Fernandes and Jasiulionis, 2019; Zhu et al., 2021). In summary, telomere-directed chromatin weakening and TPE-OLD represent a mechanistic link between telomere loss and maladaptive gene programs that mediate vascular stiffness, fibrosis, and poor contractile reserve. In the future, single-cell/spatial multi-omics, including targeting epigenetic editing (e.g., dCas9-DNMT/TET or SIRT6-enhancing) and telomere-stabilizing, redox-modulating procedures, should be compared with trial-ready endpoints to determine whether the restoration of telomere-chromatin homeostasis translates into long-term cardioprotection (Zhang et al., 2021; Gupta et al., 2025; Ng et al., 2020).

In summary, telomere-directed chromatin weakening and TPE-OLD provide a mechanistic bridge between telomere loss and maladaptive gene programs that drive vascular stiffness, fibrosis, and impaired contractile reserve. The downstream cascade linking telomere dysfunction to senescence-associated remodeling outcomes is summarized in Fig. 7.2).

7.3 Cell-type-specific senescence in the heart and vasculature

7.3.1 Cardiomyocytes and cardiac progenitor compartments

Cardiomyocytes exhibit premature postnatal exit from the cell cycle and reduced telomerase activity, and their telomeres are susceptible to oxidative damage and replication stress, which entrap long DNA damage foci. Similarly, cardiac progenitor compartments and resident stromal/progenitor populations that are limited in their renewal also accumulate telomere loss, which limits clonal proliferation and paracrine help to repair. During telomere shortening and depletion of shelterin protection, p21/p16-RB arrest signaling via ATM/ATR-p53 stabilizes cellular dysfunction, and mitochondrial dysfunction increases ROS, further enhancing guanine oxidation and maintaining telomere dysfunction foci (Aix et al., 2016; Evangelou et al., 2023). These checkpoints in cardiomyocytes consolidate permanent cell cycle withdrawal, calcium handling, mitochondrial ATP delivery degradation, and sarcomeric remodeling, which helps obstruct the contractile reserve. Simultaneously, senescent fibroblasts and progenitor cells express SASP mediators (IL-6, TGF-b, and MMPs), which stiffen the matrix, disrupt electrical continuity, and decrease angiogenic and survival signals required during postinjury recovery. Subsequent bouts of hemodynamic or ischemic stress transform adaptive hypertrophy into fibrotic remodeling, and progenitor depletion restricts de novo myocyte production and exacerbates diastolic or systolic dysfunction (Hussain et al., 2023, 2024). Together, the literature is consistent in its findings of a telomere-driven program in cardiomyocytes and progenitor niches, which links contractile loss with the inability to remodel and progress heart failure. In the future, single-cell/spatial readouts and rigorous endpoint telomere-stabilizing and redox-modulating strategies, in combination with mitophagy enhancement, SASP inhibition, and exercise-based conditioning, should be considered (Moslehi et al., 2012; Yeh et al., 2019).

7.3.2 Endothelial and vascular smooth muscle cells

Endothelial cells and vascular smooth muscle cells are loci of telomere attrition, which causes vascular dysfunction. A reduction in telomerase activity and hemodynamic stress in such cells exposes the ends of chromosomes and activates senescence programs, remodeling the tone, structure, and inflammation of these cells. Telomere uncapping promotes ATM/ATR checkpoints, stabilizes p53, and activates p21/p16, which inhibits proliferation, suppresses eNOS signaling, decreases the availability of nitric oxide, and preserves DNA damage foci. The increase in mitochondrial ROS occurs due to mitophagy impairment, oxidation of telomeric repeats, and enhancement of senescence-associated secretory output-IL-6, IL-8, MCP-1, and MMPs that attract leukocytes and weaken the subendothelial matrix. VSMCs then switch to synthetic with increased collagen synthesis, elastin degradation, and osteogenic (RUNX2) markers, resulting in intimal thickening, medial calcification, and increased pulse-wave velocity (Morgan et al., 2018; Bloom et al., 2023). Barrier leak impairment and microvascular rarefaction impair perfusion and tighten resistance beds, placing a load on the left ventricle and aggravating coronary flow reserve. In models and cohorts, shorter vascular or leukocyte telomeres correlate with endothelial dysfunction, arterial stiffness, and atherosclerotic disease progression, and cellular senescence is associated with the clinical risk. All studies indicate that endothelial and VSMCs' senescence mediated by telomeres coordinates inflammation, matrix remodeling, and hemodynamic failure. Telomere-stabilizing, redox-modulating, and mitophagy-enhancing strategies should be combined with rigorous vascular endpoints to produce sustained clinical outcomes (Tritakis, 2016).

7.3.3 Cardiac fibroblasts, SASP, and paracrine spread

Cardiac fibroblasts are focal discriminators of the myocardial extracellular matrix, and their conversion to a telomere-based senescent state essentially rewrites tissue homeostasis. In this niche, paracrine stress is propagated between cardiomyocytes, the endothelium, and immune cells by SASP and small vesicle signaling. Telomere dysfunction-induced foci are maintained by shelterin erosion and exposition of chromosome ends, ATM/ATR checkpoints, p53 stabilization, p21/p16 arrest, and mitochondrial dysfunction. Mitophagosome malfunction increases ROS levels, thereby perpetuating telomere dysfunction. These signals reactive to NF-kB and cGAS-STING pathways upregulate SASP mediators -IL-1b, IL-6, IL-8, TGF-β, MCP-1, and matrix enzyme (MMP-2/-9) counter-modulators (TIMPs) and change collagen I/III balance, cross-linking, and stiffening of the ventricular walls. Mechano-transduction through integrins and YAP/TAZ further stabilizes the senescent phenotype and enhances myofibroblast characteristics with α-SMA enrichment (Hortells et al., 2019; Rao et al., 2025). Exosomes and miRNA cargo decrease cardiomyocyte calcium handling, NO signaling in the endothelium, and inflammatory recruitment in the heart. Repeated ischemic or pressure-overload signals subsequently transform reparative scarring into diffuse fibrosis, decrease electrical continuity, and form arrhythmogenic substrates that deplete diastolic filling and contractile reserves. Experimental in vitro and in vivo data have associated shorter fibroblast or leukocyte telomeres and increased levels of SASP with fibrosis load and negative remodeling in heart failure. These findings suggest that fibroblast senescence and SASP-mediated paracrine diffusion are regulators of matrix remodeling, stiffness, and electrical heterogeneity, which deteriorate myocardial function. In the future, telomere-stabilizing/redox-modulating interventions with SASP elimination, senescence-selective elimination, and matrix-targeted therapeutics should be examined using single-cell/spatial reporters and trial-ready outcomes to project specific pathway correction to long-term clinical outcomes (Emanueli et al., 2015; Liang et al., 2024).

7.4 Telomere attrition across cardiovascular conditions

7.4.1 Atherosclerosis, endothelial dysfunction, and arterial stiffness

Atherosclerosis, endothelial dysfunction, and arterial stiffness are closely related vascular phenotypes that become more severe with age. In this triad, endothelial and smooth muscle cell telomere attrition is a more proximate event that facilitates the onset of lesions and hemodynamic impairment. Telomere uncapping and shortening engage ATM/ATR roadblocks, stabilize p53, and induce p21/p16, which inhibit endothelial growth, decrease eNOS activity, and lower the bioavailability of nitric oxide, all of which advance the adherence of leukocytes to barriers, barrier permeability, and subendothelial lipid retention. The maintenance of DNA damage foci and ROS in mitochondria supports a SASP rich in IL-6, IL-8, MCP-1, and MMPs, increasing inflammation and matrix proteolysis, which increases intimal thickening. Simultaneously, vascular smooth muscle cells switch to a contraction-to-synthetic transition, enhance collagen deposition, develop osteogenic phenotypes (e.g., RUNX2 upregulation), and promote medial calcification, elastin fragmentation, and pulse-wave velocity increments, thereby increasing systolic load and impairing coronary perfusion (Hooglugt et al., 2022; Canale and Andreassi, 2025). In cohorts of people, shorter leukocyte or arterial wall telomeres are related to impaired flow-mediated dilation, increased plaque burden, and accelerated arterial stiffening, making cellular senescence a clinical risk factor. Together, the literature suggests that telomere-mediated senescence coordinates endothelial dysfunction and matrix remodeling, which ultimately leads to atherogenesis and progressive stiffening, with direct effects on myocardial workload and events. In the future, integrated strategies are needed to stabilize telomeres, mitigate redox stress, enhance mitophagy, and manage biomechanical and metabolic risks using single-cell/spatial readouts and trial-ready endpoints (analysis of plaque biology: flow-mediated dilation, pulse-wave velocity, imaging) to transform pathway correction into vascular benefits (Bloom et al., 2022; Banerjee et al., 2023).

7.4.2 Heart failure phenotypes and cardiomyopathies

Heart failure phenotypes minimize age-related (age-associated) biology of ejection fraction (HFrEF) and conserved ejection fraction (HFpEF) mechanics. Systolic pump failure characterizes HFrEF, whereas stiff diastole characterizes

HFpEF. In both, telomere attrition is present in cardiomyocytes, fibroblasts, and vascular cells, which initiate senescence programs that control the remodeling and symptoms. Gradual telomere erosion and uncapping open ATM/ATR gates, stabilize p53, and trigger p21/p16, imposing growth arrest, which is linked to mitochondrial dysfunction, overproduction of ROS, and foci of telomere damage. Viability and contractile cardiomyocyte loss and dysfunctional calcium cycling in HFrEF reduce stroke work, and SASP pro-fibrotic fibroblasts IL-6, TGF-b, and MMPs stimulate interstitial fibrosis, contributing to systolic reserve depletion and arrhythmogenic substrates. Endothelial senescence in HFpEF reduces nitric oxide bioavailability and leads to microvascular rarefaction and titin hypophosphorylation-induced stiffness, resulting in impaired relaxation and pulmonary venous hypertension despite preserved EF (Lin et al., 2021; Ma et al., 2020). These processes include telomere-short VSMCs acceleration of matrix deposition and calcification, fibroblast acceleration, toxin-related amplification of cardiomyocyte senescence, impairment of cardiomyocyte regeneration, and propagation of maladaptive signaling. In cohorts and models, the shortest telomeres and an increased number of senescence indicators are linked to poorer ventricular mechanics, increased filling pressure, poor prognosis, and cellular aging, which correlate with clinical outcomes in both phenotypes. Overall, there is evidence that telomere-driven senescence is a common axis that enhances fibrosis, microvascular dysfunction, and contractile deterioration in HFrEF, HFpEF, and major cardiomyopathies. Combined telomere-stabilizing and redox-reducing interventions, mitophagy enhancement, SASP decrease, and rigid endpoints (LVEF, global longitudinal strain, E/e', and exercise capacity) should be included in future analyses to obtain a sustained clinical response to the intervention of corrective pathways (Yeh and Wang, 2016; Gulej et al., 2025).

7.4.3 Hypertension, pulmonary vascular disease, and arrhythmic remodeling

The interconnected syndromes of vascular and electrical aging include hypertension, pulmonary vascular disease, and arrhythmia remodeling. Under these circumstances, stress responses triggered by telomere erosion of endothelial, smooth muscle, and myocardial cells consolidate senescence and reinvigorate hemodynamic and conduction functions. The shortening and uncapping of telomeres stimulate ATM/ATR-p53 checkpoints, cause p21/p16 arrest, increase mitochondrial ROS, which inhibits endothelial nitric oxide signaling, favors vasoconstrictor bias, and enhances collagen deposition and elastin fragmentation that stiffen systemic arteries and increase afterload. Secretory, matrix-remodeling senescent VSMCs resistance in resistance vessels increases media thickness and pulse-wave velocity, which strengthens systolic hypertension and left ventricular strain (Collins and Tzima, 2011; Noureddine et al., 2011). Endothelial dysfunction in the pulmonary circulation, coupled with SASP cytokines (e.g., IL-6 and MCP-1) and protease-mediated intimal fibrosis, medial remodeling, and microvascular rarefaction, augments pulmonary vascular resistance and acclimatizes right ventricular stress. The combination of connexin-43 asplasia, calcium-handling instability, and fibroblast senescence through telomeres leads to imbalanced fibrosis, connexin-43 asplasia, atrial fibrillation, and ventricular arrhythmia in the myocardium. Models and cohort evidence of shorter telomeres being associated with increasing blood pressure, reduced vasoreactivity, pulmonary hemodynamic load, and arrhythmic events support a consistent sequence of telomere erosion and vascular electrical remodeling (Sun et al., 2022). Taken together, evidence suggests that telomere-based senescence coordinates stiffness, microvascular dysfunction, and conduction instability, which is translated into hypertensive load, increasing pulmonary pressure, and susceptibility to rhythm. In the future, telomere-stabilizing, redox-stress-reducing, mitophagy-enhancing, SASP-attenuating, and stringent regulation of biomechanical and metabolic risk-corrective pathways should be investigated with trial-ready endpoints (ambulatory BP, pulse-wave velocity, RV-PA coupling, ECG/AF burden) to convert pathway correction into sustained clinical advantages (Roger et al., 2021; Lemay et al., 2025).

7.5 Exogenous and endogenous modifiers of telomere length in CVD

7.5.1 Lifestyle and environmental stressors

Lifestyle and environmental stressors, such as smoking, air pollution, psychosocial stress, diet, and exercise, shape telomere dynamics via oxidative, inflammatory, and neuroendocrine pathways. In cardiovascular tissues, these inputs modulate mitochondrial function and DNA damage responses, steering endothelial and myocardial cells toward maintenance or senescence.

Smoking: Combustion-derived oxidants and aldehydes increase ROS, create telomere dysfunction foci, and dampen telomerase activity, which narrows the repair capacity; endothelial nitric oxide signaling decreases, leukocyte adhesion increases, and arterial remodeling accelerates (Münzel et al., 2017; Aher et al., 2025).

Air pollution: Fine/ultrafine particles and associated metals generate mitochondrial ROS and systemic inflammation, shorten leukocyte telomeres, and impair flow-mediated dilation; persistent exposure reinforces arterial stiffness via matrix remodeling and reduces the bioavailability of nitric oxide (Daiber et al., 2020; Shkirkova et al., 2020).

Psychosocial stress: Sustained hypothalamic-pituitary-adrenal and sympathetic activation elevates cortisol and catecholamine levels, promotes insulin resistance and low-grade inflammation, and is associated with shorter telomeres. Sleep disruption and circadian misalignment compound the oxidative load and replication stress (Pervanidou and Chrousos, 2012).

Diet: A high glycemic burden and proinflammatory lipids heighten oxidative flux and methylation drift that erode telomeres, whereas Mediterranean-style patterns rich in polyphenols, omega-3s, fiber, and folate support redox defenses and methyl-group balance, consistent with telomere maintenance (Polom and Boccardi, 2025; Baliou et al., 2024).

Exercise: Regular moderate-vigorous activity activates PGC-1α-driven mitochondrial biogenesis, lowers ROS levels, improves endothelial function, and is associated with longer telomere length. Excessive unperiodized training may transiently increase oxidative stress, highlighting the importance of recovery (Botta et al., 2013; El Assar et al., 2022).

Collectively, the evidence indicates that these exposures converge on ROS, DNA damage signaling, and epigenetic drift to influence telomere length, vascular stiffness, and myocardial resilience. Looking ahead, multifactorial interventions combining smoking cessation, clean air access, stress reduction, dietary optimization, and structured exercise tracked with telomere/SASP biomarkers and vascular endpoints should be prioritized to translate risk modification into durable protection against cardiovascular diseases.

7.5.2 Metabolic inflammatory states

Metabolically inflammatory states, such as obesity, insulin resistance, and diabetes, accelerate cardiovascular telomere erosion by coupling nutrient overload with oxidative stress and inflammation. These conditions reshape endothelial, smooth muscle, fibroblast, and cardiomyocyte biology through mitochondrial dysfunction and sustained DNA damage-signaling pathways.

Obesity: Adiposity elevates circulating FFAs and cytokines (TNF-α and IL-6), increases mitochondrial ROS levels, and induces telomere uncapping with ATM/ATR activation. p53-p21/p16 arrest emerges alongside endothelial nitric oxide impairment and microvascular rarefaction (Masenga et al., 2023; Ndisang et al., 2014).

Insulin resistance: Hyperinsulinemia and impaired PI3K-Akt signaling depress eNOS, increase NADPH oxidase activity, and sustain replication stress at guanine-rich telomeres. Mitophagy shortfalls allow dysfunctional mitochondria to persist, reinforcing telomeric shortening (Cree-Green et al., 2017; Khoi et al., 2024).

Diabetes: Hyperglycemia drives advanced glycation end products, activates the RAGE-NF-κB pathways, and boosts ROS that oxidize telomeric repeats; telomerase activity decreases, γH2AX/53BP1 foci persist, and SASP mediators (IL-6, MCP-1, MMPs) remodel the matrix toward stiffness and diastolic load (Chen et al., 2024; Mengstie et al., 2022).

Lipo-toxicity and ectopic fat: Cardiac and perivascular lipid accumulation produce ceramides and acylcarnitines that impair mitochondrial respiration, intensify telomere injury, and promote fibroblast and VSMC phenotype switching toward fibrosis and calcification (Park et al., 2008; Badmus et al., 2024).

Collectively, evidence links metabolic inflammatory stress to faster telomere loss, durable senescence programs, and hemodynamic compromise, which converge to atherosclerosis, heart failure, and arrhythmic risk. Future studies should test multicomponent regimens of weight reduction, insulin sensitization, glycemic control, mitophagy enhancement, and redox/telomere-stabilizing strategies with single-cell/spatial readouts and trial-ready endpoints to translate pathway correction into durable cardioprotection.

7.5.3 Genetic variation and inherited telomeropathies (TERT, TERC, POT1, shelterin genes)

TERT, TERC, POT1, and genetic variations of other shelterin genes (such as TINF2, ACD/TPP1, and TERF1/TERF2) are used to vary telomerase action or end-protection and to define the default rate of telomere shortening in cardiovascular tissues. Loss-of-function alleles or hypomorphic variants of inherited telomeropathies are associated with critically short telomeres, which result in hypodiminished endothelial, smooth muscle, fibroblast, and cardiomyocyte compartments undergoing premature senescence when exposed to hemodynamic and metabolic stress (Chakraborty and Banerjee, 2024). This is due to reduced telomerase holoenzyme activity (TERT/TERC) or defective binding of single-strand overhangs and stabilization of caps (POT1/shelterin), which reveal chromosome ends, activate ATM/ATR signaling, and enhance stable p53-p21/p16 checkpoints that change temporary arrest into permanent senescence. Mitochondrial ROS increases telomeric guanine oxidation. Nitric oxide bioavailability decreases, leukocyte adhesion increases, and atherogenesis seeds in the endothelium; medial thickening and calcification promote regeneration in vascular smooth muscle; fibroblast SASP is

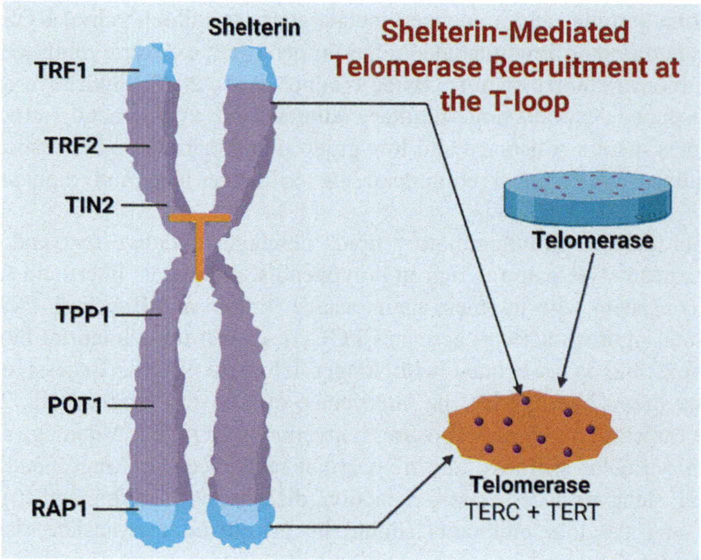

FIGURE 7.3 **Shelterin complex and telomere protection.** This image shows the protection process of T-loop ends and the attraction of telomerase by the shelterin complex. TRF1/TRF2 bind to telomeric DNA; TIN2 links the TRF proteins to the TPP1-POT1 heterodimer, which binds the single-strand overhang, whereas RAP1 is linked to TRF2. Shelterin uses the telomerase ribonucleoprotein-TERT catalytic subunit through the TPP1 TEL-patch interface, and G-rich repeats are elongated with the help of the TERC RNA template. End-to-end fusions and replication of cardiovascular tissues are prevented, and replication competence is strongly maintained by coordinated capping and lengthening.

reinforced, and fibroblast telomere damage limits regenerative capacity in the myocardium, and adverse remodeling is promoted (Guo et al., 2007; Mir et al., 2020). In cohorts, risk alleles at telomere maintenance loci and shorter leukocyte TL are related to coronary disease burden, arterial stiffness, and undesirable heart failure trajectory, and the germline architecture is linked to clinical risk through a telomere–senescence axis. These results suggest that unusual pathogenic forms and normal variants of telomere genes stratify susceptibility to vascular aging and myocardial dysfunction due to increased cellular senescence. In the future, genetics-plus-telomere phenotyping, cell-specific iPSCs, responsible consideration of telomere-stabilizing/telomerase-modulating strategies, redox, and mitophagy interventions should be compared to single-cell/spatial outcome measures and clinical endpoint results fit to test integrative genetics-plus-telomeres translational knowledge into sustained benefit (Sanchez et al., 2022; Carlquist et al., 2016). These findings suggest that both rare pathogenic variants and common polymorphisms in telomere-maintenance genes stratify susceptibility to vascular aging and myocardial dysfunction through a shared telomere–senescence axis. The molecular architecture of shelterin-mediated end protection and telomerase recruitment is summarized in Fig. 7.3.

7.6 Biomarkers and measurement for clinical and population studies

7.6.1 Assay platforms and preanalytical variables (TRF, qPCR, Flow-FISH, digital assays)

Assays and preanalytical variables define the usefulness of telomere metrics as biologically important or marginally constrained by methodology in cardiovascular research. In TRF, qPCR, Flow-FISH, and digital assays, the throughput, resolution, and cell-type specificity of each platform are traded against various sources of biases. TRF (Southern blot of terminal restriction fragments) provides length in base pairs and a distribution profile, but is sensitive to high DNA input and is susceptible to batch effects and the use of calibrators, as well as DNA quality. qPCR has rapid low-input T/S ratios in large cohorts but is sensitive to batch drift, calibrator selection, and DNA quality. Flow-FISH measures telomeres at single-cell resolution in leukocyte subsets with greater interpretability in inflammatory situations, where cell composition changes with inflammation. Digital methods, such as ddPCR absolute quantification and ultrasensitive methods, such as TeSLA shortest telomeres, provide a superior level of precision and prognosis (Wand et al., 2016; Eisenberg, 2016). Preanalytical controls are final: the choice of anticoagulants, time-to-processing, storage, freeze/thaw cycles, hemolysis, DNA extraction chemistry, and leukocyte subset distribution all change apparent length, and rigorous QC (replicates, interplate controls, z -scoring, and Bland–Altman surveillance) and reference materials are required to achieve cross-study comparability. Cardiovascular risk is best characterized by harmonized pipelines that pair platform strengths (e.g., cohort-scale qPCR anchored to TRF/Flow-FISH subsamples and digital assays of short-telomere burden). Taken together, it has been shown that platform-sensitive measurement and stringent preanalytics provide telomere measurements that monitor vascular dysfunction and remodeling. Standard reference materials and reporting should be used in future studies, shorter telomere scores should be combined with single-cell phenotyping, and harmonization of assays must be incorporated into future trials to transform telomere biology into strong cardiovascular outcomes (Bailey et al., 2015; Guder and Narayanan, 2015).

7.6.2 Circulating and tissue correlates (leukocyte TL, cfDNA TL, SASP panels)

Complementary windows to cardiovascular aging: Tissue and circulating correlates. Complementary information on cardiovascular aging, leukocyte TL, cell-free DNA telomere length (cfDNA TL), and SASP. Dynamic damage and local senescence load are measured with the help of tissue readouts and cfDNA TL. Leukocyte TL is a scalable surrogate marker for pathophysiological telomere conditions. Telomere shortening or uncapping is correlated with leukocyte TL trends, which provide cumulative estimates of oxidative and inflammatory exposures. However, because changes in leukocyte subsets and preanalytical manipulations bias estimates, cell composition normalization and constant calibrators are required. Telomere dysfunction is associated with endothelial, smooth muscle, and fibroblast senescence, which is co-localized by telomere-associated DNA damage foci (TAF) identified by immuno-FISH and released by injured vascular and myocardial cells, allowing serial and minimally invasive monitoring (Fyhrquist and Saijonmaa, 2012; Dai et al., 2024). Parallel SASP panels IL-6, IL-8, MCP-1, MMPs, GDF15, and their analogs measure paracrine activity that is associated with TL erosion and anticipates endothelial dysfunction, arterial stiffness, and unfavorable remodeling. Pathway activity (combined with vascular imaging and functional endpoints). The combination of leukocyte TL (a lifetime burden), short-term dynamics (cfDNA TL), and SASP activity scores with vascular imaging and functional endpoints using integrative models enhances risk stratification. All these convergent pieces of evidence favor the use of circulating and tissue biomarkers as consistent evidence of telomere-driven senescence and cardiovascular hazards. Future research should standardize preanalytics, shortest telomere measures, and single-cell/spatial mapping, and incorporate composite TL-SASP scores into future trials to convert marker patterns into practical cardioprotection (Fuster and Andrés, 2006; Qiao et al., 2025).

7.6.3 Prognostic value, risk stratification, and limitations of TL metrics

The prognostic value, risk stratification, and limitations of TL measurements depend on the ability of the measured TL to predict accumulated damage, present pathway activity, and impending clinical risk. In cardiovascular cohorts, shorter leukocyte or tissue TL is usually related to poor endothelial functioning, stiffening of arteries, adverse ventricular remodeling, and increased events; however, the strength and independence of such signals vary according to the assay and study design. TL adds the most to age, classic risk factors, inflammatory biomarkers, and imaging (flow-mediated dilation, pulse-wave velocity, strain), and change over time, and the burden of the shortest telomeres could be more effective than individual baseline means. Nonetheless, TL readouts are confounded by changes in leukocyte subsets, batch, DNA quality, preanalytical processing, regression dilution, and nonlinearity, which may weaken associations (Farzaneh-Far et al., 2008; Sena et al., 2022). TL is a surrogate endpoint and not a direct measure of telomeric dysfunction. Additional indicators, such as telomere-associated DNA damage foci, cfDNA TL dynamics, and SASP panels, were used to anchor biology to enhance the calibration. Comparability requires harmonized protocols, internal standards, cross-platform validation, and context-specific cutoffs that are better than universal ones. Together, these findings suggest that TL metrics can provide prognostic information on vascular dysfunction and remodeling when incorporated into multimodal models controlled for quality. Future research should prioritize the investigation of the shortest telomere, longitudinal metrics, integration of telomere length with single-cell and spatial profiling, and prespecification of trial-ready outcomes. This approach aims to determine whether telomere-targeting interventions can effectively translate prognostic signals into cardioprotective strategies (Bloom et al., 2022; Rossiello et al., 2022).

7.7 Therapeutic strategies beyond senolytics (telomere-focused and systemic)

7.7.1 Telomerase modulation and telomere-stabilizing approaches

Telomerase regulation and telomere-stabilizing methods aim to prevent cardiovascular senescence and telomere degradation. Opportunities include direct TERT/TERC augmentation and indirect stability of chromosomal ends through higher-quality control of mitochondria and redox stability. Although the shortest telomere attrition is slowed with cardiometabolic optimization (exercise, weight control, glycemic management), NAD+ replenishment, targeted mitochondrial antioxidants, and mitophagy enhancement, the theoretical results of transient telomerase upregulation are the restoration of replicative reserve, foci-reduction of telomere dysfunction, and p53-p21/p16 signaling (Li et al., 2024; Schellnegger et al., 2024). To further increase engraftment and paracrine support, experimental approaches such as ex vivo pretreatment of reparative cells, cardiac-targeted gene delivery aimed at temporary transcription, and small-molecule telomerase activators have been developed. Parallel-based end protection plans are intended to preserve shelterin activity and minimize replication stress induced by G-quadruplexes, as well as rigorous preanalytics of TL and TAF to ensure their effects on the target. The designs of trials should include dosage, duration, targeting of the tissue, and surveillance of cancer because there may be a likelihood of systemic or long-term telomerase activation that propagates the growth of premalignant clones, disrupts clonal hematopoiesis, or alters immune surveillance (Li et al., 2016; Liu et al., 2024). Paired with vascular and myocardial endpoints, pragmatic monitoring is performed using the shortest telomere burden, cfDNA TL dynamics, and a SASP panel. According to the available literature, tissue-directed telomere stabilization at a given dosage reduces DDR signaling, improves endothelial activity, and moderates the fibrotic remodeling. Short-term cardiac-targeted telomerase therapy, along with redox and mitophagy therapies, should be emphasized in the future. Mechanism correction for long-term cardioprotection should be converted using adaptive trials with single-cell/spatial readouts and preset safety gates (Fragkiadaki et al., 2025).

7.7.2 Mitochondria-targeted and redox-modulating interventions

Mitochondria-directed and redox-modulating therapies are designed to break the self-reinforcing cycle of mitochondrial dysfunction-mediated increase in ROS that weakens telomeres and telomere loss-mediated deterioration in mitochondrial quality control. The treatment targets in cardiovascular tissues to prevent the stabilization of senescence processes and reduction of DNA damage response include restoring electron-transport efficiency, enhancing mitophagy, and rebalancing redox signaling (Xu et al., 2025b; Zia et al., 2022). They encompass AMPK/SIRT1/PGC-1a activation with the help of structured exercise or small molecules to improve the oxidative capacity, NAD+ replacement to stimulate the sirtuin/PGC-1a pathways and mitochondrial biogenesis, and mitochondria-targeted antioxidants and cardiolipin stabilizers to enhance respiratory coupling. Workers on the PINK1 /Parkin pathways, including mitophagy enhancers, can be used to promote the clearance of defective organelles and reduce the persistence of ROS, and Nrf2 inducers can improve endogenous detoxification and selective inhibition of NOX, which reduces extramitochondrial ROS that contribute to telomeric guanine oxidation (Hao et al., 2025; Cantó and Auwerx, 2009). A combination of these methods works together to reduce foci of telomere-associated damage, decrease levels of p53-p21/p16 signaling, preserve endothelial nitric oxide bioavailability, and inhibit fibroblast SASP, which induces stiffness and unfavorable remodeling. SASP panels, cfDNA TL dynamics, and shortest telomere load were pragmatically studied alongside vascular and myocardial outcomes, such as diastolic metrics, pulse wave velocity, global longitudinal strain, and flow-mediated dilatation. Current research indicates that synchronous mitochondrial and redox repair may improve endothelial activity, inhibit fibrosis, and retain telomeres. To convert the mechanism of engagement into sustained cardioprotection, the focus should be on combination regimens involving exercise, specific mitochondrial therapies, and NAD+ replacement. Adaptive trials conducted on these regimens using single-cell/spatial readouts and predefined safety monitoring have been assessed (Westin et al., 2011; Fernández, 2019).

7.7.3 Exercise training, cardiometabolic optimization, and lifestyle prescriptions

Intervention on the modifiable biology that creates age-related accelerated telomere erosion and cardiovascular senescence is addressed through exercise training, cardiometabolic optimization, and lifestyle change. The aim of these interventions is to reduce the oxidative and inflammatory burden, regain mitochondrial quality control, and re-establish the hemodynamic mitochondrial homeostasis, such that the telomere-related DNA damage response is reduced instead of stabilized. AMPK-SIRT1-PGC-1a pathways, elevated mitophagy, increased endothelial bioavailability of nitric oxide, and potentially increased telomerase are activated by aerobic and combined resistance training and slow shortest telomere loss (Ludlow and Roth, 2011; Li et al., 2025). Parallel prescriptions, including weight reduction, blood pressure and lipid control, insulin sensitization, and intense glycemic regulation, inhibit NADPH oxidase-mediated ROS and reduce replication stress at guanine-rich telomeres. Unsaturated fat, fiber, and polyphenol diets improve mitochondrial efficiency and reduce the concentrations of SASP mediators. Structured sleep, stress-reduction programs, smoking cessation, and pollution avoidance further reduce redox load, which contributes to telomere erosion. Pragmatic monitoring combines telomere measures (such as shortest telomere burden and cfDNA TL dynamics) with vascular and myocardial outcomes (flow-mediated dilation, pulse wave velocity, and strain imaging) to verify these biological interactions (Sakr et al., 2023; Jomova et al., 2023). All of these studies point to the fact that integrated lifestyle and cardiometabolic interventions safeguard telomeres, enhance endothelial biofunction, and restrict fibrotic remodeling with quantifiable compliance and reserve increments. In the future, multicomponent approaches that integrate the periodization of exercise with diet quality and risk factor control with individualized redox/mitophagy support must undergo adaptive trials based on single-cell/spatial readouts and predetermined endpoints to translate mechanism correction into long-term cardioprotection (Yang et al., 2025; Semeraro et al., 2020).

7.8 Emerging technologies and translational pathways

7.8.1 Single-cell and spatial multiomics of cardiac/vessel senescence

Single-cell and spatial multi-omics can identify these senescent cell types, locate them, and determine how their signals change in their immediate tissue, allowing their identification. In combination with chromatin accessibility and pathway activity, researchers have presented telomere stress programs in healthy hearts and arteries using a combination of scRNA-seq/scATAC-seq, spatial transcriptomics, and proteome imaging. The foci detected by gH2AX/53BP1 and telomere probes in response to telomere dysfunction can be associated with the cell states of mitochondrial stress, SASP transcription, and p53-p21/p16 activity, and multiome profiles link enhancers to target genes that result in fibroblast matrix remodeling, endothelial leakage, and p53-p21/p16-induced phenotypic switching of VSMCs. RNA velocity/pseudotime predicts transits between presenescence and fixed arrest, and multiplexed RNA/protein imaging (MERFISH, IMC) and spatial barcoding (e.g., Visium, Slide-seq) determine the existence of perivascular niches where immune cells co-localize with both senescent fibroblasts and the endothelium (Chen et al., 2020; Sun, 2023). Integrative pipelines, by providing concordance between chromatin states, paracrine responses, and telomere instability, can be achieved by filtering single-cell data to the minimal telomere load (using flow-FISH or ultrasensitive telomere testing of sorted subsets). These atlases display high SASP flux microdomains, identify ligand-receptor pairings that mediate dissemination, and label patient-stratifying nodes (such as NOX, NF-kB, and TGF-b). By combining single-cell and spatial methods, mechanistic maps that combine telomere attrition, senescence programs, and tissue remodeling into testable cardiovascular disease circuits can be generated. In the future, trial-ready telomere, redox-regulating, and SASP-inhibitory therapeutic candidates should be directed by longitudinal, multimodal, cohort-based, three-dimensional combinations of telomere phenotypes with spatial multi-omics and clinically implementable results (Tao et al., 2024; Brennan et al., 2013).

7.8.2 Molecular imaging of senescence in vivo and trial-ready endpoints

Persistently growth-arrested cells are observed by molecular imaging of senescence in vivo, which are then converted into cardiovascular endpoints suitable for clinical trials. Together, they permit the noninvasive observation of the consequences of telomere damage throughout the heart and vascular system and provide measures suitable for evaluating the treatment results. The stabilization of p53-p21/p16 programs and induction of SASP offer targets for imaging agents that report SASP-reactive protease activity, p16/p21 pathway activation, or senescence-linked β-galactosidase activity (Suda et al., 2023; Ngoi et al., 2021). Translation is advancing using PET agents and nanoparticles to activate the senescent microenvironment to detect senescent load in the heart and arteries using PET/SPECT and optical reporters targeted to β-gal or galacto-conjugated reporters. Flow-mediated dilatation or perfusion MRI can be used to assess endothelial function, cardiac MRI (native/postcontrast T1, extracellular volume), and vascular imaging (carotid/femoral ultrasound, CT, or MRI calcified wall thickness) can be used to determine the relationship between regional senescent load and fibrosis, edema, and stiffness (Xiang et al., 2024; Brickute et al., 2022). Appropriate clinical trial endpoints include circulating SASP panels, dynamic aspects of cfDNA TL, safety biomarkers (NT-proBNP and high-sensitivity troponin), and vascular (flow-mediated dilatation and pulse-wave velocity) and cardiac (global longitudinal strain and diastolic indices) functions. These technologies are combined to associate structural and functional remodeling with molecular senescence, enabling mechanism-based efficacy assessments. Moreover, the alignment of composite imaging-biomarker goals and validation of senescence-targeting PET tracers with biopsy or single-cell/spatial standards should hasten the determination of telomere-stabilizing, redox-modulating, and SASP-reducing therapies (Schmitt et al., 2021; Shaddy et al., 2025).

7.8.3 Study design, regulatory considerations, and harmonized outcome measures

The capability of telomere-based techniques to generate understandable and translatable data on cardiovascular disease relies on the study design, regulatory concerns, and standardized reporting of outcomes. To guarantee the safety of the participants, sound designs should correlate biomarker enrichment with clinically significant readouts and the intent of mechanistic models with realistic goals. Adaptive phase I/II systems may be stratified using telomere measures (shortest

TABLE 7.1 Telomere-driven cardiovascular senescence: mechanism-to-translation evidence action matrix.

Mechanism	Principal mediators	Cellular effect	Cardiovascular phenotype	Translational readouts/assays	Therapeutic leverage and endpoints	References
End protection by shelterin; telomerase maintenance vs ALT	TRF1, TRF2, TIN2, TPP1, POT1, RAP1; TERT/TERC	T-loop stabilization; telomere extension; prevention of end-recognition	Preserved vascular and myocardial homeostasis	TRF/Southern, Flow-FISH, qPCR TL; TeSLA (shortest TL)	Telomere stabilization or transient telomerase up-tuning; endpoints: TL/TAF change, endothelial function	Mohammadi et al. (2025)
DNA damage response from telomere uncapping	ATM/ATR, γH2AX, 53BP1, p53, p21, p16-RB	Replicative arrest; stable senescence	Endothelial dysfunction; adverse remodeling	TAF (immuno-FISH), p21/p16, γH2AX foci	DDR quenching with upstream telomere care; endpoints: FMD, PWV, strain	Rossiello (2014), Chen et al. (2007)
Mitochondria–ROS–telomere feedback	PGC-1α/β, SIRT1, PINK1/Parkin, NOX	ROS amplification; mitophagy failure; TL erosion	Stiffness, rarefaction, contractile decline	mtDNA integrity, ROS indices; cfDNA TL dynamics; SASP panel	NAD^+ repletion, mitophagy enhancers, targeted antioxidants; endpoints: GLS, diastolic indices	Guo et al. (2025)
Epigenetic drift and TPE-OLD	H3K9me3/ H4K20me3 loss, TERRA, CTCF/ cohesin	Subtelomeric relaxation; maladaptive transcription	Fibrosis, calcification, reduced reserve	Methylome clocks, chromatin marks; single-cell/ spatial profiles	Epigenetic editing (dCas9-DNMT/TET), SIRT6 support; endpoints: ECM biomarkers, ECV (CMR)	Li and Tollefsbol (2016)
Endothelium / VSMC senescence	eNOS loss, RUNX2, MMPs	NO deficit; synthetic/ osteogenic shift	Atherosclerosis; arterial stiffness	FMD, PWV; plaque imaging; SASP cytokines	Telomere stabilization + redox control; endpoints: PWV, plaque biology	Minamino et al. (2002)
Cardiomyocyte & progenitor vulnerability	Low telomerase, Ca^{2+} handling defects, mitochondrial failure	Irreversible cell-cycle exit; reduced ATP	HFrEF/HFpEF progression	GLS, E/e′, NT-proBNP; TAF in tissue	Combined telomere/ mitochondrial therapy; endpoints: LVEF/GLS, exercise capacity	Jentzer et al. (2015)

Fibroblast SASP and paracrine spread	NF-κB, cGAS–STING, IL-6/IL-8, TGF-β, MMP-2/-9	Matrix remodeling; electrical heterogeneity	Diffuse fibrosis; arrhythmogenic substrate	ECM turnover markers; ECV (CMR)	SASP attenuation/clearance; endpoints: fibrosis regression, rhythm burden	Gagliardi et al. (2025), Loo et al. (2020)
Lifestyle and environmental modifiers	Smoking, PM2.5, psychosocial stress; diet; exercise	Oxidative/inflammatory load; telomerase modulation	Risk acceleration or mitigation	TL (leukocyte/cfDNA), SASP with vascular testing	Multicomponent lifestyle prescription; endpoints: TL trajectory, FMD, PWV	Chen et al. (2022), Zhao et al. (2021)
Metabolic-inflammatory states	Hyperglycemia, insulin resistance, FFAs, AGEs–RAGE	Replication stress; SASP amplification	Atherogenesis; HF remodeling	TL + γH2AX/53BP1; metabolic panel	Weight loss, insulin sensitizers; endpoints: TL/TAF, vascular mechanics	Wang et al. (2020), Di Lorenzo et al. (2013)
Genetic telomeropathies / TL variants	TERT, TERC, POT1; TINF2/ACD; OBFC1	Critically short TL; early senescence	Coronary burden; stiffness; HF risk	Genotyping + TL metrics	Genotype-informed selection for telomere/mitochondrial trials; safety surveillance	Revy et al. (2023)
Biomarker integration	TL (mean/shortest), cfDNA TL, TAF, SASP	Mechanism anchoring to risk	Improved stratification	Harmonized analytics; core-lab QC	Composite scores for trials; predefined thresholds	Mouliou (2023)

telomere burden, cfDNA TL dynamics, and telomere-associated damage foci) and SASP panels to address multiplicity and interim looks. They also have access to blinding, randomization, predetermined variables, and hierarchical or Bayesian decision-making. To control cardiometabolic risks and oncogenicity with telomerase-modulating medicines (e.g., serial hematology, clonal hematopoiesis screening, and imaging), safety supervision under the DSMB targets these areas of research. Assay standardization uses core lab processing, reference materials, cross-platform calibration, and data models that are compatible with the CDISC and real-world extensions (Sagris et al., 2022; Deng et al., 2022). Trial-ready outcomes combine measures of patient-centered outcomes (exercise ability and symptom rating) with assessed safety biomarkers (vascular function, flow-mediated dilation, pulse wave velocity) and cardiac mechanics (global longitudinal strain, diastolic indices, extracellular volume, using MRI). Regulatory alignment makes it possible to map risk–benefit monitoring, analytical validation, and device-drug or diagnostic-drug co-development on the already available guidelines to allow mechanisms to stage data for early efficacy. These designs can be used together to demonstrate that the normalization of telomere biology improves endothelial function, moderates fibrosis, and reduces hemodynamic loads. Integrated adaptive telomere-stabilizing, redox-modulating, and SASP-attenuating therapy assessment programs will predictably converge on consensus endpoint sets and co-valid biomarkers with spatial/single-cell assays in the future (Pelliccia et al., 2021; Biering-Sørensen et al., 2017). The major telomere-linked mechanisms, their mediators, cardiovascular phenotypes, and trial-ready translational readouts are synthesized in Table 7.1.

7.9 Conclusion and future perspectives

Telomere biology provides a consistent combination of molecular end-protection to tissue remodeling and clinical cardiovascular risk, and the data reviewed here have a chain of reaction between the protection of the shelterin and telomerase balance to enduring responses of DNA damage, mitochondrial dysfunction, microvascular rarefaction, contractile degradation, and the paracrine signaling of senescence that is orchestrated by the SASP, which connects all endothelial, vascular smooth muscle, fibroblast, and cardiomyocyte programs of Empowering telomere measurement assay standards by combining leukocyte measurements with cfDNA measurements and correlating these with tissue TAF measurements elucidates the synthesis of biomarkers into stratification, monitoring, and mechanism-anchored endpoints. The damage sensing-to-stable arrest pathway highlights the therapeutic importance of telomere-stabilizing, mitochondria-targeting, and redox-modulating interventions, enhancement of mitophagy, suppression of SASP, and intensive lifestyle interventions when examining mitophagy and fibrosis regression with the shortest-telomere burden, DDR quenching, endothelial functioning, and patient-centered outcomes. This work is novel in uniting telomere architecture, telomere maintenance, and lifespan dynamics to cell-resolved atlases and imaging to nominate trial-ready targets between molecular correction and quantifiable vascular and myocardial benefits. The scope includes the foundational biophysics of end protection, translational biomarker pipelines, and pragmatic intervention blueprints for both global and India-specific burdens. In the future, there will be a prioritization on standardizing preanalytics and reference materials, validating senescence-targeting PET tracers against either biopsy or single-cell icons, and constructing genotype-sensitive models that consider inherited telomere instability and clonal hematopoiesis during candidate selection for telomerase or mitochondrial therapy. Of equal significance are multimodal cohorts, which combine spatial multi-omics with longitudinal telomere measurements and harmonized vascular-myocardial endpoints, which make accurate go/no-go decisions in early phase trials. Together, this combination strategy can transform mechanistic understanding into long-lasting cardioprotection by decreasing the erosion of the shortest telomeres, damaging the endothelial and contractile backup, and eventually diminishing the burden of cardiovascular disease.

References

Abdel-Gabbar, M., Kordy, M.G.M., 2024. Telomere-based treatment strategy of cardiovascular diseases: imagination comes to reality. Genome Instability & Disease 5 (2), 61–75. Available at https://doi.org/10.1007/s42764-024-00123-x.

Aher, A., Doad, J., Fappi, A., Wasir, A.S., Agarwal, A., Shah, N., 2025. Pollution, climate extremes, and psychosocial stress: emerging environmental risks for cardiovascular disease: a review. International Journal of Community Medicine And Public Health 12 (10), 4806–4811. Available at https://doi.org/10.18203/2394-6040.ijcmph20253287.

Aix, E., Gutiérrez-Gutiérrez, Ó., Sánchez-Ferrer, C., Aguado, T., Flores, I., 2016. Postnatal telomere dysfunction induces cardiomyocyte cell-cycle arrest through p21 activation. Journal of Cell Biology 213 (5), 571–583. Available at https://doi.org/10.1083/jcb.201510091.

Alharbi, K.S., Almalki, W.H., Alzarea, S.I., Kazmi, I., Al-Abbasi, F.A., Afzal, O., et al., 2022a. A narrative review on the biology of piezo1 with platelet-rich plasma in cardiac cell regeneration. Chemico-Biological Interactions 363, 110011. Available at https://doi.org/10.1016/j.cbi.2022.110011.

Alharbi, K.S., Singh, Y., Afzal, O., Alfawaz Altamimi, A.S., Kazmi, I., Al-Abbasi, F.A., et al., 2022b. Molecular explanation of Wnt/βcatenin antagonist pyrvinium mediated calcium equilibrium changes in aging cardiovascular disorders. Molecular Biology Reports 49 (11), 11101–11111. Available at https://doi.org/10.1007/s11033-022-07863-7.

Alves-Fernandes, D.K., Jasiulionis, M.G., 2019. The role of SIRT1 on DNA damage response and epigenetic alterations in cancer. International Journal of Molecular Sciences 20 (13). Available at https://doi.org/10.3390/ijms20133153.

Badmus, O.O., da Silva, A.A., Li, X., Taylor, L.C., Greer, J.R., Wasson, A.R., et al., 2024. Cardiac lipotoxicity and fibrosis underlie impaired contractility in a mouse model of metabolic dysfunction-associated steatotic liver disease. FASEB BioAdvances 6 (5), 131–142. Available at https://doi.org/10.1096/fba.2023-00139.

Bailey, L.B., Stover, P.J., McNulty, H., Fenech, M.F., Gregory, J.F., Mills, J.L., et al., 2015. Biomarkers of nutrition for development—folate review. The Journal of Nutrition 145 (7), 1636S. Available at https://doi.org/10.3945/jn.114.206599.

Baliou, S., Ioannou, P., Apetroaei, M.-M., Vakonaki, E., Fragkiadaki, P., Kirithras, E., et al., 2024. The impact of the mediterranean diet on telomere biology: implications for disease management—a narrative review. Nutrients 16 (15), 2525. Available at https://doi.org/10.3390/nu16152525.

Bandaria, J.N., Qin, P., Berk, V., Chu, S., Yildiz, A., 2016. Shelterin protects chromosome ends by compacting telomeric chromatin. Cell 164 (4), 735–746. Available at https://doi.org/10.1016/j.cell.2016.01.036.

Banerjee, P., Rosales, J.E., Chau, K., Nguyen, M.T.H., Kotla, S., Lin, S.H., et al., 2023. Possible molecular mechanisms underlying the development of atherosclerosis in cancer survivors. Frontiers in Cardiovascular Medicine 10. Available at https://doi.org/10.3389/fcvm.2023.1186679.

Bhattacharyya, J., Mihara, K., Bhattacharjee, D., Mukherjee, M., 2017. Telomere length as a potential biomarker of coronary artery disease. Indian Journal of Medical Research 145 (6), 730–737. Available at https://doi.org/10.4103/0971-5916.216974.

Biering-Sørensen, T., Biering-Sørensen, S.R., Olsen, F.J., Sengeløv, M., Jørgensen, P.G., Mogelvang, R., et al., 2017. Global longitudinal strain by echocardiography predicts long-term risk of cardiovascular morbidity and mortality in a low-risk general population: the Copenhagen city heart study. Circulation: Cardiovascular Imaging 10 (3). Available at https://doi.org/10.1161/CIRCIMAGING.116.005521.

Bloom, S.I., Tucker, J.R., Lim, J., Thomas, T.G., Stoddard, G.J., Lesniewski, L.A., et al., 2022. Aging results in DNA damage and telomere dysfunction that is greater in endothelial versus vascular smooth muscle cells and is exacerbated in atheroprone regions. GeroScience 44 (6), 2741–2755. Available at https://doi.org/10.1007/s11357-022-00681-6.

Bloom, S.I., Liu, Y., Tucker, J.R., Islam, M.T., Machin, D.R., Abdeahad, H., et al., 2023. Endothelial cell telomere dysfunction induces senescence and results in vascular and metabolic impairments. Aging Cell 22 (8). Available at https://doi.org/10.1111/acel.13875.

Botta, A., Laher, I., Beam, J., DeCoffe, D., Brown, K., Halder, S., et al., 2013. Short term exercise induces PGC-1α, ameliorates inflammation and increases mitochondrial membrane proteins but fails to increase respiratory enzymes in aging diabetic hearts. PLoS One 8 (8), e70248. Available at https://doi.org/10.1371/journal.pone.0070248.

Brennan, C.W., Verhaak, R.G.W., McKenna, A., Campos, B., Noushmehr, H., Salama, S.R., et al., 2013. The somatic genomic landscape of glioblastoma. Cell 155 (2), 462–477. Available at https://doi.org/10.1016/j.cell.2013.09.034.

Brickute, D., Chen, C., Braga, M., Barnes, C., Wang, N., Allott, L., et al., 2022. Design, synthesis, and evaluation of a novel PET imaging agent targeting lipofuscin in senescent cells. RSC Advances 12 (40), 26372–26381. Available at https://doi.org/10.1039/d2ra04535d.

Canale, P., Andreassi, M.G., 2025. Targeting telomere shortening in vascular aging and atherosclerosis: therapeutic promise of Astragalus membranaceus. Journal of Cardiovascular Development and Disease 12 (9). Available at https://doi.org/10.3390/jcdd12090341.

Cantó, C., Auwerx, J., 2009. PGC-1α, SIRT1 and AMPK, an energy sensing network that controls energy expenditure. Current Opinion in Lipidology 20 (2), 98–105. Available at https://doi.org/10.1097/MOL.0b013e328328d0a4.

Cao, C., Gong, W., Shuai, Y., Rasouli, S., Ge, Q., Khan, A., et al., 2025. Canonical and non-canonical functions of the non-coding RNA component (TERC) of telomerase complex. Cell & Bioscience 15 (1). Available at https://doi.org/10.1186/s13578-025-01367-0.

Carlquist, J.F., Knight, S., Cawthon, R.M., Le, V.T., Jared Bunch, T., Horne, B.D., et al., 2016. Shortened telomere length is associated with paroxysmal atrial fibrillation among cardiovascular patients enrolled in the Intermountain Heart Collaborative Study. Heart Rhythm: the Official Journal of the Heart Rhythm Society 13 (1), 21–27. Available at https://doi.org/10.1016/j.hrthm.2015.07.032.

Chakraborty, S., Banerjee, S., 2024. Combatting cellular immortality in cancers by targeting the shelterin protein complex. Biology Direct 19 (1). Available at https://doi.org/10.1186/s13062-024-00552-4.

Chellappan, D.K., Chellian, J., Rahmah, N.S.N., Gan, W.J., Banerjee, P., Sanyal, S., et al., 2023. Hypoglycaemic molecules for the management of diabetes mellitus from marine sources. Diabetes, Metabolic Syndrome and Obesity 16, 2187–2223. Available at https://doi.org/10.2147/DMSO.S390741.

Chen, J.-H., Hales, C.N., Ozanne, S.E., 2007. DNA damage, cellular senescence and organismal ageing: causal or correlative? Nucleic Acids Research 35 (22), 7417–7428. Available at https://doi.org/10.1093/nar/gkm681.

Chen, Z., Liu, P., Xia, X., Wang, L., Li, X., 2022. The underlying mechanism of PM2.5-induced ischemic stroke. Environmental Pollution 310, 119827. Available at https://doi.org/10.1016/j.envpol.2022.119827.

Chen, C., et al., 2020. Applications of multi-omics analysis in human diseases. MedComm 4 (4), e315.

Chen, Y., et al., 2024. Advanced glycation end products and reactive oxygen species: uncovering the potential role of ferroptosis in diabetic complications. Molecular Medicine (Cambridge, Mass.) 30 (1), 141.

Chevalier, R., Murcia Pienkowski, V., Jullien, N., Caron, L., Pinard, P.V., Magdinier, F., et al., 2025. Telomere position effect-over long distances acts as a genome-wide epigenetic regulator through a common alu element. Aging Cell 24 (6). Available at https://doi.org/10.1111/acel.70027.

Collins, C., Tzima, E., 2011. Hemodynamic forces in endothelial dysfunction and vascular aging. Experimental Gerontology 46 (2-3), 185–188. Available at https://doi.org/10.1016/j.exger.2010.09.010.

Cree-Green, M., Rahat, H., Newcomer, B.R., Bergman, B.C., Brown, M.S., Coe, G.V., et al., 2017. Insulin resistance, hyperinsulinemia, and mitochondria dysfunction in nonobese girls with polycystic ovarian syndrome. Journal of the Endocrine Society 1 (7), 931–944. Available at https://doi.org/10.1210/js.2017-00192.

Dai, M., Li, K., Sacirovic, M., Zemmrich, C., Ritter, O., Bramlage, P., et al., 2024. Cell-free plasma telomere length correlated with the risk of cardiovascular events using machine learning classifiers. Scientific Reports 14 (1). Available at https://doi.org/10.1038/s41598-024-76686-2.

Daiber, A., Kuntic, M., Hahad, O., Delogu, L.G., Rohrbach, S., Di Lisa, F., et al., 2020. Effects of air pollution particles (ultrafine and fine particulate matter) on mitochondrial function and oxidative stress – implications for cardiovascular and neurodegenerative diseases. Archives of Biochemistry and Biophysics 696, 108662. Available at https://doi.org/10.1016/j.abb.2020.108662.

Deng, Y., Li, Q., Zhou, F., Li, G., Liu, J., Lv, J., et al., 2022. Telomere length and the risk of cardiovascular diseases: a Mendelian randomization study. Frontiers in Cardiovascular Medicine 9. Available at https://doi.org/10.3389/fcvm.2022.1012615.

Di Lorenzo, C., Dell'Agli, M., Colombo, E., Sangiovanni, E., Restani, P., 2013. Metabolic syndrome and inflammation: a critical review of in vitro and clinical approaches for benefit assessment of plant food supplements. Evidence-Based Complementary and Alternative Medicine 2013, 1–10. Available at https://doi.org/10.1155/2013/782461.

Doroschuk, N.A., Postnov, A.Y., Doroschuk, A.D., Ryzhkova, A.I., Sinyov, V.V., Sazonova, M.D., et al., 2021. An original biomarker for the risk of developing cardiovascular diseases and their complications: telomere length. Toxicology Reports 8, 499–504. Available at https://doi.org/10.1016/j.toxrep.2021.02.024.

Eisenberg, D.T.A., 2016. Telomere length measurement validity: the coefficient of variation is invalid and cannot be used to compare quantitative polymerase chain reaction and Southern blot telomere length measurement techniques. International Journal of Epidemiology 45 (4), 1295–1298. Available at https://doi.org/10.1093/ije/dyw191.

El Assar, M., Álvarez-Bustos, A., Sosa, P., Angulo, J., Rodríguez-Mañas, L., 2022. Effect of physical activity/exercise on oxidative stress and inflammation in muscle and vascular aging. International Journal of Molecular Sciences 23 (15), 8713. Available at https://doi.org/10.3390/ijms23158713.

Emanueli, C., Shearn, A.I.U., Angelini, G.D., Sahoo, S., 2015. Exosomes and exosomal miRNAs in cardiovascular protection and repair. Vascular Pharmacology 71, 24–30. Available at https://doi.org/10.1016/j.vph.2015.02.008.

Evangelou, K., Vasileiou, P.V.S., Papaspyropoulos, A., Hazapis, O., Petty, R., Demaria, M., et al., 2023. Cellular senescence and cardiovascular diseases: moving to the "heart" of the problem. Physiological Reviews 103 (1), 609–647. Available at https://doi.org/10.1152/physrev.00007.2022.

Farzaneh-Far, R., Cawthon, R.M., Na, B., Browner, W.S., Schiller, N.B., Whooley, M.A., 2008. Prognostic value of leukocyte telomere length in patients with stable coronary artery disease: data from the heart and soul study. Arteriosclerosis, Thrombosis, and Vascular Biology 28 (7), 1379–1384. Available at https://doi.org/10.1161/ATVBAHA.108.167049.

Fernández, E.P., 2019. The aging cardiovascular system: genetic and epigenetic determinants of vascular outcomes and cardiometabolic risk. https://repub.eur.nl/pub/120108/20-propositions.pdf.

Fragkiadaki, P., Apetroaei, M.-M., Kouvidi, E., Vakonaki, E., Renieri, E., Fragkiadoulaki, I., et al., 2025. The association between short telomere length and cardiovascular disease. Cytogenetic and Genome Research 164 (5-6), 202–210. Available at https://doi.org/10.1159/000542795.

Fuster, J.J., Andrés, V., 2006. Telomere biology and cardiovascular disease. Circulation Research 99 (11), 1167–1180. Available at https://doi.org/10.1161/01.RES.0000251281.00845.18.

Fyhrquist, F., Saijonmaa, O., 2012. Telomere length and cardiovascular aging. Annals of Medicine 44 (1), S138. Available at https://doi.org/10.3109/07853890.2012.660497.

Gadhave, D.G., Sugandhi, V.V., Jha, S.K., Nangare, S.N., Gupta, G., Singh, S.K., et al., 2024. Neurodegenerative disorders: mechanisms of degeneration and therapeutic approaches with their clinical relevance. Ageing Research Reviews 99, 102357. Available at https://doi.org/10.1016/j.arr.2024.102357.

Gagliardi, A., Migliari, S., Guercio, A., Baldari, G., Graziani, T., Cervati, V., et al., 2025. Emerging radioligands as tools to track multi-organ senescence. Diagnostics 15 (19). Available at https://doi.org/10.3390/diagnostics15192518.

Guan, J., Li, T., Ma, F., Wang, N., Zhang, H., Li, J., et al., 2025. DNA damage-dependent mechanisms of ionizing radiation-induced cellular senescence. PeerJ 13, e20087. Available at https://doi.org/10.7717/peerj.20087.

Guder, W.G., Narayanan, S., 2015. Pre-Examination Procedures in Laboratory Diagnostics: Preanalytical Aspects and Their Impact on the Quality of Medical Laboratory Result. Walter de Gruyter GmbH, Germany, pp. 1–410. Available at http://doi.org/10.1515/9783110334043.

Gulej, R., Patai, R., Ungvari, A., Kallai, A., Tarantini, S., Yabluchanskiy, A., et al., 2025. Impacts of systemic milieu on cerebrovascular and brain aging: insights from heterochronic parabiosis, blood exchange, and plasma transfer experiments. GeroScience. Available at https://doi.org/10.1007/s11357-025-01657-y.

Guo, X., Deng, Y., Lin, Y., Cosme-Blanco, W., Chan, S., He, H., et al., 2007. Dysfunctional telomeres activate an ATM-ATR-dependent DNA damage response to suppress tumorigenesis. The EMBO Journal 26 (22), 4709–4719. Available at https://doi.org/10.1038/sj.emboj.7601893.

Guo, Q.Q., Wang, S.S., Jiang, X.Y., Xie, X.C., Zou, Y., Liu, J.W., et al., 2025. Mitochondrial ROS triggers mitophagy through activating the DNA damage response signaling pathway. Proceedings of the National Academy of Sciences of the United States of America 122 (40). Available at https://doi.org/10.1073/pnas.2502841122.

Gupta, G., Samuel, V.P., M, R.M., Rani, B., Sasikumar, Y., Nayak, P.P., et al., 2025. Caspase-independent cell death in lung cancer: from mechanisms to clinical applications. Naunyn-Schmiedeberg's Archives of Pharmacology 398 (10), 13031–13048. Available at https://doi.org/10.1007/s00210-025-04149-0.

Hao, Y., Chen, F., Ren, X., Huang, X., Zhou, X., 2025. Harnessing mitochondrial biogenesis to combat acute kidney injury: current insights and futuredirections. Genes & Diseases 12 (6), 101645. Available at https://doi.org/10.1016/j.gendis.2025.101645.

Herbig, U., Jobling, W.A., Chen, B.P.C., Chen, D.J., Sedivy, J.M., 2004. Telomere shortening triggers senescence of human cells through a pathway involving ATM, p53, and p21CIP1, but Not p16INK4a. Molecular Cell 14 (4), 501–513. Available at https://doi.org/10.1016/s1097-2765(04)00256-4.

Hoffmann, J., Richardson, G., Haendeler, J., Altschmied, J., Andrés, V., Spyridopoulos, I., 2021. Telomerase as a therapeutic target in cardiovascular disease. Arteriosclerosis, Thrombosis, and Vascular Biology 41 (3), 1047–1061. Available at https://doi.org/10.1161/atvbaha.120.315695.

Hooglugt, A., Klatt, O., Huveneers, S., 2022. Vascular stiffening and endothelial dysfunction in atherosclerosis. Current Opinion in Lipidology 33 (6), 353–363. Available at https://doi.org/10.1097/MOL.0000000000000852.

Hortells, L., Johansen, A.K.Z., Yutzey, K.E., 2019. Cardiac fibroblasts and the extracellular matrix in regenerative and nonregenerative hearts. Journal of Cardiovascular Development and Disease 6 (3), 1–17. Available at https://doi.org/10.3390/jcdd6030029.

Hussain, M.S., Gupta, G., Goyal, A., Thapa, R., Almalki, W.H., Kazmi, I., et al., 2023. From nature to therapy: Luteolin's potential as an immune system modulator in inflammatory disorders. Journal of Biochemical and Molecular Toxicology 37 (11). Available at https://doi.org/10.1002/jbt.23482.

Hussain, S., Gupta, G., Shahwan, M., Bansal, P., Kaur, H., Deorari, M., et al., 2024. Non-coding RNA: a key regulator in the glutathione-GPX4 pathway of ferroptosis. Non-Coding RNA Research 9 (4), 1222–1234. Available at https://doi.org/10.1016/j.ncrna.2024.05.007.

Jentzer, J.C., Chonde, M.D., Dezfulian, C., 2015. Myocardial dysfunction and shock after cardiac arrest. BioMed Research International 2015. Available at https://doi.org/10.1155/2015/314796.

Jomova, K., Raptova, R., Alomar, S.Y., Alwasel, S.H., Nepovimova, E., Kuca, K., et al., 2023. Reactive oxygen species, toxicity, oxidative stress, and antioxidants: chronic diseases and aging. Archives of Toxicology 97 (10), 2499–2574. Available at https://doi.org/10.1007/s00204-023-03562-9.

Khoi, C.S., Lin, T.Y., Chiang, C.K., 2024. Targeting insulin resistance, reactive oxygen species, inflammation, programmed cell death, ER stress, and mitochondrial dysfunction for the therapeutic prevention of free fatty acid-induced vascular endothelial lipotoxicity. Antioxidants 13 (12). Available at https://doi.org/10.3390/antiox13121486.

Lemay, S.E., Kuwabara, Y., Bonnet, S., Potus, F., Provencher, S., Adnot, S., et al., 2025. Cellular senescence in adult pulmonary hypertension: current state and future challenges. European Respiratory Review: An Official Journal of the European Respiratory Society 34 (177). Available at https://doi.org/10.1183/16000617.0030-2025.

Li, X., Tamama, K., Xie, X., Guan, J., Hosoda, T., 2016. Improving cell engraftment in cardiac stem cell therapy. Stem Cells International 2016 (1). Available at https://doi.org/10.1155/2016/7168797.

Li, Y., Tollefsbol, T.O., 2016. Age-related epigenetic drift and phenotypic plasticity loss: implications in prevention of age-related human diseases. Epigenomics 8 (12), 1637–1651. Available at https://doi.org/10.2217/epi-2016-0078.

Li, Y., Tian, X., Luo, J., Bao, T., Wang, S., Wu, X., 2024. Molecular mechanisms of aging and anti-aging strategies. Cell Communication and Signaling 22 (1). Available at https://doi.org/10.1186/s12964-024-01663-1.

Li, X., Pang, X., Sun, H., Zhang, B., Wang, H., Wu, N., et al., 2025. Cardiac aging: molecular mechanisms and therapeutic interventions. Pharmacological Research 221, 107954. Available at https://doi.org/10.1016/j.phrs.2025.107954.

Liang, R., Abudurexiti, N., Wu, J., Ling, J., Peng, Z., Yuan, H., et al., 2024. Exosomes and miRNAs in cardiovascular diseases and transcatheter pulmonary valve replacement: advancements, gaps and perspectives. International Journal of Molecular Sciences 25 (24), 13686. Available at https://doi.org/10.3390/ijms252413686.

Lin, J., Epel, E., 2022. Stress and telomere shortening: insights from cellular mechanisms. Ageing Research Reviews 73, 101507. Available at https://doi.org/10.1016/j.arr.2021.101507.

Lin, Y., Fu, S., Yao, Y., Li, Y., Zhao, Y., Luo, L., 2021. Heart failure with preserved ejection fraction based on aging and comorbidities. Journal of Translational Medicine 19 (1). Available at https://doi.org/10.1186/s12967-021-02935-x.

Liu, R., Xu, C., Zhang, W., Cao, Y., Ye, J., Li, B., et al., 2022. FUNDC1-mediated mitophagy and HIF1α activation drives pulmonary hypertension during hypoxia. Cell Death & Disease 13 (7). Available at https://doi.org/10.1038/s41419-022-05091-2.

Liu, C., Zhang, D., Long, K., Qi, W., Pang, L., Li, J., et al., 2024. From exosomes to mitochondria and myocardial infarction: molecular insight and therapeutic challenge. Pharmacological Research 209, 107468. Available at https://doi.org/10.1016/j.phrs.2024.107468.

Loo, T.M., Miyata, K., Tanaka, Y., Takahashi, A., 2020. Cellular senescence and senescence-associated secretory phenotype via the cGAS-STING signaling pathway in cancer. Cancer Science 111 (2), 304–311. Available at https://doi.org/10.1111/cas.14266.

Ludlow, A.T., Roth, S.M., 2011. Physical activity and telomere biology: exploring the link with aging-related disease prevention. Journal of Aging Research 2011. Available at https://doi.org/10.4061/2011/790378.

Ma, C., Luo, H., Fan, L., Liu, X., Gao, C., 2020. Heart failure with preserved ejection fraction: an update on pathophysiology, diagnosis, treatment, and prognosis. Brazilian Journal of Medical and Biological Research 53 (7). Available at https://doi.org/10.1590/1414-431x20209646.

Masenga, S.K., Kabwe, L.S., Chakulya, M., Kirabo, A., 2023. Mechanisms of oxidative stress in metabolic syndrome. International Journal of Molecular Sciences 24 (9). Available at https://doi.org/10.3390/ijms24097898.

Mengstie, M.A., Chekol Abebe, E., Behaile Teklemariam, A., Tilahun Mulu, A., Agidew, M.M., Teshome Azezew, M., et al., 2022. Endogenous advanced glycation end products in the pathogenesis of chronic diabetic complications. Frontiers in Molecular Biosciences 9. Available at https://doi.org/10.3389/fmolb.2022.1002710.

Minamino, T., Miyauchi, H., Yoshida, T., Ishida, Y., Yoshida, H., Komuro, I., 2002. Endothelial cell senescence in human atherosclerosis: role of telomere in endothelial dysfunction. Circulation 105 (13), 1541–1544. Available at https://doi.org/10.1161/01.CIR.0000013836.85741.17.

Mir, S.M., Tehrani, S.S., Goodarzi, G., Jamalpoor, Z., Asadi, J., Khelghati, N., et al., 2020. Shelterin complex at telomeres: implications in ageing. Clinical Interventions in Aging 15, 827–839. Available at https://doi.org/10.2147/CIA.S256425.

Mohammadi, A., Jones, D.T., Mohammadi, S., Heidari-Bateni, G., Frishman, W.H., Aronow, W.S., 2025. Telomere dynamics in cardiovascular aging: from molecular mechanisms to precision medicine. Cardiology in Review. Available at https://doi.org/10.1097/crd.0000000000001018.

Morgan, R.G., Donato, A.J., Walker, A.E., 2018. Telomere uncapping and vascular aging. American Journal of Physiology - Heart and Circulatory Physiology 315 (1), H1. Available at https://doi.org/10.1152/ajpheart.00008.2018.

Moslehi, J., Depinho, R.A., Sahin, E., 2012. Telomeres and mitochondria in the aging heart. Circulation Research 110 (9), 1226–1237. Available at https://doi.org/10.1161/CIRCRESAHA.111.246868.

Mouliou, D.S., 2023. C-reactive protein: pathophysiology, diagnosis, false test results and a novel diagnostic algorithm for clinicians. Diseases 11 (4). Available at https://doi.org/10.3390/diseases11040132.

Münzel, T., Sørensen, M., Gori, T., Schmidt, F.P., Rao, X., Brook, F.R., et al., 2017. Environmental stressors and cardio-metabolic disease: Part II-mechanistic insights. European Heart Journal 38 (8), 557–564. Available at https://doi.org/10.1093/eurheartj/ehw294.

Ndisang, J.F., Vannacci, A., Rastogi, S., 2014. Oxidative stress and inflammation in obesity, diabetes, hypertension, and related cardiometabolic complications. Oxidative Medicine and Cellular Longevity 2014. Available at https://doi.org/10.1155/2014/506948.

Ng, P.Q., Ling, L.S.C., Chellian, J., Madheswaran, T., Panneersel-Vam, J., Kunnath, A.P., et al., 2020. Applications of nanocarriers as drug delivery vehicles for active phytoconstituents. Current Pharmaceutical Design 26 (36), 4580–4590. Available at https://doi.org/10.2174/1381612826666200610111013.

Ngoi, N.Y.L., Liew, A.Q.X., Chong, S.J.F., Davids, M.S., Clement, M.-V., Pervaiz, S., 2021. The redox-senescence axis and its therapeutic targeting. Redox Biology 45, 102032. Available at https://doi.org/10.1016/j.redox.2021.102032.

Noureddine, H., Gary-Bobo, G., Alifano, M., Marcos, E., Saker, M., Vienney, N., et al., 2011. Pulmonary artery smooth muscle cell senescence is a pathogenic mechanism for pulmonary hypertension in chronic lung disease. Circulation Research 109 (5), 543–553. Available at https://doi.org/10.1161/circresaha.111.241299.

Park, T.S., Hu, Y., Noh, H.L., Drosatos, K., Okajima, K., Buchanan, J., et al., 2008. Ceramide is a cardiotoxin in lipotoxic cardiomyopathy. Journal of Lipid Research 49 (10), 2101–2112. Available at https://doi.org/10.1194/jlr.M800147-JLR200.

Pelliccia, A., Sharma, S., Gati, S., Bäck, M., Börjesson, M., Caselli, S., et al., 2021. 2020 ESC guidelines on sports cardiology and exercise in patients with cardiovascular disease. European Heart Journal 42 (1), 17–96. Available at https://doi.org/10.1093/eurheartj/ehaa605.

Pervanidou, P., Chrousos, G.P., 2012. Metabolic consequences of stress during childhood and adolescence. Metabolism: Clinical and Experimental 61 (5), 611–619. Available at https://doi.org/10.1016/j.metabol.2011.10.005.

Polom, J., Boccardi, V., 2025. Employing nutrition to delay aging: a plant-based telomere-friendly dietary revolution. Nutrients 17 (12). Available at https://doi.org/10.3390/nu17122004.

Qiao, J., Wang, Q., Zhao, Y., Chang, M., Sun, S., Zhang, P., et al., 2025. Contribution of leukocyte telomere length to cardiovascular disease onset from genome-wide cross-trait analysis. Nature Communications 16 (1). Available at https://doi.org/10.1038/s41467-025-63707-5.

Rao, W., Li, D., Zhang, Q., Liu, T., Gu, Z., Huang, L., et al., 2025. Complex regulation of cardiac fibrosis: insights from immune cells and signaling pathways. Journal of Translational Medicine 23 (1). Available at https://doi.org/10.1186/s12967-025-06260-5.

Revy, P., Kannengiesser, C., Bertuch, A.A., 2023. Genetics of human telomere biology disorders. Nature Reviews Genetics 24 (2), 86–108. Available at https://doi.org/10.1038/s41576-022-00527-z.

Robin, J.D., Ludlow, A.T., Batten, K., Magdinier, F., Stadler, G., Wagner, K.R., et al., 2014. Telomere position effect: regulation of gene expression with progressive telomere shortening over long distances. Genes and Development 28 (22), 2464–2476. Available at https://doi.org/10.1101/gad.251041.114.

Roger, I., Milara, J., Belhadj, N., Cortijo, J., 2021. Senescence alterations in pulmonary hypertension. Cells 10 (12), 3456. Available at https://doi.org/10.3390/cells10123456.

Rossiello, F., 2014. Persistent dna damage at telomeres, caused by trf2-mediated dna repair inhibition, triggers cellular senescence and is associated with primates ageing.

Rossiello, F., Jurk, D., Passos, J.F., d'Adda di Fagagna, F., 2022. Telomere dysfunction in ageing and age-related diseases. Nature Cell Biology 24 (2), 135–147. Available at https://doi.org/10.1038/s41556-022-00842-x.

Sagris, M., Theofilis, P., Antonopoulos, A.S., Tsioufis, K., Tousoulis, D., 2022. Telomere length: a cardiovascular biomarker and a novel therapeutic target. International Journal of Molecular Sciences 23 (24), 16010. Available at https://doi.org/10.3390/ijms232416010.

Sakr, H.F., Sirasanagandla, S.R., Das, S., Bima, A.I., Elsamanoudy, A.Z., 2023. Insulin resistance and hypertension: mechanisms involved and modifying factors for effective glucose control. Biomedicines 11 (8). Available at https://doi.org/10.3390/biomedicines11082271.

Samuel, V.P., Gupta, G., Dahiya, R., Jain, D.A., Mishra, A., Dua, K., 2019. Current update on preclinical and clinical studies of resveratrol, a naturally occurring phenolic compound. Critical Reviews in Eukaryotic Gene Expression 29 (6), 529–537. Available at https://doi.org/10.1615/CritRevEukaryotGeneExpr.2019027836.

Sanchez, M., Kannengiesser, C., Hoang, S., Potier, L., Fumeron, F., Venteclef, N., et al., 2022. Leukocyte telomere length, allelic variations in related genes and risk of coronary heart disease in people with long-standing type 1 diabetes. Cardiovascular Diabetology 21 (1). Available at https://doi.org/10.1186/s12933-022-01635-0.

Sawasaki, K., Nakamura, M., Kimura, N., Kawahito, K., Yamazaki, M., Fujie, H., et al., 2024. Endothelial-derived nitric oxide impacts vascular smooth muscle cell phenotypes under high wall shear stress condition. Biochemical and Biophysical Research Communications 740, 151005. Available at https://doi.org/10.1016/j.bbrc.2024.151005.

Schellnegger, M., Hofmann, E., Carnieletto, M., Kamolz, L.-P., 2024. Unlocking longevity: the role of telomeres and its targeting interventions. Frontiers in Aging 5. Available at https://doi.org/10.3389/fragi.2024.1339317.

Schmitt, W., Rühs, H., Burghaus, R., Diedrich, C., Duwal, S., Eissing, T., et al., 2021. NT-proBNP qualifies as a surrogate for clinical end points in heart failure. Clinical Pharmacology & Therapeutics 110 (2), 498–507. Available at https://doi.org/10.1002/cpt.2222.

Semeraro, M.D., Smith, C., Kaiser, M., Levinger, I., Duque, G., Gruber, H.J., et al., 2020. Physical activity, a modulator of aging through effects on telomere biology. Impact Journals LLC, Austria Aging 12 (13), 13803–13823. Available at https://doi.org/10.18632/aging.103504.

Sena, C.M., Gonçalves, L., Seiça, R., 2022. Methods to evaluate vascular function: a crucial approach towards predictive, preventive, and personalised medicine. EPMA Journal 13 (2), 209–235. Available at https://doi.org/10.1007/s13167-022-00280-7.

Shaddy, R., Gong, J., Garito, T., Solar-Yohay, S., Zhang, S., Prescott, M.F., et al., 2025. Association between NT-proBNP changes and clinical outcomes in paediatric patients with heart failure: insights from PANORAMA-HF and PARADIGM-HF. ESC Heart Failure 12 (4), 3042–3052. Available at https://doi.org/10.1002/ehf2.15326.

Shkirkova, K., Lamorie-Foote, K., Connor, M., Patel, A., Barisano, G., Baertsch, H., et al., 2020. Effects of ambient particulate matter on vascular tissue: a review. Journal of Toxicology and Environmental Health, Part B 23 (7), 319–350. Available at https://doi.org/10.1080/10937404.2020.1822971.

Singh, N., Vishwas, S., Kaur, A., Kaur, H., Kakoty, V., Khursheed, R., et al., 2023. Harnessing role of sesamol and its nanoformulations against neurodegenerative diseases. Biomedicine & Pharmacotherapy 167, 115512. Available at https://doi.org/10.1016/j.biopha.2023.115512.

Suda, M., Paul, K.H., Minamino, T., Miller, J.D., Lerman, A., Ellison-Hughes, G.M., et al., 2023. Senescent cells: a therapeutic target in cardiovascular diseases. Cells 12 (9), 1296. Available at https://doi.org/10.3390/cells12091296.

Sun, Y., Wang, X., Liu, T., Zhu, X., Pan, X., 2022. The multifaceted role of the SASP in atherosclerosis: from mechanisms to therapeutic opportunities. Cell & Bioscience 12 (1). Available at https://doi.org/10.1186/s13578-022-00815-5.

Sun, Y., 2023. An updated landscape of cellular senescence heterogeneity: mechanisms, technologies and senotherapies. Translational Medicine of Aging 7, 46–51. Available at https://doi.org/10.1016/j.tma.2023.06.001.

Sutanto, H., Fetarayani, D., Narendra, M.R., Nasution, S.A., 2025. The role of the senescence-associated secretory phenotype in cardiovascular disease among the elderly. European Journal of Internal Medicine 141, 106488. Available at https://doi.org/10.1016/j.ejim.2025.106488.

Tan, X.W., Fong, A.Y.Y., Hwang, S.S., 2023. Telomere and telomerase biology in cardiovascular disease: a state-of-the-art review and outlook. Journal of Asian Pacific Society of Cardiology 2. Available at https://doi.org/10.15420/japsc.2023.26.

Tao, W., Yu, Z., Han, J.D.J., 2024. Single-cell senescence identification reveals senescence heterogeneity, trajectory, and modulators. Cell Metabolism 36 (5), 1126. Available at https://doi.org/10.1016/j.cmet.2024.03.009.

Thapa, R., Afzal, O., Alfawaz Altamimi, A.S., Goyal, A., Almalki, W.H., Alzarea, S.I., et al., 2023. Galangin as an inflammatory response modulator: an updated overview and therapeutic potential. Chemico-Biological Interactions 378, 110482. Available at https://doi.org/10.1016/j.cbi.2023.110482.

Tritakis, V., 2016. Association of arterial stiffness with coronary flow reserve in revascularized coronary artery disease patients. World Journal of Cardiology 8 (2), 231. Available at https://doi.org/10.4330/wjc.v8.i2.231.

Wadhwa, R., Paudel, K.R., Mehta, M., Shukla, S.D., Sunkara, K., Prasher, P., et al., 2020. Beyond the obvious: smoking and respiratory infection implications on Alzheimer's disease. CNS & Neurological Disorders - Drug Targets 19 (9), 698–708. Available at https://doi.org/10.2174/1871527319999200817112427.

Wand, T., Fang, M., Chen, C., Hardy, N., McCoy, J.P., Dumitriu, B., et al., 2016. Telomere content measurement in human hematopoietic cells: comparative analysis of qPCR and Flow-FISH techniques. Cytometry Part A 89 (10), 914–921. Available at https://doi.org/10.1002/cyto.a.22982.

Wang, Y., Sharpless, N., Chang, S., 2013. P16INK4a protects against dysfunctional telomere-induced ATR-dependent DNA damage responses. Journal of Clinical Investigation 123 (10), 4489–4501. Available at https://doi.org/10.1172/JCI69574.

Wang, H.H., Lee, D.K., Liu, M., Portincasa, P., Wang, D.Q.-H., 2020. Novel insights into the pathogenesis and management of the metabolic syndrome. Pediatric Gastroenterology, Hepatology & Nutrition 23 (3), 189. Available at https://doi.org/10.5223/pghn.2020.23.3.189.

Wei, X., Zhou, Y., Shao, E., Shi, X., Han, Y., Zhang, Y., et al., 2024. Tert promotes cardiac regenerative repair after MI through alleviating ROS-induced DNA damage response in cardiomyocyte. Cell Death Discovery 10 (1). Available at https://doi.org/10.1038/s41420-024-02135-8.

Westin, E.R., Aykin-Burns, N., Buckingham, E.M., Spitz, D.R., Goldman, F.D., Klingelhutz, A.J., 2011. The p53/p21WAF/CIP pathway mediates oxidative stress and senescence in dyskeratosis congenita cells with telomerase insufficiency. Antioxidants and Redox Signaling 14 (6), 985–997. Available at https://doi.org/10.1089/ars.2010.3444.

Xiang, X., Dong, C., Zhou, L., Liu, J., Rabinowitz, Z.M., Zhang, Y., et al., 2024. Novel PET imaging probe for quantitative detection of senescence in vivo. Journal of Medicinal Chemistry 67 (7), 5924–5934. Available at https://doi.org/10.1021/acs.jmedchem.4c00179.

Xin, H., Liu, D., Songyang, Z., 2008. The telosome/shelterin complex and its functions. Genome Biology 9 (9), 232. Available at https://doi.org/10.1186/gb-2008-9-9-232.

Xu, C., Qiu, Z., Guo, Q., Huang, Y., Zhao, Y., Zhao, R., 2025a. The role of cellular senescence in cardiovascular disease. Cell Death Discovery 11 (1). Available at https://doi.org/10.1038/s41420-025-02720-5.

Xu, X., Pang, Y., Fan, X., 2025b. Mitochondria in oxidative stress, inflammation and aging: from mechanisms to therapeutic advances. Signal Transduction and Targeted Therapy 10 (1). Available at https://doi.org/10.1038/s41392-025-02253-4.

Yang, C.H., Huang, M.L.H., Davis, W.A., Jenkins, A.J., Davis, T.M.E., 2025. Determinants of temporal change in telomere length and its associations with chronic complications and mortality in type 2 diabetes: the Fremantle diabetes study phase II. Cardiovascular Diabetology 24 (1). Available at https://doi.org/10.1186/s12933-025-02832-3.

Yeh, J.K., Wang, C.Y., 2016. Telomeres and telomerase in cardiovascular diseases. Genes 7 (9). Available at https://doi.org/10.3390/genes7090058.

Yeh, J.K., Lin, M.H., Wang, C.Y., 2019. Telomeres as therapeutic targets in heart disease. JACC: Basic to Translational Science 4 (7), 855–865. Available at https://doi.org/10.1016/j.jacbts.2019.05.009.

Zhang, N., Li, Y., Lai, T.-P., Shay, J.W., Danuser, G., 2021. Imaging assay to probe the role of telomere length shortening on telomere-gene interactions in single cells. Chromosoma 130 (1), 61–73. Available at https://doi.org/10.1007/s00412-020-00747-4.

Zhao, T., Qi, W., Yang, P., Yang, L., Shi, Y., Zhou, L., et al., 2021. Mechanisms of cardiovascular toxicity induced by PM2.5: a review. Environmental Science and Pollution Research 28 (46), 65033–65051. Available at https://doi.org/10.1007/s11356-021-16735-9.

Zhu, X., Chen, Z., Shen, W., Huang, G., Sedivy, J.M., Wang, H., et al., 2021. Inflammation, epigenetics, and metabolism converge to cell senescence and ageing: the regulation and intervention. Signal Transduction and Targeted Therapy 6 (1). Available at https://doi.org/10.1038/s41392-021-00646-9.

Zia, A., Farkhondeh, T., Pourbagher-Shahri, A.M., Samarghandian, S., 2022. The roles of mitochondrial dysfunction and reactive oxygen species in aging and senescence. Current Molecular Medicine 22 (1), 37–49. Available at https://doi.org/10.2174/1566524021666210218112616.

Chapter 8

Epigenetic modifications in cardiac aging and senescence

S. Afsar[1] and Rahamat Unissa Syed[2]

[1]*Department of Virology, Sri Venkateswara University, Tirupati, Andhra Pradesh, India,* [2]*Department of Pharmaceutics, College of Pharmacy, University of Ha'il, Hail, Saudi Arabia*

8.1 Introduction

Epigenetic modifications, which are heritable yet reversible chemical and structural changes to DNA and chromatin that modulate gene expression without altering the nucleotide sequence, provide a unifying framework for understanding how lifetime exposures, metabolic shifts, and stochastic processes reshape cardiac cell identity. This introduction defines key mechanisms, including DNA methylation, histone acetylation and methylation, chromatin remodeling, and noncoding RNAs, which together govern cardiomyocyte, fibroblast, and vascular cell phenotypes (Handy et al., 2011; Lossi et al., 2024). Cardiac aging is characterized by progressive structural, functional, and molecular deterioration of the heart, which occurs with chronological age and is accelerated by conventional cardiovascular risk factors. It manifests clinically as reduced reserve, increased interstitial fibrosis, impaired relaxation, electrophysiological instability, and a higher incidence of heart failure and ischemic events, thereby linking molecular alterations to the health consequences of the population. Cellular senescence, a state of durable cell-cycle arrest accompanied by a complex senescence-associated secretory phenotype (SASP), serves as both a hallmark and an active driver of cardiac aging by promoting paracrine inflammation, matrix remodeling, and functional decline when senescent cardiomyocytes, fibroblasts, or endothelial cells accumulate with age or after injury (Vakka et al., 2023; Triposkiadis et al., 2019). Defining these terms in sequence clarifies causality: epigenetic drift and targeted chromatin remodeling precede and contribute to the activation of senescence programs, which in turn amplify tissue-level ageing phenotypes and clinical diseases; this chronological framing guides subsequent mechanistic and translational sections. Clinically, the burden of cardiovascular disease underscores translational urgency: Cardiovascular diseases remain the leading cause of global mortality, accounting for tens of millions of deaths annually and representing approximately one-third of all global deaths in recent estimates (Kane and Sinclair, 2019; Sedivy et al., 2008). These data emphasize that molecular insights into aging and senescence have major implications for public health. In India, the public health impact is particularly acute, with cardiovascular disease constituting nearly one-third of adult deaths and driving substantial disability and premature mortality, a pattern that amplifies the importance of aging biology research tailored to regional demographics and health systems. The prevalence of heart failure and the projected rise in cardiovascular morbidity with population aging further link basic epigenetic discoveries to imminent clinical needs, as interventions that modulate chromatin marks, DNA methylation landscapes, or senescence-associated secretory phenotypes could plausibly delay functional decline and reduce the healthcare burden (Guo et al., 2022; Jan et al., 2024). Translational relevance therefore runs from molecular assays to population trials: robust, age-stratified epigenetic clocks and single-cell chromatin maps can identify causal nodes, while small molecules, epigenetic editing tools, and senolytic strategies offer therapeutic entry points whose efficacy must be evaluated in preclinical models that recapitulate human cardiac ageing and then in pragmatic clinical trials (Replogle and De Jager, 2015; Tarkhov et al., 2024). To maximize societal benefit, research priorities must include longitudinal, population-linked epigenomic studies in diverse Indian cohorts, harmonized biomarker pipelines, integration with national surveillance systems, and pragmatic trials that assess whether reversing maladaptive epigenetic states or clearing senescent cells improves functional end points; such an agenda aligns with global warnings about rising cardiovascular burden and age-related morbidity reported in contemporary world heart reports and burden-of-disease analyses. Therefore we emphasize rigorous reproducibility,

Cellular Senescence and Cardiac Aging. DOI: https://doi.org/10.1016/B978-0-443-45811-8.00009-X

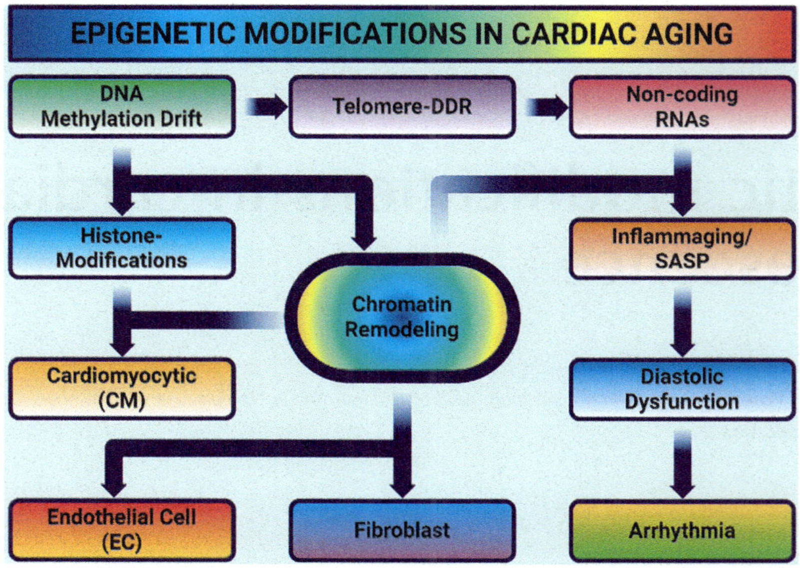

FIGURE 8.1 **Epigenetic drivers of cardiac aging.** This figure summarizes how major epigenetic mechanisms, including DNA methylation drift, histone modifications, telomere-related DNA damage signaling, and noncoding RNAs, reshape gene expression in cardiac cells during aging. These changes promote inflammation, fibrosis, endothelial dysfunction, impaired excitation–contraction coupling, diastolic dysfunction, and arrhythmia.

ethically responsible translation, and multidisciplinary collaboration spanning molecular biology, geriatrics, cardiology, and public health to ensure that the findings benefit patients across India and globally and improve population health outcomes (Rozek et al., 2014; Jin et al., 2015) (Fig. 8.1).

8.2 Molecular hallmarks of epigenetic aging in the heart

8.2.1 DNA methylation and cardiac epigenetic clocks

DNA methylation reshapes cardiac gene regulation during aging, and epigenetic clocks derived from CpG patterns estimate biological age relevant to myocardial function. Together, these measures integrate cumulative exposure to oxidative stress, inflammation, and metabolic load into quantitative readouts of the risk of remodeling. In cardiomyocytes, promoter and enhancer methylation drift alters calcium-handling, mitochondrial, and contractile programs, whereas in endothelial cells and fibroblasts, it biases angiogenesis and matrix turnover, respectively (Field et al., 2018; Reale et al., 2022). The global 5mC/5hmC balance shifts with DNMT/TET activity, NAD+-linked metabolism, and telomere–DDR signaling, reinforcing locus-specific methylation changes. Cell-free DNA methylomes and nucleosome footprints mirror tissue states in plasma, enabling minimally invasive tracking of cancer. Tissue-informed clocks adapted from multi-tissue and heart-enriched CpG panels correlate with diastolic stiffness, arrhythmic vulnerability, and recovery after pressure unloading (Reale et al., 2022; Song et al., 2025). Integration with single-cell ATAC/RNA maps anchors clock features to specific cell states and improves interpretability, while longitudinal calibration incorporates the effects of therapy from RAAS blockade, SGLT2 inhibition, exercise, and senescence-targeted regimens. Analytical rigor requires standardized preanalytics, batch correction, and external validation against imaging and proteomic end points. As a collective outcome, converging evidence supports cardiac methylation clocks as scalable biomarkers that refine prognosis and guide the timing of intervention. Future studies should harmonize heart-specific clocks, define actionable cutoffs, and embed methylation monitoring into stage-specific clinical decision pathways (Xu et al., 2021a; Ranzoni et al., 2021).

8.2.2 Histone modifications and chromatin states

Histone modifications and chromatin states encode the regulation of genes in the aging heart under metabolic and inflammatory pressure. Across cardiac cell types, shifting patterns of acetylation and methylation reshape enhancer potency, promoter repression, and the three-dimensional genome architecture. With age, H3K27ac and H3K4me1 decline in oxidative and calcium-handling programs, whereas super-enhancers erode. H3K27me3 and H3K9me3 expand over angiogenic and contractile loci to consolidate stress-adaptive and maladaptive phenotypes (Yi and Kim, 2020; Wang et al., 2013). Polycomb (PRC2) redistribution and SWI/SNF remodeling alter nucleosome positioning and loop insulation, thereby weakening promoter–enhancer coupling and narrowing transcriptional flexibility. Sirtuin activity decreases with NAD+ erosion, tipping the balance toward hyperacetylation at inflammatory modules, and BET-reader occupancy stabilizes NF-κB-centered networks that amplify SASP signaling. Metabolite availability (acetyl-CoA, SAM, α-ketoglutarate) couples the mitochondrial state to writer/eraser flux, while lamina and heterochromatin changes shift lamin-associated domains and accessible corridors (Sims et al., 2008; Yang and Li, 2020). These chromatin reconfigurations converge on impaired excitation–contraction efficiency, endothelial dysfunction, fibroblast activation, and diminished reserve under stress, and are traceable in circulating chromatin fragments and EV cargo that mirror tissue states. Therapeutically, exercise, metabolic optimization, and agents targeting HDACs, BET proteins, or PRC2, alongside mitophagy support, offer complementary leverage when deployed with cell-type precision and careful safety profiling. Collectively, these findings identify histone- and chromatin-level remodeling as a central, actionable hallmark of cardiac aging. Future studies should integrate longitudinal single-cell multi-omics with targeted epigenetic perturbations to validate biomarkers, personalize interventions, and secure durable benefits (Calandrelli et al., 2020; Rafehi et al., 2014).

8.2.3 Chromatin remodeling, nucleosome positioning, and 3D genome architecture

Chromatin remodeling, nucleosome positioning, and three-dimensional genome architecture govern how aging hearts reprogram gene expression. Across cardiomyocytes, endothelium, and fibroblasts, ATP-dependent remodelers, histone variants, and loop organizers coordinate access to the stress-response circuits. Early metabolic and mechanical stress reduces remodeler efficiency, alters nucleosome spacing at enhancers, and weakens CTCF/cohesin-anchored boundaries, diminishing promoter–enhancer communication. Polycomb redistribution and SWI/SNF retargeting rewire insulated neighborhoods, whereas lamin perturbations shift lamina-associated domains and expose previously repressed corridors (Piatti et al., 2011; Harris et al., 2025). Consequently, super-enhancers supporting oxidative metabolism erode, inflammatory and pro-fibrotic modules gain accessibility, and topologically associating domains become noisier, thereby increasing the cell-to-cell heterogeneity. Histone variant exchange of H2A.Z modifies nucleosome stability and fosters paused polymerase at stress hubs, and altered loop extrusion kinetics misalign phase-separated condensates that concentrate transcriptional machinery. Mitochondrial dysfunction further reshapes chromatin through metabolite fluxes of acetyl-CoA, SAM, and alpha-ketoglutarate, modulating writer and eraser activity, which reinforces new nucleosome landscapes (Dong et al., 2018). Single-cell and spatial assays have linked these architectural shifts to impaired excitation–contraction coupling, endothelial rarefaction, and fibroblast activation, while circulating chromatin fragments mirror nucleosome footprints and boundary erosion in the plasma. Therapeutically, exercise and metabolic optimization, along with agents targeting BET readers, HDACs, PRC2, or cohesin dynamics, may restore communication between enhancers and promoters if delivered with cell-type precision and safety monitoring. Collectively, converging studies have identified chromatin- and 3D genome-level remodeling as a central, actionable hallmark of cardiac aging. Future studies should integrate longitudinal multiomics, perturbation screens, and anchored liquid biopsy readouts to validate biomarkers, personalize interventions, and secure durable benefits (Zhuang et al., 2025; Liu et al., 2025).

8.3 Cell type-specific epigenetic remodeling

8.3.1 Cardiomyocytes: contractile dysfunction, hypertrophy, and epigenetic drift

Cardiomyocytes exhibit contractile dysfunction, hypertrophy, and epigenetic drift as they age. These interlinked changes progressively reshape excitation–contraction coupling and gene regulation throughout the lifespan. Early defects in calcium cycling and mitochondrial energetics reduce peak tension and prolong relaxation, whereas oxidative stress alters sarcomeric proteins and ion channels to amplify mechanical inefficiency. Persistent hemodynamic load and neurohormonal drive engage MAPK and calcineurin NFAT signaling, promoting concentric growth, fetal gene reexpression, and metabolic inflexibility (Howlett, 2010; Yang et al., 2018). Concurrently, epigenetic drift accumulates: DNA methylation is redistributed at promoters and enhancers; histone marks such as H3K27ac decline at oxidative genes, whereas repressive marks expand at contractile programs; chromatin architecture becomes less insulated; and noncoding RNAs shift toward pro-hypertrophic and pro-senescent profiles. Mitochondrial biogenesis wanes with reduced PGC-1α activity, sirtuin tone falls, and telomere–p53–p21 pathways reinforce a senescence-like phenotype that increases SASP phenotype mediators and paracrine crosstalk with fibroblasts and immune cells (King et al., 2016; Wang et al., 2022). Collectively, these layers weaken excitation–contraction efficiency, stiffen the ventricle through indirect pro-fibrotic signaling, and lower functional reserve under stress. Mechanistic insights nominate actionable levers: restoring Ca2+ handling, improving mitochondrial quality control, and rebalancing epigenetic enzymes (DNMT/TET, class I/II HDACs, and BET readers). Lifestyle and metabolic interventions (exercise, caloric moderation, and NAD+ repletion) can complement these agents. Collectively, converging evidence positions cardiomyocyte epigenetic remodeling as both a driver and readout of contractile aging. Future studies should integrate longitudinal multi-omics data with cell-specific perturbations to prioritize safe, durable, and stage-tailored epigenetic therapies (Luo et al., 2015; Ren and Bode, 2000).

8.3.2 Endothelium and vascular smooth muscle: vascular aging pathways

The endothelium and vascular smooth muscle cells (VSMC) jointly drive vascular aging pathways that reshape arterial tone, stiffness, and reactivity. Over time, endothelial nitric oxide signaling decreases, and VSMCs shift from a contractile to a synthetic, pro-inflammatory, and pro-calcific phenotype. Disturbed shear stress lowers KLF2/KLF4 and eNOS levels, elevates mitochondrial ROS levels, and reduces NO levels, thereby enhancing leukocyte adhesion, permeability, microvascular rarefaction, and endothelial dysfunction. Parallel telomere attrition and DNA damage activate the p53-p21 and p16-Rb checkpoints, fostering senescence and a persistent NF-κB-centered inflammatory program (Yildiz, 2007; Jaminon et al., 2019). Epigenetically, enhancer DNA methylation drifts, H3K27ac decreases at vasoprotective genes, H3K9me3 expands at angiogenic programs, and the TET/HDAC/SIRT axes remodel chromatin accessibility. Endothelial-to-mesenchymal transition is reinforced by TGF-β/SMAD and Wnt cues, while VSMCs undergo phenotypic switching and osteogenic reprogramming via RUNX2, BMP, and altered microRNAs, miR-34a and miR-21. cGAS-STING signaling cements transcriptional memory, extracellular vesicle traffic SASP factors, and matrix cross-linking increases pulse-wave velocity and systolic load (Wang et al., 2021a). These changes converge to impaired vasodilation, heightened tone, and diffuse calcification, which limit the coronary reserve and accelerate cardiac remodeling. Therapeutically, exercise, RAAS blockade, lipid lowering, NO-eNOS support, mitochondrial quality-control enhancers, Nrf2 activators, and epigenetic modulators show complementary potential. Collectively, evidence implicates epigenetically imprinted endothelial-VSMC crosstalk as the core engine of vascular aging. Future studies should combine longitudinal single-cell multi-omics with stage-specific senescence and epigenetic interventions to validate biomarkers and personalize therapy (Ma et al., 2010; Pacurari et al., 2014).

8.3.3 Cardiac fibroblasts and immune cells: fibrosis, inflammaging, and senescence-associated secretory phenotype

Cardiac fibroblasts and immune cells orchestrate fibrosis, inflammaging, and the SASP in the aging heart. Working together, they remodel the matrix and immunity, gradually reducing the diastolic reserve and reparative capacity. Mechanical load and mitochondrial stress activate fibroblasts via TGF-β/SMAD, YAP/TAZ, and angiotensin signaling, driving myofibroblast conversion, collagen I/III deposition, and lysyl-oxidase cross-linking, which stiffens the myocardium. Aged macrophages adopt pro-inflammatory states, dendritic cells resolve poorly, and T-cell repertoires contract; these shifts amplify IL-1, IL-6, and TNF waves that sustain fibroblast activation (Osorio et al., 2023; Lindner et al., 2014). Senescent stromal and immune cells release SASP factors and extracellular vesicles, spreading paracrine senescence and distorting endothelial–fibroblast–myocyte crosstalk. Epigenetically, DNA methylation drifts at the enhancers of matrix and cytokine genes, H3K27ac/H3K4me1 landscapes are rewired, heterochromatin marks expand at antifibrotic programs, and chromatin accessibility favors pro-fibrotic circuits. Inflammasome and cGAS-STING pathways interface with NF-κB-centered readers and writers to imprint chronic transcriptional memory, while noncoding RNAs bias fibroblast and macrophage phenotypes (Birch and Gil, 2020; Kim and Detmar, 2024). These layers yield persistent matrix accumulation, impaired clearance, microvascular rarefaction, and conduction heterogeneity, which together degrade performance. Therapeutically, attenuating danger sensing, rebalancing macrophage polarization, modulating TGF-β and angiotensin nodes, and pairing senescence-targeted approaches with epigenetic modulators represent converging avenues. Collectively, evidence identifies epigenetically imprinted crosstalk between fibroblasts and immune cells as the central driver of fibrotic aging. Future studies should combine single-cell and spatial multi-omics with stage-specific interventions to validate biomarkers, personalize therapy, and confirm its durability (Zhang and Ji, 2023) (Fig. 8.2).

8.4 Noncoding RNA networks in cardiac senescence

8.4.1 microRNAs and isomiRs in stress responses

MicroRNAs and their isomiRs orchestrate cardiac stress responses. By tuning translation and mRNA stability in an isoform- and context-specific manner, they integrate oxidative, mechanical, and inflammatory cues into a coordinated

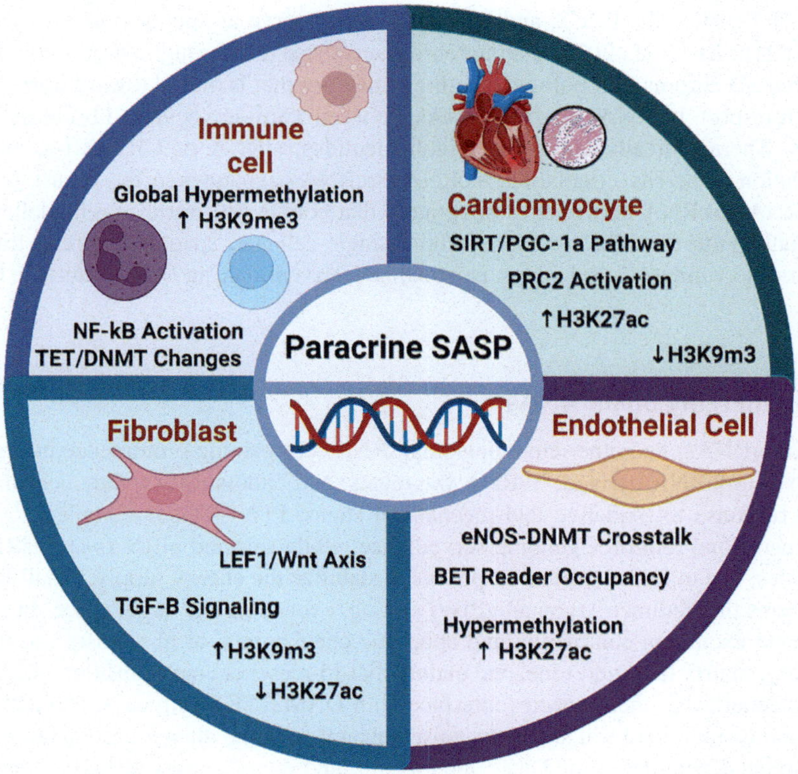

FIGURE 8.2 **Cell-specific epigenetic remodeling in the aging heart.** This quadrant diagram illustrates cell type-specific epigenetic changes in cardiomyocytes, endothelial cells, fibroblasts, and immune cells, coordinated through a central senescence-associated secretory phenotype (SASP)-mediated signaling hub. The figure highlights how these interacting pathways reinforce maladaptive remodeling, fibrosis, inflammation, and vascular dysfunction in the aging heart.

program. Stress reshapes biogenesis, altering Drosha/Dicer processing, ADAR A-to-I editing, TUT4/7 tailing, and exonuclease trimming, creating seed-shifted variants with targets and Argonaute sorting. In cardiomyocytes, miR-34a and miR-1/133 isomiRs regulate calcium handling and apoptosis, whereas in fibroblasts, miR-21 and miR-29 families calibrate TGF-β signaling and collagen turnover. Endothelial networks recruit miR-126 and stress-induced isomiRs to manage angiogenic tone, and VSMCs use miR-145/143 to guide contractile and synthetic switching (Cordeiro et al., 2025; Kura et al., 2020). Mitochondrial ROS and unfolded protein responses rewire isomiR repertoires, and extracellular vesicles shuttle cargo that propagates paracrine senescence and SASP polarization. Reciprocally, microRNAs target epigenetic regulators DNMTs, TETs, HDACs, SIRT1/6, and BET readers, while chromatin state and DNA methylation govern microRNA locus transcription, closing a feedback loop that imprints stress memory. Clinically, circulating isomiR signatures offer biomarkers, and antimiR or mimic strategies show promise when combined with cell-specific delivery and off-target safeguards. Collectively, evidence positions miRNA-isomiR plasticity as an integrator linking stress sensing to epigenetic changes and cardiac remodeling. Future studies should use single-cell small RNA profiling with CLIP-guided target maps to advance isomiR-aware therapeutics and personalized risk stratification (Lu et al., 2022; Weng et al., 2020).

8.4.2 lncRNAs and circRNAs in remodeling and repair

Long noncoding RNAs (lncRNAs) and circular RNAs (circRNAs) play important roles in cardiac remodeling and repair. As modular regulators, they integrate injury, metabolic stress, and aging signals into cell-type-specific programs. In cardiomyocytes, lncRNAs scaffold chromatin modifiers to balance hypertrophic growth and contractile recovery, whereas circRNAs adjust calcium handling and mitochondrial function. In fibroblasts, lncRNAs modulate TGF-β signaling and collagen synthesis, whereas circRNAs modulate myofibroblast conversion and matrix turnover. Endothelial cells deploy lncRNA-circRNA circuits to control angiogenic sprouting and endothelial-to-mesenchymal transition, shaping revascularization after an injury. Mechanistically, these RNAs act as guides, scaffolds, or decoys for epigenetic writers, erasers, and readers (DNMTs, TETs, HDACs, PRC2, and BET), and many serve as sponges for microRNAs and RNA-binding proteins (Wang et al., 2021b; Xie et al., 2025). Aging reprograms these loci through DNA methylation drift, histone-mark redistribution, and altered 3D chromatin topology, creating a memory that is biased toward fibrosis and incomplete repair. Extracellular vesicles transport lncRNAs and circRNAs between cardiomyocytes, fibroblasts, and endothelial cells, extending their signals. Therapeutically, antisense oligonucleotides, siRNA or CRISPRi-a, and nanoparticle delivery enable targeted modulation, whereas circulating lncRNA/circRNA signatures offer robust biomarkers. Collectively, evidence positions lncRNA-circRNA networks as integrators that couple cell-intrinsic remodeling to the epigenetic state and intercellular crosstalk. Future studies should combine single-cell and spatial long-read transcriptomics with cell-targeted RNA therapeutics to validate causal nodes, personalize interventions, and ensure durable benefits (An et al., 2025; Issa, 2014).

8.4.3 piRNAs and emerging Small RNAs

PIWI-interacting RNAs (piRNAs) and other emerging small RNAs are gaining prominence in the study of cardiac aging. Along with tRNA-derived, snoRNA-derived, Y-RNA fragments, and endo-siRNAs, they coordinate stress adaptation in multiple cell types. In response to oxidative and mechanical stress, PIWI–TDRD complexes guide the transcriptional restraint of mobile elements and reinforce genome surveillance, while selected piRNAs and tsRNAs tune mitochondrial translation, redox enzymes, and unfolded protein responses to stabilize the energy supply. Small RNA editing, tailing, and trimming generate isoforms that redirect Argonaute/PIWI sorting, expanding the target space. In cardiomyocytes, tsRNAs and endo-siRNAs regulate excitation contraction and apoptotic checkpoints; in fibroblasts and the endothelium, circuits calibrate TGF-β signaling, matrix turnover, tone, and endothelial-to-mesenchymal transition (Ahmadi Asouri et al., 2024; Ghosh et al., 2012). Epigenetically, these species interface with DNMT/TET pathways, class I/II HDACs, and PRC2 to modulate methylation and histone marks at stress-response enhancers, while altered chromatin states reshape small RNA locus expression. Age-related remodeling of Dicer, angiogenin, and exonucleases reshapes small RNA repertoires, and extracellular vesicles distribute cargo propagating paracrine senescence and SASP polarization. Clinically, circulating piRNA and tsRNA signatures show promise as robust blood biomarkers, and PIWI/tsRNA-targeted oligonucleotides or small molecules are plausible therapeutic interventions. Collectively, converging data place piRNA-centered and small RNA networks as modulators linking genome defense to remodeling. Future studies should deploy single-cell, long-read, and CLIP-guided maps to prioritize causal nodes and enable precise and durable therapy (Huang et al., 2022; Jiang et al., 2019).

8.5 Crosstalk with core aging pathways

8.5.1 Telomere biology, DNA damage response, and epigenetic drift

Telomere biology, DNA damage response, and epigenetic drift intersect to program cardiac aging. Progressive telomere attrition and shelterin erosion expose chromosome ends, priming persistent checkpoint activation in postmitotic and proliferative cardiac cells. ATM/ATR signaling, γH2AX foci, and p53-p21/p16-Rb axes enforce cell cycle arrest, whereas mitochondrial ROS and replication stress accelerate telomere fragility and inflammation. Declining TERT activity and altered POT1/TRF2 handling destabilize end protection, promoting cGAS-STING sensing of cytosolic DNA and NF-κB-driven SASP programs (Fumagalli et al., 2012; Rossiello et al., 2022). Concurrently, epigenetic drift redistributes DNA methylation at promoters and enhancers, reduces H3K9me3 heterochromatin, and perturbs PRC2-mediated H3K27me3, loosening lineage boundaries in cardiomyocytes, the endothelium, and fibroblasts. Sirtuin decline and PARP1 hyperactivation deplete NAD+, impairing repair and mitochondrial biogenesis, whereas TET/DNMT imbalance skews methylation clocks toward older age. These lesions rewire metabolic set points, attenuate proteostasis, and couple telomere–DDR signaling to chromatin readers/writers that memorialize stress, amplifying cellular senescence and paracrine fibrosis. Interventions that reduce oxidative stress, enhance mitophagy, or temper DDR signaling may preserve telomere integrity. Restoring SIRT1/6 tone, rebalancing TET/DNMT, or locus-selective epigenetic editing can stabilize youthful programs. Collectively, converging studies indicate that telomere erosion, maladaptive DDR, and epigenetic drift form a reinforcing circuit that degrades cardiac resilience and promotes remodeling. Future studies should integrate longitudinal single-cell multi-omics with senotherapeutics and epigenetic modulators to validate biomarkers, refine risk stratification, and deliver stage-appropriate, clinically durable cardioprotection (Fukuda et al., 2025; Huang et al., 2021).

8.5.2 Mitochondria–nucleus communication, sirtuins, and metabolic reprogramming

Mitochondrial dysfunction in aging cardiomyocytes initiates retrograde signaling to the nucleus through altered NAD+/NADH ratios, shifts in acetyl-CoA pools, and increased reactive oxygen species levels, which activate stress-responsive transcriptional programs. Sirtuins sense these metabolic changes and directly deacetylate histones and transcription factors to reconfigure the chromatin and transcriptional outcomes. As NAD+ availability declines with age, SIRT1 and SIRT3 activities decrease, leading to increased histone acetylation, altered chromatin accessibility, and dysregulated expression of genes governing mitochondrial biogenesis and quality control (Tocchi et al., 2015; Zhang et al., 2025). Concomitantly, metabolic reprogramming toward glycolysis and impaired fatty acid oxidation changes the levels of methyl donors and acetyl groups, which in turn influence DNA methylation and histone modification patterns, further reshaping nuclear transcription. Reduced mitonuclear communication diminishes PGC-1α-driven mitochondrial renewal and compromises mitophagy, amplifying reactive oxygen species production and sustaining maladaptive epigenetic conditions. This reinforcing loop between altered metabolism, attenuated sirtuin signaling, and chromatin erosion accelerates cardiomyocyte functional decline, fosters a pro-inflammatory secretory phenotype, and weakens contractile resilience. Interventions that restore NAD+ pools, activate sirtuins, or correct metabolic fluxes can reinstate youthful chromatin landscapes and mitochondrial function, offering mechanistic insights and actionable targets for therapeutic intervention. Such strategies hold promises for ameliorating age-related cardiac decline, enhancing resilience, and reducing the cardiovascular disease burden (Vakka et al., 2023; Yu et al., 2018).

8.5.3 Inflammatory pathways, senescence-associated secretory phenotype effectors, and epigenetic modulation

Inflammatory pathways, SASP effectors, and epigenetic modulation jointly shape the adaptation and failure of the aging heart adapts and fails. Under sterile stress, danger cues converge on NF-κB, JAK/STAT, and inflammasome circuits to set the inflammatory tone. cGAS-STING activation by cytosolic DNA sustains type I interferon and primes NLRP3, while angiotensin II and disturbed shear amplify endothelial stromal crosstalk. Senescent cardiomyocytes, fibroblasts, and immune cells release SASP mediators, including IL-6, IL-1β, TNF, TGF-β, chemokines, MMPs, and procoagulant vesicles, which propagate paracrine damage, matrix remodeling, and microvascular dysfunction (Khavinson et al., 2022; Li et al., 2023). These signals feed back onto the epigenome: DNMT/TET imbalance reshapes CpG methylation at enhancers; histone acetylation declines at vasoprotective genes as class I/II HDAC activity increases; repressive H3K9me3 spreads, H3K27me3 distribution shifts, and BET-reader occupancy entrench NF-κB programs. Sirtuins and NAD+ erosion weaken chromatin guardianship and mitochondrial quality control, whereas noncoding RNAs modulate macrophage polarization, fibroblast activation, and endothelial-to-mesenchymal transition. Together, inflammatory drivers and SASP effectors create a self-reinforcing memory that couples impaired vasodilation and diastolic stiffness with fibrotic deposition, electrical heterogeneity, and reduced reserve under stress (Roy et al., 2022). Therapeutically, multipronged strategies, such as exercise and risk factor control, RAAS and inflammasome blockade, Nrf2 activation, mitochondrial and mitophagy support, senolytics/senomorphics, and epigenetic modulators, such as HDAC or BET inhibitors, offer complementary leverage. Collectively, evidence indicates that targeting the inflammatory SASP epigenome triad can blunt cardiac remodeling and extend the healthspan. Future studies should combine longitudinal multi-omics with stage-specific combinations to validate biomarkers, personalize treatment, and secure durable benefits (Qin et al., 2022).

8.6 Environmental and lifestyle modulators of cardiac epigenetics

8.6.1 Diet, exercise, and epigenetic plasticity

Diet, exercise, and related lifestyle factors exert a powerful influence on cardiac epigenetic plasticity by modulating DNA methylation, histone modifications, and noncoding RNA networks to support healthy myocardial aging. Beginning with diet, nutritional inputs, especially bioactive compounds and methyl-donor micronutrients, can alter methylation patterns in cardiac cells, thereby regulating the expression of genes involved in metabolism, oxidative stress response, and mitochondrial function. Concurrently, regular physical activity triggers adaptive epigenetic remodeling; for example, endurance or resistance training can induce promoter hypomethylation of genes encoding antioxidant and mitochondrial regulators, increase histone acetylation at loci of cardioprotective genes, and modulate microRNA expression profiles that guard against pathological remodeling and inflammation (Ostaiza-Cardenas et al., 2025; Abraham et al., 2023). Together, diet and exercise synergistically sustain epigenetic flexibility, helping cardiomyocytes and supporting cells counteract senescence-associated epigenetic drift and maintain transcriptional homeostasis. Over time, this interplay may slow epigenetic aging in cardiac tissue, reduce maladaptive gene silencing, and preserve the functional resilience of the heart in older age. Thus lifestyle interventions act as dynamic modulators of the cardiac epigenome, offering a nonpharmacological route to influence the trajectory of cardiac aging (Wu et al., 2021).

8.6.2 Exposome factors: hypoxia, toxins, and psychosocial stress

Early-life and adult exposures within the exposome, such as intermittent or chronic hypoxia, environmental toxins, and sustained psychosocial stress, sequentially reshape the cardiac epigenetic landscape through convergent molecular mechanisms. Initially, hypoxic episodes activate hypoxia-inducible factors that recruit chromatin modifiers, producing locus-specific DNA methylation changes and histone marks that adjust the metabolic and angiogenic gene programs. Subsequent or overlapping exposure to airborne, chemical, or heavy metal toxins further perturbs methylation profiles and histone acetylation patterns while promoting oxidative DNA damage that amplifies epigenetic drift. Concurrently, prolonged psychosocial stress engages neuroendocrine signaling cascades that alter microRNA networks and long noncoding RNAs, fostering sustained inflammatory transcriptional states that predispose the myocardium to maladaptive remodeling (Baccarelli and Ghosh, 2012; de Prado-Bert et al., 2021). Over time, these ordered exposures accumulate, reinforcing senescence-associated secretory phenotypes, impairing mitochondrial function, and increasing vulnerability to fibrosis and contractile dysfunction. Importantly, some epigenetic modifications are dynamic and responsive to mitigation, reducing toxin burden, improving oxygenation, and stress-reduction interventions can partially restore epigenetic homeostasis and downstream gene expression. Therefore recognizing the sequential influence of hypoxia, toxins, and psychosocial stress on cardiac epigenetics highlights both the mechanistic links to aging phenotypes and opportunities for targeted environmental and behavioral interventions to slow cardiac senescence. Future research integrating longitudinal epigenomic profiling with exposome measurements will effectively refine intervention timing, targets, and policies (Xiao et al., 2025; Machado-Oliveira et al., 2020).

8.6.3 Sex differences and comorbidities (diabetes, hypertension)

Sex differences and metabolic comorbidities, such as diabetes and hypertension, exert profound influences on cardiac epigenetic remodeling, shaping the trajectory of cardiac aging and senescence through hormone-dependent and metabolic signaling pathways. In females, estrogen exerts cardioprotective effects by modulating DNA methylation and histone acetylation of genes that regulate oxidative stress, inflammation, and mitochondrial biogenesis. However, postmenopausal decline in estrogen levels leads to increased epigenetic silencing of cardioprotective genes and heightened vulnerability to fibrosis. In males, androgen-driven transcriptional programs interact with differential methylation patterns that may enhance susceptibility to cardiac hypertrophy and vascular dysfunction (McClements et al., 2025; Chella Krishnan et al., 2018). Diabetes induces hyperglycemia-induced oxidative stress, leading to aberrant DNA methylation of metabolic and inflammatory genes, whereas histone deacetylase activation disrupts normal chromatin accessibility. Similarly, hypertension induces sustained mechanical and oxidative stress, which alters histone methylation and noncoding RNA expression, reinforcing the maladaptive gene networks linked to hypertrophy and fibrosis. The intersection of sex-specific hormone signaling with these comorbid epigenetic perturbations accelerates cardiac aging, amplifies senescence-associated secretory phenotypes, and diminishes the regenerative potential. Importantly, emerging evidence suggests that lifestyle modification, glycemic control, and antihypertensive therapies can partially reverse maladaptive epigenetic marks, highlighting the potential for precision interventions tailored by sex and comorbidity profiles to mitigate cardiac aging (Binjawhar et al., 2023; Khullar et al., 2017).

8.7 Biomarkers and characterization technologies

8.7.1 Single-cell and spatial epigenomics (scATAC-seq, ChIP-seq, and integrative multiomics)

Single-cell and spatial epigenomics, including scATAC-seq, cell-type resolved ChIP-seq, and integrative multiomics, map chromatin regulation across cardiac cell populations. By linking open chromatin, histone marks, and transcripts to spatial niches, these platforms reveal how aging rewires regulatory circuits. scATAC-seq identifies senescence-skewed enhancer usage, diminished cardioprotective motifs (KLF2/KLF4/MEF2), and increased occupancy of inflammatory and pro-fibrotic drivers (AP-1, NF-κB, TEAD), whereas cell-state atlases connect accessibility to inflexibility and impaired calcium handling (Zhang et al., 2022; Wu et al., 2024). ChIP-seq resolves the redistribution of H3K27ac and H3K27me3 across endothelial, fibroblast, and myocyte programs, exposing super-enhancer erosion at oxidative genes and the gain of repressive marks near contractile and angiogenic loci. Spatial transcriptomics combined with chromatin profiling localizes SASP hubs at border zones, reveals endothelial-to-mesenchymal transition corridors, and links macrophage polarization to fibroblast activation gradients. Multi-omic integration with ATAC, RNA, proteome, and methylome layers reconstructs gene regulatory networks, infers transcription factor activity, and nominates ligand receptor circuits that coordinate fibrosis, microvascular rarefaction, and heterogeneity (Jiang et al., 2025; Young et al., 2011). Circulating biomarker discovery leverages cell-of-origin signatures, cell-free chromatin fragments, and EV-borne epigenetic cargo to index tissue remodeling, while machine-learning classifiers improve risk stratification. Analytical rigor hinges on standardized nuclei isolation, batch correction, peak-to-gene linkage, and validation with perturbation screens or CRISPR interference. Collectively, converging studies position single-cell and spatial epigenomics as foundational for biomarker development and target discovery in cardiac aging. Future studies should deploy longitudinal multicenter cohorts with harmonized pipelines and interventional readouts to verify causal nodes and accelerate clinically actionable stage-specific therapies (Rahat et al., 2020; De Michino et al., 2023).

8.7.2 Circulating epigenetic biomarkers and liquid biopsy platforms

Circulating epigenetic biomarkers and liquid biopsy platforms offer minimally invasive insights into cardiac aging. By sampling plasma- and serum-derived analytes, these tools capture tissue remodeling signals that evolve with fibrosis, inflammation, and senescence. Cell-free DNA (cfDNA) methylation patterns and 5-hydroxymethylcytosine profiles infer tissue-of-origin and pathway activity, whereas nucleosome footprints across promoters and enhancers reflect cardiomyocyte and stromal transcriptional states (Gai and Sun, 2019). Complementary cargo in extracellular vesicles microRNAs, lncRNAs, circRNAs, and chromatin-associated protein indexes SASP intensity and endothelial–fibroblast crosstalk. Mitochondrial cfDNA levels and fragmentation signatures track oxidative stress and impaired mitophagy, and age-associated methylation “clocks” adapted to cardiovascular tissues estimate biological age beyond chronological age. Technologically, targeted bisulfite assays, cfMeDIP-seq, and low-coverage whole-genome methylation profiling delineate methylomes at scale; small RNA sequencing and digital PCR quantify circulating RNAs; and immuno-capture of EVs enriches cell-type specificity. Analytical rigor depends on preanalytical control (fast processing, hemolysis checks, spike-in standards), batch correction, and cross-platform validation with matched tissue multiomics and imaging end points (Charidemou et al., 2025; Xu et al., 2021b). Clinically, composite panels that integrate cfDNA methylation, EV-borne ncRNAs, and standard biomarkers improve risk stratification for heart failure progression, microvascular dysfunction, and response to antifibrotic or senescence-targeted therapies. As a collective outcome, converging evidence supports circulating epigenetic signatures as scalable indicators of cardiac biological age and remodeling burden in HF. Future studies should focus on developing standardized pipelines, longitudinal cohorts, and interventional trials that coregister liquid biopsy readouts with single-cell and spatial epigenomics to verify causality, refine cutoffs, and deliver actionable, stage-specific monitoring (Bozack et al., 2024; Desiderio et al., 2024).

8.7.3 Epigenetic risk stratification and prognostic scoring systems

Epigenetic risk stratification and prognostic scoring systems translate chromatin biology into clinical predictions. By quantifying DNA methylation clocks, accessibility states, and noncoding RNA signatures, these tools estimate biological age and imminent remodeling risk beyond conventional metrics. Panels integrating cfDNA methylation, EV-borne microRNAs/lncRNAs, and histone-mark surrogates are mapped to outcomes such as heart failure progression, diastolic dysfunction, and arrhythmic events, whereas tissue-informed priors from single-cell and spatial atlases anchor model features to specific cardiac cell states (Liang et al., 2024; Benčik et al., 2025). Feature selection combines pathway-aware reduction with machine learning, followed by calibration, internal and external validation, and assessment of discrimination, reclassification, and decision-curve utility against standard clinical scores. Multimodal models fuse epigenetic indices with imaging markers (strain, ECV), proteomics (NT-proBNP, galectin-3), and genetics to improve individualized risk horizons and therapy prioritization. Longitudinal updating accommodates treatment effects of RAAS blockade, SGLT2 inhibition, exercise, or senescence-targeted regimens, while guardrails address batch effects, preanalytical variance, and demographic bias (Ji et al., 2024). Implementation requires harmonized pipelines, reference standards, and transparent reporting, with cutoffs linked to actionable thresholds that trigger surveillance intensity or the initiation of anti-fibrotic and metabolic therapies. As a collective outcome, converging evidence supports epigenetically anchored scores as scalable complements to clinical assessment that refine prognosis and guide the timing of intervention. Future studies should couple multicenter longitudinal cohorts with interventional readouts to finalize robust cutoffs, enable adaptive risk monitoring, and embed epigenetic scoring within decision pathways for stage-specific durable cardioprotection (Chybowska et al., 2024; Namli et al., 2024) (Fig. 8.3).

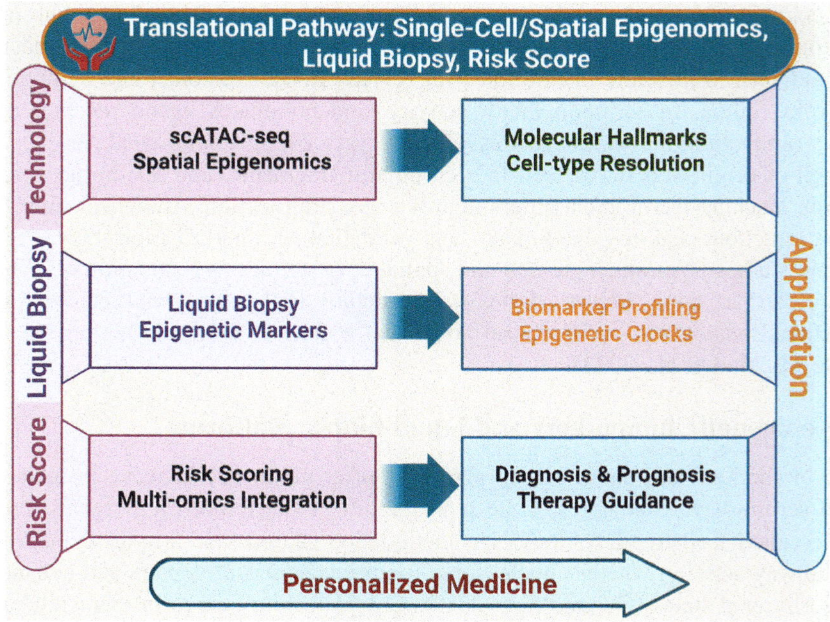

FIGURE 8.3 **Epigenomic biomarker pipeline for precision cardiology.** This figure presents a discovery-to-clinic workflow in which single-cell and spatial epigenomic technologies identify cardiac aging signatures that are translated into circulating biomarkers and epigenetic clocks. These data are integrated into multi-omics risk models that support diagnosis, prognosis, and personalized therapeutic decision-making.

8.8 Therapeutic targeting of epigenetic mechanisms

8.8.1 Small-molecule modulators (DNMTi, HDACi, BET inhibitors)

In recent years, small-molecule epigenetic modulators, such as DNMT, HDAC, and BET inhibitors, have emerged as promising strategies for reversing the aberrant epigenetic states associated with cardiac aging and senescence. DNMT inhibitors (e.g., azacitidine and decitabine) act by reducing DNA methylation at hypermethylated promoters, thereby reactivating silenced cardioprotective or stress-response genes. Following DNA demethylation, HDAC inhibitors intervene at the histone level: by inhibiting histone deacetylases, they increase histone acetylation, relax chromatin, and promote the transcription of genes that counteract senescence or enhance repair (Zhang et al., 2021; Li et al., 2024). Concurrently, bromodomain and extra-terminal motif (BET) inhibitors block the binding of BET-reader proteins to acetylated histones, thereby disrupting pro-senescence transcriptional programs. Together, these modulators produce coordinated reprogramming: DNMTi resets abnormal DNA methylation, HDACi promotes a more permissive chromatin state, and BET inhibitors prevent pathogenic gene activation despite acetylation. In preclinical studies, the combination of these small molecules has been shown to ameliorate the features of cardiac aging, reduce fibrotic remodeling, preserve mitochondrial function, and suppress the senescence-associated secretory phenotype. Thus within the therapeutic targeting of epigenetic mechanisms, small-molecule modulators represent a multilayered and interlinked approach to intervene in the epigenetic drivers of cardiac aging (Alqahtani et al., 2019; Morgado-Pascual et al., 2019).

8.8.2 RNA- and gene-editing approaches (miRNA/ASO, CRISPR/dCas9 epigenome editing)

In targeting cardiac aging and senescence, RNA-based modulators, such as miRNAs and antisense oligonucleotides (ASOs), offer early precision by restoring or suppressing specific noncoding transcripts that regulate epigenetic enzymes and pathways, thereby rebalancing the chromatin states and gene expression. Subsequently, CRISPR/dCas9-based epigenome editing tools have been developed, offering locus-specific reprogramming of DNA methylation or histone modifications without altering the underlying sequence, thus enforcing durable and selective gene regulation. First, miRNAs known to decline in aged cardiomyocytes can be supplemented or mimicked to inhibit pro-senescence factors, whereas ASOs can knock down maladaptive noncoding RNAs that drive pathological gene silencing (Lee et al., 2025; Yang et al., 2022). Guided by these shifts, dCas9 is fused to epigenetic effectors (e.g., DNMTs, TETs, HATs, and

TABLE 8.1 Epigenetic processes in cardiac aging: mediators, cellular and tissue consequences, and translational applications.

Process	Molecular mediators	Cellular effects	Tissue outcomes	Translational application	References
DNA methylation and epigenetic clocks	DNMT/TET balance; 5mC/5hmC; cfDNA methylome	Reprogrammed oxidative, calcium-handling, and mitochondrial genes	Biological-age shift; diastolic stiffness; arrhythmic vulnerability	Liquid-biopsy clocks; prognosis and timing of intervention	(Li and Ma, 2025; Horvath and Raj, 2018)
Histone modifications and chromatin states	H3K27ac/H3K4me1 loss; H3K27me3/H3K9me3 gain; PRC2; BET; sirtuins	Reduced enhancer potency; repressed contractile/angiogenic loci	Contractile inefficiency; endothelial dysfunction; fibroblast activation	HDAC/BET/PRC2 modulation; exercise-metabolic optimization	(Igolkina et al., 2019)
Chromatin remodeling and 3D genome	SWI/SNF, ISWI, CHD, INO80; CTCF/cohesin; H2A.Z; lamina	Altered nucleosome spacing; weakened boundaries; super-enhancer erosion	Increased heterogeneity; pro-inflammatory/pro-fibrotic programs	Topology-stabilizing strategies; mitophagy support	(de Bruijn, 2022)
Cardiomyocyte remodeling	PGC-1α/SIRT; PRC2; telomere–DDR (p53–p21)	Impaired Ca2+ cycling; hypertrophy; senescence-like state	Reduced reserve; load-induced stiffness	Mitochondrial QC; Ca2+ handling restoration; epigenetic enzyme rebalance	(Xiong et al., 2015)
Endothelium/VSMC vascular aging	KLF2/KLF4/eNOS; RUNX2/BMP; TET/HDAC/SIRT	Endothelial dysfunction; VSMC synthetic/osteogenic switching	Microvascular rarefaction; calcification; high tone	RAAS blockade; Nrf2 activators; NO-eNOS support; epigenetic modulators	(Sun et al., 2012)
Fibroblast/Immune SASP	TGF-β/SMAD; YAP/TAZ; NF-κB; cGAS–STING	Myofibroblast conversion; SASP amplification	Interstitial fibrosis; conduction heterogeneity	Danger-sensing attenuation; macrophage repolarization; epigenetic therapy	(Piersma et al., 2015)
microRNAs and isomiRs	miR-34a, miR-1/133, miR-21/29; editing/tailing	Target rewiring; Ca2+ handling; collagen turnover	Stress-memory imprinting; paracrine spread via EVs	Circulating signatures; antimiR/mimic therapeutics	(Nersisyan et al., 2022)
lncRNAs and circRNAs	PRC2/BET/DNMT/TET interaction; miRNA sponging	Chromatin scaffolding; mitochondrial/angiogenic tuning	Bias toward fibrosis; incomplete repair	ASO/siRNA/CRISPRi-a; long-read profiling	(Jiang et al., 2022; Liu et al., 2021)
piRNAs and small RNAs	PIWI–TDRD complexes; tsRNAs; endo-siRNAs	Genome surveillance; mitochondrial translation control	SASP polarization; ECM remodeling	Blood biomarkers; PIWI/tsRNA-targeted oligos	(Hanusek et al., 2022)
Telomere biology and DDR	ATM/ATR; γH2AX; cGAS–STING	Senescence enforcement; NAD+ depletion	Paracrine fibrosis; resilience loss	DDR tempering; SIRT1/6 support; locus-selective editing	(Ebata et al., 2022)

(Continued)

TABLE 8.1 (Continued)

Process	Molecular mediators	Cellular effects	Tissue outcomes	Translational application	References
Mitochondria–nucleus signaling	NAD+/sirtuins; acetyl-CoA/SAM flux	Histone deacetylation loss; chromatin erosion	Metabolic inflexibility; pro-inflammatory state	NAD+ repletion; sirtuin activation; flux correction	(Pirinen et al., 2012)
Inflammatory pathways and SASP	NF-κB; JAK/STAT; NLRP3	Persistent inflammatory transcriptional memory	Diastolic stiffness; fibrotic deposition	RAAS/inflammasome blockade; HDAC/BET inhibitors	(Wang et al., 2020; Chen et al., 2023)
Circulating epigenetic biomarkers	cfDNA methylation; 5hmC; EV-borne ncRNAs	Tissue-of-origin inference; pathway readout	Biological-age and remodeling index	Standardized pipelines; clinical monitoring	(Cao et al., 2020)
Risk stratification models	Epigenetic indices + imaging/proteomics/genetics	Calibrated discrimination and reclassification	Stage-tailored prognosis	Decision-support integration; adaptive updating	(Pashayan et al., 2016)
Therapeutic epigenetic targeting	DNMTi/HDACi/BETi; senolytics/senomorphics	Coordinated reprogramming of pathogenic networks	Fibrosis attenuation; function preservation	Rational combinations; safety-profiled delivery	(Bruyer et al., 2018)

SASP, senescence-associated secretory phenotype.

HDACs) that can be directed to specific promoters or enhancers, enabling controlled activation or repression of target genes implicated in mitochondrial function, fibrosis, or cell cycle arrest. In preclinical cardiac models, such editing has corrected aberrant methylation patterns, reactivated protective transcriptional programs, attenuated the senescence-associated secretory phenotype, and improved the functional metrics. Thus the combination of RNA modulation with CRISPR-based epigenome editing offers a hierarchically integrated strategy for reconfiguring epigenetic landscapes in aged hearts in a precise, sustained, and adaptive manner (Kovalev et al., 2025; Garitano et al., 2025).

8.8.3 Lifestyle and pharmacologic rejuvenation strategies (e.g., sirtuin/NAD+ pathways)

Lifestyle interventions, such as caloric restriction, exercise, and intermittent fasting, hold promise for boosting NAD^+ levels and activating sirtuins, which in turn modulate epigenetic regulators to reverse age-related chromatin drift and improve cardiac resilience. In the first instance, boosting NAD^+ via precursor supplementation (e.g., nicotinamide riboside and nicotinamide mononucleotide) replenishes the substrate pool required by sirtuin deacetylases, thereby enhancing their capacity to deacetylate histones and nonhistone proteins and tighten aberrantly loosened chromatin in aging hearts. As sirtuins regain activity, they improve mitochondrial biogenesis and antioxidant defenses, thereby reducing reactive oxygen species and DNA damage, which otherwise drive maladaptive epigenetic alterations (An et al., 2025; Kane and Sinclair, 2018). Parallel to this, pharmacologic sirtuin activators (e.g., resveratrol, SRT-1720) further potentiate the deacetylation of key histone marks, reinforcing the repression of pro-inflammatory and pro-fibrotic gene programs. Simultaneously, increased NAD^+ and sirtuin activity can engage in crosstalk with other epigenetic modifiers (e.g., influencing the activity of PARP or influencing histone methylation) to stabilize chromatin homeostasis. In preclinical cardiac aging models, these strategies have been shown to restore youthful gene expression patterns, reduce senescence marker levels, preserve contractile function, and attenuate fibrosis. Thus lifestyle modulation and NAD^+/sirtuin-targeted therapeutics synergistically induce epigenetic rejuvenation in the aging heart (Dai et al., 2018; Zhao et al., 2021) (Table 8.1).

8.9 Conclusion and future perspectives

Cardiac aging is best understood as an epigenetically orchestrated system process in which DNA methylation drift, histone code reprogramming, chromatin remodeling, and noncoding RNA circuits progressively rewire cardiomyocytes, the endothelium, fibroblasts, and immune cells, coherently linking molecular lesions to evolving tissue phenotypes. The novelty of this work is the integration of single-cell and spatial epigenomics with liquid biopsy readouts and risk scoring to deliver a stage-specific, cell-type–resolved framework for prognosis and intervention, while the scope spans molecular hallmarks, biomarker development, and therapeutic levers ranging from lifestyle and metabolic optimization to senescence-directed and epigenetic modulators. Evidence connects telomere DNA damage signaling to accessibility shifts that entrench inflammatory and pro-fibrotic programs, relates metabolite flux (acetyl-CoA, SAM, NAD+) to the balance of writers/erasers and 3D genome topology, and demonstrates how SASP-driven crosstalk accelerates matrix stiffening, microvascular rarefaction, and electrical heterogeneity. Converging datasets nominate actionable axes rebalancing TET/DNMT activity, tempering PRC2 and BET occupancy, sustaining sirtuin tone and mitophagy, and pairing RAAS or inflammasome blockade with targeted epigenetic agents, provided delivery is cell-specific and safety is rigorously profiled. Translational progress hinges on harmonized preanalytics, transparent pipelines, and external validation that coregister single-cell multiomics with imaging, proteomics, and other outcomes. Adaptive risk models must incorporate treatment effects and demographic differences. Priorities include longitudinal, multicenter cohorts that track circulating epigenetic signatures alongside tissue states, perturbation screens to confirm causal nodes, refinement of actionable cutoffs, and trials of rational combinations aligned with the disease stage, mechanism, and exposure. Collectively, an epigenetically anchored precision cardiology strategy appears feasible and is poised to extend the healthspan by slowing remodeling, restoring reserve, and guiding the timing and intensity of therapy.

References

Abraham, M.J., El Sherbini, A., El-Diasty, M., Askari, S., Szewczuk, M.R., 2023. Restoring epigenetic reprogramming with diet and exercise to improve health-related metabolic diseases. Biomolecules 13 (2).

Ahmadi Asouri, S., Aghadavood, E., Mirzaei, H., Abaspour, A., Esmaeil Shahaboddin, M., 2024. PIWI-interacting RNAs (PiRNAs) as emerging biomarkers and therapeutic targets in biliary tract cancers: a comprehensive review. Heliyon 10 (13), e33767.

Alqahtani, A., et al., 2019. Bromodomain and extra-terminal motif inhibitors: a review of preclinical and clinical advances in cancer therapy. Future Sci OA 5 (3), Fso372.

An, Y., et al., 2025. Epigenetic regulation of aging and its rejuvenation. MedComm (2020) 6 (9), e70369.

Baccarelli, A., Ghosh, S., 2012. Environmental exposures, epigenetics and cardiovascular disease. Current Opinion in Clinical Nutrition and Metabolic Care 15 (4), 323–329.

Benčik, I., Saftić Martinović, L., Mladenić, T., Ostojić, S., Dević Pavlić, S., 2025. From DNA methylation and histone modifications to non-coding RNAs: evaluating tools for epigenetic research. Applied Sciences 15 (18), 9940.

Binjawhar, D.N., Alhazmi, A.T., Bin Jawhar, W.N., MohammedSaeed, W., Safi, S.Z., 2023. Hyperglycemia-induced oxidative stress and epigenetic regulation of ET-1 gene in endothelial cells. Frontiers in Genetics 14, 1167773.

Birch, J., Gil, J., 2020. Senescence and the SASP: many therapeutic avenues. Genes & Development 34 (no. 23-24), 1565–1576.

Bozack, A.K., Navas-Acien, A., Cardenas, A., 2024. DNA methylation-based biomarkers of protein levels and cardiovascular disease risk: opportunities and challenges for precision cardiology. Circulation: Genomic and Precision Medicine 17 (2), e004571.

Bruyer, A., et al., 2018. DNMTi/HDACi combined epigenetic targeted treatment induces reprogramming of myeloma cells in the direction of normal plasma cells. British Journal of Cancer 118 (8), 1062–1073.

Calandrelli, R., et al., 2020. Stress-induced RNA-chromatin interactions promote endothelial dysfunction. Nature Communications 11 (1), 5211.

Cao, F., et al., 2020. Integrated epigenetic biomarkers in circulating cell-free DNA as a robust classifier for pancreatic cancer. Clinical Epigenetics 12 (1), 112.

Charidemou, E., Felekkis, K., Papaneophytou, C., 2025. From natriuretic peptides to microRNAs: multi-analyte liquid biopsy horizons in heart failure. Biomolecules 15 (8).

Chella Krishnan, K., Mehrabian, M., Lusis, A.J., 2018. Sex differences in metabolism and cardiometabolic disorders. Current Opinion in Lipidology 29 (5), 404–410.

Chen, Y., et al., 2023. The NLRP3 inflammasome: contributions to inflammation-related diseases. Cellular & Molecular Biology Letters 28 (1), 51.

Chybowska, A.D., et al., 2024. Epigenetic contributions to clinical risk prediction of cardiovascular disease. Circulation: Genomic and Precision Medicine 17 (1), e004265.

Cordeiro, M.A., de Carvalho, A., Spadari, R.C., 2025. MicroRNAs' impact on heart diseases. International Journal of Molecular Sciences 26 (12).

Dai, H., Sinclair, D.A., Ellis, J.L., Steegborn, C., 2018. Sirtuin activators and inhibitors: Promises, achievements, and challenges. Pharmacology & Therapeutics 188, 140–154.

de Bruijn, S., 2022. Global organization starts locally: chromatin remodelers regulate genome architecture. The Plant Cell 34 (7), 2576–2577.

Desiderio, A., et al., 2024. DNA methylation in cardiovascular disease and heart failure: novel prediction models? Clinical Epigenetics 16 (1), 115.

De Michino, S.D., et al., 2023. Simulating cell-free chromatin using preclinical models for cancer-specific biomarker discovery. bioRxiv 2023 11, 16.567416.

de Prado-Bert, P., et al., 2021. The early-life exposome and epigenetic age acceleration in children. Environment International 155, 106683.

Dong, Q., Wang, Y., Qi, S., Gai, K., He, Q., Wang, Y., 2018. Histone variant H2A.Z antagonizes the positive effect of the transcriptional activator CPC1 to regulate catalase-3 expression under normal and oxidative stress conditions. Free Radical Biology & Medicine 121, 136–148.

Ebata, H., Loo, T.M., Takahashi, A., 2022. Telomere maintenance and the cGAS-STING pathway in cancer. Cells 11 (12).

Field, A.E., Robertson, N.A., Wang, T., Havas, A., Ideker, T., Adams, P.D., 2018. DNA methylation clocks in aging: categories, causes, and consequences. Molecular Cell 71 (6), 882–895.

Fukuda, K., Shimura, C., Shinkai, Y., 2025. H3K27me3 and the PRC1-H2AK119ub pathway cooperatively maintain heterochromatin and transcriptional silencing after the loss of H3K9 methylation. Epigenetics Chromatin 18 (1), 26.

Fumagalli, M., et al., 2012. Telomeric DNA damage is irreparable and causes persistent DNA-damage-response activation. Nature Cell Biology 14 (4), 355–365.

Gai, W., Sun, K., 2019. Epigenetic biomarkers in cell-free DNA and applications in liquid biopsy. Genes (Basel) 10 (1).

Garitano, N., Aguado-Alvaro, L.P., Pelacho, B., 2025. Emerging epigenetic therapies for the treatment of cardiac fibrosis. Biomedicines 13 (5).

Ghosh, A.K., Nagpal, V., Covington, J.W., Michaels, M.A., Vaughan, D.E., 2012. Molecular basis of cardiac endothelial-to-mesenchymal transition (EndMT): differential expression of microRNAs during EndMT. Cellular Signalling 24 (5), 1031–1036.

Guo, J., et al., 2022. Aging and aging-related diseases: from molecular mechanisms to interventions and treatments. Signal Transduction and Targeted Therapy 7 (1), 391.

Handy, D.E., Castro, R., Loscalzo, J., 2011. Epigenetic modifications: basic mechanisms and role in cardiovascular disease. Circulation 123 (19), 2145–2156.

Hanusek, K., Poletajew, S., Kryst, P., Piekiełko-Witkowska, A., Bogusławska, J., 2022. piRNAs and PIWI proteins as diagnostic and prognostic markers of genitourinary cancers. Biomolecules 12 (2).

Harris, S., et al., 2025. Nucleosome repositioning in cardiac reprogramming. PLoS One 20 (1), e0317718.

Horvath, S., Raj, K., 2018. DNA methylation-based biomarkers and the epigenetic clock theory of ageing. Nature Reviews. Genetics 19 (6), 371–384.

Howlett, S.E., 2010. Age-associated changes in excitation-contraction coupling are more prominent in ventricular myocytes from male rats than in myocytes from female rats. American Journal of Physiology - Heart and Circulatory Physiology 298 (2), H659–H670.

Huang, M., Wu, Q., Jiang, Z.H., 2022. Epigenetic alterations under oxidative stress in stem cells. Oxidative Medicine and Cellular Longevity 2022, 6439097.

Huang, T.C., et al., 2021. Sex-specific chromatin remodelling safeguards transcription in germ cells. Nature 600 (7890), 737–742.

Igolkina, A.A., et al., 2019. H3K4me3, H3K9ac, H3K27ac, H3K27me3 and H3K9me3 histone tags suggest distinct regulatory evolution of open and condensed chromatin landmarks. Cells 8 (9).

Issa, J.P., 2014. Aging and epigenetic drift: a vicious cycle. The Journal of Clinical Investigation 124 (1), 24–29.

Jaminon, A., Reesink, K., Kroon, A., Schurgers, L., 2019. The role of vascular smooth muscle cells in arterial remodeling: focus on calcification-related processes. International Journal of Molecular Sciences 20 (22).

Jan, B., Dar, M.I., Choudhary, B., Basist, P., Khan, R., Alhalmi, A., 2024. Cardiovascular diseases among indian older adults: a comprehensive review. Cardiovascular Therapeutics 2024, 6894693.

Jiang, S., Postovit, L., Cattaneo, A., Binder, E.B., Aitchison, K.J., 2019. Epigenetic modifications in stress response genes associated with childhood trauma. Frontiers in Psychiatry 10, 808.

Jiang, M., Wang, Y., Wang, J., Feng, S., Wang, X., 2022. The etiological roles of miRNAs, lncRNAs, and circRNAs in neuropathic pain: a narrative review. Journal of Clinical Laboratory Analysis 36 (8), e24592.

Jiang, N., Wen, Z., Tao, H., Liao, H., 2025. Improved ChIP sequencing for H3K27ac profiling and super-enhancer analysis assisted by fluorescence-activated sorting of formalin-fixed paraffin-embedded tissues. Biological Procedures Online 27 (1), 1.

Jin, K., Simpkins, J.W., Ji, X., Leis, M., Stambler, I., 2015. The critical need to promote research of aging and aging-related diseases to improve health and longevity of the elderly population. Aging Disease 6 (1), 1–5.

Ji, Q., et al., 2024. Multimodal omics approaches to aging and age-related diseases. Phenomics 4 (1), 56–71.

Kane, A.E., Sinclair, D.A., 2019. Epigenetic changes during aging and their reprogramming potential. Critical Reviews in Biochemistry and Molecular Biology 54 (1), 61–83.

Kane, A.E., Sinclair, D.A., 2018. Sirtuins and NAD(+) in the development and treatment of metabolic and cardiovascular diseases. Circulation Research 123 (7), 868–885.

Khavinson, V., Linkova, N., Dyatlova, A., Kantemirova, R., Kozlov, K., 2022. Senescence-associated secretory phenotype of cardiovascular system cells and inflammaging: perspectives of peptide regulation. Cells 12 (1).

Khullar, M., Cheema, B.S., Raut, S.K., 2017. Emerging evidence of epigenetic modifications in vascular complication of diabetes. Frontiers in Endocrinology (Lausanne) 8, 237.

Kim, J., Detmar, M., Nov 2024. Stromal cells and epigenetics: emerging key players of chronic inflammatory skin diseases. BMB Reports 57 (11), 465–471.

King, A.D., et al., 2016. Reversible regulation of promoter and enhancer histone landscape by DNA methylation in mouse embryonic stem cells. Cell Reports 17 (1), 289–302.

Kovalev, M.A., et al., 2025. Epigenome engineering using dcas systems for biomedical applications and biotechnology: current achievements, opportunities and challenges. International Journal of Molecular Sciences 26 (13).

Kura, B., Szeiffova Bacova, B., Kalocayova, B., Sykora, M., Slezak, J., 2020. Oxidative stress-responsive microRNAs in heart injury. International Journal of Molecular Sciences 21 (1).

Lee, H.G., Kim, J., Jang, C.-Y., Shin, M., 2025. Antisense oligonucleotide therapeutics targeting age-related diseases. BioDrugs 40 (2), 237–261.

Liang, R., Tang, Q., Chen, J., Zhu, L., 2024. Epigenetic clocks: beyond biological age, using the past to predict the present and future. Aging Disease 16 (6), 3520–3545.

Lindner, D., et al., 2014. Cardiac fibroblasts support cardiac inflammation in heart failure. Basic Research in Cardiology 109 (5), 428.

Liu, Y., Dong, G., Yu, J., Liang, P., 2025. Integration of single-cell and spatial transcriptomics reveals fibroblast subtypes in hepatocellular carcinoma: spatial distribution, differentiation trajectories, and therapeutic potential. Journal of Translational Medicine 23 (1), 198.

Liu, C.G., et al., 2021. Long non-coding RNAs and circular RNAs in tumor angiogenesis: From mechanisms to clinical significance. Molecular Therapy Oncolytics 22, 336–354.

Li, T., Chen, Y., Li, S., 2024. The advances in the development of epigenetic modifications therapeutic drugs delivery systems. International Journal of Nanomedicine 19, 10623–10637.

Li, X., Li, C., Zhang, W., Wang, Y., Qian, P., Huang, H., 2023. Inflammation and aging: signaling pathways and intervention therapies. Signal Transduction and Targeted Therapy 8 (1), 239.

Li, Y., Ma, J., 2025. DNA methylation-based epigenetic clocks and cardiovascular disease: opportunities and challenges,. JACC CardioOncology 7 (6), 705–707.

Lossi, L., Castagna, C., Merighi, A., 2024. An overview of the epigenetic modifications in the brain under normal and pathological conditions. International Journal of Molecular Sciences 25 (7).

Luo, T., Kim, J.K., Chen, B., Abdel-Latif, A., Kitakaze, M., Yan, L., 2015. Attenuation of ER stress prevents post-infarction-induced cardiac rupture and remodeling by modulating both cardiac apoptosis and fibrosis. Chemico-Biological Interactions 225, 90–98.

Lu, H., et al., 2022. Mitochondrial unfolded protein response and integrated stress response as promising therapeutic targets for mitochondrial diseases. Cells 12 (1).

Machado-Oliveira, G., Ramos, C., Marques, A.R.A., Vieira, O.V., 2020. Cell senescence, multiple organelle dysfunction and atherosclerosis. Cells 9 (10).

Ma, T.K., Kam, K.K., Yan, B.P., Lam, Y.Y., 2010. Renin-angiotensin-aldosterone system blockade for cardiovascular diseases: current status. British Journal of Pharmacology 160 (6), 1273–1292.

McClements, L., Kautzky-Willer, A., Kararigas, G., Ahmed, S.B., Stallone, J.N., 2025. The role of sex differences in cardiovascular, metabolic, and immune functions in health and disease: a review for "Sex Differences in Health Awareness Day. Biology of Sex Differences 16 (1), 33.

Morgado-Pascual, J.L., Rayego-Mateos, S., Tejedor, L., Suarez-Alvarez, B., Ruiz-Ortega, M., 2019. Bromodomain and extraterminal proteins as novel epigenetic targets for renal diseases. Frontiers in Pharmacology 10, 1315.

Namli, T., et al., 2024. A scalable and transparent data pipeline for AI-enabled health data ecosystems. Frontiers in Medicine (Lausanne) 11, 1393123.

Nersisyan, S., Zhiyanov, A., Engibaryan, N., Maltseva, D., Tonevitsky, A., 2022. A novel approach for a joint analysis of isomiR and mRNA expression data reveals features of isomiR targeting in breast cancer. Frontiers in Genetics 13, 1070528.

Osorio, J.M., et al., 2023. Senescent cardiac fibroblasts: a key role in cardiac fibrosis. Biochimica Et Biophysica Acta-Molecular Basis of Disease 1869 (4), 166642.

Ostaiza-Cardenas, J., et al., 2025. Epigenetic modulation by life-style: advances in diet, exercise, and mindfulness for disease prevention and health optimization. Frontiers in Nutrition 12, 1632999.

Pacurari, M., Kafoury, R., Tchounwou, P.B., Ndebele, K., 2014. The renin-angiotensin-aldosterone system in vascular inflammation and remodeling. nternational Journal of Inflammation 2014, 689360.

Pashayan, N., Reisel, D., Widschwendter, M., 2016. Integration of genetic and epigenetic markers for risk stratification: opportunities and challenges. Per Med 13 (2), 93–95.

Piatti, P., Zeilner, A., Lusser, A., 2011. ATP-dependent chromatin remodeling factors and their roles in affecting nucleosome fiber composition. International Journal of Molecular Sciences 12 (10), 6544–6565.

Piersma, B., Bank, R.A., Boersema, M., 2015. Signaling in fibrosis: TGF-β, WNT, and YAP/TAZ converge. Frontiers in Medicine (Lausanne) 2, 59.

Pirinen, E., Lo Sasso, G., Auwerx, J., 2012. Mitochondrial sirtuins and metabolic homeostasis. Best Practice & Research. Clinical Endocrinology & Metabolism 26 (6), 759–770.

Qin, S., et al., 2022. Phytochemical activators of Nrf2: a review of therapeutic strategies in diabetes. Acta Biochimica et Biophysica Sinica (Shanghai) 55 (1), 11–22.

Rafehi, H., et al., 2014. Vascular histone deacetylation by pharmacological HDAC inhibition. Genome Research 24 (8), 1271–1284.

Rahat, B., Ali, T., Sapehia, D., Mahajan, A., Kaur, J., 2020. Circulating cell-free nucleic acids as epigenetic biomarkers in precision medicine. Frontiers in Genetics 11, 844.

Ranzoni, A.M., et al., 2021. Integrative single-cell RNA-Seq and ATAC-Seq analysis of human developmental hematopoiesis. Cell Stem Cell 28 (3), 472–487.e7.

Reale, A., Tagliatesta, S., Zardo, G., Zampieri, M., 2022. Counteracting aged DNA methylation states to combat ageing and age-related diseases. Mechanisms of Ageing and Development 206, 111695.

Ren, J., Bode, A.M., 2000. Altered cardiac excitation-contraction coupling in ventricular myocytes from spontaneously diabetic BB rats. American Journal of Physiology-Heart and Circulatory Physiology 279 (1), H238–H244.

Replogle, J.M., De Jager, P.L., 2015. Epigenomics in translational research. Translational Research: the Journal of Laboratory and Clinical Medicine 165 (1), 7–11.

Rossiello, F., Jurk, D., Passos, J.F., d'Adda di Fagagna, F., 2022. Telomere dysfunction in ageing and age-related diseases. Nature Cell Biology 24 (2), 135–147.

Roy, A., Padhi, S.S., Khyriem, I., Nikose, S., Sankar, S.H.H., Bharathavikru, R.S., 2022. Resetting the epigenome: methylation dynamics in cancer stem cells. Frontiers in Cell and Developmental Biology 10, 909424.

Rozek, L.S., Dolinoy, D.C., Sartor, M.A., Omenn, G.S., 2014. Epigenetics: relevance and implications for public health. Annual Review of Public Health 35, 105–122.

Sedivy, J.M., Banumathy, G., Adams, P.D., 2008. Aging by epigenetics–a consequence of chromatin damage? Experimental Cell Research 314 (9), 1909–1917.

Sims, H.I., Baughman, C.B., Schnitzler, G.R., 2008. Human SWI/SNF directs sequence-specific chromatin changes on promoter polynucleosomes. Nucleic Acids Research 36 (19), 6118–6131.

Song, D., Zhang, Z., Zheng, J., Zhang, W., Cai, J., 2025. 5-Hydroxymethylcytosine modifications in circulating cell-free DNA: frontiers of cancer detection, monitoring, and prognostic evaluation. Biomarker Research 13 (1), 39.

Sun, Y., et al., 2012. Smooth muscle cell-specific runx2 deficiency inhibits vascular calcification. Circulation Research 111 (5), 543–552.

Tarkhov, A.E., et al., Jun 2024. Nature of epigenetic aging from a single-cell perspective. Nature Aging 4 (6), 854–870.

Tocchi, A., Quarles, E.K., Basisty, N., Gitari, L., Rabinovitch, P.S., 2015. Mitochondrial dysfunction in cardiac aging. Biochimica et Biophysica Acta 1847 (11), 1424–1433.

Triposkiadis, F., Xanthopoulos, A., Butler, J., 2019. Cardiovascular aging and heart failure: JACC review topic of the week. Journal of the American College of Cardiology 74 (6), 804–813.

Vakka, A., Warren, J.S., Drosatos, K., 2023. Cardiovascular aging: from cellular and molecular changes to therapeutic interventions. Journal of Cardiovascular Aging 3 (3).

Wang, T.Y., Chang, M.M., Li, Y.J., Huang, T.C., Chien, S., Wu, C.C., 2021a. Maintenance of HDACs and H3K9me3 prevents arterial flow-induced venous endothelial damage. Frontiers in Cell and Developmental Biology 9, 642150.

Wang, Y., Chen, J., Cowan, D.B., Wang, D.Z., 2021b. Non-coding RNAs in cardiac regeneration: mechanism of action and therapeutic potential. Seminars in Cell & Developmental Biology 118, 150–162.

Wang, M., Yu, Q., Wang, L., Gu, H., 2013. Distinct patterns of histone modifications at cardiac-specific gene promoters between cardiac stem cells and mesenchymal stem cells. American Journal of Physiology-Cell Physiology 304 (11), C1080–C1090.

Wang, K., et al., 2022. Epigenetic regulation of aging: implications for interventions of aging and diseases. Signal Transduction and Targeted Therapy 7 (1), 374.

Wang, Z., et al., 2020. NLRP3 inflammasome and inflammatory diseases. Oxidative Medicine and Cellular Longevity 2020, 4063562.

Weng, H., et al., 2020. A new vision of mitochondrial unfolded protein response to the sirtuin family. Current Neuropharmacology 18 (7), 613–623.

Wu, G., Zhang, X., Gao, F., 2021. The epigenetic landscape of exercise in cardiac health and disease. Journal of Sport and Health Science 10 (6), 648–659.

Wu, X., et al., 2024. Single-cell sequencing to multi-omics: technologies and applications. Biomarker Research 12 (1), 110.

Xiao, J., et al., 2025. Immunoinflammation and post-translational modifications in the aging process. Journal of Translational Medicine 23 (1), 910.

Xie, X., et al., 2025. The role of long non-coding RNAs in cardiovascular diseases: a comprehensive review. Non-Coding RNA Research 11, 158–187.

Xiong, S., Patrushev, N., Forouzandeh, F., Hilenski, L., Alexander, R.W., 2015. PGC-1α modulates telomere function and DNA damage in protecting against aging-related chronic diseases. Cell Reports 12 (9), 1391–1399.

Xu, H., et al., 2021a. Integrative single-cell RNA-Seq and ATAC-Seq analysis of peripheral mononuclear cells in patients with ankylosing spondylitis. Frontiers in Immunology 12, 760381.

Xu, H., Ni, Y.Q., Liu, Y.S., 2021b. Mechanisms of action of MiRNAs and LncRNAs in extracellular vesicle in atherosclerosis. Frontiers in Cardiovascular Medicine 8, 733985.

Yang, Y., Li, G., 2020. Post-translational modifications of PRC2: signals directing its activity. Epigenetics Chromatin 13 (1), 47.

Yang, R., et al., 2018. Mitochondrial-mediated oxidative Ca(2+)/calmodulin-dependent kinase II activation induces early afterdepolarizations in guinea pig cardiomyocytes: an in silico study. Journal of the American Heart Association 7 (15), e008939.

Yang, X., et al., 2022. RNA modifications in aging-associated cardiovascular diseases. Aging (Albany NY) 14 (19), 8110–8136.

Yildiz, O., 2007. Vascular smooth muscle and endothelial functions in aging. Annals of the New York Academy of Sciences 1100, 353–360.

Yi, S.J., Kim, K., 2020. New insights into the role of histone changes in aging. International Journal of Molecular Sciences 21 (21).

Young, M.D., et al., 2011. ChIP-seq analysis reveals distinct H3K27me3 profiles that correlate with transcriptional activity. Nucleic Acids Research 39 (17), 7415–7427.

Yu, X., Ma, R., Wu, Y., Zhai, Y., Li, S., 2018. Reciprocal regulation of metabolic reprogramming and epigenetic modifications in cancer. Frontiers in Genetics 9, 394.

Zhang, S., Gong, Y., Li, C., Yang, W., Li, L., 2021. Beyond regulations at DNA levels: a review of epigenetic therapeutics targeting cancer stem cells. Cell Proliferation 54 (2), e12963.

Zhang, Y., Ji, Q., 2023. Macrophage polarization in osteoarthritis progression: a promising therapeutic target. Frontiers in Cell and Developmental Biology 11, 1269724.

Zhang, X., et al., 2025. Mitochondrial dysfunction in the regulation of aging and aging-related diseases. Cell Communication and Signaling: CCS 23 (1), 290.

Zhang, X., Qiu, H., Zhang, F., Ding, S., 2022. Advances in single-cell multi-omics and application in cardiovascular research. Frontiers in Cell and Developmental Biology 10, 883861.

Zhao, T., et al., 2021. Mechanisms of cardiovascular toxicity induced by PM2. 5: a review. Environmental Science and Pollution Research 28 (46), 65033–65051.

Zhuang, W., et al., 2025. Integrated single-cell and spatial transcriptomic analysis reveals a pathological niche formed by FAP+ fibroblasts, immune, and endothelial cells in psoriatic lesions. Clinical, Cosmetic and Investigational Dermatology 18, 2323–2340.

Chapter 9

Senescence-induced cardiac fibrosis and remodeling

Chintan Aundhia[1], Muhammad Afzal[2], Riya Mishra[3], Surya Nath Pandey[4], Md. Raihan Siddique[5], Santosh Kumar Singh[6], Neelam Singla[5], Gaurav Gupta[7,8], and Rajiv Dahiya[9]

[1]*Department of Pharmacy, Sumandeep Vidyapeeth Deemed to be University, Vadodara, Gujarat, India,* [2]*Department of Pharmaceutical Sciences, Pharmacy Program, Batterjee Medical College, Jeddah, Saudi Arabia,* [3]*Department of Pharmacy, Faculty of Pharmaceutical Science and Nursing, Vivekananda Global University, Jaipur, Rajasthan, India,* [4]*Department of Pharmacology, Teerthanker Mahaveer College of Pharmacy, Teerthanker Mahaveer University, Moradabad, Uttar Pradesh, India,* [5]*School of Pharmacy, Suresh Gyan Vihar University, Jaipur, Rajasthan, India,* [6]*ICFAI School of Pharmaceutical Sciences, The ICFAI University, Jaipur, Rajasthan, India,* [7]*Centre for Research Impact & Outcome, Chitkara College of Pharmacy, Chitkara University, Rajpura, Punjab, India,* [8]*Centre of Medical and Bio-Allied Health Sciences Research, Ajman University, Ajman, United Arab Emirates,* [9]*School of Pharmacy, Faculty of Medical Sciences, The University of the West Indies, St. Augustine, West Indies, Trinidad and Tobago*

9.1 Introduction

Cellular senescence is now being seen as the transition between microscopic injury and macroscopic cardiac remodeling; therefore, fibrosis is not merely an outcome that has to be tolerated but can be quantified, simulated, and repurposed for therapeutic use. This chapter discusses how the conversion of oxidative stress, telomere erosion, metabolic disequilibrium, and pathological overload into enduring extracellular matrix (ECM) deposition and chamber stiffening by senescent programs of cardiomyocytes, fibroblasts, endothelial cells, and immune populations (Xu et al., 2025; Li et al., 2023). We refer to senescence as a stress-induced, mostly irreversible growth arrest imposed by p53-p21 and p16INK4a-Rb checkpoints, characterized by a pro-inflammatory senescence-associated secretory phenotype (SASP) and mitochondrial dysfunction, which alters the state of neighboring cells. Fibrosis is a wound-healing reaction that becomes maladaptive when myofibroblast persistence, imbalance of protease inhibitors, and cross-linking of collagen increase filling pressures and inhibit relaxation. Remodeling refers to the concerted effects of structural and functional results, as represented by imaging, hemodynamic, and histological findings (Wagner and Wagner, 2022; Kumari and Jat, 2021). To maintain these concepts, we focus on three couplings: triggers to arrest, SASP to fibroblast activation, and matrix mechanics to transcriptional memory. We then overlaid the major SASP axes (TGF-b/SMAD, IL-6/STAT3, NF-kB), epigenetic and transcriptional locks that maintain these states, and stiffness sensing by integrins and YAP/TAZ. An echocardiography, CMR, and molecular imaging readout survey and circulating and tissue biomarker indexation of burden and progression of pathology all indicate the direct relationship between multicellular crosstalk, involving endothelial-to-mesenchymal transition, perivascular input, and immune modulation, and clinical syndromes that span HFpEF, HFrEF, postmyocardial infarction healing, and hypertensive disease (Kumari and Jat, 2021; Howes et al., 2024). To offer comparable endpoints, models, and quantitative evaluations have been integrated with histomorphometry, single-cell/spatial omics, biomechanics, 2D/3D systems, organoids, precision-cut slices, and in vivo paradigms. Only agents relevant to cardiology and endpoints that were biologically plausible were included. Senolytics, senomorphics, antifibrotics, pro-resolution, mechanotransduction, regenerative, or gene/RNA methods are combined in therapeutic sections by focusing on the timing, dose, delivery, safety, and monitoring of biomarker-based responses (Milligan et al., 2020; Atlante et al., 2025). This chapter is innovative in that it connects spatial biology to imaging and liquid biomarkers to trial design and devises stiffness sensing, metabolic constriction, and epigenetic persistence into a positive feedback loop that explains fibrosis across phenotypes. The practical objectives include standardization of language, assays side by side with clinical measures, and actionable nodes where senescence-fibrosis couplings can be broken to halt progression, reestablish compliance, and improve patient-centered outcomes. Focusing on regions of vascular indeterminism, we excluded

Cellular Senescence and Cardiac Aging. DOI: https://doi.org/10.1016/B978-0-443-45811-8.00007-6

congenital anomalies, primary infiltrative cardiomyopathies, and noncardiac fibroses, taking advantage of the imaging, hemodynamic, and biopsy correlations to obtain targets and remain focused (Mulholland and Leedham, 2024; Sofias et al., 2021). Finally, because mechanisms must be anchored to quantifiable variables to ensure that hypotheses are testable and decisions can be data-driven, we linked mechanisms to these quantifiable anchors in CMR (e.g., strain and extracellular volume), blood (e.g., collagen peptides and LOX activity), and spatial maps (e.g., single-cell states). Speculative mechanisms are only presented with evidence for convergent mechanisms and are denoted as provisional to serve as guidance in validating them rather than oversimplifying them (Tseng et al., 2016; Malhotra et al., 2013).

9.2 Cellular senescence in the cardiac microenvironment

9.2.1 Senescent cardiomyocytes, fibroblasts, endothelial, and immune cells

Senescent cardiomyocytes, fibroblasts, endothelial cells, and immune cells cooperatively drive fibrotic remodeling when cell cycle arrest converges with maladaptive signaling. Within the cardiac microenvironment, these senescent populations accumulate with age and stress, secrete SASP factors, and reprogram neighboring cells toward persistence and matrix deposition. Cardiomyocyte senescence reduces contractile reserve and calcium handling, elevates mitochondrial ROS levels, and releases DAMPs that recruit and activate stromal and immune effectors in the heart. Senescent fibroblasts resist apoptosis, sustain myofibroblast phenotypes, and increase collagen I/III synthesis with LOX-mediated cross-linking, which stiffens the myocardium (Tang et al., 2020; Wu et al., 2020). Endothelial senescence impairs nitric oxide bioavailability, promotes EndoMT, and increases leukocyte adhesion, thereby intensifying the perivascular fibrosis. Senescent immune cell subsets, including macrophages and T cells, adopt inflammatory programs that amplify TGF-β, IL-6, and chemokine networks, reinforcing fibroblast activation and ECM turnover. These cell-type-specific programs intersect through paracrine SASP cues, altered mechanotransduction, and epigenetic enforcement of the p16INK4a/p21 pathways, creating self-sustaining fibrotic niches. Concurrently, impaired protease-inhibitor balance and matricellular signaling remodel collagen architecture, elevate passive stiffness, and depress diastolic compliance, linking cellular senescence to adverse ventricular remodeling and heart failure trajectories (Hayashi et al., 2006; Ramadhiani et al., 2021). Integrated profiling across models and patient studies revealed convergent signatures of oxidative stress, mitochondrial dysfunction, and immune-fibroblast crosstalk, which nominated measurable biomarkers for stratification. Collectively, these insights connect senescent cardiomyocytes, fibroblasts, endothelium, and immune cells to the fibrotic logic underlying the stiffness and pump dysfunction. Future studies should prioritize mechanism-guided senotherapies, antifibrotic combinations, and biomarker-anchored trials to attenuate remodeling and restore myocardial resilience (Correia et al., 2024; Riou and Geny, 2025).

9.2.2 Hallmarks of senescence: cell-cycle arrest, senescence-associated secretory phenotype, and mitochondrial dysfunction

Cell cycle arrest, a pro-inflammatory SASP, and mitochondrial dysfunction are the core hallmarks of senescent biology in the heart. Within the cardiac microenvironment, these hallmarks emerge during aging, metabolic stress, ischemia, and pressure-overload, collectively initiating fibrotic remodeling. DNA damage and telomere attrition activate the p53-p21 and p16INK4a-Rb checkpoints, locking cardiomyocytes and stromal cells in growth arrest, which redirects programs toward survival. Concurrently, mitochondrial depolarization, impaired mitophagy, and reactive oxygen species amplify DNA damage and stabilize NF-κB signaling, which couples with the cGAS-STING and JAK/STAT pathways to program the SASP output (Zhao et al., 2025; Martini and Passos, 2023). The SASP encompasses TGF-β, IL-6, IL-1 cytokines, chemokines, matrix metalloproteinases, and profibrotic matricellular cues that recruit immune cells, activate fibroblasts, and reinforce myofibroblast persistence. Endothelial cells exposed to these mediators lose nitric oxide bioavailability, upregulate adhesion molecules, and undergo EndoMT, linking vascular inflammation to collagen deposition in the heart. Mechanotransduction further tightens this loop, as LOX-driven cross-linking and integrin/FAK/YAP signaling increase stiffness, which sustains SASP expression and deepens cell cycle arrest through epigenetic enforcement (Herbstein et al., 2024; Grafanaki et al., 2025). In preclinical and clinical studies, convergent signatures of checkpoint activation, redox imbalance, inflammation, and ECM remodeling have been associated with HFpEF trajectories and adverse outcomes. Together, these observations explain how cell cycle arrest, SASP programs, and mitochondrial dysfunction coevolve to entrench fibrosis, impair relaxation, and degrade the pump performance. Looking forward, mechanism-guided senotherapies paired with antifibrotic and metabolic modulators, anchored by imaging and circulating biomarkers, should be tested to determine whether interrupting these hallmarks can reverse remodeling and restore myocardial resilience (Michels da Silva et al., 2019; Yan et al., 2025).

9.2.3 Senescence triggers: oxidative stress, telomere attrition, mechanical load

Oxidative stress, telomere attrition, and mechanical load are the primary triggers for imprinting senescence programs in cardiomyocytes, fibroblasts, the endothelium, and immune cells. In the cardiac microenvironment, these stressors accumulate with aging, hypertension, ischemia, and metabolic diseases, thereby initiating maladaptive remodeling. Reactive oxygen species from mitochondria and NADPH oxidases damage DNA, proteins, and lipids, activating the p53-p21 and p16INK4a-Rb checkpoints, which enforce growth arrest. Telomere shortening, shelterin loss, and persistent DNA damage foci stabilize checkpoint signaling, promoting a SASP that releases cytokines, chemokines, and proteases (Erusalimsky, 2020; Schneider et al., 2025). Chronic pressure overload and cyclic stretch engage the integrin-FAK-YAP/TAZ pathways, unmask latent TGF-β, and stiffen the matrix through lysyl-oxidase cross-linking, thereby reinforcing fibrogenic signaling. Mitochondrial depolarization and impaired mitophagy elevate mtDNA levels and trigger the cGAS-STING and NF-κB pathways, amplifying inflammation and SASP output. Endothelial dysfunction reduces nitric oxide levels and increases adhesion molecules, fostering leukocyte recruitment and EndoMT. Mechano-inflammatory cues convert fibroblasts into apoptosis-resistant myofibroblasts. These cascades converge on ECM-remodeling, shifting collagen isoforms, skewing the MMP-MMP-TIMP balance, and increasing passive stiffness, which impairs diastolic filling (Futakuchi et al., 2018; Li et al., 2022). Across models and clinical cohorts, evidence links these triggers to HFpEF trajectories, microvascular rarefaction, and adverse outcomes. Together, these observations establish oxidative stress, telomere erosion, and mechanical load as interlocking drivers of cellular senescence and accelerated fibrotic remodeling. Looking ahead, mechanism-guided strategies combining redox control, DNA damage moderation, and mechanotransduction blockade with biomarker-anchored selection can be used to test whether the interruption of triggers can halt progression and restore myocardial compliance (Zeng and Chen, 2019; Mongirdienė et al., 2022) (Fig. 9.1).

9.3 Pathways coupling senescence to fibrosis

9.3.1 Senescence-associated secretory phenotype signaling axes: TGF-β/SMAD, IL-6/STAT3, NF-κB

TGF-β/SMAD, IL-6/STAT3, and NF-κB constitute the core SASP signaling axes that translate cellular senescence into a profibrotic action. In the senescent cardiac milieu, these axes are activated in response to DNA damage, mitochondrial stress, and mechanical load to coordinate the activation of fibroblasts and remodeling of the ECM. TGF-β from senescent cardiomyocytes, fibroblasts, and endothelium activates TGFBR1/2, phosphorylates SMAD2/3-SMAD4 complexes, drives collagen and PAI-1 transcription, and increases tension and myofibroblast persistence via the MAPK/RhoA pathway (Tominaga and Suzuki, 2019; Kojima et al., 2013). Concurrently, IL-6 engages gp130/JAK to phosphorylate STAT3, which cooperates with SMADs at profibrotic enhancers and promotes survival programs that render activated fibroblasts resistant to apoptosis. Persistent DNA damage, mitochondrial ROS, and cytosolic DNA activate NF-κB via IKK and cGAS-STING, sustaining chemokines, MMPs, and adhesion molecules that recruit leukocytes and increase TGF-β bioavailability, thereby promoting fibrosis. These pathways are mutually reinforcing: NF-κB heightens IL-6 and TGF-β expression, STAT3 loosens chromatin at SMAD targets, and TGF-β-driven stiffness feeds back via integrin-FAK-YAP to maintain the SASP output (Garbers et al., 2015; Ma et al., 2025). Endothelial dysfunction and EndoMT are integrated into this network, reducing nitric oxide levels, increasing VCAM-1/ICAM-1 levels, and seeding perivascular fibrosis, which couples microvascular rarefaction to diastolic dysfunction. Across models and patient cohorts, convergent signatures of SMAD activation, STAT3 phosphorylation, and NF-κB activity were associated with collagen cross-linking, altered MMP/TIMP balance, and impaired compliance. Together, these findings position SASP-centered TGF-β/SMAD, IL-6/STAT3, and NF-κB modules as interlocking drivers that entrench fibrosis and worsen outcomes. Future studies should test mechanism-guided combinations that co-target these axes with biomarker-anchored stratification to arrest remodeling and restore myocardial resilience (Widmer and Lerman, 2014; Cerbulescu et al., 2024).

9.3.2 Epigenetic and transcriptional reprogramming in profibrotic states

Epigenetic and transcriptional reprogramming fix senescent signals into durable profibrotic states in cardiomyocytes, fibroblasts, and endothelium. In the cardiac microenvironment, stress-activated checkpoints reshape the chromatin to favor myofibroblast priming, inflammatory tone, and ECM synthesis. DNA hypermethylation at antifibrotic loci and hypomethylation at enhancer hubs cooperate with histone marks-H3K27ac/H3K4me1 at super-enhancers and H3K27me3 redistribution to license SMAD3, NF-κB, STAT3, and TEAD/YAP programs that drive COL1A1, COL3A1, and PAI-1 transcription. SWI/SNF and other remodelers increase accessibility at mechanotransduction and TGF-β response elements, while mitochondrial dysfunction and altered metabolites (acetyl-CoA, α-KG, and NAD+) tune

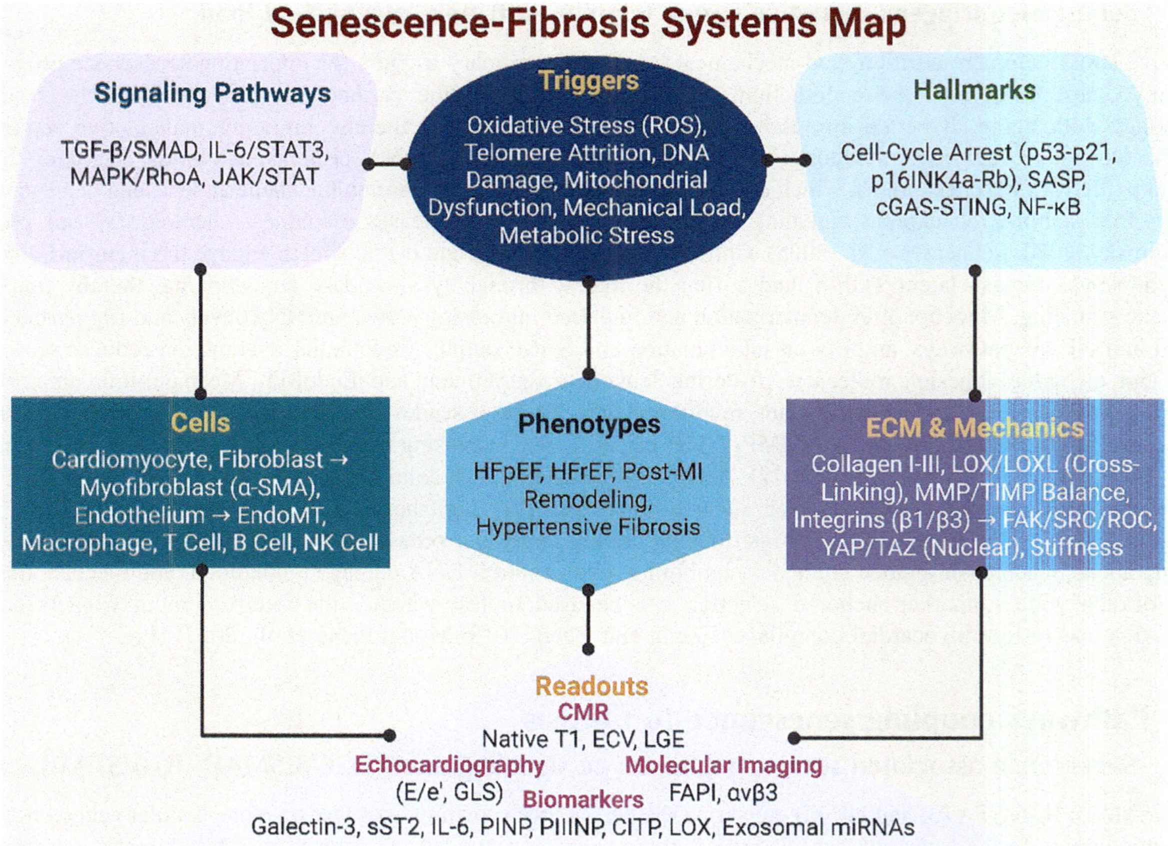

FIGURE 9.1 This figure shows a senescence-fibrosis systems map that connects the upstream triggers with the clinical outcomes. Senescence hallmarks, such as cell cycle arrest, SASP, cGAS-STING, and NF-kB, are triggered by oxidative stress, telomere attrition, DNA damage, mitochondrial dysfunction, and mechanical and metabolic stress. Signaling axes (TGF-b/SMAD, IL-6 / STAT3, MAPK/RhoA, and STAT) are core signaling axes that include cardiomyocytes, fibroblasts-myofibroblasts, endothelium-endothelial-to-mesenchymal transition, and immune cells. The activation of ECM/mechanical modules (collagen I/III, LOX/LOXL, MMP/TIMP, integrins-FAK/SRC/ROCK, and YAP/TAZ) increases stiffness, resulting in HFpEF, HFrEF, and postreadouts, including echocardiography, CMR, staging, and monitoring of molecular imaging and biomarkers circulating in the blood. *SASP*, senescence-associated secretory phenotype.

HATs, HDACs, and demethylases to reinforce these states (Peng et al., 2025; Komal et al., 2024). Noncoding RNAs further stabilize the network: miRNAs constrain the protease-inhibitor balance, and lncRNAs scaffold chromatin modifiers at profibrotic promoters, amplifying endothelial-to-mesenchymal transition and immune stromal crosstalk. Matrix stiffening via LOX cross-linking feeds back through integrin-FAK-YAP to maintain open chromatin at SASP and ECM genes, creating a memory that persists after inciting stress. Across models and patient cohorts, multi-omic signatures converge on enhancer remodeling, checkpoint activation, and collagen cross-linking, which are linked to the pathogenesis of diastolic dysfunction. Collectively, these findings explain how epigenetic and transcriptional circuits entrench fibrosis and predict its clinical course. Looking forward, rational combinations of senotherapies with epigenetic modulators (e.g., BET, EZH2, and HDAC/DNMT targeting) and mechanotransduction blockers prioritized by imaging and circulating biomarker panels should be tested to determine whether remodeling can be halted or reversed (Ratti et al., 2020).

9.3.3 Senescence-extracellular matrix feedback loops and matricellular cues

Senescence-ECM feedback loops and matricellular cues define how the aging heart converts transient stress into a state of sustained fibrosis. In this niche, senescent cardiomyocytes, fibroblasts, and endothelium are remodeled by the ECM in a self-reinforcing circuit. SASP factors elevate TGF-β activity, unmask latent ligands, and drive LOX-mediated collagen cross-linking, which stiffens the myocardium and magnifies integrin-FAK-YAP/TAZ signaling. Stiffness and altered topography, in turn, stabilize myofibroblast phenotypes, deepen p16/p21 checkpoint enforcement, and maintain open chromatin at the ECM and inflammatory gene loci. Matricellular proteins, including osteopontin, periostin, thrombospondins, and tenascin-C, act as context setters, clustering receptors, modulating latent TGF-β activation, and guiding immune stromal crosstalk, which sustains SASP output (Zhai and Sadoshima, 2024; Młynarska et al., 2025). Protease-inhibitor imbalance shifts MMP-TIMP dynamics, exposes cryptic ECM fragments, and generates DAMP signals that recruit leukocytes and amplify NF-κB signaling pathways. Endothelial dysfunction and perivascular fibrosis couple microvascular rarefaction to diastolic stiffness, while exosomal miRNAs and altered metabolites propagate fibrogenic memory across cell types, depress relaxation and contractile reserve, and worsen the efficiency of the heart. Collectively, these observations explain how matrix mechanics and matricellular signaling entrench cellular senescence and accelerate adverse remodeling in the heart. Future studies should test combined senotherapies with LOX or integrin/YAP modulators and biomarker-anchored selection to disrupt feedback, restore compliance, and improve outcomes (Raffetto and Khalil, 2008; Mohammed et al., 2015).

9.4 Multicellular crosstalk in fibrotic remodeling

9.4.1 Senescence-driven fibroblast activation and myofibroblast persistence

Senescence-driven fibroblast activation and myofibroblast persistence anchor maladaptive remodeling in the heart. In this microenvironment, senescent cardiomyocytes, endothelium, and immune cells release SASP mediators, conditioning fibroblasts for activation. The paracrine TGF-β/SMAD, IL-6/STAT3, and NF-κB pathways, reinforced by integrin-FAK-YAP signaling, induce α-SMA, drive collagen I/III production, and promote LOX cross-linking, which stiffens the myocardium. Mechanical stiffening feeds back to maintain transcriptional and epigenetic programs, protecting myofibroblasts from apoptosis and supporting matrix deposition in the heart. Endothelial dysfunction increases adhesion molecules and EndoMT, adding fibroblast-like cells, while recruited macrophages supply TGF-β, PDGF, and osteopontin, stabilizing their activation. An altered MMP/TIMP balance reduces matrix turnover, exposes cryptic fragments with DAMPs, and sustains leukocyte influx, extending SASP signaling and the fibrogenic loop (Rouhi et al., 2022; Baggett et al., 2023). Mitochondrial stress and impaired mitophagy in fibroblasts elevate ROS and cGAS-STING activity, amplifying inflammatory transcription and deepening cell cycle arrest via the p16/p21 checkpoints. Together, these processes convert transient injury signals into durable fibrotic niches, increasing passive stiffness, impairing diastolic filling, and predisposing patients to HFpEF. Across models and cohorts, convergent signatures of checkpoint activation, STAT3/SMAD phosphorylation, LOX activity, and collagen cross-linking were associated with reduced compliance and worse outcomes. Collectively, these findings position senescence-conditioned fibroblasts and persistent myofibroblasts as drivers of remodeling and potential therapeutic targets. Future studies should test biomarker-anchored combinations of senotherapies and antifibrotic or mechanotransduction modulators to dismantle persistence and improve patient outcomes (Baggett et al., 2023; Zhou et al., 2024).

9.4.2 Endothelial-to-mesenchymal and perivascular contributions

Endothelial-to-mesenchymal transition (EndoMT) and perivascular cell dynamics are conduits through which senescence programs initiate fibrosis. Within the vascular niche, senescent endothelium and pericytes/VSMCs rewire the signaling pathways that convert microvascular stress into matrix deposition. Oxidative and hemodynamic stress reduce nitric oxide levels, elevate adhesion molecules, and expose latent TGF-β, priming SMAD2/3 activation and EndoMT, whereas IL-6/STAT3 and NF-κB amplify recruitment (Peng et al., 2022; Xie and Wang, 2023). Pericyte detachment, PDGF-B/PDGFR-β signaling, and angiotensin II push perivascular cells toward fibroblast-like phenotypes, secreting collagen and LOX, and increasing the stiffness of the vascular wall. Hypoxia and mitochondrial dysfunction stabilize the HIF pathways and cGAS-STING, sustaining SASP output and inflammation and preserving endothelial-to-mesenchymal transition (EndoMT). Matrix-remodeling skews the MMP/TIMP balance, reveals fragments and DAMPs, and strengthens integrin-FAK-YAP/TAZ mechanotransduction, feeding back to maintain profibrotic transcription and epigenetic memory. Microvascular rarefaction follows, coupling impaired perfusion to diastolic dysfunction as perivascular fibrosis spreads into the interstitium (Song et al., 2005; Lin et al., 2008). Across models and cohorts, convergent signatures of SMAD/STAT3 activity, PDGFR-β engagement, LOX cross-linking, and low capillary density were associated with increased stiffness and reduced compliance. Collectively, these observations place EndoMT and perivascular remodeling at the center of senescence-associated fibrosis and adverse outcomes. Future studies should test biomarker-guided regimens targeting endothelial senescence and the PDGFR-β/TGF-β axis with mechanotransduction modulators to interrupt fibrosis and restore myocardial resilience (Glück et al., 2017; Schrimpf et al., 2014).

9.4.3 Innate and adaptive immune modulators of the fibrotic microenvironment

Innate and adaptive immune modulators orchestrate the translation of senescence into cardiac fibrosis within the fibrotic microenvironment of the heart. During aging and stress, SASP cues from senescent cells prime macrophages, neutrophils, and NK cells to sustain the profibrotic signaling. Monocyte recruitment via CCL2/CCR2 yields macrophages that release TGF-β, IL-1β, and osteopontin, thereby promoting fibroblast activation, α-smooth muscle actin induction, and LOX-dependent collagen cross-linking. Neutrophil extracellular traps and proteases expose cryptic matrix fragments that intensify DAMP-NF-κB signaling. NK and γδ T cells modulate senescent cell clearance; however, when exhausted, they permit SASP persistence. On the adaptive axis, Th2/Th17 polarization and PD-1high CD8+ dysfunction amplify IL-6/STAT3 and TGF-β/SMAD programs, whereas Tregs can limit inflammation but also promote tissue repair that consolidates scars (Bengel et al., 2024; Rurik et al., 2021). B cells contribute to cytokines and antibody complexes that engage Fc receptors, recruit myeloid cells, and reinforce chemokine loops. Endothelial activation increases VCAM-1/ICAM-1, facilitates leukocyte trafficking, promotes EndoMT, expands fibroblast-like pools, and couples perivascular fibrosis to microvascular rarefaction. Convergent immune signatures correlated with collagen cross-linking, impaired compliance, and HFpEF trajectories across all models and cohorts. Collectively, these observations define the immune senescence circuits that stabilize persistence and adverse remodeling in the heart. Future studies should evaluate combinations that rebalance innate and adaptive cue senotherapies with CCR2, STAT3, or PD-1/PD-L1 modulators to interrupt fibrosis and restore the myocardial function (van Buul et al., 2010).

9.5 Mechanics, metabolism, and matrix remodeling

9.5.1 Mechanotransduction pathways (YAP/TAZ, integrins) in stiffness sensing

YAP/TAZ and integrin-based mechanotransduction translate matrix stiffness into profibrotic gene expression in the senescent heart. In stiffening myocardium, focal adhesion signaling and nuclear YAP/TAZ drive fibroblast activation. Integrin engagement clusters focal adhesions, activates FAK/Src, and connects actin stress fibers to the nucleus via the LINC complex, increasing nuclear tension and chromatin accessibility at the COL1A1, COL3A1, and CTGF enhancers. Elevated lysyl oxidase-mediated collagen cross-linking augments rigidity, reinforcing integrin/FAK/ROCK signaling and promoting YAP/TAZ nuclear retention. YAP/TAZ cooperates with TEAD and SMAD, integrating TGF-β inputs to induce α-SMA, PAI-1, and matricellular proteins that stabilize myofibroblasts. Endothelial cells sense stiffness through β1/β3 integrins, upregulate VCAM-1/ICAM-1, and undergo EndoMT, expanding fibroblast-like pools and perivascular fibrosis (Garoffolo and Pesce, 2019; Gandin et al., 2025). Cardiomyocyte mechanosensing via costameres alters Ca2+ handling and mitochondrial demand, increasing ROS that sustains NF-κB and SASP signaling into stromal cell activation. Force-responsive enhancers gain H3K27ac and accessibility, whereas shifts in acetyl-CoA and NAD+ levels tune HAT/HDAC activity, reinforcing profibrotic programs. Across models and patient cohorts, markers of YAP/TAZ activation, integrin/FAK signaling, and LOX activity were correlated with collagen cross-linking, impaired compliance, HFpEF trajectories, adverse events, and major hospitalization risk. Collectively, these findings establish stiffness-driven mechanotransduction as a central driver that locks senescence into persistent fibrosis and worsens the outcomes. Future studies should test combinations of senotherapies with LOX inhibitors, integrin/FAK or ROCK blockers, and YAP/TAZ modulators guided by imaging and circulating biomarkers to halt remodeling and restore durable myocardial resilience (Hegyi et al., 2021; Gevaert et al., 2019).

9.5.2 Extracellular matrix turnover: collagen isoforms, cross-linking, and protease networks

ECM turnover in the senescent heart is defined by shifts in collagen isoforms, cross-linking intensity, and protease-inhibitor imbalances. These changes emerge as stress and aging reshape matrix synthesis, processing, and degradation, favoring increased stiffness and fibrotic persistence. Early after injury or hemodynamic load, fibroblasts increase COL3A1 for provisional scaffolding before transitioning toward COL1A1-rich matrices that bear long-term tension, while matricellular cues (periostin, osteopontin, and thrombospondins) orchestrate the formation of fibrils. Lysyl-oxidase (LOX/LOXL) activity increases the number of intermolecular cross-links, increases insolubility, and elevates stiffness, which feeds back through integrin-FAK-YAP. In parallel, SASP mediators and hypoxia reprogram protease networks: MMP-2/9 and MT1-MMP expand pericellular remodeling, ADAMTS variants trim proteoglycans, and imbalances with TIMPs reduce controlled turnover, expose cryptic fragments, and generate DAMPs that reinforce inflammation (Kwak, 2013; Zamilpa and Lindsey, 2010). Endothelial dysfunction and perivascular remodeling skew protease gradients, while impaired autophagy and mitochondrial ROS in fibroblasts stabilize profibrotic transcription at collagen and cross-linking genes. Over time, an increased collagen I:III ratio, advanced cross-linking (including AGEs), and reduced elastin content depress diastolic compliance, slow relaxation, and couple microvascular rarefaction to HFpEF. Across models and patient cohorts, signature-LOX activity, MMP/TIMP ratios, and C-terminal telopeptides were associated with stiffness and adverse outcomes in patients with AS. Collectively, these patterns explain how altered ECM turnover drives senescence-associated remodeling. Future studies should test biomarker-anchored strategies that combine senotherapies with LOX inhibition, targeted MMP modulation, and mechanotransduction blockers to restore compliance and improve the prognosis (Penna and Pagliaro, 2025).

9.5.3 Metabolic rewiring of senescent cells and bioenergetic constraints

Metabolic rewiring of senescent cells and subsequent bioenergetic constraints shape the way the aging heart locks into fibrosis. In the fibrotic microenvironment, cardiomyocytes, fibroblasts, and endothelium under stress adopt persistent shifts in substrate use and redox balance, reinforcing the senescence programs. Declining mitochondrial quality control and impaired mitophagy reduce the efficiency of oxidative phosphorylation, elevate reactive oxygen species levels, and activate DNA damage checkpoints that stabilize p53-p21 and p16INK4a-Rb arrest (Grootaert, 2024; Xie et al., 2023). To sustain survival and SASP output, senescent cells increase glycolytic flux and pentose-phosphate pathway activity for ATP and NADPH, while anaplerotic inputs from glutamine support TCA intermediates used in biosynthesis. Altered fatty acid handling reduces fatty acid oxidation in cardiomyocytes and lipogenic reprogramming in fibroblasts, shifting acetyl-CoA availability, changing histone acetylation at profibrotic loci, and cooperating with HIF-1α and NF-κB to maintain cytokine, chemokine, and matrix-remodeling gene expression. The accumulation of TCA cycle by-products (succinate and fumarate) and fluctuating α-ketoglutarate/NAD+ pools tune dioxygenases and sirtuins, embedding profibrotic transcription through epigenetic persistence (Lane et al., 2024). Increased ECM synthesis relies on proline and ascorbate-dependent collagen maturation, while LOX-driven cross-linking amplifies stiffness, which feeds back via integrin-FAK-YAP/TAZ to fortify the metabolic mechanical loop. Across models and patient cohorts, coherent signatures of mitochondrial dysfunction, glycolytic bias, and redox imbalance were associated with collagen cross-linking, diastolic impairment, and adverse outcomes. Collectively, these observations position metabolic constraints as drivers that entrench senescence and accelerate the remodeling processes. Future studies should test biomarker-anchored combinations that pair senotherapies with targets in mitophagy, NAD+ and redox restoration, substrate utilization, and mechanotransduction to restore myocardial efficiency and compliance (Guertl et al., 2000; Chu and Quan, 2024) (Fig. 9.2).

9.6 Clinical phenotypes, imaging, and biomarkers

9.6.1 Clinical presentations: HFpEF/HFrEF, post-MI, and hypertensive fibrosis

HFpEF, HFrEF, postmyocardial infarction remodeling, and hypertensive heart disease exhibit overlapping fibrotic phenotypes that are rooted in senescence biology. Across these presentations, sustained cell cycle arrest, SASP signaling, and mitochondrial dysfunction converge with mechanical load to regulate matrix deposition and stiffness. In HFpEF, microvascular inflammation, endothelial dysfunction, and perivascular fibrosis elevate filling pressures despite a preserved ejection fraction, whereas collagen I-rich matrices and LOX cross-linking reduce compliance (Malik et al., 2025; Lindsey et al., 2020). In HFrEF, myocyte loss after ischemic or nonischemic injury triggers replacement fibrosis with persistent myofibroblast activity, altered MMP/TIMP balance, and chamber dilation, which reduces systolic performance. After MI, infarct healing proceeds from provisional COL3A1 scaffolding to a cross-linked scar, where persistent senescence and neurohumoral drives promote adverse remodeling. In long-standing hypertension, pressure overload and concentric hypertrophy couple integrin-FAK-YAP signaling to fibroblast activation, producing diffuse interstitial fibrosis that stiffens the ventricle (Talman and Ruskoaho, 2016; Díez et al., 2020). Multimodal imaging CMR T1 mapping, extracellular volume, late gadolinium enhancement, and strain maps burden and distribution, while circulating biomarkers such as galectin-3, soluble ST2, collagen telopeptides track activity and risk, and exosomal miRNAs, collagen propeptides, and LOX activity were used. These signatures are integrated with natriuretic peptides and troponins to refine the diagnosis and stratification of HF. Collectively, convergent clinical and imaging data link senescence-anchored fibrosis to impaired compliance, progressive dysfunction, hospitalization, and increased mortality. Future studies should test biomarker-anchored trials that combine senotherapies with antifibrotic and mechanotransduction modulators to halt remodeling and improve patient-centered outcomes (Pan et al., 2020).

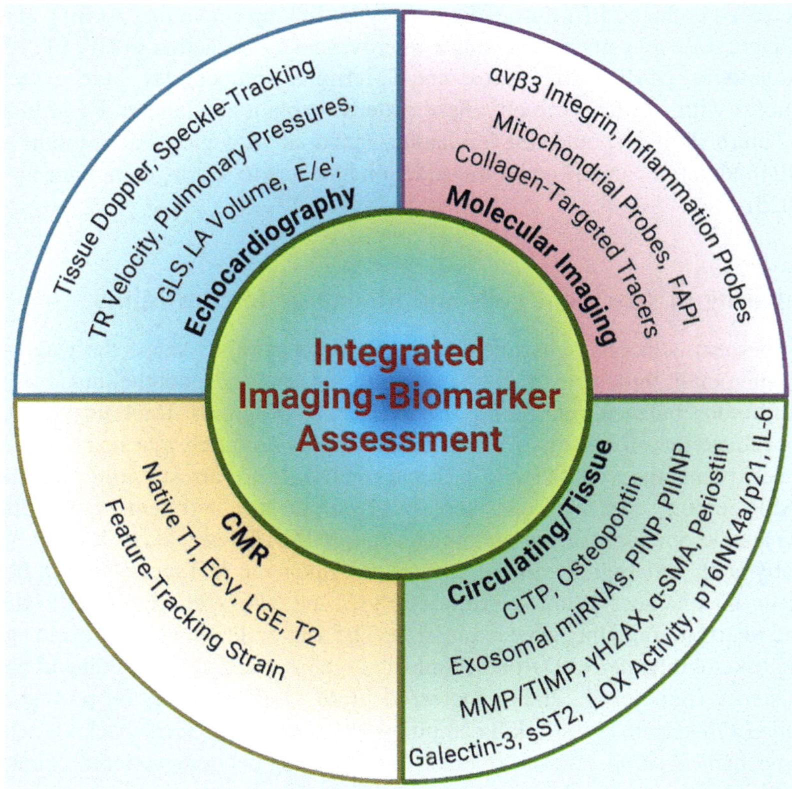

FIGURE 9.2 This radial figure is a summary of an Integrated Imaging-Biomarker Assessment of senescence-anchored cardiac fibrosis. The wheel comprises four domains: echocardiography (E/e, GLS, LA volume, TR velocity, tissue Doppler, speckle-tracking), CMR (native T1, extracellular volume, late gadolinium enhancement, T2, feature-tracking strain), Molecular Imaging (FAPI, avb3 integrin, collagen-targeted, mitochondrial, and inflammation probes), and circulating/tissue markers (galectin-3, sST2, IL-6, PINP, PII). Combined, these measurements allow the staging and stratification of risks, as well as pharmacodynamic follow-ups over the fibrosis pathways.

9.6.2 Imaging readouts (echocardiography, cardiovascular magnetic resonance, molecular imaging of fibrosis)

Imaging readouts from echocardiography, cardiovascular magnetic resonance (CMR), and molecular probes provide complementary insights into senescence-anchored fibrosis. In clinical workflows, echocardiography offers rapid hemodynamic assessment, whereas CMR and targeted tracers quantify diffuse and focal matrix remodeling with high specificity. Echocardiography tracks diastolic dysfunction via E/e′, left atrial volume, tricuspid regurgitant velocity, and pulmonary pressures, and detects early systolic impairment through global longitudinal strain that falls before ejection fraction declines; tissue Doppler and speckle-tracking further reveal load-dependent stiffening that accompanies collagen cross-linking (Klettas et al., 2022; Barton et al., 2023). CMR native T1 and extracellular volume fraction map diffuse interstitial expansion, late gadolinium enhancement delineates focal scar, and T2/T2 characterizes edema and iron that confound remodeling. Strain by feature-tracking refines mechanics, whereas parametric mapping correlates with collagen burden and outcomes. Molecular imaging extends beyond structure: fibroblast activation protein inhibitors (e.g., FAPI) highlight activated fibroblasts, αvβ3-integrin, and collagen-targeted tracers capture matrix deposition, and inflammatory or mitochondrial probes index SASP-linked biology; hybrid PET/MR integrates cellular activity with high-fidelity anatomy for longitudinal response tracking (Pan et al., 2020; Haaf et al., 2016). Convergent imaging signatures across syndromes, including HFpEF, HFrEF, postmyocardial infarction (MI), and hypertensive disease, are associated with impaired compliance, hospitalization risk, and mortality. Collectively, these readouts enable phenotyping, risk stratification, and pharmacodynamic monitoring of senescence-driven remodeling of the heart. Future studies should harmonize multiparametric thresholds, validate molecular tracers against biopsies and circulating biomarkers, and embed imaging-anchored enrichment in trials testing senotherapies and antifibrotic combinations (Lange et al., 2024).

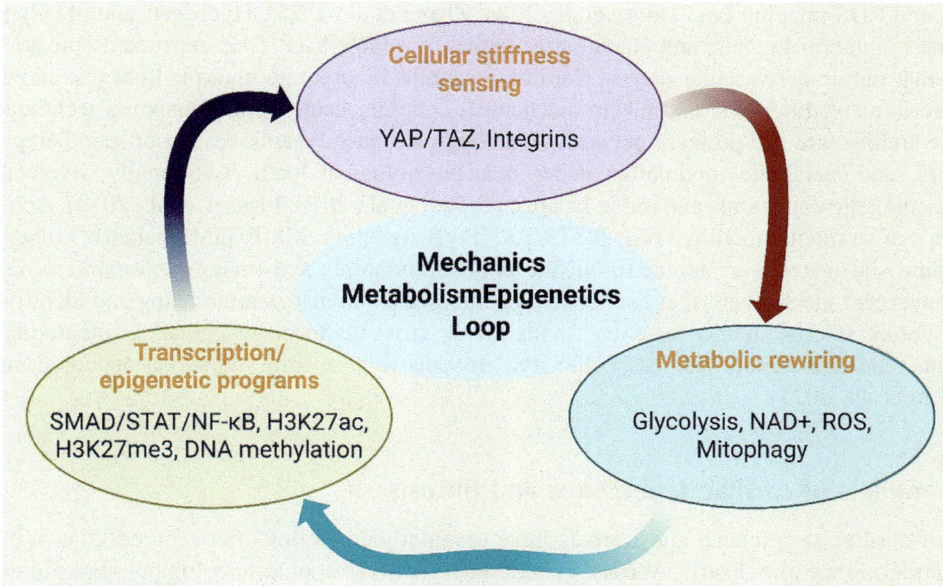

FIGURE 9.3 This figure shows a self-perpetuating cycle that maintains fibrosis, anchored on senescence. The stiffness of the cell is sensed through YAP/TAZ and integrin into nuclear signals, which signal an increase in the rigidity of the matrix. These clues stimulate metabolic rewiring, enhanced glycolysis, NAD+ balance, increased ROS, and damaged mitophagy, which enhances the stress signaling. Subsequently, transcriptional/epigenetic programs, SMAD/STAT/NF-kB activity, H3K27ac/H3K27me3 remodeling, and DNA methylation are promoted by metabolic changes that stabilize the expression of profibrotic genes. Feedback arrows suggest the strengthening of every module by others, making mechanical, metabolic, and chromatin changes a permanent remodel.

9.6.3 Circulating and tissue biomarkers of senescence and fibrotic burden

Circulating and tissue biomarkers of senescence and fibrotic burden provide complementary insights into tracking disease activity and guiding therapy. In the cardiac setting, these readouts link cellular arrest and SASP signaling to matrix remodeling and clinical risk. In the blood, galectin-3, soluble ST2, high-sensitivity IL-6, and growth-differentiation markers reflect inflammatory fibrotic tone, while collagen turnover peptides (PINP, PIIINP, CITP), C-terminal telopeptides, and lysyl-oxidase activity index synthesis and cross-linking, exosomal miRNAs, and lncRNAs add lineage-specific context and may report EndoMT or fibroblast activation. Senescence surrogates, such as circulating p16INK4a/p21 expression in relevant cell fractions, telomere attrition indices, and DNA damage markers, are aligned with SASP intensity and predict remodeling trajectories (Hu et al., 2022; Lockhart et al., 2020). In tissue, immunohistochemistry for p16/p21, γH2AX foci, and SA-β-gal surrogates co-localizes with α-SMA+ myofibroblasts, periostin/osteopontin, FAP, and a rising collagen I:III ratio; MMP/TIMP balance and advanced cross-links (including AGEs) map to stiffness, while EndoMT markers (α-SMA, vimentin, Snail/Slug) extend perivascular fibrosis into the interstitium. Multimodal integration with CMR T1/ECV or strain improves stratification, and serial sampling enables pharmacodynamic monitoring of senotherapies, antifibrotics, and mechanotransduction modulators. Collectively, convergent circulating and tissue signatures correlate with impaired compliance, HF hospitalization risk, and mortality, supporting biomarker-anchored enrichment and response assessment. Future studies should harmonize assay platforms, validate multi-analyte panels against biopsy and imaging, and develop threshold-based algorithms that personalize combination therapy to halt remodeling and restore myocardial resilience in patients with HF (Jimenez, 2013; Hua et al., 2025) (Fig. 9.3).

9.7 Models and quantitative assessment

9.7.1 In vitro and ex vivo systems (2D/3D, organoids, tissue slices)

Two-dimensional and three-dimensional in vitro systems, cardiac organoids, and precision-cut myocardial slices provide scalable platforms for investigating the mechanisms underlying senescence-driven fibrosis. These models recapitulate stromal–parenchymal interactions and permit controlled manipulation of mechanical load, oxygen tension, and cytokine gradients. In 2D cocultures, senescent cardiomyocytes or endothelium-conditioned fibroblasts are activated through SASP factors, enabling high-throughput readouts of α-SMA induction, collagen transcription, and LOX activity alongside calcium handling and ROS imaging (van Doorn et al., 2024; Zhang et al., 2025). Hydrogels and 3D bioprinted constructs add tunable stiffness, integrin ligands, and spatial constraints, revealing YAP/TAZ-dependent activation and diffusion-limited gradients that mimic perivascular niches. Cardiac organoids incorporate multiple lineages, allowing the mapping of macrophage recruitment, EndoMT, and electromechanical coupling using optical mapping techniques. Precision-cut slices preserve the architecture and protease networks, enabling pharmacodynamic testing of senotherapies, integrin/FAK or LOX inhibitors, and metabolic modulators under near-physiological load. Additionally, live-cell metabolic flux analyses calibrate energetic constraints and redox balance (Caliari et al., 2016; Idaszek et al., 2019). Across platforms, the quantitative metrics of strain/contractility, TGF-β/STAT3/NF-κB signaling, MMP/TIMP balance, collagen I:III ratio, and extracellular volume surrogates were aligned with the clinical endpoints and imaging biomarkers. Collectively, these systems reveal convergent mechanisms that anchor cellular senescence to matrix remodeling and identify potential targets for intervention. Future studies should focus on standardizing cross-platform benchmarks, integrating bioreactors and single-cell readouts, and embedding patient-specific iPSC models to accelerate biomarker-anchored mechanism-guided therapy design (Liu et al., 2025).

9.7.2 In vivo models of cardiac senescence and fibrosis

In vivo models of cardiac senescence and fibrosis have revealed causal links between cellular aging programs and maladaptive remodeling of the heart. Across genetic, surgical, and pharmacological paradigms, these models reproduce human phenotypes and enable a quantitative assessment of the fibrotic burden. Aged or telomerase-deficient mice accumulate p16INK4a/p21 expression, γH2AX foci, and SASP signaling, which precede collagen deposition and diastolic dysfunction. Progeroid Ercc1-/Δ strains and pressure overload (transverse aortic constriction) accelerate checkpoint activation, EndoMT, and myofibroblast persistence in the heart. Angiotensin II infusion, DOCA-salt hypertension, and postmyocardial infarction healing add neurohumoral drivers that expand interstitial and perivascular fibrosis (Luan et al., 2024; Osorio et al., 2023). Readouts included echocardiographic E/e′, strain, and left atrial size; invasive pressure-volume loops for relaxation and stiffness constants; and tissue metrics, including picrosirius red area, hydroxyproline, second-harmonic generation, LOX activity, collagen I:III ratio, and MMP/TIMP balance. Molecular imaging with FAPI or integrin-targeted tracers localizes activated fibroblasts, whereas reporter lines (p16-3MR, INK-ATTAC) map the dynamics and test clearance of senescent cells. Intervention studies with senolytics, CCR2 or STAT3 blockade, LOX inhibition, and YAP/TAZ modulation established the mechanisms and pharmacodynamic windows. Collectively, convergent findings link the intensity of senescence to stiffness, impaired compliance, and other clinical risk surrogates. Future studies should harmonize endpoints across models and embed imaging-anchored, biomarker-guided trials to translate mechanism-based combinations (Haschak, 2021; Chaher et al., 2023).

9.7.3 Quantification: histomorphometry, single-cell/spatial omics, biomechanics

The quantification of senescence-linked fibrosis integrates histomorphometry, single-cell/spatial omics, and biomechanics to convert biological data into metrics that can be compared. In practice, analyses progress from tissue architecture to cellular programs and, finally, to the material properties that govern cardiac function. Histomorphometry maps the collagen burden using picrosirius red or second-harmonic generation, distinguishes interstitial from perivascular fibrosis, and derives collagen I:III ratios, fiber thickness, alignment, and lacunarity. Adjunct measures capture LOX activity, MMP/TIMP balance, and periostin/osteopontin distribution in the Single-cell RNA/ATAC sequencing resolves senescent cardiomyocytes, fibroblasts, endothelium, and immune subsets, whereas spatial transcriptomics and multiplex proteomics localize SASP cytokines, SMAD/STAT/NF-κB activity, EndoMT markers, and myofibroblast persistence within stiffness hotspots (Ghazal et al., 2025; Liu et al., 2022). Biomechanical assays, such as atomic force microscopy, microindentation, tensile testing, and traction force microscopy, quantify the elastic modulus, viscoelasticity, and stress relaxation, which correlate with integrin-FAK-YAP/TAZ signaling and diastolic dysfunction. Image-based strain mapping links these properties to regional mechanics. Cross-modal registration aligns omics patches with fiber orientation and stiffness fields, enabling the prediction of extracellular volume and strain from the tissue features. Collectively, these pipelines yield reproducible endpoints that track the severity of remodeling, therapeutic response, and risk. Future studies should harmonize acquisition and analysis standards, build reference atlases spanning species and disease states, and pair multimodal readouts with biomarker-anchored trials to guide the development of senotherapies, antifibrotics, and mechanotransduction modulators (Weber et al., 2022; Li et al., 2018).

9.8 Therapeutic strategies and translational pathways

9.8.1 Senolytics and senomorphics: targets, combinations, and dosing paradigms

Senolytics and senomorphics provide complementary routes to disarm senescence-driven fibrosis by selectively clearing damaged cells and downregulating the SASP output. In the cardiovascular context, target selection, rational combinations, and dosing paradigms determine whether remodeling is halted or merely delayed. Senolytics that inhibit pro-survival networks (e.g., BCL-2/BCL-xL, PI3K/AKT) use intermittent hit-and-run schedules for pulse clearance while sparing regeneration. Senomorphics that modulate mTOR/FOXO, JAK/STAT, or NF-κB act as chronic controllers to reduce the levels of cytokines, chemokines, and matrix enzymes. In fibroblasts and endothelium, combining SASP dampening with antifibrotic agents (TGF-β pathway, LOX cross-linking) and mechanotransduction blockers (integrin/FAK/ROCK, YAP/TAZ) addresses both the drivers and enablers of persistence (Zhang et al., 2023a). Postmyocardial infarction (MI) or pressure-overload models suggest therapeutic windows: early modulation to temper inflammatory priming, followed by intermittent senolytic cycles as myofibroblasts stabilize, and finally, maintenance senomorphics to prevent re-accumulation. Biomarker-anchored titration guided by circulating IL-6, galectin-3, collagen propeptides, and imaging ECV or strain links the dose to the pharmacodynamic response, while safety monitors track cytopenias, hepatic signals, and QT risk. Drug-drug interactions, age-related pharmacokinetics, and comorbidity profiles shape selection and interval spacing, and device-enabled local delivery may reduce systemic exposure. Collectively, convergent preclinical and early clinical data indicate that mechanism-guided combination regimens can reduce fibrosis, improve compliance, and enhance the functional capacity. Future studies should standardize endpoints, embed adaptive dosing informed by multi-omics and imaging readouts, and test pragmatic schedules that maximize efficacy while minimizing toxicity (Yong et al., 2025; Saxena et al., 2016).

9.8.2 Antifibrotic and pro-resolution approaches (TGF-β modulators, extracellular matrix enzymatics)

Antifibrotic and pro-resolution approaches aim to interrupt profibrotic signaling while restoring a balanced matrix turnover. Within this framework, TGF-β modulators and ECM enzymes converge to reduce stiffness, resolve inflammation, and normalize the healing dynamics. TGF-β targeting spans ligand traps and neutralizing antibodies that curb bioactive ligands, small-molecule ALK5/TGFBR1 inhibitors that limit SMAD2/3 transcriptional drive, and strategies that block latent TGF-β activation via αv-integrins or thrombospondin, thereby diminishing myofibroblast persistence and PAI-1-linked matrix accumulation. Because global TGF-β blockade risks impaired repair, selective or context-dependent inhibition and short pulses aligned with postinjury phases are prioritized (Li et al., 2024; Sgonc and Wick, 2008). ECM-directed interventions act downstream: LOX/LOXL inhibition restrains collagen cross-linking and passive stiffness; careful MMP/TIMP rebalancing enhances controlled collagen turnover; and targeted enzymatic or peptide tools disrupt maladaptive cross-links while sparing the essential architecture of the ECM. Pro-resolution mediators, including specialized lipid mediators, temper neutrophil-macrophage signaling and reduce SASP amplification, creating a biochemical environment permissive for matrix normalization. Device-enabled or catheter-based delivery, injectable hydrogels, and affinity-guided nanoparticles concentrate these agents in fibrotic territories and lower systemic exposure, while combination regimens couple TGF-β modulation with mechanotransduction blockers (integrin/FAK/ROCK, YAP/TAZ) or senotherapies for durable benefits (Sethi et al., 2011; Cabral-Pacheco et al., 2020). Biomarker-anchored monitoring of galectin-3, soluble ST2, collagen propeptides, LOX activity, and imaging readouts (CMR T1/ECV, strain, and fibroblast activation tracers) guide the timing, dose, and response assessment. Collectively, emerging data indicate that targeted, time-sensitive TGF-β control combined with ECM enzymatics can reduce stiffness, improve diastolic function, and reduce adverse events. Future studies should refine patient selection, validate pharmacodynamic thresholds, and test adaptive combination schedules that maximize efficacy while safeguarding physiological repair processes (Shah et al., 2020; Patel et al., 2020).

9.8.3 Regenerative and reprogramming strategies (cell, gene, and RNA therapies)

Regenerative and reprogramming strategies, including cell, gene, and RNA therapies, aim to rebuild function while dismantling senescence-anchored fibrosis. In the remodeling heart, these approaches pair the renewal of cardiomyocytes with the reeducation of fibroblasts and the endothelium to restore matrix balance. Cell therapies using cardiac progenitors, cardiosphere-derived cells, or mesenchymal stromal cells deliver paracrine cues that temper TGF-β/IL-6 signaling, recruit reparative macrophages, and provide aligned matrices with engineered patches that reduce wall stress (Ahmad et al., 2025;

TABLE 9.1 Senescence-driven cardiac fibrosis: mechanisms, readouts, and therapeutic targets.

Core keywords	Mechanistic node	Cellular/matrix consequence	Clinical phenotype links	Diagnostics/quantitative metrics	References
Oxidative stress; telomere attrition; DNA damage; metabolic/mechanical load	Checkpoint arrest (p53-p21; p16^INK4a–Rb); SASP amplification	Fibroblast activation; EndoMT; immune crosstalk; ECM synthesis	HFpEF, HFrEF, post-MI, hypertensive fibrosis	Echo/CMR framework; biomarkers overview	Opresko et al. (2025); Zhang et al. (2023b)
TGF-β/SMAD; IL-6/STAT3; NF-κB; cGAS-STING	Interlocking SASP axes; SMAD–STAT cooperation	α-SMA⁺ myofibroblast persistence; collagen I/III ↑	Compliance ↓; diastolic dysfunction	Chemokines/MMPs; VCAM-1/ICAM-1	Lu et al. (2012)
Integrins (β1/β3); FAK/Src/ROCK; YAP/TAZ	Stiffness sensing → nuclear YAP/TAZ; TEAD/SMAD cooperation	CTGF/PAI-1 induction; apoptosis resistance; perivascular fibrosis	Stiffness ↑; hospitalization risk ↑	YAP/TAZ, FAK signaling; LOX activity	Longmate et al. (2024)
Collagen I/III; LOX/LOXL; MMP/TIMP; ADAMTS	Isoform shift; cross-linking ↑; protease imbalance	Insoluble collagen; DAMP exposure; inflammatory reinforcement	Diffuse/interstitial fibrosis	CITP, PINP/PIIINP; LOX activity; elastin ↓	Cox et al. (2013)
Cardiomyocyte; Fibroblast→Myofibroblast; Endothelium→EndoMT; Macrophage/T, B, NK cells	SASP paracrine reprogramming; EndoMT/immune recruitment	Matrix deposition; microvascular rarefaction	HFpEF trajectories; adverse remodeling	Collagen ratio; LOX; endothelial activation	Goldsmith et al. (2014)
CCL2/CCR2; IL-1β; PD-1^high CD8⁺; Tregs; NETs	Myeloid/lymphoid reinforcement of TGF-β/IL-6 programs	α-SMA induction; cross-linking ↑; chemokine loops	Compliance ↓; risk ↑	Immune signatures; fibrosis linkage	O'Connor et al. (2015)
Echocardiography	Hemodynamic/ systolic–diastolic assessment	Load-dependent stiffening; early systolic impairment	Phenotyping; risk	E/e′; GLS; LA volume; TR velocity; tissue Doppler	Dini et al. (2024)
CMR	Diffuse vs focal fibrosis quantification	Interstitial expansion; scar delineation	Prognostic mapping	Native T1; ECV; LGE; feature-tracking strain	Karamitsos et al. (2020); Backhaus et al. (2020)
Molecular imaging	Activated fibroblast/matrix activity	Cellular activity + anatomy (hybrid PET/MR)	Response tracking	FAPI; αvβ3; collagen/inflammation/ mitochondrial tracers	Rauber et al. (2023)
Circulating/tissue biomarkers	Inflammation/turnover; senescence burden	Trajectory and pharmacodynamics	Risk-stratification	Galectin-3; sST2; IL-6; PINP/PIIINP/CITP; LOX; MMP/TIMP; p16/p21; γH2AX; α-SMA; periostin; exosomal miRNAs	Mansfield et al. (2024)
Quantification pipelines	Histomorphometry; scRNA/ ATAC; Spatial proteomics; AFM/microindentation	Cross-modal linkage of cell states, fiber architecture, stiffness	Severity, response, risk	Collagen I:III; fiber alignment; modulus/ viscoelasticity; strain mapping	Ranzoni et al. (2021)
Senolytics/senomorphics	Intermittent clearance; SASP modulation	Senescent cell load ↓; cytokine output ↓	Compliance ↑; function ↑	Dosing by ECV/GLS; IL-6; galectin-3; collagen peptides; safety panels	Robbins et al. (2021)
TGF-β modulation; αv-integrins; LOX inhibition	Context-selective pathway/ matrix targeting	Turnover rebalance; stiffness feedback ↓	Remodeling attenuation	Imaging response (T1/ECV/strain); LOX activity	Munger and Sheppard (2011)
Integrin/FAK/ROCK; YAP/ TAZ; AAV; siRNA/ASO; miRNA; iCM	Mechanotransduction blockade; regeneration/ reprogramming	Weaken stiffness loop; re-educate stroma; rebuild function	Outcome improvement	Biomarker-anchored dosing; device-enabled delivery	Hicks-Berthet and Varelas (2017)

He et al., 2022). Gene modalities delivered by AAV or nonviral vectors can inhibit profibrotic drivers (SMAD signaling, CTGF), enhance mitophagy and calcium handling, and express antifibrotic decoys that limit myofibroblast persistence in the heart. In vivo reprogramming converts resident fibroblasts to induced cardiomyocyte-like cells with factor cocktails (e.g., Gata4/Mef2c/Tbx5) and cooperates with epigenetic modulators to bypass chromatin barriers. RNA therapeutics, such as siRNA, antisense RNA, and miRNA mimics/inhibitors, silence SASP mediators and cross-linking enzymes, thereby inhibiting SASP. Exosome nanoparticles enable fibroblast targeting via FAP- or integrin-directed ligands (Wang et al., 2021; Saadat et al., 2020). Because immune activation, arrhythmia risk, and off-target remodeling remain concerns, dosing paradigms favor phased regimens: anti-inflammatory priming, targeted gene/RNA delivery during scar maturation, and maintenance cycles tuned by imaging and circulating collagen propeptides. Device-assisted delivery and transient gene switching enhance precision and safety. Across models and early trials, these platforms improved contractile reserve, reduced stiffness, and shrank the fibrotic area while preserving perfusion. Looking forward, the convergence of single-cell maps, vector design, and combinations with senotherapies or mechanotransduction blockers should enable the reversal of remodeling and translate into clinical benefits for patients with HFpEF (Chen et al., 2024) (Table 9.1).

9.9 Conclusion and future perspectives

Cellular senescence emerges as a programmable driver linking molecular injury to ECM accumulation, myocardial stiffness, and clinically recognizable remodeling of the heart. Triggers, including oxidative stress, telomere attrition, mitochondrial dysfunction, and pathological load, checkpoint enforcement, and the SASP converge to activate fibroblasts and sustain persistent myofibroblasts. Endothelial dysfunction, immune amplification, and matrix mechanotransduction further hard-wire these states, coupling perivascular fibrosis with impaired relaxation and adverse effects. Our novelty lies in unifying stiffness sensing, metabolic constraint, and epigenetic persistence into a self-reinforcing loop and mapping a practical measurement-to-therapy continuum that ties spatial biology to imaging and circulating biomarkers. Within this scope, we connected mechanistic nodes to clinical syndromes spanning HFpEF, HFrEF, postmyocardial infarction remodeling, and hypertensive disease, and aligned assays from histomorphometry to single-cell and spatial omics with echocardiography, CMR, and molecular imaging techniques. We also frame quantitative biomarker panels—galectin-3, soluble ST2, collagen propeptides, MMP/TIMP ratios, lysyl-oxidase activity, and selected exosomal microRNAs—as anchors for risk-stratification and pharmacodynamic tracking. Taken together, convergent evidence across models and cohorts links senescence intensity with collagen cross-linking, extracellular volume expansion, reduced compliance, hospitalization risk, and mortality. On this foundation, therapeutic opportunities center on senolytics to reduce the senescent cell burden, senomorphics to temper SASP signals, antifibrotic and pro-resolution agents to rebalance turnover, mechanotransduction modulators to weaken stiffness feedback, and regenerative, gene, or RNA platforms to rebuild function. Timing, dosing, delivery, and safety require careful tuning, favoring intermittent clearance cycles coupled with maintenance controllers, image-guided localization, and biomarker-anchored titrations. Looking ahead, priorities include harmonized endpoints across preclinical and clinical studies, validated thresholds that integrate imaging and liquid biomarkers, and adaptive trial designs that match the mechanisms to patient phenotypes. Equally important, single-cell and spatial atlases should guide target selection, while advances in vector engineering, delivery devices, and combination scheduling should improve precision and reduce off-target risks. Collectively, the field now has a coherent blueprint that links mechanisms to measurable outcomes and points toward therapies capable of preventing disease progression, restoring compliance, and improving the quality of life. Our perspective is to keep recommendations anchored to reproducible metrics and transparent uncertainty so that the evidence remains actionable, cumulative, and ready for clinical application.

References

Ahmad, W., et al., 2025. In vivo targeted reprogramming of cardiac fibroblasts for heart regeneration: advances and therapeutic potential. Bioengineering (Basel) 12 (9), 940. Available at https://doi.org/10.3390/bioengineering12090940.

Atlante, S., et al., 2025. Senolytic therapies for cardiovascular aging: tackling fibrosis and metabolic dysfunction. European Journal of Internal Medicine 140, 106413.

Backhaus, S.J., et al., 2020. Real-time cardiovascular magnetic resonance T1 and extracellular volume fraction mapping for tissue characterisation in aortic stenosis. Journal of Cardiovascular Magnetic Resonance: Official Journal of the Society for Cardiovascular Magnetic Resonance 22 (1), 46.

Baggett, B.C., et al., 2023. Myofibroblast senescence promotes arrhythmogenic remodeling in the aged infarcted rabbit heart. Elife 12, e84088. Available at https://doi.org/10.7554/eLife.84088.

Barton, A.K., et al., 2023. Emerging molecular imaging targets and tools for myocardial fibrosis detection. European Heart Journal - Cardiovascular Imaging 24 (3), 261–275.

Bengel, F., et al., 2024. Linking immune modulation to cardiac fibrosis. Nature Cardiovascular Research 3 (4), 414–419.

Cabral-Pacheco, G.A., et al., 2020. The roles of matrix metalloproteinases and their inhibitors in human diseases. International Journal of Molecular Sciences 21 (24), 9739. Available at https://doi.org/10.3390/ijms21249739.

Caliari, S.R., et al., 2016. Dimensionality and spreading influence MSC YAP/TAZ signaling in hydrogel environments. Biomaterials 103, 314–323.

Cerbulescu, T., et al., 2024. The impact of matrix metalloproteinases and their tissue inhibitors in patients with chronic glaucoma - a literature review. Romanian Journal of Morphology and Embryology = Revue Roumaine de Morphologie et Embryologie 65 (4), 557–565.

Chaher, N., Phinikaridou, A., Botnar, R.M., 2023. Developing a Collagen III Targeted MRI Probe for Molecular Imaging of Cardiac Fibrosis.

Chen, R., et al., 2024. Macrophages in cardiovascular diseases: molecular mechanisms and therapeutic targets. Signal Transduction and Targeted Therapy 9 (1), 130.

Chu, C.Q., Quan, T., 2024. Fibroblast Yap/Taz signaling in extracellular matrix homeostasis and tissue fibrosis. Journal of Clinical Medicine 13 (12), 3358. Available at https://doi.org/10.3390/jcm13123358.

Correia, S.P., et al., 2024. Quantitative proteomics of patient fibroblasts reveal biomarkers and diagnostic signatures of mitochondrial disease. JCI Insight 9 (20), e178645. Available at https://doi.org/10.1172/jci.insight.178645.

Cox, T.R., et al., 2013. LOX-mediated collagen crosslinking is responsible for fibrosis-enhanced metastasis. Cancer Research 73 (6), 1721–1732.

Díez, J., González, A., Kovacic, J.C., 2020. Myocardial interstitial fibrosis in nonischemic heart disease, part 3/4: JACC focus seminar. Journal of the American College of Cardiology 75 (17), 2204–2218.

Dini, F.L., et al., 2024. Echocardiography in the assessment of heart failure patients. Diagnostics (Basel) 14 (23), 2730. Available at https://doi.org/10.3390/diagnostics14232730.

Erusalimsky, J.D., 2020. Oxidative stress, telomeres and cellular senescence: What non-drug interventions might break the link? Free Radical Biology and Medicine 150, 87–95.

Futakuchi, A., et al., 2018. YAP/TAZ are essential for TGF-β2-mediated conjunctival fibrosis. Investigative Ophthalmology and Visual Science 59 (7), 3069–3078.

Gandin, A., et al., 2025. YAP/TAZ mechanotransduction is a fast cellular response to subnuclear adhesions revealed by mechanically dynamic substrates. bioRxiv 2025 06. 04.656127.

Garbers, C., Aparicio-Siegmund, S., Rose-John, S., 2015. The IL-6/gp130/STAT3 signaling axis: recent advances towards specific inhibition. Current Opinion in Immunology 34, 75–82.

Garoffolo, G., Pesce, M., 2019. Mechanotransduction in the cardiovascular system: from developmental origins to homeostasis and pathology. Cells 8 (12), 1607. Available at https://doi.org/10.3390/cells8121607.

Gevaert, A.B., et al., 2019. Heart failure with preserved ejection fraction: a review of cardiac and noncardiac pathophysiology. Frontiers in Physiology 10, 638.

Ghazal, R., et al., 2025. Cardiac fibrosis in the multi-omics era: implications for heart failure. Circulation Research 136 (7), 773–802.

Glück, S., et al., 2017. Innate immune sensing of cytosolic chromatin fragments through cGAS promotes senescence. Nature Cell Biology 19 (9), 1061–1070.

Goldsmith, E.C., et al., 2014. Myocardial fibroblast-matrix interactions and potential therapeutic targets. Journal of Molecular and Cellular Cardiology 70, 92–99.

Grafanaki, K., et al., 2025. Fibrotic disease of the skin and lung: shared pathways, environmental drivers, and therapeutic opportunities in a changing climate. International Journal of Molecular Sciences 26 (17), 8394. Available at https://doi.org/10.3390/ijms26178394.

Grootaert, M.O.J., 2024. Cell senescence in cardiometabolic diseases. NPJ Aging 10 (1), 46.

Guertl, B., Noehammer, C., Hoefler, G., 2000. Metabolic cardiomyopathies. International Journal of Experimental Pathology 81 (6), 349–372.

Haaf, P., et al., 2016. Cardiac T1 mapping and extracellular volume (ECV) in clinical practice: a comprehensive review. Journal of Cardiovascular Magnetic Resonance: Official Journal of the Society for Cardiovascular Magnetic Resonance 18 (1), 89.

Haschak, M., 2021. Age-Related Remodeling is Associated with Altered Cardiac Resident and Circulation-derived Macrophage Phenotype and Function. University of Pittsburgh.

Hayashi, T., et al., 2006. Endothelial cellular senescence is inhibited by nitric oxide: implications in atherosclerosis associated with menopause and diabetes. Proceedings of the National Academy of Sciences of the United States of America 103 (45), 17018–17023.

He, X., et al., 2022. Advances in cellular reprogramming-based approaches for heart regenerative repair. Cells 11 (23), 3914. Available at https://doi.org/10.3390/cells11233914.

Hegyi, B., et al., 2021. Cardiomyocyte Na(+) and Ca(2+) mishandling drives vicious cycle involving CaMKII, ROS, and ryanodine receptors. Basic Research in Cardiology 116 (1), 58.

Herbstein, F., et al., 2024. The SASP factor IL-6 sustains cell-autonomous senescent cells via a cGAS-STING-NFκB intracrine senescent noncanonical pathway. Aging Cell 23 (10), e14258.

Hicks-Berthet, J., Varelas, X., 2017. Integrin-FAK-CDC42-PP1A signaling gnaws at YAP/TAZ activity to control incisor stem cells. Bioessays: News and Reviews in Molecular, Cellular and Developmental Biology 39 (10). Available at https://doi.org/10.1002/bies.201700116.

Howes, A.M., et al., 2024. Fibroblast senescence-associated extracellular matrix promotes heterogeneous lung niche. APL Bioengineering 8 (2), 026119.

Hu, C., et al., 2022. Cellular senescence in cardiovascular diseases: a systematic review. Aging Disease 13 (1), 103–128.

Hua, K., Hong, H., Wang, X., 2025. Biomarker-guided adaptive enrichment design with threshold detection for clinical trials with time-to-event outcome. Journal of Biopharmaceutical Statistics 35 (6), 1209–1226. Available at https://doi.org/10.1080/10543406.2025.2489291.

Idaszek, J., et al., 2019. 3D bioprinting of hydrogel constructs with cell and material gradients for the regeneration of full-thickness chondral defect using a microfluidic printing head. Biofabrication 11 (4), 044101.

Jimenez, S.A., 2013. Role of endothelial to mesenchymal transition in the pathogenesis of the vascular alterations in systemic sclerosis. ISRN Rheumatology 2013, 835948. Available at https://doi.org/10.1155/2013/835948.

Karamitsos, T.D., et al., 2020. Myocardial tissue characterization and fibrosis by imaging. JACC: Cardiovascular Imaging 13 (5), 1221–1234.

Klettas, D., et al., 2022. Echocardiographic and cardiac magnetic resonance imaging-derived strains in relation to late gadolinium enhancement in hypertrophic cardiomyopathy. The American Journal of Cardiology 171, 132–139.

Kojima, H., et al., 2013. IL-6-STAT3 signaling and premature senescence. Jakstat 2 (4), e25763.

Komal, S., et al., 2024. Epigenetic regulation in myocardial fibroblasts and its impact on cardiovascular diseases. Pharmaceuticals (Basel) 17 (10), 1353. Available at https://doi.org/10.3390/ph17101353.

Kumari, R., Jat, P., 2021. Mechanisms of cellular senescence: cell cycle arrest and senescence associated secretory phenotype. Frontiers in Cell and Developmental Biology 9, 645593.

Kwak, H.B., 2013. Aging, exercise, and extracellular matrix in the heart. Journal of Exercise Rehabilitation 9 (3), 338–347.

Lane, A.N., Higashi, R.M., Fan, T.W., 2024. Challenges of spatially resolved metabolism in cancer research. Metabolites 14 (7), 383. Available at https://doi.org/10.3390/metabo14070383.

Lange, T., et al., 2024. CMR-based cardiac phenotyping in different forms of heart failure. The International Journal of Cardiovascular Imaging 40 (7), 1585–1596.

Li, D., et al., 2023. Cellular senescence in cardiovascular diseases: from pathogenesis to therapeutic challenges. Journal of Cardiovascular Development and Disease 10 (10), 439. Available at https://doi.org/10.3390/jcdd10100439.

Li, H., et al., 2022. Extracellular matrix stiffness and TGFβ2 regulate YAP/TAZ activity in human trabecular meshwork cells. Frontiers in Cell and Developmental Biology 10, 844342.

Li, M., et al., 2018. Atomic force microscopy studies on cellular elastic and viscoelastic properties. Science China Life Sciences 61 (1), 57–67.

Li, M., et al., 2024. Exploring the antifibrotic mechanisms of ghrelin: modulating TGF-β signalling in organ fibrosis. Expert Reviews in Molecular Medicine 27, e8. Available at https://doi.org/10.1017/erm.2024.38.

Lin, S.L., et al., 2008. Pericytes and perivascular fibroblasts are the primary source of collagen-producing cells in obstructive fibrosis of the kidney. The American Journal of Pathology 173 (6), 1617–1627.

Lindsey, M.L., et al., 2020. Focusing heart failure research on myocardial fibrosis to prioritize translation. Journal of Cardiac Failure 26 (10), 876–884.

Liu, H., et al., 2022. Dynamic and static biomechanical traits of cardiac fibrosis. Frontiers in Bioengineering and Biotechnology 10, 1042030.

Liu, J., Yu, H., Xu, Y., 2025. Targeting cellular senescence: pathophysiology in multisystem age-related diseases. Biomedicines 13 (7), 1727. Available at https://doi.org/10.3390/biomedicines13071727.

Lockhart, S.M., Saudek, V., O'Rahilly, S., 2020. GDF15: a hormone conveying somatic distress to the brain. Endocrine Reviews 41 (4), bnaa007. Available at https://doi.org/10.1210/endrev/bnaa007.

Longmate, W.M., et al., 2024. Keratinocyte integrin α3β1 induces expression of the macrophage stimulating factor, CSF-1, through a YAP/TEAD-dependent mechanism. Matrix Biology: Journal of the International Society for Matrix Biology 127, 48–56.

Lu, P., et al., 2012. Role of TGF-β1/Smad3 signaling pathway in secretion of type I and III collagen by vascular smooth muscle cells of rats undergoing balloon injury. Journal of Biomedicine and Biotechnology 2012, 965953.

Luan, Y., et al., 2024. Cardiac cell senescence: molecular mechanisms, key proteins and therapeutic targets. Cell Death Discovery 10 (1), 78.

Ma, X., et al., 2025. Molecular mechanisms in liver repair and regeneration: from physiology to therapeutics. Signal Transduction and Targeted Therapy 10 (1), 63.

Malhotra, A., et al., 2013. HypothesisFinder: a strategy for the detection of speculative statements in scientific text. PLoS Computational Biology 9 (7), e1003117.

Malik, M.K., et al., 2025. Evolving role of myocardial fibrosis in heart failure with preserved ejection fraction. Frontiers in Cardiovascular Medicine 12, 1573346.

Mansfield, L., et al., 2024. Emerging insights in senescence: pathways from preclinical models to therapeutic innovations. NPJ Aging 10 (1), 53.

Martini, H., Passos, J.F., 2023. Cellular senescence: all roads lead to mitochondria. The FEBS Journal 290 (5), 1186–1202.

Michels da Silva, D., Langer, H., Graf, T., 2019. Inflammatory and molecular pathways in heart failure-ischemia, HFpEF and transthyretin cardiac amyloidosis. International Journal of Molecular Sciences 20 (9), 2322. Available at https://doi.org/10.3390/ijms20092322.

Milligan, D.A., Tyler, E.J., Bishop, C.L., 2020. Tissue engineering to better understand senescence: organotypics come of age. Mechanisms of Ageing and Development 190, 111261.

Młynarska, E., et al., 2025. Inflammaging and senescence-driven extracellular matrix remodeling in age-associated cardiovascular disease. Biomolecules 15 (10), 1452.

Mohammed, S.F., et al., 2015. Coronary microvascular rarefaction and myocardial fibrosis in heart failure with preserved ejection fraction. Circulation 131 (6), 550–559.

Mongirdienė, A., et al., 2022. Reactive oxygen species induced pathways in heart failure pathogenesis and potential therapeutic strategies. Biomedicines 10 (3), 602. Available at https://doi.org/10.3390/biomedicines10030602.

Mulholland, E.J., Leedham, S.J., 2024. Redefining clinical practice through spatial profiling: a revolution in tissue analysis. Annals of the Royal College of Surgeons of England 106 (4), 305–312.

Munger, J.S., Sheppard, D., 2011. Cross talk among TGF-β signaling pathways, integrins, and the extracellular matrix. Cold Spring Harbor Perspectives in Biology 3 (11), a005017.

O'Connor, T., Borsig, L., Heikenwalder, M., 2015. CCL2-CCR2 signaling in disease pathogenesis. Endocrine, Metabolic and Immune Disorders Drug Targets 15 (2), 105–118.

Opresko, P.L., Sanford, S.L., De Rosa, M., 2025. Oxidative stress and DNA damage at telomeres. Cold Spring Harbor Perspectives in Biology 17 (6), a041707. Available at https://doi.org/10.1101/cshperspect.a041707.

Osorio, J.M., et al., 2023. Senescent cardiac fibroblasts: a key role in cardiac fibrosis. Biochimica et Biophysica Acta - Molecular Basis of Disease 1869 (4), 166642.

Pan, J.A., Kerwin, M.J., Salerno, M., 2020. Native T1 mapping, extracellular volume mapping, and late gadolinium enhancement in cardiac amyloidosis. A meta-analysis. JACC Cardiovascular Imaging 13 (6), 1299–1310.

Patel, D.M., et al., 2020. Association of plasma-soluble ST2 and galectin-3 with cardiovascular events and mortality following cardiac surgery. American Heart Journal 220, 253–263.

Peng, Q., et al., 2022. The role of endothelial-to-mesenchymal transition in cardiovascular disease. Cells 11 (11), 1834. Available at https://doi.org/10.3390/cells11111834.

Peng, W.G., Getachew, A., Zhou, Y., 2025. Decoding the epigenetic and transcriptional basis of direct cardiac reprogramming. Stem Cells 43 (3), sxaf002. Available at https://doi.org/10.1093/stmcls/sxaf002.

Penna, C., Pagliaro, P., 2025. Endothelial dysfunction: redox imbalance, NLRP3 inflammasome, and inflammatory responses in cardiovascular diseases. Antioxidants (Basel) 14 (3), 256. Available at https://doi.org/10.3390/antiox14030256.

Raffetto, J.D., Khalil, R.A., 2008. Matrix metalloproteinases and their inhibitors in vascular remodeling and vascular disease. Biochemical Pharmacology 75 (2), 346–359.

Ramadhiani, R., et al., 2021. Endothelial cell senescence exacerbates pulmonary fibrosis potentially through accelerated endothelial to mesenchymal transition. The Kobe Journal of Medical Sciences 67 (3), E84–e91.

Ranzoni, A.M., et al., 2021. Integrative single-cell RNA-Seq and ATAC-Seq analysis of human developmental hematopoiesis. Cell Stem Cell 28 (3), 472–487.e7.

Ratti, M., et al., 2020. MicroRNAs (miRNAs) and long non-coding RNAs (lncRNAs) as new tools for cancer therapy: first steps from bench to bedside. Targeted Oncology 15 (3), 261–278.

Rauber, S., et al., 2023. Molecular imaging with fibroblast activation protein tracers depicts inflammatory joint damage and its transition to resolution of inflammation. bioRxiv p. 2023.01. 17.524425.

Riou, M., Geny, B., 2025. Mitochondrial dysfunction and oxidative stress: emerging insights in muscle and cardiovascular disease mechanisms. Antioxidants (Basel) 14 (8), 902. Available at https://doi.org/10.3390/antiox14080902.

Robbins, P.D., et al., 2021. Senolytic drugs: reducing senescent cell viability to extend health span. Annual Review of Pharmacology and Toxicology 61, 779–803.

Rouhi, L., et al., 2022. Deletion of the Lmna gene in fibroblasts causes senescence-associated dilated cardiomyopathy by activating the double-stranded DNA damage response and induction of senescence-associated secretory phenotype. Journal of Cardiovascular Aging 2 (3), 30. Available at https://doi.org/10.20517/jca.2022.14.

Rurik, J.G., Aghajanian, H., Epstein, J.A., 2021. Immune cells and immunotherapy for cardiac injury and repair. Circulation Research 128 (11), 1766–1779.

Saadat, S., et al., 2020. Pivotal role of TGF-β/Smad signaling in cardiac fibrosis: non-coding RNAs as effectual players. Frontiers in Cardiovascular Medicine 7, 588347.

Saxena, A., Russo, I., Frangogiannis, N.G., 2016. Inflammation as a therapeutic target in myocardial infarction: learning from past failures to meet future challenges. Translational Research: the Journal of Laboratory and Clinical Medicine 167 (1), 152–166.

Schneider, S.E., et al., 2025. Mechanical stress triggers premature senescence in cardiac fibroblasts. Advanced Science (Weinh) 12 (47), e13314. Available at https://doi.org/10.1002/advs.202513314.

Schrimpf, C., et al., 2014. The role of pericyte detachment in vascular rarefaction. Journal of Vascular Research 51 (4), 247–258.

Sethi, A., et al., 2011. Transforming growth factor-beta induces extracellular matrix protein cross-linking lysyl oxidase (LOX) genes in human trabecular meshwork cells. Investigative Ophthalmology and Visual Science 52 (8), 5240–5250.

Sgonc, R., Wick, G., 2008. Pro- and anti-fibrotic effects of TGF-beta in scleroderma. Rheumatology (Oxford) 47 (Suppl 5), v5–v7.

Shah, N.N., et al., 2020. Galactin-3 and soluble ST2 as complementary tools to cardiac MRI for sudden cardiac death risk stratification in heart failure: a review. JRSM Cardiovascular Disease 9 p. 2048004020957840.

Sofias, A.M., et al., 2021. Therapeutic and diagnostic targeting of fibrosis in metabolic, proliferative and viral disorders. Advanced Drug Delivery Reviews 175, 113831.

Song, S., et al., 2005. PDGFRbeta+ perivascular progenitor cells in tumours regulate pericyte differentiation and vascular survival. Nature Cell Biology 7 (9), 870–879.

Talman, V., Ruskoaho, H., 2016. Cardiac fibrosis in myocardial infarction-from repair and remodeling to regeneration. Cell and Tissue Research 365 (3), 563–581.

Tang, X., Li, P.H., Chen, H.Z., 2020. Cardiomyocyte senescence and cellular communications within myocardial microenvironments. Frontiers in Endocrinology 11, 280.

Tominaga, K., Suzuki, H.I., 2019. TGF-β signaling in cellular senescence and aging-related pathology. International Journal of Molecular Sciences 20 (20), 5002. Available at https://doi.org/10.3390/ijms20205002.

Tseng, W.Y., Su, M.Y., Tseng, Y.H., 2016. Introduction to cardiovascular magnetic resonance: technical principles and clinical applications. Acta Cardiologica Sinica 32 (2), 129–144.

van Buul, J.D., et al., 2010. ICAM-1 clustering on endothelial cells recruits VCAM-1. Journal of Biomedicine and Biotechnology 2010, 120328.

van Doorn, E.C.H., et al., 2024. Advancing 3D engineered in vitro models for heart failure research: key features and considerations. Bioengineering (Basel) 11 (12), 1220. Available at https://doi.org/10.3390/bioengineering11121220.

Wagner, K.D., Wagner, N., 2022. The senescence markers p16INK4A, p14ARF/p19ARF, and p21 in organ development and homeostasis. Cells 11 (12), 1966. Available at https://doi.org/10.3390/cells11121966.

Wang, J., et al., 2021. Non-viral gene delivery systems for treatment of myocardial infarction: targeting strategies and cardiac cell modulation. Pharmaceutics 13 (9), 1520. Available at https://doi.org/10.3390/pharmaceutics13091520.

Weber, A., Benitez, R., Toca-Herrera, J.L., 2022. Measuring biological materials mechanics with atomic force microscopy - determination of viscoelastic cell properties from stress relaxation experiments. Microscopy Research and Technique 85 (10), 3284–3295.

Widmer, R.J., Lerman, A., 2014. Endothelial dysfunction and cardiovascular disease. Global Cardiology Science and Practice 2014 (3), 291–308.

Wu, C.M., et al., 2020. The emerging role of cell senescence in atherosclerosis. Clinical Chemistry and Laboratory Medicine: CCLM / FESCC 59 (1), 27–38.

Xie, J.G., Wang, D.P., 2023. The role of endothelial-to-mesenchymal transition (EndoMT) in organ fibrosis. American Journal of Medical and Clinical Research & Reviews 2 (11), 1–8.

Xie, S., et al., 2023. Metabolic landscape in cardiac aging: insights into molecular biology and therapeutic implications. Signal Transduction and Targeted Therapy 8 (1), 114.

Xu, C., et al., 2025. The role of cellular senescence in cardiovascular disease. Cell Death Discovery 11 (1), 431.

Yan, X., et al., 2025. The mechanisms associated with inflammation and coronary microvascular dysfunction in heart failure with preserved ejection fraction. Medical Principles and Practice: International Journal of the Kuwait University, Health Science Centre 35 (2), 101–113. Available at https://doi.org/10.1159/000548233.

Yong, J., et al., 2025. Post-myocardial infarction cardiac remodeling: multidimensional mechanisms and clinical prospects of stem cell therapy. Stem Cell Reviews and Reports 21 (5), 1369–1427.

Zamilpa, R., Lindsey, M.L., 2010. Extracellular matrix turnover and signaling during cardiac remodeling following MI: causes and consequences. Journal of Molecular and Cellular Cardiology 48 (3), 558–563.

Zeng, H., Chen, J.X., 2019. Microvascular rarefaction and heart failure with preserved ejection fraction. Frontiers in Cardiovascular Medicine 6, 15.

Zhai, P., Sadoshima, J., 2024. Cardiomyocyte senescence and the potential therapeutic role of senolytics in the heart. Journal of Cardiovascular Aging 4 (2), 18. Available at https://doi.org/10.20517/jca.2024.06.

Zhang, H., et al., 2025. Application of human cardiac organoids in cardiovascular disease research. Frontiers in Cell and Developmental Biology 13, 1564889.

Zhang, J.Q., et al., 2023. Cellular senescence of renal tubular epithelial cells in renal fibrosis. Frontiers in Endocrinology 14, 1085605.

Zhang, L., et al., 2023a. Targeting cellular senescence with senotherapeutics: senolytics and senomorphics. The FEBS Journal 290 (5), 1362–1383.

Zhao, S., et al., 2025. Cellular senescence as a key player in chronic heart failure pathogenesis: Unraveling mechanisms and therapeutic opportunities. Progress in Biophysics and Molecular Biology 196, 8–18.

Zhou, X., et al., 2024. Mitophagy and cGAS-STING crosstalk in neuroinflammation. Acta Pharmaceutica Sinica B 14 (8), 3327–3361.

Chapter 10

Metabolic disorders, obesity, and accelerated cardiac senescence

Ayesha Sultana[1], Nasreen Sulthana[2], Pathakota Roja[2], Amita Joshi Rana[3], Md Sadique Hussain[4,5], and Bindu Madhavi Boddupalli[6]

[1]*Department of Pharmaceutics, Yenepoya Pharmacy College & Research Centre, Yenepoya (Deemed to be University), Mangalore, Karnataka, India,* [2]*Department of Pharmacology, Sree Dattha Institute of Pharmacy, Sheriguda, Ibrahimpatnam, Hyderabad, Telangana, India,* [3]*College of Pharmacy, Graphic Era Hill University, Bhimtal, Uttarakhand, India,* [4]*Uttaranchal Institute of Pharmaceutical Sciences, Uttaranchal University, Dehradun, Uttarakhand, India,* [5]*School of Pharmaceutical Sciences, Lovely Professional University, Phagwara, Punjab, India,* [6]*Department of Pharmaceutics, School of Pharmacy, Mount Kenya University, Thika, Kenya*

10.1 Introduction

Metabolic disorders are categorized as acquired and congenital metabolic disorders. The most common acquired metabolic disorders include obesity, diabetes, dyslipidemia, hyperthyroidism, Cushing syndrome, hyperuricemia, and haemochromatosis (Garg et al., 2022). These metabolic abnormalities are mostly caused by stress and a westernized lifestyle. Due to stress, superoxide production, lipid metabolism, rise in insulin resistance, metabolic syndrome is becoming more complex every day (Jakubiak et al., 2021). These metabolic disorders contribute to heart diseases and cause millions of deaths annually (Eroglu et al., 2024).

The breakdown of cell-specific biochemical functions in metabolic diseases impairs the capacity to efficiently absorb and use vital nutrients such as proteins, lipids, and carbohydrates. These processes impact energy utilization, oxidation of fatty acids, protein to amino acid conversion, and fat retention (Amatruda et al., 2019). There is an existing interplay between all the diseases occurring due to metabolic disorders. Obesity, diabetes, and cardiac senescence are closely associated (Marino et al., 2022). The factors interlinking these three disorders suggest high body mass index, abdominal obesity, increased hypertension, elevated lipid profiles, leading to type 2 diabetes, plaque formation, persistent inflammation, risk of coronary diseases, strokes, heart failure, and increased cardiovascular mortality (Pakhare and Anjankar, 2024).

Research shows that chronological aging is related to age-related conditions like heart failure, diabetes, and atherosclerotic disease, and the pathophysiology of these conditions is driven by senescent cell accumulation. A physiological condition known as senescence occurs when cells cease to divide yet remain alive and metabolically active. Senescent cell accumulation is due to aging (Majewska and Krizhanovsky, 2025). Changes in the structure of cells, gene regulation, and activity are its most distinctive features. Although we still fail to comprehend, it is essential to the development of the detrimental effects of aging (Katsuumi et al., 2018).

Three types of cell senescence may be distinguished as stress-induced senescence (caused by oxidative stress or anticancer drugs), replicative senescence (shortening of telomeres), and oncogene-driven senescence (caused by activation of oncogene) (González-Gualda et al., 2021; Hussain et al., 2025). One of the main reasons for cell aging has been suggested to be oxidative damage, which may be brought on by an excess of ROS (reactive oxygen species) or a lack of antioxidant defense (Colavitti and Finkel, 2005). Metabolic inducers of cell senescence cause sirtuin to decrease its activity when NAD is lost. Sirtuins have several anti-senescent properties because they regulate energy metabolism, encourage DNA damage repair, and reduce inflammation (Grabowska et al., 2017). Oxidative stress promotes cell senescence by damaging DNA and accelerating mitochondrial malfunction due to increased ROS production and oxidative mitochondrial DNA damage. Defective autophagy and mitophagy can result in cell senescence (Grootaert, 2024).

Cellular Senescence and Cardiac Aging. DOI: https://doi.org/10.1016/B978-0-443-45811-8.00017-9

Research shows that "vascular senescence" or "cellular senescence" in the vasculature is a major contributor to the advancement of cardiovascular diseases (CVDs) and metabolic disorders. Atherosclerosis, systolic heart failure, and systemic metabolic dysfunction are all linked to vascular senescence (Fig. 10.1) (Gogiraju et al., 2015; Yoshida et al., 2015; Yokoyama et al., 2014; Wang and Bennett, 2012). Obese or heart failure patients have greater amounts of p53 in their visceral fat, aging arteries, and failing hearts (Gutiérrez-Cuevas et al., 2021). p53-induced cell death is bad for aging and age-related diseases such as plaque, diabetes, heart failure, and obesity (Vaddavalli and Schumacher, 2022). According to research, senescent cells may induce tissue transformation in age-associated disorders, indicating that cell senescence suppression or elimination may be a viable next-generation therapeutic option. A recent study found that genetic engineering might prevent age-associated deterioration in the kidneys and heart of mice by removing senescent cells (Baker et al., 2016).

Cellular senescence, oxidative stress, and metabolic dysregulation are all linked to inflammation (Leyane et al., 2022), which affects heart function. Inflammation is a significant issue in obesity and diabetes, triggered by NF-κB and NLRP3 inflammasome activity, resulting in the release of inflammatory mediators and profibrotic cytokines (Rohm et al., 2022). Because inflammation plays a part in cardiac aging, cytokines may help slow the heart's aging process. However, it is uncertain how well they can prevent coronary artery disease. Alterations in epigenetics are another indicator of aging that is regulated by metabolism. Epigenetic clocks can use DNA methylation at specific genomic locations to determine age in biology (Horvath and Raj, 2018). Histone variations play particular roles in genomic stability and transcriptional control (Ferrand et al., 2020). Research on cardiac aging is expected to grow significantly, given the present illness patterns in our aging population. Knowing the causes of cardiac aging that cause cardiac failure while preserving the stable ejection rate is essential to avoid this condition. Finding triggers, pathogenic nodes, and increasing the aged myocardium's vulnerability to other cardiac disorders depend on understanding the molecular processes driving cardiac aging (Derumeaux et al., 2024).

Age-related cardiac failure is caused by extensive exposure to various stressors, including sporogenous, inherited, reactive, autophagic, and inflammatory factors, along with the accumulation of senescent cells (Fang et al., 2025). Cardiac aging is a natural process that causes cardiovascular activity to deteriorate, increasing vulnerability, and vascular remodeling, which is exacerbated by metabolic, inflammatory, and senescence factors. Exercise and calorie restriction slow it down, but obesity and diabetes accelerate it. Developing preventive and therapeutic strategies requires understanding heart aging (Hu et al., 2025). The progressive deterioration of heart shape and functionality brought on by aging cells and molecules is known as cardiac senescence. It is distinguished by hypertrophy of cardiomyocytes,

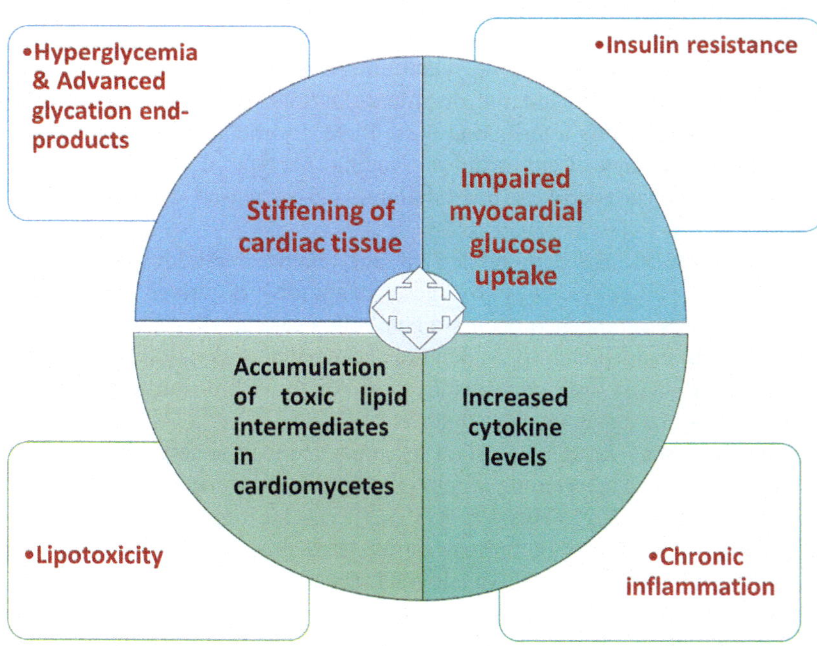

FIGURE 10.1 **llustrates how senescence can be activated by insulin resistance, hyperglycemia, lipotoxicity, chronic inflammation, and the changes occurring.** Senescence in cardiac tissues is triggered by metabolic factors such as insulin resistance, hyperglycemia, lipotoxicity, and chronic inflammation. This figure visualizes molecular pathways, including oxidative stress, impaired mitophagy, and changes in sirtuin activity, that accelerate cardiac aging. It highlights how these interconnected disturbances promote cellular senescence, fibrosis, and decline in cardiac function.

remodeling of the extracellular matrix and fibrosis, mitochondrial dysfunction, telomere shortening, elevated levels of indicators of senescence (like $p16^{INK4a}$ and $p21^{CIP1}$), and altered proteostasis and autophagy (Mehdizadeh et al., 2022). The detrimental factors for CVD include metabolic conditions such as non-insulin-dependent diabetes mellitus, obesity, and insulin resistance. There is growing evidence that these disorders accelerate cardiac aging (cardiac senescence) as well as circulatory dysfunction. The effects of aging on metabolism are significant. For example, obesity and diabetes exacerbate mitochondrial dysfunction, a crucial aging characteristic. Mitochondrial DNA damage, p53-linked senescence, and alterations in fission/fusion all prevent mitochondrial turnover in aging and metabolic diseases. This may change cellular metabolism, hinder autophagy, encourage apoptosis, and cause senescence. Mitochondrial dysfunction and its consequences are crucial targets in reducing cardiac aging. Furthermore, senescent pathways are activated by persistent DNA damage in telomere areas without a change in telomere length, which induces the cyclin-dependent kinase inhibitors $p16^{Ink4a}$, $p15^{Ink4b}$, and $p21^{CIP}$ (Anderson et al., 2019; Martini et al., 2021) (Fig. 10.1).

10.2 Pathways linking metabolic disorders and cardiovascular senescence

The main cause of metabolic illnesses is deficiencies in the enzyme needed to convert one metabolite into another. The abnormalities or indicators of metabolic illnesses are caused by either an excess of one metabolite or a deficiency of one or more metabolites. Metabolic syndrome, prevalent in the context of the obesity and diabetes epidemic, greatly increases the risk of CVDs and diabetic complications (Fig. 10.2) (Nagamani et al., 2019). The ability of a cell to perform its original role is compromised as it becomes senescent, which may result in a reduction in the overall function of the tissue. This might be partially caused by significant changes in gene expression; some of them could be exclusive to a particular cell type, while others might be confined to senescence. Furthermore, it has been proposed that aging cells display indiscriminate gene regulation because of the random activation of genes brought on by continuous chromatin remodeling (Burton and Faragher, 2018). Research also indicates that senescent cells in pancreatic beta cells may contribute to diabetes pathogenesis due to their role in insulin production. Dysfunction in these cells can impair insulin secretion, influenced by factors such as inflammation, autoimmunity, oxidative stress, and ER stress (Murakami et al., 2022).

Heart aging is closely linked to non-insulin-dependent diabetes mellitus through several pathways: (i) High blood sugar causes the formation of advanced glycation end products (AGEs), which harden the cardiac matrix and decrease contractility. (ii) Insulin resistance drives the myocardium to prefer fatty acids over glucose, leading to inefficient energy utilization and heightened oxidative stress. (iii) Cardiomyocyte death is the outcome of a mitochondrial malfunction, which raises the concentration of reactive oxygen species and lowers ATP synthesis; a decrease in sex steroids, growth-stimulating hormone (GH), and insulin-like growth factor-1 (IGF-1) is also reported (Rezuş et al., 2020).

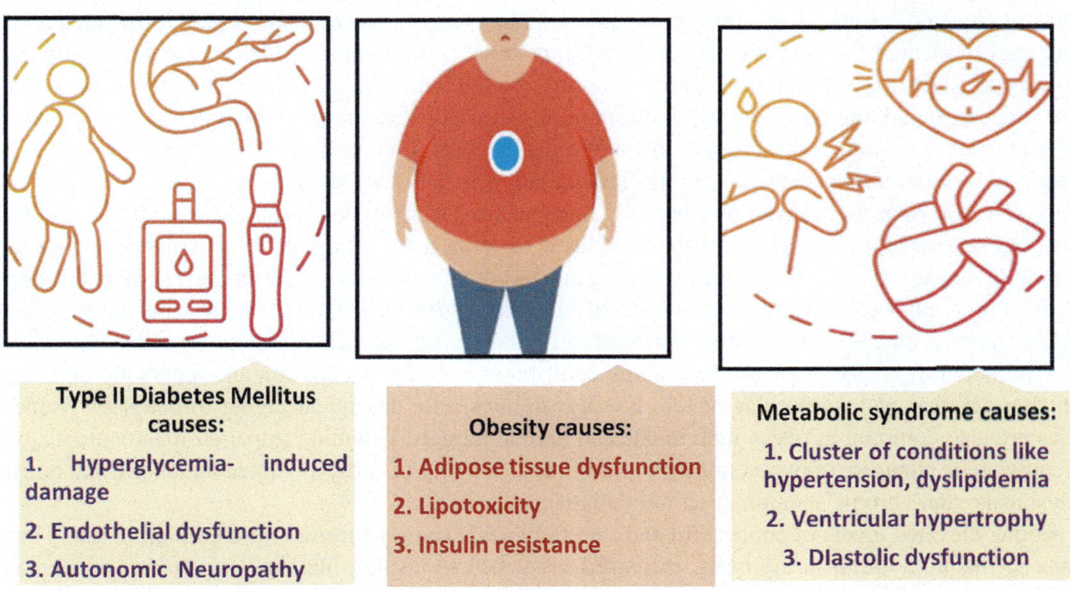

FIGURE 10.2 **Metabolic disorders, obesity, and accelerated cardiac senescence cause metabolic syndrome.** This figure demonstrates how metabolic syndrome arises from obesity and metabolic disorders, and its consequences on cardiovascular aging. Key components such as hyperglycemia, insulin resistance, and dyslipidemia interact to drive chronic inflammation and cardiac remodeling. It emphasizes the vicious cycle where cardiac senescence further exacerbates metabolic dysfunction and disease progression.

Hyperinsulinemia and related insulin resistance (IR) have been linked to cardiac senescence in human adipose and liver cells, according to recent research. Their interconnectedness is demonstrated by the fact that elevated cardiac senescence also stimulates cellular IR (Spinelli et al., 2023). Diabetes also raises the chance of adverse consequences, such as diabetic retinal damage, peripheral neuropathy, and heart failure, by causing microvascular and macrovascular damage. These disorders impair movement, impair sensory and motor abilities, and have a detrimental impact on general well-being (Zhang et al., 2023).

Apart from insulin resistance, dyslipidemia (lipotoxicity) is a condition where excess circulating lipids can enter the heart, which can result in chronic low-grade inflammation, increased risk of mitochondrial damage, and cardiomyocyte death (Mann et al., 2024). Dyslipidemia raises the risk of CVDs and causes thrombosis. It includes excessive cholesterol, hypertriglyceridemia, and increased low-density lipoproteins (LDLs) (Nakamura, 2024). Meta-inflammation and chronic low-grade inflammatory states are frequently accompanied by metabolic alterations that are directly associated with the onset of CVDs. Because of oxidative stress and persistently inflamed fat tissue, meta-inflammation is characterized by high blood levels of cytokines that trigger inflammation, including IL-1β, IL-6, and TNF-α (Russo et al., 2021). The ROS intensifies this stress process, which causes posttranslational modifications in genetic material, fatty acids, and peptides. AGEs, which are connected to dysfunction in cells and CVDs, are formed when oxidative stress is aggravated by hyperglycemia (Silveira Rossi et al., 2022).

Aging significantly affects cardiac health by impairing metabolic sensing mechanisms, specifically through the roles of AMPK and sirtuins. These key intracellular sensors, essential for functions like mitochondrial health and oxidative balance, become less effective with age, resulting in elevated oxidative damage and impaired mitochondrial activity. This dysfunction contributes to the progression of heart-related conditions. One significant stress sensor is the AMPK pathway, frequently downregulated in metabolic disorders, impairing mitochondrial biogenesis. Sirtuins, notably SIRT1 and SIRT3, are downregulated in obesity/diabetes, guarding against inflammation, malfunction of mitochondrial activity, and DNA, which is frequent in metabolic diseases and aging, leading to bioenergetic failure (Barcena et al., 2025). As the heart ages, autophagy is also affected due to downregulation of autophagy genes, increased mTOR activity, mitochondrial impairment, and contractile dysfunction.

10.3 Obesity and cardiac senescence

Obesity affects the overall cells, tissues, and organs, contributing to cardiac aging. Strong clinical and experimental data point to a direct molecular connection between aging and obesity in the etiology of several age-related CVDs (Ruperez et al., 2025). The adipose tissue is essential to sustaining blood glucose, fat, and energy regulation, the greatest energy reserve, and an endocrine organ; nevertheless, these metabolic processes deteriorate with aging and obesity (Liu et al., 2020). It leads to elevated insulin levels, fat accumulation, and a changed biochemical profile, downregulating cognitive capacity and cardiovascular risk factors. These conditions cause arrhythmias, autonomic dysfunction, and increase demand for oxygen and rise ventricular filling pressure (Volpe and Gallo, 2023). Due to impaired metabolism, inflammatory processes, and altered adipocytokine formation, fat tissue senescence, which is common in overweight and aging individuals, causes insulin resistance and metabolic dysfunction. It contributes to the development of diabetes and other metabolic deteriorations by causing chronic inflammation that affects metabolic organs (Zhao and Yue, 2024). Various kinds of cells, such as immune system cells and mature fat cells, experience senescence at different rates in distinct fat depots, emphasizing their complexity. In addition to being a sign of aging, mitochondrial dysfunction is also a result of obesity (Cojocaru et al., 2023). Obesity results in heart dysfunction and ubiquitinated protein buildup (Ye et al., 2024). Both systemic and local consequences of obesity lead to heart aging. Even though it has been reported that fat tissue comprises more aged cells, the curative potential of regulating senescence in metabolic and CVDs associated with obesity is yet unknown. It is also unknown how aging contributes to heart dysfunction and metabolic diseases caused by obesity. At the cellular and molecular level, obesity mimics the biological aging processes. Numerous aging characteristics are also common in obese individuals and animal models, including impaired macroautophagy, decreased proteostasis, epigenetic changes, impaired mitochondrial function, aging of cells, disrupted metabolic and nutrient-sensing pathways, dysbiosis, and chronic inflammation (Abdellatif et al., 2023).

In obese people, elevated levels of cholesterol and free fatty acids cause abnormal lipid buildup in many organs, which can contribute to plaque buildup in the heart, increased resistance to insulin, metabolic disorders, diabetes, high blood pressure, and heart disease (Henning, 2021). Dysfunction of visceral white fat cells aggravates these problems by causing unfavorable changes in the gut bacteria, activating the renin-angiotensin-aldosterone pathway, increasing inflammation-promoting substances, and promoting oxidative stress (Gutiérrez-Cuevas et al., 2021; Xin et al., 2019; Yoosuf et al., 2021).

TABLE 10.1 Accelerated cardiac senescence pathophysiology. This table summarizes the mechanisms involved in the accelerated cardiac senescence and the factors causing it.

S. no.	Accelerated cardiac senescence causes	Mechanism
1.	Mitochondrial dysfunction and cellular stress	• Free radicals hinder the synthesis of ATP and alter mitochondrial DNA. • Dysfunctional mitochondria promote apoptosis and senescence-associated secretory phenotype (SASP).
2.	Cell senescence	• Among cardiac fibroblasts and cardiomyocytes, permanent cell cycle disruption occurs. • Senescent cells exacerbate tissue remodeling through the release of growth regulators, matrix-degrading enzymes, and cytokines associated with inflammation.
3.	Epigenetic changes	• Aging phenotypes are caused by changes in gene expression patterns brought on by metabolic stress-related DNA methylation and histone alterations.
4.	Telomere shortening	• In cardiac cells, telomere attrition is accelerated by obesity and metabolic dysfunction, which lowers the cells' ability for repair and proliferation.

Obesity elevates leptin and resistin levels, contributing to endothelial dysfunction and fibrosis (Engin, 2024). Low adiponectin levels reduce (Fig. 10.2) the myocardium's antiinflammatory and metabolic functions. Additionally, chronic inflammation in cardiac cells leads to the senescence-associated secretory phenotype (SASP) (Khavinson et al., 2023). It can induce tissue damage, cellular dysfunction, persistent inflammation, organ aging, suppression of stem cells, and imbalance (Wang et al., 2024). Epicardial adipose tissue affects the myocardium and coronary arteries by releasing cytokines like TNF-α and IL-6 (Konwerski et al., 2022), which contribute to atrial fibrillation and myocardial fibrosis. Increased sympathetic tone and volume overload elevate myocardial stress, leading to hypertrophy and heart failure.

Contemporary research indicates that leptin, a hormone produced by adipocytes, is linked to heart disease, metabolic syndrome, and obesity (Vilariño-García et al., 2024). Under normal conditions, leptin plays a role in regulating pressure and volume; however, persistent resistance to leptin and elevated levels of hyperleptinemia may be linked to heart and kidney disorders as well as hypertension (Poetsch et al., 2020). Future studies will explore whether reducing leptin levels or blocking its actions can protect heart and kidney health in obese individuals diagnosed with metabolic syndrome. Obesity-related inflammatory signals activate serine kinases, which inhibit insulin action (Saltiel and Olefsky, 2017).

Age-associated cardiac failure is a condition where individuals develop heart failure without having any known history of detrimental conditions, such as high blood pressure, obesity, insulin resistance, or plaque-associated heart diseases. It seems that the physiological aging process renders people more susceptible to the progression of cardiac failure, given that the prevalence of cardiac disease rises with age. A major contributing element to the pathophysiology of heart failure is cellular senescence, as several studies have shown (Shih et al., 2011; Dhalla et al., 2012). Vascular endothelial cell (EC) senescence is a condition brought on by aging and is typified by an increase in ROS, p53/p21, p16, SA-β-Gal, and a response to inflammation, as well as a decrease in eNOS/NO levels (Bu et al., 2023). The development of diabetes and CVDs is significantly influenced by EC senescence (Marino et al., 2022). Physical activity may prevent ECs from senescing (Meng et al., 2023).

Aging is linked to arterial remodeling that increases the intima/media thickness ratio, a characteristic of aged arteries, which can predispose individuals to CVDs (Castelli et al., 2023). Aging is a normal process, but a rise in the intima/media thickness ratio is a feature of aging arteries, which has been demonstrated to increase two to three times between the ages of 20 and 90 (Lakatta, 2013; Lakatta and Levy, 2003). Key mechanisms are summarized in Table 10.1.

10.4 Factors impacted by metabolic disorders and cardiac senescence

Structural and functional changes due to aging cause the "contractile" characteristic of vascular smooth cells in muscles to change to a "synthetic" type as people age. This transition causes intimal thickening, which raises artery permeability and causes atherosclerotic disease. Furthermore, the arterial media widens and lacks cellularity with age (van der Laan et al., 2025). The aorta's length and circumference expand as individuals age due to elevated collagen synthesis and a decline in elastin concentration (Mauriello et al., 1992). Age-related changes in artery stiffness, elasticity, and compliance lead to a drop in diastolic blood pressure and a spike in systolic blood pressure (Herzog et al., 2025; Samuel et al., 2019; Shaik Mohamed Sayed et al., 2023).

Arterial diseases and vascular senescence have particular characteristics of aging arteries; it is the buildup of senescent vascular smooth muscle cells (VSMCs), aged fat-laden cells, and old vascular EC. Increased levels of free radicals, p53/p21, p16, SA-β-Gal, inflammation, gradual shortening of telomere, and decreased levels of eNOS/NO are linked to the accumulation of these aging cells. The senescence of these cells supports damaging changes in arteries, causing hardening and the onset of hypertension (Morgan et al., 2013; Ramanunny et al., 2022; Reshma et al., 2023).

ECs, which are essential for preserving vascular equilibrium and biological activities, including clotting, blood pressure control, angiogenesis, and basic metabolic processes, exhibit senescence in arterial diseases. Aging ECs display dysfunctional phenotypes, including reduced migration, decreased angiogenic molecule expression, and poor nitric oxide production, which promotes vasodilatation. Defective ECs often decrease NOS activity, impairing Endothelium-dependent dilatation (EDD), a predictor of CVDs, and contribute to the development of pro-oxidant, pro-inflammatory, vasoconstrictor, and prothrombotic characteristics. Atherosclerotic plaque has been discovered to contain senescent endothelium cells (Minamino et al., 2002). Aging is linked to elevated levels of pro-inflammatory cytokines, including interleukin-6, tumor necrosis factor alpha, and monocyte chemoattractant protein-1, which have been identified as significant indicators of inflammatory response (Ungvari et al., 2010; Chan et al., 2020; Clarence et al., 2022).

Due to aging senescence of vascular smooth muscle cells in arterial disorders, transition from contractile to synthetic due to disruption in TGF-β signaling, resulting in elevated expression of ICAM-1, angiotensinogen, and improved inducible NOS (iNOS) activity (Faggiotto et al., 1984; Lakatta, 2015). As individuals age, their intima thickens due to a rise in collagen synthesis and a decline in elastin levels. VSMCs exhibit dysregulated synthesis of growth regulators, extracellular matrix modulators, and cytokines that are pro-inflammatory. When the cells are exposed to free radicals and angiotensin II, this dysregulation speeds up the remodeling of the vascular system (Herbert et al., 2008; Zhao et al., 2015).

The process of immune cell senescence in arterial disorders has been demonstrated to be associated with the length of the telomere in human vascular cells. A robust correlation has been observed between telomere shortening and the development of diabetes and CVDs (D'Mello et al., 2015; Haycock et al., 2014). The mortality rate is greater among older people with shorter telomeres in their leukocytes, as heart disease causes more deaths (Cawthon et al., 2003). MCP-1, IL-6, IL-1β, and TNF-α are among the cytokines that are pro-inflammatory and free radicals that are produced in greater quantities in individuals with atherosclerosis (Calvert et al., 2011).

It has been shown that systemic metabolic conditions and organ dysfunction are caused by the senescence of ECs in metabolic syndrome, which stems from vascular disruption, whereas capillary networks are essential for evolution and regulation (Ferrara, 2004; Shimizu et al., 2014). Although skeletal muscle regulation is essential for systemic metabolic function as it facilitates the removal of glucose, metabolic stress can result in insulin resistance and fat accumulation (Gilbert, 2021). It was shown that systemic glucose intolerance results from the suppression of skeletal muscle metabolism caused by EC senescence. Dietary obesity-induced metabolic stress raises the expression of p53 inside the vascular endothelium (Yokoyama et al., 2014; Hussain et al., 2021; Kaur et al., 2025).

The development of insulin resistance and the advancement of metabolic diseases have been linked to the effects of metabolic disorder and vascular senescence, which cause persistent inflammation in visceral adipose tissue. This condition is especially noticeable in people with diabetes, obesity, and heart failure. Reduced insulin signaling and increased MAPK signaling in vascular cells are hallmarks of vascular senescence, which is linked to diabetes and obesity. This results in specific insulin resistance, which causes vascular remodeling by cell migration, proliferation, and vasoconstriction (Hill et al., 2021).

Mitochondrial impairments during cardiac aging predominantly affect interfibrillar mitochondria (IFM), which are crucial for maintaining cardiac function (Hinton et al., 2024). Unlike subsarcolemmal mitochondria (SSM), IFM decline correlates with age-related cardiac dysfunction. Studies utilizing transmission electron microscopy (TEM) showed that aging significantly reduces the inner mitochondrial membrane (IMM) in cardiomyocytes, though cristae morphology remains unchanged. The impaired renewal and clearance of IFM contribute to abnormal mitochondrial accumulation and increased free radical species, exacerbating cellular stress and promoting aging phenotypes. Notably, aging hearts exhibit swelling mitochondria with damaged cristae, indicating that mitochondrial integrity is compromised early in the process, preceding cardiac hypertrophy and fibrosis (Xie et al., 2023). Considering hypoxia-regulated proteins in relation to aging, such as peroxisome proliferator-activated receptors (PPARs) and hypoxia-inducible factor-1 (HIF-1), is essential since hypertrophy is a frequent sign of cardiac aging and can interfere with oxygen delivery, resulting in hypoxia (Belanger et al., 2002). HIF-1 is essential for controlling the heart's energy metabolism (Cerychova and Pavlinkova, 2018). HIF-1α facilitates the metabolic transition from oxidative to glycolytic pathways in hypoxic environments (Kierans and Taylor, 2021; Knutson et al., 2021; Ng et al., 2019; Pandey et al., 2022).

10.5 Conclusion and future perspective

Through systemic inflammation, oxidative stress, and metabolic disturbances, metabolic diseases and obesity hasten cardiac senescence, making tailored therapies necessary to enhance cardiovascular health in the elderly. Focusing on inflammatory pathways, epigenetic modulation, and mitochondrial health may provide novel strategies to slow the heart's aging caused by metabolic diseases. Exercise and diet can boost mitochondrial biogenesis, reduce obesity, and increase insulin sensitivity. There is evidence that limiting calories and fasting for an intermittent period can postpone the aging of the heart. Interventions for positive aging and reducing the risk of metabolic conditions can be enhanced by encouraging stress management, having access to mental health care, and healthy lifestyle modifications.

Human health is positively impacted by a variety of dietary components, including complex carbs, omega-3 fatty acids, fermented foods, prebiotics, polyphenols, and spices. They lower the risk of gastrointestinal diseases and metabolic dysfunction by acting as efficient remedies. Consuming sufficient quantities regularly aids in the restoration of a healthy gut microbiota, which enhances insulin signaling, physiological functions, and immunological responses while reducing chronic inflammation. A pharmacological drug like Metformin is a medication that activates AMPK, enhances mitochondrial function, and has antisenescent properties. SGLT2 inhibitors have been shown to improve cardiac outcomes in both patients with and without diabetes. Senolytics, including Dasatinib and Quercetin, are new medications that eliminate senescent cells. To delay cardiac aging, several therapies focusing on metabolism and mitochondria are recommended, including antioxidants, mitophagy inhibitors, Nampt supplementation, and sirtuin stimulation.

References

Abdellatif, M., Rainer, P.P., Sedej, S., Kroemer, G., 2023. Hallmarks of cardiovascular ageing. Nature Reviews Cardiology 20 (11), 754–777. Available at https://doi.org/10.1038/s41569-023-00881-3.

Amatruda, M., Ippolito, G., Vizzuso, S., Vizzari, G., Banderali, G., Verduci, E., 2019. Epigenetic effects of n-3 LCPUFAs: a role in pediatric metabolic syndrome. International Journal of Molecular Sciences 20 (9), 2118. Available at https://doi.org/10.3390/ijms20092118.

Anderson, R., Lagnado, A., Maggiorani, D., Walaszczyk, A., Dookun, E., Chapman, J., et al., 2019. Length-independent telomere damage drives post-mitotic cardiomyocyte senescence. EMBO Journal 38 (5). Available at https://doi.org/10.15252/embj.2018100492.

Baker, D.J., Childs, B.G., Durik, M., Wijers, M.E., Sieben, C.J., Zhong, J., et al., 2016. Naturally occurring p16 Ink4a-positive cells shorten healthy lifespan. Nature 530 (7589), 184–189. Available at https://doi.org/10.1038/nature16932.

Barcena, M.L., Aslam, M., Norman, K., Ott, C., Ladilov, Y., 2025. Role of AMPK and sirtuins in aging heart: basic and translational aspects. International Society on Aging and Disease, Germany Aging and Disease 16 (5). Available at https://doi.org/10.14336/AD.2024.1216.

Belanger, A.J., Lu, H., Date, T., Liu, L.X., Vincent, K.A., Akita, G.Y., et al., 2002. Hypoxia up-regulates expression of peroxisome proliferator-activated receptor γ angiopoietin-related gene (PGAR) in cardiomyocytes: role of hypoxia inducible factor 1α Journal of Molecular and Cellular Cardiology 34 (7), 765–774. Available at https://doi.org/10.1006/jmcc.2002.2021.

Bu, L.L., Yuan, H.H., Xie, L.L., Guo, M.H., Liao, D.F., Zheng, X.L., 2023. New dawn for atherosclerosis: vascular endothelial cell senescence and death. International Journal of Molecular Sciences 24 (20). Available at https://doi.org/10.3390/ijms242015160.

Burton, D.G.A., Faragher, R.G.A., 2018. Obesity and type-2 diabetes as inducers of premature cellular senescence and ageing. Biogerontology 19 (6), 447–459. Available at https://doi.org/10.1007/s10522-018-9763-7.

Calvert, P.A., Liew, T.V., Gorenne, I., Clarke, M., Costopoulos, C., Obaid, D.R., et al., 2011. Leukocyte telomere length is associated with high-risk plaques on virtual histology intravascular ultrasound and increased proinflammatory activity. Arteriosclerosis, Thrombosis, and Vascular Biology 31 (9), 2157–2164. Available at https://doi.org/10.1161/ATVBAHA.111.229237.

Castelli, R., Gidaro, A., Casu, G., Merella, P., Profili, N.I., Donadoni, M., et al., 2023. Aging of the arterial system. International Journal of Molecular Sciences 24 (8), 6910. Available at https://doi.org/10.3390/ijms24086910.

Cawthon, R.M., Smith, K.R., O'Brien, E., Sivatchenko, A., Kerber, R.A., 2003. Association between telomere length in blood and mortality in people aged 60 years or older. Lancet 361 (9355), 393–395. Available at https://doi.org/10.1016/S0140-6736(03)12384-7.

Cerychova, R., Pavlinkova, G., 2018. HIF-1, metabolism, and diabetes in the embryonic and adult heart. Frontiers in Endocrinology 9. Available at https://doi.org/10.3389/fendo.2018.00460.

Chan, Y., Ng, S.W., Xin Tan, J.Z., Gupta, G., Tambuwala, M.M., Bakshi, H.A., et al., 2020. Emerging therapeutic potential of the iridoid molecule, asperuloside: a snapshot of its underlying molecular mechanisms. Chemico-Biological Interactions 315, 108911. Available at https://doi.org/10.1016/j.cbi.2019.108911.

Clarence, D.D., Paudel, K.R., Manandhar, B., Singh, S.K., Devkota, H.P., Panneerselvam, J., et al., 2022. Unravelling the therapeutic potential of nano-delivered functional foods in chronic respiratory diseases. Nutrients 14 (18). Available at https://doi.org/10.3390/nu14183828.

Cojocaru, K.A., Luchian, I., Goriuc, A., Antoci, L.M., Ciobanu, C.G., Popescu, R., et al., 2023. Mitochondrial dysfunction, oxidative stress, and therapeutic strategies in diabetes, obesity, and cardiovascular disease. Antioxidants 12 (3). Available at https://doi.org/10.3390/antiox12030658.

Colavitti, R., Finkel, T., 2005. Reactive oxygen species as mediators of cellular senescence. IUBMB Life 57 (4-5), 277–281. Available at https://doi.org/10.1080/15216540500091890.

Derumeaux, G., Sawaki, D., Czibik, G., 2024. Trajectories in cardiac aging: journey to the crossroads of inflammation and metabolism. Circulation 149 (25), 1935–1937. Available at https://doi.org/10.1161/CIRCULATIONAHA.123.065116.

Dhalla, N.S., Rangi, S., Babick, A.P., Zieroth, S., Elimban, V., 2012. Cardiac remodeling and subcellular defects in heart failure due to myocardial infarction and aging. Heart failure reviews 17 (4-5), 671–681. Available at https://doi.org/10.1007/s10741-011-9278-7.

D'Mello, M.J.J., Ross, S.A., Briel, M., Anand, S.S., Gerstein, H., Paré, G., 2015. Association between shortened leukocyte telomere length and cardiometabolic outcomes: systematic review and meta-analysis. Circulation: Cardiovascular Genetics 8 (1), 82–90. Available at https://doi.org/10.1161/CIRCGENETICS.113.000485.

Engin, A., 2024. Endothelial dysfunction in obesity and therapeutic targets. In: ENGIN, A.B., ENGIN, A. (Eds.), Obesity and Lipotoxicity. Advances in Experimental Medicine and Biology, vol 1460. Springer, Cham, pp. 489–538. Available at https://doi.org/10.1007/978-3-031-63657-8_17.

Eroglu, T., Capone, F., Schiattarella, G.G., 2024. The evolving landscape of cardiometabolic diseases. EBioMedicine 109. Available at https://doi.org/10.1016/j.ebiom.2024.105447.

Faggiotto, A., Ross, R., Harker, L., 1984. Studies of hypercholesterolemia in the nonhuman primate. I. Changes that lead to fatty streak formation. Arteriosclerosis 4 (4), 323–340. Available at https://doi.org/10.1161/01.atv.4.4.323.

Fang, Z., Raza, U., Song, J., Lu, J., Yao, S., Liu, X., et al., 2025. Systemic aging fuels heart failure: molecular mechanisms and therapeutic avenues. ESC Heart Failure 12 (2), 1059–1080. Available at https://doi.org/10.1002/ehf2.14947.

Ferrand, J., Rondinelli, B., Polo, S.E., 2020. Histone variants: guardians of genome integrity. Cells 9 (11). Available at https://doi.org/10.3390/cells9112424.

Ferrara, N., 2004. Vascular endothelial growth factor: basic science and clinical progress. Endocrine Reviews 25 (4), 581–611. Available at https://doi.org/10.1210/er.2003-0027.

Garg, C., Khan, H., Kaur, A., Singh, T.G., Sharma, V.K., Singh, S.K., 2022. Therapeutic implications of sonic hedgehog pathway in metabolic disorders: novel target for effective treatment. Pharmacological Research 179, 106194. Available at https://doi.org/10.1016/j.phrs.2022.106194.

Gilbert, M., 2021. Role of skeletal muscle lipids in the pathogenesis of insulin resistance of obesity and type 2 diabetes. Journal of Diabetes Investigation 12 (11), 1934–1941. Available at https://doi.org/10.1111/jdi.13614.

Gogiraju, R., Xu, X., Bochenek, M.L., Steinbrecher, J.H., Lehnart, S.E., Wenzel, P., et al., 2015. Endothelial p53 deletion improves angiogenesis and prevents cardiac fibrosis and heart failure induced by pressure overload in mice. Journal of the American Heart Association 4 (2). Available at https://doi.org/10.1161/jaha.115.001770.

González-Gualda, E., Baker, A.G., Fruk, L., Muñoz-Espín, D., 2021. A guide to assessing cellular senescence in vitro and in vivo. The FEBS journal 288 (1), 56–80. Available at https://doi.org/10.1111/febs.15570.

Grabowska, W., Sikora, E., Bielak-Zmijewska, A., 2017. Sirtuins, a promising target in slowing down the ageing process. Biogerontology 18 (4), 447–476. Available at https://doi.org/10.1007/s10522-017-9685-9.

Grootaert, M.O.J., 2024. Cell senescence in cardiometabolic diseases. Nature Research, Belgium npj Aging 10 (1). Available at https://doi.org/10.1038/s41514-024-00170-4.

Gutiérrez-Cuevas, J., Sandoval-Rodriguez, A., Meza-Rios, A., Monroy-Ramírez, H.C., Galicia-Moreno, M., García-Bañuelos, J., et al., 2021. Molecular mechanisms of obesity-linked cardiac dysfunction: an up-date on current knowledge. Cells 10 (3), 629. Available at https://doi.org/10.3390/cells10030629.

Haycock, P.C., Heydon, E.E., Kaptoge, S., Butterworth, A.S., Thompson, A., Willeit, P., 2014. Leucocyte telomere length and risk of cardiovascular disease: systematic review and meta-analysis. British Medical Journal 349 (jul08 3), g4227. Available at https://doi.org/10.1136/bmj.g4227.

Henning, 2021. Obesity and obesity-induced inflammatory disease contribute to atherosclerosis: a review of the pathophysiology and treatment of obesity. American Journal of Cardiovascular Disease 11 (4), 504–529.

Herbert, K.E., Mistry, Y., Hastings, R., Poolman, T., Niklason, L., Williams, B., 2008. Angiotensin II-mediated oxidative DNA damage accelerates cellular senescence in cultured human vascular smooth muscle cells via telomere-dependent and independent pathways. Circulation Research 102 (2), 201–208. Available at https://doi.org/10.1161/CIRCRESAHA.107.158626.

Herzog, M.J., Müller, P., Lechner, K., Stiebler, M., Arndt, P., Kunz, M., et al., 2025. Arterial stiffness and vascular aging: mechanisms, prevention, and therapy. Signal Transduction and Targeted Therapy 10 (1). Available at https://doi.org/10.1038/s41392-025-02346-0.

Hill, M.A., Yang, Y., Zhang, L., Sun, Z., Jia, G., Parrish, A.R., et al., 2021. Insulin resistance, cardiovascular stiffening and cardiovascular disease. Metabolism: Clinical and Experimental 119, 154766. Available at https://doi.org/10.1016/j.metabol.2021.154766.

Hinton, A., Claypool, S.M., Neikirk, K., Senoo, N., Wanjalla, C.N., Kirabo, A., et al., 2024. Mitochondrial structure and function in human heart failure. Circulation Research 135 (2), 372–396. Available at https://doi.org/10.1161/CIRCRESAHA.124.323800.

Horvath, S., Raj, K., 2018. DNA methylation-based biomarkers and the epigenetic clock theory of ageing. Nature Reviews Genetics 19 (6), 371–384. Available at https://doi.org/10.1038/s41576-018-0004-3.

Hu, L., Li, J., Tang, Z., Gong, P., Chang, Z., Yang, C., et al., 2025. How does biological age acceleration mediate the associations of obesity with cardiovascular disease? Evidence from international multi-cohort studies. Cardiovascular Diabetology 24 (1). Available at https://doi.org/10.1186/s12933-025-02770-0.

Hussain, M.S., Agrawal, N., Ilma, B., Rekha, M.M., Nayak, P.P., Kaur, M., et al., 2025. Autophagy and cellular senescence in Alzheimer's disease: key drivers of neurodegeneration. CNS Neuroscience and Therapeutics 31 (7). Available at https://doi.org/10.1111/cns.70503.

Hussain, S., Khan, M.S., Jamali, M.C., Siddiqui, A.N., Gupta, G., Hussain, M.S., et al., 2021. Impact of bariatric surgery in reducing macrovascular complications in severely obese T2DM patients. Obesity Surgery 31 (5), 1929–1936. Available at https://doi.org/10.1007/s11695-020-05155-2.

Jakubiak, G.K., et al., 2021. Obesity and insulin resistance" is the component of the metabolic syndrome most strongly associated with oxidative stress. Antioxidants (Basel) 11 (1).

Katsuumi, G., Shimizu, I., Yoshida, Y., Minamino, T., 2018. Vascular senescence in cardiovascular and metabolic diseases. Frontiers in Cardiovascular Medicine 5. Available at https://doi.org/10.3389/fcvm.2018.00018.

Kaur, P., Khan, H., Kaur Grewal, A., Dua, K., Kumar Singh, S., Gupta, G., et al., 2025. Exploring therapeutic strategies: the relationship between metabolic disorders and FOXO signalling in Alzheimer's disease. CNS & Neurological Disorders - Drug Targets 24 (3), 196–207. Available at https://doi.org/10.2174/0118715273321002240919102841.

Khavinson, V., Linkova, N., Dyatlova, A., Kantemirova, R., Kozlov, K., 2023. Senescence-associated secretory phenotype of cardiovascular system cells and inflammaging: perspectives of peptide regulation. Cells 12 (1), 106. Available at https://doi.org/10.3390/cells12010106.

Kierans, S.J., Taylor, C.T., 2021. Regulation of glycolysis by the hypoxia-inducible factor (HIF): implications for cellular physiology. The Journal of Physiology 599 (1), 23–37. Available at https://doi.org/10.1113/jp280572.

Knutson, A.K., Williams, A.L., Boisvert, W.A., Shohet, R.V., 2021. HIF in the heart: development, metabolism, ischemia, and atherosclerosis. Journal of Clinical Investigation 131 (17). Available at https://doi.org/10.1172/JCI137557.

Konwerski, M., Gąsecka, A., Opolski, G., Grabowski, M., Mazurek, T., 2022. Role of epicardial adipose tissue in cardiovascular diseases: a review. Biology 11 (3), 355. Available at https://doi.org/10.3390/biology11030355.

Lakatta, E.G., 2013. The reality of aging viewed from the arterial wall. Artery Research 7 (2), 73–80. Available at https://doi.org/10.1016/j.artres.2013.01.003.

Lakatta, E.G., 2015. So! What's aging? Is cardiovascular aging a disease? Journal of Molecular and Cellular Cardiology 83, 1–13. Available at https://doi.org/10.1016/j.yjmcc.2015.04.005.

Lakatta, E.G., Levy, D., 2003. Arterial and cardiac aging: major shareholders in cardiovascular disease enterprises: Part II: the aging heart in health: links to heart disease. Circulation 107 (2), 346–354. Available at https://doi.org/10.1161/01.CIR.0000048893.62841.F7.

Leyane, T.S., Jere, S.W., Houreld, N.N., 2022. Oxidative stress in ageing and chronic degenerative pathologies: molecular mechanisms involved in counteracting oxidative stress and chronic inflammation. International Journal of Molecular Sciences 23 (13). Available at https://doi.org/10.3390/ijms23137273.

Liu, Z., Wu, K.K.L., Jiang, X., Xu, A., Cheng, K.K.Y., 2020. The role of adipose tissue senescence in obesity- and ageing-related metabolic disorders. Clinical Science 134 (2), 315–330. Available at https://doi.org/10.1042/cs20190966.

Majewska, J., Krizhanovsky, V., 2025. Immune surveillance of senescent cells in aging and disease. Nature Aging 5 (8), 1415–1424. Available at https://doi.org/10.1038/s43587-025-00910-5.

Mann, V., Sundaresan, A., Shishodia, S., 2024. Overnutrition and lipotoxicity: impaired efferocytosis and chronic inflammation as precursors to multifaceted disease pathogenesis. Biology 13 (4). Available at https://doi.org/10.3390/biology13040241.

Marino, F., Scalise, M., Salerno, N., Salerno, L., Molinaro, C., Cappetta, D., et al., 2022. Diabetes-induced cellular senescence and senescence-associated secretory phenotype impair cardiac regeneration and function independently of age. Diabetes 71 (5), 1081–1098. Available at https://doi.org/10.2337/db21-0536.

Martini, H., Lefevre, L., Sayir, S., Itier, R., Maggiorani, D., Dutaur, M., et al., 2021. Selective cardiomyocyte oxidative stress leads to bystander senescence of cardiac stromal cells. International Journal of Molecular Sciences 22 (5), 2245. Available at https://doi.org/10.3390/ijms22052245.

Mauriello, A., Orlandi, A., Palmieri, G., Spagnoli, L.G., Oberholzer, M., Christen, H., 1992. Age-related modification of average volume and anisotropy of vascular smooth muscle cells. Pathology - Research and Practice 188 (4-5), 630–636. Available at https://doi.org/10.1016/s0344-0338(11)80070-1.

Mehdizadeh, M., Aguilar, M., Thorin, E., Ferbeyre, G., Nattel, S., 2022. The role of cellular senescence in cardiac disease: basic biology and clinical relevance. Nature Reviews Cardiology 19 (4), 250–264. Available at https://doi.org/10.1038/s41569-021-00624-2.

Meng, J., Geng, Q., Jin, S., Teng, X., Xiao, L., Wu, Y., et al., 2023. Exercise protects vascular function by countering senescent cells in older adults. Frontiers in Physiology 14. Available at https://doi.org/10.3389/fphys.2023.1138162.

Minamino, T., Miyauchi, H., Yoshida, T., Ishida, Y., Yoshida, H., Komuro, I., 2002. Endothelial cell senescence in human atherosclerosis: role of telomere in endothelial dysfunction. Circulation 105 (13), 1541–1544. Available at https://doi.org/10.1161/01.CIR.0000013836.85741.17.

Morgan, R.G., Ives, S.J., Lesniewski, L.A., Cawthon, R.M., Andtbacka, R.H.I., Noyes, R.D., et al., 2013. Age-related telomere uncapping is associated with cellular senescence and inflammation independent of telomere shortening in human arteries. American Journal of Physiology - Heart and Circulatory Physiology 305 (2). Available at https://doi.org/10.1152/ajpheart.00197.2013 H251.

Murakami, T., Inagaki, N., Kondoh, H., 2022. Cellular senescence in diabetes mellitus: distinct senotherapeutic strategies for adipose tissue and pancreatic β cells. Frontiers in Endocrinology 13. Available at https://doi.org/10.3389/fendo.2022.869414.

Nagamani, S., Sahoo, R., Muneeswaran, G., Narahari Sastry, G., 2019. Data science driven drug repurposing for metabolic disorders. Silico Drug Design: Repurposing Techniques and Methodologies. Elsevier, India, pp. 191–227. Available at 10.1016/B978-0-12-816125-8.00007-9.

Nakamura, M., 2024. Lipotoxicity as a therapeutic target in obesity and diabetic cardiomyopathy. Journal of Pharmacy & Pharmaceutical Sciences 27. Available at https://doi.org/10.3389/jpps.2024.12568.

Ng, S.W., Chan, Y., Chellappan, D.K., Madheswaran, T., Zeeshan, F., Chan, Y.L., et al., 2019. Molecular modulators of celastrol as the keystones for its diverse pharmacological activities. Biomedicine and Pharmacotherapy 109, 1785–1792. Available at https://doi.org/10.1016/j.biopha.2018.11.051.

Pakhare, M., Anjankar, A., 2024. Critical correlation between obesity and cardiovascular diseases and recent advancements in obesity. Cureus 16 (1). Available at https://doi.org/10.7759/cureus.51681.

Pandey, N.K., Singh, S.K., Kumar, B., Gulati, M., Vishwas, S., Khursheed, R., et al., 2022. Expanding arsenal against diabetes mellitus through nanoformulations loaded with glimepiride and simvastatin: a comparative study. Environmental Science and Pollution Research 29 (34), 51976–51988. Available at https://doi.org/10.1007/s11356-022-19371-z.

Poetsch, M.S., Strano, A., Guan, K., 2020. Role of leptin in cardiovascular diseases. Frontiers in Endocrinology 11. Available at https://doi.org/10.3389/fendo.2020.00354.

Ramanunny, A.K., Wadhwa, S., Gulati, M., Vishwas, S., Khursheed, R., Paudel, K.R., et al., 2022. Journey of Alpinia galanga from kitchen spice to nutraceutical to folk medicine to nanomedicine. Journal of Ethnopharmacology 291, 115144. Available at https://doi.org/10.1016/j.jep.2022.115144.

Reshma, A., Tamilanban, T., Chitra, V., Subramaniyan, V., Gupta, G., Fuloria, N.K., et al., 2023. Anti-obesity effects of olivetol in adult zebrafish model induced by short-term high-fat diet. Scientific Reports 13 (1). Available at https://doi.org/10.1038/s41598-023-44462-3.

Rezuş, E., Burlui, A., Cardoneanu, A., Rezuş, C., Codreanu, C., Pârvu, M., et al., 2020. Inactivity and skeletal muscle metabolism: a vicious cycle in old age. International Journal of Molecular Sciences 21 (2), 592. Available at https://doi.org/10.3390/ijms21020592.

Rohm, T.V., Meier, D.T., Olefsky, J.M., Donath, M.Y., 2022. Inflammation in obesity, diabetes, and related disorders. Immunity 55 (1), 31–55. Available at https://doi.org/10.1016/j.immuni.2021.12.013.

Ruperez, C., Madeo, F., de Cabo, R., Kroemer, G., Abdellatif, M., 2025. Obesity accelerates cardiovascular ageing. European Heart Journal 46 (23), 2161–2185. Available at https://doi.org/10.1093/eurheartj/ehaf216.

Russo, S., Kwiatkowski, M., Govorukhina, N., Bischoff, R., Melgert, B.N., 2021. Meta-inflammation and metabolic reprogramming of macrophages in diabetes and obesity: the importance of metabolites. Frontiers in immunology 12. Available at https://doi.org/10.3389/fimmu.2021.746151.

Saltiel, A.R., Olefsky, J.M., 2017. Inflammatory mechanisms linking obesity and metabolic disease. Journal of Clinical Investigation 127 (1), 1–4. Available at https://doi.org/10.1172/JCI92035.

Samuel, V.P., Gupta, G., Dahiya, R., Jain, D.A., Mishra, A., Dua, K., 2019. Current update on preclinical and clinical studies of resveratrol, a naturally occurring phenolic compound. Critical Reviews in Eukaryotic Gene Expression 29 (6), 529–537. Available at https://doi.org/10.1615/CritRevEukaryotGeneExpr.2019027836.

Shaik Mohamed Sayed, U.F., Moshawih, S., Goh, H.P., Kifli, N., Gupta, G., Singh, S.K., et al., 2023. Natural products as novel anti-obesity agents: insights into mechanisms of action and potential for therapeutic management. Frontiers in Pharmacology 14. Available at https://doi.org/10.3389/fphar.2023.1182937.

Shih, H., Lee, B., Lee, R.J., Boyle, A.J., 2011. The aging heart and post-infarction left ventricular remodeling. Journal of the American College of Cardiology 57 (1), 9–17. Available at https://doi.org/10.1016/j.jacc.2010.08.623.

Shimizu, I., Aprahamian, T., Kikuchi, R., Shimizu, A., Papanicolaou, K.N., MacLauchlan, S., et al., 2014. Vascular rarefaction mediates whitening of brown fat in obesity. Journal of Clinical Investigation 124 (5), 2099–2112. Available at https://doi.org/10.1172/jci71643.

Silveira Rossi, J.L., Barbalho, S.M., Reverete de Araujo, R., Bechara, M.D., Sloan, K.P., Sloan, L.A., 2022. Metabolic syndrome and cardiovascular diseases: going beyond traditional risk factors. Diabetes/Metabolism Research and Reviews 38 (3). Available at https://doi.org/10.1002/dmrr.3502.

Spinelli, R., Baboota, R.K., Gogg, S., Beguinot, F., Blüher, M., Nerstedt, A., et al., 2023. Increased cell senescence in human metabolic disorders. Journal of Clinical Investigation 133 (12). Available at https://doi.org/10.1172/jci169922.

Ungvari, Z., Kaley, G., De Cabo, R., Sonntag, W.E., Csiszar, A., 2010. Mechanisms of vascular aging: new perspectives. Journals of Gerontology - Series A Biological Sciences and Medical Sciences 65 (10), 1028–1041. Available at https://doi.org/10.1093/gerona/glq113.

Vaddavalli, P.L., Schumacher, B., 2022. The p53 network: cellular and systemic DNA damage responses in cancer and aging. Trends in Genetics 38 (6), 598–612. Available at https://doi.org/10.1016/j.tig.2022.02.010.

van der Laan, K.W.F., Neutel, C.H.G., Pencheva, M.G., Wesley, C.D., Krüger, D.N., Schalkwijk, C.G., et al., 2025. Vascular aging impairs active modulation of murine aortic stiffness by smooth muscle cells. American Journal of Physiology - Cell Physiology 329 (5), C1398. Available at https://doi.org/10.1152/ajpcell.00387.2025.

Vilariño-García, T., Polonio-González, M., Pérez-Pérez, A., Ribalta, J., Arrieta, F., Aguilar, M., et al., 2024. Role of leptin in obesity, cardiovascular disease, and type 2 diabetes. International Journal of Molecular Sciences 25 (4), 2338. Available at https://doi.org/10.3390/ijms25042338.

Volpe, M., Gallo, G., 2023. Obesity and cardiovascular disease: an executive document on pathophysiological and clinical links promoted by the Italian Society of Cardiovascular Prevention (SIPREC). Frontiers in Cardiovascular Medicine 10. Available at https://doi.org/10.3389/fcvm.2023.1136340.

Wang, B., Han, J., Elisseeff, J.H., Demaria, M., 2024. The senescence-associated secretory phenotype and its physiological and pathological implications. Nature Reviews Molecular Cell Biology 25 (12), 958–978. Available at https://doi.org/10.1038/s41580-024-00727-x.

Wang, J.C., Bennett, M., 2012. Aging and atherosclerosis: mechanisms, functional consequences, and potential therapeutics for cellular senescence. Circulation Research 111 (2), 245–259. Available at https://doi.org/10.1161/CIRCRESAHA.111.261388.

Xie, S., Xu, S.-C., Deng, W., Tang, Q., 2023. Metabolic landscape in cardiac aging: insights into molecular biology and therapeutic implications. Signal Transduction and Targeted Therapy 8 (1). Available at https://doi.org/10.1038/s41392-023-01378-8.

Xin, G.L.L., Khee, Y.P., Ying, T.Y., Chellian, J., Gupta, G., Kunnath, A.P., et al., 2019. Current status on immunological therapies for type 1 diabetes mellitus. Current Diabetes Reports 19 (5). Available at https://doi.org/10.1007/s11892-019-1144-3.

Ye, B., Chen, Y., Chen, X., Xu, D., Jiang, Y., Lin, W., et al., 2024. Deubiquitinase USP25 alleviates obesity-induced cardiac remodeling and dysfunction by downregulating TAK1 and reducing TAK1-mediated inflammation. JACC: Basic to Translational Science 9 (11), 1287–1304. Available at https://doi.org/10.1016/j.jacbts.2024.06.001.

Yokoyama, M., Okada, S., Nakagomi, A., Moriya, J., Shimizu, I., Nojima, A., et al., 2014. Inhibition of endothelial p53 improves metabolic abnormalities related to dietary obesity. Cell Reports 7 (5), 1691–1703. Available at https://doi.org/10.1016/j.celrep.2014.04.046.

Yoosuf, S., Gupta, G., Bhargava, R., Kumar, N., Ranjan, P., Pandey, R.M., et al., 2021. Magnitude of psychiatric comorbidity in patients with obesity in Northern Indian population. Diabetes & Metabolic Syndrome: Clinical Research & Reviews 15 (5), 102270. Available at https://doi.org/10.1016/j.dsx.2021.102270.

Yoshida, Y., Shimizu, I., Katsuumi, G., Jiao, S., Suda, M., Hayashi, Y., et al., 2015. p53-Induced inflammation exacerbates cardiac dysfunction during pressure overload. Journal of Molecular and Cellular Cardiology 85, 183–198. Available at https://doi.org/10.1016/j.yjmcc.2015.06.001.

Zhang, K., Ma, Y., Luo, Y., Song, Y., Xiong, G., Ma, Y., et al., 2023. Metabolic diseases and healthy aging: identifying environmental and behavioral risk factors and promoting public health. Frontiers in Public Health 11. Available at https://doi.org/10.3389/fpubh.2023.1253506.

Zhao, W., Zheng, X.-L., Peng, D.-Q., Zhao, S.-P., 2015. Myocyte enhancer factor 2A regulates hydrogen peroxide-induced senescence of vascular smooth muscle cells via microRNA-143. Journal of Cellular Physiology 230 (9), 2202–2211. Available at https://doi.org/10.1002/jcp.24948.

Zhao, Y., Yue, R., 2024. Aging adipose tissue, insulin resistance, and type 2 diabetes. Biogerontology 25 (1), 53–69. Available at https://doi.org/10.1007/s10522-023-10067-6.

Chapter 11

Autophagy, mitophagy, and senescence in the heart

Chintan Aundhia[1], Muhammad Afzal[2], Surya Nath Pandey[3], Md Raihan Siddique[4], Riya Thapa[5], Asif Ahmad Bhat[5], Doli Rani Das[6], Neelam Singla[4], Santosh Kumar Singh[7], and Gaurav Gupta[8,9]

[1]*Department of Pharmacy, Sumandeep Vidyapeeth Deemed to be University, Vadodara, Gujarat, India,* [2]*Department of Pharmaceutical Sciences, Pharmacy Program, Batterjee Medical College, Jeddah, Saudi Arabia,* [3]*Department of Pharmacology, Teerthanker Mahaveer College of Pharmacy, Teerthanker Mahaveer University, Moradabad, Uttar Pradesh, India,* [4]*School of Pharmacy, Suresh Gyan Vihar University, Jaipur, Rajasthan, India,* [5]*Faculty of Pharmacy, AIMST University, Bedong, Kedah, Malaysia,* [6]*Shree Krishna College of Pharmacy, Sitapur, Uttar Pradesh, India,* [7]*ICFAI School of Pharmaceutical Sciences, The ICFAI University, Jaipur, Rajasthan, India,* [8]*Centre for Research Impact & Outcome, Chitkara College of Pharmacy, Chitkara University, Rajpura, Punjab, India,* [9]*Centre of Medical and Bio-Allied Health Sciences Research, Ajman University, Ajman, United Arab Emirates*

11.1 Introduction

Cardiac aging indicates that proteostasis and mitochondrial quality control decrease with age; thus we initially defined autophagy as the lysosome-mediated degradation of damaged proteins and organelles, mitophagy as the selective degradation of dysfunctional mitochondria by PINK1-Parkin and receptor-mediated pathways, including BNIP3, NIX, and FUNDC1, and cellular senescence as long-lasting cell cycle arrest sustained by p16INK4a and p21CIP1. The combination of these programs dictates either the recovery of stressed cardiomyocytes, fibroblasts, and endothelial cells or the development of adverse remodeling (Ghosh et al., 2020; Yun et al., 2025). Mechanistic crosstalk: ULK1–Beclin-1-ATG-driven autophagy suppresses proteotoxic stress, mitophagy suppresses reactive oxygen species (ROS) and mitochondrial DNA leakage, and both inhibit cGAS-STING-NF-kB activation; however, established senescence inhibits lysosome biogenesis and autophagic flux, which builds upon itself. These processes are important because cardiovascular diseases are a major global health concern (Li et al., 2025). In 2022, the World Health Organization estimated 19.8 million CVD deaths, accounting for approximately 32% of all deaths, and 85% of all cardiovascular diseases are caused by heart attack and stroke. The new Global Burden of Disease estimates show that cardiovascular disease has become the leading cause of approximately one in three deaths in 2023 worldwide and is estimated to account for approximately 437 million disability-adjusted life-years of the aging population, despite declining age-standardized rates. In India, modern sources have recorded that cardiovascular disease is the underlying cause of nearly one-third of all deaths in the country, and a recent countrywide heart failure registry recorded a 1-year mortality rate of almost 22% in over ten thousand individuals with an average age of approximately 60 years (Lindstrom et al., 2022; Piñeiro et al., 2023; Mendis et al., 2022). This epidemiology is relevant to myocardial remodeling; in autophagy initiation, which is equalized to lysosomal capacity, and mitophagy, which is optimally balanced, ATP supply, calcium processing, endothelial bioavailability of nitric oxide, and matrix plasticity are maintained. In the case of initiation failure or poor lysosomal acidification, stalled flux is represented by LC3-II and p62, dysfunctional mitochondria persist, inflammatory signaling is activated by danger-associated molecular patterns, and tissue stiffness and electrical heterogeneity arise (Nàger et al., 2025; Zhang et al., 2025a). Senescence conceptually combines irreversible DNA damage, mitochondrial pathology, and oxidative stress into organ-scale responses, and autophagy and Mitophagy provide cues that maladaptation occurs when reversible stress becomes too intense, making them appealing therapeutic targets in load, metabolic strain, and ischemia-reperfusion models. This chapter is novel in the integration of molecular circuitry, which includes macroautophagy, PINK1-Parkin and receptor-mediated mitophagy, and chaperone-mediated autophagy with spatial and single-cell biomarker systems such as LC3 and p62 dynamics, phospho-ubiquitin, mt-Keima, and p16 or p21, using SASP panels (Stojanovic et al., 2025; Cui et al.,

Cellular Senescence and Cardiac Aging. DOI: https://doi.org/10.1016/B978-0-443-45811-8.00001-5

2012). We also compared these measurements with several translational levers, such as mTOR and AMPK modulators, mitophagy enhancers, senotherapeutics, gene- and cell-based vectors, and structured lifestyle programs, on an integrated map that connects the measurable flux to functional endpoints, such as myocardial strain, endothelial reactivity, and diastolic compliance. Its purpose and objectives are to integrate the processes into the modern world and Indian epidemiology, recognize actionable nodes in which autophagy activation can be integrated into lysosomal and mitochondrial capacity without loss of organelle bulk, and develop trial-ready strategies and composite biomarkers of stratification and tracking. All evidence confirms a practical hypothesis: the recovery of autophagy–mitophagy flux in the entire repertoire of cardiac cell types can limit the burden of senescence, limit inflammatory remodeling, enhance age-related and comorbidity-driving resilience, and enhance long-term clinical translation (Korolchuk et al., 2017; Alam et al., 2020).

11.2 Canonical autophagy pathways in cardiac cells

11.2.1 Initiation and autophagosome biogenesis (ULK1–Beclin-1–ATG axis)

Cardiac cell energetic or oxidative stress inhibits mTORC1 and/or activates AMPK, inducing the release of the ULK1 complex (ULK1–ATG13–FIP200–ATG101) to phosphorylate its downstream targets, which initiates isolation-membrane nucleation. Signaling by ULK1 subsequently forms the Beclin-1-VPS34 class III PI3K core, which contains ATG14L and VPS15, creating PI3P microdomains at the ER-mitochondria points of contact, which are vacuoles that recruit DFCP1 and WIPI2 to initiate phagocytosis. Membranes are provided by ATG9-positive vesicles and the ATG2-WIPI axis, and rim expansion and curvature stabilization are mediated by two ubiquitin-like conjugation systems: ATG12–ATG5–ATG16L1 and LC3/ATG8 lipidation mediated by ATG7/ATG3 and ATG4 recycling, respectively (Ganley et al., 2009; Gómez-Virgilio et al., 2022). The nascent autophagosome is aided by ESCRT-III elements (e.g., CHMP2A) and VPS4 to close the vesicle and convert it into a cargo-competent vesicle, which is then transported for fusion with lysosomes. In cardiomyocytes, it induces optimal initiation of proteotoxic stress and protects mitochondria against secondary damage in response to ischemia-reperfusion or pressure overload. In fibroblasts and endothelium, it facilitates matrix homeostasis and regulates vascular tone. Upon maintaining DNA-damage signaling through telomere attrition, the strain on this axis increases; insufficient biogenesis or blockage of flux results in p62, ROS, and cGAS-STING-NF-κB signaling, enriching the SASP, which facilitates antisurvival remodeling. Together, these results indicate that an optimally adjusted ULK1–Beclin-1–ATG program protects against proteostasis, inhibits senescence, and improves contractile reserves in aging hearts. Future studies should focus on mapping ULK1 substrate networks in cardiac cells, generating selective Vps34/Beclin-1 modulators and TFEB-coupled modulators, and validating flux biomarkers (LC3 turnover, p62 clearance, and PI3P reporters) to inform patient stratification and therapy (Rabinovich-Nikitin and Kirshenbaum, 2021; Maejima et al., 2016).

11.2.2 Cargo recognition and lysosomal degradation

In cardiac autophagy, the selection of cargo is initiated when ubiquitinated damaged proteins and organelles are identified by stress-responsive E3-ubiquitinases. This process facilitates the binding of receptors, including p62/SQSTM1, NBR1, OPTN, NDP52, and TAX1BP1, to the cargo. These receptors interact with LC3/GABARAP-decorated phagophores through LC3-interacting regions. Additionally, the ATG12–ATG5–ATG16L1 system is employed to oligomerize substrates within the phagophores. Upon closure, the autophagosomes are trafficked to lysosomes by Rab7 and the HOPS tethering complex. v-ATPase acidification of lysosomes activates cathepsins and other hydrolytic enzymes, and recycling of degradation products helps to maintain cardiomyocyte energetics. Telomere loss increases DNA damage and ROS loads in the aging heart, ubiquitination, and dependence on specific recognition and lysosomal competence (Jeong et al., 2019; Riley and Tait, 2016). Deteriorating ubiquitinated cargo and lysosome competence occur under impaired LAMP2, TFEB/TFE3 signaling, acidification, and profibrotic SASP summative adverse remodeling. Healthy cargo–receptor interactions and autolysosomal degradation include proteotoxicity, mitochondrial quality, and excitation–contraction coupling among cardiomyocytes, fibroblasts, and the endothelium. Altogether, evidence suggests that the preservation of efficient cargo recognition along with lysosomal activity inhibits the burden of senescence, preserves the myocardial structure, and enhances diastolic compliance. Next-generation views emphasize standardized flux biomarkers (LC3 turnover and p62 clearance), methods that re-establish lysosomal acidity and cathepsin activity, and TFEB-based plans consistent with cell type-specific delivery to increase clearance without excessive self-digestion (Ren et al., 2024; Shyam et al., 2021).

11.2.3 Chaperone-mediated autophagy in myocardial homeostasis

Chaperone-mediated autophagy (CMA) helps maintain myocardial proteostasis by identifying cytosolic proteins with KFERQ-like motifs, binding them with Hsc70/HSPa8 at the microtubules and co-chaperones, transporting the substrates to lysosomal LAMP2A, and transmembrane degradation of unfolded polypeptides using luminal lys-Hsc70. CMA is rapidly stimulated in response to oxidative or metabolic stress in cardiomyocytes to replenish important metabolic enzymes, redox regulators, and contractile modulators; maintain ATP production; restrict ROS growth; and stabilize calcium homeostasis (Robert et al., 2019; Valdor and Martinez-Vicente, 2024). CMA suppresses profibrotic and pro-inflammatory signaling in cardiac fibroblasts and endothelial cells, which would otherwise cause stiffening of the matrix and inhibition of NO bioavailability. Telomere loss enhances DNA damage signatures and oxidative stress and increases CMA activity, but aging lowers LAMP2A levels, modifies lysosomal membrane structure and acidity, and decreases CMA flux, which allows the build-up of oxidized proteins, perturbs mitochondrial homeostasis, stimulates cGAS-STING and NF-kB, and condenses an SASP that drives deleterious remodeling (Bachmann et al., 2022). Crosstalk with macroautophagy provides complementary clearance, for which CMA is insufficient, proteotoxic stress increases, mitophagy is overfertilized, and senescence programs stabilize among cardiomyocytes, fibroblasts, and endothelium. Collectively, these studies show that preserved CMA inhibits telomere-associated senescence, maintains ventricular compliance, and is procontractile in aging hearts. Future research should focus on flux-specific biomarkers, measures that can reestablish LAMP2A expression and lysosomal competence, and cell type-specific CMA modulators, in line with TFEB-regulated lysosome biogenesis, to increase clearance without losing or exhausting any essential regulatory proteins or disrupting myocardial energetics (Kmiotek-Wasylewska et al., 2023) (Fig. 11.1).

11.3 Mitophagy mechanisms relevant to the heart

11.3.1 PINK1/Parkin-dependent mitophagy

PINK1/Parkin-dependent mitophagy is activated in cases of mitochondrial membrane potential loss, which permits PINK1 to accumulate on the outer membrane, phosphorylate ubiquitin at Ser65, and recruit cytosolic Parkin, which preevolves K63-linked ubiquitin chains on VDAC and Mfn1/2 and other substrates to tag organelles as targets. Phospho-ubiquitin and Parkin construct a signal that attracts OPTN, NDP52, and TBK1, which phosphorylate the receptors to enhance LC3-binding, and cytoskeleton-motor position-tagged mitochondria to be packaged within LC3-decorated phagophores, which fuse with lysosomes to degrade and recycle metabolites. In a model of cardiomyocytes in ischemia-reperfusion, pressure overload, or metabolic stress, early mitophagy suppresses ROS, reduces the release of mtDNA, maintains ATP delivery, and inhibits cGASP-NF-κB phosphorylation, leading to a profibrotic SASP (Bhattacharya et al., 2022; Onishi et al., 2021). Damaged mitochondria are tolerated when flux is damaged by defective PINK1 kinase activity, parkin insufficiency, or lysosome dysfunction and induce Ca2+ mishandling, permeability transition, and senescence signals that impair excitation–contraction coupling. In contrast, mitophagy is dangerous because of overconsumption, indicating the necessity of balanced regulation based on cell type and stress duration. Together, these studies indicate that the preservation of PINK1/Parkin signaling conserves mitochondrial fitness and mitigates inflammatory remodeling and contractile resilience during heart aging. Future studies are needed to confirm flux biomarkers, selective activators, or gene therapies to enhance the PINK1/Parkin and TBK1-OPTN axes in combination with lysosome-restoring and mitochondria-targeting redox therapies.

11.3.2 Receptor-mediated mitophagy (BNIP3/NIX, FUNDC1)

Mitochondrial outer membrane receptors that regulate receptor-mediated mitophagy in the heart have LC3-interacting regions and are transcriptionally controlled by HIF-1a and FOXO3 under hypoxia and oxidative stress, BNIP3 and NIX/BNIP3L, and BNIP5 through the activation of phosphorylation switches that favor LC3-binding (PGAM5-mediated dephosphorylation at Ser13) and inhibit activity (CK2-phosphorylation at Ser). The fission affected by DRP1 isolates dysfunctional fragments of mitochondria, exposing cardiolipin and lowering the membrane potential so that BNIP3/NIX and FUNDC1 can bind to LC3/GABARAP and recruit autophagosomal membranes to undergo encapsulation and degradation in lysosomes. In cardiomyocytes, ROS is suppressed by appropriate receptor signaling, mitDNA leakage, cGAS-STING activation is prevented, and ATP supply is preserved during ischemia-reperfusion, pressure overload, and metabolic stress (Zhu et al., 2013; Möbus et al., 2025). In endothelial cells and fibroblasts, receptor signaling maintains the barrier and inhibits profibrotic signaling, which would otherwise stiffen the myocardium. Telomere-capped DNA damage signaling enhances oxidative load and dependence on receptor mitophagy; however, chronic BNIP3 overexpression or

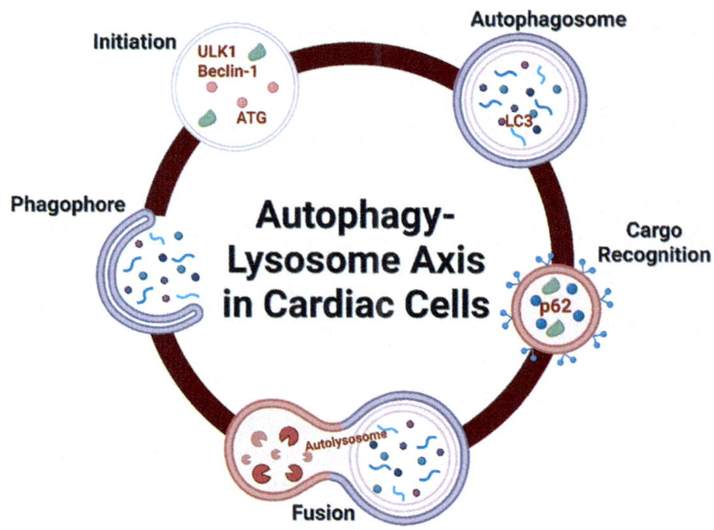

FIGURE 11.1 **The autophagy–lysosomal axis: ULK1–Beclin-1–ATG-mediated quality control loop in cardiac cells.** Summary of the autophagy–lysosome axis of the cardiac cell as a process that is in a cycle. A stress signal that activates ULK1–Beclin-1–ATG processes will trigger the formation of phagophores. Membrane expansion produces an LC3-decorated autophagosome that traps cytoplasmic cargo. Selective receptors, such as p62, identify ubiquitinated content and conjugate it with LC3. This is followed by fusion of the autophagosome with a lysosome to form an autolysosome, in which the contents are broken down by hydrolases and metabolites recycled to restore proteostasis and inhibit senescence caused by mitochondrial damage. The arrows indicate the continuity of this quality control loop.

sustained FUNDC1 stimulation may cause excessive clearance, permeability transition, and cell death, highlighting the importance of controlled flux and lysosomal competence. Together, the evidence suggests that the narrowest ranges of BNIP3/NIX and FUNDC1 activity can provide mitochondrial quality, prevent inflammatory programs of senescence, protect contractile resilience in aging hearts, and that receptor dysregulation promotes poor remodeling and the risk of heart failure. Future studies should focus on receptor-selective flux signatures (e.g., phospho-FUNDC1 status and BNIP3/NIX reporters) and cell type-specific delivery methods coupled with lysosome-rescuing and redox-balancing solutions to improve mitochondrial stability (Bloom et al., 2023).

11.3.3 Mitochondrial dynamics, reactive oxygen species, and damage signaling

The fusion–fission equilibrium, where MFN1/2 and OPA1 maintain the continuity of the network and cristae architecture, fission segregation of dysfunctional segments through DRP1, and protein degradation of OPA1 to downregulate membrane potential and promote ROS to damage lipids, proteins, and BtDNA, determines mitochondrial quality in the heart. PINK1 is stabilized by loss of potential, and parkin is recruited to damaged organelles. However, receptor pathways, including those involving BNIP3, NIX, and FUNDC1, are activated by hypoxia and long-term ROS to link fragmented mitochondria to LC3/GABARAP-labeled autophagosomes. Cristae remodeling and permeability transition release or exposure of all mitochondrial DNA and cardiolipin, initiating TLR9 and cGAS-STING pathways, enhancing NF-kB activity, and causing paracrine signals to activate fibroblasts and inhibit endothelial nitric oxide bioavailability (Adebayo et al., 2021; Ong et al., 2017). Competent mitophagy and lysosome flux, both mediated by TFEB/TFE3 transcriptional programs, recycle substrates to maintain ATP synthesis and restrict calcium mishandling in cardiomyocytes. Failure to maintain lysosomal acidification and cargo movement results in lysosome overload and accumulation of LC3-II and p62, persistence of damaged mitochondria, and senescence via p16^INK4a/p21 CIP1 and a SASP that stiffens Calibrated fission with maintained OPA1 processing and timely activation of PINK1/Parkin and BNIP3/FUNDC1, on the other hand, replenishes redox balance, suppresses inflammatory signaling, and helps to maintain contractile reserve. Taken together, the literature suggests that the ability to tune mitochondria to mitophagy helps maintain myocardial resilience and prevents the development of diastolic dysfunction. Future studies should establish cell type-specific set points for DRP1 and OPA1, biomarker-validated interactions between dynamics and mitophagy flux, and precision strategies, such as TFEB activation and selective PINK1/Parkin modulators and mitochondria-targeted redox therapies, to repair quality control with intact.

11.4 Cellular senescence in the myocardium

11.4.1 Triggers: DNA damage, mitochondrial dysfunction, and oxidative stress

Telomere shortening, replication stress, and reactive aldehyde damage DNA in cardiomyocytes, triggering ATM/ATR-CHK signaling that stabilizes p53 and works with p16 64a-RB to stabilize cell cycle arrest, but induces metabolic rewiring. Subsequently, increased mitochondrial damage (impaired electron transport via complexes I/III, dysregulated Ca2+ processing, and loss of membrane potential) exacerbates this damage response, inhibiting biogenesis mediated by PGC-1a, thus impairing antioxidant defenses. Oxidative stress also alters both the nuclear and mitochondrial genomes and maintains gH2AX and 53BP1 foci and cytosolic DNA leakage, leading to cGAS-STING-NF-κB and p38 MAPK/C/EBPb convergence to entrain a stable SASP that propagates inflammation, fibroblast activation, and microvascular dysfunction (Aix et al., 2016; Linders et al., 2024). The first line of defense against these insults includes autophagy and mitophagy to remove damaged proteins and depolarize mitochondria. However, long-term stress, lysosomal dysfunction, or faulty PINK1/Parkin and BNIP3/FUNDC1 signaling lead to cargo perfection, chronic ROS, and irreversible senescence. The paracrine cues involved in endothelial and fibroblast senescence are feed-forward mechanisms that stiffen the matrix, impair perfusion, and disrupt excitation-contraction coupling, thereby accelerating diastolic dysfunction and adverse remodeling. Taken together, our results show that highly synchronized control of DNA damage, mitochondrial quality maintenance, and oxidative stress restraint decreases the burden of senescence and safeguards myocardial architecture and functioning. DDR modulators, mitophagy enhancers, and mitochondria-targeted redox therapies could be incorporated into future studies, along with the validation of flux and senescence biomarkers, to inform patient stratification and responses to therapy (Albano et al., 2023; Zuo et al., 2020).

11.4.2 Effector programs: p16INK4a/p21CIP1, senescence-associated secretory phenotype, and tissue remodeling

DNA damage and mitochondrial stress stabilize p53 to trigger p21CIP1, increase p16INK4a, and enforce RB-dependent cell cycle arrest, heterochromatin formation, and metabolic rewiring, which increase mTOR tone and silence NAD +/SIRT1 signaling. Autophagy defects allow GATA4 buildup, which collaborates with cGAS-STING and NF-kB to produce a SASP full of IL-6, IL-1β, TNF-α, TGF-b, chemokines, and matrix-active factors. This milieu inhibits cardiomyocyte contraction and calcium handling and promotes myofibroblast differentiation, collagen I/III deposition, lysyl oxidase cross-linking, and titin-stiffening. It suppresses nitric oxide bioavailability and capillary density and increases susceptibility to ischemia-reperfusion in the endothelium. This reciprocal feedback mechanism of matrix remodeling via integrin and TGF-β signaling into p16/p21 programs, electrical heterogeneity, and fibro-fatty expansion forms an arrhythmogenic substrate (Mijit et al., 2020; Abuetabh et al., 2022; O'Sullivan et al., 2024). In the case of competent mitophagy and macroautophagy, SASP strength is maintained, and tissue mechanics are plastic. In the case of flux curtailment, ongoing LC3-II/p62 build-up supports inflammatory defense and solidifies adverse remodeling. Overall, research suggests that the reduction of tuning effector nodes, that is, the attenuation of maladaptive responses of SASP without destabilizing cytostatic arrest, can be used to reduce fibrosis, enhance diastolic compliance, and stabilize conduction. Future projections focus on composite biomarkers that combine p16/p21 activity with SASP and matrix turnover, cell type-specific therapy involving senomorphics, intermittent senolysis, and mitophagy, and a study that integrates SASP reprogramming with proteostasis and vascular function restoration to delay cardiac aging (O'Sullivan et al., 2024; Lim et al., 2011).

11.4.3 Senescence across cardiac cell types (cardiomyocytes, fibroblasts, endothelium)

Senescence manifests heterogeneously across cardiac cell types yet converges on persistent DNA damage signaling, mitochondrial dysfunction, and disturbed proteostasis. In cardiomyocytes, telomere attrition and oxidative load stabilize p53/p21 and p16INK4a/RB programs, remodel chromatin, depress PGC-1α-driven biogenesis, and impair calcium cycling; insufficient autophagy and mitophagy allow damaged mitochondria and proteotoxic aggregates to persist, amplifying ROS and weakening contractile reserve. Senescent fibroblasts arise through TGF-β and mechanotransduction cues, adopting a myofibroblast phenotype that secretes collagen I/III, periostin, and lysyl oxidase, crosslinks the extracellular matrix, and stiffens the myocardium. Their SASP enriches IL-6, IL-1β, and CCL2, which propagate inflammation and recruit immune cells, further reinforcing the senescence circuits. Endothelial cells undergo senescence under shear stress imbalance, hyperglycemia, and hypoxia, with eNOS uncoupling, diminished nitric oxide bioavailability, and episodic EndoMT reducing capillary density and barrier integrity; defective mitophagy and lysosomal flux increase mtDNA release and cGAS-STING activation, intensifying paracrine injury to myocytes and interstitial cells (Li et al., 2021; Herbstein et al., 2024). Crosstalk among these compartments creates a feed-forward loop: cardiomyocyte DAMPs and mitochondrial peptides activate fibroblasts, a stiff matrix perturbs myocyte mechanosignaling, and endothelial dysfunction limits perfusion, drives diastolic dysfunction, and arrhythmogenic remodeling. Collectively, studies indicate that preserving autophagy–mitophagy competence and dampening SASP across cell types curbs the senescence burden, protects ventricular compliance, and maintains microvascular function. Future perspectives emphasize composite biomarkers that couple p16/p21 activity with mitophagy flux and matrix turnover, cell-type-selective senomorphics combined with intermittent senolysis, and targeted restoration of lysosomal competence and mitochondrial quality to slow cardiac aging while preserving essential homeostatic signaling (Li et al., 2025; Colliva et al., 2020).

11.5 Crosstalk between autophagy/mitophagy and senescence

11.5.1 Autophagy insufficiency: triggers of senescence and functional decline

Senescence occurs heterogeneously among cell types in the heart but is focused on continued DNA damage signaling, mitochondrial dysfunction, and disrupted proteostasis. Telomere shortening and oxidative stress stabilize the p53/p21 and p16INK4a/RB programs, chromatin remodeling, suppression of PGC-1a-mediated biogenesis, and damage to mitochondria and protein aggregates in cardiomyocytes, which reduces the contractile reserve after insufficient autophagy and mitophagy. TGF-β and mechanotransduction induce senescent fibroblasts into a myofibroblast phenotype, secreting collagen I/III, periostin, and lysyl oxidase that cross-link the matrix and stiffen the myocardium, and induce IL-6, IL-1b, and CCL2 SASP, which enhances inflammation and attracts immune cells to augment senescence circuits (Dixit et al., 2025). Shear stress imbalance, hyperglycemia, and hypoxia cause endothelial cell senescence, uncoupling of eNOS, depletion of nitric oxide bioavailability, and periodic EndoMT, decreasing capillary density and barrier integrity. Mitophagy and lysosomal flux defects augment the release of mitochondrial DNA and cGAS-STING activation, worsening the paracrine injury of myocytes and interstitial cells. Interference between such compartments generates a feed-forward loop; fibroblast activation by cardiomyocyte DAMPs and mitochondrial peptides stiff matrix mechanosignals myocytes and results in endothelial dysfunction that restricts perfusion, leading to diastolic dysfunction and arrhythmogenic remodeling. Collectively, the literature shows that the maintenance of autophagy–mitophagy capacity and suppression of SASP in different cell types prevents the excessive burden of senescence, preserves ventricular compliance, and retains microvascular viability. Future studies should include composite biomarkers that integrate p16/p21 activity with mitophagy flux and matrix turnover, cell type-specific senomorphics alongside intermittent senolysis, and specific restoration of lysosomal competence and mitochondrial quality to slow the aging of the heart while maintaining critical homeostatic signaling (Zhang et al., 2018a; Janaszak-Jasiecka et al., 2023; Bharath et al., 2014).

11.5.2 Mitophagy as a brake on senescence and inflammaging

Mitophagy suppresses senescence and inflammation of the heart by selectively eliminating depolarized or oxidatively damaged mitochondria before their dysfunction can cause cellular stress. Loss of membrane potential stabilizes PINK1, which triggers Parkin to form phospho-ubiquitin scars, and hypoxia and persistent ROS activate BNIP3/NIX and FUNDC1 receptors, which recruit LC3/GABARAP membranes that present damaged organelles to lysosomes to undergo degradation and meet on-time clearance, inhibit mitochondrial ROS, cardiolipin, and mtDNA exposure, and cGAS-STING-NF-kB activation, preventing the enhancement of p16INK4a/p21 programs and profibrotic SASP. Competent mitophagy protects ATP, calcium regulation, excitation-contraction coupling in cardiomyocytes, nitric oxide bioavailability, myofibroblast transition in endothelial and fibroblast compartments, microvascular integrity, and matrix homeostasis (Zhou et al., 2024). In the case of loss of lysosomal capacity or PINK1/Parkin and receptor signaling, dysfunctional mitochondria survive, increase ROS and DAMP signals, and stabilize irreversible senescence circuits in cell types. Taken together, the results of these studies have shown that the maintenance of mitophagy flux suppresses inflammatory remodeling, reduces diastolic stiffening, and maintains contractile reserves in cardiac aging. Future directions include composite biomarkers between phospho-ubiquitin and mt-Keima reporters, SASP reporters, TFEB-targeted restoration of lysosomal competence, calibrated PINK1/Parkin or FUNDC1 activation, and cell type-selective approaches that integrate mitophagy enhancers with mitochondria-targeted redox therapies to preserve quality control without disrupting bioenergetics (Quinn et al., 2020; Ivankovic et al., 2016).

11.5.3 Situational dependency: protective versus maladaptive responses

Cross-bridging activation or inhibition of autophagy–mitochondrial crosstalk with stress intensity and duration has beneficial and maladaptive effects. Brief energetic or oxidative stress triggers AMPK, suppresses mTORC1, and activates ULK1–Beclin-1 to degrade damaged proteins, whereas PINK1/Parkin and BNIP3/FUNDC1 delete depolarized mitochondria, restore ATP, and inhibit ROS. In a sustained injury, this can result in the inability to keep pace with lysosomal capacity and TFEB/TFE3 programs, cargo retention, and cGAS-STING-NF-κB signaling that stabilizes p16INK4a/p21CIP1 arrest and a profibrotic SASP. Flux promotes calcium–calcium interactions and contractile reserves in cardiomyocytes, dampens myofibroblast transition in fibroblasts, and increases the bioavailability of nitric oxide in the endothelium. However, when mitophagy is excessive, or mTOR is inhibited for a long time, mitochondrial mass can be lost, and bioenergetics can be compromised (Lee et al., 2012; Desideri et al., 2025). Stalled flux results in LC3-II/p62 accumulation, increased fission, and OPA1 dysprocessing, intensifying oxidative stress and senescence. Thus the context of cytoprotective or maladaptive autophagy is dependent on whether it is ischemia-reperfusion or pressure overload, young or aged lysosomes, and cardiomyocytes or stromal compartments. Together, research shows that calibrated activation in parity with lysosomal competence prevents inflammaging, lessens diastolic stiffening, and maintains ventricular function and vascularity. Future studies should create composite biomarkers of flux, experiment with TFEB-directed restoration of lysosomal capacity, selective PINK1/Parkin or FUNDC1 tuning, and implement senomorphics using mitochondrial-targeted redox therapies to maintain quality control without impairing bioenergetics (Bhattacharya et al., 2022) (Fig. 11.2).

11.6 Physiologic and pathologic modulators in cardiac aging

11.6.1 Hemodynamic load, metabolic stress, and ischemia-reperfusion

Cardiac aging is shaped by the sequential activation of quality-control and stress-response pathways: pressure/volume load induces wall stress and neurohormonal tone, which activates AMPK, and episodic inhibition of mTORC1 initiates ULK1–Beclin-1–ATG-mediated autophagosome biogenesis; parallel mitochondrial strain causes one to induces DRP1-mediated fission, and the drop in membrane potential triggers the PINK1/Parkin and receptor (BNIP3/NIX); lipotoxic and glucotoxic metabolites increase ROS and block calcium processing during nutrient excess or insulin unresponsiveness that initially enhances compensatory autophagy but progressively overwhelms lysosomal capacity regulated by TFEB/TFE3 such that flux decelerates and p62 and damaged mitochondria accumulate (Barcena et al., 2025; Qiu et al., 2023). To prevent ROS-induced mitochondrial injury, abrupt reoxygenation in I/R spikes the permeability transition pore, thereby activating the cGAS-STING-NF-kB axis, which stalls mitochondrial flux and preserves the mitochondrial genome, and pro-fibrotic SASP, which stiffens the myocardium and decreases microvascular integrity, respectively. Calibrated mitophagy-autophagy sustains ATP supply, nitric oxide bioavailability, and matrix homeostasis across cardiomyocytes, fibroblasts, and the endothelium, whereas excessive mitophagy leads to mass depletion and energy defects, and insufficient clearance leads to inflammation and diastolic dysfunction (Kent et al., 2021; Zhang et al., 2025b). All these studies suggest a possible role for matching autophagy initiation to lysosomal competence in mitigating remodeling, improving contractile reserve, and reducing arrhythmic susceptibility to load, metabolic stress, and I/R. Future research will involve the determination of composite flux biomarkers for patient stratification, trial of TFEB-directed lysosome repair and selective PINK1/Parkin or FUNDC1 priming, and trial of mitochondria-targeted redox and Ca2+-handling therapies to maintain quality (Yun et al., 2025; Schwemmlein et al., 2022).

11.6.2 Comorbid drivers: obesity, diabetes, hypertension

Obesity, diabetes, and hypertension converge on the aging heart by elevating nutrient and mechanical stress, which dysregulates autophagy–mitophagy and accelerates senescence. Adiposity promotes lipotoxic intermediates and low-grade inflammation, which increases ROS production and decreases the mitochondrial membrane potential. Diabetes induces hyperglycemia-driven AGE/RAGE and PKC signaling, which impairs calcium handling and lysosomal acidity. Hypertension imposes wall stress with RAAS and sympathetic activation, which augments the DRP1-skewed fission and proteotoxic load. As the AMPK-mTORC1 balance tilts, ULK1–Beclin-1 initiation becomes intermittent, ATG conjugation and LC3 lipidation become inefficient, and PINK1/Parkin or BNIP3/FUNDC1-mediated mitophagy cannot keep pace, allowing depolarized mitochondria and ubiquitinated aggregates to persist in the cells (Nag et al., 2024; Salvatore et al., 2021). mtDNA leakage then engages cGAS-STING and NF-κB, amplifying p16^INK4a/p21^CIP1 programs and a SASP that drives fibroblast activation, collagen cross-linking, endothelial nitric oxide loss, and

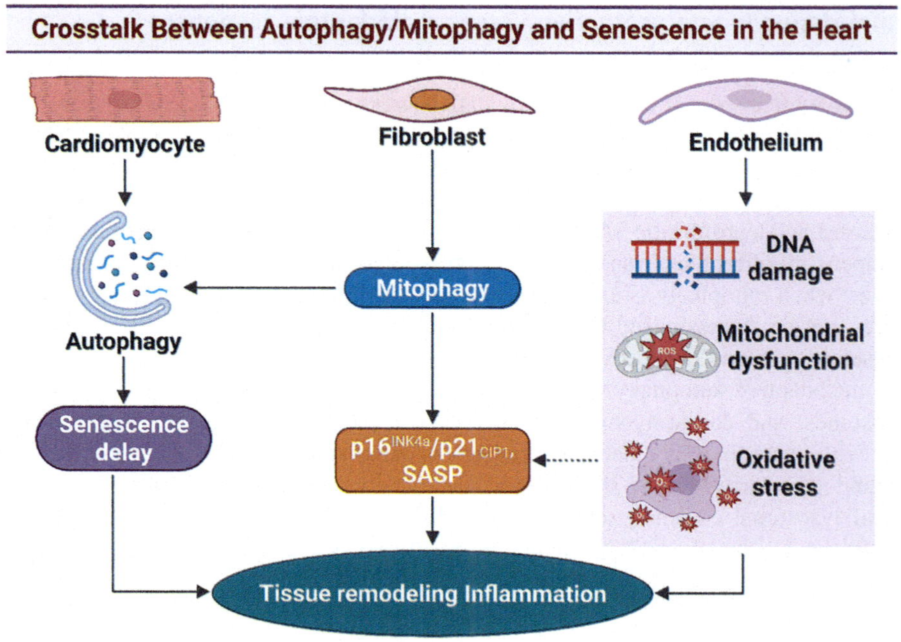

FIGURE 11.2 **Cell type-specific autophagy–mitophagy–senescence crosstalk in cardiac tissue: cardiomyocytes, fibroblasts, and endothelium.** The schematic represents crosstalk between autophagy, mitophagy, and senescence among cardiac cell types. Autophagy in cardiomyocytes minimizes proteotoxic stress and senescence. Mitochondrial removal in fibroblasts eliminates dysfunctional mitochondria, lowers them, and suppressing p16 -INK4a/p21CIP1 signals and SASP. Prosenescent cues are reciprocated by endothelial injury, DNA damage, mitochondrial dysfunction, and oxidative stress. The activation of SASP factors and inflammation are responsible for quality control pathway failures, which cause tissue remodeling and inflammation when activated by SASP factors and DAMP signals. On the other hand, autophagy-mitochondrial flux maintains mitochondrial fitness, inhibits maladaptive remodeling, and enhances diastolic compliance. *SASP*, senescence-associated secretory phenotype.

microvascular rarefaction, culminating in diastolic stiffening and arrhythmogenic remodeling. Collectively, studies indicate that these comorbid drivers synergistically blunt clearance pathways, intensify inflammaging, and hasten functional decline, whereas restoring flux preserves the ventricular compliance and microvascular integrity. Future studies should prioritize composite biomarkers linking autophagy/mitophagy activity to SASP and matrix turnover, integration of metabolic therapies (weight reduction, SGLT2 inhibitors, GLP-1 receptor agonists) with TFEB-guided lysosome restoration, and selective activation of PINK1/Parkin or FUNDC1 along with mitochondria-targeted redox support to re-establish quality control without compromising bioenergetics (Du et al., 2022; Guo et al., 2020).

11.6.3 Lifestyle and nutritional modifiers (e.g., caloric restriction mimetics)

Lifestyle and nutritional interventions modulate cardiac aging by rebalancing the nutrient-sensing and stress-response pathways that govern autophagy, mitophagy, and cellular senescence. Caloric restriction and exercise elevate AMP/ATP and NAD^+/NADH ratios, activating AMPK and SIRT1, while attenuating mTORC1 to release ULK1–Beclin-1 initiation and expand LC3-dependent autophagosome formation. In parallel, mitochondrial biogenesis and antioxidant capacity increase through SIRT1/PGC-1α signaling, and mitophagy is reinforced via the PINK1/Parkin and receptor pathways (BNIP3/NIX, FUNDC1), thereby culling depolarized mitochondria, lowering ROS, and stabilizing calcium handling. Caloric restriction mimetics, including metformin, resveratrol, spermidine, and ketone bodies, recapitulate these nodes by promoting AMPK activation, sirtuin engagement, TFEB-driven lysosomal biogenesis, and improved lysosomal acidification, all of which together enhance cargo recognition, autolysosomal degradation, and metabolic recycling. As proteotoxic and mitochondrial stress decline, DNA-damage signaling and cGAS-STING-NF-κB activation wane, p16INK4a/p21CIP1 programs are contained, and SASP intensity diminishes across cardiomyocytes, fibroblasts, and endothelium, easing matrix stiffness and preserving microvascular function (De-Leon-Covarrubias et al., 2024; Hussain et al., 2025; Pandey et al., 2022). Collectively, studies indicate that coordinated lifestyle measures and mimetics sustain autophagy mitophagy flux, limit inflammaging, and maintain diastolic compliance and contractile reserves in the aging heart. Future perspectives should establish composite biomarkers linking flux with SASP and matrix turnover, define dose timing windows and cell-type-specific responses, and test combinations of exercise, dietary modulation, and mimetics alongside lysosome-restoring and mitochondria-targeted redox strategies to prolong myocardial resilience without compromising bioenergetics.

11.7 Assessment and mapping technologies

11.7.1 Autophagy/mitophagy readouts (LC3, p62, mt-Keima, mitophagy reporters)

Assessment of autophagy/mitophagy in the heart relies on complementary flux readouts that track cargo passage rather than static abundance, beginning with LC3 conversion, where LC3-I lipidation to LC3-II marks autophagosome membranes. True flux was inferred by LC3-II turnover in the presence or absence of lysosomal inhibitors (e.g., bafilomycin A1 or chloroquine) alongside puncta quantification by immunostaining or tandem reporters. p62/SQSTM1 provides a reciprocal signal that decreases with efficient degradation and accumulation when biogenesis outpaces lysosomal clearance. As p62 is transcriptionally regulated by stress pathways, paired LC3 measurements and inhibitor controls are essential to avoid misclassification (Roychan, 2025; Baudot et al., 2015). For mitophagy, pH-sensitive reporters enable organelle-specific flux mapping. mt-Keima shifts excitation upon delivery to the acidic lysosome, supporting in vivo quantification in cardiomyocytes, whereas dual-fluorophore systems (e.g., mito-QC or mCherry-GFP fusions to outer-membrane anchors) distinguish autophagosomes from autolysosome pools by selective GFP quenching. These tools can be combined with parkin translocation, phospho-ubiquitin Ser65 immunodetection, TOMM20 loss assays, or mitochondrial DNA copy analyses to strengthen the assignment. In cardiac tissues, they benefit from AAV9 delivery or transgenic reporters to overcome motion and autofluorescence (McWilliams et al., 2016; Wang, 2020). Rigor requires time-matched controls, lysosomal competency checks, and cell-type resolved analyses across cardiomyocytes, fibroblasts, and endothelium. Collectively, studies show that multiplexed LC3/p62 dynamics with mt-Keima or mito-QC resolve autophagy–mitophagy bottlenecks that align with senescence and remodeling severity. Future studies should focus on standardizing cardiac flux panels, integrating single-cell and spatial imaging with longitudinal in vivo reporters, and developing clinically translatable biomarkers that link mitophagy flux to senescence signatures for patient stratification and therapy monitoring (Korolchuk et al., 2017; Gupta et al., 2025a; Hussain et al., 2024).

11.7.2 Senescence biomarkers (p16/p21, senescence-associated beta-galactosidase, senescence-associated secretory phenotype panels) and spatial readouts

Senescence assessment in the heart integrates cell-cycle arrest markers, lysosomal activity, secretory profiles, and spatial context to distinguish stable senescence from transient stress responses; accordingly, p16INK4a and p21CIP1 immunodetection and transcript levels indicate RB-dependent arrest, while colocalization with γH2AX/53BP1 telomere-associated foci strengthens attribution to DNA damage, and lipofuscin or GL13 complements senescence specificity in postmitotic cardiomyocytes. senescence-associated beta-galactosidase (SA-β-gal) provides a practical readout for lysosomal β-galactosidase activity at pH 6.0. However, high basal lysosomal content in cardiomyocytes and macrophages necessitates control and parallel markers to avoid false positives (Kumari and Jat, 2021). SASP panels measured by multiplex ELISA, proteomics, or spatial transcriptomics quantify inflammatory and matrix-active mediators (e.g., IL-6, IL-1β, TNF-α, TGF-β, MMPs, TIMPs, and CCL2), and coupling these signatures to cGAS-STING pathway activation helps to link mitochondrial DNA leakage to inflammaging. Spatial readouts, such as multiplex immunofluorescence, imaging mass cytometry, RNAscope, and spatial RNA/proteomics, map p16/p21 and SASP to cardiomyocytes, fibroblasts, and endothelium, align senescent niches with fibrosis and microvascular rarefaction, and enable coregistration with autophagy–mitophagy flux reporters (LC3/p62 dynamics, phospho-ubiquitin, and mt-Keima) to identify clearance bottlenecks (Xia et al., 2025; Lopes-Paciencia et al., 2019). Longitudinal AAV9-delivered reporters and single-cell multiomics resolve heterogeneity and track responses to load, metabolic stress, and ischemia-reperfusion. Collectively, studies indicate that composite biomarker panels integrating p16/p21, refined SA-β-gal, and SASP with spatial mapping best quantify senescence burden and its paracrine footprint, correlate with diastolic stiffening, and predict remodeling severity. Future studies should standardize cardiac-specific panels, harmonize spatial pipelines with flux metrics for patient stratification, and develop clinically translatable assays, such as circulating SASP multiplexes and imaging surrogates, to monitor senescence alongside autophagy–mitophagy restoration (Song et al., 2024; Zhu et al., 2024).

11.7.3 Imaging and functional assays for cardiac quality control

Imaging of autophagy–mitophagy, mito-KEMA in vivo, and visualization of ULK1–Beclin-1 puncta and autophagosome-lysosome contacts were combined with imaging biomarkers of cardiac quality control. Tandem mCherry-GFP-LC3 distinguishes between autophagosomes and autolysosomes, mito-QC as a reporter of mitophagy quantifies mitophagy in vivo, and super-resolution or two-photon microscopy visualizes U Contractile mechanics (speckle-tracking echocardiography, strain imaging, pressure-volume), and/or oxidative phosphorylation and coupling (high-resolution respirometry or Seahorse assays) were reported in intact hearts, whereas flux status (longitudinal studies across cardiomyocytes, fibroblasts, endothelium) was measured using AAV9-delivered reporters (McWilliams and Ganley, 2019; Montava-Garriga et al., 2020). The combination of these modalities is associated with stalled LC3/p62 turnover or low mitochondrial Keima signaling with impaired respiration, diastolic stiffening, enhanced inflammatory readouts, and recovered lysosomal acidity and mitophagy with better strain and endothelial performance. Together, research points out that multimodal, flux-anchored imaging in connection with functional analyses is the most powerful method for detecting bottlenecks between defective clearance and the severity of remodeling. Future efforts should involve the standardization of cardiac flux panels, the development of longitudinal reporters in vivo that can be used with clinical imaging surrogates, and the validation of translational endpoints that can be linked to restoring mitophagy to better mechanics and less inflammaging (Kelly et al., 2024; Pugsley, 2017; Gupta et al., 2025b; Prasher et al., 2023) (Fig. 11.3).

11.8 Therapeutic strategies targeting quality control and senescence

11.8.1 mTOR/AMPK modulators, mitophagy enhancers (rapalogs, spermidine, urolithin A)

Modulating mTOR/AMPK and improving mitophagy to reestablish cardiac quality control by coordinating the initiation of autophagy with lysosomal and mitochondrial capacity, transient inhibition of mTORC1 by rapalogs liberates ULK1–Beclin-1 to increase LC3-dependent biogenesis and promotes TFEB/TFE3 programs to enhance lysosomal acidity and degradative flux, and activation of AMPK in response to energy stress further phosphorylates ULK1 and suppresses these nodes. These are complemented by mitophagy-directed agents: spermidine prevents EP300-mediated acetylation to derepress autophagy genes and enhance proteostasis, and urolithin A elicits a conserved mitophagy response culling depolarized mitochondria and promoting biogenesis via PGC-1a/NRF signaling, reducing ROS, preventing the release of mitochondrial DNA, and preventing cGAS-STING-NF-κB activation (Zhang et al., 2018b; Li et al., 2014). Coordinated autophagy–mitophagy maintains ATP, calcium dynamics, and excitation-contraction coupling in cardiomyocytes, attenuates myofibroblast transition, and nitric oxide bioavailability maintains matrix rigidity and microvascular rarefaction in fibroblasts and endothelium, respectively. Normalization of flux results in reduced p164a/p214CIP1 pressure and inhibition of SASP strength, which is reflected in increased diastolic compliance and decreased load-induced arrhythmogenic remodeling, metabolic strain, and ischemia-reperfusion. Together, these studies suggest that any additive combination, including mTOR/AMPK modulators with mitophagy enhancers, can restore the loss of clearance without reducing the mass of mitochondria and without disturbing the bioenergetics. Future clinical efforts should establish dose-timing ranges for lysosomal competence, prove composite biomarkers to bind LC3/p62 dynamics to phospho-ubiquitin and mt-Keima radiation for patient stratification, and design cell type-specific delivery with mitochondria-targeted redox support to maintain long-term resilience with minimum off-target effects (Li et al., 2014; Bartolomé et al., 2017; Carosi et al., 2022).

11.8.2 Senotherapeutics: senolytics and senomorphics in cardiac contexts

Heart senotherapeutics aims to interrupt the negative cardiac consequences of chronic DNA damage and mitochondrial stress by eliminating senescent cells (senolytics) or reprogramming the senescent cell secretory phenotype (senomorphics). Thus BCL-2 family inhibitors and combinations, such as tyrosine kinase inhibitors and flavonoids, preferentially eliminate p164a/p21-high cells that promote fibrosis, whereas agents that suppress NF-κB/JAK-STAT or reprogram autophagy-lysosomes are not effective. Oxidative and metabolic stress increase the load of pro-inflammatory and matrix-remodeling cells. Senomorphics of rapalogs, metformin, spermidine, or SIRT1 activators enhance proteostasis, mitophagy, and endothelial nitric oxide bioavailability, ease the load of cardiac diastolic stiffness, and stabilize excitation-contraction coupling (Park and Shin, 2022; Suda et al., 2024; Goyal et al., 2025; Samuel et al., 2024). Owing to the potential for impaired repair in cases of excessive clearance or general pathway inhibition, dosing is usually intermittent and progressively accompanied by strategies that correspond to autophagy initiation to lysosomal capacity and minimize off-target effects through cell type-selective delivery. Overall, literature shows that sensitive senotherapeutic approaches

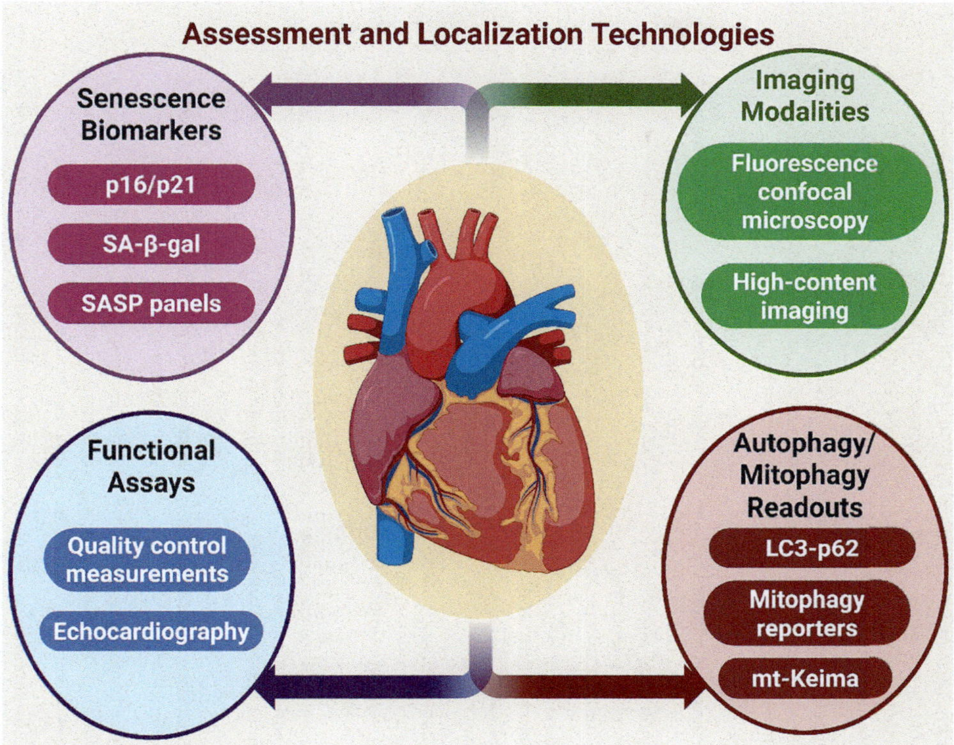

FIGURE 11.3 **Integrated assessment framework for cardiac quality control: spatial mapping of senescence biomarkers and autophagy–mitophagy flux with functional endpoints.** Overview of the assessment and localization of cardiac quality control. The senescence biomarkers in the left panel, including p16/p21, SA-β-gal, and SASP panels, were used to measure cell cycle arrest and inflammatory secretomes. The right panel shows the readout of autophagy/mitophagy, the dynamics of the products of autophagy (LC3-p62), and the reporters of mitophagy (mitochondrial turnover: mt-Keima), which were used to measure the flux. The high green circle represents imaging modalities, such as fluorescence confocal microscopy and high-content imaging, which are used for spatial mapping. The bottom blue circle depicts functional assays, quality control measurements, and echocardiography connecting molecular flux to organ-level performance in the center of the heart image. *SASP*, senescence-associated secretory phenotype.

have anti-SASP signaling, decreased collagen deposition, and maintained microvascular integrity in cases of mitochondrial quality control compatibility. Future studies should focus on the use of composite biomarkers to connect p16/p21 activity to SASP panels and mitophagy flux, precision schedules involving a combination of senomorphics and mitophagy enhancers, TFEB- directing lysosome restoration, and rigorous and cell-type-resolved trials to achieve long-lasting functional improvements with few side effects (Bonam et al., 2021; Raffaele and Vinciguerra, 2022; Bhat et al., 2024; Thapa et al., 2024b).

11.8.3 Gene/cell-based and lifestyle interventions: translation pathways

Gene/cell-based and lifestyle intervention translation pathways intersect on the restoration of cardiac quality regulation and the moderation of senescence; thus initial attempts should focus on defining targets (ULK1–Beclin-1 initiation, TFEB/TFE3 lysosome biogenesis, PINK1/Parkin or BNIP3/FUNDC1 mitophagy, and LAMP2A to CMA) and refining delivery by AAV9 or lipid nanoparticles to deliver CRISPRa/i, RNAi, or cDNA payload Cardiomyocytes reprogrammed into induced pluripotent stem cells, engineered mesenchymal stromal cells, exosomes that are enriched with autophagy/mitophagy-promoting miRNAs and proteins as cardiomyocyte paracrine modulators; cardiac-tropic capsids, immune and off-target mitigated through dose fractionation, and promoters regulable are engineered as paracrine modulators; select cardiomyocyte paracrine signaling pathways. Lifestyle translation occurs via formal exercise, food adjustments, and dietary caloric-restriction-mimetic protocols to activate AMPK-SIRT1, diminish mTORC1, improve TFEB programs, and fix mitophagy without diminishing mitochondrial mass, cardiac rehabilitation models to enforce adherence, telemetry, and metabolic luminance (Vásquez-Trincado et al., 2016; Roberts and Markby, 2021). The development based on the

TABLE 11.1 Autophagy–mitophagy–senescence landscape in the aging heart.

Mechanism	Key proteins/nodes	Cellular context	Functional outcome	Biomarker/readout	References
Macroautophagy initiation	ULK1–ATG13–FIP200–ATG101; Beclin-1–VPS34–ATG14L; ATG12–ATG5–ATG16L1; LC3 lipidation	Cardiomyocyte, Fibroblast, Endothelium	Proteostasis maintenance; protection during ischemia-reperfusion and pressure overload	LC3-I→LC3-II turnover (± lysosomal inhibitors); p62/SQSTM1 clearance	Gammoh et al. (2013)
Cargo selection and trafficking	p62/SQSTM1, NBR1, OPTN, NDP52, TAX1BP1; Rab7–HOPS; STX17–SNAP29–VAMP8; v-ATPase	Cardiomyocyte	Targeted removal of ubiquitinated cargo; efficient autolysosomal degradation	LC3 puncta with receptor colocalization; lysosomal acidification assays	Zellner and Behrends (2021)
Chaperone-mediated autophagy	LAMP2A; Hsc70/HSPA8; lys-Hsc70; KFERQ-like motifs	Cardiomyocyte, Fibroblast, Endothelium	Renewal of metabolic/redox proteins; restraint of senescence-associated secretory phenotype (SASP) signaling	LAMP2A abundance; substrate uptake assays; CMA reporter constructs	Losmanová et al. (2020)
PINK1/parkin-dependent mitophagy	PINK1; Parkin; phospho-ubiquitin (Ser65); OPTN/NDP52; TBK1	Cardiomyocyte under I/R, overload, metabolic stress	Removal of depolarized mitochondria; ROS reduction; ATP preservation	Parkin translocation; p-Ub(Ser65) immunodetection; mt-Keima flux	Heo et al. (2015); Li et al. (2022)
Receptor-mediated mitophagy	BNIP3, NIX/BNIP3L, FUNDC1; DRP1-mediated fission; PGAM5/CK2/ULK1 phospho-switches	Cardiomyocyte, Endothelium, Fibroblast	Quality control under hypoxia/oxidative stress; prevention of mtDNA leakage	Receptor phosphorylation status (e.g., FUNDC1-Ser17); mito-QC reporters	Liu et al. (2014), (2022)
Mitochondrial dynamics and damage signaling	MFN1/2, OPA1, DRP1; cGAS-STING; NF-κB	Whole myocardium	Balance of fusion–fission coupled to mitophagy; limitation of inflammaging	OPA1 processing; DRP1 activation; cGAS-STING activation markers	Lin et al. (2022)
Senescence effectors	p16INK4a, p21CIP1, p53; SASP (IL-6, IL-1β, TNF-α, TGF-β, MMPs)	Cardiomyocyte, Fibroblast, Endothelium	Cell-cycle arrest; matrix remodeling; microvascular dysfunction	p16/p21 expression; SA-β-gal with controls; SASP multiplex panels	Lemay et al. (2025)
Physiologic/pathologic modulators	Hemodynamic load; metabolic stress; ischemia-reperfusion	Whole myocardium	Flux mismatch → diastolic stiffening, arrhythmogenic remodeling	Strain imaging; respirometry; endothelial function assays	Buja, (2023)
Therapeutic modulators	Rapalogs; AMPK activators; Spermidine; Urolithin A; Senolytics/Senomorphics	Translational context	Restoration of flux; SASP attenuation; preserved compliance	Composite panels linking LC3/p62, p-Ub, mt-Keima, SASP	Barkovskaya et al. (2025)

biomarker is then connected to the LC3/p62 dynamics, phospho-ubiquitin, and lysosomal competence with the circulating SASP panels, endothelial functionality, and strain imaging to select patients and monitor the response, scale-up of manufacturing hits GMP production, stability, and release requirements of vectors and cell products. Together, these findings suggest that gene or cell therapy in combination with lifestyle interventions can restore normal autophagy—mitophagy flux, reduce inflammation, and enhance diastolic compliance and contractile reserves. Future goals include cell type-selective delivery and control of its reversibility, Europeanization of composite flux-senescence biomarkers in stratified trials, and long-term safety follow-up to ensure durable myocardial protection and reduce the drawbacks of arrhythmic recontrol deimmunization and off-target liability (Bashir et al., 2025; Thapa et al., 2024a) (Table 11.1).

11.9 Conclusion and future perspectives

Autophagy, mitophagy, and cellular senescence interact to establish boundaries between the adapting and aging hearts. Initially, the synthesis highlighted that ULK1–Beclin-1–ATG-autophagy and PINK1/pinkin-mediated mitophagy (BNIP3/NIX, FUNDC1) determine the levels of autophagy and mitophagy, which bead levels off ROS, mtDNA leakage, and cGAS-STING-NF-kB signaling that stabilizes p16CIP1 ink4a programs. In cardiomyocytes, fibroblasts, and the endothelium, continuous flux maintains ATP delivery, calcium trafficking, bioavailability of nitric oxide, and plasticity of the matrix, where initiation failure or loss of lysosomal competence results in LC3-II/p62 build-up, facilitating the spread of mitochondrial dysfunction, leading to diastolic stiffness and arrhythmogenic remodeling. This study is novel in combining molecular circuitry (macroautophagy/mitophagy/CMA), spatial and single-cell biomarker models (LC3/p62 dynamics, phospho-ubiquitin, mt-Keima, and p16/p21 with SASP panels), and actionable levers (mTOR/AMPK modulators, mitophagy enhancers, senotherapeutics, and gene/cell-based or lifestyle programs) into a single translational map, linking flux to quantifiable functional endpoints. The mechanisms, assessment technologies, and intervention strategies are covered, with a focus on the balanced initiation of autophagy and lysosomal capacity, and the withdrawal of mitophagy without the loss of mitochondrial mass. In the future, it is anticipated that composite cardiac-specific biomarker sets should be prioritized to couple autophagy-mitochondrial flux and senescence, cell-type-resolved trials should be prioritized to align TFEB-guided lysosome restoration or calibrated PINK1/Parkin or FUNDC1 activation, and schedules need to be precise to couple senomorphics or intermittent senolysis with mitochondria-targeted redox support. Potential areas of mechanistic improvement include ULK1 substrate mapping in the human myocardium; factors that mediate OPA1 processing, which connect dynamics with mitophagy; and causal models that allow the association of spatial senescence niches with perfusion and conduction defects. Stratified deployment may be allowed with clinical AAV9 or LNP delivery using regulable promoters, rehabilitation-attached lifestyle programs, and standard imaging-circulating readouts (strain, endothelial function, and SASP multiplexes). Taken together, all evidence allows for the development of a practical hypothesis: the reinstatement of quality-control flux in all cardiac cell types may suppress senescence load, alleviate adverse remodeling, and enhance aging and comorbidity-stress resilience.

References

Abuetabh, Y., Wu, H.H., Chai, C., Al Yousef, H., Persad, S., Sergi, C.M., et al., 2022. DNA damage response revisited: the p53 family and its regulators provide endless cancer therapy opportunities. Experimental and Molecular Medicine 54 (10), 1658–1669. Available at https://doi.org/10.1038/s12276-022-00863-4.

Adebayo, M., Singh, S., Singh, A.P., Dasgupta, S., 2021. Mitochondrial fusion and fission: the fine-tune balance for cellular homeostasis. The FASEB Journal 35 (6). Available at https://doi.org/10.1096/fj.202100067r.

Aix, E., Gutiérrez-Gutiérrez, Ó., Sánchez-Ferrer, C., Aguado, T., Flores, I., 2016. Postnatal telomere dysfunction induces cardiomyocyte cell-cycle arrest through p21 activation. Journal of Cell Biology 213 (5), 571–583. Available at https://doi.org/10.1083/jcb.201510091.

Alam, S., Abdullah, C.S., Aishwarya, R., Morshed, M., Bhuiyan, M.S., 2020. Molecular perspectives of mitochondrial adaptations and their role in cardiac proteostasis. Frontiers in Physiology 11. Available at https://doi.org/10.3389/fphys.2020.01054.

Albano, G.D., Montalbano, A.M., Gagliardo, R., Profita, M., 2023. Autophagy/mitophagy in airway diseases: impact of oxidative stress on epithelial cells. Biomolecules 13 (8). Available at https://doi.org/10.3390/biom13081217.

Bachmann, J.C., Baumgart, S.J., Uryga, A.K., Bosteen, M.H., Borghetti, G., Nyberg, M., et al., 2022. Fibrotic signaling in cardiac fibroblasts and vascular smooth muscle cells: the dual roles of fibrosis in HFpEF and CAD. Cells 11 (10). Available at https://doi.org/10.3390/cells11101657.

Barcena, M.L., Aslam, M., Norman, K., Ott, C., Ladilov, Y., 2025. Role of AMPK and Sirtuins in aging heart: basic and translational aspects. Aging and Disease 16 (5). Available at https://doi.org/10.14336/AD.2024.1216.

Barkovskaya, A., Brauning, A., Thambala, A., Bhardwaj, K., Chamoli, M., Rane, A., et al., 2025. Mitigating pro-inflammatory SASP and DAMP with urolithin A: a novel senomorphic strategy. Aging Cell 24 (11), 1474–9718. Available at https://doi.org/10.1111/acel.70237.

Bartolomé, A., García-Aguilar, A., Asahara, S.-I., Kido, Y., Guillén, C., Pajvani, U.B., et al., 2017. MTORC1 regulates both general autophagy and mitophagy induction after oxidative phosphorylation uncoupling. Molecular and Cellular Biology 37 (23), 1098–5549. Available at https://doi.org/10.1128/mcb.00441–17.

Bashir, B., Gulati, M., Vishwas, S., Gupta, G., Dhanasekaran, M., Paudel, K.R., et al., 2025. Bridging gap in the treatment of Alzheimer's disease via postbiotics: current practices and future prospects. Ageing Research Reviews 105, 102689. Available at https://doi.org/10.1016/j.arr.2025.102689.

Baudot, A.D., Haller, M., Mrschtik, M., Tait, S.W.G., Ryan, K.M., 2015. Using enhanced-mitophagy to measure autophagic flux. Methods (San Diego, Calif.) 75, 105–111. Available at https://doi.org/10.1016/j.ymeth.2014.11.014.

Bharath, L.P., Mueller, R., Li, Y., Ruan, T., Kunz, D., Goodrich, R., et al., 2014. Impairment of autophagy in endothelial cells prevents shear-stress-induced increases in nitric oxide bioavailability1. Canadian Journal of Physiology and Pharmacology 92 (7), 605–612. Available at https://doi.org/10.1139/cjpp-2014-0017.

Bhat, A.A., Moglad, E., Afzal, M., Thapa, R., Almalki, W.H., Kazmi, I., et al., 2024. Therapeutic approaches targeting aging and cellular senescence in Huntington's disease. CNS Neuroscience and Therapeutics 30 (10). Available at https://doi.org/10.1111/cns.70053.

Bhattacharya, S., Yin, J., Huo, W., Chaum, E., 2022. Modeling of mitochondrial bioenergetics and autophagy impairment in MELAS-mutant iPSC-derived retinal pigment epithelial cells. Stem Cell Research and Therapy 13 (1). Available at https://doi.org/10.1186/s13287-022-02937-6.

Bloom, S.I., Liu, Y., Tucker, J.R., Islam, M.T., Machin, D.R., Abdeahad, H., et al., 2023. Endothelial cell telomere dysfunction induces senescence and results in vascular and metabolic impairments. Aging Cell 22 (8). Available at https://doi.org/10.1111/acel.13875.

Bonam, S.R., Tranchant, C., Muller, S., 2021. Autophagy-lysosomal pathway as potential therapeutic target in Parkinson's disease. Cells 10 (12). Available at https://doi.org/10.3390/cells10123547.

Buja, L.M., 2023. Pathobiology of myocardial ischemia and reperfusion injury: models, modes, molecular mechanisms, modulation, and clinical applications. Cardiology in Review 31 (5), 252–264. Available at https://doi.org/10.1097/CRD.0000000000000440.

Carosi, J.M., Fourrier, C., Bensalem, J., Sargeant, T.J., 2022. The mTOR–lysosome axis at the centre of ageing. FEBS Open Bio 12 (4), 739–757. Available at https://doi.org/10.1002/2211-5463.13347.

Colliva, A., Braga, L., Giacca, M., Zacchigna, S., 2020. Endothelial cell–cardiomyocyte crosstalk in heart development and disease. The Journal of Physiology 598 (14), 2923–2939. Available at https://doi.org/10.1113/jp276758.

Cui, H., Kong, Y., Zhang, H., 2012. Oxidative stress, mitochondrial dysfunction, and aging. Journal of Signal Transduction 2012, 1–13. Available at https://doi.org/10.1155/2012/646354.

De-Leon-Covarrubias, U.E., Perez-Trujillo, J.J., Villa-Cedillo, S.A., Martinez-Perez, A.G., Montes-de-Oca-Saucedo, C.R., Loera-Arias, Md.J., et al., 2024. Unlocking the potential: caloric restriction, caloric restriction mimetics, and their impact on cancer prevention and treatment. Metabolites 14 (8). Available at https://doi.org/10.3390/metabo14080418.

Desideri, E., Castelli, S., Ciriolo, M.R., 2025. MAPK signaling in the interplay between oxidative stress and autophagy. Antioxidants 14 (6). Available at https://doi.org/10.3390/antiox14060662.

Dixit, V., Chaubey, K.K., Dayal, D., Bajrang Chole, P., Rakesh Kumar Bachheti, M.B.T., Gupta, P.C., et al., 2025. Natural products in aging: cellular mechanisms and emerging therapeutics for age-related disorders. Journal of Aging Research 2025 (1). Available at http://doi.org/10.1155/jare/6868732.

Du, Y., Zhang, H., Nie, X., Qi, Y., Shi, S., Han, Y., et al., 2022. Link between sterile inflammation and cardiovascular diseases: focus on cGAS-STING pathway in the pathogenesis and therapeutic prospect. Frontiers in Cardiovascular Medicine 9. Available at https://doi.org/10.3389/fcvm.2022.965726.

Gammoh, N., Florey, O., Overholtzer, M., Jiang, X., 2013. Interaction between FIP200 and ATG16L1 distinguishes ULK1 complex–dependent and –independent autophagy. Nature Structural and Molecular Biology 20 (2), 144–149. Available at https://doi.org/10.1038/nsmb.2475.

Ganley, I.G., Lam, D.H., Wang, J., Ding, X., Chen, S., Jiang, X., 2009. ULK1·ATG13·FIP200 complex mediates mTOR signaling and is essential for autophagy. Journal of Biological Chemistry 284 (18), 12297–12305. Available at http://doi.org/10.1074/jbc.M900573200.

Ghosh, R., Vinod, V., Symons, J.D., Boudina, S., 2020. Protein and mitochondria quality control mechanisms and cardiac aging. Cells 9 (4), 933. Available at https://doi.org/10.3390/cells9040933.

Gómez-Virgilio, L., Silva-Lucero, M.-del-C., Flores-Morelos, D.-S., Gallardo-Nieto, J., Lopez-Toledo, G., Abarca-Fernandez, A.-M., et al., 2022. Autophagy: a key regulator of homeostasis and disease: an overview of molecular mechanisms and modulators. Cells 11 (15), 2262. Available at https://doi.org/10.3390/cells11152262.

Goyal, K., Afzal, M., Babu, M.A., Roopashree, R., Bansal, P., Mishra, S., et al., 2025. Icariin as a therapeutic agent in breast cancer: modulating apoptosis and suppressing proliferation. Journal of Complementary and Integrative Medicine 22 (3), 417–426. Available at https://doi.org/10.1515/jcim-2024-0405.

Guo, Y., Gu, R., Gan, D., Hu, F., Li, G., Xu, G., 2020. Mitochondrial DNA drives noncanonical inflammation activation via cGAS–STING signaling pathway in retinal microvascular endothelial cells. Cell Communication and Signaling 18 (1). Available at https://doi.org/10.1186/s12964-020-00637-3.

Gupta, S., Afzal, M., Agrawal, N., Almalki, W.H., Rana, M., Gangola, S., et al., 2025a. Harnessing the FOXO-SIRT1 axis: insights into cellular stress, metabolism, and aging. Biogerontology 26 (2). Available at https://doi.org/10.1007/s10522-025-10207-0.

Gupta, G., Samuel, V.P., Rani, R.M.M.B., Sasikumar, Y., Nayak, P.P., et al., 2025b. Caspase-independent cell death in lung cancer: from mechanisms to clinical applications. Naunyn-Schmiedeberg's Archives of Pharmacology 398 (10), 13031–13048. Available at https://doi.org/10.1007/s00210-025-04149-0.

Heo, J.M., Ordureau, A., Paulo, J.A., Rinehart, J., Harper, J.W., 2015. The PINK1-PARKIN mitochondrial ubiquitylation pathway drives a program of OPTN/NDP52 recruitment and TBK1 activation to promote mitophagy. Molecular Cell 60 (1), 7–20. Available at https://doi.org/10.1016/j.molcel.2015.08.016.

Herbstein, F., Sapochnik, M., Attorresi, A., Pollak, C., Senin, S., Gonilski-Pacin, D., et al., 2024. The SASP factor IL-6 sustains cell-autonomous senescent cells via a cGAS-STING-NFκB intracrine senescent noncanonical pathway. Aging Cell 23 (10). Available at https://doi.org/10.1111/acel.14258.

Hussain, M.S., Agrawal, N., Ilma, B., Rekha, M.M., Nayak, P.P., Kaur, M., et al., 2025. Autophagy and cellular senescence in Alzheimer's disease: key drivers of neurodegeneration. CNS Neuroscience and Therapeutics 31 (7). Available at https://doi.org/10.1111/cns.70503.

Hussain, M.S., Moglad, E., Afzal, M., Sharma, S., Gupta, G., Sivaprasad, G.V., et al., 2024. Autophagy-associated non-coding RNAs: unraveling their impact on Parkinson's disease pathogenesis. CNS Neuroscience and Therapeutics 30 (5). Available at https://doi.org/10.1111/cns.14763.

Ivankovic, D., Chau, K.-Y., Schapira, A.H.V., Gegg, M.E., 2016. Mitochondrial and lysosomal biogenesis are activated following PINK 1/parkin-mediated mitophagy. Journal of Neurochemistry 136 (2), 388–402. Available at https://doi.org/10.1111/jnc.13412.

Janaszak-Jasiecka, A., Płoska, A., Wierońska, J.M., Dobrucki, L.W., Kalinowski, L., 2023. Endothelial dysfunction due to eNOS uncoupling: molecular mechanisms as potential therapeutic targets. Cellular and Molecular Biology Letters 28 (1). Available at https://doi.org/10.1186/s11658-023-00423-2.

Jeong, S.J., Zhang, X., Rodriguez-Velez, A., Evans, T.D., Razani, B., 2019. P62/SQSTM1 and selective autophagy in cardiometabolic diseases. Antioxidants and Redox Signaling 31 (6), 458–471. Available at https://doi.org/10.1089/ars.2018.7649.

Kelly, G., Kataura, T., Panek, J., Ma, G., Salmonowicz, H., Davis, A., et al., 2024. Suppressed basal mitophagy drives cellular aging phenotypes that can be reversed by a p62-targeting small molecule. Developmental Cell 59 (15), 1924. Available at https://doi.org/10.1016/j.devcel.2024.04.020.

Kent, A.C., El Baradie, K.B.Y., Hamrick, M.W., 2021. Targeting the mitochondrial permeability transition pore to prevent age-associated cell damage and neurodegeneration. Oxidative Medicine and Cellular Longevity 2021. Available at https://doi.org/10.1155/2021/6626484.

Kmiotek-Wasylewska, K., Bobis-Wozowicz, S., Karnas, E., Orpel, M., Woźnicka, O., Madeja, Z., et al., 2023. Anti-inflammatory, anti-fibrotic and pro-cardiomyogenic effects of genetically engineered extracellular vesicles enriched in miR-1 and miR-199a on human cardiac fibroblasts. Stem Cell Reviews and Reports 19 (8), 2756–2773. Available at https://doi.org/10.1007/s12015-023-10621-2.

Korolchuk, V.I., Miwa, S., Carroll, B., von Zglinicki, T., 2017. Mitochondria in cell senescence: is mitophagy the weakest link? EBioMedicine 21, 7–13. Available at https://doi.org/10.1016/j.ebiom.2017.03.020.

Kumari, R., Jat, P., 2021. Mechanisms of cellular senescence: cell cycle arrest and senescence associated secretory phenotype. Frontiers in Cell and Developmental Biology 9. Available at https://doi.org/10.3389/fcell.2021.645593.

Lee, J., Giordano, S., Zhang, J., 2012. Autophagy, mitochondria and oxidative stress: cross-talk and redox signalling. Biochemical Journal 441 (2), 523–540. Available at http://doi.org/10.1042/BJ20111451.

Lemay, S.E., Kuwabara, Y., Bonnet, S., Potus, F., Provencher, S., Adnot, S., et al., 2025. Cellular senescence in adult pulmonary hypertension: current state and future challenges. European Respiratory Review: An Official Journal of the European Respiratory Society 34 (177). Available at https://doi.org/10.1183/16000617.0030-2025.

Li, Y., Chen, C., Yao, F., Su, Q., Liu, D., Xue, R.G., et al., 2014. AMPK inhibits cardiac hypertrophy by promoting autophagy via mTORC1. Archives of Biochemistry and Biophysics 558, 79–86. Available at https://doi.org/10.1016/j.abb.2014.06.023.

Li, A., Gao, M., Liu, B., Qin, Y., chen, L., Liu, H., et al., 2022. Mitochondrial autophagy: molecular mechanisms and implications for cardiovascular disease. Cell Death and Disease 13 (5). Available at https://doi.org/10.1038/s41419-022-04906-6.

Li, Y., Kračun, D., Dustin, C.M., El Massry, M., Yuan, S., Goossen, C.J., et al., 2021. Forestalling age-impaired angiogenesis and blood flow by targeting NOX: Interplay of NOX1, IL-6, and SASP in propagating cell senescence. Proceedings of the National Academy of Sciences 118 (42). Available at https://doi.org/10.1073/pnas.2015666118.

Li, X., Pang, X., Sun, H., Zhang, B., Wang, H., Wu, N., et al., 2025. Cardiac aging: molecular mechanisms and therapeutic interventions. Pharmacological Research 221, 107954. Available at https://doi.org/10.1016/j.phrs.2025.107954.

Lim, J., Kim, H.W., Youdim, M.B.H., Rhyu, I.J., Choe, K.M., Oh, Y.J., 2011. Binding preference of p62 towards LC3-II during dopaminergic neurotoxin-induced impairment of autophagic flux. Autophagy 7 (1), 51–60. Available at https://doi.org/10.4161/auto.7.1.13909.

Lin, J., Duan, J., Wang, Q., Xu, S., Zhou, S., Yao, K., 2022. Mitochondrial dynamics and mitophagy in cardiometabolic disease. Frontiers in Cardiovascular Medicine 9. Available at https://doi.org/10.3389/fcvm.2022.917135.

Linders, A.N., Dias, I.B., López Fernández, T., Tocchetti, C.G., Bomer, N., Van der Meer, P., 2024. A review of the pathophysiological mechanisms of doxorubicin-induced cardiotoxicity and aging. npj Aging 10 (1). Available at https://doi.org/10.1038/s41514-024-00135-7.

Lindstrom, M., DeCleene, N., Dorsey, H., Fuster, V., Johnson, C.O., LeGrand, K.E., et al., 2022. Global burden of cardiovascular diseases and risks collaboration, 1990-2021. Journal of the American College of Cardiology 80 (25), 2372–2425. Available at https://doi.org/10.1016/j.jacc.2022.11.001.

Liu, L., Sakakibara, K., Chen, Q., Okamoto, K., 2014. Receptor-mediated mitophagy in yeast and mammalian systems. Cell Research 24 (7), 787–795. Available at https://doi.org/10.1038/cr.2014.75.

Liu, H., Zang, C., Yuan, F., Ju, C., Shang, M., Ning, J., et al., 2022. The role of FUNDC1 in mitophagy, mitochondrial dynamics and human diseases. Biochemical Pharmacology 197, 114891. Available at https://doi.org/10.1016/j.bcp.2021.114891.

Lopes-Paciencia, S., Saint-Germain, E., Rowell, M.-C., Ruiz, A.F., Kalegari, P., Ferbeyre, G., 2019. The senescence-associated secretory phenotype and its regulation. Cytokine 117, 15–22. Available at https://doi.org/10.1016/j.cyto.2019.01.013.

Losmanová, T., Janser, F.A., Humbert, M., Tokarchuk, I., Schläfli, A.M., Neppl, C., et al., 2020. Chaperone-mediated autophagy markers LAMP2A and HSC70 are independent adverse prognostic markers in primary resected squamous cell carcinomas of the lung. Oxidative Medicine and Cellular Longevity 2020, 1–12. Available at https://doi.org/10.1155/2020/8506572.

Maejima, Y., Isobe, M., Sadoshima, J., 2016. Regulation of autophagy by Beclin 1 in the heart. Journal of Molecular and Cellular Cardiology 95, 19–25. Available at https://doi.org/10.1016/j.yjmcc.2015.10.032.

McWilliams, T.G., Ganley, I.G., 2019. Investigating mitophagy and mitochondrial morphology in vivo using mito-QC: a comprehensive guide. Methods in Molecular Biology 1880, 621–642. Available at https://doi.org/10.1007/978-1-4939-8873-0_41.

McWilliams, T.G., Prescott, A.R., Allen, G.F.G., Tamjar, J., Munson, M.J., Thomson, C., et al., 2016. Mito-QC illuminates mitophagy and mitochondrial architecture in vivo. Journal of Cell Biology 214 (3), 333–345. Available at https://doi.org/10.1083/jcb.201603039.

Mendis, S., Graham, I., Narula, J., 2022. Addressing the global burden of cardiovascular diseases; need for scalable and sustainable frameworks. Global Heart 17 (1). Available at https://doi.org/10.5334/gh.1139.

Mijit, M., Caracciolo, V., Melillo, A., Amicarelli, F., Giordano, A., 2020. Role of p53 in the regulation of cellular senescence. Biomolecules 10 (3), 420. Available at https://doi.org/10.3390/biom10030420.

Möbus, L., Ylä-Outinen, L., Mannino, L., Migliaccio, G., Kosunen, K., D'Alessandro, N., et al., 2025. Endothelial sensitivity to pro-fibrotic signals links systemic exposure to pulmonary fibrosis. Cell Death and Disease 16 (1). Available at https://doi.org/10.1038/s41419-025-07824-5.

Montava-Garriga, L., Singh, F., Ball, G., Ganley, I.G., 2020. Semi-automated quantitation of mitophagy in cells and tissues. Mechanisms of Ageing and Development 185, 111196. Available at https://doi.org/10.1016/j.mad.2019.111196.

Nag, S., Mitra, O., Maturi, B., Kaur, S.P., Saini, A., Nama, M., et al., 2024. Autophagy and mitophagy as potential therapeutic targets in diabetic heart condition: Harnessing the power of nanotheranostics. Asian Journal of Pharmaceutical Sciences 19 (3), 100927. Available at https://doi.org/10.1016/j.ajps.2024.100927.

Nàger, M., Calvoli, M., Larsen, K.B., Birgisdottir, A.B., 2025. The multifaceted role of autophagy and mitophagy in cardiovascular health and disease. Autophagy Reports 4 (1). Available at https://doi.org/10.1080/27694127.2025.2572511.

Ong, S.B., Kalkhoran, S.B., Hernández-Reséndiz, S., Samangouei, P., Ong, S.G., Hausenloy, D.J., 2017. Mitochondrial-shaping proteins in cardiac health and disease – the long and the short of it!. Cardiovascular Drugs and Therapy 31 (1), 87–107. Available at https://doi.org/10.1007/s10557-016-6710-1.

Onishi, M., Yamano, K., Sato, M., Matsuda, N., Okamoto, K., 2021. Molecular mechanisms and physiological functions of mitophagy. The EMBO Journal 40 (3). Available at https://doi.org/10.15252/embj.2020104705.

O'Sullivan, E.A., Wallis, R., Mossa, F., Bishop, C.L., 2024. The paradox of senescent-marker positive cancer cells: challenges and opportunities. npj Aging 10 (1). Available at https://doi.org/10.1038/s41514-024-00168-y.

Pandey, S.N., Singh, G., Semwal, B.C., Gupta, G., Alharbi, K.S., Almalki, W.H., et al., 2022. Therapeutic approaches of nutraceuticals in the prevention of Alzheimer's disease. Journal of Food Biochemistry 46 (12). Available at https://doi.org/10.1111/jfbc.14426.

Park, J., Shin, D.W., 2022. Senotherapeutics and their molecular mechanism for improving aging. Biomolecules and Therapeutics 30 (6), 490–500. Available at https://doi.org/10.4062/biomolther.2022.114.

Piñeiro, D., Narula, J., Pervan, B., Hadeed, L., 2023. World heart day 2023: knowing your heart. Argentina Indian Journal of Medical Research 158 (3), 213–215. Available at https://doi.org/10.4103/ijmr.ijmr_1689_23.

Prasher, P., Sharma, M., Singh, S.K., Gulati, M., Chellappan, D.K., Rajput, R., et al., 2023. Spermidine as a promising anticancer agent: recent advances and newer insights on its molecular mechanisms. Frontiers in Chemistry 11. Available at https://doi.org/10.3389/fchem.2023.1164477.

Pugsley, H.R., 2017. Assessing autophagic flux by measuring LC3, p62, and LAMP1 co-localization using multispectral imaging flow cytometry. Journal of Visualized Experiments 2017 (125). Available at http://doi.org/10.3791/55637.

Quinn, P.M.J., Moreira, P.I., Ambrósio, A.F., Alves, C.H., 2020. PINK1/PARKIN signalling in neurodegeneration and neuroinflammation. Acta Neuropathologica Communications 8 (1), 189. Available at https://doi.org/10.1186/s40478-020-01062-w.

Qiu, Z., Li, Y., Fu, Y., Yang, Y., 2023. Research progress of AMP-activated protein kinase and cardiac aging. Open Life Sciences 18 (1). Available at https://doi.org/10.1515/biol-2022-0710.

Rabinovich-Nikitin, I., Kirshenbaum, L.A., 2021. GSK-3β mediates cardiac senescence through inhibition of ULK1 directed autophagy. Journal of Cardiovascular Aging 1 (1). Available at https://doi.org/10.20517/jca.2021.24.

Raffaele, M., Vinciguerra, M., 2022. The costs and benefits of senotherapeutics for human health. The Lancet Healthy Longevity 3 (1), e67. Available at https://doi.org/10.1016/S2666-7568(21)00300-7.

Ren, X., Wang, J., Wei, H., Li, X., Tian, Y., Wang, Z., et al., 2024. Impaired TFEB-mediated autophagy-lysosome fusion promotes tubular cell cycle G2/M arrest and renal fibrosis by suppressing ATP6V0C expression and interacting with SNAREs. International Journal of Biological Sciences 20 (5), 1905–1926. Available at https://doi.org/10.7150/ijbs.91480.

Riley, J.S., Tait, S.W.G., 2016. Mechanisms of mitophagy: putting the powerhouse into the doghouse. Biological Chemistry 397 (7), 617–635. Available at https://doi.org/10.1515/hsz-2016-0137.

Robert, G., Jacquel, A., Auberger, P., 2019. Chaperone-mediated autophagy and its emerging role in hematological malignancies. Cells 8 (10). Available at https://doi.org/10.3390/cells8101260.

Roberts, F.L., Markby, G.R., 2021. New insights into molecular mechanisms mediating adaptation to exercise; a review focusing on mitochondrial biogenesis, mitochondrial function, mitophagy and autophagy. Cells 10 (10). Available at https://doi.org/10.3390/cells10102639.

Roychan, R., 2025. Comparing p62 and LC3 as autophagy biomarkers: insights into cellular homeostasis. The Plymouth Student Scientist 18 (1). Available at https://doi.org/10.70156/1754-2383.1510.

Salvatore, T., Pafundi, P.C., Galiero, R., Albanese, G., Di Martino, A., Caturano, A., et al., 2021. The diabetic cardiomyopathy: the contributing pathophysiological mechanisms. Frontiers in Medicine 8. Available at https://doi.org/10.3389/fmed.2021.695792.

Samuel, V.P., Moglad, E., Afzal, M., Kazmi, I., Alzarea, S.I., Ali, H., et al., 2024. Exploring ubiquitin-specific proteases as therapeutic targets in Glioblastoma. Pathology - Research and Practice 260, 155443. Available at https://doi.org/10.1016/j.prp.2024.155443.

Schwemmlein, J., Maack, C., Bertero, E., 2022. Mitochondria as therapeutic targets in heart failure. Current Heart Failure Reports 19 (2), 27–37. Available at https://doi.org/10.1007/s11897-022-00539-0.

Shyam, R., Ogando, D.G., Choi, M., Liton, P.B., Bonanno, J.A., 2021. Mitochondrial ROS induced lysosomal dysfunction and autophagy impairment in an animal model of congenital hereditary endothelial dystrophy. Investigative Opthalmology and Visual Science 62 (12), 15. Available at https://doi.org/10.1167/iovs.62.12.15.

Song, Y., Spurlock, B., Liu, J., Qian, L., 2024. Cardiac aging in the multi-omics era: high-throughput sequencing insights. Cells 13 (20), 1683. Available at https://doi.org/10.3390/cells13201683.

Stojanovic, B., Jovanovic, I., Dimitrijevic Stojanovic, M., Stojanovic, B.S., Kovacevic, V., Radosavljevic, I., et al., 2025. Oxidative stress-driven cellular senescence: mechanistic crosstalk and therapeutic horizons. Antioxidants 14 (8), 987. Available at https://doi.org/10.3390/antiox14080987.

Suda, M., Paul, K.H., Tripathi, U., Minamino, T., Tchkonia, T., Kirkland, J.L., 2024. Targeting cell senescence and senolytics: novel interventions for age-related endocrine dysfunction. Endocrine Reviews 45 (5), 655–675. Available at https://doi.org/10.1210/endrev/bnae010.

Thapa, R., Moglad, E., Afzal, M., Gupta, G., Bhat, A.A., Almalki, W.H., et al., 2024a. The role of sirtuin 1 in ageing and neurodegenerative disease: a molecular perspective. Ageing Research Reviews 102, 102545. Available at https://doi.org/10.1016/j.arr.2024.102545.

Thapa, R., Moglad, E., Afzal, M., Gupta, G., Bhat, A.A., Almalki, W.H., et al., 2024b. ncRNAs and their impact on dopaminergic neurons: autophagy pathways in Parkinson's disease. Ageing Research Reviews 98, 102327. Available at https://doi.org/10.1016/j.arr.2024.102327.

Valdor, R., Martinez-Vicente, M., 2024. The role of chaperone-mediated autophagy in tissue homeostasis and disease pathogenesis. Biomedicines 12 (2). Available at https://doi.org/10.3390/biomedicines12020257.

Vásquez-Trincado, C., García-Carvajal, I., Pennanen, C., Parra, V., Hill, J.A., Rothermel, B.A., et al., 2016. Mitochondrial dynamics, mitophagy and cardiovascular disease. The Journal of Physiology 594 (3), 509–525. Available at https://doi.org/10.1113/jp271301.

Wang, C., 2020. A sensitive and quantitative mkeima assay for mitophagy via FACS. Current Protocols in Cell Biology 86 (1). Available at https://doi.org/10.1002/cpcb.99.

Xia, L., Yan, X., Zhang, H., 2025. Mitochondrial DNA-activated cGAS-STING pathway in cancer: mechanisms and therapeutic implications. Biochimica et Biophysica Acta (BBA) - Reviews on Cancer 1880 (1), 189249. Available at https://doi.org/10.1016/j.bbcan.2024.189249.

Yun, H.R., Singh, M.K., Han, S., Ranbhise, J.S., Ha, J., Kim, S.S., et al., 2025. Roles of autophagy and oxidative stress in cardiovascular disease. Antioxidants 14 (10). Available at https://doi.org/10.3390/antiox14101263.

Zellner, S., Behrends, C., 2021. Autophagosome content profiling reveals receptor-specific cargo candidates. Autophagy 17 (5), 1281–1283. Available at https://doi.org/10.1080/15548627.2021.1909410.

Zhang, H., Liu, B., Li, T., Zhu, Y., Luo, G., Jiang, Y., et al., 2018a. AMPK activation serves a critical role in mitochondria quality control via modulating mitophagy in the heart under chronic hypoxia. International Journal of Molecular Medicine 41 (1), 69–76. Available at https://doi.org/10.3892/ijmm.2017.3213.

Zhang, J.X., Qu, X.L., Chu, P., Xie, D.J., Zhu, L.L., Chao, Y.L., et al., 2018b. Low shear stress induces vascular eNOS uncoupling via autophagy-mediated eNOS phosphorylation. Biochimica et Biophysica Acta - Molecular Cell Research 1865 (5), 709–720. Available at https://doi.org/10.1016/j.bbamcr.2018.02.005.

Zhang, X., He, J., Xu, Z., Yang, Y., 2025a. Naringin regulates the cGAS-STING pathway to improve mitochondrial dysfunction and ferroptosis after myocardial ischemia–reperfusion injury. Cytotechnology 77 (3). Available at https://doi.org/10.1007/s10616-025-00762-2.

Zhang, X., Shao, S., Li, Q., Wang, Y., Kong, M., Zhang, C., 2025b. Roles of autophagy, mitophagy, and mitochondria in left ventricular remodeling after myocardial infarction. Reviews in Cardiovascular Medicine 26 (3). Available at https://doi.org/10.31083/rcm28195.

Zhou, X., Wang, J., Yu, L., Qiao, G., Qin, D., Yuen-Kwan Law, B., et al., 2024. Mitophagy and cGAS–STING crosstalk in neuroinflammation. Acta Pharmaceutica Sinica B 14 (8), 3327–3361. Available at https://doi.org/10.1016/j.apsb.2024.05.012.

Zhu, Y., Anastasiadis, Z.P., Espindola Netto, J.M., Evans, T., Tchkonia, T., Kirkland, J.L., 2024. Past and future directions for research on cellular senescence. Cold Spring Harbor Perspectives in Medicine 14 (2), a041205. Available at https://doi.org/10.1101/cshperspect.a041205.

Zhu, Y., Massen, S., Terenzio, M., Lang, V., Chen-Lindner, S., Eils, R., et al., 2013. Modulation of serines 17 and 24 in the LC3-interacting region of Bnip3 determines pro-survival mitophagy versus apoptosis. Journal of Biological Chemistry 288 (2), 1099–1113. Available at https://doi.org/10.1074/jbc.m112.399345.

Zuo, Z., Jing, K., Wu, H., Wang, S., Ye, L., Li, Z., et al., 2020. Mechanisms and functions of mitophagy and potential roles in renal disease. Frontiers in Physiology 11. Available at https://doi.org/10.3389/fphys.2020.00935.

Chapter 12

Senolytics and senomorphics: emerging therapeutic strategies

Rabab Fatima[1], Mudasir Maqbool[2], Amita Joshi Rana[3,4], and Md Sadique Hussain[5,6]
[1]*School of Health Science & Technology, UPES, Dehradun, Uttarakhand, India,* [2]*Department of Pharmacology, Government Medical College Baramulla, Baramulla, Jammu and Kashmir, India,* [3]*College of Pharmacy, Graphic Era Hill University, Bhimtal, Uttarakhand, India,* [4]*Centre for Promotion of Research, Graphic Era (Deemed to be) University, Dehradun, Uttarakhand, India,* [5]*Uttaranchal Institute of Pharmaceutical Sciences, Uttaranchal University, Dehradun, Uttarakhand, India,* [6]*School of Pharmaceutical Sciences, Lovely Professional University, Phagwara, Punjab, India*

12.1 Introduction

Aging is a major predisposing factor in the occurrence of cardiovascular diseases (CVDs), the most common cause of fatality all over the globe (López-Otín et al., 2023). As the world demographic is shifting toward an aging population, there has been an increasing need to find new interventions that can target the molecular causes of cardiac aging (Moskalev et al., 2022). Another defining characteristic of aging that contributes to cardiovascular pathophysiology includes cell senescence, whereby senescent cells experience a stable cell cycle halt with alterations in phenotype characterized by chromatin modifications, metabolic dysregulation, and the acquisition of a pro-inflammatory secretome known as the senescence-associated secretory phenotype (SASP) (Huang et al., 2022; Alqahtani et al., 2025).

Senescent cells in the cardiovascular system build up in the vascular endothelium, cardiac fibroblasts, smooth muscle cells, and cardiomyocytes, where they persistently reduce cardiovascular tissue functioning and regenerative potential (Mehdizadeh et al., 2022). Such cells, initially protective as they pause the reproduction of damaged or mutated cells or cancerous cells, can turn treacherous when they do not disappear as they should, and they are not eliminated by the immune system (Luan et al., 2024). Their SASP also promotes chronic sterile inflammation, extracellular matrix remodeling, and paracrine senescence, which further accelerates the onset of the abnormalities in cardiac dysfunction with age and disease (Khavinson et al., 2023; Salminen et al., 2012). In addition, senescent cells exhibit apoptosis resistance caused by upregulation of antiapoptotic proteins, especially of the BCL-2 family (Hu et al., 2022), and are thus potential candidates for targeted destruction or inhibition of activity (Hu et al., 2022).

As a response to the adverse influence of senescent cells, there has been the development of two distinct therapeutic agents (Zhang et al., 2023): senolytics, a potential treatment aimed at selectively killing senescent cells through apoptosis, and also senomorphics (sometimes referred to as senostatics), which aim to create treatments that will decline the SASP or otherwise mitigate the actions of the senescent cells without the need to kill them (Zhang et al., 2023; Hussain et al., 2025). All these approaches can best be categorized as senotherapeutics and have demonstrated successful preclinical studies in cardiac aging mitigation, enhanced vascular integrity, and general cardiovascular performance (Gasek et al., 2021; Suda et al., 2023).

Senolytics like dasatinib, quercetin, fisetin, and navitoclax act on the survival pathways used by the senescent cells, notably PI3K/AKT, BCL-2/BCL-XL and other kinase pathways (Kirkland and Tchkonia, 2020). The resultant activation of apoptosis in senescent cells at the expense of these pathways is how senolytics alleviate SASP-associated tissue dysfunction. Particularly, intermittent use of senolytics has been found useful in animal models of vascular aging, atherosclerosis, and cardiac failure without being toxic to normal proliferating cells to a significant degree (Atlante et al., 2025). The best-known and most-studied senolytic combination, Dasatinib and quercetin (D+Q), has demonstrated the ability to also enhance endothelial functioning and decrease arterial stiffness in old mice, an effect that will meet clinically relevant cardiovascular outcomes (Nieto et al., 2024).

Cellular Senescence and Cardiac Aging. DOI: https://doi.org/10.1016/B978-0-443-45811-8.00002-7

Senomorphics are an alternative approach that is more subtle since senescent cells are kept alive, but their dangerous secretory capacity is modulated (Kaur et al., 2024). Other agents like rapamycin, metformin, resveratrol, and kaempferol have been shown to downregulate SASP components through inhibition of major regulatory pathways like mTOR, NF-kappaB, and JAK/STAT (Tang et al., 2025; Feng et al., 2021). This therapeutic modulation may improve inflammation, fibrosis, and endothelial dysfunction, the key characteristics of cardiac aging, as well as preserve tissue homeostasis and reduce the chances of overstimulation of the immune activities related to senescent cell removal (Gonçalinho et al., 2023). In addition, senomorphics can provide an alternative that can be safer in older and multimorbid patients, and ablation of senescent cells can also be dangerous in the case of off-target toxicities or the loss of positive effects of senescent cells and their associated functions (Martel et al., 2020).

The focus of senescent cell targeting in cardiovascular aging is based on the accumulating evidence that senescence can be involved in nearly all aspects of age-related pathology that contribute to CVDs, including atherosclerosis, myocardial infarction, and heart failure with preserved ejection fraction (HFpEF), and vascular calcification (Chen et al., 2022a). Senotherapeutics could have the possible benefits of restoring regenerative signaling, enhancing vascular compliance, and ameliorating myocardial fibrosis by clearing or controlling senescent cells or their secretome. These interventions are expected not only to be used to treat overt disease but have the potential to slow or prevent the development of the onset of age-related cardiac dysfunction and potentially prolong cardiovascular healthspan (Robbins et al., 2021).

The objectives of this chapter are to investigate the mechanistic rationale, preclinical, and clinical progress and translational prospects of senolytic and senomorphic approaches toward cardiac aging. We commenced with explaining the molecular signaling underpinning senescence and the weaknesses that render senescent cells therapeutically approachable. Then, we get into the different but complementary actions of senolytics and senomorphics, with particular agents being tested and their readings of interest to CVDs. These strategies are compared, and a discussion of their safety profile, their possible synergistic combinations, and their prospects is given. Last, we discuss the situation with clinical trials, regulatory aspects, and the new paradigm of a personalized senotherapy informed by profiling of biomarkers. The combination of these insights will give an overview of how targeting cell senescence will be a result in the fight against cardiovascular aging, with its slow acceptance into the clinical practice of cardiac pathology, which is bound to revolutionize the field of age-related conditions, including cardiac diseases.

12.2 Mechanistic basis for targeting senescent cells

12.2.1 Senescence-associated signaling pathways

Cellular senescence is primarily orchestrated by the interconnected molecular circuits involving critical tumor suppressor pathways. The p53/p21 axis mainly serves as a primary DNA damage response mechanism, where p53 activation leads to transcriptional overexpression of p21, causing G1/S cell cycle halt. A prolonged overexpression of p53, p16INK4A, pRB, and p21WAF1/CIP1 is sufficient to induce senescence (Jan et al., 2025). The p16INK4a pathway operates through cyclin-dependent kinase inhibition, specifically targeting the CDK4/6 complex and preserving retinoblastoma protein (pRb) in its hypophosphorylated, growth-suppressive state. NF-κB cascade represents a significant inflammatory mediator in senescent cells, influencing the SASPs through transcriptional regulation of inflammatory mediators, chemokines, and matrix-degrading enzymes (Kumar and Thirumurugan, 2023). This persistent inflammatory activation led to a feed-forward cycle that reinforces a senescent state while simultaneously promoting tissue dysfunction.

12.2.2 Vulnerabilities of senescent cells

Senescent cells show a paradoxical survival advantage despite their growth-arrested state, primarily through over-expression of BCL-2 proteins (Hu et al., 2022). Senescent cells upregulated the antiapoptotic proteins (BCL-W & BCL-XL), creating dependencies that can be therapeutically exploited. This antiapoptotic network collectively suppressed the mitochondrial-mediated apoptosis by preventing cytochrome c release and caspase triggering (Huang et al., 2022). The therapeutic vulnerability lies in this heightened dependence on pro-survival signals. Senescent cells are sensitive to downregulation of Bcl-xl, Bcl-w, and Bcl-2, making them selectively targetable through BH3-mimetic compounds, including navitoclax (ABT-263) and specific BCL-XL inhibitors (Davan-Wetton et al., 2021). This selectivity arises because normal, proliferating cells maintain multiple survival pathways, whereas senescent cells become increasingly reliant on BCL-2 family proteins for survival.

12.2.3 Senescence-associated secretory phenotype modulation in cardiovascular pathology

Senescent cells lack proliferative ability and release SASP, which can worsen several CVDs. The SASP represents a double-edged mechanism in CVDs progression and includes IL-6, IL-1β, TNF-α, and matrix metalloproteinases (MMPs), which collectively promote vascular inflammation, endothelial dysfunction, and arterial stiffening (Xu et al., 2025). In cardiovascular contexts, SASP modulation offers therapeutic potential through multiple mechanisms (Khavinson et al., 2023; Banerjee et al., 2021). Pro-inflammatory SASP factors accelerate atherosclerosis by promoting macrophage recruitment, foam cell formation, and plaque instability (Shang et al., 2023). Matrix-remodeling enzymes within the SASP contribute to vascular remodeling and hypertension development, where SASP-mediated paracrine senescence is reported to increase the burden of dysfunctional cells within cardiovascular cells (Liu et al., 2025). The various therapeutic strategies targeting SASP are direct cytokine inhibition, NF-κB pathway modulation, and upstream senescence prevention (Sahu and Satapathy, 2025), where the selective removal of senescent cells through senolytics represents the most direct approach, precisely removing the cellular source of pathogenic SASP factors while preserving functional tissue architecture and regenerative capacity (Chen et al., 2022b). Cellular senescence drives cardiovascular dysfunction through paracrine and autocrine effects of SASP (Mehdizadeh et al., 2022; Zhao et al., 2025), leading to inflammation, fibrosis, and impaired vascular function in multiple cardiac cell types (Fig. 12.1).

12.3 Senolytic agents: overview and cardiovascular applications

12.3.1 Small molecule senolytics

Senolytics selectively remove the senescent cells by exploiting their antiapoptotic dependencies. The first senolytic compounds, Dasatinib, Quercetin, Fisetin, and Navitoclax, were discovered using a hypothesis-driven strategy, targeting vulnerabilities in the senescent cell survival pathway, where Dasatinib and Quercetin constituted the most widely researched senolytic combination (Kirkland and Tchkonia, 2020). Dasatinib is a multikinase inhibitor originally developed for chronic myeloid leukemia, which targets the senescent cells through inhibition of survival kinases, including SRC family kinases. On the other hand, quercetin, a natural flavonoid, disrupts the antiapoptotic signaling through multiple pathways, including PI3K/AKT modulation. The doses of both the drugs used were between 5–50 μM and 0.5–1.5 nM, respectively, with an intermittent dosing protocol reducing the systemic toxicity concern. Navitoclax (ABT-263) functions as a selective BCL-2/BCL-XL inhibitor, directly targeting the heightened antiapoptotic dependence atypical of senescent cells (Zhu et al., 2016). This led to the identification of quercetin and dasatinib as senolytics, agents that target some of these routes, inducing apoptosis preferentially in senescent cells. Fisetin represents another natural compound with senolytic features, demonstrating efficacy through multiple antiapoptotic pathway disruption (Spiegel, 2025; Fatima et al., 2025).

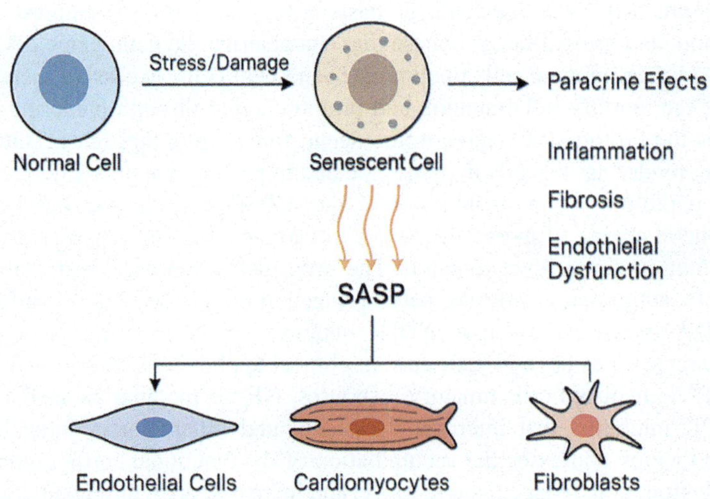

FIGURE 12.1 **Pathophysiology of cellular senescence in cardiac aging.** Stress or damage induces a transition from normal to senescent cells, which secrete pro-inflammatory SASP factors. These SASP molecules act on endothelial cells, cardiomyocytes, and fibroblasts, contributing to hallmarks of cardiac aging.

12.3.2 Cardiovascular-specific impacts

Senolytic therapies can manage a diverse array of CVDs and age-related CVDs where senescence is known to offer a pathogenic function, including arterial dysfunction and arterial stiffening correlated with severe hypertension, highlighting their broad therapeutic potential. A single dose of senolytics may enhance the vasomotor function, but the structural and functional impact of long-term senolytic treatment is still unknown, emphasizing the need for a chronic treatment protocol. Senolytic intervention restored the endothelial nitric oxide (NO) bioavailability, reducing the inflammatory mediator expression, and improved the acetylcholine-mediated vasodilation in aged and atherosclerotic models. The mechanism is mainly the elimination of senescent endothelial cells that exhibited impaired nitric oxide production, increased oxidative stress, and enhanced inflammatory signaling (Atlante et al., 2025).

Cellular senescence has been implicated in several age-related disorders, as well as metabolic syndrome, establishing the broad relevance to cardiac pathophysiology. The potential therapeutic mechanism observed is a marked reduction of the SASP-mediated pro-fibrotic signaling, improving cardiomyocyte contractility, and enhancing cardiac regenerative potential (Dookun et al., 2022). Senescent cardiac fibroblasts secreted excessive collagen and MMPs, contributing to both fibrosis and matrix degradation that characterized the pathological remodeling (Picos et al., 2025). Preclinical heart failure studies demonstrated significant therapeutic potential for senolytic intervention. The senescence of vascular cells promoted the development of age-related CVDs, while inhibition of vascular cell senescence ameliorated the phenotypic features of aging (Shimizu and Minamino, 2019). Senolytics demonstrated induction of apoptosis of senescent cardiovascular cells and alleviated the CVD onset and growth, with marked improvement in left ventricular function, and reduced cardiac fibrosis in vivo (Okuno et al., 2020).

12.4 Senomorphic agents: overview and utility in cardiac aging

In the process of cardiac aging, senomorphic therapy addresses the accumulation of dysfunctional cardiomyocytes and cardiac fibroblasts that lead to diastolic dysfunction, reduced contractility, and increased arrhythmogenicity. The therapeutic rationale involves modulating SASP-mediated paracrine effects that promote cardiac fibrosis, endothelial dysfunction, and impaired regenerative capacity. Senomorphics may preserve cardiac reserve while reducing age-related inflammatory burden, potentially slowing progression of heart failure with preserved ejection fraction and other age-related cardiac pathologies.

Senomorphic agents are also called senostatics, or senomorphics, and mainly function by inhibiting the SASP by targeting JAK–STAT, NF-κB, mTOR, p38 MAPK, all participated in the secretion of SASPs. This approach preserves potentially beneficial aspects of cellular senescence while mitigating the inflammatory damage (Zhang et al., 2023). Unlike senolytics that eliminate senescent cells, senomorphics offer sustained modulation of harmful SASP components while maintaining cellular integrity. The mechanistic basis positions them to suppress pro-inflammatory cytokine production, MMPs production, and growth factor release that characterize the pathogenic SASP profile. While SASP can serve a beneficial role by attracting immune cells to eliminate senescent cells, excessive secretion, particularly for TGF-β family proteins and VEGF, can amplify inflammation and promote aging through paracrine signaling (Stojanović et al., 2025). Rapamycin stands as the prototypical senomorphic agent, functioning through mTOR pathway inhibition, where mTOR drives its conversion, rendering the cells hypertrophic and hyperfunctional (SASPs), which ultimately leads to the progression of age-related pathology. It is an inhibitor of the mTOR cascade and has been demonstrated to extend longevity and reduce epigenetic ageing (Tang et al., 2025). In cardiac contexts, rapamycin reduced the SASP-mediated inflammation while preserving cardiomyocyte function. The drug metformin exhibited senomorphic properties through AMPK activation and mTOR suppression, offering cardioprotective effects beyond its antidiabetic actions. Rapamycin and metformin are two FDA-approved, effective mTOR inhibitors proposed for this purpose, exhibiting significant anticancer and antiaging characteristics beyond their current clinical applications. Additional senomorphic agents include JAK inhibitors targeting STAT-mediated inflammatory responses, NF-κB modulators such as aspirin and certain natural compounds, and p38 MAPK inhibitors that interrupt stress-activated inflammatory cascades (Zhang et al., 2023). In cardiac aging, senomorphic therapy addresses the accumulation of dysfunctional cardiomyocytes and cardiac fibroblasts that contribute to diastolic dysfunction, reduced contractility, and increased arrhythmogenicity. The presence of SASP and a slight inflammatory response characteristic of the inflammaging process forms the prerequisite for CVDs. The therapeutic rationale reported was modulating SASP-mediated paracrine effects that promoted cardiac fibrosis, endothelial dysfunction, and impaired regenerative capacity (Birch and Gil, 2020). Kaempferol inhibited the SASPs by blocking NF-κB p65 functioning and IκBζ expression via the IRAK1/IκBα axis (Feng et al., 2021). This lowered the IL-1α and IL-1β in senescent cells and aging rat kidneys. In ovariectomized rats on a high-fat diet, kaempferol (50–100 mg/kg) reduced aortic lipid deposition and lesion size in a dose-dependent manner, likely through GPER-mediated stimulation of the PI3K/AKT/Nrf2 cascade, indicating antiinflammatory and antioxidant potential (Dawson et al., 2022). Colchicine (0.5 mg/day) significantly reduced low-attenuation atherosclerotic plaque volume in ACS patients over 12 months. Though total plaque volume did not differ significantly, a decrease in other coronary CTA markers and LDL levels suggested antiatherosclerotic effects, likely via inflammation modulation. In a double-blind clinical trial involving 40 postmyocardial infarction Caucasian individuals, daily supplementation with 10 mg of resveratrol for 90 days significantly enhanced left ventricular diastolic function and enhanced flow-mediated vasodilation compared to the placebo (Teimouri et al., 2022). Similarly, in another randomized clinical trial, a daily dose of 500 mg resveratrol for a month led to a rise in circulating SIRT1 levels and a reduction in plasma noradrenaline, indicating its potential to suppress inflammation and ameliorate endothelial dysfunction (Gonçalinho et al., 2023).

12.5 Comparative analysis: senolytics versus senomorphics

The innovation in the field of senotherapeutics has given rise to two primary approaches for targeting senescent cells: senolytics, which selectively trigger apoptosis in these cells, and senomorphics, which mitigate harmful features of the senescent phenotype, such as SASP, without eliminating the cells themselves. The divergent yet complementary mechanisms of senolytics and senomorphics are illustrated in, outlining how each class modulates senescence through distinct molecular targets. Senolytics operate through the selective elimination of senescent cells by exploiting their heightened antiapoptotic dependencies (Fig. 12.2).

These agents target BCL-2 family proteins, survival kinases, and cellular stress pathways that senescent cells mainly rely upon for survival (Rad and Grillari, 2024). The therapeutic goal is to remove the cellular source of inflammatory SASP factors while potentially restoring the tissue regenerative capacity. Preclinical data indicated that most of the senolytics mitigate the diseased conditions in several organs, improve physical function and resilience, and reduce all causes of mortality, even if administered to the aged. Senomorphics, on the other hand, functioned by suppressing pathological SASP secretion without cell elimination. Their targets are cascades associated with the SASP expressions, including the PI3k/Akt, p38 MAPK, and mTOR, achieving therapeutic benefit through inflammatory modulation while preserving cellular integrity (Grootaert, 2024). Senomorphics indirectly disrupts the senescence procedure, acts as SASP inhibitors, offering sustained antiinflammatory effects without triggering apoptosis. Succinctly, it presents a comprehensive layout of the senolytics and senomorphics discussed above, with their mechanism of action (Table 12.1).

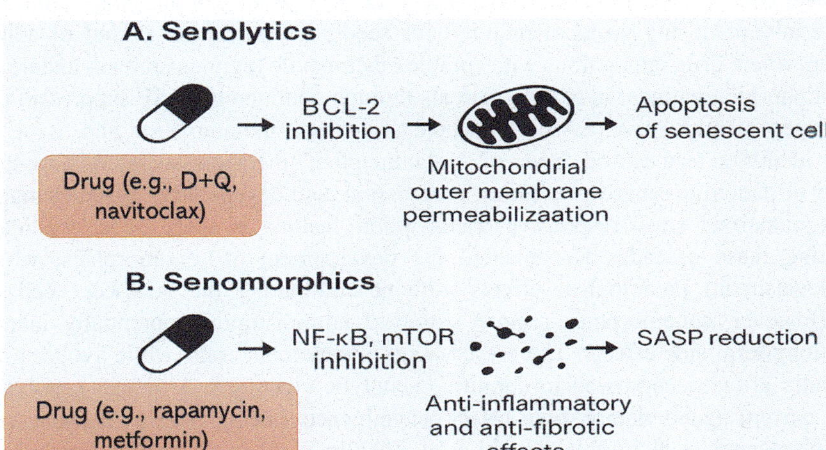

FIGURE 12.2 Mechanisms of action of senotherapeutics. Panel A: Senolytics (e.g., dasatinib + quercetin, navitoclax) inhibit BCL-2 family proteins, disrupt mitochondrial integrity, and induce apoptosis in senescent cells. Panel B: Senomorphics (e.g., rapamycin, metformin) suppress SASP by inhibiting NF-κB and mTOR pathways, leading to antiinflammatory and antifibrotic outcomes.

TABLE 12.1 An overlay of senolytics and senomorphics and their mechanism of action in different diseases.

Agent	Type	Mechanism of action	Diseases/Conditions targeted
Kaempferol	Senomorphic	Inhibits NF-κB p65 and IκBζ via the IRAK1/IκBα cascade, reducing SASP expression	Aging, atherosclerosis, inflammation, renal aging
Resveratrol	Senomorphic	Activates SIRT1, suppresses NF-κB, lowers SASP (e.g., TNF-α, IL-1β)	Cardiovascular diseases, stroke, myocardial injury, aging
Colchicine	Senomorphic-like	Antiinflammatory, affects neutrophil activation and IL-1β signaling	Atherosclerosis, coronary artery disease
Statins	Senomorphic-like	Lipid-lowering indirectly reduces SASP via antiinflammatory and endothelial repair	Atherosclerosis, ischemic heart disease
PCSK9 inhibitors	Senomorphic-like	Reduce LDL cholesterol and inflammation; plaque regression	Hyperlipidemia, atherosclerosis
Dasatinib + quercetin	Senolytic	Promote apoptosis in senescent cells by suppressing antiapoptotic routes	Idiopathic pulmonary fibrosis, an aging-related disease
Fisetin	Senolytic	Facilitates senescent cell apoptosis via modulation of ROS and survival pathways	Age-related dysfunction,
Navitoclax (ABT-263)	Senolytic	Suppresses BCL-2/BCL-xL, triggering apoptosis in senescent cells	Cancer, fibrosis, aging-related pathologies
Curcumin	Senomorphic	Antioxidant and antiinflammatory; modulates SASP and NF-κB	Aging, inflammation

12.6 Risk–benefit profiles in elderly or multimorbid patients

The differential safety profiles of senolytics and senomorphics present critical considerations for the elderly population. Senomorphic, which suppresses the pathological SASP production without killing the senescent cells, could be a safer alternative and as effective as compared to the senolytic agent. This safety advantage stems from avoiding potential complications related to widespread cellular elimination in multimorbid patients. Senolytics mainly carry theoretical risks in elderly patients, such as potential depletion of cells with residual beneficial functions, disruption of tissue architecture, and temporary inflammatory response observed during cell clearance (Martel et al., 2020; Alharbi et al., 2022a; Thapa et al., 2023). The intermittent dosing protocol required for senolytics may pose a kind of adherence challenge in a multimorbid population, where drug interactions with complex therapeutic regimens remain underexplored. Senomorphics offers an advantage for chronic treatment in elderly patients through continuous SASP suppression with potentially fewer acute toxicities. When persistently active, SASP contributed to chronic inflammation and tissue dysfunction (McHugh et al., 2025). It is now identified as a central driver of "inflammaging," the age-associated chronic inflammatory state, as well as a key promoter of paracrine senescence, tumor progression, and degenerative tissue changes. SASP expression is tightly modulated by numerous stress-responsive transcription factors as well as intracellular signaling cascades. Therapeutically targeting these cascades has enabled the development of senomorphics, which reduce the SASP production and its downstream pathological effects without eliminating the senescent cells themselves (Ya and Bayraktutan, 2024). However, senomorphics require sustained administration, potentially increasing the cumulative toxicity concerns and long-term side effects. The preservation of senescent cells, while avoiding acute elimination risk, may limit the therapeutic efficacy compared to definitive senolytic clearance. Although senolytics selectively remove senescent cells, many current agents demonstrate off-target influence, posing risks to nonsenescent cells and essential immune populations (Narayanan et al., 2023; Alharbi et al., 2022b; Singh et al., 2019). For instance, navitoclax, while effective at removing senescent cells, is closely related to dose-limiting thrombocytopenia due to its inhibition of BCL-xL in platelets (McCorkle et al., 2024). Similarly, broad-spectrum kinase inhibitors and several pro-apoptotic therapeutics can cause cytotoxicity, especially with chronic use or in association with other therapies (Revskij et al., 2024).

These safety concerns are especially critical in older adults, who are also the most vulnerable to adverse effects due to common comorbidities, increased frailty, and the prevalence of polypharmacy. Therefore, enhancing the specificity and delivery mechanisms of senotherapeutics is a pressing challenge. Promising strategies include the development of tissue-targeted prodrugs, antibody—drug conjugates, and nanoparticle-based delivery systems, all aimed at improving therapeutic precision while minimizing collateral damage. For a multimorbid elderly patient, a combination approach or sequential therapy may help optimize the therapeutic outcome while minimizing the individual drug limitations.

12.7 Combination therapies and synergistic strategies

12.7.1 Rationale for combining senolytics with senomorphics

Combining the senomorphic and senolytic therapies gives a potentially effective way of preventing the multifold effects of cellular senescence on aging and age-related diseases. Senolytics are substances that specifically remove senescent cells and lead to their apoptosis, accumulating with age and causing various pathologies, such as neurodegeneration, CVDs, osteoarthritis, and frailty (Zhang et al., 2023). Senescent cells are reported to secrete pro-inflammatory factors and disrupt tissue homeostasis, and removal of such cells is seen as an effective therapeutic method to hinder the associated diseases (Lagoumtzi and Chondrogianni, 2021). Senolytics carry significant health benefits in that they remove these dysfunctional cells, but, because they also have the potential to cause unwanted inflammatory responses and tissue damage, they can be problematic in people who are already in decline or have immune dysfunction. Senomorphics, on the other hand, and in contrast to senolytics, do not cause the death of senescent cells per se but alter the SASPs (de Magalhães, 2024). The SASP is a mixture of a number of pro-inflammatory cytokines, growth factors, and proteases in which senescent cells contribute to the enhancement of inflammation and tissue degeneration (Huang et al., 2022). By acting on the SASP, senomorphics have the effect of dampening the harmful role of senescent cells without killing them directly, thereby avoiding the inflammation and chronic destruction typically produced by senescent cells (Thompson et al., 2022). It is critical that this strategy is particularly relevant in some tissues, such as in the brain, where senescent cells, including neurons, are important in the tissue structure and functioning. The logic behind combining the senolytics and senomorphics is the complementary nature of their mechanism (Zhang et al., 2022). By getting rid of the senescent cells, mainly in an effective way by senolytics, and by reducing the effects of the SASP, mainly being effectively dealt with by senomorphics, inflammation and tissue damage decrease (Li et al., 2024). Such a synergistic approach would not only enhance the efficiency of senotherapeutic drugs but would also help to minimize their risks, making the therapeutic technology safer and more balanced (Robbins et al., 2021). An example is that dasatinib and quercetin (D+Q) are common senolytic agents, and their combination has proven to be extremely effective in preclinical models of age-related disorders (Power et al., 2023; Altamish et al., 2020; Singh et al., 2022). Whereas dasatinib will interfere with the pro-survival dogmas of senescent cells, quercetin will also influence the SASP, which will reduce the inflammation and enhance regeneration of the tissues. Collectively, D+Q has been found to have beneficial outcomes on numerous aging-related conditions, including Alzheimer's disease, frailty, and bone disease (Gasek et al., 2021).

12.7.2 Dual-targeting of senescence-associated secretory phenotype + senescent cell clearance

The concept of dual-targeting, that is, eliminating the senescent cells and regulating their secretory phenotype (SASP), appears as a convincing step forward in aging and age-related diseases therapy (Ogrodnik et al., 2021). With age, senescent cells accumulate as a result of DNA damage, telomere shortening, and other stresses and contribute to fibrosis, inflammation, and loss of tissue regeneration ability (Baker et al., 2011). On the one hand, senolytic agents, which include dasatinib and quercetin, are developed with the intent of selectively destroying senescent cells, but in practice, the result can be that the deleterious consequences of SASP are still present, which can, in turn, promote chronic inflammation and tissue dysfunction (He and Sharpless, 2017). This renders the dual-targeting approach especially alluring, since it hits both the source of such cells and the inflammatory surroundings in which senescent cells thrive. Senolytics (e.g., dasatinib) and senomorphic agents (e.g., quercetin) have demonstrated an important therapeutic potential in preclinical trials in combination (Lucas et al., 2023; Babu et al., 2025; Kakkar et al., 2025). Dasatinib is effective in the elimination of senescent cells, whereas quercetin affects the factors of SASP, decreasing the inflammatory process caused by the senescence of cells (Zhao et al., 2024). In the models of Alzheimer's disease, for instance, the combination of D+Q has been found to always enhance brain functioning, diminish neuroinflammation, and boost memory function and cognitive understanding. Likewise, in the conditions of frailty and osteoarthritis, the combination had a better result in physical function and reduced disease progression (Muñoz-Espín and Serrano, 2014). Therefore, senescence should be targeted in two ways: effectively removing senescent cells and inhibiting the deleterious SASP, which will have a more comprehensive effect in treating aging-related pathologies and leading to better therapeutic results and fewer chances of side effects (Schafer et al., 2018).

12.7.3 Role in multimodal cardiac aging interventions

Multimodal treatments that address cardiac aging using the combination of senolytic and senomorphic therapies are also coming up as potential ways of combating age-related CVDs. Cellular senescence is vital in the aging of the heart, and it promotes the occurrence of heart failure, myocardial infarction, plus atherosclerosis (Okuno et al., 2020). Senescence accumulation of cells is observed during aging in the heart, and it is involved in inflammation, fibrosis, and inefficient tissue regenerative potential, all of which worsen cardiac dynamics (Li et al., 2024). Although senescent cells have deleterious immunological effects, therapies that specifically eliminate senescent cells, senolytic therapies, are highly effective in lowering the burden of senescent cells in the heart (Zhang et al., 2023). As an example, preclinical trials with dasatinib and quercetin have discovered that clearing senescent cells leads to improvement of heart functions, decreased inflammation, and tissue repair following cardiac injury (Muñoz-Espín and Serrano, 2014). Nevertheless, whereas senolytics would be useful in removing the senescent cells, they do not bear directly on the inflammatory milieu formed by the senescent cells, especially the SASP, which would enhance tissue damage (Power et al., 2023). This is where senomorphic treatments are involved. Senomorphics regulate SASP, which helps in suppressing the secretion of pro-inflammatory cytokines and other substances that contribute to chronic inflammation and fibrosis (Baker et al., 2011). Through action on the inflammatory mediators of senescence, senomorphics overcome senescence dissemination and subsequent tissue degeneration. Senolytic and senomorphic therapies also have a combined solution potential as they not only eradicate senescent cells but also substantially reduce the chronic inflammation that those cause (Zhang et al., 2023). This two-sided targeting promotes cardiac regeneration, tissue homeostasis restoration, and better heart activity. Additionally, these therapies could also be combined with the more traditional cardiac-aging therapeutics, like statins, which have also displayed senomorphic activity, which could, in turn, add beneficial effects to the overall cardiac aging treatment (Muñoz-Espín and Serrano, 2014).

12.8 Safety, toxicology, and regulatory concerns

12.8.1 Risk of off-target effects and unintended tissue damage

The possibility of off-target effects and unintended tissue destruction cannot be unacknowledged as a serious obstacle on the way to the creation of an effectively functioning senolytic and senomorphic treatment. Dasatinib and navitoclax, which are senolytic agents, destroy senescent cells, and they do it by interfering with the important survival functions, including BCL-2 proteins (Schmitt et al., 2022; Balaraman et al., 2025; Gupta et al., 2018). Such disparate targeting has the side effect of (unintentionally) killing nonsenescent cells. As an example, the BCL-2 family inhibitor navitoclax causes thrombocytopenia and neutropenia due to disruption of BCL-xL in platelets and neutrophils. The off-target effects make senolytics a problematic treatment to administer, especially in populations at risk of multimorbidity and frailty, as most elderly would fall in this category (Miller et al., 2023). Similarly, senomorphic treatments, which have an effect on the SASP without killing senescent cells, can also have pleiotropic effects. These agents have nonspecific effects that can affect other cellular pathways outside SASP, and this could lead to unwanted effects like immune suppression or even interference in tissue homeostasis (Imb et al., 2024). Additionally, there are chances that the long-term intake of senomorphics will result in a pharmacodynamic tolerance, necessitating dose increases or combination treatments, which further adds to the risk of toxicity. Research is also focused on making senotherapeutics more specific to mitigate these risks (Zheng et al., 2024). Examples offered to reduce the undesirable effects arising as a result of off-targets include tissue-targeted delivery systems, nanoparticles, and prodrugs that respond to tissue-specific enzyme activation and thereby inhibit the unwanted effects (Zhang and Zhou, 2024).

12.8.2 Immunogenicity and inflammation risk with senescent cell clearance

The risk of inflammation and immunogenicity of clearance of senescent cells is a key issue in the strategies of application of senolytic therapies. The senescent cells, even though they are the cause of aging pathologies, produce a variety of pro-inflammatory cocktails referred to as the SASPs (Miller et al., 2023). Such molecules are cytokines, chemokines, and proteins that degrade the extracellular matrix and cause a low-grade chronic inflammation that is denoted as inflammaging. Senolytic therapies could help to reverse these inflammatory complications by clearing the senescence burden load; dasatinib or navitoclax may be used as senolytic therapies to achieve this effect (Gasek et al., 2021). But on the other side, this destruction can also cause immune events and lead to the further development of inflammation in the tissue as a result of the pro-inflammatory cues carried by the senescent cells being released in the course of their removal (de Magalhães, 2024). Moreover, it is established that senescent cells do not undergo immune surveillance due to a number of factors, one of which is the upregulation of immune checkpoint proteins, such as PD-L1, that suppresses the action of T cells and natural killer (NK) cells (Lagoumtzi and Chondrogianni, 2021). The immune evasion complicates the immune response toward the removal of senescent cells, and it may complicate the efficiency of therapies targeting the clearance of such cells. Also, there is the possibility of collateral tissue damage through the off-target effect of immune activation once senescent cells are removed, or the inflammatory conditions that are already present may be worsened (Li et al., 2024; Zheng et al., 2024). An instance is where the intervention of senescent cells in relative tissues, such as the cardiovascular system or the brain, might result in unnecessary inflammation or even activation of the immune system. Such potential risk-reduction measures include combination regimens of senolytics with antiinflammatory reagents or senomorphic treatment, which can clear SASP signals without cell death, both to offer a less dangerous, narrower interface with senescent cell clearance (Robbins et al., 2021; Zhao et al., 2024).

12.8.3 Clinical translation challenges

There are many issues connected with clinical translation of senolytic and senomorphic-based therapy concerning dose, duration, and appropriate biomarkers of responsiveness. Establishing the optimal dose and treatment duration is vital, since not only senolytics but also senomorphics have off-target effects and toxicity potential, especially in extended treatments (Zhang and Zhou, 2024). Senolytic medicines, including dasatinib and quercetin, have been shown to improve therapy when used intermittently in preclinical trials, although it is unknown what dosing should be used to treat human patients. Immune suppression or cardiovascular toxicity can be an adverse result of continuous or high-dose administration (Imb et al., 2024). Striking the right balance between efficacy and safety is thus a pertinent problem in the production of such therapies to become clinically viable. The other problem is finding trustworthy biomarkers that could measure the therapeutic response and evaluate the effectiveness of the senotherapeutic interventions (Zhang et al., 2022). Senescence markers, including SASP molecules, p16INK4A as a mark of senescence, are being studied, although their applicability and value in the clinical trials of humans are currently unclear (Miller et al., 2023). Moreover, these biomarkers do not necessarily correlate with functional outcomes, such as enhanced physical ability or better thoughts, and it is hard to gauge how well the treatment is going in clinical practice. To address these hurdles, active clinical trials are targeting optimal dosing regimens, tracking adverse effects, and optimizing serum levels of biomarkers of senescence and therapy responses to enhance the translation of senolytic and senomorphic approaches to people (Gasek et al., 2021).

12.9 Clinical pipeline and trial readiness

12.9.1 Snapshot of senescence-targeted drugs in cardiovascular-related trials

Multiple senescence-driven drugs have already entered cardiovascular-related clinical trials, which is an indication of the accumulating interest in this field as a remedy to age-related CVDs. The most promising approach in the trials is the use of senolytic agents that specifically kill senescent cells (Chen et al., 2022a). The most investigated combination is dasatinib and quercetin (D+Q), which has been positively reported in model studies of heart failure, atherosclerosis, and myocardial infarction. The mixture has proved to have a reduction in senescent cell burden and cardiac improvements through improvement in the left ventricular ejection fraction (LVEF) and regeneration of tissues (Owens et al., 2021). Other candidates slated to be tested in patients include UBX0101, which is a senolytic drug that was tested to target senescent cells in patients with osteoarthritis. Though the Phase II clinical trial of osteoarthritis was ineffective in terms of main outcomes, it is believed that there is a certain cardiovascular aspect in the application of the drug since it also has affinity in diabetic macular edema (Sweeney et al., 2023). Scientists are still investigating its success in alternative models of disease with special reference to cardiovascular health. There is also research into the development of drugs such as ABT-263 (Navitoclax) that have been developed to help kill senescent cells by blocking the BCL-2 family proteins in attempts to find a way of treating the disease (Suda et al., 2023). The compound has demonstrated effectiveness in preclinical tests to eliminate senescent cells and limit fibrosis in heart failure and ischemic heart disease experiments. Further, less-destructive agents in the SASP, for example, rapamycin and resveratrol, are being examined in clinical studies as possible drugs to lower inflammation and cardiovascular system fibrosis (Lee et al., 2021). Such compounds have been demonstrated to mitigate the harmful consequences of senescence-related inflammation of heart failure, which in turn may increase the overall heart functioning. Moreover, cardiovascular trials are also being carried out on drugs, such as metformin, which is already known to have some antiaging effects, to determine senolytic and antiinflammatory effects (Jia et al., 2019). Metformin is effective in preclinical models to help decrease the burden of senescence and cardiovascular health, and therefore, it is the main drug in clinical trials to target senescence in aging hearts (De Bartolo et al., 2025).

12.9.2 Regulatory perspectives on aging interventions

The regulatory attitudes toward aging interventions, especially those of the FDA and EMA, have their peculiarities, as intervening in cellular senescence is a new effort. Both agencies have the problem of defining surrogate endpoints related to aging that have been complex because classic clinical trial designs get in the way of reflecting the full potential of antiaging drugs (Okuno et al., 2020; de Magalhães, 2024). The process of aging is multifactorial and not that quick, which is why it is hard to specify the specific biomarkers of aging and senescence. In many current clinical trials, there are no standardized methods of measuring the outcome of aging, including biological age, frailty, or resilience, and so it is difficult to determine the effectiveness of senotherapeutics (Huang et al., 2022; Schafer et al., 2018). Furthermore, aging interventions are being proposed to be used as a primary preventive that not only aims to improve age-related disorders once a disease has developed but also to retard the development of age-related functional impairment in people who are still symptom-free (Thompson et al., 2022). Such a transition makes risk–benefit analysis and the length of follow-up in clinical trials difficult to determine. In these challenges, regulatory agencies like the FDA and EMA are considering adaptive trial designs like umbrella and platform trials that provide flexibility in conducting testing of multiple agents or dosing approaches in aging populations (Owens et al., 2021). Moreover, the increase in the market of unregulated antiaging products and the exaggeration of the claims made is threatening the credibility of the people and the integrity of science. Regulatory agencies are being admonished to partner with geroscience researchers to come up with coherent, clinically translatable guidelines and make sure that aging interventions have an aggressive regulation process before entering the marketplace (Imb et al., 2024).

12.9.3 Role of compassionate use and expanded access protocols

Expanded access and compassionate use procedures are also crucial in making access to senotherapeutics easier prior to finalization of regulatory approvals by authorities like the FDA and EMA. These programs aim to ensure that patients caught in bad or life-threatening situations receive investigational therapies, as a last resort when no acceptable alternative treatment is available (Robbins et al., 2021). Since cellular senescence is a hallmark of aging and a driving factor of disease burden due to chronic disease, including CVDs, the promise of senotherapies is of great interest. Nevertheless, therapies such as senolytics and senomorphics are still undergoing clinical trials in the first stage, and their safety is not fully determined (Miller et al., 2023). Expanded access programs grant patients, particularly those with severe heart failure or other heart diseases, the chance to access these potential treatments that are otherwise not available in controlled clinical trials (Lucas et al., 2023). One example is that not all patients would be in a position to participate in clinical trials that have rigorous inclusion criteria and therefore would ensure that they have access to senolytic therapies such as dasatinib and quercetin that have been found to lower levels of senescent cell populations and improve cardiac capabilities (He and Sharpless, 2017). Although compassionate use is very important, it has also given rise to an ethical concern of the posed risk of exposing patients to unapproved drug treatments, which may have negative effects (Zhang et al., 2022). This necessitates corrective measures to be taken by regulatory agencies in order to strike a balance between access of patients and strict surveillance and safety data gathering in order to introduce such therapies with control and sobriety (Suda et al., 2023).

12.10 Future outlook

A potential strategy of senescence-targeting therapy is personalized senotherapy, wherein the strategy is differentiated according to personal biomarkers. As oncology has shifted to a personalized approach, involving tumor genomics and molecular diagnostics to help inform treatment choices, senotherapy in the future may also lead to the use of a patient-specific senescence profile to make personalized choices (Zhang and Zhou, 2024). The increased multiomics technologies, such as transcriptomics, proteomics, and metabolomics, are making it possible to extensively map the senescence signatures across tissues and individuals. Such instruments are starting to build the foundation of senescence atlases, which can lead to the identification of biomarkers and creation of personalized treatments (Lee et al., 2021). One of the big secrets behind this personalized strategy is the introduction of certain biomarkers related to cellular senescence. Existing markers to detect senescent cells include SA-β-gal, p16INK4A, and CD9, and further studies on the topic are currently related to identifying more specific biomarkers to start specific treatment in targeted therapy (Ogrodnik et al., 2021). Moreover, artificial intelligence and machine learning are also being applied to the multiomics data analysis, making it faster to discover new senescence-related pathways and better to stratify patients. Employing such technologies in practice will enable targeting the senescent cells more efficiently with fewer systemic adverse effects and thus a more positive effect on therapies (Muñoz-Espín and Serrano, 2014). Personalized senotherapy has the potential, therefore, to offer a significant advance over current levels of treatment of age-related diseases by delivering more accurate and safe interventions that are individualized to the biology of the aging process (Zhao et al., 2024).

12.10.1 Integration with artificial intelligence-based drug discovery

The implementation of AI-driven drug discovery in senotherapy has great potential in the development of the area of senescence-related therapy. With AI, machine learning models, researchers will be able to enhance the speed at which new senotherapeutic agents may be identified, and with even greater specificity of targeting drugs (i.e., targeting only disease-relevant cells), which can potentially improve outcomes (Wan et al., 2023). The processing of multiple omics data, such as genomic, transcriptomic, and proteomic data, using AI has the potential to identify (specific) biomarkers related to senescent cells and identify possible therapeutic targets. The design of drug delivery vehicles, for example, nanoparticles or prodrugs that are selective to senescent cells, can be optimized as well using AI-driven methods to keep off-target effects to a minimum (Khosla, 2024). As another example, the AI models may help to calculate the effectiveness of various targeting ligands or delivery platforms to activate the drug more accurately in senescent tissues. Also, AI could be applied to increase the creation of senolytic prodrugs that activate in senescent cells and, thereby, win safer profiles and decreased systemic hazard (Shen et al., 2024). The ongoing development of AI is expected to result in more efficient and custom senotherapeutic agents being built into the drug development pipelines soon. These tools will enable new senolytic and senomorphic drugs to be identified and the clinical trial designs to be optimized, by making it possible to monitor patient responses in real-time depending on AI-analyzed biomarkers (Chang et al., 2024).

12.10.2 Role in preventive cardiology and healthspan extension

The possible preventive cardiology and healthspan extension avenue of senotherapy is an area of research that is gathering traction and being realized by the hopeful connotation it may have to aging and the occurrence of CVDs (Jia et al., 2019). Senescence of cells, the programmed loss of cell division capacity without a change in metabolism, is further evident with age and is a leading mechanism of cardiovascular impairment, inflammation, and fibrosis (Power et al., 2023). Endothelial cells and smooth muscle cells are senescent cells in the heart and contribute to the pathogenesis of such diseases as atherosclerosis, heart failure, and myocardial infarction. Senolytics and senomorphics are two categories or sets of interventions that target senescence, and they have much potential in preventive cardiology (Schmitt et al., 2022). Such induced elimination of senescent cells by senolytics can improve the dysfunctional cardiovascular status through chronic inflammation and fibrosis. As an example, dasatinib and quercetin (D+Q) were demonstrated as a potential treatment to slow or prevent heart failure since they are capable of enhancing cardiac regenerative capacity and left ventricular ejection fraction in aged mice models (Zheng et al., 2024). A complementary strategy is senomorphics, the drugs can alter the deleterious impact of the SASPs, without killing the senescent cells. Such treatments may help ameliorate the inflammation and tissue destruction through SASP factors and improve heart health and integrity. Such use of senotherapeutics in preventive cardiology may, therefore, expand healthspan due to fewer CVDs with advanced age and enhanced cardiovascular resilience in general (de Magalhães, 2024; De Bartolo et al., 2025).

12.11 Conclusion

The presence of senescent cells in cardiovascular tissues has been established to play a central role in the age-related cardiac dysfunction, vascular remodeling, and chronic inflammation. The current development of senotherapeutic interventions, notably senolytics and senomorphics, provides a paradigm shift in the field of addressing causative mechanisms at the core of cardiovascular aging as opposed to smoking out the symptoms. Senolytics, by selectively destroying senescent cells through induction of apoptosis, hold promise for rejuvenation of both tissue and endothelial integrity and for rejuvenating myocardial fibrosis via preclinical testing. Instead, the safer option is senomorphics because these drugs do not alter cellular viability or increase tissue degeneration by blocking the SASPs.

Each type of therapeutic modality has several significant strengths and weaknesses. Although a more conclusive option is provided by senolytics via cell clearance, its off-target effects (such as cytotoxicity and immune activation) raise areas of concern, especially underrepresented categories of individuals, for example, frailty or multimorbid conditions. The more appropriate long-term management would be through senomorphics, as they had a favorable safety profile in long-term use and when used as adjuncts to current cardiovascular management. Combinations of these strategies, either sequentially or in combination regimens, might have synergy and allow the beneficial effects of the senescence of cells to appear and the therapeutic advantage to take its maximum.

With senescence-targeting agents currently in the early phase of clinical trials, there is also a pivotal lack of biomarker demonstration, dose-guiding schedules, and safety assessments in the long term. Novel fabrications of drug delivery, like tissue-selective nanoparticles or prodrugs, or their further implementations in senescence profiling based on omics and AI-based systems, will enhance the precision and efficacy of these interventions. In the future, senotherapeutics have great potential not only to cure or alleviate pathological diseases of the cardiovascular system, but also to slow down cardiac aging itself. Their possible inclusion into prevention cardiology paradigms may alter the view of age-associated cardiovascular management, focusing on resilience, regeneration, and extension of healthspan. Future translational research and clinical verification are essential to achieve the maximum therapeutic effect of senolytics and senomorphics in the treatment of cardiovascular senescence.

References

Alharbi, K.S., Almalki, W.H., Alzarea, S.I., Kazmi, I., Al-Abbasi, F.A., Afzal, O., et al., 2022a. A narrative review on the biology of piezo1 with platelet-rich plasma in cardiac cell regeneration. Chemico-Biological Interactions 363, 110011. Available at https://doi.org/10.1016/j.cbi.2022.110011.

Alharbi, K.S., Singh, Y., Afzal, O., Alfawaz Altamimi, A.S., Kazmi, I., Al-Abbasi, F.A., et al., 2022b. Molecular explanation of Wnt/βcatenin antagonist pyrvinium mediated calcium equilibrium changes in aging cardiovascular disorders. Molecular Biology Reports 49 (11), 11101–11111. Available at https://doi.org/10.1007/s11033-022-07863-7.

Alqahtani, S., Alqahtani, T., Venkatesan, K., Sivadasan, D., Ahmed, R., Sirag, N., et al., 2025. SASP modulation for cellular rejuvenation and tissue homeostasis: therapeutic strategies and molecular insights. Cells 14 (8), 608. Available at https://doi.org/10.3390/cells14080608.

Altamish, M., Samuel, V.P., Dahiya, R., Singh, Y., Deb, P.K., Bakshi, H.A., et al., 2020. Molecular signaling of G-protein-coupled receptor in chronic heart failure and associated complications. Drug Development Research 81 (1), 23–31. Available at https://doi.org/10.1002/ddr.21627.

Atlante, S., Gottardi Zamperla, M., Cis, L., Farsetti, A., Gaetano, C., 2025. Senolytic therapies for cardiovascular aging: tackling fibrosis and metabolic dysfunction. European Journal of Internal Medicine 140, 106413. Available at https://doi.org/10.1016/j.ejim.2025.07.009.

Babu, M.A., Jyothi S, R., Kaur, I., Kumar, S., Sharma, N., Kumar, M.R., et al., 2025. The role of GATA4 in mesenchymal stem cell senescence: a new frontier in regenerative medicine. Regenerative Therapy 28, 214–226. Available at https://doi.org/10.1016/j.reth.2024.11.017.

Baker, D.J., Wijshake, T., Tchkonia, T., Lebrasseur, N.K., Childs, B.G., Van De Sluis, B., et al., 2011. Clearance of p16 Ink4a-positive senescent cells delays ageing-associated disorders. Nature 479 (7372), 232–236. Available at https://doi.org/10.1038/nature10600.

Balaraman, A.K., Altamimi, A.S.A., Babu, M.A., Goyal, K., PadmaPriya, G., Bansal, P., et al., 2025. The interplay of senescence and MMPs in myocardial infarction: implications for cardiac aging and therapeutics. Biogerontology 26 (1). Available at https://doi.org/10.1007/s10522-025-10190-6.

Banerjee, P., Kotla, S., Reddy Velatooru, L., Abe, R.J., Davis, E.A., Cooke, J.P., et al., 2021. Senescence-associated secretory phenotype as a hinge between cardiovascular diseases and cancer. Frontiers in Cardiovascular Medicine 8. Available at https://doi.org/10.3389/fcvm.2021.763930.

Birch, J., Gil, J., 2020. Senescence and the SASP: many therapeutic avenues. Genes and Development 34 (23-24), 1565–1576. Available at https://doi.org/10.1101/gad.343129.120.

Chang, M., Dong, Y., Cruickshank-Taylor, A.B., Gnawali, G., Bi, F., Wang, W., 2024. Senolytic prodrugs: a promising approach to enhancing senescence-targeting intervention. Chembiochem: A European Journal of Chemical Biology 25 (22). Available at https://doi.org/10.1002/cbic.202400355.

Chen, M.S., Lee, R.T., Garbern, J.C., 2022a. Senescence mechanisms and targets in the heart. Cardiovascular Research 118 (5), 1173–1187. Available at https://doi.org/10.1093/cvr/cvab161.

Chen, B., Zhu, R., Hu, H., Zhan, M., Wang, T., Huang, F., et al., 2022b. Elimination of senescent cells by senolytics facilitates bony endplate microvessel formation and mitigates disc degeneration in aged mice. Frontiers in Cell and Developmental Biology 10. Available at https://doi.org/10.3389/fcell.2022.853688.

Davan-Wetton, C.S.A., Pessolano, E., Perretti, M., Montero-Melendez, T., 2021. Senescence under appraisal: hopes and challenges revisited. Cellular and Molecular Life Sciences 78 (7), 3333–3354. Available at https://doi.org/10.1007/s00018-020-03746-x.

Dawson, L.P., Lum, M., Nerleker, N., Nicholls, S.J., Layland, J., 2022. Coronary atherosclerotic plaque regression: JACC state-of-the-art review. Journal of the American College of Cardiology 79 (1), 66–82. Available at https://doi.org/10.1016/j.jacc.2021.10.035.

De Bartolo, A., Angelone, T., Rocca, C., 2025. Elucidating emerging signaling pathways driving endothelial dysfunction in cardiovascular aging. Vascular Pharmacology 158, 107462. Available at https://doi.org/10.1016/j.vph.2025.107462.

Dookun, E., Passos, J.F., Arthur, H.M., Richardson, G.D., 2022. Therapeutic potential of senolytics in cardiovascular disease. Cardiovascular Drugs and Therapy 36 (1), 187–196. Available at https://doi.org/10.1007/s10557-020-07075-w.

Fatima, R., Soni, P., Sharma, M., Prasher, P., Kaverikana, R., Mangalpady, S.S., et al., 2025. Fisetin as a chemoprotective and chemotherapeutic agent: mechanistic insights and future directions in cancer therapy. Medical Oncology 42 (4). Available at https://doi.org/10.1007/s12032-025-02664-x.

Feng, Z., Wang, C., Jin, Y., Meng, Q., Wu, J., Sun, H., 2021. Kaempferol-induced GPER upregulation attenuates atherosclerosis via the PI3K/AKT/Nrf2 pathway. Pharmaceutical Biology 59 (1), 1106–1116. Available at https://doi.org/10.1080/13880209.2021.1961823.

Gasek, N.S., Kuchel, G.A., Kirkland, J.L., Xu, M., 2021. Strategies for targeting senescent cells in human disease. Nature Aging 1 (10), 870–879. Available at https://doi.org/10.1038/s43587-021-00121-8.

Gonçalinho, G.H.F., Kuwabara, K.L., Faria, N.Fd.O., Goes, M.Fd.S., Roggerio, A., Avakian, S.D., et al., 2023. Sirtuin 1 and vascular function in healthy women and men: a randomized clinical trial comparing the effects of energy restriction and resveratrol. Nutrients 15 (13). Available at https://doi.org/10.3390/nu15132949.

Grootaert, M.O.J., 2024. Cell senescence in cardiometabolic diseases. npj Aging 10 (1). Available at https://doi.org/10.1038/s41514-024-00170-4.

Gupta, G., De Jesus Andreoli Pinto, T., Chellappan, D.K., Mishra, A., Malipeddi, H., Dua, K., 2018. A clinical update on metformin and lung cancer in diabetic patients. Panminerva Medica 60 (2), 70–75. Available at https://doi.org/10.23736/S0031-0808.18.03394-3.

He, S., Sharpless, N.E., 2017. Senescence in health and disease. Cell 169 (6), 1000–1011. Available at https://doi.org/10.1016/j.cell.2017.05.015.

Hu, L., Li, H., Zi, M., Li, W., Liu, J., Yang, Y., et al., 2022. Why senescent cells are resistant to apoptosis: an insight for senolytic development. Frontiers in Cell and Developmental Biology 10. Available at https://doi.org/10.3389/fcell.2022.822816.

Huang, W., Hickson, L.T.J., Eirin, A., Kirkland, J.L., Lerman, L.O., 2022. Cellular senescence: the good, the bad and the unknown. Nature Reviews Nephrology 18 (10), 611–627. Available at https://doi.org/10.1038/s41581-022-00601-z.

Hussain, M.S., Agrawal, N., Ilma, B., Rekha, M.M., Nayak, P.P., Kaur, M., et al., 2025. Autophagy and cellular senescence in Alzheimer's disease: key drivers of neurodegeneration. CNS Neuroscience and Therapeutics 31 (7). Available at https://doi.org/10.1111/cns.70503.

Imb, M., Véghelyi, Z., Maurer, M., Kühnel, H., 2024. Exploring senolytic and senomorphic properties of medicinal plants for anti-aging therapies. International Journal of Molecular Sciences 25 (19), 10419. Available at https://doi.org/10.3390/ijms251910419.

Jan, N., Sofi, S., Mir, A.A., Masoodi, G., Mir, M.A., 2025. Balancing senescence and apoptosis: therapeutic insights into aging and cancer. Molecular and Cellular Biochemistry 480 (11), 5821–5850. Available at https://doi.org/10.1007/s11010-025-05355-3.

Jia, G., Aroor, A.R., Jia, C., Sowers, J.R., 2019. Endothelial cell senescence in aging-related vascular dysfunction. Biochimica et Biophysica Acta (BBA) - Molecular Basis of Disease 1865 (7), 1802–1809. Available at https://doi.org/10.1016/j.bbadis.2018.08.008.

Kakkar, C., Sharma, V., Mannan, A., Gupta, G., Singh, S., Kumar, P., et al., 2025. Diabetic cardiomyopathy: an update on emerging pathological mechanisms. Current Cardiology Reviews 21 (2). Available at https://doi.org/10.2174/011573403x331870241025094307.

Kaur, G., Sohanur Rahman, M., Shaikh, S., Panda, K., Chinnapaiyan, S., Santiago Estevez, M., et al., 2024. Emerging roles of senolytics/senomorphics in HIV-related co-morbidities. Biochemical Pharmacology 228, 116179. Available at https://doi.org/10.1016/j.bcp.2024.116179.

Khavinson, V., Linkova, N., Dyatlova, A., Kantemirova, R., Kozlov, K., 2023. Senescence-associated secretory phenotype of cardiovascular system cells and inflammaging: perspectives of peptide regulation. Cells 12 (1), 106. Available at https://doi.org/10.3390/cells12010106.

Khosla, S., 2024. A pipeline for senolytics. Journal of Clinical Investigation 134 (9). Available at https://doi.org/10.1172/jci180558.

Kirkland, J.L., Tchkonia, T., 2020. Senolytic drugs: from discovery to translation. Journal of Internal Medicine 288 (5), 518–536. Available at https://doi.org/10.1111/joim.13141.

Kumar, A., Thirumurugan, K., 2023. Understanding cellular senescence: pathways involved, therapeutics and longevity aiding. Cell Cycle (Georgetown, Tex.) 22 (20), 2324–2345. Available at https://doi.org/10.1080/15384101.2023.2287929.

Lagoumtzi, S.M., Chondrogianni, N., 2021. Senolytics and senomorphics: natural and synthetic therapeutics in the treatment of aging and chronic diseases. Free Radical Biology and Medicine 171, 169–190. Available at https://doi.org/10.1016/j.freeradbiomed.2021.05.003.

Lee, S., Wang, E.Y., Steinberg, A.B., Walton, C.C., Chinta, S.J., Andersen, J.K., 2021. A guide to senolytic intervention in neurodegenerative disease. Mechanisms of Ageing and Development 200, 111585. Available at https://doi.org/10.1016/j.mad.2021.111585.

Li, B., Lyu, P., Tang, J., Li, J., Ouchi, T., Fan, Y., et al., 2024. The potential role and therapeutic relevance of cellular senescence in skeletal pathophysiology. The Journals of Gerontology, Series A: Biological Sciences and Medical Sciences 79 (4). Available at https://doi.org/10.1093/gerona/glae037.

Liu, J., Yu, H., Xu, Y., 2025. Targeting cellular senescence: pathophysiology in multisystem age-related diseases. Biomedicines 13 (7). Available at https://doi.org/10.3390/biomedicines13071727.

López-Otín, C., Blasco, M.A., Partridge, L., Serrano, M., Kroemer, G., 2023. Hallmarks of aging: an expanding universe. Cell 186 (2), 243–278. Available at https://doi.org/10.1016/j.cell.2022.11.001.

Luan, Y., Zhu, X., Jiao, Y., Liu, H., Huang, Z., Pei, J., et al., 2024. Cardiac cell senescence: molecular mechanisms, key proteins and therapeutic targets. Cell Death Discovery 10 (1). Available at https://doi.org/10.1038/s41420-023-01792-5.

Lucas, V., Cavadas, C., Aveleira, C.A., 2023. Cellular senescence: from mechanisms to current biomarkers and senotherapies. Pharmacological Reviews 75 (4), 675–713. Available at https://doi.org/10.1124/pharmrev.122.000622.

de Magalhães, J.P., 2024. Cellular senescence in normal physiology. Science (New York, N.Y.) 384 (6702), 1300–1301. Available at https://doi.org/10.1126/science.adj7050.

Martel, J., Ojcius, D.M., Wu, C.-Y., Peng, H.-H., Voisin, L., Perfettini, J.-L., et al., 2020. Emerging use of senolytics and senomorphics against aging and chronic diseases. Medicinal Research Reviews 40 (6), 2114–2131. Available at https://doi.org/10.1002/med.21702.

McCorkle, J.R., Ahn, R., Cao, C.D., Hill, K.S., Dietrich, C.S., Kolesar, J.M., 2024. Antineoplastic drug synergy of artesunate with navitoclax in models of high-grade serous ovarian cancer. Cancers 16 (7). Available at https://doi.org/10.3390/cancers16071321.

McHugh, D., Durán, I., Gil, J., 2025. Senescence as a therapeutic target in cancer and age-related diseases. Nature Reviews Drug Discovery 24 (1), 57–71. Available at https://doi.org/10.1038/s41573-024-01074-4.

Mehdizadeh, M., Aguilar, M., Thorin, E., Ferbeyre, G., Nattel, S., 2022. The role of cellular senescence in cardiac disease: basic biology and clinical relevance. Nature Reviews Cardiology 19 (4), 250–264. Available at https://doi.org/10.1038/s41569-021-00624-2.

Miller, S.J., Darji, R.Y., Walaieh, S., Lewis, J.A., Logan, R., 2023. Senolytic and senomorphic secondary metabolites as therapeutic agents in Drosophila melanogaster models of Parkinson's disease. Frontiers in Neurology 14. Available at https://doi.org/10.3389/fneur.2023.1271941.

Moskalev, A., Guvatova, Z., De, I., Lopes, A., Beckett, C.W., Kennedy, B.K., et al., 2022. Targeting aging mechanisms: pharmacological perspectives. Trends in Endocrinology & Metabolism 33 (4), 266–280. Available at https://doi.org/10.1016/j.tem.2022.01.007.

Muñoz-Espín, D., Serrano, M., 2014. Cellular senescence: from physiology to pathology. Nature Reviews. Molecular Cell Biology 15 (7), 482–496. Available at https://doi.org/10.1038/nrm3823.

Narayanan, J., Tamilanban, T., Kumar, P.S., Guru, A., Muthupandian, S., Kathiravan, M.K., et al., 2023. Role and mechanistic actions of protein kinase inhibitors as an effective drug target for cancer and COVID. Archives of Microbiology 205 (6). Available at https://doi.org/10.1007/s00203-023-03559-z.

Nieto, M., Konigsberg, M., Silva-Palacios, A., 2024. Quercetin and dasatinib, two powerful senolytics in age-related cardiovascular disease. Biogerontology 25 (1), 71–82. Available at https://doi.org/10.1007/s10522-023-10068-5.

Ogrodnik, M., Evans, S.A., Fielder, E., Victorelli, S., Kruger, P., Salmonowicz, H., et al., 2021. Whole-body senescent cell clearance alleviates age-related brain inflammation and cognitive impairment in mice. Aging Cell 20 (2). Available at https://doi.org/10.1111/acel.13296.

Okuno, K., Cicalese, S., Elliott, K.J., Kawai, T., Hashimoto, T., Eguchi, S., 2020. Targeting molecular mechanism of vascular smooth muscle senescence induced by angiotensin II, a potential therapy via senolytics and senomorphics. International Journal of Molecular Sciences 21 (18), 6579. Available at https://doi.org/10.3390/ijms21186579.

Owens, W.A., Walaszczyk, A., Spyridopoulos, I., Dookun, E., Richardson, G.D., 2021. Senescence and senolytics in cardiovascular disease: promise and potential pitfalls. Mechanisms of Ageing and Development 198, 111540. Available at https://doi.org/10.1016/j.mad.2021.111540.

Picos, A., Seoane, N., Campos-Toimil, M., Viña, D., 2025. Vascular senescence and aging: mechanisms, clinical implications, and therapeutic prospects. Biogerontology 26 (3). Available at https://doi.org/10.1007/s10522-025-10256-5.

Power, H., Valtchev, P., Dehghani, F., Schindeler, A., 2023. Strategies for senolytic drug discovery. Aging Cell 22 (10). Available at https://doi.org/10.1111/acel.13948.

Rad, A.N., Grillari, J., 2024. Current senolytics: mode of action, efficacy and limitations, and their future. Mechanisms of Ageing and Development 217, 111888. Available at https://doi.org/10.1016/j.mad.2023.111888.

Revskij, D., Woitas, A., Kölle, B., Umstätter, C., Zechner, D., Khan, F.M., et al., 2024. Effects of triggers of senescence and senolysis in murine pancreatic cancer cells. Hepatobiliary & Pancreatic Diseases International 23 (6), 628–637. Available at https://doi.org/10.1016/j.hbpd.2024.06.001.

Robbins, P.D., Jurk, D., Khosla, S., Kirkland, J.L., Lebrasseur, N.K., Miller, J.D., et al., 2021. Senolytic drugs: reducing senescent cell viability to extend health span. Annual Review of Pharmacology and Toxicology 61, 779–803. Available at https://doi.org/10.1146/annurev-pharmtox-050120-105018.

Sahu, P., Satapathy, T., 2025. Immunopharmacology of senescence: targeting the senescence-associated secretory phenotype (SASP)—a mechanism-based review. Inflammopharmacology 33 (8), 4291–4310. Available at https://doi.org/10.1007/s10787-025-01867-y.

Salminen, A., Kauppinen, A., Kaarniranta, K., 2012. Emerging role of NF-κB signaling in the induction of senescence-associated secretory phenotype (SASP). Cellular Signalling 24 (4), 835–845. Available at https://doi.org/10.1016/j.cellsig.2011.12.006.

Schafer, M.J., Haak, A.J., Tschumperlin, D.J., LeBrasseur, N.K., 2018. Targeting senescent cells in fibrosis: pathology, paradox, and practical considerations. Current Rheumatology Reports 20 (1). Available at https://doi.org/10.1007/s11926-018-0712-x.

Schmitt, C.A., Wang, B., Demaria, M., 2022. Senescence and cancer — role and therapeutic opportunities. Nature Reviews Clinical Oncology 19 (10), 619–636. Available at https://doi.org/10.1038/s41571-022-00668-4.

Shang, D., Liu, H., Tu, Z., 2023. Pro-inflammatory cytokines mediating senescence of vascular endothelial cells in atherosclerosis. Fundamental and Clinical Pharmacology 37 (5), 928–936. Available at https://doi.org/10.1111/fcp.12915.

Shen, M., Fu, J., Zhang, Y., Chang, Y., Li, X., Cheng, H., et al., 2024. A novel senolytic drug for pulmonary fibrosis: BTSA1 targets apoptosis of senescent myofibroblasts by activating BAX. Aging Cell 23 (9). Available at https://doi.org/10.1111/acel.14229.

Shimizu, I., Minamino, T., 2019. Cellular senescence in cardiac diseases. Journal of Cardiology 74 (4), 313–319. Available at https://doi.org/10.1016/j.jjcc.2019.05.002.

Singh, Y., Samuel, V.P., Dahiya, S., Gupta, G., Gillhotra, R., Mishra, A., et al., 2019. Combinational effect of angiotensin receptor blocker and folic acid therapy on uric acid and creatinine level in hyperhomocysteinemia-associated hypertension. Biotechnology and Applied Biochemistry 66 (5), 715–719. Available at https://doi.org/10.1002/bab.1799.

Singh, Y., Fuloria, N.K., Fuloria, S., Subramaniyan, V., Almalki, W.H., Al-abbasi, F.A., et al., 2022. A European pharmacotherapeutic agent roflumilast exploring integrated preclinical and clinical evidence for SARS CoV-2 mediated inflammation to organ damage. British Journal of Clinical Pharmacology 88 (8), 3562–3565. Available at https://doi.org/10.1111/bcp.15328.

Spiegel, M., 2025. Fisetin as a blueprint for senotherapeutic agents – elucidating geroprotective and senolytic properties with molecular modeling. Chemistry – A European Journal 31 (16). Available at https://doi.org/10.1002/chem.202403755.

Stojanović, S.D., Thum, T., Bauersachs, J., 2025. Anti-senescence therapies: a new concept to address cardiovascular disease. Cardiovascular Research 121 (5), 730–747. Available at https://doi.org/10.1093/cvr/cvaf030.

Suda, M., Paul, K.H., Minamino, T., Miller, J.D., Lerman, A., Ellison-Hughes, G.M., et al., 2023. Senescent cells: a therapeutic target in cardiovascular diseases. Cells 12 (9), 1296. Available at https://doi.org/10.3390/cells12091296.

Sweeney, M., Cook, S.A., Gil, J., 2023. Therapeutic opportunities for senolysis in cardiovascular disease. FEBS Journal 290 (5), 1235–1255. Available at https://doi.org/10.1111/febs.16351.

Tang, H., Zhang, X., Hu, S., Song, Y., Jin, W., Zou, J., et al., 2025. Natural products acting as senolytics and senomorphics alleviate cardiovascular diseases by targeting senescent cells. Targets 3 (3), 23. Available at https://doi.org/10.3390/targets3030023.

Teimouri, M., Homayouni-Tabrizi, M., Rajabian, A., Amiri, H., Hosseini, H., 2022. Anti-inflammatory effects of resveratrol in patients with cardiovascular disease: a systematic review and meta-analysis of randomized controlled trials. Complementary Therapies in Medicine 70, 102863. Available at https://doi.org/10.1016/j.ctim.2022.102863.

Thapa, R., Afzal, O., Alfawaz Altamimi, A.S., Goyal, A., Almalki, W.H., Alzarea, S.I., et al., 2023. Galangin as an inflammatory response modulator: an updated overview and therapeutic potential. Chemico-Biological Interactions 378, 110482. Available at https://doi.org/10.1016/j.cbi.2023.110482.

Thompson, E.L., Pitcher, L.E., Niedernhofer, L.J., Robbins, P.D., 2022. Targeting cellular senescence with senotherapeutics: development of new approaches for skin care. Plastic & Reconstructive Surgery 150 (2), 12S. Available at https://doi.org/10.1097/prs.0000000000009668.

Wan, R., Srikaram, P., Guntupalli, V., Hu, C., Chen, Q., Gao, P., 2023. Cellular senescence in asthma: from pathogenesis to therapeutic challenges. eBioMedicine 94, 104717. Available at https://doi.org/10.1016/j.ebiom.2023.104717.

Xu, C., Qiu, Z., Guo, Q., Huang, Y., Zhao, Y., Zhao, R., 2025. The role of cellular senescence in cardiovascular disease. Cell Death Discovery 11 (1). Available at https://doi.org/10.1038/s41420-025-02720-5.

Ya, J., Bayraktutan, U., 2024. Senolytics and senomorphics targeting p38MAPK/NF-κB pathway protect endothelial cells from oxidative stress-mediated premature senescence. Cells 13 (15). Available at https://doi.org/10.3390/cells13151292.

Zhang, L., Pitcher, L.E., Prahalad, V., Niedernhofer, L.J., Robbins, P.D., 2023. Targeting cellular senescence with senotherapeutics: senolytics and senomorphics. The FEBS Journal 290 (5), 1362–1383. Available at https://doi.org/10.1111/febs.16350.

Zhang, L., Pitcher, L.E., Yousefzadeh, M.J., Niedernhofer, L.J., Robbins, P.D., Zhu, Y., 2022. Cellular senescence: a key therapeutic target in aging and diseases. Journal of Clinical Investigation 132 (15). Available at https://doi.org/10.1172/jci158450.

Zhang, Y., Zhou, Y., 2024. Advances in targeted therapies for age-related osteoarthritis: a comprehensive review of current research. Biomedicine & Pharmacotherapy 179, 117314. Available at https://doi.org/10.1016/j.biopha.2024.117314.

Zhao, H., Liu, Z., Chen, H., Han, M., Zhang, M., Liu, K., et al., 2024. Identifying specific functional roles for senescence across cell types. Cell 187 (25), 7314. Available at https://doi.org/10.1016/j.cell.2024.09.021.

Zhao, S., Zhang, Y., Zhao, Y., Lu, X., 2025. Cellular senescence as a key player in chronic heart failure pathogenesis: Unraveling mechanisms and therapeutic opportunities. Progress in Biophysics and Molecular Biology 196, 8–18. Available at https://doi.org/10.1016/j.pbiomolbio.2025.02.002.

Zheng, L., He, S., Wang, H., Li, J., Liu, Y., Liu, S., 2024. Targeting cellular senescence in aging and age-related diseases: challenges, considerations, and the emerging role of senolytic and senomorphic therapies. Aging and Disease 15 (6), 2554–2594. Available at https://doi.org/10.14336/AD.2024.0206.

Zhu, Y., Tchkonia, T., Fuhrmann-Stroissnigg, H., Dai, H.M., Ling, Y.Y., Stout, M.B., et al., 2016. Identification of a novel senolytic agent, navitoclax, targeting the Bcl-2 family of anti-apoptotic factors. Aging Cell 15 (3), 428–435. Available at https://doi.org/10.1111/acel.12445.

Chapter 13

Biomarkers of senescence in cardiovascular diseases

Bharathi Kalidass[1], Prasanna Srinivasan Ramalingam[1], Gothandam Kodiveri Muthukaliannan[1], Purushothaman Balakrishnan[2], Janaki Ramaiah Mekala[1], Sivakumar Arumugam[1], and Subhajit Dutta[3,4]

[1]*School of Biosciences and Technology, Vellore Institute of Technology, Vellore, Tamil Nadu, India,* [2]*Department of Biomaterials, Saveetha Dental College and Hospitals, SIMATS, Saveetha University, Chennai, Tamil Nadu, India,* [3]*Functional Genomics and Metabolism Research Unit, Department of Biochemistry and Molecular Biology, University of Southern Denmark, Odense, Denmark,* [4]*Center for Functional Genomics and Tissue Plasticity (ATLAS), University of Southern Denmark, Odense, Denmark*

13.1 Introduction

13.1.1 Overview of cellular senescence in the cardiovascular context

Cellular senescence is a complex, multifactorial, and stress-induced cellular event. It is defined by stable cell cycle arrest in response to various stimuli such as intrinsic and extrinsic stressors, including telomere attrition, epigenetic changes, radiation, oxidative and genetic stressors, perturbed proteostasis, chromatin disorganization, mitotic signals, oncogenic activation, inflammation or tissue damage signals, and nutrient deprivation (Kumari and Jat, 2021). In the cardiovascular system, senescence is characterized by amplified pacing frequency, irregular shortening, dysregulated contraction, endothelial cells, and SASP factor production, which results in the accumulation of senescent cells over time and leads to the development and progression of cardiovascular diseases such as cardiac fibrosis, vascular stiffening, atherosclerosis, and heart failure (Luan et al., 2024).

The cardiac environment contains different types of cells, such as 70% of noncardiac cells, which encompass fibroblasts, pericytes, smooth muscle cells, and 30% of cardiomyocytes, which are grouped into atrial and ventricular myocytes. The orchestration of these cells facilitates the integral functions of the heart (Grandi et al., 2023). The existence of different cell environments, the cardiac senescence is characterized into cardiomyocyte senescence and endothelial cell senescence. In cardiomyocyte senescence, which is mediated by DNA damage, oxidative stress, and stimulation of $p16^{INK4a}$/RB and p53-p21cascades, results in mitochondrial dysfunction, impaired contractility, and secretion of pro-inflammatory SASP regulators (Camacho-Encina et al., 2024). In cardiac endothelial cells, senescence is induced by metabolic imbalance, oxidative stress, and epigenetic modifications like miR-217 upregulation and SIRT1/6 downregulation, significantly impairing vascular integrity by diminishing nitric oxide bioavailability and enhancing endothelia-1 expression (Zhang et al., 2017a; Scioli et al., 2020). These maladaptive cascades trigger endothelial dysfunction, regulating vascular inflammation and contributing to the disease progression of heart failure, atherosclerosis, and atrial fibrillation.

The progressive accumulation of senescent cells in the cardiac environment impedes intracellular communication, facilitates sterile inflammation, and leads to adverse remodeling. In aged people, the senescent-derived microenvironment plays a crucial role in cardiovascular pathologies (Song et al., 2020). Consequently, emerging therapeutic strategies, including senomorphics and senolytics, have exhibited a promising therapeutic avenue for cardiovascular diseases (Saliev and Singh, 2025). A comprehensive understanding and integration of biomarker technology would be a potent framework for early detection and characterization of the disease.

13.1.2 Importance of identifying biomarkers for early detection, diagnosis, and prognosis

According to the National Cancer Institute, biomarkers are formulated as a "biological molecule present in blood, other body fluids, or tissue that is an indicator of a healthy or pathological homeostasis or the presence of a disease condition"

Cellular Senescence and Cardiac Aging. DOI: https://doi.org/10.1016/B978-0-443-45811-8.00010-6

(https://www.cancer.gov/publications/dictionaries/cancer-terms/def/biomarker). Moreover, biomarkers are employed to measure the pharmacological response to the treatments by evaluating the treatment efficacy. The detection of a potential biomarker is crucial for early detection, accurate diagnosis, prognosis, and treatment monitoring (Ahmad et al., 2023). Biomarkers are a potent tool to distinguish pathological environments from healthy, with their genetic variability, transcriptional dysregulation, and posttranslational modifications. The spectrum of biomarkers encompasses diverse molecular entities, including nucleic acids, proteins, antibodies, and peptides, and often exhibits as complex molecular fingerprints like transcriptomic, proteomic, and metabolic profiles (López et al., 2025). Early detection using specific and sensitive biomarkers enables timely therapeutics, which is critical in preventing disease progression and diminishing the morbidity and mortality (Deulkar et al., 2024). The detection techniques include invasive (high-resolution imaging and tissue biopsy) and noninvasive approaches (samplings of biofluids such as blood, urine, cerebrospinal fluid (CSF), saliva, and sweat). The invasive techniques offer site-specific insights and fail to enclose the systemic dynamics, whereas noninvasive methods provide greater insights into systemic dynamics but are constrained by biomarker dilution, instability, and background interferences (Diaz et al., 2023). Therefore, robust evaluation of biomarker specificity, sensitivity, and clinical applicability is crucial for their effective administration into diagnostic and therapeutic approaches.

They provide diagnostic specificity by delineating disease subtype and its key underlying mechanisms, which dissect the potent part for tailoring the specific and accurate therapeutic strategies for the disease. In terms of prognosis, biomarkers predict disease progression, recurrence risk, and treatment susceptibility. This chapter delves into the current understanding of biomarkers in cardiovascular diseases by intersecting cellular senescence and its role in detecting cardiac diseases.

13.2 Cellular senescence: general concepts

13.2.1 Definition and features of cellular senescence

Senescence is a Latin word derived from "senex," meaning old. In biology, cellular senescence is a complex stress-regulated process characterized by a persistent and significantly irreversible arrest of cell cycle progression, typically at the G1 phase, despite continued metabolic activity (Ajoolabady et al., 2025). In contrast to transient quiescence, which is a reversible cell cycle arrest that can be returned to proliferation with an appropriate mitogenic stimulus, whereas, even under mitogenic stimulus, cellular senescence is sustained by the activation of cell cycle arrest, mediated by the upregulation of cell cycle inhibitory pathways (Marescal and Cheeseman, 2020). The senescent phenotypes are heterogeneous and microenvironment-dependent and are characterized by several cellular and molecular profiles. It represents a polygenic biological process of aging, characterized by the progressive decline of cellular processes, ultimately impairing tissue homeostasis and organismal function (Hussain et al., 2025). It is typically concomitant with a declined reproductive capacity and an elevated risk of mortality (McHugh and Gil, 2018). The heterogenic phenotypic nature of senescent cells typically involves cell cycle arrest via the upregulation of cyclin-dependent kinase (CDK) inhibitors, apoptosis resistance, mediated by improved expression of antiapoptotic proteins like BCL-2, metabolic reprogramming, which includes mitochondrial dysfunction via reactive oxygen species (ROS) production, and modified oxidative phosphorylation process, epigenetic remodeling, such as generation of senescence-associated heterochromatin foci (SAHF), which downregulate proliferation-associated genes, and lastly lysosomal expansion and enhanced SA-ß-gal regulation (Huang et al., 2022).

13.2.2 Mechanisms and triggers

The key underlying mechanisms of cellular senescence and aging levels are not fully understood. It can be triggered by diverse upstream stimuli that intersect on two canonical effector mechanisms, including $p16^{INK4a}$/RB and $p53\text{-}p21^{WAF/CIP1}$ cascades, as shown in (Roger et al., 2021). Activation of these cascades enables a stable, persistent proliferative arrest, serving as a critical barrier against uncontrolled cell division. Telomeres are repetitive DNA sequences (tandem TTAGGG repeats) present at the ends of linear chromosomes, associated with several proteins that form a cluster known as "shelterin" (Tardat and Déjardin, 2018). Telomeres are composed of a C-lagging strand and a G-leading strand, with the G-rich strand ending in a 3' single-stranded (ss) overhang composed of repetitive sequences. This overhang is postulated to fold back and invade the adjacent double-stranded (ds) telomeric DNA, forming a looped architecture, termed a t-loop (Chang et al., 2022). The shelterin complex comprises six core protein components, such as protection of telomeres 1 (POT1), adrenocortical dysplasia protein homolog (TPP1), telomeric repeat-binding factor 1 (TRF1), telomeric repeat-binding factor 2 (TRF2), TRF2-interacting protein (RAP1), and TRF1-interacting nuclear factor 2 (TIN2) (Bandaria et al., 2016).

TRF1 and TRF2 couple to ds telomeric DNA, facilitating structural stability of the telomeric duplex. Whereas POT1 peculiarly identifies and binds to the ss 3'G-rich overhang, hence preventing the activation of DNA damage signaling pathways and maintaining the t-loop structure. TPP1 and TIN2 act as the scaffolding and regulatory subunits, driving interprotein interactions and safeguarding the expression of telomeric length (Kalal et al., 2023; Xin et al., 2008; Nakamura et al., 2008) (Fig. 13.1).

In most somatic cells, the lack of telomerase activity provokes telomeres' susceptibility to progressive shortening via the end-replication problem, which originates from the inherent biochemical constraint of DNA polymerase that cannot replicate the terminal region of the lagging strand. Due to the inherent limitation of lagging-strand synthesis, the removal of the terminal RNA primers creates an unrepeated gap that DNA polymerase cannot extend, resulting in irreversible telomere shortening. This end-replication problem fundamentally constrains cellular proliferative capacity and governs replicative senescence (Srinivas et al., 2020). Telomere shortening or uncapping serves as a crucial molecular stimulus for cellular senescence by activating a persistent DNA damage response (DDR) that the cell senses as irreversible genomic lesions (Roger et al., 2021). This chronic DDR activation is primarily mediated by ATM/ATR kinases and stabilization of p53, which facilitates precise stimulation of CDK inhibitors p53-$p21^{WAF/CIP1}$, regulating a sustained cell cycle arrest. Conversely, this pathway can lead to aging and tissue loss functionality by enhancing the accumulation of senescent cells, even though the mechanism efficiently regulates the spread of damaged cells and assists in the prevention of genomic integrity (Burgess et al., 2014; Abuetabh et al., 2022). Additionally, telomere-induced senescence is stimulus-dependent, which indicates that telomere structure can interact distinctly according to cellular environment, DNA repair, and the role of epigenetics in modulating telomere sensitivity and functions. Therefore, telomere-mediated senescence exhibits a duality and balance between suppression of carcinogenesis and contribution to organismal aging, which highlights the dual nature of this pathway as related to health and disease (Zhu et al., 2021).

Nuclear DNA is steadily exposed to damage from both endogenous metabolic by-products and exogenous environmental factors. Disruption of the double helix skeleton, either ss or ds DNA, triggers DDR, a conserved and intricately regulated signaling network that is fundamental to maintaining genomic stability (Zhu et al., 2021). DDR activation is not only a significant repair mechanism but also a decisive determinant of cell fate, with persistent or

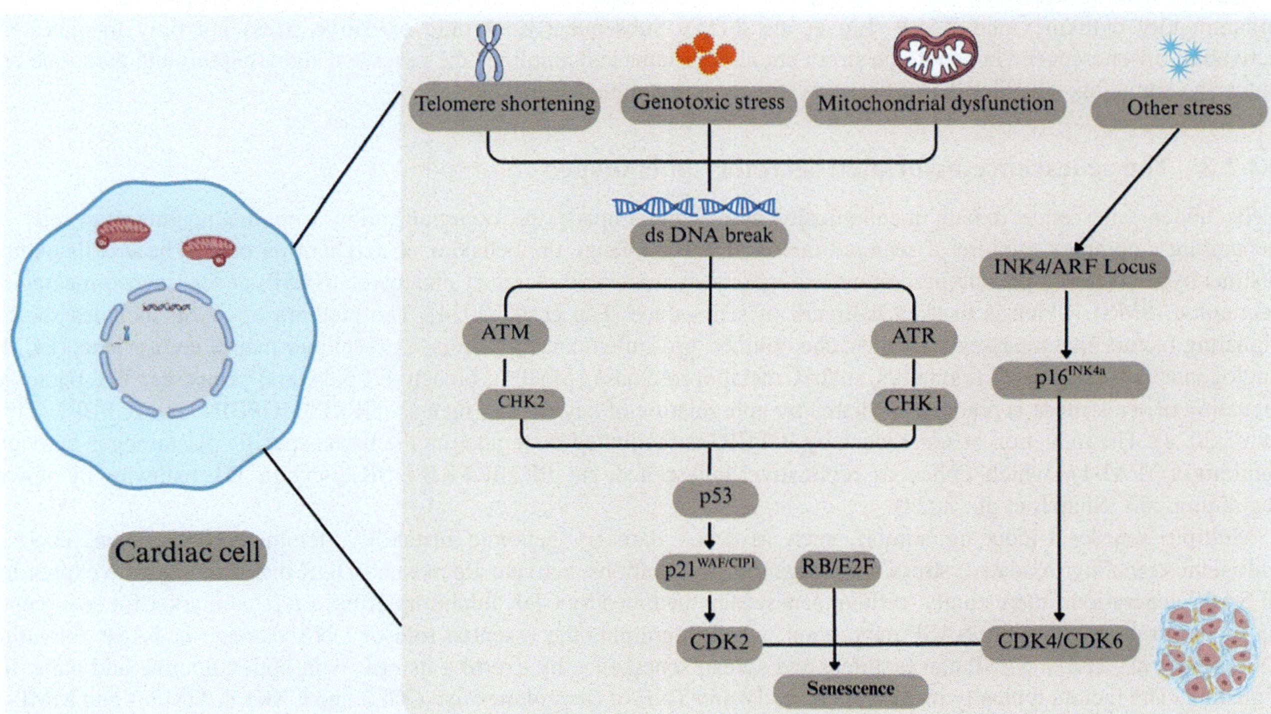

FIGURE 13.1 Pathways leading to cardiac cell senescence. Multiple stressors including telomere shortening, genotoxic stress, mitochondrial dysfunction, and other cellular insults converge on the DNA damage response and INK4/ARF locus to activate downstream signaling p53/ $p21^{CIP1/Waf1}$- and $p16^{INK4a}$- associated pathways, driving stable cell cycle arrest and the senescent phenotype in cardiac cells.

irreversible damage serving as a primary trigger for cellular senescence. The key senescence-associated damages include unresolved ds-breaks, dysregulated telomeres, and oxidative DNA modifications (Nakamura et al., 2008; Rossiello et al., 2014).

The DDR functions through an orchestrated cascade comprising damage recognition, signal transduction, and effector engagement. Generally, the MRN complex (MRE11-RAD50-NBS1) acts as a primary detector of ds-breaks, mediating the recruitment and upregulation of ataxia-telangiectasia (ATM) kinase (Syed and Tainer, 2018). Conversely, poly(ADP-ribose) polymerase 1 (PARP1) shows a rapid kinetic binding profile to ss-breaks, triggering base excision repair through the recruitment of downstream repair proteins. For nonhomologous end joining (NHEJ), the Ku70/Ku80 heterodimer exhibits a strong binding interaction to DNA termini, offering the accumulation of NHEJ machinery (Wang et al., 2006; De Lorenzo et al., 2013). This stimulus activates the apical DDR kinases ATM, ataxia-telangiectasia and Rad3-related (ATR), and DNA-dependent protein kinase catalytic subunit (DNA-PKcs) and phosphorylates a wide range of substrates integral to damage signaling. Among these, histone H2AX, a variant of the histone H2A protein, rapidly phosphorylated at the site of DNA damage, especially ds-breaks, forming γH2AX, serves as a chromatin marker that delineates surrounding domains of ds-break (Helt et al., 2005). Moreover, the tumor suppressor p53 phosphorylates and stabilizes its function in mediating DNA repair, cell cycle arrest, or apoptosis. The checkpoint kinases CHK1 and CHK2 are stimulated to reinforce cell cycle arrest, providing the temporal coordination between the DNA repair cascade and cell cycle progression. The phosphorylation process serves a dual function via amplifying the initial recognition signal and constructing a barrier that permits repair to occur (Vlatkovic et al., 2022). The DDR signaling cascade can initiate permanent cell cycle arrest (senescence) or programmed cell death (apoptosis) when there is unsuccessful repair and irreversible damage.

Oxidative stress plays a crucial role in the stimulation and maintenance of cellular senescence by regulating molecular damage and activating stress response pathways, resulting from an imbalance between ROS generation and the capacity of free radical scavenging systems (Dash et al., 2025). Increased ROS levels induce oxidative damage to macromolecules such as DNA, proteins, and lipids, thereby disrupting cellular integrity and functions. ROS is vulnerable to telomeres, which accelerates telomere attrition and promotes telomere dysfunction. This dysregulation generates a persistent DDR, activating key signaling cascades that enforce cell cycle arrest (Barnes et al., 2019). Further, ROS influence redox-sensitive signaling pathways such as NF-κB, p38 MAPK, and AP-1, facilitating the transcriptional activation of pro-inflammatory cytokines and SASP (Liu et al., 2025). Subsequently, chronic oxidative stress not only mediates the activation of senescence via genotoxic stress but also sustains and amplifies the senescent signaling through paracrine and autocrine signaling, contributing to tissue dysfunction and aging.

13.2.3 The senescence-associated secretory phenotype

Cells under senescence remain metabolically active and impede the communication with the neighboring cells by generating a complex mixture of secreted factors that can change the behavior of neighboring cells. These cells exhibit distinct hypersecretory phenotypes known as senescence-associated secretory phenotype (SASP) or senescence-messaging secretome (SMS), which is the key hallmark of senescence (Fan et al., 2024). The plethora of SASP includes soluble signaling factors encompasses cytokines, chemokines, pro-inflammatory factors, extracellular matrix components (ECM), angiogenic factors, growth regulators, matrix metalloproteinase (MMPs), bioactive lipids, and proteases. The paracrine signaling of senescence is potently mediated by upregulation of key SASPs such as IGFBP3, IGFBP4, and IGFBP7 (Zhao et al., 2024). The induction of senescence by IGFBP3 is facilitated by regulating the tissue-specific plasminogen activator inhibitor-1 (PAI-1), which enhances replicative senescence via PI(3)K-PKB-GSK3β-cyclin D1 pathway by down-regulating p53 (Shaikh et al., 2024).

Multiple senescent-inducing stimuli, such as DNA damage, genomic instability, telomere dysfunction, aberrant mitogenic signaling, oxidative stress, and epigenomic alterations, activate a persistent DDR that promotes the expression of SASP generation. Interestingly, cellular senescence mediated by CDK inhibitors offers a typical marker for senescence but does not upregulate the SASP expression, which highlights the essential role of DNA damage in SASP activation (Nadeem et al., 2025). SASP can facilitate and spread senescence by exerting its effects in both autocrine and paracrine signaling. The factors typically include IL-6, IL-1ß and TNF-α, chemokines like CCL2 and CXCL8, MMP-1 and MMP-3, growth factors comprise TGF-ß and VEGF, and extracellular vesicles. The SASP is critically orchestrated at the transcriptional level by signaling pathways involving NF-κB, upregulation of kinases like p38 MAPK, mTOR, and C/EDPß, which integrate stress and damage signals to regulate SASP expression (Quintero-Fabián et al., 2019; Wang et al., 2025).

The SASP carries out several mechanistic roles impacting both senescent cells and their local tissue microenvironment. Autocrine SASP factors enhance the expression of CDK inhibitors and sustain DNA damage response signaling via positive feedback mechanisms, stabilizing and reinforcing the senescence growth arrest. Paracrine SASP factors induce senescence in nearby cells. This process is called "bystander senescence" and increases the senescence population in the tissue (Liao et al., 2021). The SASP also alters immune surveillance by recruiting many types of immune effector cells to facilitate the clearance of senescent cells and tissue remodeling and repair. However, chronic expression of the SASP can impact immune function, leading to immunosenescence or immune suppression, allowing accumulation of senescent cells and consequently contributing to tissue dysfunction. SASP-associated proteases remodel the ECM and alter tissue structure and function, while pro-inflammatory cytokines establish a persistent low-grade inflammatory state, "inflammaging", which has been implicated in age-related pathologies including cardiovascular disease, fibrosis, and cancer (Alqahtani et al., 2025). Through paracrine effects, SASP factors can alter the behavior of adjacent non-senescent cells, which includes angiogenesis, epithelial-to-mesenchymal transition (EMT), and stemness.

13.3 Types of senescence biomarkers

The identification of reliable biomarkers for cellular senescence is crucial for a comprehensive understanding of aging biology, age-associated disease, and therapeutic strategies for targeting senescent cells, as shown in Table 13.1. According to their characteristic, biomarkers are classified into molecular, cellular, and tissue level, circulating versus tissue-specific biomarkers.

13.3.1 Molecular biomarkers

Molecular biomarkers are measured at the molecular entities of proteomic and genomic levels, such as nucleic acids, proteins, and their modifications, which aid in capturing both upregulated activation and downregulated manifestation of senescent cells. These biomarkers are functionally at the fundamental bimolecular levels and are characterized into DNA-based, RNA-based, and Protein-based cohorts, providing distinct mechanistic insights into the molecular level (Piñero et al., 2023).

13.3.1.1 DNA-based biomarkers

At the genome level, DNA-based biomarkers encompass sequence changes (qualitative) and copy number alterations (quantitative), as well as measuring epigenetic modifications that regulate gene expression. It captures the germline and somatic mutations, copy number variations (CNVs), single-nucleotide polymorphisms (SNPs), insertions/deletions (indels), and structural rearrangements. In epigenetic modification, especially DNA methylation-derived aging clocks at CpG islands are a potent biomarker (Ren et al., 2019). During the epigenetic mechanism, DNA methylation (5-methylcytosine) is generated by the insertion of a methyl group to the C-5 position of the cytosine ring, which is catalyzed by methyltransferase (DNMT). There are three classes of DNMT, including DNMT1, DNMT3A, and DNMT3B, where DNMT1 declined with aging and decreased DNA methylation (Cui and Xu, 2018). Inversely, during the aging process in mammalian cells, upregulated expression of DNMT3A and DNMT3B is responsible for the de novo methylation of CpG islands. As an epigenetic biomarker, DNA methylation has been utilized to detect chronological and biological aging. Likewise, histone modification undergoes the modification of chromatin accessibility and gene expression without altering the nucleotide sequence (Vaidya et al., 2025).

13.3.1.2 RNA-based biomarkers

RNA acts as an essential mediator for genetic information transfer and is critically involved in transcription and posttranslational modifications. Specifically, RNA is chemically liable under alkaline conditions, they can be accurately detected and quantified at minimal abundance using molecular techniques (Liu et al., 2024a). Due to their dynamic expression profiles and availability of accurate amplifying techniques, compared to protein biomarkers, RNA biomarkers offer enhanced sensitivity and specificity. RNA biomarkers include messenger RNAs (mRNAs), which are products of transcription, noncoding RNAs (microRNAs, long noncoding RNAs, circular RNAs) that act through posttranscriptional regulation, and alternative splicing variants that may potentially change the profile of protein isoforms (Liu et al., 2024a). Transcriptomic studies through RNA-seq or microarrays can indicate targets of a dysregulated pathway, but RNA molecules are very labile and are easily degraded through handling, causing them to have a degree of preanalytical variability (Gallego Romero et al., 2014). RNA expression changes should be interpreted cautiously because they could reflect secondary effects instead of the primary driving factor of the disease.

TABLE 13.1 Major biomarkers associated with cellular senescence and their relevance to cardiovascular pathologies. Key biomarkers such as p16^{INK4a}, SA-ß-gal, P53, CD8^{+}, CD57^{+} - T cells, and SASP provide molecular insights into mechanisms of vascular aging, atherosclerosis, and heart failure.

Biomarker	Type	Detection method	Major cardiovascular disease associations	Advantages	Limitations	References
SA-ß-gal	Histochemical enzyme activity	Histochemical staining	Age-related Myocardial dysfunction	Widely used marker; visually localizes senescent cells in tissue context	Susceptible to false detection in cells contains high lysosomal contents	Lyu et al. (2018)
p16^{INK4a}	Cell cycle inhibitor proteins	Immunohistochemistry, gene expression	Antifibrosis and improves heart dysfunction	Reflects activation of senescence pathways; widely studied	Expression is stimulation-dependent and may occur in non-senescent inflammatory cells	Zhao et al. (2021)
SASP	Circulating protein biomarkers	ELISA	Irreversible Pulmonary Arterial Hypertension	Circulating levels give noninvasive insight into senescent burden and risk prediction of CVDs	High inter-individual variability, poor tissue origin specificity	Van Der Feen et al. (2020)
CD8^{+}, CD57^{+} - T cells	Adaptive immune effectors	Flow cytometry	Cardiac mortality, Arterial Stiffness	Well-characterized subset and measurable in circulation	Widely used marker for inflammatory response.	Tae Yu et al. (2015), Yu et al., 2017
P53	Tumor suppressor/ cell cycle pathway	Western blot, IHC, gene expression assays	Atherosclerotic vascular disease, Age-related myocardial dysfunction, age-related heart failure, oxidative stress, and endothelial dysfunction	Key upstream regulator; indicates senescence pathway activation	p53 is involved in many stress pathways beyond senescence, limiting specificity	Gevaert et al. (2017), Wang et al. (2018), Min et al. (2007)

13.3.1.3 Protein-based biomarkers

Protein-based biomarkers of cellular senescence include a diverse array of nonsecretory and secretory proteins, whose altered abundance, localization, or activity represent the complex and context-dependent biology of the senescent phenotype. While cell cycle inhibitors such as $p16^{INK4a}$ (CDKN2A) and $p21^{Cip1/Waf1}$ (CDKN1A) are often considered canonical senescence markers mediating irreversible G1 arrest via inhibition of CDK4/6 and hypophosphorylation of the retinoblastoma protein (pRB) (p16) and transcriptional activation through p53 (p21), the expression of CDKN2A and CDKN1A is neither ubiquitous across senescence subtypes nor fully specific, since they can also be induced in response to transient stress, or through differentiation (Wagner and Wagner, 2022). $p27^{Kip1}$ (CDKN1B) also contributes to reinstating arrest in some models but is not widely generalizable. Persistent DDR proteins such as γH2AX and 53BP1, at DNA segments with chromatin changes that can reinforce senescence (DNA-SCARS), and chromatin remodelers like macroH2A within SAHF, provide structural and mechanistic evidence of genome instability, but could be elevated in other DNA damage–initiated pathologies as well (Bencivenga et al., 2025; Liu et al., 2024b).

The SASP includes proteins such as IL-6, IL-8, IL-1α/β, CCL2, CXCL1, VEGF, amphiregulin, and MMP family members, provides a measurable extracellular signature, but its composition varies widely depending on the cell type, senescence-trigger, and time since induction, while also overlapping with inflammation and injury responses. For example, markers of lysosomal biogenesis like SA-β-gal, cathepsin B, and cathepsin D may reflect greater lysosomal biogenesis, but the extent of their functional relevance in terms of promoting senescence maintenance is contentious, while they are also known to be false positives in non-senescent lysosome-rich cells (Valieva et al., 2022). Markers of metabolism and structure (e.g., GLUT1, vimentin, fibronectin, laminin, and other integrin subtypes) reflect the changing landscapes of cellular architecture, bioenergetics, etc., but these hyperstructured metabolically active changes are not limited to senescence (Tomecka et al., 2024). Mitochondrial proteins, including SOD2, PGC-1α, and TOM20 as indicators of oxidative stress and metabolic stress, are also ubiquitous among other recovery and adaptation to stress programs (Rius-Pérez et al., 2020). Importantly there is no specific protein or protein marker with absolute specificity or sensitivity to senescence; this means that a more rigorous multiparametric approach that combines cell cycle regulators, DDR proteins, SASP factors, and metabolic markers, and validates all factors at some point within the time-based spectrum and context (i.e. tissue)-specific, is essential to validate true senescence and distinguish it from similar, phenotypically, but possibly reversible stress states.

13.3.2 Cellular and tissue-level markers

Cellular and tissue-level markers of senescence include structural, phenotypic, and histological changes that occur at the single-cellular or multicellular tissue level. These biomarkers offer the cumulative downregulation effects of epigenetic reprogramming, genomic instability, proteostatic imbalance, and chronic stress signaling that determine the senescent phenotype. While the molecular biomarkers provide the molecular level resolution, cellular and tissue markers illustrate the integrated, morpho-functional implications of the senescence process within the microenvironment (Low et al., 2021). Cellular-level biomarkers of senescence contain various structural, enzymatic, and chromatin-mediated modifications that collectively depict the irreversible cell cycle arrest and phenotypic remodeling of senescent cells. Senescent cells exhibit altered morphology, including flattened and enlarged phenotypes, a change in the ratio of cytoplasm to nucleus, heterogeneous organelle distribution, and irregular nuclear envelope morphology, which are mainly caused by cytoskeletal reorganization, enhanced lysosomal expansion, and aberrant accumulation of oxidatively damaged biomolecules (Song et al., 2020). SA-β-gal activity is an example of a commonly employed enzymatic marker of senescence, which is histochemically detectable at pH 6.0, resulting in increased lysosomal mass. However, it is important to note that typical SA-β-gal activity is not completely specific, as it can also happen in non-senescent cells, activated macrophages, or cells maintained in culture for extended periods (de Mera-Rodríguez et al., 2021).

Chromatin and DNA damage-mediated biomarkers provide further structural and functional evidence of senescence. SAHF exhibits as DAPI-dense nuclear domains, enriched in transcriptional repressive histone modifications, especially trimethylated histones H3 at lysine 9 (H3K9me3), enabling stable silencing of E2F targets, which are proliferation-promoting genes (Kumar et al., 2013). These structural modifications of the chromatin will be visualized using fluorescence microscopy, utilizing antibodies specific to the modified histones. By colocalization of γ-H2AX and 53BP1 in DNA damage foci, with a subset localizing to the telomeric region as telomere-associated foci (TAF), we capture the persistent DNA-damage signaling using immunofluorescence and high-content imaging, which results in the existence of irreversible ds-break. The SASP triggers detectable extracellular modifications, including deposition of MMPs, ECM remodeling, and disrupted cell-cell communication dynamics, which lead to modifications in the tissue microenvironment and immune cells' requirements (Kleiner et al., 2015).

Tissue-level biomarkers of senescence measure the spatial distribution, structural modifications, and microenvironmental changes triggered by the accumulation of senescent cells within the organs or tissues, but their interpretations are constrained by overlapping and context-derived variability with other pathological states (Di Micco et al., 2021). By applying immunohistochemistry and RNA in situ hybridization techniques in the tissue microenvironment, the regions of chronic injury, fibrotic remodeling, or aging-associated degeneration could be detected using specific markers such as $p16^{INK4a}$, $p21^{Cip1/Waf}$, and SASP expressed on the accumulated senescent cells. However, the existence of these markers is not completely specific to senescence, as transient stress-related expression and persistent expression in differentiation states can generate a similar pattern of expression (Zhang et al., 2017b). The stains, including Masson's trichrome or picrosirius red and haematoxylin and eosin (H&E), are employed to quantify the structural modifications comprising ECM disorganization, deuteriation of parenchymal architecture, and basement membrane thickening. These histopathological signatures are the key characteristic features of non-senescent fibrotic or inflammatory conditions, limiting their ability to independently provide diagnostic values (López De Padilla et al., 2021).

Microenvironment changes caused by chronic SASP signaling include stable infiltrates of immune cells such as macrophages, T lymphocytes, and natural killer (NK) cells into the tissue microenvironment, which can be detected by immunolabeling with lineage-specific markers such as CD3, CD68, and NKp46 (Kale et al., 2020). While the recruitment of immune effector cells can certainly be part of a senescence surveillance mechanism, an immune profile can be observed at any time by tissue infiltration during an infection, autoimmune-mediated disease, or in response to injury, regardless of the senescent state. As a result of SASP-mediated secretion of pro-fibrotic cytokines, growth factors, and matrix-modifying enzymes, ECM remodeling and fibrotic deposition can also result in tissue microenvironment changes, specifically, as evidenced by increased collagen crosslinking, elastin fragmentation, and stiffening of the matrix (Mebratu et al., 2023). These ECM changes can be critical indicators of the long-term presence of senescent cells, can be evaluated by biochemical measures (eg, tissue collagen content and level of oncogenesis, elastic fiber total area, elastic fibers quantity) or by AC staining, and should always be examined in combination with additional biomarkers that will help distinguish senescence-associated fibrosis from fibrosis of other etiology (Blokland et al., 2020). Overall, a critical multiparametric approach is therefore needed to rigorously and meaningfully evaluate tissue-level senescence signatures in both experimental and clinical settings.

13.3.3 Circulating versus tissue-specific biomarkers

Senescence biomarkers can be widely distinguished into circulating and tissue-specific classes based on their origin, detection method, and applicability as represented in Table 13.2. Circulating biomarkers are released into systemic circulation from various tissues and can be measured in relatively noninvasive biofluids such as blood, urine, and cerebrospinal fluid. The ease of collection and analysis makes circulating biomarkers appealing for early disease detection, monitoring patients over time, and large-scale screening programs, but their specificity is often diminished due to the potential biomarker's source (e.g., circulating tumor DNA, C-reactive protein, microRNAs) (Donatti et al., 2020). Conversely, tissue-specific biomarkers are only produced from a specific cell type, tissue, or organ (i.e., HER2 for breast cancer or amyloid-β plaque imaging in the brains of patients with Alzheimer's), allowing for high specificity and historical details about the local pathological mechanism and consequence. However, more often than not, the evaluation of tissue-specific biomarkers requires invasive sampling (e.g., biopsy), limiting their frequency in longitudinal planning or monitoring (Chung et al., 2013). Circulating biomarkers allow for dynamic systemic assessment, while tissue-specific biomarkers allow for fine detail disease characterization, and both have advantages and challenges. Circulating biomarkers allow for accessibility and ease of longitudinal monitoring, while tissue-specific biomarkers provide unmatched local accuracy, at the expense of invasiveness and logistical sampling constraints.

13.4 Key biomarkers of senescence in cardiovascular diseases

Senescence of cardiovascular tissues is mechanistically important in the development of atherosclerosis, vascular stiffening, myocardial remodeling, and heart failure (Katsuumi et al., 2018). Telomere length is one of the well-studied biomarkers, where consistently shorter telomeres have been monitored in leukocytes, endothelial cells, and vascular smooth muscle cells in a diverse array of cardiovascular diseases. Although methods such as qPCR, terminal restriction fragment (TRF) analysis, and fluorescence in situ hybridization (FISH) make telomere length quantification possible, researchers note limitations in specificity due to variations between methods, sensitivity to the quality of the sample, or possible influences of genetic and lifestyle factors on telomere length (Butt et al., 2010). SA-β-gal remains a prevalent lysosomal marker in tissue biopsies, especially for assessing senescence in endothelial and myocardial cells (Ogrodnik et al., 2024). However, its

TABLE 13.2 The principal characteristics of senescence biomarkers detected in circulation and within tissue, focusing on their biological origin. Circulating markers offer systemic, minimally invasive monitoring, whereas tissue markers enable mechanistic insights at the cellular level.

Biomarker	Circulating or tissue-specific	Sample type	Clinical relevance	Utility in CVD	Comments
SA-β-galactosidase	Tissue-specific	Myocardial biopsies, vascular tissue sections	Histological hallmark of senescence; indicates lysosomal expansion	Detects senescent cells in plaques, fibrotic myocardium	Gold-standard histological assay, but not applicable for routine clinical use due to invasiveness
$p16^{INK4a}$ / $p21^{CIP1}$	Tissue-specific	Cardiac tissue biopsies, vascular tissue	Reflects irreversible cell cycle arrest; associated with endothelial dysfunction and vascular stiffness	Identifies localized senescent burden in atherosclerosis or heart failure	Requires invasive sampling; expression may occur in inflammation without senescence
SASP factors (IL-6, IL-1β, IL-8, MMPs, GDF-15, TNFR1/2, MCP-1)	Circulating	Plasma, serum	Indicate systemic pro-inflammatory senescence signaling	Predicts HF, CAD progression, and mortality risk	Noninvasive; lacks tissue origin specificity; influenced by comorbidities
Multiplex proteomic panels (e.g., GDF-15, MMP-1, TNFR2)	Circulating	Plasma, serum	Improves prognostic accuracy beyond age/sex	Stratifies CVD risk, especially in elderly cohorts	Requires advanced analytical platforms; expensive
Soluble ST2	Circulating	Serum	Prognostic for HF and post-ACS outcomes	Supports risk stratification in HF and ACS patients	Not senescence-specific; reflects myocardial stress
p53	Tissue-specific (with possible circulating fragments in EVs)	Tissue biopsies, isolated EVs	Master regulator of DNA-damage-induced senescence	Links oxidative/genotoxic stress to remodeling	Lack of absolute specificity due to roles in apoptosis/repair
Extracellular vesicle (EV) cargo (miRNAs, proteins)	Circulating	Plasma-derived EVs	Reflects intercellular senescence communication	Potential early detection tool for vascular aging	Experimental; no clinical cut-offs established
$CD4^{+}$ T lymphocytes	Circulating and tissue-specific	Peripheral blood, vascular/lymphoid tissue	Mediate immune-driven vascular remodeling and stiffness	Potential immune marker of CVD progression	Strong preclinical evidence; human causality studies are limited
CD55 (Decay-accelerating factor)	Circulating and tissue-specific	Plasma, endothelial cell membranes	Regulates complement activation; implicated in atherogenesis modulation	May modulate vascular inflammation and plaque stability	Role in stiffness/cardiac death not fully characterized

usefulness in saliva, blood, or plasma is limited, and the possibility of false positives can arise in scenarios involving macrophage activation or lysosomal storage disorders.

$p16^{INK4a}$ and $p21^{CIP1}$ expression are hallmark features in senescent cardiovascular cells; however, upregulation of $p16^{INK4a}$ and $p21^{CIP1}$ can also occur in transient stress responses, induced pluripotent stem cell (iPS) differentiation states, or standard activation, so it is not exclusive to senescence. While markers of DNA damage response, such as γ-H2AX, have been strongly linked with both atherosclerosis and heart failure, their occurrence has been more localized to DNA damage accompanying senescence when found in association with more persistent double-strand breaks in senescent TAF (Grootaert, 2024). DNA damage can also be present as a result of ischemic injury or oxidative stress, and thus is not linked to senescence. Circulating SASP markers such as IL-6, IL-1β, and MCP-1 offer a systemic and minimally invasive index of senescence-associated inflammation, but such elevation is widely reported in a range of states, including infection, autoimmune disease, and general inflammatory states, which can complicate clinical utility or interpretation (Picos et al., 2025). Circulating microRNAs (e.g., miR-34a, miR-146a) and exosome-associated patterns suggest promising noninvasive biomarkers of senescence, but these are also still early in the validation process, with unknowns including their tissue origin, stability, and standard detection protocols (Picos et al., 2025). To develop a valid diagnostic strategy for cardiovascular senescence, therefore will need a multiparametric approach incorporating both tissue-regional and circulating markers, including appropriate demarcation of biological and methodological confounders.

13.5 Techniques for detecting senescence biomarkers

The précised identification of senescence requires analytical techniques that can monitor molecular, cellular, and histological modifications with great sensitivity and accurate specificity. The selection of an appropriate biomarker detection technique is based on its class (DNA, RNA, functional phenotype, or proteins), biological matrix (cultured cells, fresh or fixed tissue, or body fluids), and its application (mechanistic, clinical diagnosis, or longitudinal monitoring) (Picos et al., 2025). The detection of senescence biomarkers comprises molecular, histological, and omics-based techniques. Immunohistochemistry (IHC) continues to be one of the most popular techniques for recognizing senescence-associated proteins, including $p16^{INK4a}$, $^{p21CIP1/WAF1}$, and γ-H2AX, without sacrificing tissue architecture, while also enabling spatial and contextual measurements of the location of senescent cells. However, its accessibility is limited by factors including variability in antibody specificity, antigen retrieval, and fixation, as well as by the presence of background nonspecific staining and variability in scoring caused by inter-observer differences that introduce added subjectivity, all of which detracts from reproducibility in research and in clinical practice (Giatromanolaki et al., 2020).

Quantitative PCR (qPCR) permits high-sensitivity, high-throughput quantification of senescence-related transcripts, including cell cycle inhibitors, and SASP mediators. However, qPCR carries an analytical bias that depends on the quality of the RNA sample, the design of the primers, and normalized stability to reference genes (Sanders et al., 2014). Finally, a key limitation of transcriptional changes measured by qPCR is that they do not always correspond with stable, or functional senescence, meaning that qPCR results alone may be of limited diagnostic utility.

The enzyme-linked immunosorbent assay (ELISA) is a powerful, highly reproducible approach for absolute quantification of soluble SASP factors in biological fluids or conditioned media. However, standard ELISA formats cannot detect multiple analytes simultaneously, restricting the detection ability to explore SASP complexity. Multiplex bead-based immunoassays allow for the detection of greater numbers of analytes but add cost and potential experimental complications by virtue of increased cross-reactivity within assays (Samiminemati et al., 2024).

Flow cytometry allows for multiparametric, single-cell resolution analysis of senescence-associated markers across heterogeneous populations of cells. This comprises discovering surface antigens, marking DNA damage, and metabolic activity alterations. However, during the dissociation steps, fragile senescent cells can undergo phenotypic changes, and permeabilization steps for intracellular staining can impact antigen integrity and signal quality and reproducibility (Adewoye et al., 2020).

Next-generation sequencing (NGS) based methods, such as RNA sequencing or whole genome sequencing, provide unparalleled resolution for dissecting differences in transcriptomes, genomic instability, and telomere shortening in the context of senescence. While NGS-based methods can produce very deep and comprehensive datasets, they come with many limitations, including high costs, specialized bioinformatic expertise, and susceptibility to errors resulting from nucleic acid degradation, particularly when formalin-fixed or formalin-archived samples are used (Qin, 2019). These limitations are what make NGS not currently scalable for routine clinical diagnostic purposes.

13.6 Clinical relevance of senescence biomarkers in cardiovascular diseases

Cardiovascular diseases, including heart failure, atherosclerosis, and hypertension, are tightly regulated by the aberrant accumulation of senescent cells in myocardial, vascular, and immune compartments. Senescence biomarkers provide a potential for early detection, staging, and therapeutic monitoring in these environments, as shown in Fig. 13.2.

13.6.1 Correlation with disease stage/severity

Atherosclerotic lesions reliably express canonical cell cycle inhibitors, such as $p16^{INK4a}$ and $p21^{CIP1/WAF1}$, in vascular smooth muscle cells (VSMCs) and endothelial cells, as well as persistent markers of DNA damage response such as γ-H2AX and TAF. These changes demonstrate markers of stress-related senescent and replicative senescent cells and reflect a dramatic shift in cellular phenotype to a pro-inflammatory and matrix-degrading phenotype. The SASP in VSMCs, which is characterized by the secretion of MMPs and inflammatory cytokines, accelerates degradation of the extracellular matrix and of the fibrous cap as well as recruits macrophages into the core of the lesion (Bennett et al., 2016). These pro-inflammatory changes reflect the process of destabilization, which increases the risk of plaque rupture and ultimately, myocardial infarction, stroke, and other adverse vascular events. Notably, these biomarkers increase spatially and temporally as lesions progress and transition from early, lipid-rich fatty streaks to advanced fibroatheromas with high rupture potential, potentially providing a means to stratify cardiovascular risk.

In heart failure, pathways of systemic and myocardial senescence reveal both diagnostic and mechanistic relevance. Circulating SASP factors, such as IL-6, TNF-α, and some abbreviated relevant MMPs, among others, exhibit proportional increases in their levels based on increasing severity of ventricular dysfunction and correlate closely with echocardiographic measures of adverse remodeling based on ejection fraction and left ventricular mass index (Khavinson et al., 2023). In the heart, senescence of myofibroblasts leads to maladaptive deposition of extracellular matrix components, namely collagen type I and collagen type III, which promote myocardial stiffening and contribute to diastolic dysfunction. Biomarkers of fibrotic remodeling, such as PAI-1, are elevated in both the tissue and circulation to illuminate fibrotic burden (Nikolov and Popovski, 2022). Importantly, though these biomarkers have prognostic value based on staffing experience metrics capturing care processes, they also lack absolute specificity as centrally located by profile identification

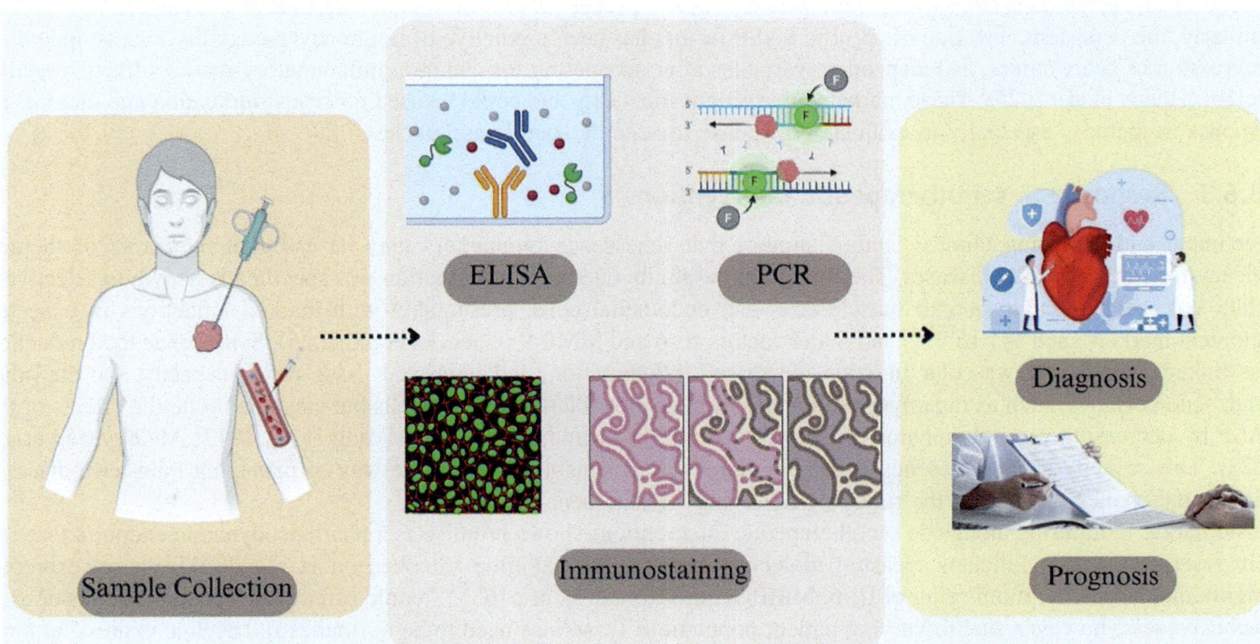

FIGURE 13.2 Clinical applications of senescence biomarkers in cardiovascular disease. Patient-derived blood or tissue samples can be analyzed for senescence markers using assays such as ELISA, PCR, and immunostaining. These biomarkers provide insights into cellular aging processes within the cardiovascular system and may support early diagnosis, risk stratification, and prognosis in patients with heart disease.

could be activated through nonsenescent fibrotic processes, hence, the need for multimarker panels to enhance classification accuracy.

Associative vascular aging in hypertension is closely linked with endothelial senescence as exhibited through diminished expression of endothelial nitric oxide synthase (eNOS) and increased measures of oxidative stress and its by-products, that is, 8-oxo-2'-deoxyguanosine (8-oxo-dG), which visibly represent cumulative DNA oxidation. Physiologically, the presence of senescence produces a decline in endothelial function as the potential for vasodilation decreases; induction of pro-thrombotic and pro-inflammatory vascular phenotypes may increase (Hernandez-Navarro et al., 2024). These oxidative stress and senescence behaviors correlate with elevated levels of such biochemical markers that affirm greater vascular structural alterations and calcification, an increase in vascular stiffness and wall thickness. In clinical practice, there is emerging evidence that such molecular markers correlate with functional measures such as pulse wave velocity, which measures arterial stiffness and correlates with long-term cardiovascular morbidity (Gopcevic et al., 2021). Whereas oxidative stress biomarkers directly correlate to vascular remission, therefore must consider other diagnoses of acute inflammatory response where structural changes are relatively gained with mild oxidative stress markers, without a proper clinical context.

13.6.2 Prognostic versus diagnostic value

Several circulating senescence-associated biomarkers exhibit remarkable potential for the detection of the disease as early as it precedes structural or functional changes in the cardiovascular system. MicroRNAs detectable in the peripheral blood, such as miR-34a and miR-146a, were found in the early stages of endothelial senescence and may enable identification of vascular aging prior to progression of atherosclerotic plaque or stiffening of the arteries. $p16^{INK4a}$ in leukocytes is also an early indicator of subclinical vascular aging, indicative of cumulative cellular stress and replicative senescence. These markers would potentially aid early intervention, especially in populations at high-risk with no cardiovascular disease clinical manifestations. However, their diagnostic specificity may be limited by their involvement in other age-related or inflammatory conditions, necessitating integration with complementary biomarkers and imaging modalities for precise clinical application (Kong et al., 2025).

Alternatively, some senescence-associated biomarkers have better predictive value than others for future cardiovascular events. Although shortened leukocyte telomere length has consistently been associated with increased rates of myocardial infarction, stroke, and cardiovascular mortality, it is also a proper prognostic factor independent of standard risk scores. Similarly, the persistent elevation of specific SASP factors has been predictive of future adverse cardiac events, including progression of heart failure, as independent variables after accounting for common inflammatory markers like C-reactive protein (Huang et al., 2025). These markers have prognostic utility and could be used for risk stratification and monitoring response to therapy, especially to evaluate a response to senotherapeutic approaches.

13.6.3 Response to senotherapeutic interventions

Preclinical and emerging clinical studies suggest that senescence biomarkers may be essential indicators of therapy efficacy for cardiovascular diseases. The senolytics dasatinib–quercetin combination was specifically shown to selectively ablate senescent vascular smooth muscle cells and endothelial cells, presumably with marked reductions in expected canonical markers such as $p16^{INK4a}$ and SASP factors IL-6 and MMP-9 (Gadecka et al., 2025). Senescence load reduction was linked to improved vascular function, decreased inflammation, and improved tissue repair capacity. On the other hand, senomorphics, such as rapamycin and metformin, cannot ablate senescent cells but can ameliorate the effects of the SASP by decreasing pro-inflammatory cytokines and matrix-degrading enzymes (Wang et al., 2017; Abdelgawad et al., 2023). This strategy to modify senescence phenotype is a reasonable and sensible way of managing long-term diseases without the chance of injuring the tissue by removing all senescent cells at once.

Biomarker monitoring alongside senotherapeutic interventions shows promise as a pharmacodynamic endpoint that can help researchers and clinicians measure molecular changes in real time (Stephenson et al., 2021). As an example, longitudinal biomarker monitoring of IL-6, MMP-9, and expression of $p16^{INK4a}$ could offer quantitative measures of drug responsiveness; however, due to varying patient populations (baselines need to be normalized), baseline expression must be considered (Mariani et al., 2021). Also, in the emerging field of regenerative cardiovascular medicine, preimplantation screening of autologous stem or progenitor cell populations for markers of senescence has been shown to improve posttransplant survival, functional recovery, engraftment into host tissue, and decrease posttransplant graft failure by preexisting senescent cell populations (Deng et al., 2025). Overall, these observations suggest that senescence biomarkers can act as both a therapeutic target and a dynamic readout of success for proactive intervention in regard to cardiovascular disease.

13.7 Limitations and future directions

13.7.1 Specificity and sensitivity issues

A primary limitation is a lack of absolute specificity. Most markers of senescence, especially those attached to SASP, are not specific to senescence, but are secreted and released under circumstances of acute inflammation, infection, and other metabolic reactions (Schmitt et al., 2022). Likewise, DNA damage markers and telomere shortening are not specific to senescence, which are generated from transient stress response. The spatial heterogeneity of senescence cell accumulation further constrained the sensitivity of the marker. During biopsies and imaging studies, sampling errors are potentially limited by depicting focal regions within tissues that may harbor a high senescence burden while adjacent areas remain unaffected (Klepacki et al., 2025).

As well, circulating biomarkers lack sensitivity in part due to dilution and degradation when entering the circulatory system, limiting their ability to detect low-abundance signals from localized senescent cell populations (Wang and Dreesen, 2018). Inter-individual variability due to age, genetic background, lifestyle, and comorbidities adds another layer of complexity in interpreting biomarker levels (Castruita et al., 2022).

13.7.2 Multimarker panels and omics approaches

The utilization of a simple biomarker for detecting or monitoring senescence is necessarily limited due to the substantial heterogeneity of senescent cell phenotypes, context-dependent functions of senescent cells, and the overlap between markers of pathological aging and physiological aging (Shah et al., 2025). Single-molecule approaches typically fail to detect transient, reversible, or cell-type-specific aspects of senescence, which ultimately yields poor diagnostic and prognostic performance. Composite biomarker panels that integrate molecular markers (such as $p16^{INK4a}$ and telomere length), SASP, cytokine profiling, and imaging-derived changes related to tissue architecture can offer a more complete assessment that can increase both sensitivity and specificity (Niazi, 2025; Basisty et al., 2020a). However, the use of composite panels is not without challenges, which include inter-individual variability (or difference in cellular response to exposures), difficulties with technical standardization, and the potential for confounding due to comorbidities or systemic inflammation, which is unrelated to the senescence process.

Multi-omics approaches, such as transcriptomics, proteomics, metabolomics, and epigenomics, may allow for the detection of systems-level signatures of senescence (Chen et al., 2023). The potential to use machine learning-based integration of multi-omics datasets to identify new predictive biomarkers and therapeutic targets is powerful but does have drawbacks. High-dimensional data requires large, well-annotated cohorts to minimize the risk of overfitting; batch effects or other issues related to sample quality can lead to misleading results (Athieniti and Spyrou, 2023). Also of note, while advanced technologies such as spatial transcriptomics and single-cell multi-omics offer enhanced resolution for mapping the cellular niches of senescent cells, they require a considerable commitment of time, equipment, and expertise, and are impractical for immediate clinical use (Sartori et al., 2025). The successful translation of these innovative approaches of analyzing senescence will rely on maintaining a balance between the analytic complexity and the realities of clinical useability, reproducibility, and scalability.

13.7.3 Emerging biomarkers and technological advances

Novel biomarkers present a unique opportunity to improve upon current challenges of senescence detection and monitoring. Senescence-associated extracellular vesicles (saEVs), enriched with microRNAs, proteins, and lipids specific to the originating cell type, are a noninvasive method of tracing senescent cells' tissue of origin and functional status. DNA methylation clocks that are modified to the epigenetic profile associated with senescence provide a quantitative, temporal measurement of biological age that informs about aging through mechanisms that are not solely based on chronological clock time (Basisty et al., 2020b; He et al., 2021). Senescence-targeted PET tracers will also allow in vivo imaging of senescent cell burden and provide a means to spatially and temporally track senescent cells across tissues by allowing studies in heterogeneous diseases such as atherosclerosis or heart failure (Xiang et al., 2024).

Regarding the technological side of the modifications, microfluidic single-cell platforms are increasingly available to enable high-resolution SASP profiling, and there is a clear capability to detect intercellular heterogeneity that can be obscured in bulk assays (Erfanian et al., 2023). Technologies such as Raman spectroscopy offer label-free imaging that can produce a molecular fingerprint of a senescent cell without the use of an exogenous label, while CRISPR-based biosensors would enable real-time, cell-specific monitoring of senescence-associated gene expression (Lyu et al., 2018). In the translational arena, the development of miniaturized, high-throughput, point-of-care diagnostic tools will someday give potential for integration of these technologies toward rapid, noninvasive assays for senescence profiling. If such favorable technologies are standardized and validated, as pointed out above, this would also facilitate precision-guided delivery of senotherapeutics and more longitudinal monitoring of therapeutic efficacy in the clinical setting.

13.8 Conclusion

Biomarkers of senescence are central to elucidating the contribution of cellular senescence to the onset, progression, and mortality of cardiovascular diseases. However, no single biomarker currently provides sufficient diagnostic or prognostic specificity. Pathological and physiological senescence may share overlapping molecular features, underscoring the need for multimarker panels that integrate cell cycle inhibitors (e.g., $p16^{INK4a}$, $p21^{CIP1/WAF1}$), DNA damage markers (e.g., γ-H2AX), telomere length, SASP cytokines (e.g., IL-6), and circulating microRNAs.

Emerging multi-omics approaches (transcriptomics, proteomics, metabolomics, and epigenomics), combined with computational modeling, hold promise for improving both the specificity and sensitivity of senescence biomarkers by capturing the multifaceted nature of senescence. Clinically, biomarker-guided strategies such as senolytics and senomorphics may allow for more precise targeting of senescent burden, while regenerative therapies can benefit from preimplantation screening of senescence outcomes.

At the same time, advances in technology platforms, including single-cell SASP profiling, spatial transcriptomics, senescence-specific PET imaging, and microfluidic assays, offer unprecedented resolution for biomarker discovery and validation. Looking ahead, point-of-care assays may enable rapid, noninvasive assessment of senescent burden, improving risk stratification and therapeutic decision-making. Future priorities should include cross-cohort validation studies, harmonization and standardization of assays, and the development of biomarker-guided clinical algorithms to ensure adoption, scalability, and cost-effectiveness in precision cardiovascular medicine.

References

Abdelgawad, I.Y., Agostinucci, K., Sadaf, B., Grant, M.K.O., Zordoky, B.N., 2023. Metformin mitigates SASP secretion and LPS-triggered hyper-inflammation in doxorubicin-induced senescent endothelial cells. Frontiers in Aging 4. Available at https://doi.org/10.3389/fragi.2023.1170434.

Abuetabh, Y., Wu, H.H., Chai, C., Al Yousef, H., Persad, S., Sergi, C.M., et al., 2022. DNA damage response revisited: the p53 family and its regulators provide endless cancer therapy opportunities. Experimental and Molecular Medicine 54 (10), 1658–1669. Available at https://doi.org/10.1038/s12276-022-00863-4.

Adewoye, A.B., Tampakis, D., Follenzi, A., Stolzing, A., 2020. Multiparameter flow cytometric detection and quantification of senescent cells in vitro. Biogerontology 21 (6), 773–786. Available at https://doi.org/10.1007/s10522-020-09893-9.

Ahmad, A., Imran, M., Ahsan, H., 2023. Biomarkers as biomedical bioindicators: approaches and techniques for the detection, analysis, and validation of novel biomarkers of diseases. Pharmaceutics 15 (6), 1630. Available at https://doi.org/10.3390/pharmaceutics15061630.

Ajoolabady, A., Pratico, D., Bahijri, S., Eldakhakhny, B., Tuomilehto, J., Wu, F., et al., 2025. Hallmarks and mechanisms of cellular senescence in aging and disease. Cell Death Discovery 11 (1). Available at https://doi.org/10.1038/s41420-025-02655-x.

Alqahtani, S., Alqahtani, T., Venkatesan, K., Sivadasan, D., Ahmed, R., Sirag, N., et al., 2025. SASP modulation for cellular rejuvenation and tissue homeostasis: therapeutic strategies and molecular insights. Cells 14 (8), 608. Available at https://doi.org/10.3390/cells14080608.

Athieniti, E., Spyrou, G.M., 2023. A guide to multi-omics data collection and integration for translational medicine. Computational and Structural Biotechnology Journal 21, 134–149. Available at https://doi.org/10.1016/j.csbj.2022.11.050.

Bandaria, J.N., Qin, P., Berk, V., Chu, S., Yildiz, A., 2016. Shelterin protects chromosome ends by compacting telomeric chromatin. Cell 164 (4), 735–746. Available at https://doi.org/10.1016/j.cell.2016.01.036.

Barnes, R.P., Fouquerel, E., Opresko, P.L., 2019. The impact of oxidative DNA damage and stress on telomere homeostasis. Mechanisms of Ageing and Development 177, 37–45. Available at https://doi.org/10.1016/j.mad.2018.03.013.

Basisty, N., Kale, A., Hee Jeon, O., Kuehnemann, C., Payne, T., Rao, C., et al., 2020a. A proteomic atlas of senescence-associated secretomes for aging biomarker development. PLOS Biology 18 (1), e3000599. Available at https://doi.org/10.1371/journal.pbio.3000599.

Basisty, N., Kale, A., Patel, S., Campisi, J., Schilling, B., 2020b. The power of proteomics to monitor senescence-associated secretory phenotypes and beyond: toward clinical applications. Expert Review of Proteomics 17 (4), 297–308. Available at https://doi.org/10.1080/14789450.2020.1766976.

Bencivenga, D., Stampone, E., Azhar, J., Parente, D., Ali, W., Del Vecchio, V., et al., 2025. p27Kip1 and tumors: characterization of CDKN1B variants identified in MEN4 and breast cancer. Cells 14 (3), 188. Available at https://doi.org/10.3390/cells14030188.

Bennett, M.R., Sinha, S., Owens, G.K., 2016. Vascular smooth muscle cells in atherosclerosis. Circulation Research 118 (4), 692–702. Available at https://doi.org/10.1161/CIRCRESAHA.115.306361.

Blokland, K.E.C., Pouwels, S.D., Schuliga, M., Knight, D.A., Burgess, J.K., 2020. Regulation of cellular senescence by extracellular matrix during chronic fibrotic diseases. Clinical Science 134 (20), 2681–2706. Available at https://doi.org/10.1042/CS20190893.

Burgess, R.C., Burman, B., Kruhlak, M.J., Misteli, T., 2014. Activation of DNA damage response signaling by condensed chromatin. Cell Reports 9 (5), 1703–1717. Available at https://doi.org/10.1016/j.celrep.2014.10.060.

Butt, H.Z., Atturu, G., London, N.J., Sayers, R.D., Bown, M.J., 2010. Telomere length dynamics in vascular disease: a review. European Journal of Vascular and Endovascular Surgery 40 (1), 17–26. Available at https://doi.org/10.1016/j.ejvs.2010.04.012.

Camacho-Encina, M., Booth, L.K., Redgrave, R.E., Folaranmi, O., Spyridopoulos, I., Richardson, G.D., 2024. Cellular senescence, mitochondrial dysfunction, and their link to cardiovascular disease. Cells 13 (4). Available at https://doi.org/10.3390/cells13040353.

Castruita, P.A., Piña-Escudero, S.D., Rentería, M.E., Yokoyama, J.S., 2022. Genetic, social, and lifestyle drivers of healthy aging and longevity. Current Genetic Medicine Reports 10 (3), 25–34. Available at https://doi.org/10.1007/s40142-022-00205-w.

Chang, T.R., Long, X., Shastry, S., Parks, J.W., Stone, M.D., 2022. Single-molecule mechanical analysis of strand invasion in human telomere DNA. Biochemistry 61 (15), 1554–1560. Available at https://doi.org/10.1021/acs.biochem.1c00448.

Chen, C., Wang, J., Pan, D., Wang, X., Xu, Y., Yan, J., et al., 2023. Applications of multi-omics analysis in human diseases. MedComm 4 (4). Available at https://doi.org/10.1002/mco2.315.

Chung, L., Shibli, S., Moore, K., Elder, E.E., Boyle, F.M., Marsh, D.J., et al., 2013. Tissue biomarkers of breast cancer and their association with conventional pathologic features. British Journal of Cancer 108 (2), 351–360. Available at https://doi.org/10.1038/bjc.2012.552.

Cui, D., Xu, X., 2018. DNA methyltransferases, DNA methylation, and age-associated cognitive function. International Journal of Molecular Sciences 19 (5), 1315. Available at https://doi.org/10.3390/ijms19051315.

Dash, U.C., Bhol, N.K., Swain, S.K., Samal, R.R., Nayak, P.K., Raina, V., et al., 2025. Oxidative stress and inflammation in the pathogenesis of neurological disorders: mechanisms and implications. Acta Pharmaceutica Sinica B 15 (1), 15–34. Available at https://doi.org/10.1016/j.apsb.2024.10.004.

De Lorenzo, S.B., Patel, A.G., Hurley, R.M., Kaufmann, S.H., 2013. The elephant and the blind men: making sense of PARP inhibitors in homologous recombination deficient tumor cells. Frontiers in Oncology 3. Available at https://doi.org/10.3389/fonc.2013.00228.

de Mera-Rodríguez, J.A., Álvarez-Hernán, G., Gañán, Y., Martín-Partido, G., Rodríguez-León, J., Francisco-Morcillo, J., 2021. Is senescence-associated β-galactosidase a reliable in vivo marker of cellular senescence during embryonic development? Frontiers in Cell and Developmental Biology 9. Available at https://doi.org/10.3389/fcell.2021.623175.

Deng, S., Xie, H., Xie, B., 2025. Cell-based regenerative and rejuvenation strategies for treating neurodegenerative diseases. Stem Cell Research & Therapy 16 (1). Available at https://doi.org/10.1186/s13287-025-04285-7.

Deulkar, P., Singam, A., Jain, A., 2024. A comprehensive review of the role of biomarkers in the early detection of endocrine disorders in critical illnesses. Cureus 16 (5). Available at https://doi.org/10.7759/cureus.61409.

Di Micco, R., Krizhanovsky, V., Baker, D., d'Adda di Fagagna, F., 2021. Cellular senescence in ageing: from mechanisms to therapeutic opportunities. Nature Reviews Molecular Cell Biology 22 (2), 75–95. Available at https://doi.org/10.1038/s41580-020-00314-w.

Diaz, P.M., Leehans, A., Ravishankar, P., Daily, A., 2023. Multiomic approaches for cancer biomarker discovery in liquid biopsies: advances and challenges. Biomarker Insights 18. Available at https://doi.org/10.1177/11772719231204508.

Donatti, A., Canto, A.M., Godoi, A.B., da Rosa, D.C., Lopes-Cendes, I., 2020. Circulating metabolites as potential biomarkers for neurological disorders—metabolites in neurological disorders. Metabolites 10 (10), 1–32. Available at https://doi.org/10.3390/metabo10100389.

Erfanian, N., Heydari, A.A., Feriz, A.M., Iañez, P., Derakhshani, A., Ghasemigol, M., et al., 2023. Deep learning applications in single-cell genomics and transcriptomics data analysis. Biomedicine & Pharmacotherapy 165, 115077. Available at https://doi.org/10.1016/j.biopha.2023.115077.

Fan, H., Qiao, Z., Li, J., Shang, G., Shang, C., Chen, S., et al., 2024. Recent advances in senescence-associated secretory phenotype and osteoporosis. Heliyon 10 (4), e25538. Available at https://doi.org/10.1016/j.heliyon.2024.e25538.

Gadecka, A., Nowak, N., Bulanda, E., Janiszewska, D., Dudkowska, M., Sikora, E., et al., 2025. The senolytic cocktail, dasatinib and quercetin, impacts the chromatin structure of both young and senescent vascular smooth muscle cells. GeroScience 47 (3), 3907–3925. Available at https://doi.org/10.1007/s11357-024-01504-6.

Gallego Romero, I., Pai, A.A., Tung, J., Gilad, Y., 2014. RNA-seq: impact of RNA degradation on transcript quantification. BMC Biology 12. Available at https://doi.org/10.1186/1741-7007-12-42.

Gevaert, A.B., Shakeri, H., Leloup, A.J., Van Hove, C.E., De Meyer, G.R.Y., Vrints, C.J., et al., 2017. Endothelial senescence contributes to heart failure with preserved ejection fraction in an aging mouse model. Lippincott Williams and Wilkins, Belgium Circulation: Heart Failure 10 (6). Available at https://doi.org/10.1161/CIRCHEARTFAILURE.116.003806.

Giatromanolaki, A., Kouroupi, M., Balaska, K., Koukourakis, M.I., 2020. Immunohistochemical detection of senescence markers in human sarcomas. Pathology - Research and Practice 216 (2), 152800. Available at https://doi.org/10.1016/j.prp.2019.152800.

Gopcevic, K.R., Gkaliagkousi, E., Nemcsik, J., Acet, Ö., Bernal-Lopez, M.R., Bruno, R.M., et al., 2021. Pathophysiology of circulating biomarkers and relationship with vascular aging: a review of the literature from VascAgeNet group on circulating biomarkers, European Cooperation in Science and Technology Action 18216. Frontiers in Physiology 12. Available at https://doi.org/10.3389/fphys.2021.789690.

Grandi, E., Navedo, M.F., Saucerman, J.J., Bers, D.M., Chiamvimonvat, N., Dixon, R.E., et al., 2023. Diversity of cells and signals in the cardiovascular system. Journal of Physiology 601 (13), 2547–2592. Available at https://doi.org/10.1113/JP284011.

Grootaert, M.O.J., 2024. Cell senescence in cardiometabolic diseases. npj Aging 10 (1). Available at https://doi.org/10.1038/s41514-024-00170-4.

He, X., Liu, J., Liu, B., Shi, J., 2021. The use of DNA methylation clock in aging research. Experimental Biology and Medicine 246 (4), 436–446. Available at https://doi.org/10.1177/1535370220968802.

Helt, C.E., Cliby, W.A., Keng, P.C., Bambara, R.A., O'Reilly, M.A., 2005. Ataxia telangiectasia mutated (ATM) and ATM and Rad3-related protein exhibit selective target specificities in response to different forms of DNA damage. Journal of Biological Chemistry 280 (2), 1186–1192. Available at https://doi.org/10.1074/jbc.M410873200.

Hernandez-Navarro, I., Botana, L., Diez-Mata, J., Tesoro, L., Jimenez-Guirado, B., Gonzalez-Cucharero, C., et al., 2024. Replicative endothelial cell senescence may lead to endothelial dysfunction by increasing the BH2/BH4 ratio induced by oxidative stress, reducing BH4 availability, and decreasing the expression of eNOS. International Journal of Molecular Sciences 25 (18), 9890. Available at https://doi.org/10.3390/ijms25189890.

Huang, W., Hickson, L.T.J., Eirin, A., Kirkland, J.L., Lerman, L.O., 2022. Cellular senescence: the good, the bad and the unknown. Nature Reviews Nephrology 18 (10), 611–627. Available at https://doi.org/10.1038/s41581-022-00601-z.

Huang, X., Huang, L., Lu, J., Cheng, L., Wu, D., Li, L., et al., 2025. The relationship between telomere length and aging-related diseases. Clinical and Experimental Medicine 25 (1). Available at https://doi.org/10.1007/s10238-025-01608-z.

Hussain, M.S., Agrawal, N., Ilma, B., M M, R., Nayak, P.P., Kaur, M., et al., 2025. Autophagy and cellular senescence in Alzheimer's disease: key drivers of neurodegeneration. CNS Neuroscience & Therapeutics 31 (7). Available at https://doi.org/10.1111/cns.70503.

Kalal, A.A., Shetty, R.A., Manjappa, A.B., Kulkarni, N.V., Shetty, P., 2023. Prognostic significance of dysregulation of shelterin complex and its correlation with telomere length and cytogenetics in multiple myeloma. Journal of Genetic Engineering and Biotechnology 21 (1). Available at https://doi.org/10.1186/s43141-023-00504-x.

Kale, A., Sharma, A., Stolzing, A., Stolzing, A., Desprez, P.Y., Desprez, P.Y., et al., 2020. Role of immune cells in the removal of deleterious senescent cells. Immunity and Ageing 17 (1). Available at https://doi.org/10.1186/s12979-020-00187-9.

Katsuumi, G., Shimizu, I., Yoshida, Y., Minamino, T., 2018. Vascular senescence in cardiovascular and metabolic diseases. Frontiers in Cardiovascular Medicine 5. Available at https://doi.org/10.3389/fcvm.2018.00018.

Khavinson, V., Linkova, N., Dyatlova, A., Kantemirova, R., Kozlov, K., 2023. Senescence-associated secretory phenotype of cardiovascular system cells and inflammaging: perspectives of peptide regulation. Cells 12 (1), 106. Available at https://doi.org/10.3390/cells12010106.

Kleiner, R.E., Verma, P., Molloy, K.R., Chait, B.T., Kapoor, T.M., 2015. Chemical proteomics reveals a γH2AX-53BP1 interaction in the DNA damage response. Nature Chemical Biology 11 (10), 807–814. Available at https://doi.org/10.1038/nchembio.1908.

Klepacki, H., Kowalczuk, K., Łepkowska, N., Hermanowicz, J.M., 2025. Molecular regulation of SASP in cellular senescence: therapeutic implications and translational challenges. Cells 14 (13). Available at https://doi.org/10.3390/cells14130942.

Kong, M., Li, J., Tong, N., 2025. The role of peripheral blood microRNAs in the pathogenesis and treatment response of age-related macular degeneration. Future Science OA 11 (1). Available at https://doi.org/10.1080/20565623.2025.2482499.

Kumar, R., Horikoshi, N., Singh, M., Gupta, A., Misra, H.S., Albuquerque, K., et al., 2013. Chromatin modifications and the DNA damage response to ionizing radiation. Frontiers in Oncology 2. Available at https://doi.org/10.3389/fonc.2012.00214.

Kumari, R., Jat, P., 2021. Mechanisms of cellular senescence: cell cycle arrest and senescence associated secretory phenotype. Frontiers in Cell and Developmental Biology 9. Available at https://doi.org/10.3389/fcell.2021.645593.

Liao, Z., Yeo, H.L., Wong, S.W., Zhao, Y., 2021. Cellular senescence: mechanisms and therapeutic potential. Biomedicines 9 (12). Available at https://doi.org/10.3390/biomedicines9121769.

Liu, S., Liu, J., Wang, Y., Deng, F., Deng, Z., 2025. Oxidative stress: signaling pathways, biological functions, and disease. MedComm 6 (7). Available at https://doi.org/10.1002/mco2.70268.

Liu, W.W., Zheng, S.Q., Li, T., Fei, Y.F., Wang, C., Zhang, S., et al., 2024a. RNA modifications in cellular metabolism: implications for metabolism-targeted therapy and immunotherapy. Signal Transduction and Targeted Therapy 9 (1). Available at https://doi.org/10.1038/s41392-024-01777-5.

Liu, Y., Lomeli, I., Kron, S.J., 2024b. Therapy-induced cellular senescence: potentiating tumor elimination or driving cancer resistance and recurrence? Cells 13 (15). Available at https://doi.org/10.3390/cells13151281.

López, R., Pereira, F.J., Aller, A.J., 2025. An appraisal at a glance of metallome and disease biomarkers. European Journal of Medicinal Chemistry 297, 117931. Available at https://doi.org/10.1016/j.ejmech.2025.117931.

López De Padilla, C.M., Coenen, M.J., Tovar, A., De la Vega, R.E., Evans, C.H., Müller, S.A., 2021. Picrosirius red staining: revisiting its application to the qualitative and quantitative assessment of collagen type I and Type III in tendon. Journal of Histochemistry and Cytochemistry 69 (10), 633–643. Available at https://doi.org/10.1369/00221554211046777.

Low, E., Alimohammadiha, G., Smith, L.A., Costello, L.F., Przyborski, S.A., von Zglinicki, T., et al., 2021. How good is the evidence that cellular senescence causes skin ageing? Ageing Research Reviews 71. Available at https://doi.org/10.1016/j.arr.2021.101456.

Luan, Y., Zhu, X., Jiao, Y., Liu, H., Huang, Z., Pei, J., et al., 2024. Cardiac cell senescence: molecular mechanisms, key proteins and therapeutic targets. Cell Death Discovery 10 (1). Available at https://doi.org/10.1038/s41420-023-01792-5.

Lyu, G., Guan, Y., Zhang, C., Zong, L., Sun, L., Huang, X., et al., 2018. TGF-β signaling alters H4K20me3 status via miR-29 and contributes to cellular senescence and cardiac aging. Nature Communications 9 (1). Available at https://doi.org/10.1038/s41467-018-04994-z.

Marescal, O., Cheeseman, I.M., 2020. Cellular mechanisms and regulation of quiescence. Developmental Cell 55 (3), 259–271. Available at https://doi.org/10.1016/j.devcel.2020.09.029.

Mariani, N., Cattane, N., Pariante, C., Cattaneo, A., 2021. Gene expression studies in depression development and treatment: an overview of the underlying molecular mechanisms and biological processes to identify biomarkers. Translational Psychiatry 11 (1). Available at https://doi.org/10.1038/s41398-021-01469-6.

McHugh, D., Gil, J., 2018. Senescence and aging: causes, consequences, and therapeutic avenues. Journal of Cell Biology 217 (1), 65–77. Available at https://doi.org/10.1083/jcb.201708092.

Mebratu, Y.A., Soni, S., Rosas, L., Rojas, M., Horowitz, J.C., Nho, R., 2023. The aged extracellular matrix and the profibrotic role of senescence-associated secretory phenotype. American Journal of Physiology - Cell Physiology 325 (3), C565. Available at https://doi.org/10.1152/ajpcell.00124.2023.

Min, L., Mogi, M., Iwanami, J., Li, J., Sakata, A., Fujita, T., et al., 2007. Cross-talk between aldosterone and angiotensin II in vascular smooth muscle cell senescence. Cardiovascular Research 76 (3), 506–516. Available at https://doi.org/10.1016/j.cardiores.2007.07.008.

Nadeem, J., Sultana, R., Parveen, A., Kim, S.Y., 2025. Recent advances in anti-aging therapeutic strategies targeting DNA damage response and senescence-associated secretory phenotype-linked signaling cascade. Cell Biochemistry and Function 43 (3). Available at https://doi.org/10.1002/cbf.70046.

Nakamura, A.J., Chiang, Y.J., Hathcock, K.S., Horikawa, I., Sedelnikova, O.A., Hodes, R.J., et al., 2008. Both telomeric and non-telomeric DNA damage are determinants of mammalian cellular senescence. Epigenetics & Chromatin 1 (1). Available at https://doi.org/10.1186/1756-8935-1-6.

Niazi, S.K., 2025. Telomere-targeted medicine: bridging molecular mechanisms and clinical applications in age-related diseases. Life Sciences 379, 123904. Available at https://doi.org/10.1016/j.lfs.2025.123904.

Nikolov, A., Popovski, N., 2022. Extracellular matrix in heart disease: focus on circulating collagen type I and III derived peptides as biomarkers of myocardial fibrosis and their potential in the prognosis of heart failure: a concise review. Metabolites 12 (4), 297. Available at https://doi.org/10.3390/metabo12040297.

Ogrodnik, M., Carlos Acosta, J., Adams, P.D., d'Adda di Fagagna, F., Baker, D.J., Bishop, C.L., et al., 2024. Guidelines for minimal information on cellular senescence experimentation in vivo. Cell 187 (16), 4150–4175. Available at https://doi.org/10.1016/j.cell.2024.05.059.

Picos, A., Seoane, N., Campos-Toimil, M., Viña, D., 2025. Vascular senescence and aging: mechanisms, clinical implications, and therapeutic prospects. Biogerontology 26 (3). Available at https://doi.org/10.1007/s10522-025-10256-5.

Piñero, J., Fraga, P.S.R., Valls-Margarit, J., Ronzano, F., Accuosto, P., Jane, R.L., et al., 2023. Genomic and proteomic biomarker landscape in clinical trials. Computational and Structural Biotechnology Journal 21, 2110–2118. Available at https://doi.org/10.1016/j.csbj.2023.03.014.

Qin, D., 2019. Next-generation sequencing and its clinical application. Cancer Biology and Medicine 16 (1), 4–10. Available at https://doi.org/10.20892/j.issn.2095-3941.2018.0055.

Quintero-Fabián, S., Arreola, R., Becerril-Villanueva, E., Torres-Romero, J.C., Arana-Argáez, V., Lara-Riegos, J., et al., 2019. Role of matrix metalloproteinases in angiogenesis and cancer. Frontiers in Oncology 9. Available at https://doi.org/10.3389/fonc.2019.01370.

Ren, Y., Huang, S., Dai, C., Xie, D., Zheng, L., Xie, H., et al., 2019. Germline predisposition and copy number alteration in pre-stage lung adenocarcinomas presenting as ground-glass nodules. Frontiers in Oncology 9. Available at https://doi.org/10.3389/fonc.2019.00288.

Rius-Pérez, S., Torres-Cuevas, I., Millán, I., Ortega, Á.L., Pérez, S., Sandhu, M.A., 2020. PGC-1 α, inflammation, and oxidative stress: an integrative view in metabolism. Oxidative Medicine and Cellular Longevity 2020. Available at https://doi.org/10.1155/2020/1452696.

Roger, L., Tomas, F., Gire, V., 2021. Mechanisms and regulation of cellular senescence. International Journal of Molecular Sciences 22 (23), 13173. Available at https://doi.org/10.3390/ijms222313173.

Rossiello, F., Herbig, U., Longhese, M.P., Fumagalli, M., d'Adda di Fagagna, F., 2014. Irreparable telomeric DNA damage and persistent DDR signalling as a shared causative mechanism of cellular senescence and ageing. Current Opinion in Genetics & Development 26, 89–95. Available at https://doi.org/10.1016/j.gde.2014.06.009.

Saliev, T., Singh, P.B., 2025. Targeting senescence: a review of senolytics and senomorphics in anti-aging interventions. Biomolecules 15 (6). Available at https://doi.org/10.3390/biom15060860.

Samiminemati, A., Aprile, D., Siniscalco, D., Di Bernardo, G., 2024. Methods to Investigate the Secretome of Senescent Cells. Methods and Protocols 7 (4). Available at https://doi.org/10.3390/mps7040052.

Sanders, R., Mason, D.J., Foy, C.A., Huggett, J.F., 2014. Considerations for accurate gene expression measurement by reverse transcription quantitative PCR when analysing clinical samples. Analytical and Bioanalytical Chemistry 406 (26), 6471–6483. Available at https://doi.org/10.1007/s00216-014-7857-x.

Sartori, F., Codicè, F., Caranzano, I., Rollo, C., Birolo, G., Fariselli, P., et al., 2025. A comprehensive review of deep learning applications with multi-omics data in cancer research. Genes 16 (6), 648. Available at https://doi.org/10.3390/genes16060648.

Schmitt, C.A., Wang, B., Demaria, M., 2022. Senescence and cancer — role and therapeutic opportunities. Nature Reviews Clinical Oncology 19 (10), 619–636. Available at https://doi.org/10.1038/s41571-022-00668-4.

Scioli, M.G., Storti, G., D'Amico, F., Rodríguez Guzmán, R., Centofanti, F., Doldo, E., et al., 2020. Oxidative stress and new pathogenetic mechanisms in endothelial dysfunction: potential diagnostic biomarkers and therapeutic targets. Journal of Clinical Medicine 9 (6), 1995. Available at https://doi.org/10.3390/jcm9061995.

Shah, J., Al-Hashimi, A., Benedetto, M., Ruchaya, P.J., 2025. From bench to bedside: the critical need for standardized senescence detection. Archives of Cardiovascular Diseases 118 (3), 205–211. Available at https://doi.org/10.1016/j.acvd.2024.12.008.

Shaikh, S.B., Balaya, R.D.A., Dagamajalu, S., Bhandary, Y.P., Unwalla, H., Prasad, T.S.K., et al., 2024. A signaling pathway map of plasminogen activator inhibitor-1 (PAI-1/SERPINE-1): a review of an innovative frontier in molecular aging and cellular senescence. Cell Communication and Signaling 22 (1). Available at https://doi.org/10.1186/s12964-024-01910-5.

Song, P., Zhao, Q., Zou, M.-H., 2020. Targeting senescent cells to attenuate cardiovascular disease progression. Ageing Research Reviews 60, 101072. Available at https://doi.org/10.1016/j.arr.2020.101072.

Song, S., Lam, E.W.F., Tchkonia, T., Kirkland, J.L., Sun, Y., 2020. Senescent cells: emerging targets for human aging and age-related diseases. Trends in Biochemical Sciences 45 (7), 578–592. Available at https://doi.org/10.1016/j.tibs.2020.03.008.

Srinivas, N., Rachakonda, S., Kumar, R., 2020. Telomeres and telomere length: a general overview. Cancers 12 (3), 558. Available at https://doi.org/10.3390/cancers12030558.

Stephenson, D., Badawy, R., Mathur, S., Tome, M., Rochester, L., Mirelman, A., et al., 2021. Digital progression biomarkers as novel endpoints in clinical trials: a multistakeholder perspective. Journal of Parkinson's Disease 11 (s1), S103. Available at https://doi.org/10.3233/jpd-202428.

Syed, A., Tainer, J.A., 2018. The MRE11-RAD50-NBS1 complex conducts the orchestration of damage signaling and outcomes to stress in DNA replication and repair. Annual Review of Biochemistry 87, 263–294. Available at https://doi.org/10.1146/annurev-biochem-062917-012415.

Tae Yu, H., Youn, J.C., Lee, J., Park, S., Chi, H.S., Lee, J., et al., 2015. Characterization of CD8+ CD57+ T cells in patients with acute myocardial infarction. Cellular and Molecular Immunology 12 (4), 466–473. Available at https://doi.org/10.1038/cmi.2014.74.

Tardat, M., Déjardin, J., 2018. Telomere chromatin establishment and its maintenance during mammalian development. Chromosoma 127 (1), 3–18. Available at https://doi.org/10.1007/s00412-017-0656-3.

Tomecka, P., Kunachowicz, D., Górczyńska, J., Gebuza, M., Kuźnicki, J., Skinderowicz, K., et al., 2024. Factors determining epithelial-mesenchymal transition in cancer progression. International Journal of Molecular Sciences 25 (16), 8972. Available at https://doi.org/10.3390/ijms25168972.

Vaidya, H., Jelinek, J., Issa, J.P.J., 2025. DNA methylation, aging, and cancer. Epigenomes 9 (2). Available at https://doi.org/10.3390/epigenomes9020018.

Valieva, Y., Ivanova, E., Fayzullin, A., Kurkov, A., Igrunkova, A., 2022. Senescence-associated β-galactosidase detection in pathology. Diagnostics 12 (10), 2309. Available at https://doi.org/10.3390/diagnostics12102309.

Van Der Feen, D.E., Bossers, G.P.L., Hagdorn, Q.A.J., Moonen, J.R., Kurakula, K., Szulcek, R., et al., 2020. Cellular senescence impairs the reversibility of pulmonary arterial hypertension. Science Translational Medicine 12 (554). Available at https://doi.org/10.1126/SCITRANSLMED.AAW4974.

Vlatkovic, T., Veldwijk, M.R., Giordano, F.A., Herskind, C., 2022. Targeting cell cycle checkpoint kinases to overcome intrinsic radioresistance in brain tumor cells. Cancers 14 (3). Available at https://doi.org/10.3390/cancers14030701.

Wagner, K.D., Wagner, N., 2022. The senescence markers p16INK4A, p14ARF/p19ARF, and p21 in organ development and homeostasis. Cells 11 (12). Available at https://doi.org/10.3390/cells11121966.

Wang, A.S., Dreesen, O., 2018. Biomarkers of cellular senescence and skin aging. Frontiers in Genetics 9. Available at https://doi.org/10.3389/fgene.2018.00247.

Wang, H., Yu, Y., Li, R., Zhang, H., Chen, Z.-sheng, Sun, C., et al., 2025. Immunoregulatory mechanisms in the aging microenvironment: targeting the senescence-associated secretory phenotype for cancer immunotherapy. Acta Pharmaceutica Sinica B 15 (9), 4476–4496. Available at https://doi.org/10.1016/j.apsb.2025.07.022.

Wang, M., Wu, W., Wu, W., Rosidi, B., Zhang, L., Wang, H., et al., 2006. PARP-1 and Ku compete for repair of DNA double strand breaks by distinct NHEJ pathways. Nucleic Acids Research 34 (21), 6170–6182. Available at https://doi.org/10.1093/nar/gkl840.

Wang, R., Yu, Z., Sunchu, B., Shoaf, J., Dang, I., Zhao, S., et al., 2017. Rapamycin inhibits the secretory phenotype of senescent cells by a Nrf2-independent mechanism. Aging Cell 16 (3), 564–574. Available at https://doi.org/10.1111/acel.12587.

Wang, Y.C., Lee, A.S., Lu, L.S., Ke, L.Y., Chen, W.Y., Dong, J.W., et al., 2018. Human electronegative LDL induces mitochondrial dysfunction and premature senescence of vascular cells in vivo. Aging Cell 17 (4). Available at https://doi.org/10.1111/acel.12792.

Xiang, X., Dong, C., Zhou, L., Liu, J., Rabinowitz, Z.M., Zhang, Y., et al., 2024. Novel PET imaging probe for quantitative detection of senescence in vivo. Journal of Medicinal Chemistry 67 (7), 5924–5934. Available at https://doi.org/10.1021/acs.jmedchem.4c00179.

Xin, H., Liu, D., Songyang, Z., 2008. The telosome/shelterin complex and its functions. Genome Biology 9 (9), 232. Available at https://doi.org/10.1186/gb-2008-9-9-232.

Yu, Y., Kim, S., Park, K., et al., 2017. Arterial stiffness is associated with cytomegalovirus-specific senescent CD8+ T cells. Journal of the American Heart Association 6. Available at https://doi.org/10.1161/JAHA.117.006535.

Zhang, W., Huang, Q., Zeng, Z., Wu, J., Zhang, Y., Chen, Z., et al., 2017a. Sirt1 inhibits oxidative stress in vascular endothelial cells. Oxidative Medicine and Cellular Longevity 2017 (1). Available at https://doi.org/10.1155/2017/7543973.

Zhang, Y., Alexander, P.B., Wang, X.-F., 2017b. TGF-β family signaling in the control of cell proliferation and survival. Cold Spring Harbor Perspectives in Biology 9 (4), a022145. Available at https://doi.org/10.1101/cshperspect.a022145.

Zhao, J., He, X., Zuo, M., Li, X., Sun, Z., 2021. Anagliptin prevented interleukin 1β (IL-1β)-induced cellular senescence in vascular smooth muscle cells through increasing the expression of sirtuin1 (SIRT1). Bioengineered 12 (1), 3968–3977. Available at https://doi.org/10.1080/21655979.2021.1948289.

Zhao, S., Qiao, Z., Pfeifer, R., Pape, H.-C., Mao, K., Tang, H., et al., 2024. Modulation of fracture healing by senescence-associated secretory phenotype (SASP): a narrative review of the current literature. European Journal of Medical Research 29 (1), 38. Available at https://doi.org/10.1186/s40001-023-01604-7.

Zhu, X., Chen, Z., Shen, W., Huang, G., Sedivy, J.M., Wang, H., et al., 2021. Inflammation, epigenetics, and metabolism converge to cell senescence and ageing: the regulation and intervention. Signal Transduction and Targeted Therapy 6 (1). Available at https://doi.org/10.1038/s41392-021-00646-9.

Chapter 14

Noninvasive techniques for tracking cardiac aging

Chintan Aundhia[1], Muhammad Afzal[2], Surya Nath Pandey[3], Md. Raihan Siddique[4], Riya Thapa[5], Asif Ahmad Bhat[5], Md. Sadique Hussain[6,7], Neelam Singla[4], Santosh Kumar Singh[8], and Gaurav Gupta[9,10]

[1]*Department of Pharmacy, Sumandeep Vidyapeeth Deemed to be University, Vadodara, Gujarat, India,* [2]*Department of Pharmaceutical Sciences, Pharmacy Program, Batterjee Medical College, Jeddah, Saudi Arabia,* [3]*Department of Pharmacology, Teerthanker Mahaveer College of Pharmacy, Teerthanker Mahaveer University, Moradabad, Uttar Pradesh, India,* [4]*School of Pharmacy, Suresh Gyan Vihar University, Jaipur, Rajasthan, India,* [5]*Faculty of Pharmacy, AIMST University, Bedong, Kedah, Malaysia,* [6]*Uttaranchal Institute of Pharmaceutical Sciences, Uttaranchal University, Dehradun, Uttarakhand, India,* [7]*School of Pharmaceutical Sciences, Lovely Professional University, Phagwara, Punjab, India,* [8]*ICFAI School of Pharmaceutical Sciences, The ICFAI University, Jaipur, Rajasthan, India,* [9]*Centre for Research Impact & Outcome, Chitkara College of Pharmacy, Chitkara University, Rajpura, Punjab, India,* [10]*Centre of Medical and Bio-Allied Health Sciences Research, Ajman University, Ajman, United Arab Emirates*

14.1 Introduction

Cardiovascular research is a global and interdisciplinary effort. Prominent research hubs in the United Kingdom, Europe, the United States, the Middle East, Asia, and Australia are leading the charge to enhance our understanding of this field. These institutions combine expertise in clinical cardiology, imaging science, genetics, computational biology, and public health to propel progress in cardiovascular studies (Lauer, 2012). The involvement of well-known organizations, such as the William Harvey Research Institute, Broad Institute, Massachusetts General Hospital, and Baker Heart and Diabetes Institute, is an indicator of how current research incorporates the most recent biomedical science using data to guide the study (Parvin et al., 2025). In addition to conventional medicine, computer science, occupational medicine, and space studies have contributed to the importance of environmental, occupational, and computational perspectives in understanding complex cardiovascular processes. The breadth of this method provides a strong scientific base for research on cardiovascular aging, which also provides the necessary ethical and regulatory rigor. This is done by ensuring a high level of compliance with the existing standards of human and animal research, patient consent policies, and international regulations, such as those promulgated by the FDA and EMA (Pavez Loriè et al., 2021). The specified commitments help increase the reliability, accountability, and translational relevance of the research. Simultaneously, collaborative projects are placed against a rapidly changing demographic environment. Within the 21st century, demographic aging has been the order of the day: by 2020, the number of persons aged 60 years and above had matched the number of children under five years of age, and in the coming years, the population aged 80 years and above is set to increase by almost threefold. Although this demographic change suggests more years of life, it is also causing acute problems for healthcare systems, economic systems, and societies as a whole (Ismail et al., 2021; Alharbi et al., 2022a,b). The key to resolving these issues is a healthy lifespan, whereby long years are spent in good health and with freedom of choice, instead of the burden of chronic disease.

Biologically, aging is a nonlinear cumulative damage to the body in terms of molecules and cells, resulting in a decline in the resiliency of physiological systems within the body. Compared to chronological age, biological aging differs significantly among individuals and is influenced by genes, life choices, and environmental exposure. Such heterogeneity is the reason why some people might have cardiovascular health at a very old age, while others suffer from early degradation (Dutta et al., 2024). Cardiovascular disease is the most prevalent cause of morbidity and mortality in the elderly, making it necessary to understand the mechanisms underlying age-related changes in the heart and vessels in more

Cellular Senescence and Cardiac Aging. DOI: https://doi.org/10.1016/B978-0-443-45811-8.00005-2

detail. Recent developments in imaging modalities and molecular diagnostics have made it easier to noninvasively evaluate the cardinal aspects of cardiovascular aging (vascular stiffness, impaired endothelial homeostasis, impaired mitochondrial activity, and low-grade systemic inflammation) (Wargocka-Matuszewska et al., 2023). These tools, in combination with computational modeling and large-scale data integration, create new opportunities to trace biological trajectories, stratify risk, and identify early intervention points. With the incorporation of mechanistic knowledge in the clinical setting, the combination of state-of-the-art imaging methods, biomarker analysis, and massive data analytics is set to transform the study, prevention, and treatment of cardiovascular aging (Loewe et al., 2025; Altamish et al., 2020; Babu et al., 2025). This background statement frames the aging of the cardiovascular system as both a scientific problem and a relevant social issue. Through international cooperation, technological advancement, and a focus on interventions that are more personalized, the field aims to not only increase the lifespan of the individual but also improve the quality of the overall aging process.

14.2 Understanding biological and healthful aging

14.2.1 The certainty of physiological senescence

The modern discussion on the imminence of senescence as a physiological phenomenon has advanced significantly. Conventional theoretical models propose that senescence is an irreversible, time-dependent process of dysfunctional progression. However, recent empirical studies have shown that biological senescence does not constitute an automatic operation of chronological age; rather, it is significantly regulated by a combination of determinants (Dodig et al., 2019). Cellular senescence, which was once accepted as an inevitable outcome of temporal development, has now become an extrinsic stressor with the potential to cause senescence, including oxidative insults and inflammation (Liao et al., 2021; Dahiya et al., 2025). Mitochondrial dysfunction is not only an amplifier of biological decay, but it is also a potential therapeutic target, which blurs the line between natural decay and opportunities that can be inhibited or salvaged (Tian et al., 2023).

Genetic and epigenetic influences also complicate the senescence phenotype. The role of heritable factors in lifespan extension, as observed in centenarian cohorts, along with an environmentally controlled set of epigenetic processes, explains how experience determinants influence the course of biological aging. The dualistic effect is argued to support the idea that senescence has a genetic basis that is also predetermined and flexible in the phenotype. Empirical evidence proves that cellular age can be reprogrammed with the help of chemical treatments, and lifestyle changes, including caloric intake and exercise, seem to extend health span (Morris et al., 2019). These findings challenge the belief that senescence is an inevitable process; rather, the concept can be modeled as a process that can be manipulated.

However, it is difficult to separate normative and pathologic senescence. Gradual deterioration in physiological functioning is expected and may be compared to sudden deterioration, which is more typical of the underlying pathology. The most crucial determinant is the rate of senescence, which is used as an index to distinguish between healthy and diseased pathways. This rate offers a methodological basis on which people who are at high risk are categorized, preemptive interventions are provided to prevent the manifestation of diseases, and strategies to extend the lifespan in optimum health are informed. These views, as a result, describe senescence as not a predetermined fate, but as something that can be changed depending on biological, environmental, and timely therapeutic factors.

14.2.2 Disease-driven modulation of aging patterns

Aging is still among the strongest risk factors for chronic disease and disability, but modern-day data show that the association does not merely exist due to chronological time but is also mediated by the rate of biological aging. Large population-based studies indicate that individuals with signs of faster biological aging (especially when accompanied by adverse lifestyle habits) are more vulnerable to cardiovascular disease, neoplasia, and premature death. This supports the idea that biological aging is a focal mechanistic pathway by which external exposures are converted into poor health (Bland, 2018). One of the key pathways of aging to disease is the sustained low-grade inflammatory environment, sometimes called inflammaging, along with the resultant immune dysfunction, called immunosenescence. Although these processes were previously considered harmful, they may have adaptive roles when maintained within the physiological limits. A controlled state of proinflammatory activity can be used as a protective response to enhance resistance to external insults and delay disease development (Fulop et al., 2018). However, the disruption of this balance activates the aging curbing process in their biological pathways, resulting in the earlier expression of chronic diseases. It is worth noting that there is a reciprocal, cyclical relationship between aging and pathology: biological senescence increases susceptibility to disease, and disease increases biological aging, so that a feedback loop occurs, becoming more risky as age advances

(Bachmann et al., 2020; Balaraman et al., 2025; Gupta et al., 2018). The rising toll of multimorbidity, which refers to the presence of multiple chronic diseases in an individual, is already based on this vicious cycle and is a significant risk factor for the health of the population. Aging has been acknowledged as a disease risk factor, which has highlighted the urgency of interventions that will have a parallel impact on biological aging processes as well as disease processes, which would break the vicious cycle and increase healthy lifespan.

14.3 Predicted age-related adaptations in the cardiovascular system

14.3.1 Cellular and molecular markers of cardiovascular senescence

Cardiovascular aging is a consequence of the convergence of basic processes that become less functional with age, impairing cardiac and vascular function. The main pointers, including impaired autophagy, disturbed proteostasis, genomic instability (including clonal hematopoiesis), and massive epigenetic changes, directly erode tissue integrity. These changes disrupt the homeostatic ability of the cardiovascular system to bear physiological loads (Pacinella et al., 2022). In contrast to these drivers, the antagonistic hallmarks are adaptive responses, which gradually become maladaptations. The main pathways are mitochondrial dysfunction, neurohormonal imbalance, and cellular senescence, the last of which is often enhanced by telomere loss. Initially, the perturbations play compensatory roles, but with the passage of time, they play a role in increased oxidative stress, inefficiencies, and increased inflammatory signaling. All these processes lead to the eventual failure of myocardial contractility, arterial elasticity, and endothelial resilience (Zheng et al., 2024; Bhat et al., 2024). These hallmarks are integrated to cause an accumulation of exponentially growing organ dysfunction. Old age makes the cardiovascular system more vulnerable to physiological stress factors, thus increasing the prevalence of age-related pathologies, including heart failure, arrhythmias, and vascular stiffening. These molecular insights provide a mechanistic explanation for why aging has become the strongest risk factor for cardiovascular disease and form the basis for future therapeutic approaches that may be used to slow or even reverse this gradual loss (Wang et al., 2024).

14.4 Senescence-associated remodeling patterns in cardiovascular structures

14.4.1 Cardiac myofibers

Myocardial remodeling is an age-related process mediated by changes at the cellular and tissue levels. Progressive senescence is also linked to fewer cardiomyocytes, compensated by hypertrophy, leading to a myocardial makeup of fewer but larger myocytes (Werbner et al., 2023). Increased fibroblast activity and extracellular matrix reorganization enable myocardial fibrosis and augment myocardial stiffness, which in turn reduces ventricular compliance and predisposes the myocardium to diastolic dysfunction, which is a prelude to age-associated heart failure (Xiao et al., 2025). Another detrimental impact is that senescent cardiomyocytes release paracrine mediators that trigger cell cycle arrest in neighboring cells and promote the spread of inflammatory signaling cascades. At the molecular level, a reduction in autophagic activity impairs the main quality control system of the myocardium, which involves the recycling of impaired organelles and macromolecules, thus triggering dysfunction of the cellular machinery and consequent cell death. Similarly, oxidative damage to proteins, nucleic acids, and lipids caused by the growth of reactive oxygen species undermines myocardial integrity (Peng et al., 2024).

Although the association between oxidative stress and cardiovascular pathology cannot be denied, the use of antioxidant interventions in clinical practice to reduce age-induced cardiac disease has produced mixed results. Other molecular determinants include telomere attrition and IGF-1 dysregulation, which cause cellular senescence and structural remodeling (Silva Menezes Júnior et al., 2024). Notably, these secondary changes are enhanced by gradual increases in afterload, which can be attributed to age-related changes in vascular and valvular stiffness. The aggregate changes in the histological and molecular remodeling of the myocardium weaken its functional reserve and increase the predisposition to heart failure, arrhythmias, and ischemic insult in old age.

14.4.2 Valves

Old age causes significant changes in the structural integrity and work capacity of cardiac valves, which are dependent on the synchronization of valve endothelial cells (VECs) and interstitial cells (VICs) to maintain diastolic and systolic competence of the heart. As they grow older, VECs show reduced proliferative ability, reduced nitric oxide production, and a loss of their ability to repair stretch-induced injury (Anstine et al., 2016). Simultaneously, oxidative stress intensifies due to the excess production of reactive oxygen species, which further weakens cell protection mechanisms. The subsequent loss of endothelial cell density interferes with intercellular communication, increases paracellular permeability, and weakens the interface between the endothelial and interstitial layers of the cornea. These perturbations in cellular status affect intercellular communication between VECs and VICs and trigger maladaptive remodeling reactions (Cooper et al., 2002).

Interstitial cells become more proliferative and migratory, thus favoring extracellular matrix remodeling with over-deposition of collagenous and elastic fibers. Dysfunctional VECs may induce VICs to deviate into a pro-calcific fate over time and create a pathogenic pathway leading to progressive valve calcification (Rodriguez and Masters, 2009). The overall effect is progressive thickening, stiffening, and degeneration of valve structures, which have a reduced capacity to repair. These pathophysiological mechanisms explain the high incidence of valvular heart disease in older adults. Aortic stenosis, mitral regurgitation, and aortic regurgitation are the most common clinical manifestations that constitute a significant proportion of cardiovascular morbidity in the elderly population.

14.4.3 Myocardial electrophysiological circuitry

With advancing age, the cardiac conduction system undergoes profound structural and functional remodeling, compromising its ability to maintain rhythmic control. In the sinoatrial (SA) node, age-related changes include increased interstitial fibrosis, reduced expression of the gap junction protein connexin-43, and diminished ionic currents, such as L- and T-type calcium currents and pacemaker If currents. These alterations collectively slow diastolic depolarization and action potential conduction across the node, which manifests as reduced intrinsic and maximum heart rates (Mesquita et al., 2025). A direct outcome of these changes is the increased prevalence of sick sinus syndrome in older adults. In the atrioventricular (AV) node, aging prolongs the effective refractory period, impairing conduction between the atria and ventricles. Structural degeneration extends further into the His-Purkinje system, where replacement fibrosis in the His bundle and left bundle branch, coupled with repolarization abnormalities in the Purkinje fibers, contribute to conduction block and arrhythmic risk (Andersen et al., 1998). The culmination of these age-related modifications is a markedly higher

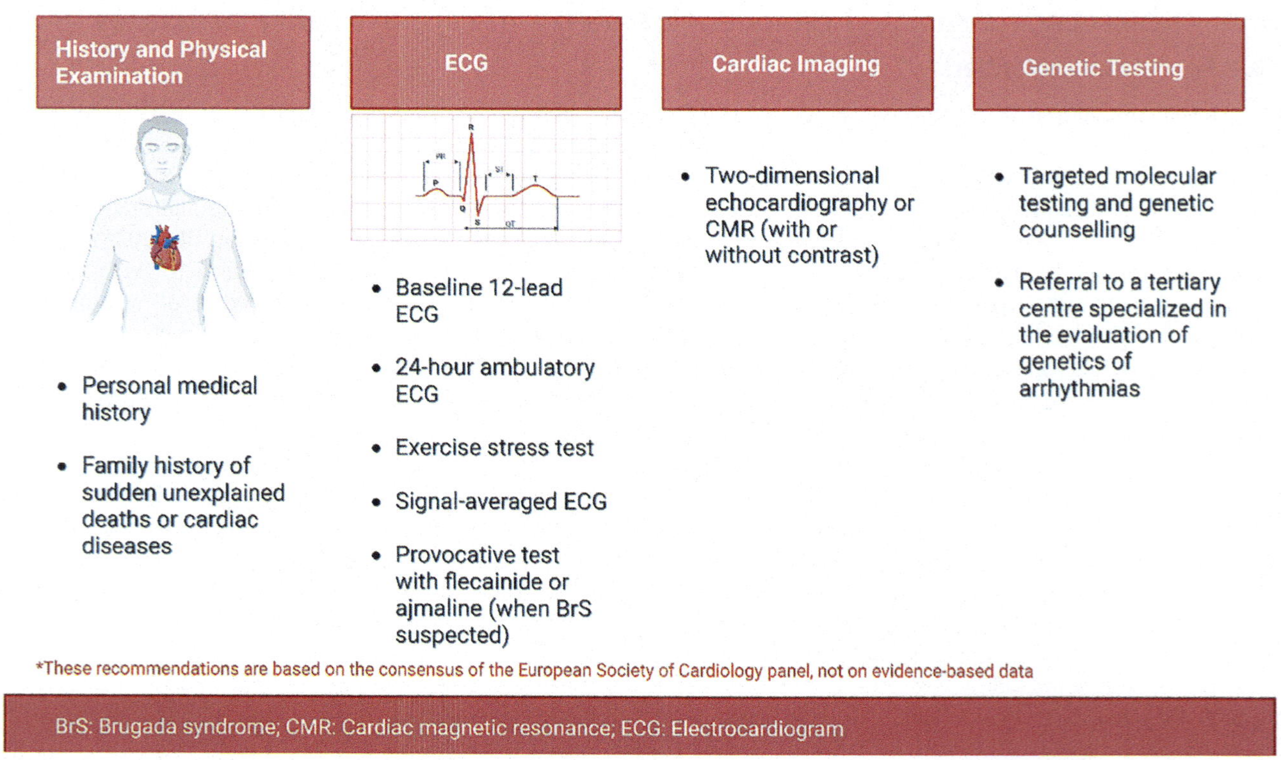

FIGURE 14.1 **Multimodal assessment of age-related cardiac conduction disorders.** Multimodal evaluation of age-related conduction disorders. Progressive remodeling of the SA/AV nodes and His-Purkinje system increases bradyarrhythmias and conduction blocks in older adults. The panel summarized a pragmatic assessment pathway: detailed history and examination (including familial sudden death), electrocardiography (baseline 12-lead, ambulatory/holter, exercise testing, signal-averaged ECG, and drug provocation when Brugada syndrome is suspected), cardiac imaging with echocardiography or CMR, and targeted genetic testing with counseling when inherited arrhythmia is possible. This integrated approach guides diagnosis, risk stratification, and pacemaker- or disease-specific therapy.

incidence of bradyarrhythmias and conduction disturbances in the elderly, as reflected in the substantially increased need for pacemaker implantation in aging populations (Fig. 14.1).

14.4.4 Angio-vascular system

Vascular aging is most evident in the arteries, where structural and functional decline progressively alter the hemodynamic parameters and increase disease susceptibility. In the aorta, age-related changes primarily involve the intima, leading to reduced elasticity and altered vessel diameter. These changes contribute to higher systolic and pulse pressures and diminished compliance. Arterial enlargement is common in muscular arteries, such as the carotid and femoral arteries. Medial hypertrophy is a major predictor of age-related increases in intima-media thickness (Xu et al., 2017; Singh et al., 2022, 2020). Two processes, namely arterial stiffening and endothelial dysfunction, underlie the degradation of vascular function. Disruption of elastin and deposition of fibrous tissues lead to stiffening of the arteries; however, this process occurs more in the central arteries than in the peripheral vessels. This stiffness increases the velocity of pulse waves and systolic blood pressure, whereas the previous reflection of the wave increases the central pressure and cardiac afterload (Cocciolone et al., 2018). The dampening of central relative to peripheral arterial pressure amplification adversely affects microvascular perfusion, predisposing tissues to ischemic harm. As a prominent example of biological aging, endothelial dysfunction is triggered by cardiovascular risk determinants, at least to some extent, but can also be caused by age-related, age-specific processes. In addition to the onset of atherogenesis, endothelial impairment alters vasomotor homeostasis, inhibits vascular proliferation, and predisposes the endothelium to thrombosis and inflammation (Young et al., 2021). These phenomena together explain the role of vascular remodeling in the age-related increase in cardiovascular morbidity.

14.5 Gender-dependent patterns of age-associated cardiac remodeling

Sex-based differences have a salient effect on cardiovascular remodeling development with senescence, which is a process controlled by endocrine conditions and particular exposure to cardiovascular risk factors. Male patients tend to have left ventricular hypertrophy with myocardial fibrosis, which presents as eccentric remodeling of the heart muscle (Kruszewska et al., 2022). Concentric hypertrophy, on the other hand, is more common in females and is a remodeling phenotype that is closely related to heart failure with preserved ejection fraction in the elderly population. Major sex-related differences can also be identified in vascular and valvular remodeling. Male participants always have a greater degree of arterial rigidity and rate of calcific aortic valve disease during their lifetime (Kaur and Lau, 2022). Women are especially vulnerable to mitral valve pathology and show a rapid increase in the rate of arterial stiffening after menopause, a process that has been directly linked to the dampening of estrogen-induced protective signals (Ryczkowska et al., 2023). Collectively, these data indicate that cardiovascular aging is not a homogenous process across sexes but a pattern of remodeling of a unique type. Defining sex-specific patterns is crucial for creating personalized diagnostic recommendations and preventive interventions designed to reduce the influence of sex-related risks in the aging population.

14.6 Biological age assessment through noninvasive tools

Reorganization of the cardiovascular system during aging has become a key issue in surveillance using noninvasive methodologies. The characteristics of structural and functional alterations defining physiological aging are becoming increasingly accurately defined through imaging modalities as the development of structural and functional degradation proceeds. An example is the use of echocardiography to identify variations in ventricular geometry, wall thickness, and cardiac function during diastole. Cardiac magnetic resonance imaging also provides a high-resolution evaluation of tissue composition, myocardial fibrosis, and remodeling of chambers (Pender et al., 2025; Saini et al., 2024). Vascular imaging techniques, such as carotid ultrasonography and computed tomography angiography, determine the stiffness of arteries, intima-media thickness, and calcification; thus, they are indicators of vascular aging at the system level (Stein et al., 2008). In addition to structural visualization, modern modalities, such as positron emission tomography (PET) and molecular imaging, enable an overall view of inflammatory processes, metabolic signatures, and microvascular integrity, thus making it possible to make a conceptual transition between anatomical and biologically inclined analyses. These methods go beyond conventional imaging, incorporating functional and metabolic values that frame anatomical findings. The modalities described above are supplementary devices to standard imaging that provide functional and metabolic data to complement and elaborate on the fixed anatomical measurements and, therefore, improve the clinical decision-making process and the quality of diagnosis (Phelps, 2000). All of these noninvasive modalities provide objective indices that

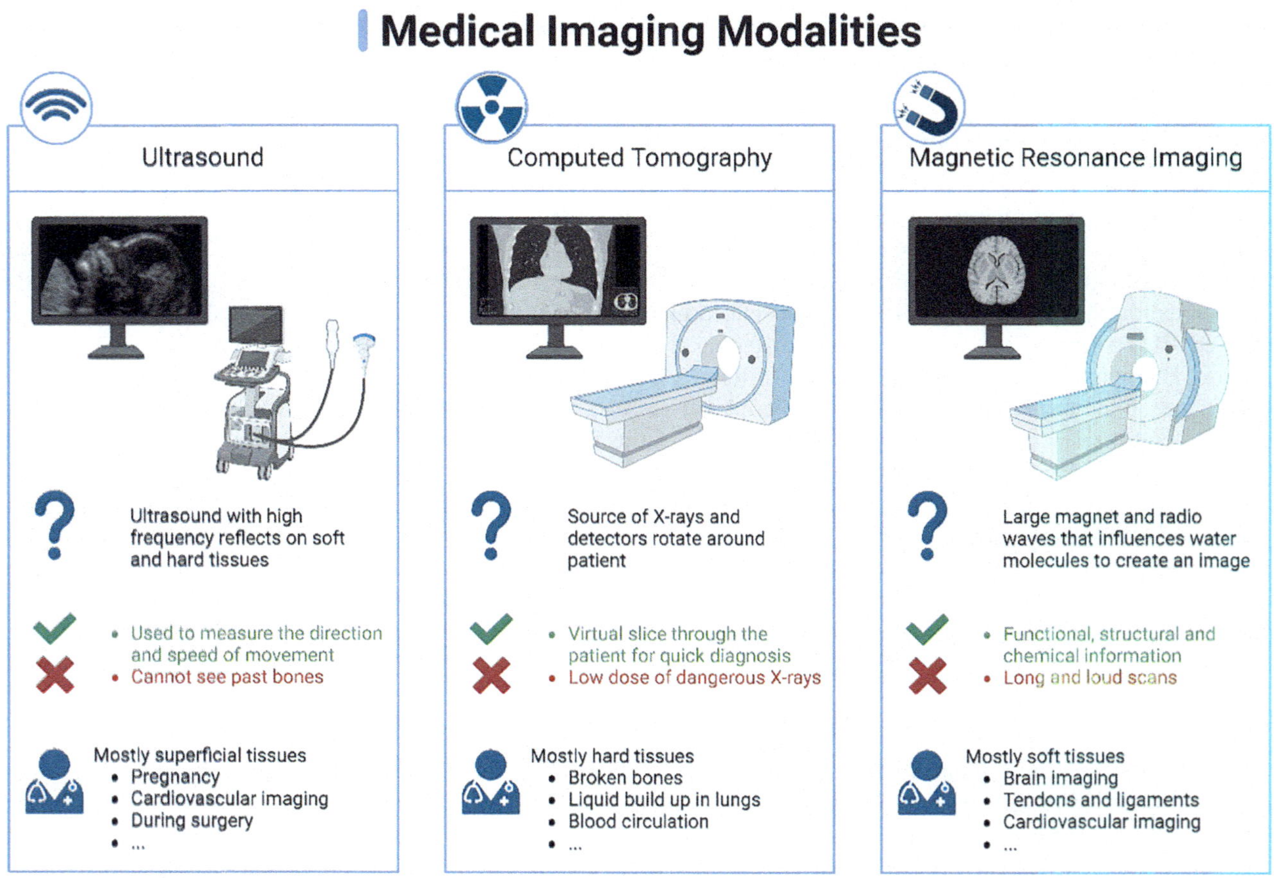

FIGURE 14.2 **Noninvasive imaging toolkit for cardiovascular aging.** Noninvasive imaging toolkit for tracking cardiovascular aging. Echocardiography (ultrasound) provides a bedside assessment of chamber geometry, wall thickness, diastolic filling, strain, and valvular function, which are central for detecting hypertrophy, volumetric remodeling, and regurgitation. Computed tomography (CT) offers high-resolution vascular and valvular anatomy, quantifies coronary/valvular calcium and plaque, and supports the angiographic evaluation of arterial stiffness and luminal narrowing. Cardiac magnetic resonance (CMR/MRI) integrates precise volumetrics with tissue characterization (late gadolinium enhancement, native T1/T2, and extracellular volume) to map diffuse fibrosis, myocardial remodeling, and flow dynamics. Together, these complementary modalities yield objective indices that distinguish physiological from pathological remodeling, refine risk stratification, and guide therapy across aging-related phenotypes (hypertrophy, fibrosis, chamber resizing, valvular disease, and vascular calcification).

define the course of cardiovascular senescence, distinguish between normative and pathological remodeling, and identify individuals at risk of accelerated decay (Fig. 14.2).

14.6.1 Hypertrophy

Left ventricular hypertrophy can be easily measured using echocardiography or cardiac magnetic resonance imaging; thus, it has become a central noninvasive measure in the study of age-related cardiovascular remodeling. Traditionally, LV hypertrophy was considered a harmless consequence of senescence, but longitudinal studies have indicated that the majority of hypertrophic deterioration is due to the accumulated risk factors of vascular disease in the elderly, but not the aging process itself (Frenzel et al., 2024). In contrast, patients with low-risk factors are those who retain a stable myocardial mass between early adulthood and advanced age, suggesting that pathological drivers rather than the process of aging itself are the major determinants of hypertrophic remodeling (Shih et al., 2011).

Physiological myocardial aging is defined by the loss of cardiomyocytes, which is compensated for by the hypertrophy of other surviving cells. This balance often maintains a steady or relatively low myocardial mass compared with extensive hypertrophic remodeling. However, this homeostatic progression may be abrogated by pathological conditions (He et al., 2022). A good example is transthyretin amyloidosis, a disease commonly observed in the geriatric population and associated with significant left-ventricular wall thickening and extracellular matrix proliferation, which can be easily detected using cardiac magnetic resonance imaging or amyloid-specific imaging agents (Porcari et al., 2022). It has been demonstrated that left ventricular hypertrophy is not an inevitable sign of aging. Instead, it is a combination of physiological old age, exposure to vascular risk factors, and disease-specific pathways. Therefore, the sound use of hypertrophy as a surrogate biomarker of cardiovascular aging is necessary.

14.6.2 Tissue fibrogenesis

As patients age, the number of cardiomyocytes gradually decreases, the extracellular matrix becomes more deposed, and myocardial fibrosis becomes more prominent, leading to the tightening of the ventricles and cardiac impairment. Cardiac remodeling is a characteristic feature of cardiac senescence and is inherently associated with diastolic dysfunction and high rates of heart failure in the aging population (Frantz et al., 2022). Cardiac magnetic resonance (CMR) imaging is a noninvasive imaging modality that can be used to study fibrotic changes in the myocardium. Such techniques include native T1 mapping and extracellular volume (ECV) quantification, which permit the identification of diffuse myocardial fibrosis even when cardiovascular disease is not present. Large population-based studies have shown that high myocardial T1 values have prognostic value and can identify patients at high risk of experiencing adverse cardiovascular events well before the development of clinical disease (Marques et al., 2022). Fibrosis does not involve the ventricles alone, and fibrotic remodeling of the left atrium with age has also been reported. Direct quantification remains difficult; however, surrogate markers such as atrial strain or functional parameters may be helpful (Karakasis et al., 2025). Collectively, these imaging methods underscore the idea that noninvasive fibrosis indicators can be used as early predictors of physiological aging and to rank cardiovascular risks throughout the lifespan.

14.6.3 Decrease of ventricular volumes

A decrease in ventricular volume in the elderly is a slow but steady process and a well-known age-related physiological change. Transthoracic echocardiography remains the main modality for assessing these volumetric changes; however, cardiac magnetic resonance imaging offers greater accuracy with better definition of endocardial boundaries and full coverage of the entire heart. Reductions in the sizes of the left ventricular chambers are accompanied by the maintenance of myocardial mass, thus ushering in a process of remodeling that predisposes to a more concentric myocardial architecture. This remodeling process occurs more strongly in women, whereby the maintenance of myocardial mass coupled with a decrease in cavity size intensifies the tendency of concentric left-ventricular adaptation (Haberka et al., 2024). The detected changes represent the most complicated interplay between structural remodeling and functional reserve and, as a result, can be used as a noninvasive biomarker of cardiovascular senescence to monitor longitudinally to differentiate between physiological and pathological remodeling.

14.6.4 Systolic and diastolic function

The effective work of the left ventricle is closely connected to the specific architecture of its myofibers, where the layers of spiral fibers produce rhythmic relaxation and contraction. With age, this finely tuned system undergoes measurable decline, particularly in diastolic function. Impaired myocardial relaxation, calcium handling abnormalities, increased fibrosis, and greater ventricular stiffness all contribute to the development of diastolic dysfunction (Sengupta et al., 2007). Echocardiography remains the most practical modality for detecting age-related changes using surrogate markers of left ventricular filling pressure and their hemodynamic effects. Assessing systolic function during aging is a complex process. Compensation between myofiber layers can mask early deficits, often resulting in preserved or even "supranormal" ejection fraction despite underlying dysfunction (Ganau et al., 2024). Although LVEF is widely used, it fails to capture subtle fiber-level alterations and may overestimate the function in older adults. Alternatives, such as the LV global function index, which adjusts for ventricular structure, offer a more accurate reflection of health and demonstrate linear associations with outcomes in large cohorts (Nwabuo et al., 2019).

More advanced techniques, particularly strain imaging, provide deeper insights into myocardial mechanics by quantifying fiber-specific deformations. Longitudinal strain has emerged as the most sensitive marker for early dysfunction, often declining before changes in LVEF become apparent. Speckle-tracking echocardiography and cardiac MRI-based approaches (such as tagging and feature tracking) extend this capability, although the variability across methods highlights the need for standardization (Echeverría et al., 2022). Similarly, myocardial torsion, which reflects the twisting and untwisting of the ventricle, shows promise as an early marker of cardiac aging but requires further validation (Zlibut et al., 2025). Together, these imaging tools reveal that while diastolic decline is a consistent hallmark of aging, the systolic reserve can remain deceptively preserved until later stages. The comprehensive use of strain and torsion imaging, alongside traditional indices, is crucial for tracking physiological aging with greater precision.

14.6.5 Left atrial volumes and function

Noninvasive imaging is an important method for explaining ontogenetic changes in the left atrial (LA) architecture and performance throughout life. The most common modalities used to measure atrial dimensions are transthoracic echocardiography and cardiac magnetic resonance imaging (CMR). These two methods constantly indicate age-related changes that increase the left atrial volume with a gradual decrease in functional indices. These data indicate a reduction in atrial adaptive capacity in accordance with changing hemodynamic needs (Kuchynka et al., 2015). The reservoir and conduit stages are among the functional elements that are most prone to dysfunction, hence limiting the blood storage and forward transfer ability of the atrium. Instead of using traditional volumetric evaluation, developing tools, especially atrial strain images, have higher sensitivity in the detection of early dysfunction (Gaynor et al., 2005). The measurement of strain not only outlines the existence of minor impairment of atrial performance but also provides information on diastolic left ventricular function that is directly related to atrial remodeling. Volumetric and strain-based measures of the left atrium are powerful, noninvasive cardiovascular senescence biomarkers that can be utilized in the future to improve risk stratification and preventive treatment (Samiei et al., 2021; Singh et al., 2019; Solanki et al., 2024).

14.6.6 Right atrial structure and function

Right atrial (RA) remodeling has recently been recognized as an age-related contributory factor to cardiovascular aging and arrhythmogenesis. One of the key structural changes involves a shift in the orientation of the caval veins: in younger individuals, the superior and inferior caval vein outlets face each other, whereas with aging, they progressively lateralize. This alteration disrupts normal hemodynamics and changes the RA inflow pattern (Wehrum et al., 2018). In youth, the right atrial blood flow typically forms a smooth clockwise helical rotation with minimal turbulence, promoting efficient filling. In contrast, in older adults, this organized rotation often disappears and is replaced by a turbulent flow. These hemodynamic disturbances are accompanied by measurable declines in the mean and peak systolic caval vein blood flow, reflecting impaired venous return and atrial function (Wehrum et al., 2018). Structurally, the RA undergoes gradual dilation with age, and its geometry becomes increasingly spherical. These morphological adaptations are particularly pronounced in individuals who progress from healthy aging to paroxysmal atrial fibrillation and eventually to persistent atrial fibrillation. Thus, the noninvasive assessment of the RA size, shape, and inflow dynamics provides valuable insights into physiological aging and early arrhythmogenic remodeling.

14.6.7 Valve structure and function

Valvular aging is a central component of cardiovascular remodeling, and noninvasive imaging systems, such as echocardiography, cardiac magnetic resonance (CMR), and cardiac computed tomography (CCT), are increasingly used to assess valvular aging. These methods provide comprehensive structural and functional data that, in turn, make it possible to identify degenerative changes that often occur prior to the onset of an overt clinical disease (Mathew et al., 2018). Progressive leaflet thickening, calcification, loss of leaflet mobility, changes in flow dynamics, and annular geometry changes are some of the most egregious imaging biomarkers of valvular aging. Echocardiography is the crucial modality for diagnosing early functional impairment, whereas cardiac magnetic resonance imaging (CMR) is used to supplement the examination by measuring the volume of regurgitant and tissue structure. Cardiac computed tomography (CCT), which is granted by its high spatial resolution, is particularly useful in defining the burden of calcification and pre-therapeutic planning of interventions, including transcatheter valve replacement (Shiota, 2017; Thapa et al., 2023). All of these imaging biomarkers support the progressive nature of valvular dysfunction with age. Multimodal imaging, which can differentiate the physiological aging process and pathological degeneration in older adults, is not only useful in the stratification of risk but also allows the timely initiation of treatment interventions.

14.6.8 Valvular thickening

Valve thickening is another typical feature of aging remodeling, and as time progresses, the valves lose their pliability and integrity. Noninvasive imaging modalities, particularly echocardiography, cardiac magnetic resonance (CMR), and cardiac computed tomography (CCT), are often used to detect and quantitatively assess the degree of thickening (Izquierdo-Gómez et al., 2017). The older a person is, the less elastic the valve wall, and thus, the fewer valves are capable of withstanding the hemodynamic forces on the valves during systole and diastole. This gradual hardening results in two significant functional sequelae. The former is valvular incompetence, which is defined as incomplete closure of the leaflets that allows retrograde blood flow. The latter is stenosis, which is caused by thickening and limited mobility of the leaflets, thus hindering forward flow. Both negative conditions damage cardiac performance and increase susceptibility to symptomatic valvular disease among the elderly. In turn, the early detection of patients at risk and the timely administration of interventions that would preserve cardiac function require meticulous imaging-based evaluation of valvular thickening (Czarny and Resar, 2014).

14.6.9 Calcification

Valvular calcification is one of the best-known attributes of cardiovascular aging. The gradual deposition of calcium on the leaflets of the valves results in a decrease in elasticity, thus retarding the normal opening and closing abilities.

This stiffening is closely associated with impaired function and leads to stenosis, deviant hemodynamics, and modified preload. Modern imaging technologies have helped identify and measure such changes accurately (Chen et al., 2019). Echocardiography can easily visualize the presence of calcific deposits and the potential resultant effects of this on the motion of the leaflets, but cardiac computed tomography (CCT) can provide a precise assessment of the calcium load (Faggiano et al., 2021). The Agatston scoring system, which is used to measure coronary calcium, has been extensively used to measure valvular disease as an objective measure of the level of calcification. Calcification is not only an indicator of senescence but also a prognostic indicator of the development of clinically important valvular pathology, making it a central noninvasive biomarker. Monitoring its course provides an overall evaluation of the severity of the disease, its prognostic value, and the chances of it being resolved therapeutically (Obisesan et al., 2021).

14.6.10 Flow changes

The aging process has a significant effect on valvular hemodynamics, as confirmed by structural changes, including leaflet thickening and fusion of the commissures, which result in measurable changes in flow dynamics. Aortic valve remodeling is most pronounced with age and is becoming increasingly associated with structural calcification and functional losses that impair exercise capacity (Nilsson, 2014). A major observation is that it develops turbulent oscillatory flow at the valve leaflet level. This abnormal flow pattern promotes inflammation and triggers valvular interstitial cells, accelerating the calcification process. The eccentricity of flow, which is measured by the magnitude of systolic movement, progressively increases with age in the ascending aorta. Such increments show a positive relationship with chronological age and act as independent predictors of decreased peak oxygen consumption, thereby acting as a measure of low exercise tolerance (Zhao et al., 2023). The consequences of these changes include increased energy loss in the circulatory system, which increases the inefficient nature of the aging cardiovascular system. Therefore, the study of flow dynamics explains not only the mechanistic basis of valvular pathology but also the functional connection between aging, inflammation, calcification, and progressive deterioration of cardiovascular performance.

14.6.11 Valvular stenosis

Thickening and calcification of the valve leaflets are progressive changes that occur in the cardiovascular system, making valvular stenosis an age-related pathology. These structural distortions limit the opening of the valve and prevent the forward flow of blood, which makes the heart work harder by producing increased pressure (Stewart et al., 2019). Consequently, the velocity of blood flowing through the constricted opening increases, hemodynamic turbulence occurs, and pressure differences that can be measured accurately appear. Noninvasive imaging is used to measure the severity of stenosis (Mynard et al., 2020). Echocardiography is the foundation of valve area and transvalvular gradient evaluation, whereas cardiovascular magnetic resonance (CMR) and cardiac computed tomography (CCT) are used to provide additional information on valve anatomy, calcium load, and dynamic flow patterns. In addition to the valve itself, CMR tissue mapping methods provide the opportunity to assess myocardial remodeling, such as hypertrophy and fibrosis, secondary to chronic pressure syndrome (Manzo et al., 2023). This combination of modalities not only defines the extent of valvular obstruction but also helps monitor the systemic consequences of valvular stenosis, which provides a complete mode of defining the effect of valvular stenosis in the environment of heart failure that occurs with age.

14.6.12 Valvular regurgitation

Cardiovascular aging often causes valvular regurgitation, which gradually impairs the type of fibrotic remodeling and subsequent cellular tissue density reduction. These changes in the structure of the leaflets weaken their integrity, making them more vulnerable to tether or rupture, and eventually leading to valve incompetence and retrograde blood flow

(Cawley and Otto, 2009). This regurgitation during this age causes chronic volume overload of the myocardium and aggravates remodeling processes that predispose individuals to heart failure. Diagnostic evaluation and longitudinal monitoring are mainly conducted using noninvasive imaging techniques. The issue of transthoracic echocardiography is the most widely used, as the most common method of measuring the extent of regurgitation, but the use of cardiovascular magnetic resonance (CMR) has also been shown to be a highly sensitive modality that can accurately determine the absolute regurgitant volume and its fraction (Chatrath and Papadakis, 2022). Notably, cardiac magnetic resonance imaging not only improves the diagnostic accuracy but also provides prognostic information, which proves that the degree of regurgitation is associated with unfavorable left ventricular remodeling and dysfunction (Hagendorff et al., 2024; Pfaller et al., 2021). These findings indicate that valvular regurgitation, which is age-dependent, is not only a structural but also a dynamic process with far-reaching hemodynamic and prognostic consequences. The diagnosis of this condition in aging populations is based on the latest imaging methods, which greatly contribute to early diagnosis and enhance long-term results.

14.6.13 Vasculature

Vascular aging is a complex process that can now be described in detail with the help of multimodal imaging. These techniques can identify the structural, geometric, and functional changes that occur with age; hence, they provide evidence of the actual processes that can result in arterial stiffening and remodeling (Ranard et al., 2023). The gradual reduction of elastin, along with increased deposition of collagen in arterial walls are powerful signs of vascular senescence, which is accompanied by a progressive loss of arterial elasticity. This augmented stiffness not only disrupts blood pressure control but also predisposes patients to a high cardiovascular risk (Ruiz et al., 2015). Various methods have been developed to measure arterial stiffness; however, their application in clinical practice is still limited by technical variability and the effect of arterial pressure (Shehab et al., 2019). The use of imaging modalities, especially cardiovascular magnetic resonance (CMR) and vascular ultrasound, is also considered an option. These methods can measure pulse-wave propagation in the aorta and carotid femoral arteries, and can use more modern methods, such as deformation imaging and shearwave elastography. All modalities have unique benefits: shear-wave imaging allows the determination of localized arterial stiffness at any point in the cardiac cycle, and distensibility-based measures cover the entire cardiac cycle (Vermes et al., 2022).

As a person ages, the use of vascular ultrasonography shows a progressive increase in the thickness of the intimal and medial layers, which is further amplified in patients with hypertension or atherosclerosis. The most commonly used surrogate marker of vascular aging is carotid intima-media thickness (Vinco et al., 2024). In addition to thickening, aging causes the build-up of atheromatous plaque and produces changes in the composition of the plaque, with older people having a higher number of calcified, brittle, or negatively remodeled plaques (Chen et al., 2023). Atherosclerotic plaque burden and morphology can be accurately evaluated using complex imaging tools, such as coronary computed tomography (CCT), intravascular ultrasound, and optical coherence tomography (Clemente et al., 2020). Vascular calcification is a critical process involved in aging. The amount of calcium deposition, which is measured by coronary artery calcium scoring, has a close relationship with plaque burden and cardiovascular events (Gui et al., 2024). In fact, calcium scoring is steadily becoming a measure of vascular age, providing prognostic information that is not dependent on chronological age. (Leopold, 2015; Quaglino et al., 2020). All these improvements in imaging prove that the process of vascular aging may now be tracked with increased accuracy, which preconditions better risk stratification and opens possibilities for earlier intervention to guarantee vascular health maintenance.

14.6.14 Electrocardiography

ECG is a straightforward but potent instrument for monitoring age-dependent modifications in cardiac conduction, most of which have been identified. However, these changes are not distributed equally, with their frequency and intensity varying across sexes and ethnicities, demonstrating the role of both biological and demographic factors in cardiovascular aging (Haas and Deserno, 2025). In the past several years, the introduction of artificial intelligence has enabled novel opportunities to utilize ECG as a biological aging marker. Deep learning algorithms currently predict chronological age directly using an ECG trace, and the difference between the predicted and actual ages is a powerful biomarker of physiological aging (Lv and Liu, 2025). Increased age, as predicted by ECG, compared to chronological age, is strongly associated with poor cardiovascular events, such as lower left ventricular ejection fraction, heart failure, hypertension, coronary artery disease, atrial fibrillation, stroke, and all-cause mortality (Pal et al., 2024). These suggestions indicate that AI-based ECG may be a noninvasive and highly accessible alternative to cardiac imaging. Combined with structural and functional measurements, it provides a feasible method for biological age estimation and those at high risk, and thus a promising instrument in preventive cardiology and population healthcare (Swetha et al., 2024) (Fig. 14.2).

Noninvasive imaging toolkit for tracking cardiovascular aging. Echocardiography (ultrasound) provides a bedside assessment of chamber geometry, wall thickness, diastolic filling, strain, and valvular function, which are central for detecting hypertrophy, volumetric remodeling, and regurgitation. Computed tomography (CT) offers high-resolution vascular and valvular anatomy, quantifies coronary/valvular calcium and plaque, and supports the angiographic evaluation of arterial stiffness and luminal narrowing. Cardiac magnetic resonance (CMR/MRI) integrates precise volumetrics with tissue characterization (late gadolinium enhancement, native T1/T2, and extracellular volume) to map diffuse fibrosis, myocardial remodeling, and flow dynamics. Together, these complementary modalities yield objective indices that distinguish physiological from pathological remodeling, refine risk stratification, and guide therapy across aging-related phenotypes (hypertrophy, fibrosis, chamber resizing, valvular disease, and vascular calcification).

14.7 The role of "big data" in understanding aging

14.7.1 Big data approaches for biological heart-age estimation

The emergence of a new generation of biobanks incorporating molecular profiling, imaging, and electronic health record data has made it possible to develop a new generation of big data approaches for biological aging. In cardiology, these resources are already being used to predict the biological age of the heart, a new health and resiliency concept that transcends chronological age (Arai et al., 2023). Machine learning models that predict chronological age using biomarkers, such as ECG signals and imaging measures, as well as biochemical and physiological characteristics, are typically used to estimate the biological heart age (Arsenault et al., 2024). The resulting age gap (biological age minus chronological age) can be used as an indicator of either rapid or durable cardiovascular aging. A positive age gap is a sign of an older-than-expected heart, which has been associated with increased risks of cardiovascular and metabolic diseases and mortality (Généreux et al., 2023).

The effectiveness of such a methodological solution is justified by the overall examination of CMR-produced features and ECG data of approximately 40,000 participants in the UK Biobank cohort. Disparities in cardiac measures using these models by age had a statistically significant correlation with genetic polymorphisms and exposure to environmental factors, with obesity and poor lipid profiles being the most common cardiovascular aging accelerators (Gumpangseth et al., 2024). In addition, similar models based on the use of electrocardiographic data alone have shown that even elementary and broadly available data can be powerful predictors of biological age. In addition to traditional biomarkers, more advanced imaging methods are widening the scope of cardiac age (Iakunchykova et al., 2023). The radiomics of CMR, which isolates texture and morphological changes in pixelated areas of routine cardiac imaging, has been proven to identify subtle temporal variations in ventricular structure and myocardial tissue properties (Otto et al., 2024). Radiometric cardiac age models can be used as predictors of intrinsic senescence and risk factors that can be modified, making them sensitive in detecting premature biological aging of the myocardium (Raisi-Estabragh et al., 2022; Salih et al., 2023). These big data plans explain why machine learning methodologies combined with multimodal biomarker profiles can potentially transform cardiovascular risk assessment. The measurement of biological cardiac age can provide clinicians with more accurate and personalized measurements that could guide preventive measures, preemptive curative measures, and longitudinal monitoring of cardiovascular disease (Vandenberk et al., 2022).

14.7.2 DNA methylation

DNA methylation has proven to be one of the strongest biomarkers of biological aging, forming the basis of epigenetic timekeeping systems. By applying machine learning to genome-wide methylation profiles, researchers can estimate the "epigenetic age" and quantify the rate of aging at the molecular level (Cao et al., 2021). When epigenetic age exceeds chronological age, it is consistently linked to adverse outcomes, including cancer, dementia, cardiovascular disease, and higher all-cause mortality rates. Importantly, lifestyle and health behaviors exert measurable influences on methylation-based aging trajectories (Desiderio et al., 2024). Poor diet quality and a higher body mass index are associated with greater age acceleration, whereas healthier dietary patterns, normal weight, and overall good cardiovascular health are associated with slower epigenetic aging and reduced cardiovascular mortality. This highlights the potential of interventions targeting lifestyle factors to favorably modify biological aging (Jain and Epstein, 2024). Recent studies have also demonstrated that subclinical atherosclerosis, even in individuals without overt cardiovascular disease, is strongly associated with accelerated epigenetic aging (Liu et al., 2023). This finding bridges molecular biology and imaging, underscoring the value of integrating methylation data with cardiovascular phenotyping to refine risk prediction (Lu et al., 2022).

Collectively, these insights show how big data approaches, particularly epigenetic profiling, can deepen our understanding of cardiovascular aging and open new pathways for personalized prevention strategies.

14.7.3 Telomere length

Telomeres, protective nucleoprotein caps at the ends of chromosomes, play a fundamental role in maintaining genomic stability during repeated cell divisions. With each somatic division, telomeres progressively shorten, eventually triggering cellular senescence and loss of regenerative capacity. This procedure determines telomere length as a broadly accepted biomarker of biological aging (Argentieri et al., 2024). Leukocyte telomere length (LTL) has since become a convenient surrogate endpoint of telomere dynamics in the system, as it is correlated with telomere length in a variety of tissues. In addition, the accumulated effects of genetic determinants and environmental impact (such as oxidative stress and inflammatory events) on LTL shortening highlight the role of not only chronological aging but also the effects these elements have on telomere erosion (Aung et al., 2023). It is important to note that abbreviated leukocyte telomere length (LTL) has been linked to a high risk of coronary artery disease, heart failure, and harmful left ventricular remodeling in large-scale population studies that utilized biobank and genomic data. In addition, Mendelian randomization studies support a causal relationship between telomere attrition and atherogenesis, thus supporting the contributory role of telomeres in the pathogenesis of cardiovascular diseases (Burren et al., 2024). By adding telomere length to the data analytics of large populations, researchers can better measure the biological age of populations, identify the people most at risk, and understand the genetic and environmental factors of cardiovascular senescence (Butts et al., 2024). This highlights LTL as a geriatric biomarker and a potentially unmet point of access for precision medicine initiatives for the risk stratification of cardiovascular disease.

14.7.4 Clonal hematopoiesis of indeterminate potential

Clonal hematopoiesis of indeterminate potential (CHIP) has emerged as a striking example of how large-scale genomic data can reshape our understanding of aging and cardiovascular risk. Population-level sequencing studies have shown that approximately 1 in 10 individuals over the age of 70 years harbor subclinical preleukemic mutations in circulating blood cells, despite showing no overt signs of hematologic malignancy. These mutations commonly affect genes that regulate epigenetic processes, such as DNMT3A, TET2, and ASXL1 (Gumuser et al., 2023). Although originally linked to leukemia risk, CHIP is now recognized as an independent driver of cardiovascular disease. Observational and mechanistic studies have demonstrated a strong association between atherosclerotic cardiovascular disease and heart failure, with the magnitude of risk varying according to the specific mutated gene (Lin et al., 2024). The underlying mechanisms appear to involve chronic inflammatory activation and enhanced cardiac fibrosis, which together accelerate vascular and myocardial aging (Marnell et al., 2021; Niroula et al., 2021). By uncovering these hidden genomic patterns, big data approaches have revealed how somatic mutations substantially influence cardiovascular health trajectories in older adults (Sikking et al., 2023). This recognition highlights CHIP as both a biomarker of biological aging and a potential therapeutic target for reducing age-related cardiovascular risk.

14.7.5 Proteomics and omics

Proteomics and metabolomics provide powerful insights into the translation of genetic information into functional phenotypes, offering unique insights into the biology of aging. The aging proteome is shaped by multiple processes, including altered protein translation, accumulation of post-translational modifications, aggregation, and impaired degradation. Together, these shifts contribute to functional decline in tissues such as the heart and vasculature (Bhatia et al., 2022). Importantly, key signaling pathways, including insulin-like growth factor regulation and hypoxia-inducible factor signaling, show age-dependent changes, making them potential biomarkers for tracking biological aging in humans. Despite these advances, cardiovascular-specific proteomic signatures remain relatively unexplored (Ghazal et al., 2025). A more precise understanding of the proteomic changes linked to cardiovascular disease could identify early markers of dysfunction and provide therapeutic entry points (Gu et al., 2022). This field has already begun to inform the development of novel interventions. The development of senolytic therapies, which selectively clear senescent cells, is a promising avenue for slowing or reversing biological aging (Oh et al., 2025). Similarly, improvements in amyloid typing have demonstrated how proteomic analysis can directly guide the development of targeted therapies (Oh et al., 2023). By mapping the molecular fingerprints of aging, proteomics, and related omics technologies, researchers have the potential to not only refine diagnostics but also accelerate personalized interventions, paving the way for therapies that extend the health span while reducing the cardiovascular burden in older populations.

14.8 Conclusions and future work

Promoting healthy cardiovascular aging has emerged as a pressing priority in global public health. Biological aging, with its systemic molecular and cellular changes, drives the gradual loss of function and predisposes individuals to a spectrum of age-related diseases. Within the cardiovascular system, these alterations manifest in predictable yet variable ways that can be tracked using advanced imaging techniques and noninvasive biomarkers. Such tools not only enable the monitoring of physiological aging but also help identify deviations that may signal early disease or accelerated decline. The integration of cardiovascular imaging with biomarker-derived estimates of "biological heart age" offers a promising framework for understanding the aging trajectory. These approaches allow researchers to disentangle the roles of genetic predisposition, environmental exposure, and lifestyle factors in shaping cardiovascular health across the lifespan. However, the predictive models vary depending on the population studied, the markers chosen, and the statistical methods applied.

Looking forward, one of the key challenges will be determining whether biological age metrics, such as "delta age," provide incremental clinical value beyond chronological age and traditional risk factors. As larger high-quality datasets become available and computational tools continue to evolve, this field is likely to expand rapidly in the future. Multimodal cardiovascular imaging will remain central to these efforts, advancing mechanistic insights and supporting the development of individualized and population-based strategies aimed at extending the health span. By bridging biology, technology, and clinical translation, future research has the potential to reshape our understanding of cardiovascular aging, its monitoring, and its management.

References

Alharbi, K.S., Almalki, W.H., Alzarea, S.I., Kazmi, I., Al-Abbasi, F.A., Afzal, O., et al., 2022a. A narrative review on the biology of piezo1 with platelet-rich plasma in cardiac cell regeneration. Chemico-Biological Interactions 363, 110011. Available at https://doi.org/10.1016/j.cbi.2022.110011.

Alharbi, K.S., Singh, Y., Afzal, O., Alfawaz Altamimi, A.S., Kazmi, I., Al-Abbasi, F.A., et al., 2022b. Molecular explanation of Wnt/βcatenin antagonist pyrvinium mediated calcium equilibrium changes in aging cardiovascular disorders. Molecular Biology Reports 49 (11), 11101–11111. Available at https://doi.org/10.1007/s11033-022-07863-7.

Altamish, M., Samuel, V.P., Dahiya, R., Singh, Y., Deb, P.K., Bakshi, H.A., et al., 2020. Molecular signaling of G-protein-coupled receptor in chronic heart failure and associated complications. Drug Development Research 81 (1), 23–31. Available at https://doi.org/10.1002/ddr.21627.

Andersen, H.R., Nielsen, J.C., Thomsen, P.E.B., Thuesen, L., Vesterlund, T., Pedersen, A.K., et al., 1998. Atrioventricular conduction during long-term follow-up of patients with sick sinus syndrome. Circulation 98 (13), 1315–1321. Available at https://doi.org/10.1161/01.CIR.98.13.1315.

Anstine, L.J., Bobba, C., Ghadiali, S., Lincoln, J., 2016. Growth and maturation of heart valves leads to changes in endothelial cell distribution, impaired function, decreased metabolism and reduced cell proliferation. Journal of Molecular and Cellular Cardiology 100, 72–82. Available at https://doi.org/10.1016/j.yjmcc.2016.10.006.

Arai, H., Kawakubo, M., Kadokami, T., 2023. Editorial for "Image-Based Biological Heart Age Estimation Reveals Differential Aging Patterns Across Cardiac Chambers" Journal of Magnetic Resonance Imaging 58 (6), 1813–1814. Available at https://doi.org/10.1002/jmri.28687.

Argentieri, M.A., Xiao, S., Bennett, D., Winchester, L., Nevado-Holgado, A.J., Ghose, U., et al., 2024. Proteomic aging clock predicts mortality and risk of common age-related diseases in diverse populations. Nature Medicine 30 (9), 2450–2460. Available at https://doi.org/10.1038/s41591-024-03164-7.

Arsenault, B.J., Loganath, K., Girard, A., Botezatu, S., Zheng, K.H., Tzolos, E., et al., 2024. Lipoprotein(a) and calcific aortic valve stenosis progression: a systematic review and meta-analysis. JAMA Cardiology 9 (9), 835–842. Available at https://doi.org/10.1001/jamacardio.2024.1882.

Aung, N., Wang, Q., van Duijvenboden, S., Burns, R., Stoma, S., Raisi-Estabragh, Z., et al., 2023. Association of longer leukocyte telomere length with cardiac size, function, and heart failure. JAMA Cardiology 8 (9), 808. Available at https://doi.org/10.1001/jamacardio.2023.2167.

Babu, M.A., Jyothi S, R., Kaur, I., Kumar, S., Sharma, N., Kumar, M.R., et al., 2025. The role of GATA4 in mesenchymal stem cell senescence: a new frontier in regenerative medicine. Regenerative Therapy 28, 214–226. Available at https://doi.org/10.1016/j.reth.2024.11.017.

Bachmann, M.C., Bellalta, S., Basoalto, R., Gómez-Valenzuela, F., Jalil, Y., Lépez, M., et al., 2020. The challenge by multiple environmental and biological factors induce inflammation in aging: their role in the promotion of chronic disease. Frontiers in Immunology 11. Available at https://doi.org/10.3389/fimmu.2020.570083.

Balaraman, A.K., Altamimi, A.S.A., Babu, M.A., Goyal, K., PadmaPriya, G., Bansal, P., et al., 2025. The interplay of senescence and MMPs in myocardial infarction: implications for cardiac aging and therapeutics. Biogerontology 26 (1). Available at https://doi.org/10.1007/s10522-025-10190-6.

Bhat, A.A., Moglad, E., Afzal, M., Thapa, R., Almalki, W.H., Kazmi, I., et al., 2024. Therapeutic approaches targeting aging and cellular senescence in Huntington's disease. CNS Neuroscience and Therapeutics 30 (10). Available at https://doi.org/10.1111/cns.70053.

Bhatia, H.S., Brunner, A.D., Öztürk, F., Kapoor, S., Rong, Z., Mai, H., et al., 2022. Spatial proteomics in three-dimensional intact specimens. Cell 185 (26), 5040. Available at https://doi.org/10.1016/j.cell.2022.11.021.

Bland, J.S., 2018. Age as a modifiable risk factor for chronic disease. Integrative Medicine (Boulder) 17 (4), 16–19.

Burren, O.S., Dhindsa, R.S., Deevi, S.V.V., Wen, S., Nag, A., Mitchell, J., et al., 2024. Genetic architecture of telomere length in 462,666 UK Biobank whole-genome sequences. Nature Genetics 56 (9), 1832–1840. Available at https://doi.org/10.1038/s41588-024-01884-7.

Butts, B., Hope, C., Herring, C., Mueller, K., Gary, R.A., 2024. The effects of exercise on telomere length in persons with heart failure. Journal of Cardiovascular Nursing 39 (3), E86. Available at https://doi.org/10.1097/jcn.0000000000001044.

Cao, J., Wu, Q., Huang, Y., Wang, L., Su, Z., Ye, H., 2021. The role of DNA methylation in syndromic and non-syndromic congenital heart disease. Clinical Epigenetics 13 (1). Available at https://doi.org/10.1186/s13148-021-01077-7.

Cawley, P.J., Otto, C.M., 2009. Valvular regurgitation: does cardiovascular magnetic resonance provide additional information compared to echocardiography? Minerva Cardioangiologica 57 (4), 521–535.

Chatrath, N., Papadakis, M., 2022. Physical activity and exercise recommendations for patients with valvular heart disease. Heart (British Cardiac Society) 108 (24), 1938–1944. Available at https://doi.org/10.1136/heartjnl-2021-319824.

Chen, H.Y., Engert, J.C., Thanassoulis, G., 2019. Risk factors for valvular calcification. Lippincott Williams and Wilkins, Canada Current Opinion in Endocrinology, Diabetes and Obesity 26 (2), 96–102. Available at https://doi.org/10.1097/MED.0000000000000471.

Chen, Z., Sun, X., Li, X., Liu, N., 2023. Oleoylethanolamide alleviates hyperlipidaemia-mediated vascular calcification via attenuating mitochondrial DNA stress triggered autophagy-dependent ferroptosis by activating PPARα Biochemical Pharmacology 208, 115379. Available at https://doi.org/10.1016/j.bcp.2022.115379.

Clemente, A., Traghella, I., Mazzone, A., Sbrana, S., Vassalle, C., 2020. Vascular and valvular calcification biomarkers. Advances in Clinical Chemistry 95, 73–103. Available at https://doi.org/10.1016/bs.acc.2019.08.002.

Cocciolone, A.J., Hawes, J.Z., Staiculescu, M.C., Johnson, E.O., Murshed, M., Wagenseil, J.E., 2018. Elastin, arterial mechanics, and cardiovascular disease. American Journal of Physiology - Heart and Circulatory Physiology 315 (2), H189. Available at https://doi.org/10.1152/ajpheart.00087.2018.

Cooper, D., Stokes, K.Y., Tailor, A., Granger, D.N., 2002. Oxidative stress promotes blood cell-endothelial cell interactions in the microcirculation. Cardiovascular Toxicology 2 (3), 165–180. Available at https://doi.org/10.1007/s12012-002-0002-7.

Czarny, Resar, 2014. Diagnosis and management of valvular aortic stenosis. Clinical Medicine Insights. Cardiology. 8 (Suppl 1), 15–24.

Dahiya, R., et al., 2025. Cellular senescence in multiple sclerosis (MS). Cellular Senescence and Brain Aging: Mechanisms, Therapeutic Strategies, and Implications for Neurodegenerative Diseases. Springer, pp. 207–225.

Desiderio, A., Pastorino, M., Campitelli, M., Longo, M., Miele, C., Napoli, R., et al., 2024. DNA methylation in cardiovascular disease and heart failure: novel prediction models? Clinical Epigenetics 16 (1). Available at https://doi.org/10.1186/s13148-024-01722-x.

Dodig, S., Čepelak, I., Pavić, I., 2019. Hallmarks of senescence and aging. Biochem Medica (Zagreb) 29 (3), 030501. Available at 10.11613/BM.2019.030501.

Dutta, S., Goodrich, J.M., Dolinoy, D.C., Ruden, D.M., 2024. Biological aging acceleration due to environmental exposures: an exciting new direction in toxicogenomics research. Genes 15 (1), 16. Available at https://doi.org/10.3390/genes15010016.

Echeverría, L.E., Rojas, L.Z., Rueda-Ochoa, O.L., Gómez-Ochoa, S.A., Mayer, M.A., Becerra-Motta, L.P., et al., 2022. Longitudinal strain by speckle tracking and echocardiographic parameters as predictors of adverse cardiovascular outcomes in chronic Chagas cardiomyopathy. International Journal of Cardiovascular Imaging 38 (6), 1245–1255. Available at https://doi.org/10.1007/s10554-021-02508-5.

Faggiano, A., Santangelo, G., Carugo, S., Pressman, G., Picano, E., Faggiano, P., 2021. Cardiovascular calcification as a marker of increased cardiovascular risk and a surrogate for subclinical atherosclerosis: role of echocardiography. Journal of Clinical Medicine 10 (8), 1668. Available at https://doi.org/10.3390/jcm10081668.

Frantz, S., Hundertmark, M.J., Schulz-Menger, J., Bengel, F.M., Bauersachs, J., 2022. Left ventricular remodelling post-myocardial infarction: pathophysiology, imaging, and novel therapies. European Heart Journal 43 (27), 2549–2561. Available at https://doi.org/10.1093/eurheartj/ehac223.

Frenzel, S., Bülow, R., Dörr, M., Felix, S.B., Friedrich, N., Völzke, H., et al., 2024. Left ventricular hypertrophy as a risk factor for accelerated brain aging: results from the study of health in pomerania. Human Brain Mapping 45 (3). Available at https://doi.org/10.1002/hbm.26567.

Fulop, T., Larbi, A., Dupuis, G., Le Page, A., Frost, E.H., Cohen, A.A., et al., 2018. Immunosenescence and inflamm-aging as two sides of the same coin: friends or foes? Frontiers in Immunology 8. Available at https://doi.org/10.3389/fimmu.2017.01960.

Ganau, A., Orrù, M., Floris, M., Saba, P.S., Loi, F., Sanna, G.D., et al., 2024. Echocardiographic heart ageing patterns predict cardiovascular and non-cardiovascular events and reflect biological age: the SardiNIA study. European Journal of Preventive Cardiology 31 (6), 677–685. Available at https://doi.org/10.1093/eurjpc/zwad254.

Gaynor, S.L., Maniar, H.S., Prasad, S.M., Steendijk, P., Moon, M.R., 2005. Reservoir and conduit function of right atrium: Impact on right ventricular filling and cardiac output. American Journal of Physiology - Heart and Circulatory Physiology 288 (5), H2140. Available at https://doi.org/10.1152/ajpheart.00566.2004.

Généreux, P., Sharma, R.P., Cubeddu, R.J., Aaron, L., Abdelfattah, O.M., Koulogiannis, K.P., et al., 2023. The mortality burden of untreated aortic stenosis. Journal of the American College of Cardiology 82 (22), 2101–2109. Available at https://doi.org/10.1016/j.jacc.2023.09.796.

Ghazal, R., Wang, M., Liu, D., Tschumperlin, D.J., Pereira, N.L., 2025. Cardiac fibrosis in the multi-omics era: implications for heart failure. Circulation Research 136 (7), 773–802. Available at https://doi.org/10.1161/circresaha.124.325402.

Gu, Y., Zhou, Y., Ju, S., Liu, X., Zhang, Z., Guo, J., et al., 2022. Multi-omics profiling visualizes dynamics of cardiac development and functions. Cell Reports 41 (13), 111891. Available at https://doi.org/10.1016/j.celrep.2022.111891.

Gui, Z., Shao, C., Zhan, Y., Wang, Z., Li, L., 2024. Vascular calcification: high incidence sites, distribution, and detection. Cardiovascular Pathology 72, 107667. Available at https://doi.org/10.1016/j.carpath.2024.107667.

Gumpangseth, T., Komutrattananont, P., Palee, P., Lekawanvijit, S., Kanchai, C., Prasitwattanaseree, S., et al., 2024. Collagen and elastic fibers assessment of the human heart valves for age estimation in Thais using image analysis. Forensic Science, Medicine and Pathology 20 (3), 920–932. Available at https://doi.org/10.1007/s12024-023-00775-3.

Gumuser, E.D., Schuermans, A., Cho, S.M.J., Sporn, Z.A., Uddin, M.M., Paruchuri, K., et al., 2023. Clonal hematopoiesis of indeterminate potential predicts adverse outcomes in patients with atherosclerotic cardiovascular disease. Journal of the American College of Cardiology 81 (20), 1996–2009. Available at https://doi.org/10.1016/j.jacc.2023.03.401.

Gupta, G., De Jesus Andreoli Pinto, T., Chellappan, D.K., Mishra, A., Malipeddi, H., Dua, K., 2018. A clinical update on metformin and lung cancer in diabetic patients. Panminerva Medica 60 (2), 70–75. Available at https://doi.org/10.23736/S0031-0808.18.03394-3.

Haas, P., Deserno, T.M., 2025. Signal quality of smart shirts recording electrocardiography. Studies in Health Technology and Informatics 329, 1920–1921. Available at https://doi.org/10.3233/SHTI251279.

Haberka, M., Starzak, M., Smolka, G., Wojakowski, W., Gąsior, Z., 2024. Echocardiography and cardiac magnetic resonance in the assessment of left-ventricle remodeling: differences implying clinical decision. Journal of Clinical Medicine 13 (6), 1620. Available at https://doi.org/10.3390/jcm13061620.

Hagendorff, A., Helfen, A., Brandt, R., Knebel, F., Altiok, E., Ewers, A., et al., 2024. Expert proposal to analyze the combination of aortic and mitral regurgitation in multiple valvular heart disease by comprehensive echocardiography. Clinical Research in Cardiology 113 (3), 393–411. Available at https://doi.org/10.1007/s00392-023-02227-y.

He, X., Du, T., Long, T., Liao, X., Dong, Y., Huang, Z.-P., 2022. Signaling cascades in the failing heart and emerging therapeutic strategies. Signal Transduction and Targeted Therapy 7 (1). Available at https://doi.org/10.1038/s41392-022-00972-6.

Iakunchykova, O., Schirmer, H., Vangberg, T., Wang, Y., Benavente, E.D., van Es, R., et al., 2023. Machine-learning-derived heart and brain age are independently associated with cognition. European Journal of Neurology 30 (9), 2611–2619. Available at https://doi.org/10.1111/ene.15902.

Ismail, Z., Ahmad, W.I.W., Hamjah, S.H., Astina, I.K., 2021. The impact of population ageing: a review. Iranian Journal of Public Health 50 (12), 2451–2460.

Izquierdo-Gómez, M.M., Hernández-Betancor, I., García-Niebla, J., Marí-López, B., Laynez-Cerdeña, I., Lacalzada-Almeida, J., 2017. Valve calcification in aortic stenosis: etiology and diagnostic imaging techniques. BioMed Research International 2017. Available at https://doi.org/10.1155/2017/5178631.

Jain, R., Epstein, J.A., 2024. Epigenetics. In: Rickert-Sperling, S., Kelly, R.G., Haas, N. (Eds.), Congenital Heart Diseases: The Broken Heart. Advances in Experimental Medicine and Biology, vol 1441. Springer, Cham, pp. 341–364. Available at https://doi.org/10.1007/978-3-031-44087-8_18.

Karakasis, P., Theofilis, P., Vlachakis, P.K., Korantzopoulos, P., Patoulias, D., Antoniadis, A.P., et al., 2025. Atrial fibrosis in atrial fibrillation: mechanistic insights, diagnostic challenges, and emerging therapeutic targets. International Journal of Molecular Sciences 26 (1), 209. Available at https://doi.org/10.3390/ijms26010209.

Kaur, G., Lau, E., 2022. Sex differences in heart failure with preserved ejection fraction: From traditional risk factors to sex-specific risk factors. Women's Health 18. Available at https://doi.org/10.1177/17455057221140209.

Kruszewska, J., Cudnoch-Jedrzejewska, A., Czarzasta, K., 2022. Remodeling and fibrosis of the cardiac muscle in the course of obesity—pathogenesis and involvement of the extracellular matrix. International Journal of Molecular Sciences 23 (8), 4195. Available at https://doi.org/10.3390/ijms23084195.

Kuchynka, P., Podzimkova, J., Masek, M., Lambert, L., Cerny, V., Danek, B., et al., 2015. The role of magnetic resonance imaging and cardiac computed tomography in the assessment of left atrial anatomy, size, and function. BioMed Research International 2015, 1–13. Available at https://doi.org/10.1155/2015/247865.

Lauer, M.S., 2012. Advancing cardiovascular research. Chest 141 (2), 500–505. Available at https://doi.org/10.1378/chest.11-2521.

Leopold, J.A., 2015. Vascular calcification: mechanisms of vascular smooth muscle cell calcification. Trends in Cardiovascular Medicine 25 (4), 267–274. Available at https://doi.org/10.1016/j.tcm.2014.10.021.

Liao, Z., Yeo, H.L., Wong, S.W., Zhao, Y., 2021. Cellular senescence: mechanisms and therapeutic potential. Biomedicines 9 (12), 1769. Available at https://doi.org/10.3390/biomedicines9121769.

Lin, A.E., Bapat, A.C., Xiao, L., Niroula, A., Ye, J., Wong, W.J., et al., 2024. Clonal hematopoiesis of indeterminate potential with loss of Tet2 enhances risk for atrial fibrillation through Nlrp3 inflammasome activation. Circulation 149 (18), 1419–1434. Available at https://doi.org/10.1161/CIRCULATIONAHA.123.065597.

Liu, J., Wu, Y., Sun, H., Liu, X., Gu, X., Zhang, Y., et al., 2023. Association between placental DNA methylation and fetal congenital heart disease. Molecular Genetics and Genomics 298 (1), 243–251. Available at https://doi.org/10.1007/s00438-022-01944-9.

Loewe, A., Hunter, P.J., Kohl, P., 2025. Computational modelling of biological systems now and then: revisiting tools and visions from the beginning of the century. Philosophical Transactions of the Royal Society A: Mathematical, Physical and Engineering Sciences 383 (2296). Available at https://doi.org/10.1098/rsta.2023.0384.

Lu, A.T., Binder, A.M., Zhang, J., Yan, Q., Reiner, A.P., Cox, S.R., et al., 2022. DNA methylation GrimAge version 2. Aging 14 (23), 9484–9549. Available at https://doi.org/10.18632/aging.204434.

Lv, H., Liu, M., 2025. Overestimated myocardial ischemia. Annals of Noninvasive Electrocardiology: The Official Journal of the International Society for Holter and Noninvasive Electrocardiology, Inc 30 (3), e70064.

Manzo, R., Ilardi, F., Nappa, D., Mariani, A., Angellotti, D., Immobile Molaro, M., et al., 2023. Echocardiographic evaluation of aortic stenosis: a comprehensive review. Diagnostics 13 (15), 2527. Available at https://doi.org/10.3390/diagnostics13152527.

Marnell, C.S., Bick, A., Natarajan, P., 2021. Clonal hematopoiesis of indeterminate potential (CHIP): linking somatic mutations, hematopoiesis, chronic inflammation and cardiovascular disease. Journal of Molecular and Cellular Cardiology 161, 98–105. Available at https://doi.org/10.1016/j.yjmcc.2021.07.004.

Marques, M.D., Weinberg, R., Kapoor, S., Ostovaneh, M.R., Kato, Y., Liu, C.Y., et al., 2022. Myocardial fibrosis by T1 mapping magnetic resonance imaging predicts incident cardiovascular events and all-cause mortality: the multi-ethnic study of atherosclerosis. European Heart Journal Cardiovascular Imaging 23 (10), 1407–1416. Available at https://doi.org/10.1093/ehjci/jeac010.

Mathew, R.C., Löffler, A.I., Salerno, M., 2018. Role of cardiac magnetic resonance imaging in valvular heart disease: diagnosis, assessment, and management. Current Cardiology Reports 20 (11). Available at https://doi.org/10.1007/s11886-018-1057-9.

Silva Menezes Júnior, A.da, França-e-Silva, A.L.G.de, Oliveira, J.M.de, Silva, D.M.da, 2024. Developing pharmacological therapies for atrial fibrillation targeting mitochondrial dysfunction and oxidative stress: a scoping review. International Journal of Molecular Sciences 25 (1), 535. Available at https://doi.org/10.3390/ijms25010535.

Mesquita, T., Miguel-Dos-Santos, R., Cingolani, E., 2025. Aging and sinus node dysfunction: mechanisms and future directions. Clinical Science 139 (11), 577–593. Available at https://doi.org/10.1042/CS20231025.

Morris, B.J., Willcox, B.J., Donlon, T.A., 2019. Genetic and epigenetic regulation of human aging and longevity. Biochimica et Biophysica Acta - Molecular Basis of Disease 1865 (7), 1718–1744. Available at https://doi.org/10.1016/j.bbadis.2018.08.039.

Mynard, J.P., Kondiboyina, A., Kowalski, R., Cheung, M.M.H., Smolich, J.J., 2020. Measurement, analysis and interpretation of pressure/flow waves in blood vessels. Frontiers in Physiology 11. Available at https://doi.org/10.3389/fphys.2020.01085.

Nilsson, P.M., 2014. Hemodynamic aging as the consequence of structural changes associated with early vascular aging (EVA). Aging and Disease 5 (2), 109–113. Available at https://doi.org/10.14336/AD.2014.0500109.

Niroula, A., Sekar, A., Murakami, M.A., Trinder, M., Agrawal, M., Wong, W.J., et al., 2021. Distinction of lymphoid and myeloid clonal hematopoiesis. Nature Medicine 27 (11), 1921–1927. Available at https://doi.org/10.1038/s41591-021-01521-4.

Nwabuo, C.C., Moreira, H.T., Vasconcellos, H.D., Mewton, N., Opdahl, A., Ogunyankin, K.O., et al., 2019. Left ventricular global function index predicts incident heart failure and cardiovascular disease in young adults: the coronary artery risk development in young adults (CARDIA) study. European Heart Journal Cardiovascular Imaging 20 (5), 533–540. Available at https://doi.org/10.1093/ehjci/jey123.

Obisesan, O.H., Osei, A.D., Uddin, S.M.I., Dzaye, O., Blaha, M.J., 2021. An update on coronary artery calcium interpretation at chest and cardiac CT. Radiology: Cardiothoracic Imaging 3 (1), e200484. Available at https://doi.org/10.1148/ryct.2021200484.

Oh, H.S.-H., Rutledge, J., Nachun, D., Pálovics, R., Abiose, O., Moran-Losada, P., et al., 2023. Organ aging signatures in the plasma proteome track health and disease. Nature 624 (7990), 164–172. Available at https://doi.org/10.1038/s41586-023-06802-1.

Oh, H.S.-H., Le Guen, Y., Rappoport, N., Urey, D.Y., Farinas, A., Rutledge, J., et al., 2025. Plasma proteomics links brain and immune system aging with healthspan and longevity. Nature Medicine 31 (8), 2703–2711. Available at https://doi.org/10.1038/s41591-025-03798-1.

Otto, C.M., Newby, D.E., Hillis, G.S., 2024. Calcific aortic stenosis: a review. JAMA: The Journal of the American Medical Association 332 (23), 2014–2026. Available at https://doi.org/10.1001/jama.2024.16477.

Pacinella, G., Ciaccio, A.M., Tuttolomondo, A., 2022. Endothelial dysfunction and chronic inflammation: the cornerstones of vascular alterations in age-related diseases. International Journal of Molecular Sciences 23 (24), 15722. Available at https://doi.org/10.3390/ijms232415722.

Pal, A., Rai, H.M., Agarwal, S., Agarwal, N., 2024. Advanced noise-resistant electrocardiography classification using hybrid wavelet-median denoising and a convolutional neural network. Sensors 24 (21), 7033. Available at https://doi.org/10.3390/s24217033.

Parvin, N., Joo, S.W., Jung, J.H., Mandal, T.K., 2025. Multimodal AI in biomedicine: pioneering the future of biomaterials, diagnostics, and personalized healthcare. Nanomaterials 15 (12), 895. Available at https://doi.org/10.3390/nano15120895.

Pavez Loriè, E., Baatout, S., Choukér, A., Buchheim, J.-I., Baselet, B., Dello Russo, C., et al., 2021. The future of personalized medicine in space: from observations to countermeasures. Frontiers in Bioengineering and Biotechnology 9. Available at https://doi.org/10.3389/fbioe.2021.739747.

Pender, A., Lewis-Owona, J., Ekiyoyo, A., Stoddard, M., 2025. Echocardiography and heart failure: an echocardiographic decision aid for the diagnosis and management of cardiomyopathies. Current Cardiology Reports 27 (1). Available at https://doi.org/10.1007/s11886-025-02194-y.

Peng, Y., Tao, Y., Liu, L., Zhang, J., Wei, B., 2024. Crosstalk among reactive oxygen species, autophagy and metabolism in myocardial ischemia and reperfusion stages. Aging and Disease 15 (3), 1075–1107. Available at https://doi.org/10.14336/AD.2023.0823-4.

Pfaller, B., Dave Javier, A., Grewal, J., Gabarin, N., Colman, J., Kiess, M., et al., 2021. Risk associated with valvular regurgitation during pregnancy. Journal of the American College of Cardiology 77 (21), 2656–2664. Available at https://doi.org/10.1016/j.jacc.2021.03.327.

Phelps, M.E., 2000. Positron emission tomography provides molecular imaging of biological processes. Proceedings of the National Academy of Sciences of the United States of America 97 (16), 9226–9233. Available at https://doi.org/10.1073/pnas.97.16.9226.

Porcari, A., Fontana, M., Gillmore, J.D., 2022. Transthyretin cardiac amyloidosis. Cardiovascular Research 118 (18), 3517–3535. Available at https://doi.org/10.1093/cvr/cvac119.

Quaglino, Boraldi, Lofaro, 2020. The biology of vascular calcification. International Review of Cell and Molecular Biology 354, 261–353.

Raisi-Estabragh, Z., Salih, A., Gkontra, P., Atehortúa, A., Radeva, P., Boscolo Galazzo, I., et al., 2022. Estimation of biological heart age using cardiovascular magnetic resonance radiomics. Scientific Reports 12 (1). Available at https://doi.org/10.1038/s41598-022-16639-9.

Ranard, L.S., Bonow, R.O., Nishimura, R., Mack, M.J., Thourani, V.H., Bavaria, J., et al., 2023. Imaging methods for evaluation of chronic aortic regurgitation in adults: JACc state-of-the-art review. Journal of the American College of Cardiology 82 (20), 1953–1966. Available at https://doi.org/10.1016/j.jacc.2023.08.051.

Rodriguez, K.J., Masters, K.S., 2009. Regulation of valvular interstitial cell calcification by components of the extracellular matrix. Journal of Biomedical Materials Research - Part A 90 (4), 1043–1053. Available at https://doi.org/10.1002/jbm.a.32187.

Ruiz, C.E., Kliger, C., Perk, G., Maisano, F., Cabalka, A.K., Landzberg, M., et al., 2015. Transcatheter therapies for the treatment of valvular and paravalvular regurgitation in acquired and congenital valvular heart disease. Journal of the American College of Cardiology 66 (2), 169–183. Available at https://doi.org/10.1016/j.jacc.2015.05.034.

Ryczkowska, K., Adach, W., Janikowski, K., Banach, M., Bielecka-Dabrowa, A., 2023. Menopause and women's cardiovascular health: is it really an obvious relationship? Archives of Medical Science 19 (2), 458–466. Available at https://doi.org/10.5114/aoms/157308.

Saini, M., Singh, S.K., Yadav, H.K., Dave, P., Gupta, M., Bhat, A.A., et al., 2024. Role of synbiotics in cardiovascular diseases. Synbiotics in Human Health: Biology to Drug Delivery. Springer Singapore, India, pp. 137–153. Available at https://doi.org/10.1007/978-981-99-5575-6_6.

Salih, A.M., Pujadas, E.R., Campello, V.M., McCracken, C., Harvey, N.C., Neubauer, S., et al., 2023. Image-based biological heart age estimation reveals differential aging patterns across cardiac chambers. Journal of Magnetic Resonance Imaging 58 (6), 1797–1812. Available at https://doi.org/10.1002/jmri.28675.

Samiei, N., Abbasi, F., Shojaeifard, M., Parsaee, M., Hosseini, S., Rezaei, Y., et al., 2021. The role of left atrial strain in detecting left ventricular diastolic dysfunction: comparison between the 2009 and 2016 recommendations. Journal of Tehran University Heart Center 16 (2), 58–67. Available at https://doi.org/10.18502/jthc.v16i2.7386.

Sengupta, P.P., Krishnamoorthy, V.K., Korinek, J., Narula, J., Vannan, M.A., Lester, S.J., et al., 2007. Left ventricular form and function revisited: applied translational science to cardiovascular ultrasound imaging. Journal of the American Society of Echocardiography 20 (5), 539–551. Available at https://doi.org/10.1016/j.echo.2006.10.013.

Shehab, S., Allida, S.M., Newton, P.J., Robson, D., Macdonald, P.S., Davidson, P.M., et al., 2019. Valvular regurgitation in a biventricular mock circulatory loop. ASAIO Journal 65 (6), 551–557. Available at https://doi.org/10.1097/mat.0000000000000852.

Shih, H., Lee, B., Lee, R.J., Boyle, A.J., 2011. The aging heart and post-infarction left ventricular remodeling. Journal of the American College of Cardiology 57 (1), 9–17. Available at https://doi.org/10.1016/j.jacc.2010.08.623.

Shiota, T., 2017. Role of echocardiography for catheter-based management of valvular heart disease. Journal of Cardiology 69 (1), 66–73. Available at https://doi.org/10.1016/j.jjcc.2016.09.015.

Sikking, M.A., Stroeks, S.L.V.M., Waring, O.J., Henkens, M.T.H.M., Riksen, N.P., Hoischen, A., et al., 2023. Clonal hematopoiesis of indeterminate potential from a heart failure specialist's point of view. Journal of the American Heart Association 12 (15). Available at https://doi.org/10.1161/JAHA.123.030603.

Singh, Y., Samuel, V.P., Dahiya, S., Gupta, G., Gillhotra, R., Mishra, A., et al., 2019. Combinational effect of angiotensin receptor blocker and folic acid therapy on uric acid and creatinine level in hyperhomocysteinemia-associated hypertension. Biotechnology and Applied Biochemistry 66 (5), 715–719. Available at https://doi.org/10.1002/bab.1799.

Singh, Y., Gupta, G., Mishra, A., Chellappan, D.K., Dua, K., 2020. Gender and age differences reveal risk patterns in COVID-19 outbreak. Alternative Therapies in Health and Medicine 26, 54–55.

Singh, Y., Fuloria, N.K., Fuloria, S., Subramaniyan, V., Almalki, W.H., Al-abbasi, F.A., et al., 2022. A European pharmacotherapeutic agent roflumilast exploring integrated preclinical and clinical evidence for SARS CoV-2 mediated inflammation to organ damage. British Journal of Clinical Pharmacology 88 (8), 3562–3565. Available at https://doi.org/10.1111/bcp.15328.

Solanki, N., Gupta, G., Chellappan, D.K., Singh, S.K., Gulati, M., Paudel, K.R., et al., 2024. Boswellic acids: a critical appraisal of their therapeutic and nutritional benefits in chronic inflammatory diseases. Endocrine, Metabolic & Immune Disorders - Drug Targets 24 (1), 116–129. Available at https://doi.org/10.2174/1871530323666230512154634.

Stein, J.H., Korcarz, C.E., Hurst, R.T., Lonn, E., Kendall, C.B., Mohler, E.R., et al., 2008. Use of carotid ultrasound to identify subclinical vascular disease and evaluate cardiovascular disease risk: a consensus statement from the american society of echocardiography carotid intima-media thickness task force endorsed by the society for vascular medicine. Journal of the American Society of Echocardiography 21 (2), 93–111. Available at https://doi.org/10.1016/j.echo.2007.11.011.

Stewart, G.C., Sun, Y.-P., O'Gara, P.T., 2019. Valvular heart disease. In: Toth, P.P., Cannon, C.P. (Eds.), Comprehensive Cardiovascular Medicine in the Primary Care Setting. Springer International Publishing, Cham, pp. 391–408.

Swetha, K., Reddy, B.S., Shobhamani, B., Sivajothi, S., 2024. Clinical findings, laboratory results, electrocardiography and echocardiography findings in dairy buffaloes with theileriosis. Acta Parasitologica 69 (4), 1896–1910. Available at https://doi.org/10.1007/s11686-024-00927-7.

Thapa, R., Afzal, O., Alfawaz Altamimi, A.S., Goyal, A., Almalki, W.H., Alzarea, S.I., et al., 2023. Galangin as an inflammatory response modulator: an updated overview and therapeutic potential. Chemico-Biological Interactions 378, 110482. Available at https://doi.org/10.1016/j.cbi.2023.110482.

Tian, Q., Lee, P.R., Walker, K.A., Ferrucci, L., 2023. Energizing mitochondria to prevent mobility loss in aging: rationale and hypotheses. Exercise and Sport Sciences Reviews 51 (3), 96–102. Available at https://doi.org/10.1249/jes.0000000000000315.

Vandenberk, B., Lei, L.Y., Ballantyne, B., Vickers, D., Liang, Z., Sheldon, R.S., et al., 2022. Cardioneuroablation for vasovagal syncope: a systematic review and meta-analysis. Heart Rhythm: the Official Journal of the Heart Rhythm Society 19 (11), 1804–1812. Available at https://doi.org/10.1016/j.hrthm.2022.06.017.

Vermes, E., Iacuzio, L., Levy, F., Bohbot, Y., Renard, C., Gerber, B., et al., 2022. Role of cardiovascular magnetic resonance in native valvular regurgitation: a comprehensive review of protocols, grading of severity, and prediction of valve surgery. Frontiers in Cardiovascular Medicine 9. Available at https://doi.org/10.3389/fcvm.2022.881141.

Vinco, G., Porto, M.D., Demattè, C., Giovanelli, C., Caruso, F., Marinetti, A., et al., 2024. Role of cardiovascular magnetic resonance in the assessment of native aortic regurgitation with insights on mixed and multiple valvular heart disease: a narrative review. Echocardiography (Mount Kisco, N.Y.) 41 (12). Available at https://doi.org/10.1111/echo.70045.

Wang, B., et al., 2024. AGE-RAGE axis and cardiovascular diseases: pathophysiologic mechanisms and prospects for clinical applications. Cardiovascular Drugs and Therapy 39, 1489–1506. Available at https://doi.org/10.1007/s10557-024-07639-0.

Wargocka-Matuszewska, W., Uhrynowski, W., Rozwadowska, N., Rogulski, Z., 2023. Recent advances in cardiovascular diseases research using animal models and pet radioisotope tracers. International Journal of Molecular Sciences 24 (1), 353. Available at https://doi.org/10.3390/ijms24010353.

Wehrum, T., Lodemann, T., Hagenlocher, P., Stuplich, J., Ngo, B.T.T., Grundmann, S., et al., 2018. Age-related changes of right atrial morphology and inflow pattern assessed using 4D flow cardiovascular magnetic resonance: results of a population-based study. Journal of Cardiovascular Magnetic Resonance 20 (1), 38. Available at https://doi.org/10.1186/s12968-018-0456-9.

Werbner, B., Tavakoli-Rouzbehani, O.M., Fatahian, A.N., Boudina, S., 2023. The dynamic interplay between cardiac mitochondrial health and myocardial structural remodeling in metabolic heart disease, aging, and heart failure. Journal of Cardiovascular Aging 3 (1). Available at https://doi.org/10.20517/jca.2022.42.

Xiao, F., Qi, J., Ma, S., Sun, L., Sun, Y., 2025. Research progress on the role and mechanism in the change of cardiac structure and function of cardiac fibrosis in the elderly. Cardiology in Review. Available at https://doi.org/10.1097/crd.0000000000000911.

Xu, X., Wang, B., Ren, C., Hu, J., Greenberg, D.A., Chen, T., et al., 2017. Age-related impairment of vascular structure and functions. Aging and Disease 8 (5), 590–610. Available at https://doi.org/10.14336/AD.2017.0430.

Young, A., Garcia, M., Sullivan, S.M., Liu, C., Moazzami, K., Ko, Y.-A., et al., 2021. Impaired peripheral microvascular function and risk of major adverse cardiovascular events in patients with coronary artery disease. Arteriosclerosis, Thrombosis, and Vascular Biology 41 (5), 1801–1809. Available at https://doi.org/10.1161/atvbaha.121.316083.

Zhao, X., Garg, P., Assadi, H., Tan, R.-S., Chai, P., Yeo, T.J., et al., 2023. Aortic flow is associated with aging and exercise capacity. European Heart Journal Open 3 (4). Available at https://doi.org/10.1093/ehjopen/oead079.

Zheng, L., He, S., Wang, H., Li, J., Liu, Y., Liu, S., 2024. Targeting cellular senescence in aging and age-related diseases: challenges, considerations, and the emerging role of senolytic and senomorphic therapies. Aging and Disease 15 (6), 2554–2594. Available at https://doi.org/10.14336/AD.2024.0206.

Zlibut, A., Muresan, I.D., Bietenbeck, M., Radu, A.D., Agoston-Coldea, L., 2025. CMR-derived strain and torsion reveal subclinical dysfunction in hypertrophic cardiomyopathy: a prospective case–control study. Biomedicines 13 (8), 1986. Available at https://doi.org/10.3390/biomedicines13081986.

Chapter 15

Clinical trials and translational advances targeting cardiac aging

Yumna Khan[1], Amna Khan[2], Hamza Khan[3], Lina Eltaib[4], and Md Sadique Hussain[5,6]

[1]*Saveetha Institute of Basic Medical Sciences (SIBMS), Saveetha Institute of Medical and Technical Sciences (SIMATS), Saveetha University, Thandalam, Chennai, Tamil Nadu, India,* [2]*Abbottabad International Medical Institute, Abbottabad, Khyber Pakhtunkhwa, Pakistan,* [3]*Department of Medicine, Khyber Medical College, Peshawar, Khyber Pakhtunkhwa, Pakistan,* [4]*Department of Pharmaceutics, College of Pharmacy, Northern Border University, Rafha, Saudi Arabia,* [5]*Uttaranchal Institute of Pharmaceutical Sciences, Uttaranchal University, Dehradun, Uttarakhand, India,* [6]*School of Pharmaceutical Sciences, Lovely Professional University, Phagwara, Punjab, India*

15.1 Introduction

Structural, functional, and biomolecular alterations in the structure of the circulatory system are the primary characteristics of the complex nature of cardiovascular system aging. In middle-aged and older people, the condition predisposes to cardiovascular disease, decreased physical capacity, and enhanced mortality (Pietri and Stefanadis, 2021). Signs of a senescent heart are diastolic dysfunction, augmented myocardial fibrosis, decreased capillary density, left ventricular hypertrophy, and changes in the extracellular matrix composition. All these lead to greater susceptibility of the senescent heart to arrhythmias, ischemic events, and heart failure (North and Sinclair, 2012; Xie et al., 2023). This is due to the different features at the cellular level, among them the presence of genomic instability, dysfunctional mitochondria, shortening of telomere, and most importantly, cellular senescence (López-Otín et al., 2023). Cellular senescence is a condition in which the cell's capacity to divide is permanently blocked, and it can be caused by both oxidative damage and telomere shortening, as well as the DNA damage-response process (Hussain et al., 2025). With advancing age, the number of senescent cells, which release a pro-inflammatory secretory phenotype termed the senescence-associated secretory phenotype (SASP), grows exponentially. These cells are endothelial cells, cardiomyocytes, and cardiac progenitor cells (CPCs). The phenotype also participates in tissue structure, changes angiogenesis, facilitates fibrosis, and suppresses tissue regeneration (Lewis-McDougall et al., 2019; Afzal et al., 2025; Thapa et al., 2024a, 2024b). Preclinical studies have shown that senolytic drugs such as dasatinib and quercetin may be employed to eradicate senescent cells in old animals to restore cardiac function, reduce fibrosis, and improve the body's regenerative capacity following myocardial infarction (Salerno et al., 2022; Zhang et al., 2022a).

Recent findings have created an increasing desire to use our knowledge of senescence biology in therapeutic applications of benefit. Senotherapeutic strategies, including modalities to eliminate cellular senescence like metabolic rejuvenators, senolytics, senomorphics, and stem cell rejuvenation, have created an opportunity to achieve a promising strategy to treat age-related cardiovascular disorders (Rad and Grillari, 2024). Interestingly, other agents such as ABT263 (navitoclax), fisetin, and agents administered through cardiac-specific nanoparticles are also under investigation in the laboratory and preliminary human trials in order to determine how well they would reverse or mitigate cardiac aging (Lee et al., 2021; Lérida-Viso et al., 2022). Nevertheless, the process of translating these findings into clinical practice is complicated and multifaceted. There is a lack of clinical trials on antisenescence interventions to cardiac aging, most of which are exploratory or feasibility studies. For instance, the REVERSE-LVH trial revealed that sacubitril /valsartan may have an effect on pathways associated with senescence, which reduce myocardial fibrosis and improve the heart structure of patients with left ventricular hypertrophy (Lee et al., 2025). Similarly, senescence pathways in a cardiovascular setting can be suppressed by antiinflammatory drugs such as ruxolitinib, SIRT activators, and rapamycin analogs (Yang et al., 2024; Bashir et al., 2025; Thapa et al., 2025). Translational strategies have also found their way to regenerative technologies. Trials such as RIMECARD and HUC-HEART have demonstrated that mesenchymal stem cell treatment can

Cellular Senescence and Cardiac Aging. DOI: https://doi.org/10.1016/B978-0-443-45811-8.00016-7

be used to enhance the functioning of the heart in patients with heart failure, suggesting potential improvements in the clinical benefits of aging cardiac stem cells (Ulus et al., 2020; Jihwaprani et al., 2025). A clinical trial is shown to have potential in repairing aging hearts that could not repair themselves with the help of cardiomyocytes produced by human induced pluripotent stem cells (hiPSC-CMs) (Ulus et al., 2020; Clancy and Santana, 2024).

Nonetheless, significant difficulties with the clinical implementation of these treatments exist. Trials in cardiovascular studies involving younger patients do not necessarily involve older individuals, thus making it hard to generalize results to the older population (Caughey et al., 2020). Patient selection and monitoring responses are further complicated by the lack of standardized markers for cardiac senescence. While a concise panel has not yet been developed, markers such as p16INK4a, IL-6, IGFBP7, and GDF-15 are currently under investigation (Hastings et al., 2024; di Candia et al., 2021; Kyryk and Parkhomenko, 2025). Moreover, not all forms of senescence are harmful; in some cases, senescent cells may aid in tissue repair and prevent tumors (Giannoula et al., 2023). Therefore senolytic therapies must be specific to avoid unintended or off-target effects. Concerns about long-term toxicity, tissue-specific differences, and immune response clearance also influence clinical trial designs. Alongside nonpharmacological methods, such as intense exercise, reduced food intake, and multimodal rehabilitation programs, have shown promise in reversing signs of cardiovascular aging. REHAB-HF and NCT02039154 studies emphasize the ability of lifestyle changes to modulate cardiac output, vascular stiffness, and functional ability in elderly people, and frequently to equal or to surpass the gains made by medications (Reeves et al., 2017; Howden et al., 2018; Hieda et al., 2020).

The role of cardiac aging as a potential target of therapeutic intervention continues to increase, and this is attributed to the subsequent incorporation of senotherapeutics into clinical practice that will necessitate a unified system of translational studies, validation of secondary endpoints, inclusiveness in design, and long-term data safety studies. This chapter covers in-depth assessments of the field, current clinical trials, translational research on senolytic and senomorphic therapies, regenerative approaches like stem cell interventions and induced pluripotent stem cells (iPSCs), significant challenges in biomarker development, patient stratification, and future directions for such trials. In order to maximize the therapeutic potential of older adults at risk of cardiovascular disease, this chapter suggests that, to accelerate the evidence-based intervention development process in the area of cardiac aging, it is necessary to map the existing scientific advances onto defining the bottlenecks at the translational levels. shows the multifactorial effects of cardiac aging, starting with ischemic damage and necrotic or apoptotic cell accumulation, which ultimately lead to postischemic cardiac fibrosis and progressive myocardial dysfunction (Fig. 15.1).

15.2 Landscape of clinical trials in cardiac aging

15.2.1 Design considerations and challenges in cardiac aging trials

The variability of health status, increased vulnerability, and low cardiac efficiency that characterize older persons complicates the standard trial designs. Enhancements in the heart structure, quality of life, and physical capabilities are the results that should be deemed acceptable in older adults and are frequently more desirable than death indicators. This kind of research requires methods that are flexible and rely on low-risk drugs, a consistent assessment method, and consider the physical constraints of the subjects involved.

Direct cardiac examinations can be successfully substituted by noninvasive surrogate measures, such as measurements of arterial stiffness and estimated heart age. The cardiovascular system ageing in middle-aged individuals without atherosclerosis has been studied using this method, and future research that concentrates on practical outcomes should be based on it. Another major example is from the EVIDENT trial (NCT01083082), which used these measurements to investigate cardiovascular aging in a nonatherosclerotic middle-aged group (García-Ortiz et al., 2014; Bhat et al., 2024a; Thapa et al., 2024a,b).

There are three key considerations that must be made when carrying out research in the area of cardiac aging. To begin with, it is important to choose suitable surrogates and functional endpoints that are relevant to the age group under study so as to have an accurate depiction of the common outcomes that older people undergo (Li et al., 2020). Secondly, the safety of the participants must be considered in the selection process, and the selected group should be representative of the general health problems that older adults encounter, while at the same time, the risks should be minimal (Caughey et al., 2020). Third, ethical protections should be incorporated to overcome typical cognitive, sensory, and physical impairments experienced by aging individuals (Altawalbeh et al., 2020). By targeting these aspects, the clinical trials will have a chance to maintain the scientific validity and be more relevant to the elderly subjects.

These methods have been utilized effectively in a range of clinical trials, such as cardiac aging rehabilitation and regenerative therapies. The necessity to design therapies based on age was also identified by the REHAB-HF trial

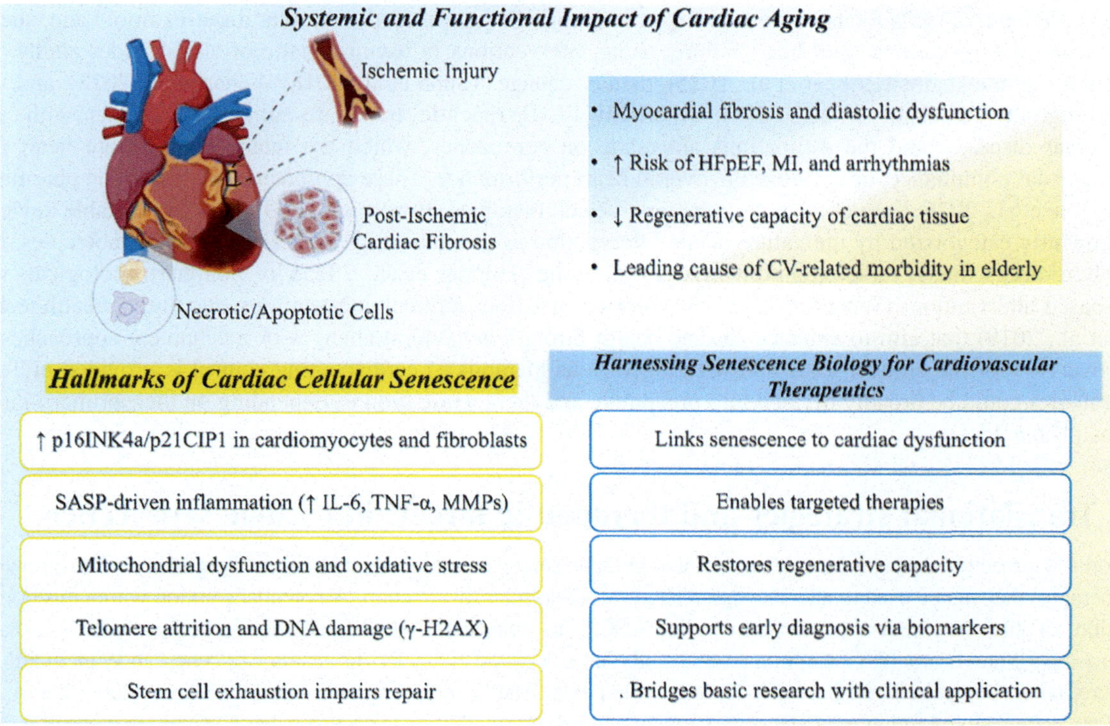

FIGURE 15.1 **Molecular drivers, systemic impact, hallmarks, and therapeutic relevance of cardiac cellular senescence in aging.** It illustrates the multifactorial effects of cardiac aging, beginning with ischemic damage and the accumulation of necrotic or apoptotic cells. This process leads to postischemic cardiac fibrosis and progressive myocardial dysfunction, impacting heart function. The schematic highlights molecular drivers, systemic impact, and therapeutic relevance of cardiac cellular senescence in aging.

(Kitzman et al., 2021) and NCT02039154 study (Howden et al., 2018) in heart failure patients of the older age group, aiming at functional outcomes of physical functioning and quality of life. Likewise, the regenerative trials RIMECARD (Bartolucci et al., 2017), HUC-HEART (Ulus et al., 2020), and C-CURE (Bartunek et al., 2013) imposed strong inclusion criteria to advance the safety and stability of the stem cell delivery to patients with hemodynamically stable conditions. Other advanced approaches have followed these preliminary methods to investigate the therapeutic potential of hiPSC-derived cardiomyocytes as a therapeutic treatment against cardiac aging. Research using these hiPSC-derived cardiomyocytes, such as NCT04982081 (Andrasch et al., 2025) and NCT03763136 (Zhang et al., 2022b), underscores the importance of thorough long-term safety assessment of aging hearts, particularly when it comes to the possibility of malignancy or immune rejection.

Regardless of these new changes in research methodology because of technology, however, there exists a problem of underrepresentation of aging people in cardiovascular research. Research findings indicate that more than half of clinical studies of ischemic heart disease (IHD) do not include people living over 75 years of age (and above) due to arbitrary age criteria. As a result, the generalizability of the findings to the most impacted groups concerning heart conditions with age is reduced (Bourgeois et al., 2017). In addition, practical barriers can also be encountered by the researchers in terms of recruiting older people through their lack of knowledge about the study protocols, recall of complex treatment regimes, and support of valid informed consent, especially in cases of sensory loss or cognitive deficit (Altawalbeh et al., 2020; Bhat et al., 2024b; Subedi et al., 2025). To overcome such challenges, future research must follow patient-centered strategies that address the realities of aging, create procedures to address the needs of aging subjects, and employ more inclusive age criteria. This would enhance study results regarding practical relevance and scientific value.

15.2.2 Classification of trials targeting cardiac aging: preventive versus therapeutic approaches

Clinical trials with the primary objective of treating cardiac aging are typically grouped based on their primary objective, intervention type, and stage of clinical development, one of the simplest distinctions between preventive and therapeutic approaches. Preventive trials focus on postponing or reducing the risk of cardiovascular diseases in old age, such as high blood pressure, atherosclerosis, and heart failure, using interventions in asymptomatic or at-risk older adults. Common interventions include statins (Hopper et al., 2025), dietary changes (Shah et al., 2018; Weber et al., 2012), and structured exercise programs (Chokshi et al., 2018). Alternatively, therapeutic trials are aimed at patients with diagnosed cardiovascular diseases, and the estimations are made on approaches, which are intended to restore heart functions, improve vascular compliance, or optimize the overall heart performance. These approaches could include pharmacological therapies (Lee et al., 2025; Kato et al., n.d.), regenerative cell therapies (Borow et al., 2019), or implantable devices. Trials are subsequently categorized by the nature of the intervention as pharmacological, including small molecules, biologics, and senolytic drugs that reverse or inhibit myocardial aging (Hopper et al., 2025), or nonpharmacological, which are lifestyle-based interventions (Weber et al., 2012), caloric restriction, surgical interventions, and digital health technologies (Schulz et al., 2019) that aim to enhance cardiac health through behavioral changes or mechanical approaches. A more detailed overview of the clinical trials investigating various methods of cardiac aging control is provided in Table 15.1. These strategies could be broadly divided into preventive and therapeutic groups, depending on their primary intervention objectives (Table 15.1).

15.3 Translational strategies and therapeutic targets in cardiac senescence

Translational approaches that can be used to combat cellular senescence are oriented to translate mechanistic knowledge into clinical therapies that can be used to alleviate age-related cardiovascular dysfunction. A potential solution within this domain is the rejuvenation of the senescent cardiac stem cells (CSCs) to reinstate their regenerative functions. CSCs isolated from decompensated human hearts (E-CSCs) in a pivotal study have depicted characteristic features of senescence, including increased IL-1b secretion and impaired repair ability. It was found that AMPK signaling was reactivated and downstream molecular dysfunctions were restored in the presence of rapamycin and resveratrol. When the rejuvenated CSCs were injected into infarcted mouse hearts, the cell death was greatly reduced, and the myocardial regeneration was improved, which represents the translational potential of molecularly enhanced, patient-derived CSC therapies (Avolio et al., 2014; Goyal et al., 2024; Jha et al., 2024). Developing on the cellular method, pharmacological destruction of senescent cells by senolytic drugs has proven to be very promising in the aging-associated cardiovascular disease models. Research was carried out where older mice of the female gender were given dasatinib and quercetin (D + Q) after they had a heart attack. Through this therapy, senescent cardiomyocytes and dysfunctional progenitor cells were found and eliminated, which resulted in improved ventricular activity and increased regenerative potential (Salerno et al., 2022). Such results highlight the potential of the application of senolytic treatments to older adults with cardiovascular issues. The induction of innate antisenescence responses has now been shown to be of significant clinical value. In older mice, it was shown that increased expression of the nuclear sirtuin SIRT6 in their hearts led to lower levels of classic senescence-related markers, as well as decreased fibrosis, mitochondrial DNA damage, and hypertrophy (Pillai et al., 2021). These results show that SIRT6 is a prospective therapeutic agent for preventing or reversing age-related cardiac remodeling.

Moreover, translational studies have also concentrated on molecular regulators and genetic factors of cardiac senescence, providing new therapeutic opportunities. Patient-centered research was conducted to identify elevated insulin-like growth factor-binding protein 7 (IGFBP7) levels in heart failure patients correlated with high levels of expression in p16, p21, and p53, and suggested that IGFBP7 could serve as a senescence-related biomarker. Genetic silencing and neutralizing antibodies of IGFBP7 in mouse models resulted in the TOO-IGFBP7 restoration of FOXO3a signaling, cardiac structural improvement, and alleviation of senescence-related problems, which confirms the suitability of IGFBP7 as a potential target to rejuvenate the heart (Zhang et al., 2022a). Further research involving Ulk1S913A knock-in mice illustrated that dysfunctional autophagy accelerates cardiac aging by increasing the burden of senescence and causing contractile dysfunction. These deleterious changes were reversed by treatment with the senolytic compound, ABT263, which gives strong preclinical evidence that autophagy modulation can be used as a therapeutic strategy (Zhai et al., 2025). Moreover, a novel approach involving the investigation of the Selenoprotein T (SELENOT)/CD36 signaling pathway indicated that the synthetic peptide PSELT alleviated DNA damage, oxidative stress, and endoplasmic reticulum stress in cardiomyocytes with doxorubicin (Hussain et al., 2025; De Bartolo et al., 2025; Gupta et al., 2025). This finding indicates a novel molecular target that has great potential to combat senescence-related heart dysfunction.

15.3.1 Challenges and limitations in translating antisenescence therapies

Senescent cell-based therapeutic approach has the potential to treat cardiac aging, but there are significant difficulties in translating the laboratory study findings to clinical applications. The main challenge is to recapitulate the complex process of human cardiac senescence in the preclinical models. For example, researchers examined myocardial biopsies of patients

TABLE 15.1 Classification of clinical trials targeting cardiac aging: preventive and therapeutic approaches.

Study name/ Trial identifiers	Intervention approach	Clinical and demographic details	Sample size	Study status	Key findings	References
PROMISE (NCT01174550)	Preventive, nonpharmacological	Biomarker-based risk stratification (IL-6, hs-cTn), CT angiography imaging in 1,796 participants with stable chest pain (mean age: 60.2 ± 8.0 years; 47.5% men); evaluated for CAD using imaging and biomarker profiling.	1,796	Completed	Higher IL-6 and hs-cTn levels were associated with greater CAD burden and MACEs. IL-6 remained an independent predictor of MACEs after adjusting for imaging markers. In nonobstructive CAD, elevated IL-6 and hs-cTn showed a stronger risk of MACEs (HR: 2.5–2.7), supporting their role in early preventive risk stratification.	Ferencik et al. (2022)
DREAM-HF (NCT02032004)	Therapeutic, pharmacological (cell-based immunotherapy)	Allogeneic mesenchymal precursor cells for advanced chronic HFrEF; includes adaptive enrichment design and a joint frailty model for recurrent events.	≥270 (147 NYHA class III + 123 class II)	Ongoing	Evaluates whether mesenchymal precursor cells reduce recurrent HF-related major adverse events and delay terminal HF. Uses adaptive enrichment for broader treatment evaluation in aging HF populations.	Borow et al. (2019)
NCT02556450	Preventive, pharmacological	Spironolactone 50 mg/day in 527 participants (median age: 73 years; 26% women) with or at high risk of CAD and elevated BNP; followed for up to 9 months.	527	Completed	Significant reductions in type-I collagen biomarkers, blood pressure, NT-proBNP, and left atrial volume. No interaction with galectin-3 levels. Supports use in preventing HF progression in aging populations.	Cleland et al. (2021)
STAREE-HEART (NCT04536870)	Preventive, pharmacological	Atorvastatin vs. placebo in 369 healthy participants aged ≥70 years (mean age: 73.0 ± 3.4 years); echocardiography, ECG, and biomarkers assessed at baseline and year 3.	369	Ongoing	Assesses the effect of statins on LV function (GLS). Baseline data showed preserved LV function and low AF prevalence (4.5%). It may prevent myocardial aging and heart failure.	Hopper et al. (2025)
REVERSE-LVH (NCT03553810)	Therapeutic, pharmacological	Sacubitril/valsartan vs. valsartan in 78 patients with essential hypertension and LVH; fibrosis assessed via cardiac MRI.	78	Completed (Phase 2)	Sacubitril/valsartan reduced interstitial volume more than valsartan (P=.006), and improved LV mass, atrial volume, filling pressure, and cardiac biomarkers. Suggests antifibrotic potential in aging hearts.	Lee et al. (2025)

(Continued)

TABLE 15.1 (Continued)

Study name/ Trial identifiers	Intervention approach	Clinical and demographic details	Sample size	Study status	Key findings	References
Lp(a) HORIZON (NCT04023552)	Therapeutic, pharmacological	Pelacarsen vs. placebo in 8,323 patients with CVD and Lp(a) ≥70 mg/dL; 1:1 randomization; standard therapy background.	8,323	Ongoing (Phase 3)	Evaluates whether lowering Lp (a) reduces MACE. Primary endpoint: CV death, nonfatal MI/stroke, or urgent revascularization. Subgroup analysis for Lp(a) ≥90 mg/dL.	Cho et al. (2025)
ABLATE Versus PACE (NCT04906668)	Therapeutic, nonpharmacological	Cryoballoon PVI vs. pacemaker + AVN ablation in elderly patients (≥75 years) with persistent AF and preserved EF; multicenter randomized trial.	196	Ongoing	Evaluates whether pacemaker + AVN ablation improves symptom control vs. PVI without increasing adverse outcomes.	Boehmer et al. (2024)
Elderly ACS 2 (NCT01777503)	Therapeutic, pharmacological	Clopidogrel vs. prasugrel in patients >74 years with STEMI or NSTEMI undergoing PCI across 32 Italian centers.	1,443	Completed	Prasugrel (5 mg/day) did not reduce ischemic events compared to clopidogrel but had similar bleeding rates. 58.9% of CV deaths occurred within 30 days.	Crimi et al. (2019)
ENGAGE AF–TIMI 48 (NCT00781391)	Therapeutic, pharmacological	Edoxaban vs. warfarin in elderly AF patients (≥75 years; $CHADS_2$ ≥2); median follow-up 2.8 years; global randomized trial.	8,474 (≥75 subgroup)	Completed	Edoxaban reduced major bleeding vs. warfarin (HR: 0.83 [0.70–0.99]) with similar stroke prevention (HR: 0.83 [0.66–1.04]); prevented 82 major bleeds per 10,000 pt-years.	Kato et al. (n.d.)
PHARM-CHF (NCT01692119)	Therapeutic, nonpharmacological	Pharmacy-based medication review, counseling, and regular dose dispensing in elderly CHF patients (mean age: 74 years; 62% male; 52% NYHA Class III).	237 (110 intervention, 127 control)	Completed	Improved adherence (+5.7%, P=.007), increased adherent patients (OR 2.9, P=.005), and better QoL. No significant changes in hospitalization/ mortality.	Schulz et al. (2019)
ACTIVE REWARD (NCT02531022)	Preventive, nonpharmacological	Wearable tracker + loss-framed financial incentives for 105 IHD patients (mean age: 60±11 years; 70% male); home-based 24-week program.	105	Completed	Incentive group showed significant step increases during ramp-up (+1061), maintenance (+1368), and follow-up (+1154) vs. control (P<.01).	Chokshi et al. (2018)

EVADE CAD (NCT02135939)	Preventive, nonpharmacological	Vegan diet vs. AHA diet in 100 CAD patients over 8 weeks; dietary counseling and groceries provided.	100	Completed	Vegan diet led to a 32% reduction in hsCRP (P=.02); no significant changes in BMI, glycemic markers, or LDL. Suggests antiinflammatory effects in CAD patients.	Shah et al. (2018)
Brazilian Cardioprotective Diet Program (NCT01453166)	Preventive, nonpharmacological	Brazilian cardioprotective diet vs. standard guideline-based diet in patients aged ≥45 years with atherothrombotic CVD.	117	Completed	Greater reductions in SBP, DBP, glucose, and BMI vs. the control group, though not statistically significant due to pilot study design.	Weber et al. (2012)
BIOSCIENCE (NCT01443104)	Therapeutic, pharmacological	BP-SES vs. DP-EES in 2,119 patients with minimal exclusion criteria; includes STEMI subgroup; 2-year follow-up.	2,119	Completed	No difference in TLF (10.5% vs. 10.4%, RR: 1.00); STEMI subgroup showed lower risk with BP-SES (RR: 0.48, P=.043).	Zbinden et al. (2016)

with severe cardiotoxicity caused by doxorubicin and found that there was a marked elevation in senescence markers in the left ventricular tissues. To recreate these conditions in vitro, dynamic engineered heart tissues (dyn-EHTs) and cardiomyocytes derived from human pluripotent stem cells were exposed to repeated, clinically relevant doses of doxorubicin. Such experimental models were able to recapitulate key features of cardiac aging, such as tissue expansion, troponin release, and declining contractile capability. Despite the introduction of senomorphic agents like 5-aminoimidazole-4-carboxamide ribonucleotide and resveratrol, the models have failed to reinstate cardiac function. This emphasizes the existence of a significant difference between the alleviation of senescence characteristics and the restoration of functionality (Linders et al., 2023).

In addition to the modeling limitations, more general scientific and physiological issues pose an additional problem to clinical translation. A significant issue is that there are no highly specific and validated biomarkers of senescence, and thus it is challenging to accurately determine and select specific therapies (Shah et al., 2025). The difficulty in specifically killing senescent cells and leaving healthy cells intact is further complicated by the fact that the senescent cells in different tissues and disease conditions are heterogeneous due to their differences (Mansfield et al., 2024). Also, long-term use has safety concerns due to the low selectivity and potential toxicity of various senolytic drugs (Rad and Grillari, 2024). The ability of the immune system to clear senescent cells has been found to be effective in younger or healthy mice but is less effective in older or immunocompromised human populations, which can reduce the effectiveness of treatment (Giannoula et al., 2023). Collectively, these issues highlight the pressing need for improved human-relevant models, better biomarker panels, and safer, more targeted antisenescence therapies to facilitate successful clinical translation.

15.3.2 Pharmacological interventions

Pharmacological approaches have demonstrated great potential in translating mechanistic understanding into therapeutic approaches to address cardiac aging. One of the key approaches is based on the application of ABT263, a Bcl-2/Bcl-xL inhibitor, using PLGA-based nanoparticles, which selectively destroy the stress-induced senescent cells (SISCs) in the heart. The strategy not only reduces remodeling of the heart, but also enhances recovery of the heart function after ischemia-reperfusion (IR) injury, with minimal side effects to the body (Lee et al., 2021). Similarly, Navitoclax therapy in models of doxorubicin-induced cardiomyopathy was effective in senescent cardiomyocytes, fibrosis, and ventricular function restoration (Lérida-Viso et al., 2022). oxorubicin-induced cardiomyopathy was effective in senescent cardiomyocytes, fibrosis, and ventricular function restoration (Pillai et al., 2021). Dasatinib and quercetin (D + Q) is another senolytic combination that has been studied most often, and which was able to remove senescent cardiomyocytes and CPCs in aged female mice following a myocardial infarction. This led to enhancement of tissue regeneration, reduction in the pro-fibrotic and inflammatory markers, and overall cardiac performance (Salerno et al., 2022). Along with senolytics, senomorphic and metabolic agents are also important in the context of cardiac aging. JAK1/2 inhibitor ruxolitinib has been reported to improve mitochondrial structure and function in models of septic cardiomyopathy. This enhancement was achieved by silencing the SASP and enhancing the antiapoptotic pathways (Yang et al., 2024). In addition, another study found that resveratrol provides antiaging effects on cardiac aging caused by D-galactose. It did so by reducing the expression of the NF-kB and Notch1, recovering the expression of short-chain fatty acids (SCFAs), and altering the gut-heart axis (Wang et al., 2024). Moreover, supplementation with NAD+ precursors, specifically nicotinamide riboside (NR) and dihydronicotinamide riboside (NRH), has been successful in rejuvenating the mitochondrion in older hearts by restoring NAD, activating SIRT1/SIRT6 pathways, repairing DNA, and improving diastolic function. Remarkably, NRH has been found to be more effective in age-related cardiac dysfunction reversal (Huang et al., 2025). Together, these findings highlight the growing potential of pharmacological regimens, such as senolytics, senomorphics, and mitochondrial therapy, as useful interventions to slow or reverse cardiac senescence and to keep the myocardium healthy during the aging process. A comprehensive overview of these therapies alongside their translational limitations is presented in Fig. 15.2.

15.4 Regenerative and nonpharmacological therapies for cardiac aging

New treatment strategies are increasingly focusing on repairing structural and functional loss of aging of the heart, especially on regenerative treatment. Interestingly, stem cell treatments proved to be useful in the repair of myocardial damage, repair enhancement, and improving outcomes in aging-related heart conditions such as heart failure with reduced ejection fraction (HFrEF) and ischemic cardiomyopathy. In the experiment of the RIMECARD, they studied the intravenous infusion of umbilical cord-derived mesenchymal stem cells (UC-MSCs) in patients with stable HFrEF. Without producing any known immune-related side effects, these stem cells have been demonstrated to greatly increase

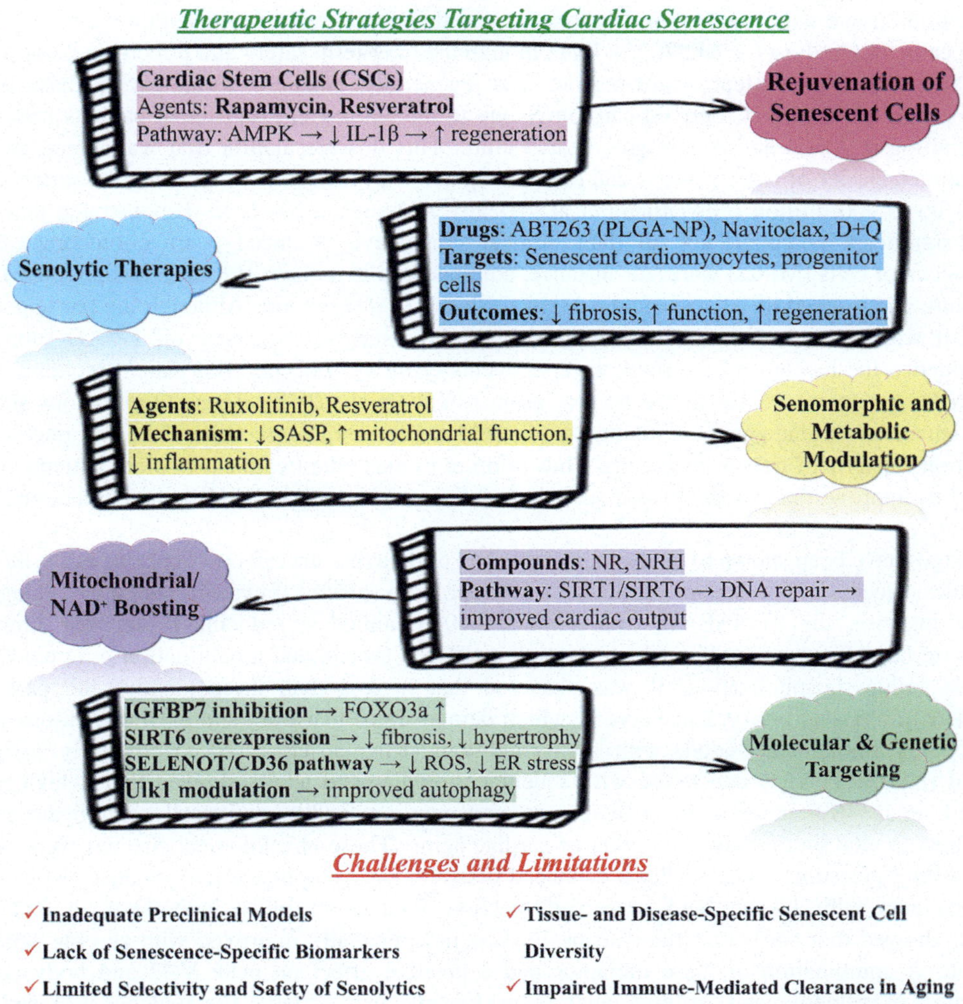

FIGURE 15.2 **Therapeutic strategies targeting cardiac senescence and associated challenges.** It depicts various therapeutic strategies targeting cardiac senescence, including pharmacological and nonpharmacological interventions. The figure summarizes challenges in translation, such as drug delivery, patient selection, and biomarkers. It emphasizes senolytic, senomorphic, metabolic, and regenerative treatments aimed at improving cardiac aging outcomes.

left ventricular ejection fraction (LVEF) and overall quality of life due to their high amounts of hepatocyte growth factor and unique immunoprivileged features (Bartolucci et al., 2017). Another significant study, the HUC-HEART study, attempted to inject human umbilical cord mesenchymal stem cells into the heart muscle during bypass graft surgery. There were no treatment side effects and a marked increase in heart function after 12 months. This comprised an increase of 5.4% in left ventricular ejection fraction, better stroke volume, and less damaged heart tissue (Ulus et al., 2020). The C-CURE trial (NCT00810238), in which autologous cardiogenically conditioned mesenchymal stem cells were used to treat chronic ischemic heart failure, showed significant improvement in the impairment of cardiac function with aging. These included improved performance in the 6-minute walk test, a reduction in left ventricular end-systolic volume, and an increase in ejection fraction (Bartunek et al., 2013). Besides mesenchymal stem cells, the ESCORT study also analyzed cardiovascular progenitors with human embryonic stem cells in fibrin patches. In patients with severe ischemic heart failure, this not only functioned but also improved the mobility of the heart muscle during bypass surgery, proving the effectiveness of using early developmental cell types to repair age-associated heart muscle injury (Menasché et al., 2018).

Permanent loss of heart muscle cells with the aging of individuals is a significant issue, particularly with severe heart failure. To tackle this, a recent Phase I study (NCT04982081) tested the direct injection of heart muscle cells made from human induced pluripotent stem cells (hiPSC-CMs), specifically known as HiCM-188. To restore the aging and nonfunctional cardiac tissue with functioning cardiomyocytes, these cells were implanted directly into the aging cardiac tissue through a catheter (Andrasch et al., 2025). Even though the long-term safety and effectiveness of this treatment are yet to be evaluated, it is a great step toward personalized regenerative medicine. Similarly, a Phase I/IIa clinical trial (NCT03763136) assessed the hiPSC-derived cardiomyocyte to the epicardium transfer in patients undergoing coronary artery bypass grafting (CABG) and progressive heart failure. This dose-escalation trial was aimed at overcoming the limited regeneration capacity of elderly hearts and revitalizing the heart functions by showing the safety and feasibility of the provision of up to 400 million cells (Zhang et al., 2022b). Although it has been demonstrated that stem cell-based therapies have significant potential, not all their clinical trials have produced positive outcomes. Intramyocardial mesenchymal precursor cells (MPCs) were, for instance, administered in a Phase II trial (NCT02362646) in patients with severe heart failure and in need of left ventricular assist device (LVAD) support. Although the participants tolerated the treatment well, it was not linked to the main objective of the temporary weaning off LVAD support and did not significantly influence the incidence of serious adverse events (70.9 vs. 78.7 per 100 patient-months) or mortality rate (14.2% in the MPC group vs. 15.1% in the control group) (Yau et al., 2019). These observations show regenerative mechanisms of advanced cardiac aging are facing the two-fold challenge of overcoming the aging microenvironment and reduced integration capacity of transplanted cells. Thus in order to make the treatment of older patients more effective, it may be required to improve the process of cell selection, delivery schemes, and add tissue engineering or gene editing options.

Daily life changes have been shown to be useful in reversing or delaying the effects of cardiac aging through improving cardiac mechanics, decreasing vascular stiffness, and improving overall physical resilience alongside cell-based approaches. For instance, the REHAB-HF trial (NCT02196038) aimed at reducing frailty and diminished physical abilities, which are key effects of cardiac aging, through a tailored transitional rehabilitation program during or shortly after acute heart failure hospitalization. It was observed that in less than three months, the participants became significantly better in physical performance scores, which demonstrates that even shorter interventions can significantly improve functionality among older adults with heart conditions (Kitzman et al., 2021). Similarly, a significant 2-year randomized trial (NCT02039154) conducted with sedentary middle-aged adults showed that a high-intensity aerobic exercise resulted in an 18% increase in peak oxygen consumption, and a statistically significant reduction in left ventricular stiffness, a vital biomechanical marker of cardiac aging. These physiological changes were observed without increased heart filling pressures, which reflects enhanced diastolic and compliance performance (Howden et al., 2018). Moreover, in patients with obesity and preserved ejection fraction in older individuals, a 2x2 factorial study (NCT00959660) showed that cardiorespiratory capacity was independently improved with aerobic activity and calorie intake restriction. A combination of these therapies had a positive effect on peak VO2 and body composition, two important indicators of metabolic and cardiovascular aging (Kitzman et al., 2016). The evidence provided by such studies is sufficient to suggest that carefully designed and multimodal lifestyle interventions can not only delay but even reverse most of the pathophysiological symptoms of aging of the heart.

15.5 Clinical biomarkers and surrogate endpoints

The significant causes of progressive physiological, biochemical, and structural alterations that comprise cardiac aging include metabolic dysfunction, cellular senescence, chronic inflammation, and changes in the characteristics of myocardial tissue. It has been found that biomarkers and surrogate endpoints that are not only indicative of the presence of these pathological changes but also provide important information about the process and reversibility of cardiac aging in many studies. Senescence of CPCs is one of the key characteristics of cardiac aging that decreases the regenerative capacity of the heart. A study that involved 119 patients with cardiovascular disease, aged 32–86, revealed that more than half of the CPCs in individuals aged over 74 years exhibited signs of senescence, as indicated by high levels of p16INK4a, SA-β-gal, and γH2AX, and shorter telomeres, indicating that cardiovascular disease is highly correlated with senescence of cardiovascular cells. These defective CPCs also led to myocardial deterioration caused by the reduced self-renewal and differentiation potential of these cells (Lewis-McDougall et al., 2019). Emphasizing the importance of cellular pathways, the study showed that the introduction of Pim-1 kinase into human CPCs facilitated cell proliferation, temporarily increased the telomeres, and induced telomerase, which is a molecular reversal of aging phenotypes. The regeneration of Pim-1 overcomes the shortening of telomeres and replicative fatigue associated with age, which emphasizes that basal cellular deficiencies cause cardiac aging and may act as therapeutic targets (Cottage et al., 2012).

Chronic systemic inflammation and metabolic disruption are all causes of inflammaging that can be evaluated by various circulating biomarkers. Higher amounts of IL-6 (≥3.3 pg/mL) were significantly associated with cardiovascular mortality (9.0% vs. 0.5%, P=.001) in a 6-year study of 322 patients with acute coronary syndrome (ACS), which suggests that it is a key element in the pathophysiology of inflammatory processes involved in biological aging of the heart. The IL-6 showed more predictive accuracy (AUC=0.72) and an adjusted hazard ratio of 8.60 than the hsCRP (Gager et al., 2020). Subsequent research based on a deep-learning-based inflammatory aging clock (iAge) of 1,001 individuals revealed CXCL9 as a major effector of vascular dysfunction and adverse cardiac remodeling. The endothelial aging process was reversed by silencing CXCL9 in human and murine models, which is translational evidence on the role of inflammaging in myocardial health (Sayed et al., 2021). Furthermore, in PREDICTOR, incorporating 2,001 individuals, high levels of IGFBP7 and GDF-15 were independently linked to diastolic dysfunction, atrial enlargement, and enhanced mortality, regardless of NT-proBNP and hsTnT (HR: 2.13 and 2.03), therefore supporting their usefulness in cardiac aging phenotyping (Meessen et al., 2021).

Another aspect of cardiac aging is the slow deterioration of the structure and activity of the myocardium, monitored successfully with the help of noninvasive imaging. Several imaging biomarkers, including cardiac magnetic resonance (CMR), echocardiography, and CT-derived radiomic features, have demonstrated the ability to assess age-related changes. For instance, a study based on CMR data of 172 healthy individuals aged 21–94 years reported a significant age-related decrease in native myocardial T1 periods, especially in men. This reduction indicated that myocardial remodeling and progressive interstitial fibrosis were present, which were two critical markers of cardiac aging (Cadour et al., 2025). Additional research employing a radiomics study with 14,902 individuals from the UK Biobank showed that aging hearts typically have lower ventricular diameters and more homogenous myocardial texture patterns. These changes included a decrease in structural variability and compliance, which are indicative of the mechanical deterioration frequently seen in aging heart tissue (Raisi-Estabragh et al., 2021).

15.6 Future directions and recommendations

An increasing collection of senotherapeutic and metabolic reagents is undergoing preclinical development with apparent cardiac aging implications. The small-molecule senolytics navitoclax (ABT263), a Bcl-2/Bcl-xL inhibitor, and dasatinib-quercetin have uniformly decreased senescent cell burden, myocardial fibrosis, and diastolic dysfunction in aged models or pressure-overload models and have shown enhanced cardiac performance in short-term paradigms of treatment (Jia et al., 2020). Recent studies in mouse models with advanced atherosclerosis and in large-animal models indicate that navitoclax can stabilize plaques and decrease the size of the lesion, which highlights the more general cardiovascular applicability of navitoclax (Karnewar et al., 2024). Nonetheless, dose-limiting thrombocytopenia and off-target toxicity are major impediments. Galacto-conjugated navitoclax, along with other pro-drugs that capitalize on senescent cell-specific β-galactosidase activity, are under development to achieve better selectivity and low hematologic toxicity, and can be considered an exciting next generation of senolytics with translational potential to older cardiovascular disease patients (Stojanović et al., 2025).

Alongside, senomorphic approaches to block the SASPs without necessarily destroying the senescent cells are becoming increasingly popular. SASP signaling, mitochondrial dysfunction, and oxidative stress in aged/stressed cardiomyocytes are mitigated by low-dose resveratrol and JAK6/STAT pathway inhibitors like ruxolitinib and result in a better systolic performance of the heart in experimental septic cardiomyopathy and other models (Yang et al., 2024). Such observations reinforce the idea that intermittent senomorphic conditioning could be added to conventional therapies aimed at targeting heart failure with inflammatory and metabolic processes that promote heart aging.

Another fast-growing field is the modulation of cellular energetics using nicotinamide adenine dinucleotide (NAD+) boosters. Nicotinamide riboside (NR) and nicotinamide mononucleotide (NMN) protect mitochondrial homeostasis, decrease ROS, and inhibit fatigue to heart failure in several preclinical paradigms of pressure overload, ischemia-reperfusion, and cardiomyopathy (Yu et al., 2025). Recent reviews note that findings with NAD+ precursors are mounting to have general cardiometabolic advantages in age-related disease and are already at an early trial phase in heart failure, making them the near-term prospects of applying geroscience-oriented cardiovascular trials (Marín-Blázquez et al., 2024; Shi et al., 2024). Lastly, molecular amelioration of distinct senescence pathways such as IGFBP7/FOXO3a signaling, SELENOT CD36 oxidative stress, and autophagy modulators like Ulk1 is promising in preclinical models of cardiac aging and remodeling (Pal, 2025). These approaches are going to be difficult to translate without a lot of focus on toxicology, biomarker selection, patient use, and combination therapy with well-established cardiovascular agents, to reduce risks yet enhance geroprotective effects.

15.7 Integrating artificial intelligence and machine learning for trial prediction and recruitment

Artificial intelligence (AI) and machine learning (ML) are on the verge of changing the lifecycle of cardiovascular trials, either in the protocol design or in long-term follow-up. The most current reviews report on how ML can be used to maximize site selection, recreate sample-size and event-rate configurations, and enhance the likelihood of trial success, and does so with an ability to adapt quickly to the growing data (Cunningham et al., 2024; Olawade et al., 2026). The use of AI-enabled tools in cardiovascular medicine, in particular, has already been utilized to improve risk prediction, patient stratification, and therapy optimization in heart failure and ischemic heart disease (Armoundas et al., 2024; Averbuch et al., 2022). Such capabilities can be reapplied to distinguish patients with increased rates of cardiovascular aging, based on clinical, imaging, and biomarker measurements to enrich trials of senescence-targeted interventions (Suleimenova et al., 2025).

The use of AI to screen electronic medical records and wearable-generated data based on predictive threats can significantly enhance routine recruitment, especially due to the underrepresented older adults, who often fail to be included in conventional trials. Digital health and AI research in clinical studies on cardiovascular disease reveal that automated prescreening and virtual recruiting platforms can significantly enhance enrollment rates and diversity and decrease costs and timeliness to completion (Lampreia et al., 2024). In aging-related cardiac trials, algorithm selection could be biased on candidates with phenotypes indicative of cardiac senescence, for example, HFpEF in multimorbidity, chronic ischemic remodeling, or with a high senescence biomarker signature, which is more likely to yield a gerotherapeutic effect.

In addition to the recruitment aspect, AI may aid in adaptive monitoring as well as individualized dosing. Constant telemetry, wearable devices, and implantable sensors produce high-rate data streams that can be analyzed by ML models to produce early warning signs of toxicity or early functional status improvement to adjust the dose in real time and to build adaptive trial designs (Costa et al., 2025). Notably, open-ended model development, external validation, and focus on algorithmic bias are essential activities needed to prevent systematically screening out or misclassifying frail, multimorbid, or socioeconomically disadvantaged elders (Cunningham et al., 2024; Armoundas et al., 2024).

15.8 Conclusions

Cardiac aging represents a major clinical and biological issue due to the buildup of genetic and cellular changes, notably the rise in senescent cells within myocardial and vascular tissues. In recent years, the understanding of the processes that cause cellular senescence has significantly increased, as well as the associated immune reactions that are associated with the SASP, mitochondrial dysfunction, and telomere shortening. This knowledge forms the basis of the translation work focused on changing or reducing the size of senescent populations to improve cardiac performance and resilience in the elderly.

This chapter outlines the most recent developments in the research of clinical and translational studies dealing with cardiac senescence, highlighting current progress in the measurement of metabolic interventions, senolytic and senomorphic therapies, and regenerative therapies, such as stem cell and iPSC-based treatments of cardiomyocytes. Even though preclinical studies have indicated great potential, the clinical implementation is still in its nascent stages. Most current clinical research only focuses on biomarker endpoints, safety, and feasibility, without considering direct cardiac outcomes.

Some critical issues remain regarding the design and implementation of the trials, especially concerning the various elderly patient groups, reproducibility of senescence biomarkers, and the poor knowledge of the long-term effects of clearing senescent cells. However, the therapeutic translation landscape is changing, and the new technologies of drug delivery, tissue therapy, patient stratification, including AI-enhanced trial design and multiomic biomarker platforms, are contributing to its improvement. The approaches to addressing these issues might include a coordinated initiative that integrates essential research, clinical trials, and regulatory measures to successfully implement the senescence-targeted therapies into cardiovascular care. With a mechanistic conception aligned to patient-focused innovations, the treatment of cardiac diseases associated with aging has the potential to transform and enhance health span in aging populations.

References

Afzal, M., Rekha, M.M., Sahoo, S., Pandey, S.N., Maji, C., Goyal, K., et al., 2025. Targeting the senescence-associated secretory phenotype to modify osteoarthritis in aging. Inflammopharmacology 33 (11), 6555–6575. Available at https://doi.org/10.1007/s10787-025-02001-8.

Altawalbeh, S.M., Alkhateeb, F.M., Attarabeen, O.F., 2020. Ethical issues in consenting older adults: academic researchers and community perspectives. Journal of Pharmaceutical Health Services Research 11 (1), 25–32. Available at https://doi.org/10.1111/jphs.12327.

Andrasch, Y., Ireri, M.M., Gander, J., Timm, A.-E.S., Chennappan, S., Fidan, M., et al., 2025. Impaired MC3T3-E1 osteoblast differentiation triggered by oncogenic HRAS is rescued by the farnesyltransferase inhibitor Tipifarnib. Scientific Reports 15 (1). Available at https://doi.org/10.1038/s41598-025-91592-x.

Armoundas, A.A., Narayan, S.M., Arnett, D.K., Spector-Bagdady, K., Bennett, D.A., Celi, L.A., et al., 2024. Use of artificial intelligence in improving outcomes in heart disease: a scientific statement from the American Heart Association. Circulation 149 (14). Available at https://doi.org/10.1161/CIR.0000000000001201 E1028,.

Averbuch, T., Sullivan, K., Sauer, A., Mamas, M.A., Voors, A.A., Gale, C.P., et al., 2022. Applications of artificial intelligence and machine learning in heart failure. European Heart Journal - Digital Health 3 (2), 311–322. Available at https://doi.org/10.1093/ehjdh/ztac025.

Avolio, E., Gianfranceschi, G., Cesselli, D., Caragnano, A., Athanasakis, E., Katare, R., et al., 2014. Ex vivo molecular rejuvenation improves the therapeutic activity of senescent human cardiac stem cells in a mouse model of myocardial infarction. Stem Cells 32 (9), 2373–2385. Available at https://doi.org/10.1002/stem.1728.

Bartolucci, J., Verdugo, F.J., González, P.L., Larrea, R.E., Abarzua, E., Goset, C., et al., 2017. Safety and efficacy of the intravenous infusion of umbilical cord mesenchymal stem cells in patients with heart failure: a phase 1/2 randomized controlled trial (RIMECARD trial [Randomized clinical trial of intravenous infusion umbilical cord mesenchymal stem cells on cardiopathy]). Circulation Research 121 (10), 1192–1204. Available at https://doi.org/10.1161/CIRCRESAHA.117.310712.

Bartunek, J., Behfar, A., Dolatabadi, D., Vanderheyden, M., Ostojic, M., Dens, J., et al., 2013. Cardiopoietic stem cell therapy in heart failure: the C-CURE (cardiopoietic stem cell therapy in heart failURE) multicenter randomized trial with lineage-specified biologics. Journal of the American College of Cardiology 61 (23), 2329–2338. Available at https://doi.org/10.1016/j.jacc.2013.02.071.

Bashir, B., Gulati, M., Vishwas, S., Gupta, G., Dhanasekaran, M., Paudel, K.R., et al., 2025. Bridging gap in the treatment of Alzheimer's disease via postbiotics: current practices and future prospects. Ageing Research Reviews 105, 102689. Available at https://doi.org/10.1016/j.arr.2025.102689.

Bhat, A.A., Moglad, E., Afzal, M., Thapa, R., Almalki, W.H., Kazmi, I., et al., 2024. Therapeutic approaches targeting aging and cellular senescence in Huntington's disease. CNS Neuroscience and Therapeutics 30 (10), e70053. Available at https://doi.org/10.1111/cns.70053.

Bhat, A.A., Moglad, E., Goyal, A., Afzal, M., Thapa, R., Almalki, W.H., et al., 2024. Nrf2 pathways in neuroprotection: alleviating mitochondrial dysfunction and cognitive impairment in aging. Life Sciences 357, 123056. Available at https://doi.org/10.1016/j.lfs.2024.123056.

Boehmer, A.A., Kaess, B.M., Ruckes, C., Meyer, C., Metzner, A., Rillig, A., et al., 2024. Pulmonary vein isolation or pace and ablate in elderly patients with persistent atrial fibrillation (ABLATE Versus PACE)—rationale, methods, and design. Canadian Journal of Cardiology 40 (12), 2429–2440. Available at https://doi.org/10.1016/j.cjca.2024.07.021.

Borow, K.M., Yaroshinsky, A., Greenberg, B., Perin, E.C., 2019. Phase 3 DREAM-HF trial of mesenchymal precursor cells in chronic heart failure. Circulation Research 125 (3), 265–281. Available at https://doi.org/10.1161/circresaha.119.314951.

Bourgeois, F.T., Orenstein, L., Ballakur, S., Mandl, K.D., Ioannidis, J.P.A., 2017. Exclusion of elderly people from randomized clinical trials of drugs for ischemic heart disease. Journal of the American Geriatrics Society 65 (11), 2354–2361. Available at https://doi.org/10.1111/jgs.14833.

Cadour, F., Caudron, J., Gillibert, A., Normant, S., Dacher, J.-N., 2025. Normal variations of myocardial T1, T2 and T2* values at 1.5 T cardiac MRI in sex-matched healthy volunteers. Diagnostic and Interventional Imaging 106 (6), 217–225. Available at https://doi.org/10.1016/j.diii.2025.01.005.

Caughey, G.E., Inacio, M.C., Simon Bell, J., Vitry, A.I., Shakib, S., 2020. Inclusion of older people reflective of real-world clinical practice in cardiovascular drug trials. American Heart Association Inc., Australia. Journal of the American Heart Association 9 (21). Available at https://doi.org/10.1161/JAHA.120.016936.

Cho, L., Nicholls, S.J., Nordestgaard, B.G., Landmesser, U., Tsimikas, S., Blaha, M.J., et al., 2025. Design and rationale of Lp(a)HORIZON trial: assessing the effect of lipoprotein(a) lowering with pelacarsen on major cardiovascular events in patients with CVD and elevated Lp(a). American Heart Journal 287, 1–9. Available at https://doi.org/10.1016/j.ahj.2025.03.019.

Chokshi, N.P., Adusumalli, S., Small, D.S., Morris, A., Feingold, J., Ha, Y.P., et al., 2018. Loss-framed financial incentives and personalized goal-setting to increase physical activity among ischemic heart disease patients using wearable devices: the active reward randomized trial. Journal of the American Heart Association 7 (12). Available at https://doi.org/10.1161/jaha.118.009173.

Clancy, C.E., Santana, L.F., 2024. Advances in induced pluripotent stem cell-derived cardiac myocytes: technological breakthroughs, key discoveries and new applications. Journal of Physiology 602 (16), 3871–3892. Available at https://doi.org/10.1113/JP282562.

Cleland, J.G.F., Ferreira, J.P., Mariottoni, B., Pellicori, P., Cuthbert, J., Verdonschot, J.A.J., et al., 2021. The effect of spironolactone on cardiovascular function and markers of fibrosis in people at increased risk of developing heart failure: The heart 'OMics' in AGEing (HOMAGE) randomized clinical trial. European Heart Journal 42 (6), 684–696. Available at https://doi.org/10.1093/eurheartj/ehaa758.

Costa, F., Gomez Doblas, J.J., Díaz Expósito, A., Adamo, M., D'Ascenzo, F., Kołtowski, L., et al., 2025. Artificial intelligence in cardiovascular pharmacotherapy: applications and perspectives. European Heart Journal 46 (37), 3616–3627. Available at https://doi.org/10.1093/eurheartj/ehaf474.

Cottage, C.T., Neidig, L., Sundararaman, B., Din, S., Joyo, A.Y., Bailey, B., et al., 2012. Increased mitotic rate coincident with transient telomere lengthening resulting from Pim-1 overexpression in cardiac progenitor cells. Stem Cells 30 (11), 2512–2522. Available at https://doi.org/10.1002/stem.1211.

Crimi, G., Morici, N., Ferrario, M., Luca Ferri, A., Piatti, L., Grosseto, D., et al., 2019. Time course of ischemic and bleeding burden in elderly patients with acute coronary syndromes randomized to low-dose prasugrel or clopidogrel. Journal of the American Heart Association 8 (2). Available at https://doi.org/10.1161/jaha.118.010956.

Cunningham, J.W., Abraham, W.T., Bhatt, A.S., Dunn, J., Felker, G.M., Jain, S.S., et al., 2024. Artificial intelligence in cardiovascular clinical trials. Journal of the American College of Cardiology 84 (20), 2051–2062. Available at https://doi.org/10.1016/j.jacc.2024.08.069.

De Bartolo, A., Rago, V., Romeo, N., De Cicco, M., Lefranc, B., Leprince, J., et al., 2025. Unveiling selenoprotein T as a novel regulator of cardiomyocyte senescence: pivotal role of the CD36 receptor in AC16 human cardiomyocytes. GeroScience 48, 1751–1770. Available at https://doi.org/10.1007/s11357-025-01759-7.

di Candia, A.M., de Avila, D.X., Moreira, G.R., Villacorta, H., Maisel, A.S., 2021. Growth differentiation factor-15, a novel systemic biomarker of oxidative stress, inflammation, and cellular aging: Potential role in cardiovascular diseases. American Heart Journal Plus: Cardiology Research and Practice 9, 100046. Available at https://doi.org/10.1016/j.ahjo.2021.100046.

Ferencik, M., Mayrhofer, T., Lu, M.T., Bittner, D.O., Emami, H., Puchner, S.B., et al., 2022. Coronary atherosclerosis, cardiac troponin, and interleukin-6 in patients with chest pain. JACC: Cardiovascular Imaging 15 (8), 1427–1438. Available at https://doi.org/10.1016/j.jcmg.2022.03.016.

Gager, G.M., Biesinger, B., Hofer, F., Winter, M.-P., Hengstenberg, C., Jilma, B., et al., 2020. Interleukin-6 level is a powerful predictor of long-term cardiovascular mortality in patients with acute coronary syndrome. Vascular Pharmacology 135, 106806. Available at https://doi.org/10.1016/j.vph.2020.106806.

García-Ortiz, L., Recio-Rodríguez, J.I., Schmidt-Trucksäss, A., Puigdomenech-Puig, E., Martínez-Vizcaíno, V., Fernández-Alonso, C., et al., 2014. Relationship between objectively measured physical activity and cardiovascular aging in the general population – The EVIDENT trial. Atherosclerosis 233 (2), 434–440. Available at https://doi.org/10.1016/j.atherosclerosis.2014.01.021.

Giannoula, Y., Kroemer, G., Pietrocola, F., 2023. Cellular senescence and the host immune system in aging and age-related disorders. Biomedical Journal 46 (3), 100581. Available at https://doi.org/10.1016/j.bj.2023.02.001.

Goyal, K., Afzal, M., Altamimi, A.S.A., Babu, M.A., Ballal, S., Kaur, I., et al., 2024. Chronic kidney disease and aging: dissecting the p53/p21 pathway as a therapeutic target. Biogerontology 26 (1), 32. Available at 10.1007/s10522-024-10173-z.

Gupta, S., Afzal, M., Agrawal, N., Almalki, W.H., Rana, M., Gangola, S., et al., 2025. Harnessing the FOXO-SIRT1 axis: insights into cellular stress, metabolism, and aging. Biogerontology 26 (2). Available at https://doi.org/10.1007/s10522-025-10207-0.

Hastings, M.H., Zhou, Q., Wu, C., Shabani, P., Huang, S., Yu, X., et al., 2024. Cardiac aging: from hallmarks to therapeutic opportunities. Cardiovascular Research 121 (10), 1474–1488. Available at 10.1093/cvr/cvae124.

Hieda, M., Sarma, S., Hearon, C.M., Dias, K.A., Martinez, J., Samels, M., et al., 2020. Increased myocardial stiffness in patients with high-risk left ventricular hypertrophy: the hallmark of stage-B heart failure with preserved ejection fraction. Circulation 141 (2), 115–123. Available at https://doi.org/10.1161/CIRCULATIONAHA.119.040332.

Hopper, I., Marwick, T.H., Wolfe, R., Amerena, J., Curtis, A.J., Freedman, B., et al., 2025. STAREE-HEART: a randomized placebo-controlled trial of atorvastatin effects on a marker of cardiac aging in older individuals without prior cardiovascular disease events: protocol and baseline description of participants. American Heart Journal 290, 140–152. Available at https://doi.org/10.1016/j.ahj.2025.06.004.

Howden, E.J., Sarma, S., Lawley, J.S., Opondo, M., Cornwell, W., Stoller, D., et al., 2018. Reversing the cardiac effects of sedentary aging in middle age-a randomized controlled trial: implications for heart failure prevention. Circulation 137 (15), 1549–1560. Available at https://doi.org/10.1161/CIRCULATIONAHA.117.030617.

Huang, Q., Wang, Y., Farias, K., Zeng, X., Chen, Q., Cheng, H., et al., 2025. NAD^+ replenishment mitigates cardiomyocyte senescence and corrects heart failure with preserved ejection fraction in aged mice. bioRxiv [Preprint]. Available at 10.1101/2025.07.15.664959.

Hussain, M.S., Agrawal, N., Ilma, B., Rekha, M.M., Nayak, P.P., Kaur, M., et al., 2025. Autophagy and cellular senescence in Alzheimer's disease: key drivers of neurodegeneration. CNS Neuroscience and Therapeutics 31 (7). Available at https://doi.org/10.1111/cns.70503.

Jha, S.K., De Rubis, G., Devkota, S.R., Zhang, Y., Adhikari, R., Jha, L.A., et al., 2024. Cellular senescence in lung cancer: molecular mechanisms and therapeutic interventions. Ageing Research Reviews 97, 102315. Available at https://doi.org/10.1016/j.arr.2024.102315.

Jia, K., Dai, Y., Liu, A., Li, X., Wu, L., Lu, L., et al., 2020. Senolytic agent navitoclax inhibits angiotensin II-induced heart failure in mice. Journal of Cardiovascular Pharmacology 76 (4), 452–460. Available at https://doi.org/10.1097/fjc.0000000000000878.

Jihwaprani, M.C., Charbat, M.A., Sula, I., Haider, K.H., 2025. Routes of administration of mesenchymal stem cell therapy in cardiac therapy. Handbook of Regenerative Medicine: Stem Cell-based Approach. CRC Press, Saudi Arabia, pp. 211–232. Available at http://doi.org/10.1201/9781003464594-14.

Karnewar, S., Karnewar, V., Shankman, L.S., Owens, G.K., 2024. Treatment of advanced atherosclerotic mice with ABT-263 reduced indices of plaque stability and increased mortality. JCI Insight 9 (2). Available at https://doi.org/10.1172/jci.insight.173863.

Kato, E.T., et al., 2026. Efficacy and safety of edoxaban in elderly patients with atrial fibrillation in the ENGAGE AF-TIMI 48 trial. Journal of the American Heart Association 5 (5), e003432. Available at 10.1161/JAHA.116.003432.

Kitzman, D.W., Brubaker, P., Morgan, T., Haykowsky, M., Hundley, G., Kraus, W.E., et al., 2016. Effect of caloric restriction or aerobic exercise training on peak oxygen consumption and quality of life in obese older patients with heart failure with preserved ejection fraction: a randomized clinical trial. JAMA - Journal of the American Medical Association 315 (1), 36–46. Available at https://doi.org/10.1001/jama.2015.17346.

Kitzman, D.W., Whellan, D.J., Duncan, P., Pastva, A.M., Mentz, R.J., Reeves, G.R., et al., 2021. Physical rehabilitation for older patients hospitalized for heart failure. New England Journal of Medicine 385 (3), 203–216. Available at https://doi.org/10.1056/NEJMoa2026141.

Kyryk, V., Parkhomenko, O., 2025. Potential prognostic markers of cellular senescence in age-associated cardiovascular pathology. Ukraine Fiziologichnyi Zhurnal 71 (2), 93–105. Available at https://doi.org/10.15407/fz71.02.093.

Lampreia, F., Madeira, C., Dores, H., 2024. Digital health technologies and artificial intelligence in cardiovascular clinical trials: a landscape of the European space. Digital Health 10. Available at https://doi.org/10.1177/20552076241277703.

Lee, J.R., Park, B.W., Park, J.H., Lim, S., Kwon, S.P., Hwang, J.W., et al., 2021. Local delivery of a senolytic drug in ischemia and reperfusion-injured heart attenuates cardiac remodeling and restores impaired cardiac function. Acta Biomaterialia 135, 520–533. Available at https://doi.org/10.1016/j.actbio.2021.08.028.

Lee, V., Dalakoti, M., Zheng, Q., Toh, D.-F., Boubertakh, R., Bryant, J.A., et al., 2025. Effects of sacubitril/valsartan on hypertensive heart disease: the REVERSE-LVH randomized phase 2 trial. Nature Communications 16, 6981. Available at https://doi.org/10.1038/s41467-025-62203-0.

Lérida-Viso, A., Estepa-Fernández, A., Morellá-Aucejo, Á., Lozano-Torres, B., Alfonso, M., Blandez, J.F., et al., 2022. Pharmacological senolysis reduces doxorubicin-induced cardiotoxicity and improves cardiac function in mice. Pharmacological Research 183, 106356. Available at https://doi.org/10.1016/j.phrs.2022.106356.

Lewis-McDougall, F.C., Ruchaya, P.J., Domenjo-Vila, E., Shin Teoh, T., Prata, L., Cottle, B.J., et al., 2019. Aged-senescent cells contribute to impaired heart regeneration. Aging Cell 18 (3), e12931. Available at https://doi.org/10.1111/acel.12931.

Li, L., Su, C., Chen, X., Wang, Q., Jiao, W., Luo, H., et al., 2020. Chlorogenic acids in cardiovascular disease: a review of dietary consumption, pharmacology, and pharmacokinetics. Journal of Agricultural and Food Chemistry 68 (24), 6464–6484. Available at https://doi.org/10.1021/acs.jafc.0c01554.

Linders, A.N., Dias, I.B., Ovchinnikova, E.S., Vermeer, M.C.S.C., Hoes, M.F., Markousis Mavrogenis, G., et al., 2023. Evaluation of senescence and its prevention in doxorubicin-induced cardiotoxicity using dynamic engineered heart tissues. JACC: CardioOncology 5 (3), 298–315. Available at https://doi.org/10.1016/j.jaccao.2023.03.012.

López-Otín, C., Blasco, M.A., Partridge, L., Serrano, M., Kroemer, G., 2023. Hallmarks of aging: an expanding universe. Cell 186 (2), 243–278. Available at https://doi.org/10.1016/j.cell.2022.11.001.

Mansfield, L., Ramponi, V., Gupta, K., Stevenson, T., Mathew, A.B., Barinda, A.J., et al., 2024. Emerging insights in senescence: pathways from preclinical models to therapeutic innovations. npj Aging 10 (1). Available at https://doi.org/10.1038/s41514-024-00181-1.

Marín-Blázquez, M., Rovira, J., Ramírez-Bajo, M.J., Zapata-Pérez, R., Rabadán-Ros, R., 2024. NAD(+) enhancers as therapeutic agents in the cardiorenal axis. Cell Communication and Signaling: CCS 22 (1), 537. Available at 10.1186/s12964-024-01903-4.

Meessen, J.M.T.A., Cesaroni, G., Mureddu, G.F., Boccanelli, A., Wienhues-Thelen, U.-H., Kastner, P., et al., 2021. IGFBP7 and GDF-15, but not P1NP, are associated with cardiac alterations and 10-year outcome in an elderly community-based study. BMC Cardiovascular Disorders 21 (1). Available at https://doi.org/10.1186/s12872-021-02138-8.

Menasché, P., Vanneaux, V., Hagège, A., Bel, A., Cholley, B., Parouchev, A., et al., 2018. Transplantation of human embryonic stem cell–derived cardiovascular progenitors for severe ischemic left ventricular dysfunction. Journal of the American College of Cardiology 71 (4), 429–438. Available at https://doi.org/10.1016/j.jacc.2017.11.047.

North, B.J., Sinclair, D.A., 2012. The intersection between aging and cardiovascular disease. Circulation Research 110 (8), 1097–1108. Available at https://doi.org/10.1161/CIRCRESAHA.111.246876.

Olawade, D.B., Fidelis, S.C., Marinze, S., Egbon, E., Osunmakinde, A., Osborne, A., 2026. Artificial intelligence in clinical trials: a comprehensive review of opportunities, challenges, and future directions. International Journal of Medical Informatics 206. Available at https://doi.org/10.1016/j.ijmedinf.2025.106141.

Pal, A., 2025. Cellular senescence and senotherapeutics in cardiovascular diseases. Advances in Pharmacology 104, 313–349. Available at https://doi.org/10.1016/bs.apha.2025.01.019.

Pietri, P., Stefanadis, C., 2021. Cardiovascular aging and longevity: JACC state-of-the-art review. Journal of the American College of Cardiology 77 (2), 189–204. Available at https://doi.org/10.1016/j.jacc.2020.11.023.

Pillai, V.B., Samant, S., Hund, S., Gupta, M., Gupta, M.P., 2021. The nuclear sirtuin SIRT6 protects the heart from developing aging-associated myocyte senescence and cardiac hypertrophy. Aging 13 (9), 12334–12342. Available at https://doi.org/10.18632/aging.203027.

Rad, A.N., Grillari, J., 2024. Current senolytics: mode of action, efficacy and limitations, and their future. Mechanisms of Ageing and Development 217, 111888. Available at https://doi.org/10.1016/j.mad.2023.111888.

Raisi-Estabragh, Z., Jaggi, A., Gkontra, P., McCracken, C., Aung, N., Munroe, P.B., et al., 2021. Cardiac magnetic resonance radiomics reveal differential impact of sex, age, and vascular risk factors on cardiac structure and myocardial tissue. Frontiers in Cardiovascular Medicine 8. Available at https://doi.org/10.3389/fcvm.2021.763361.

Reeves, G.R., Whellan, D.J., Duncan, P., O'Connor, C.M., Pastva, A.M., Eggebeen, J.D., et al., 2017. Rehabilitation therapy in older acute heart failure patients (REHAB-HF) trial: design and rationale. American Heart Journal 185, 130–139. Available at https://doi.org/10.1016/j.ahj.2016.12.012.

Salerno, N., Marino, F., Scalise, M., Salerno, L., Molinaro, C., Filardo, A., et al., 2022. Pharmacological clearance of senescent cells improves cardiac remodeling and function after myocardial infarction in female aged mice. Mechanisms of Ageing and Development 208, 111740. Available at https://doi.org/10.1016/j.mad.2022.111740.

Sayed, N., Huang, Y., Nguyen, K., Krejciova-Rajaniemi, Z., Grawe, A.P., Gao, T., et al., 2021. An inflammatory aging clock (iAge) based on deep learning tracks multimorbidity, immunosenescence, frailty and cardiovascular aging. Nature Aging 1 (7), 598–615. Available at https://doi.org/10.1038/s43587-021-00082-y.

Schulz, M., Griese-Mammen, N., Anker, S.D., Koehler, F., Ihle, P., Ruckes, C., et al., 2019. Pharmacy-based interdisciplinary intervention for patients with chronic heart failure: results of the PHARM-CHF randomized controlled trial. European Journal of Heart Failure 21 (8), 1012–1021. Available at https://doi.org/10.1002/ejhf.1503.

Shah, B., Newman, J.D., Woolf, K., Ganguzza, L., Guo, Y., Allen, N., et al., 2018. Anti-inflammatory effects of a vegan diet versus the american heart association–recommended diet in coronary artery disease trial. Journal of the American Heart Association 7 (23). Available at https://doi.org/10.1161/JAHA.118.011367.

Shah, J., Al-Hashimi, A., Benedetto, M., Ruchaya, P.J., 2025. From bench to bedside: the critical need for standardized senescence detection. Archives of Cardiovascular Diseases 118 (3), 205–211. Available at https://doi.org/10.1016/j.acvd.2024.12.008.

Shi, C., Wen, Z., Yang, Y., Shi, L., Liu, D., 2024. NAD+ metabolism and therapeutic strategies in cardiovascular diseases. Atherosclerosis Plus 57, 1–12. Available at https://doi.org/10.1016/j.athplu.2024.06.001.

Stojanović, S.D., Thum, T., Bauersachs, J., 2025. Anti-senescence therapies: a new concept to address cardiovascular disease. Cardiovascular Research 121 (5), 730–747. Available at https://doi.org/10.1093/cvr/cvaf030.

Subedi, S., Guntipally, M., Suwal, N., Thapa, R., Bashyal, S., Panth, N., et al., 2025. Cellular senescence in chronic obstructive pulmonary disease: molecular mechanisms and therapeutic interventions. Ageing Research Reviews 110, 102813. Available at https://doi.org/10.1016/j.arr.2025.102813.

Suleimenova, M., Abzaliyev, K., Mansurova, M., Abzaliyeva, S., Kurmanova, A., Tokhtakulinova, G., et al., 2025. A predictive model of cardiovascular aging by clinical and immunological markers using machine learning. Diagnostics 15 (7), 850. Available at https://doi.org/10.3390/diagnostics15070850.

Thapa, R., Gupta, S., Gupta, G., Bhat, A.A., Smriti, Singla, M., et al., 2024. Epithelial-mesenchymal transition to mitigate age-related progression in lung cancer. Ageing Research Reviews 102, 102576. Available at 10.1016/j.arr.2024.102576.

Thapa, R., Moglad, E., Afzal, M., Gupta, G., Bhat, A.A., Hassan almalki, W., et al., 2024. The role of sirtuin 1 in ageing and neurodegenerative disease: a molecular perspective. Ageing Research Reviews 102, 102545. Available at https://doi.org/10.1016/j.arr.2024.102545.

Thapa, R., Marianesan, A.B., Rekha, A., Ganesan, S., Kumari, M., Bhat, A.A., et al., 2025. Hypoxia-inducible factor and cellular senescence in pulmonary aging and disease. Biogerontology 26 (2). Available at https://doi.org/10.1007/s10522-025-10208-z.

Ulus, A.T., Mungan, C., Kurtoglu, M., Celikkan, F.T., Akyol, M., Sucu, M., et al., 2020. Intramyocardial transplantation of umbilical cord mesenchymal stromal cells in chronic ischemic cardiomyopathy: a controlled, randomized clinical trial (HUC-HEART Trial). International Journal of Stem Cells 13 (3), 364–376. Available at https://doi.org/10.15283/IJSC20075.

Wang, L.-F., Li, W.-J., Zhang, X.-Y., Zhang, Y.-C., Chen, G.-F., Zhou, X.-Y., et al., 2024. Resveratrol prevents age-related heart impairment through inhibiting the Notch/ NF-κB pathway. Food Science & Nutrition 12 (2), 1035–1045. Available at https://doi.org/10.1002/fsn3.3817.

Weber, B., Galante, A.P., Bersch-Ferreira, A.C., Torreglosa, C.R., Carvalho, V.O., da Silva Victor, E., et al., 2012. Effects of Brazilian cardioprotective diet program on risk factors in patients with coronary heart disease: a Brazilian cardioprotective diet randomized pilot trial. Clinics (Sao Paulo, Brazil) 67 (12), 1407–1414. Available at https://doi.org/10.6061/clinics/2012(12)10.

Xie, S., Xu, S.-C., Deng, W., Tang, Q., 2023. Metabolic landscape in cardiac aging: insights into molecular biology and therapeutic implications. Signal Transduction and Targeted Therapy 8 (1). Available at https://doi.org/10.1038/s41392-023-01378-8.

Yang, B., Li, T., Wang, Z., Zhu, Y., Niu, K., Hu, S., et al., 2024. Ruxolitinib-based senomorphic therapy mitigates cardiomyocyte senescence in septic cardiomyopathy by inhibiting the JAK2/STAT3 signaling pathway. International Journal of Biological Sciences 20 (11), 4314–4340. Available at https://doi.org/10.7150/ijbs.96489.

Yau, T.M., Pagani, F.D., Mancini, D.M., Chang, H.L., Lala, A., Woo, Y.J., et al., 2019. Intramyocardial injection of mesenchymal precursor cells and successful temporary weaning from left ventricular assist device support in patients with advanced heart failure: a randomized clinical trial. American Medical Association, Canada JAMA - Journal of the American Medical Association 321 (12), 1176–1186. Available at https://doi.org/10.1001/jama.2019.2341.

Yu, F., Zhao, H., Luo, L., Wu, W., 2025. Nicotinamide adenine dinucleotide supplementation to alleviate heart failure: a mitochondrial dysfunction perspective. Nutrients 17 (11), 1855. Available at https://doi.org/10.3390/nu17111855.

Zbinden, R., Piccolo, R., Heg, D., Roffi, M., Kurz, D.J., Muller, O., et al., 2016. Ultrathin strut biodegradable polymer sirolimus-eluting stent versus durable-polymer everolimus-eluting stent for percutaneous coronary revascularization: 2-year results of the BIOSCIENCE trial. Journal of the American Heart Association 5 (3). Available at https://doi.org/10.1161/jaha.116.003255.

Zhai, P., Sung, E.-A., Takayama, K., Ikeda, S., Matsushita, M., Nakada, Y., et al., 2025. S913 phosphorylation of Ulk1 protects the heart from aging through inhibition of cardiac senescence. Cardiovascular Research 121 (10), 1609–1622. Available at https://doi.org/10.1093/cvr/cvaf129.

Zhang, L., Smyth, D., Al-Khalaf, M., Blet, A., Du, Q., Bernick, J., et al., 2022a. Insulin-like growth factor-binding protein-7 (IGFBP7) links senescence to heart failure. Nature Cardiovascular Research 1 (12), 1195–1214. Available at https://doi.org/10.1038/s44161-022-00181-y.

Zhang, H., Xue, Y., Pan, T., Zhu, X., Chong, H., Xu, C., et al., 2022b. Epicardial injection of allogeneic human-induced-pluripotent stem cell-derived cardiomyocytes in patients with advanced heart failure: protocol for a phase I/IIa dose-escalation clinical trial. BMJ Open 12 (5), e056264. Available at https://doi.org/10.1136/bmjopen-2021-056264.

Chapter 16

Gender differences and ethnic disparities in cardiac aging

Rabab Fatima[1], Mokhtar Rejili[2], Yumna Khan[3], Lina Eltaib[4], Mudasir Maqbool[5], Amita Joshi Rana[6,7], Shahin Fatima[8], Md. Sadique Hussain[9,10], and Muhammad Tufail[11]

[1]*School of Health Science & Technology, UPES, Dehradun, Uttarakhand, India,* [2]*Department of Biology, College of Sciences, Imam Mohammad Ibn Saud Islamic University (IMSIU), Riyadh, Saudi Arabia,* [3]*Saveetha Institute of Basic Medical Sciences (SIBMS), Saveetha Institute of Medical and Technical Sciences (SIMATS), Saveetha University, Thandalam, Chennai, Tamil Nadu, India,* [4]*Department of Pharmaceutics, College of Pharmacy, Northern Border University, Rafha, Saudi Arabia,* [5]*Department of Pharmacology, Government Medical College Baramulla, Baramulla, Jammu and Kashmir, India,* [6]*College of Pharmacy, Graphic Era Hill University, Bhimtal, Uttarakhand, India,* [7]*Centre for Promotion of Research, Graphic Era (Deemed to be) University, Dehradun, Uttarakhand, India,* [8]*Department of Computer Science and Engineering, Koneru Lakshmaiah Education Foundation, Hyderabad, Telangana, India,* [9]*Uttaranchal Institute of Pharmaceutical Sciences, Uttaranchal University, Dehradun, Uttarakhand, India,* [10]*School of Pharmaceutical Sciences, Lovely Professional University, Phagwara, Punjab, India,* [11]*Department of Oral and Maxillofacial Surgery, Center of Stomatology, Xiangya Hospital, Central South University, Changsha, Hunan, P.R. China*

16.1 Introduction

In the global burden of diseases (GBDs), cardiovascular diseases (CVDs) are the primary cause of death with estimates of 20.5 million deaths in 2021, and projections suggest that it will continue to grow due to ageing populations (Tan et al., 2025). CVDs substantially impact global mortality and incur excessive healthcare costs. Notwithstanding comprehensive preventative initiatives and advancements in treatment, detrimental health behaviors and environmental variables have compromised these improvements, leading to a persistent increase in overall CVD mortality over the past three decades (Roth et al., 2020). The worldwide epidemic of metabolic disorders (Chew et al., 2022; Lin et al., 2023) and the underlying social determinants of health have common pathobiological mechanisms that exacerbate CVDs, resulting in significant social and economic consequences globally (Chong et al., 2023). In the future, males are anticipated to exhibit significantly inferior cardiovascular outcomes in comparison to females (Chong et al., 2025). The most prevalent nonmodifiable risk factor of most CVDs, including coronary artery disease, heart failure, arrhythmias, and valvular disease, is aging. Older adults, despite having no conventional risk factors, eventually experience structural and functional cardiac alterations, which include greater myocardial stiffness, loss of diastolic relaxation, mitochondrial malfunction, and increased pro-inflammatory signaling, which worsen over age and predispose the elderly to CVD (North and Sinclair, 2012; Lin and Kerkelä, 2020). Demographic changes at the population level, with an increase in the proportion of older age populations, will presumably proportionately predispose the CVD burden, especially in low- and middle-income countries with rapidly shifting epidemiology (Chong et al., 2025). Knowledge of cardiac aging consequently forms fundamental objectives in worldwide cardiovascular health and prevention models.

Aging has universal effects on cardiovascular structure and function, but is neither biologically nor clinically homogenous, regardless of population (Hussain et al., 2024). Gender- and sex-related distinctions have been acknowledged as important variables of cardiovascular aging patterns. Among others, women are characterized by a different or slower rate of myocardial remodeling, slower age-related arterial stiffening, as well as an increased lifetime risk of heart failure with preserved ejection fraction (HFpEF) (Coppi et al., 2025), and men are typified by earlier systolic dysfunction and a greater burden of coronary atherosclerosis (Ji et al., 2022; Gu et al., 2017). These variations indicate some sophisticated interplay between sex hormones, genetic factors, immune processes, cardiometabolic factors, and sociocultural determinants of cardiovascular aging throughout the life cycle.

Cellular Senescence and Cardiac Aging. DOI: https://doi.org/10.1016/B978-0-443-45811-8.00018-0

These biological differences are further aggravated by ethnic and racial differences. There is substantial evidence that people with African, South Asian, and Hispanic ancestry experience a faster progression of vascular aging, an increased prevalence of hypertension, an earlier occurrence of heart failure, and increased CVD mortality as opposed to White populations (Reeves and Odden, 2025; Sudre et al., 2022). Notably, these differences cannot be explained by genetics only; they are the products of overlapping forces such as the varying exposure to environmental stress, social determinants of health, epigenetic changes, access to healthcare, and differences in the control of risk factors (Carnethon et al., 2017). The existence of ethnicity-specific cardiac age correlates, like greater hypertrophy of the left ventricle in African Americans or higher risk of cardiometabolic diseases in South Asians, indicates the need to target preventive and treatment interventions to different populations (Basavarajaiah et al., 2008). Gender and ethnicity dependence contribute to heterogeneous cardiovascular aging pathways, which affect the onset, presentation, progression, and outcome of the disease. To attain accuracy in cardiology and mitigate inequities in universal cardiovascular wellbeing, inquiry into these disparities is vital.

The purpose of this chapter is to give an in-depth discussion of biological, epidemiological, and sociocultural determinants of gender difference and ethnic disparity in cardiac aging. It combines existing data on the sex-specific physiology, hormonal control, genetic inclinations, and lifestyle alterations that moderate cardiovascular aging in both men and women. Simultaneously, it investigates ethnic and racial differences in cardiac structure, cardiac function, aging of vascular, and CVD risk, and it takes into consideration the communities that have an excessive burden of premature cardiovascular morbidity. The emerging knowledge of genomics, epigenetics, and precision-medicine strategies, which could contribute to reducing these disparities, is also mentioned in the chapter. Through the combination of population research findings, research based on mechanisms, and clinical records, the chapter aims to identify gaps in knowledge, offer future studies, and emphasize the importance of embracing inclusive and equitable cardiovascular aging studies.

16.2 Biological basis of gender differences in cardiac aging

Estrogen has significant antisenescence effects in cardiomyocytes via several pathways that involve estrogen receptor-a (ER-a) signaling that increases telomerase activity and inhibits senescence pathways mediated by p53/p21 (Luo and Kim, 2016). Estrogen improved the expression of endothelial nitric oxide synthase (eNOS) to allow vasodilation and decrease oxidative stress by elevating the activity of superoxide dismutase. This is partially due to the mechanisms that justify the reduced incidence of CVDs in premenopausal women as opposed to that of age-matched men (Kypreos et al., 2014). Progesterone exhibited complicated CVD effects with nuclear progesterone receptors and membrane-bound receptors. It suppressed the production of inflammatory cytokines and regulated the activity of matrix metalloproteinases, and this lowered the extracellular matrix remodeling (Raghupathy and Szekeres-Bartho, 2022). Nonetheless, a few positive estrogenic actions on the vascular functioning may be counteracted by progesterone, especially when it is taken in synthetic progestin (Shah et al., 2019). Natural progesterone seems to be antiinflammatory and promotes the quiescence of the vascular smooth muscle cells.

Androgens have paradoxical cardiac tissue effects. Testosterone either stimulates hypertrophy of cardiomyocytes at the androgen receptor-mediated upregulation of IGF-1 and may increase the rate of cellular senescence by enhancing reactive oxygen species (ROS) generation (Schafstedde and Nordmeyer, 2023). Dihydrotestosterone stimulates fibroblast growth and collagen production, which leads to cardiac fibrosis. On the other hand, the physiological levels of testosterone favor mitochondrial biogenesis and endothelial functionality by nongenomic signal transduction (Yin et al., 2021).

Knowledge of the interaction of sex hormones with cardiac ageing pathways presents therapeutic interventions to age-associated cardiovascular deterioration. The estrogen-progesterone modulatory interactions and androgenic effects of the hormones on the cardiovascular system highlight the significance of hormonal balance in preserving the cardiovascular system during ageing (Xu et al., 2025).

Menopausal transition triggers cardiovascular remodeling that is critical, and women are at risk of heart failure with preserved ejection fraction (HFpEF) two- to threefold more after menopause. The withdrawal of estrogen removes ER-a-mediated cardioprotection, leading to endothelial dysfunction, arterial stiffening, and poor ventricular-arterial coupling (Ying et al., 2021). Women in postmenopause have less bioavailability of nitric oxide, more endothelin-1 expression, and more inflammatory signaling by NF-kB activation. The sudden decrease in estradiol is associated with a faster process of myocardial fibrosis, which is mediated by transforming growth factor-b increase and overproduction of collagen. Epidemiological research showed that women make up 60%–70% of HFpEF, and the prevalence peaks 5–10 years after menopause, implying that there is a definite hormonal window to work with (Raj et al., 2023).

Studies are showing a growing disparity between the male and female heart in relation to not only size and geometry, but also cellular stress responses, mitochondrial performance, and the induction of senescence pathways. These differences have important critical implications for the formation of heart failure subtypes, arrhythmogenic risk, and therapeutic responsiveness. In a study, sexual dimorphism of the left ventricle (LV) was demonstrated, which, when indexed by body surface area, men revealed a bigger LV mass and more wall thickness. It is in part due to androgen-mediated enlargements of myocardial fibers and an increased hemodynamic load of lean body mass (Bairey Merz et al., 2020). Women, on the other hand, showed a more concentric pattern of hypertrophic remodeling in response to pathological stimuli like hypertension; however, men more frequently suffer eccentric hypertrophy with chamber dilation. Large cohort studies such as the Framingham Heart Study and the MESA (Multi-Ethnic Study of Atherosclerosis) indicated a steeper age-associated increase in LV mass index in women, although at a lower absolute baseline. The sex-based difference in myocardial stiffness is mainly based on the differences in collagen turnover, cross-linking of the extracellular matrix (ECM), and fibroblast activity (Liu et al., 2017). There is increased collagen deposition and collagen I:III ratio in women with aging, which results in greater passive stiffness of the myocardium. Menopause-induced estrogen deficiency increased the rate of ECM remodeling, which was dependent on increased matrix metalloproteinase (MMPs) and tissue inhibitor of metalloproteinase (TIMPs) activity (Lam et al., 2009). Men, on the other hand, have earlier ECM degradation and loss of myocytes, which causes chamber dilation instead of stiffening. The overall implication was that heart failure with preserved ejection fraction (HFpEF) is more common in women, and reduced ejection fraction (HFrEF) is more common in men (Galis and Khatri, 2002).

The sex differences in diastolic physiology are seen at an early age in adulthood and increase with age. The women tend to have increased E/e ratio, decreased ventricular compliance, and slower relaxation rates. These are in part explained by a diminished ventricular cavity size, exaggerated stiffness of the myocardium, and an enhanced likelihood of microvascular dysfunction (Robinson et al., 2024). The reduced bioavailability of nitric oxide (NO) and endothelial responsiveness increased with age in postmenopausal women, which also adds to the diastolic dysfunction (Moreau et al., 2020). Men have stronger early systolic impairment as they experience more cumulative myocyte apoptosis and an earlier development of atherosclerosis, which, in turn, influences the diastolic filling (Choi et al., 2009).

Canonical senescence pathways such as p16 INK4a, p53, and telomerase activity are greatly influenced by sex hormones. It is normal that female cardiomyocytes show reduced p16INK4a and p53 basal expression in reproductive years since estrogen increases the DNA repair and antiapoptotic signatures (Mijit et al., 2020). The longer preservation of telomerase activity in women can guard against the shortening of telomeres and cell fatigue. The loss of estrogen after menopause, but not before, leads to faster upregulation of senescence markers, leading to convergence to male-like senescent phenotypes. Men show more p53- and p21-dependent pathways earlier, which is an indication of more oxidative stress exposure throughout the lifespan (Westin et al., 2011). The disfavor of endogenous estrogens in senescence signaling, which is immediately regained in ovariectomized females, can be supported by the use of animal models (Maldonado et al., 2023). Oxidative stress is one of the key determinants of myocardial aging and pathological remodeling, and its control in males and females is markedly different. Estrogen increased the antioxidant ability by increasing superoxide dismutase (SOD), glutathione peroxidase, and catalase, and decreasing NADPH oxidase activity. This rendered the women who have greater redox homeostasis premenopausal (Allegra et al., 2023). Increased mitochondrial production of reactive oxygen species (ROS) is also found in men in part because of androgen-induced increases in metabolic rate and decreased antioxidant buffering. Increased ROS in the male hearts facilitated an earlier diabetes of DNA damage, lipid peroxidation, and activation of the inflammatory cascade. As women age, they tend to have a more acute increase in oxidative stress after estrogen loss, which leads to microvascular inflammation and fibrosis associated with HFpEF.

Autophagy is important in cardiomyocyte proteostasis and the removal of damaged organelles. Autophagy is significantly sex differentiated yet not well known in its mechanism. The female hearts have stronger basal autophagic flux, which is enhanced by estrogen through AMPK activation and mTOR inhibition (Gatica et al., 2015). The mechanisms protect the quality of mitochondria and decrease the number of misfolded proteins. Conversely, male hearts experience earlier age reduction in autophagy, causing an increased vulnerability to protein aggregation and metabolic stress. Nevertheless, estrogen-lowering after menopause reduces autophagic capacity in women, which might be one of the reasons why they become more susceptible to HFpEF and microvascular impairments in old age (Ren et al., 2018). Structural remodeling, senescence activation, oxidative stress, mitochondrial activity, and autophagy interacted to provide specific aging pathways in the male and female hearts. Preserved mitochondrial efficiency and less oxidative burden in early life keep systolic performance in women, whereas progressive stiffening caused by ECM remodeling and inflammation by senescence favors diastolic dysfunction with age (Pagan et al., 2022). Systolic impairment and eccentric remodeling patterns are more common in men, who are exposed to increased lifetime oxidative stress, and their mitochondrial function and autophagy were already impaired. Such sex-specific pathways explained the dissimilar epidemiology of HFpEF and HFrEF and had an implication on precision medicine (Martinez et al., 2024).

There are differences between women and men with regard to left ventricular hypertrophy and diastolic dysfunction. Women usually exhibit concentric remodeling of proportional thickening of the walls and reduced cavity size without loss of ejection fraction and at the expense of diastolic filling. The pattern is linked to more myocardial stiffness due to higher phosphorylation and higher collagen 1:3 ratios (Martin and Leinwand, 2024). Eccentric hypertrophy is mostly observed in men, and it is characterized by chamber dilatation and preserved compliance to a relatively small extent (Kessler et al., 2019). Echocardiographic reports suggested that women possess greater E/e' ratios and high left atrial pressures during identical hemodynamic stress, which represented an increased intrinsic myocardial stiffness. The diastolic dysfunction progression was faster in postmenopausal women, and tissue Doppler imaging revealed lower early diastolic velocities, as well as longer isovolumetric relaxation time than age-matched men (D'Andrea et al., 2018).

The sexually dimorphic patterns of cardiac senescence markers play an important role in this regard. During the reproductive years, women have a low baseline p16 INK4a expression in the cardiomyocytes, and it is up-regulated dramatically after menopause (Xu et al., 2022). Estrogen represses p164INK4a by way of epigenetic changes, for example, elevated histone acetylation of the CDKN2A locus (Schlecht et al., 2015). Men, on the other hand, showed earlier p53 buildup under oxidative stress, which is associated with testosterone-increased production of mitochondrial ROS (Darbandi et al., 2018). The female cardiomyocytes have a higher telomerase activity up to menopause, when the telomere loss rate is increased. There is a 40%–50% decline in telomerase reverse transcriptase (TERT) expression in postmenopausal women during a period of 5 years. Men show slower telomere erosion and earlier senescence-related secretory phenotype, which is marked by increased IL-6 and IL-1b secretion of senescent cardiac fibroblasts (Razgonova et al., 2020).

There are strong sex differences in the responses to oxidative stress. Estrogen promotes the expression of antioxidant enzymes such as superoxide dismutase-2, catalase, and glutathione peroxidase, which offer excellent protection against ROS in premenopausal women (Ramírez-Expósito et al., 2014). Mitochondrial performance is sexually dimorphic, and female cardiomyocytes have a higher oxidative phosphorylation with higher stability of mitochondrial membrane potential. Autophagy flux is regulated by estrogen: PINK1/Parkin-mediated mitophagy is stimulated by estrogen to ensure quality control of mitochondria, whereas androgens counteractively inhibit the formation of autophagosomes by activating mTOR (Jin and Youle, 2012). Mitochondrial dysfunction and senescence are closely connected to each other and are also the hallmarks of ageing. It is defined as reduced respiratory capacity per mitochondrion, leading to increased generation of oxygen-free radicals. Mitochondrial dysfunction of postmenopausal women is mainly characterized by low activity of complex I and high levels of mitochondrial DNA damage, which primarily points to male senescence patterns (Miwa et al., 2022).

16.3 Ethnic disparities in cardiac aging

Inequalities in diagnosis, treatment, and monitoring of coronary artery disease (CAD) among ethnic minority groups, and especially the South Asians, have been reported since the 1980s. The observation that CAD was more prevalent among South Asians (hazard ratio [HR] 1.35, 95% CI 1.30–1.40) but had a better prognosis (HR 0.78, 95% CI 0.74–0.82) than in Caucasians was an indication that primary prevention is the key in reducing the disparities between South Asian and White populations in the context of morbidity and mortality. Despite the possibility of etiological, preventive, and therapeutic implications of similar research in an individual and a population, other CVDs and noncommunicable diseases have not been investigated as well (Banerjee, 2020). Traditionally marginalized racial, ethnic, and Indigenous groups are more likely to have CVD and risk factors. Such groups of people also tend to have more obstacles to cardiovascular health care access, which further adds to the health equity gap. To combat the issue of inequalities and disparities in cardiovascular health outcomes, the Lancet Regional Health-Europe organized professionals to assess the existing knowledge on inequalities and disparities in cardiovascular health outcomes among marginalized populations and suggest methods of combating these inequalities. A study was conducted to review disparities in CVD with reference to coronary heart disease and stroke, in terms of race, ethnicity, ancestry, and indigeneity, with the intersection of these factors with sex, gender, and socioeconomic status (SES) in Europe and North America. The selection of these regions was based on the fact that they possess well-established healthcare systems and have long-standing, and even increasing, inequalities in cardiovascular health and outcomes (Anand et al., 2025).

The African population represented the highest prevalence of hypertension in the world (46% among the African Americans compared to 28% among the Caucasians), and was associated with an earlier onset and a higher severity. This is related to twofold incidence of heart failure, mainly of HFpEF phenotypes, and higher left ventricular hypertrophy at younger ages. The South Asians are disproportionately burdened with coronary artery disease, and atherosclerosis develops 5–10 years earlier than other ethnicities despite their less favorable traditional risk factor profiles, a condition known as the South Asian paradox (Sucato et al., 2023). Hypertension causes a significant cardiovascular renal morbidity and mortality, which is also a very critical issue among African Americans; approximately 30% of all mortalities can be linked to hypertension. Hypertension among African Americans is more common, with an earlier onset, more severe, and more frequently linked to the presence of target organ damage like left ventricular hypertrophy and other cardiovascular issues, as compared to whites. It is only a quarter of all African Americans with hypertension and half of those under drug therapy who reach a blood pressure below 140/90 mm Hg. These are relatively lower control rates than among white Americans. Better risk factor control strategies should be based on the increased awareness and knowledge of the epidemiologic trends of hypertension and other cardiovascular risk factors, the rate of risk factor control, and influences on these control rates. This would probably lead to a decrease in complications of CVDs (Flack et al., 2003).

16.3.1 Genetic determinants and epigenetic determinants

CVDs are one of the leading causes of death in the global society, but the genetic and epigenetic pathways are not fully understood. Atherosclerosis contributes significantly and mostly causes ischemic heart disease and stroke (Libby et al., 2019). Numerous studies were associated with the genetic factors that relate to the development of atherosclerosis, but few of them are properly characterized. A study examined the correlation between TIMP-1 (tissue inhibitors of metalloproteinases) and ESRa (estrogen receptor a) gene methylation and the polymorphisms of MTHFR (5,10-methylenetetrahydrofolate reductase) in cardio-vascular disease (CVD) and ageing. Two groups were examined: a case group of 37 old patients with CVDs and a control group of 25 young/normal subjects. Genetic and epigenetic studies were conducted by the use of MS-PCR (Methylation Specific Polymerase Chain Reaction) and PCR-RFLP (restriction fragment length polymorphism). The level of methylation of the two genes of interest and the proportion of MTHFR polymorphisms were also found to be greater in the case group. Thus, these conditions were associated with CVD/ageing well (Huica et al., 2011). Topriceanu et al. (2023), in this context, investigated the relationship between biological ageing in terms of DNA-methylation clocks and 1946 birth cohort ECG-based heart age in Britain. They assessed four ages of DNAm and acceleration values of DNAm and two cardiac age scores based on ECG and two scores based on disease risk. It was discovered that faster ageing of DNAm (especially PhenoAge and GrimAge) was always associated with older, weaker, and more electrically impaired hearts at age 60–64. Therapeutic analyses revealed that AgeAccel of DNAm was associated with 10%–40% of the detrimental impact of cardiometabolic risk factors, including diabetes, elevated cholesterol, and prior CVDs, on cardiac ageing. These results thus indicated a biological interaction between epigenetic ageing and cardiovascular health and indicate that modulating epigenetic processes could provide new cardioprotective interventions (Topriceanu et al., 2023).

Genetic polymorphisms that are ethnically specific tend to have a large impact on the process of cardiac ageing (Niu et al., 2024). Individual responses to drugs can be determined by genetic polymorphisms that are randomly found in populations. By examining such variants, which are frequently analysed using Mendelian randomization, one can identify biological pathways of disease. In the past ten years, genetic variation studies have identified some of the major LDL-cholesterol-related genes, such as PCSK9, whose loss-of-function mutations were initially used to propose the benefit of attacking this pathway with drug therapy. Inhibitors of the PCSK9 reduced LDL-C by blocking the activity of LDL receptors, also decreased lipoprotein(a), a novel cardiovascular risk factor. Due to accumulating evidence that an increasing number of patients are either not responding to or responding less to the use of PCSK9 inhibitors, there has been an increase in interest in pharmacogenomics as well. Genetics-based, larger prospective studies should be conducted to completely assess the gene-guided PCSK9 inhibitor therapy (Krittanawong et al., 2022). Human beings have a wide range of sodium sensitivity, with some individuals responding with a marked rise in blood pressure with increased intake of sodium, in part because of genetic differences in the renal sodium-handling pathways (Mutchler et al., 2021). One of the key elements of such regulation is the epithelial sodium channel (ENaC), which is an apical heteromeric channel, made of a, b, and g subunits in the apical membrane of the distal nephron cortical collecting ducts. ENaC, being the last sodium reabsorption regulator, is an end-effector of the renin-angiotensin-aldosterone system and plays a critical role in regulating extracellular fluid volume and blood pressure in the long term. ENaC dysregulation played a direct role in sodium reabsorption and hypertension by showing gain-of-function ENaC mutations that lead to Liddle syndrome, a rare monogenic form of severe hypertension. Human and animal studies indicated that improper renal sodium retention by ENaC further promoted salt-sensitive hypertension (Sun et al., 2011).

Genetic tendencies are highly adjusted to environmental conditions, which form ethnic differences (Jelenkovic et al., 2016). The highly refined carbohydrate and saturated fat contents of traditional South Asian diets contribute significantly to the genetic predisposition to metabolic syndrome (Misra et al., 2018). On the other hand, some populations with dietary habits of the Mediterranean provide cardioprotective effects by providing favorable lipid profiles and lowering of inflammation (Pant et al., 2024). The socioeconomic status is in a critical overlap with the biology: the pro-inflammatory states are disproportionately in minority groups due to a chronic psychosocial stress caused by discrimination and a lower socioeconomic position. Biological vulnerabilities are enhanced by disparities in access to healthcare. African Americans and Hispanics have delayed diagnosis, adherence to drugs because of cost barriers, and access to advanced cardiac therapies. Exposures to the environment, such as air pollution, neighborhood food environments, and occupational hazards, are not evenly spread across ethnic groups, which causes cumulative disadvantage (Hajat et al., 2015). Allostatic load as a biological manifestation of social disadvantage accelerates the process of cardiac senescence, and this process causes self-perpetuating cycles of environmental stressors, increasing the genetic susceptibility of cardiac ageing, eventually resulting in ethnic differences in cardiac ageing outcomes (Teferi et al., 2026). An investigation by Waziri et al. (2025) in this context revealed a novel biomarker for vascular-associated biological stress as an indicator of the physiological burden of a multiplicity of vascular conditions prevalent with ageing. They used the 550 adults of the MIDUS refresher sample to test the hypothesis that this biomarker, DNA methylation-based biological ageing (GrimAge2), and allostatic load mediated the relationship between vascular disease and cognition. It was found that the impact of vascular disease on global cognition (0.60, P=.01) and executive function (0.53, P=.002) was significantly mediated by vascular-related biological stress, but not episodic memory (1.1, P=.308). There was also the mediation of global cognition (0.27, P=.002) and executive functioning (0.20, P=.002) by DNA methylation (GrimAge2), but a smaller effect on episodic memory (0.39, P=0.102). Allostatic load as a mode of mediation which was found to be a small and insignificant mode of mediation across domains. The findings, in general, indicated that cumulative vascular biological stress coupled with epigenetic ageing mechanisms played an important role in cognitive impairment related to vascular disease (Waziry et al., 2025).

16.4 Intersection of gender and ethnicity in cardiac aging

Cardiac aging is a multifaceted process and depends on various intrinsic biological interferences, where sex and ethnicity determine the course of structural and functional deterioration. The cellular and molecular differences between males and females have been demonstrated to play a role in the unique patterns of myocardial ageing. For example, males had shorter telomeres, lower mitochondrial activity, and were more prone to deleterious somatic mutations than females, suggesting that inherent cellular processes could contribute to the faster cardiac aging in men (Podolskiy et al., 2016). In support of these results, experimental research done on rats showed that the female heart and the ventricular myocytes had higher fractional shortening and slow relaxation as compared to males. Transcriptomic studies showed about 600 sexually dimorphic expressed genes with more protein kinase A activity in female myocytes in favor of baseline molecular signaling in sex-specific cardiac functionality (Trexler et al., 2017). Moreover, in mice, males had earlier aortic stiffness, vascular fibrosis, and angiotensin II-mediated vasoconstriction, and females had delayed and steeper cardiovascular function declines. Sex-specific mechanisms that lead to myocardial aging were further demonstrated through differences in perivascular fibrosis and MR-regulated profibrotic genes (DuPont et al., 2021).

Ethnicity is another important determinant of myocardial aging patterns that tends to interplay with sex to create population-specific susceptibilities. For instance, the South Asians showed a greater age-related improvement in diastolic dysfunction, as indicated by the E/e′ ratio, relative to the African Caribbeans, despite having reduced blood pressure, weight index, and left ventricular mass. South Asian ethnicity was independently predictive of higher E/e′ and mortality risk, indicating that ethnic factors can independently accelerate cardiac aging independent of predictable risk factors (Shantsila et al., 2016). Similar trends were confirmed by large cohort studies based in the United States. Of 5110 CVD-free participants, non-Hispanic Blacks adults were highest in mean excess heart age (13.0 years), followed by Mexican-Americans (10.5 years) and non-Hispanic Whites (8.5 years). Over 10 years of excess heart age increased the risk of cardiovascular mortality by 65%, and the association was most significant in non-Hispanic Whites (HR 1.83) (Hirsch et al., 2018).

In addition to these biological determinants, the interplay of social and environmental factors with sex and ethnicity also influences cardiac aging and thus combined biological and social determinants. A study demonstrated that vascular aging measures (carotid to femoral pulse velocity) in Black women (9.34 ± 1.91 m/s) had higher values than those in White women (8.84 ± 1.64 m/s), which showed that sex and race were both significant contributors to the process of arterial stiffening (Baldo et al., 2017). Epigenetic effects, such as FKBP5 demethylation in blacks, were associated with accelerated DNA methylation aging and increased cardiometabolic risk, which proves that early-life stress may increase biological susceptibility (Beach et al., 2022). At the population level, it was found that non-Hispanic Black men and women had greater predicted heart ages (58.7 and 58.9 years) than non-Hispanic White men and women (55.3 and 52.5 years), with regional and sociodemographic factors also contributing to excess heart age, especially in Massachusetts (men: 10.1 years; women: 9.1 years) (Yang et al., 2015). Furthermore, adults of New York City who were non-Latino Blacks exhibited the greatest excess heart age (8.4 years), and non-Latino White women the lowest (2.3 years), which supports the cumulative nature of biological, ethnic, and environmental risk factors (Tabaei et al., 2019).

The recent breakthroughs in the field of artificial intelligence (AI) have offered the means to measure the synergistic impact of biological, ethnic, and social determinants on cardiac aging. AI-based electrocardiography assessments in 280,323 patients revealed that financial strain (β=–1.02) and food insecurity (β=–0.74) were the variables most strongly correlated with cardiac aging, relative to comorbidities and demographic factors, and sex-stratified analysis confirmed the findings (Rajai et al., 2025). Furthermore, AI-based Heart Delta Age (HDA) was associated with hypertension, diabetes, myocardial infarction, and heart failure, and increased HDA was positively associated with all-cause (HR 1.62) and cardiovascular mortality (HR 2.12) (Mossavarali et al., 2025). In order to provide useful information on risk stratification and targeted therapies, the technologies offer objective measurements to assess the overlap in biological and social drivers of cardiovascular aging.

16.5 Multifactorial mechanisms contributing to cardiac aging disparities

Cardiac aging is affected by a series of interconnected mechanisms, such as senescence, oxidative stress, mitochondrial dysfunction, telomere attrition, and epigenetic remodeling (Fig. 16.1). These processes are not only associated with functional deterioration with age but also with differences between sex and racial/ethnic groups. The interrelationship of these processes is important to explain the multifactorial nature of discrepancies in cardiac aging. Age-related cardiac remodeling and dysfunction are central processes that are regulated by senescence and senescence-associated secretory phenotype (SASP). A large-scale proteomics analysis was performed on 2248 participants in the Penn Heart Failure Study (PHFS) with the SOMAScan V4 assay to measure plasma concentrations of 25 SASP proteins. Two SASP factors were discovered and had a significant correlation with old age, higher NYHA class, and decreased renal functioning. Notably, these variables had independent predictive ability of all-cause mortality, independent of traditional risk scores and NT-proBNP. Proteomic pathway analysis also demonstrated an association with more than a thousand proteins implicated in coagulation, complement activation, acute phase response, and metabolism regulation, demonstrating their role in senescence-related dysfunction of the heart (Salman et al., 2024). Based on these findings, there was a study of sex-specific cardiac aging, which revealed that female hearts showed a lower level of Sirt1 and Sirt3 expression, impairment of mitochondrial biogenesis, accumulation of cardiac macrophages, heightened levels of pro-inflammatory cytokines, and NF-κB activation. These age-dependent changes were highly correlated with SASP, but the male hearts lacked any significant changes, highlighting sex-specific vulnerability to senescence-driven cardiac dysfunction (Barcena de Arellano et al., 2019).

Oxidative stress (OS) has been recognized as another key determinant of disparities in cardiac aging, acting upstream of vascular and myocardial dysfunction. OS markers, 8-hydroxy-2'-deoxyguanosine (8-OHdG), and 8-isoprostane, were found to correlate with 24-hour blood pressure responses and 24-hour blood pressure in a cohort of 137 high-cardiovascular

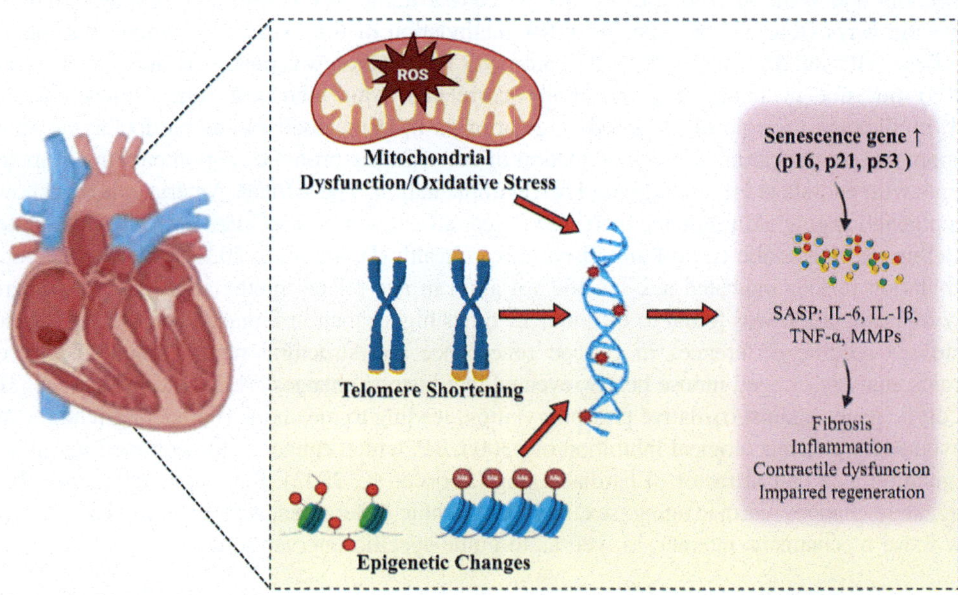

FIGURE 16.1 Drivers of cardiomyocyte senescence and cardiac aging. Schematic overview of the major molecular and cellular drivers of cardiomyocyte senescence and cardiac aging. Mitochondrial dysfunction and increased reactive oxygen species (ROS) promote telomere shortening, DNA damage, and epigenetic alterations, which activate senescence-associated pathways (e.g., p16, p21, p53) and the senescence-associated secretory phenotype (SASP), ultimately contributing to fibrosis, hypertrophy, and functional decline of the aging heart.

risk young adults. Interestingly, 8-OHdG was linked to systolic BP measurements of the European Americans and not African Americans, but on the other hand, 8-isoprostane was linked to reactive SBP and nighttime diastolic BP only in African Americans, demonstrating race-specific effects of OS on cardiovascular function (Kapuku et al., 2017). Likewise, the African Americans in the META-Health cohort had reduced glutathione plasma and a more oxidized glutathione redox potential in comparison to the Whites, despite traditional risk factors and metabolic syndrome. This increased oxidative stress was associated with endothelial dysfunction, vascular inflammation, and atherosclerosis, which offers a population-specific mechanism of accelerated cardiac aging (Morris et al., 2012). Corroborating this, a study on patients with type 2 diabetes mellitus observed that White patients were found to have greater arterial stiffness (via pulse wave velocity), increased urinary 8-OHdG, and reduced antioxidant defenses (GPx-3, SOD, selenium) than black patients. The regression analysis revealed that oxidative DNA damage was an independent predictor of vascular stiffness in White but not Black patients, suggesting ethnic variations in the role of oxidative stress in the disparity in cardiovascular aging (Zitouni et al., 2025).

Mitochondrial dysfunction further amplifies disparities in cardiac aging by compromising energy production and functional capacity. Mitochondrial dysfunction also increases differences in cardiac aging by decreasing energy generation and functional capacity. The contribution of mitochondrial dysfunction in cardiovascular aging was shown by a population-level study of more than 27,000 individuals that showed that lower blood-derived mitochondrial DNA copy number (mtDNA CN) was a predictor of greater risk of incident coronary heart disease (CHD). Though Mendelian randomization did not reveal a direct causal role of mtDNA CN, high LDL cholesterol was revealed to be a causal factor that reduced mitochondrial aging by mitigating lipid-induced metabolic stress on the mitochondria (Liu et al., 2023). Additionally, Self-identified Black participants in the older adult age group were also found to have significantly lower maximal muscle mitochondrial respiration respiration (Max OXPHOS: 50.8 vs. 60.9 pmol/(s·mg), P=.0002) and cardiorespiratory fitness (VO2 peak: 1391 vs. 1566 mL/min, P=.007) than was observed in similar White participants, regardless of their differing socioeconomic status, BMI, and physical activity. Such deficits reduced walking speed, which underscores the importance of mitochondrial dysfunction as one of the main biological factors of racial disparities in cardiac aging (Coen et al., 2025).

Another mechanism of sex- and ethnicity-specific differences in cardiac aging is telomere attrition. Shorter leukocyte telomere length (LTL) was found to be strongly linked with a higher risk and mortality of CHD in White women (a 1-kb decrease in LTL was associated with an additional 50% risk) and not in African American women, despite their longer LTL. These results suggest that ethnicity alters the influence of telomere shortening on cardiovascular outcomes (Carty et al., 2015). Similarly, shorter LTL was highly correlated with elevated pulse pressure and aortic pulse wave velocity, which is an indication of large artery stiffness, in men, but not in women, who had longer telomeres and no significant impact on arterial stiffness. These findings indicate accelerated biological aging among men and indicate sex differences in cardiovascular aging (Benetos et al., 2001).

Epigenetic alterations add to the molecular differences in cardiac aging by controlling the expression of the genes in age-, sex-, and race-specific ways. Age and sex-specific DNA methylation of the PLA2G7 promoter was also related to CHD risks; in a study of 36 CHD and 36 controls, PLA2G7 promoter methylation was positively associated with CHD in females (adjusted P=0.003) but not in males, and positively correlated with increased total cholesterol, triglycerides and apolipoprotein B, implicating the role of epigenetic regulation in lipid metabolism and cardiac aging (Jiang et al., 2013). Furthermore, Genome-wide myocardial DNA methylation of race-specific promoter region patterns of metabolic genes was observed in patients with end-stage HF undergoing LVAD implantation. The African American hearts showed the evidence of faster cardiac epigenetic aging alongside the increased 2-year all-cause mortality after LVAD, and the interaction between epigenetic remodeling and metabolic control and disparities in health (Pepin et al., 2021). A large meta-analysis of whole-blood DNA methylation further indicated sex-specific variation in age-related methylation patterns, where in males, age-related variability of methylation was found to be about 15 times higher than in females, and that biological aging in males may contribute to sex-specific differences in cardiac senescence and structural remodeling (Yusipov et al., 2020). In addition, multiomics analysis of aged mouse hearts revealed that histone changes (H3K27ac, H3K27me3, H3K4me1) were activated by glycolytic genes against oxidative phosphorylation, leading to metabolic remodeling features of cardiac ageing. This shift was reversed by pharmacological inhibition of p300/CBP, which enhanced the contractility, indicating the role of epigenetic dysregulation as a causal factor of cardiac aging (Serio et al., 2023). Since no studies directly investigated the interaction of epigenetic changes, cardiac aging, sex, and racial/ethnic differences, additional research is urgently required to comprehend how these mechanisms interact, as well as to guide specific interventions.

16.6 Clinical implications

The cardiovascular changes that come with aging do not take place consistently across the populations; gender and ethnicity interplay give rise to a considerable degree of variation in the disease consciousness, risk anticipation, therapeutic responses, and health care outcomes. These differences are critical to comprehend how to use precision cardiology models, which can reduce disparity and enhance long-term cardiovascular outcomes in different aging populations.

16.6.1 Diagnosis and risk stratification

Cardiac biomarkers are the foundational instruments in the diagnosis of heart failure, myocardial injury, and cardiometabolic risk, and have a risk of varying, respectively, wide baseline levels and diagnostic cutoff points by gender and ethnicity. As an illustration, high-sensitivity cardiac troponin (hs-cTn) levels are always lower in women than in men, even in healthy groups, which implies a sex-specific cutoff to prevent underdiagnosis due to older women having a myocardial infarction (High sensitivity cardiac troponin and the under-diagnosis of myocardial infarction in women: prospective cohort study, 2016; Lobo and Jaffe, 2023). Biomarker interpretation that accounts for gender differences has been found to enhance the quality of diagnosis and clinical decision-making by adopting sex-specific troponin thresholds (Lobo and Jaffe, 2023).

In the same manner, natriuretic peptides (BNP and NT-proBNP) are gender-specifically varied: the levels are higher in women compared to men in part owing to hormonal factors and reliance on the myocardial wall stress responses (Lam et al., 2011). Natriuretic peptide physiology is also altered by ethnicity. Research has shown that the baseline levels of BNP and NT-proBNP tend to be lower among Blacks as compared to Whites, a factor that can cause under-diagnosis of early heart failure among elderly black adults when standard thresholds are applied (Gupta et al., 2020). These ethnic variations are also caused by variability in adiposity, renal function, and genetic factors, including corin variants (Gupta et al., 2020; Rame et al., 2009). Demographic variance is also observed in other aging-related myocardial remodeling biomarkers, including galectin-3, soluble ST2, and inflammatory cytokines (Kalogeropoulos et al., 2010). Black and Hispanic adults often show increased levels of inflammatory markers indicative of increased load of cardiometabolic stressors and social determinants of health (Carroll et al., 2009; Toussaint et al., 2022). Achieving proper risk stratification in aging populations would thus rely on the incorporation of gender- and ethnicity-specific ranges of biomarkers into risk algorithms in cardiovascular risk.

16.6.2 Therapeutic responses

Gender disparities are significant in the therapeutic effectiveness, side effects, and pharmacodynamics of cardiovascular drugs. Women, by contrast, tend to respond better to thiazide diuretics and calcium channel blockers than men, and men can usually have a better blood-pressure response with ACE inhibitors or β-blockers (Tamargo et al., 2017). Female patients are also affected more frequently by adverse effects like cough when using the ACE inhibitors; edema when using the calcium channel blockers, which are sex differences in drug metabolism and vascular physiology (Tamargo et al., 2017). Statins result in a similar LDL-lowering effect in both sexes in the management of lipids, but women may experience more statin-related muscle symptoms, which might affect long-term compliance (Myerson, 2025). Estrogen signaling alterations in the postmenopause also affect statin pharmacokinetics and have the potential to alter the antiinflammatory effects of treatment (Wakatsuki et al., 2002). New senotherapeutic agents, including senolytic to cellular senescence pathways, are also potentially gender selective because of sex differences in immune aging, mitochondrial stress responses, and senescent cellular burdens of the immune system. There is limited evidence at the very early stages in relation to women's tolerance to senolytics, including dasatinib and quercetin, and these initial events indicate that women might respond differently to inflammatory/metabolic cycles (Lagoumtzi and Chondrogianni, 2021).

Furthermore, ethnicity is a significant aspect of pharmacogenomic and therapeutic variation, which affects the safety and effectiveness of cardiovascular drugs in older adults. An example is that ACE inhibitors have been known to be less effective in reducing blood pressure in Black patients (Peck et al., 2013), who usually respond better to thiazide diuretics or calcium channel blockers because of lower renin levels and different physiology of sodium handling (Gupta et al., 2010). APOL1 variants are also genetic polymorphisms that affect treatment reactions and renal side effects in Black populations (Cerdeña et al., 2021). There is great ethnic variability in statin metabolism. To attain similar levels of LDL reduction, East Asian populations need lower doses of rosuvastatin and simvastatin because plasma drug concentrations are higher than in Caucasian populations because of interactions with variations in SLCO1B1 and other transporter genes (Zhang et al., 2020). Instead, South Asians demonstrate a significant increase in the number of dyslipidemia and insulin resistance, which can lead to the necessity of more vigorous lipid-lowering measures (Misra and Shrivastava, 2013). Recently developed heart failure medications like sacubitil/valsartan are also not reliable in causing the same variability in blood-pressure response across ethnic groups, and thus demand population-specific dosage and monitoring programs (Ibrahim et al., 2020). Collectively, these results demonstrate the significance of the inclusion of pharmacogenomics, demographic data, and social determinants in the therapeutic decision-making of aging people of varying backgrounds.

16.7 Future directions

Future studies in cardiac aging need to increasingly adopt the idea of personalized and precision medicine, in which gender and ethnicity have a deep-seated impact on cardiovascular physiology, biomarker phenotypes, and therapeutic reactions. By incorporating sex-specific genomic, hormonal, and immunometabolic signatures with ethnicity-informed pharmacogenomic measures, incorporating ethnicity-informed prevention and treatment approaches is necessary to enhance outcomes in various aging populations. The cover of diversity in clinical trials is equally important since women and ethnic minorities continue to be systematically underrepresented, hampering the externalization of the existing evidence and closing the cardiovascular care gaps. Providing fair access to the educational process and tailoring the research in such a way that it can identify gender- and ethnicity-specific variations will help to bring mechanistic knowledge to clinical practice much faster. The new development of senodulders in senotherapeutics, such as senolytics and senomorphics, is promising in reducing the effects of age on cardiovascular systems, but a rigorous inquiry into whether these compounds have disparate efficacy or toxicity, based on genders and ethnicities, is necessary. Politically and globally speaking, timely improvements to structural inequities, access to state-of-the-art diagnostics and treatments, as well as the introduction of culturally sensitive preventive interventions, will become inseparable in reducing cardiac aging disparities. Taken together, these future directions highlight the necessity of the interdisciplinary and equity-oriented approach that cuts across the biological discovery, clinical innovation, and health policy to advance cardiovascular aging research and care to all populations.

16.8 Conclusion

The phenomenon known as cardiac aging is global but heterogeneous and is influenced by a complex combination of biological gender, genetic lineage, exposure factors, and social–cultural determinants. In this chapter, evidence synthesis has revealed that men and women experience different myocardial remodeling, vascular stiffening, and heart failure patterns, in part, due to the variability in hormonal signaling, immune ageing patterns, and cardiometabolic risk profiles. Such sex-specific routes overlap with ethnic inequalities that further adjust the way cardiovascular aging progresses. African, South Asian, and Hispanic people are disproportionately aging faster in the vascular system, with a higher prevalence rate of hypertension, earlier development of heart failure, and a higher rate of CVD mortality, which are affected by biological vulnerabilities as well as structural disparities in health access, environmental pressures, and risk-factor management.

Senescence biology, mitochondrial impairment, epigenetic remodeling, and proteomic studies reveal that these differences develop on both biological and social pathways. An example is how sex, race, environment, and social adversity have come together to increase cardiovascular vulnerability in Black women and regionally higher excess heart age. Innovations in biomarkers of cardiac age that are products of artificial intelligence only highlight the scale of social determinants of unfavorable cardiac aging processes even more, namely, financial strain and food insecurity.

In the future, sex-specific molecular signatures in combination with ethnicity-informed genomic, pharmacogenomic, and clinical risk profiles will be a foundation for designing applicable strategic interventions in preventing, detecting, and treating diseases and disorders fairly. Enhancing patient access to the most recent diagnostics, advancing diversity in clinical trials, and reducing structural inequalities in cardiovascular treatment are key translational priorities. Senotherapeutic interventions based on the use of emerging sources hold good promises but need to be evaluated carefully in various populations since they may vary in terms of their efficacy and tolerability according to sex and ethnicity.

References

Allegra, A., Caserta, S., Genovese, S., Pioggia, G., Gangemi, S., 2023. Gender differences in oxidative stress in relation to cancer susceptibility and survival. Antioxidants 12 (6), 1255. Available at https://doi.org/10.3390/antiox12061255.

Anand, S.S., Kandasamy, S., Marchand, M., Kavousi, M., Gulati, M., Deanfield, J., et al., 2025. Reducing inequalities in cardiovascular disease: focus on marginalized populations considering ethnicity and race. The Lancet Regional Health - Europe 56, 101371. Available at https://doi.org/10.1016/j.lanepe.2025.101371.

Bairey Merz, C.N., Nelson, M.D., Cheng, S., Wei, J., 2020. Sex differences and the left ventricle: morphology matters. European Heart Journal - Cardiovascular Imaging 21 (9), 991–993. Available at https://doi.org/10.1093/ehjci/jeaa195.

Baldo, M.P., Cunha, R.S., Ribeiro, A.L.P., Lotufo, P.A., Chor, D., Barreto, S.M., et al., 2017. Racial differences in arterial stiffness are mainly determined by blood pressure levels: results from the ELSA-Brasil study. Journal of the American Heart Association 6 (6). Available at https://doi.org/10.1161/JAHA.117.005477.

Banerjee, A., 2020. Ethnicity, heart failure and the prevention continuum: time to act. Heart (British Cardiac Society) 106 (9), 631–633. Available at https://doi.org/10.1136/heartjnl-2019-316293.

Barcena de Arellano, M.L., Pozdniakova, S., Kühl, A.A., Baczko, I., Ladilov, Y., Regitz-Zagrosek, V., 2019. Sex differences in the aging human heart: decreased sirtuins, pro-inflammatory shift and reduced anti-oxidative defense. Aging 11 (7), 1918–1933. Available at https://doi.org/10.18632/aging.101881.

Basavarajaiah, S., Boraita, A., Whyte, G., Wilson, M., Carby, L., Shah, A., et al., 2008. Ethnic differences in left ventricular remodeling in highly-trained athletes. relevance to differentiating physiologic left ventricular hypertrophy from hypertrophic cardiomyopathy. Journal of the American College of Cardiology 51 (23), 2256–2262. Available at https://doi.org/10.1016/j.jacc.2007.12.061.

Beach, S.R.H., Ong, M.L., Lei, M.K., Carter, S.E., Simons, R.L., Gibbons, F.X., et al., 2022. Methylation of FKBP5 is associated with accelerated DNA methylation ageing and cardiometabolic risk: replication in young-adult and middle-aged Black Americans. Epigenetics: Official Journal of the DNA Methylation Society 17 (9), 982–1002. Available at https://doi.org/10.1080/15592294.2021.1980688.

Benetos, A., Okuda, K., Lajemi, M., Kimura, M., Thomas, F., Skurnick, J., et al., 2001. Telomere length as an indicator of biological aging the gender effect and relation with pulse pressure and pulse wave velocity. Hypertension 37 (2), 381–385. Available at https://doi.org/10.1161/01.hyp.37.2.381.

Carnethon, M.R., Pu, J., Howard, G., Albert, M.A., Anderson, C.A.M., Bertoni, A.G., et al., 2017. Cardiovascular health in African Americans: a scientific statement from the American Heart Association. Circulation 136 (21), e393. Available at https://doi.org/10.1161/CIR.0000000000000534.

Carroll, J.F., Fulda, K.G., Chiapa, A.L., Rodriquez, M., Phelps, D.R., Cardarelli, K.M., et al., 2009. Impact of race/ethnicity on the relationship between visceral fat and inflammatory biomarkers. Obesity 17 (7), 1420–1427. Available at https://doi.org/10.1038/oby.2008.657.

Carty, C.L., Kooperberg, C., Liu, J., Herndon, M., Assimes, T., Hou, L., et al., 2015. Leukocyte telomere length and risks of incident coronary heart disease and mortality in a racially diverse population of postmenopausal women. Arteriosclerosis, Thrombosis, and Vascular Biology 35 (10), 2225–2231. Available at https://doi.org/10.1161/ATVBAHA.115.305838.

Cerdeña, J.P., Tsai, J., Grubbs, V., 2021. APOL1, black race, and kidney disease: turning attention to structural racism. American Journal of Kidney Diseases 77 (6), 857–860. Available at https://doi.org/10.1053/j.ajkd.2020.11.029.

Chew, N.W.S., Zhang, A., Kong, G., Lee, K.L., Ng, C.H., Chong, B., et al., 2022. Prognostically distinct phenotypes of metabolic health beyond obesity in aortic stenosis. American Journal of Cardiology 178, 112–118. Available at https://doi.org/10.1016/j.amjcard.2022.05.018.

Choi, Y.H., Cowan, D.B., Moran, A.M., Colan, S.D., Stamm, C., Takeuchi, K., et al., 2009. Myocyte apoptosis occurs early during the development of pressure-overload hypertrophy in infant myocardium. Journal of Thoracic and Cardiovascular Surgery 137 (6), 1356. Available at https://doi.org/10.1016/j.jtcvs.2008.12.020.

Chong, B., Kong, G., Shankar, K., Chew, H.S.J., Lin, C., Goh, R., et al., 2023. The global syndemic of metabolic diseases in the young adult population: a consortium of trends and projections from the Global Burden of Disease 2000–2019. Metabolism: Clinical and Experimental 141. Available at https://doi.org/10.1016/j.metabol.2023.155402.

Chong, B., Jayabaskaran, J., Jauhari, S.M., Chan, S.P., Goh, R., Kueh, M.T.W., et al., 2025. Global burden of cardiovascular diseases: projections from 2025 to 2050. European Journal of Preventive Cardiology 32 (11), 1001–1015. Available at https://doi.org/10.1093/eurjpc/zwae281.

Coen, P.M., Yi, F., Lui, L.Y., Distefano, G., Ramos, S.V., Goodpaster, B.H., et al., 2025. Muscle mitochondrial respiration and cardiorespiratory fitness contribute to slower walking speed of older individuals who identify as black. Aging Cell 24 (6). Available at https://doi.org/10.1111/acel.70040.

Coppi, F., Pagnoni, G., Grossule, F., Nassar, A., Maini, A., Masaracchia, G., et al., 2025. Gender-specific differences in diastolic dysfunction and HFpEF: pathophysiology, diagnosis, and therapeutic strategies. Journal of Cardiovascular Development and Disease 12 (6), 213. Available at https://doi.org/10.3390/jcdd12060213.

D'Andrea, A., Vriz, O., Ferrara, F., Cocchia, R., Conte, M., Di Maio, M., et al., 2018. Reference ranges and physiologic variations of left E/e' ratio in healthy adults: clinical and echocardiographic correlates. Journal of Cardiovascular Echography 28 (2), 101. Available at https://doi.org/10.4103/jcecho.jcecho_57_17.

Darbandi, M., Darbandi, S., Agarwal, A., Sengupta, P., Durairajanayagam, D., Henkel, R., et al., 2018. Reactive oxygen species and male reproductive hormones. Reproductive Biology and Endocrinology 16 (1). Available at https://doi.org/10.1186/s12958-018-0406-2.

DuPont, J.J., Kim, S.K., Kenney, R.M., Jaffe, I.Z., 2021. Sex differences in the time course and mechanisms of vascular and cardiac aging in mice: role of the smooth muscle cell mineralocorticoid receptor. American Journal of Physiology - Heart and Circulatory Physiology. 320 (1), H169. Available at https://doi.org/10.1152/AJPHEART.00262.2020.

Flack, J.M., Ferdinand, K.C., Nasser, S.A., 2003. Epidemiology of hypertension and cardiovascular disease in African Americans. Journal of Clinical Hypertension (Greenwich, Conn.) 5 (1), 5–11. Available at https://doi.org/10.1111/j.1524-6175.2003.02152.x.

Galis, Z.S., Khatri, J.J., 2002. Matrix metalloproteinases in vascular remodeling and atherogenesis: the good, the bad, and the ugly. Circulation Research 90 (3), 251–262. Available at https://doi.org/10.1161/res.90.3.251.

Gatica, D., Chiong, M., Lavandero, S., Klionsky, D.J., 2015. Molecular mechanisms of autophagy in the cardiovascular system. Circulation Research 116 (3), 456–467. Available at https://doi.org/10.1161/CIRCRESAHA.114.303788.

Gu, H., Gao, Y., Wang, H., Hou, Z., Han, L., Wang, X., et al., 2017. Sex differences in coronary atherosclerosis progression and major adverse cardiac events in patients with suspected coronary artery disease. Journal of Cardiovascular Computed Tomography 11 (5), 367–372. Available at https://doi.org/10.1016/j.jcct.2017.07.002.

Gupta, A.K., Poulter, N.R., Dobson, J., Eldridge, S., Cappuccio, F.P., Caulfield, M., et al., 2010. Ethnic differences in blood pressure response to first and second-line antihypertensive therapies in patients randomized in the ASCOT trial. American Journal of Hypertension 23 (9), 1023–1030. Available at https://doi.org/10.1038/ajh.2010.105.

Gupta, D.K., Walford, G.A., Ma, Y., Jarolim, P., Wang, T.J., Passino, C., 2020. Racial/ethnic differences in circulating natriuretic peptide levels: the diabetes prevention program. PLoS One 15 (2), e0229280. Available at https://doi.org/10.1371/journal.pone.0229280.

Hajat, A., Hsia, C., O'Neill, M.S., 2015. Socioeconomic disparities and air pollution exposure: a global review. Current Environmental Health Reports 2 (4), 440–450. Available at https://doi.org/10.1007/s40572-015-0069-5.

Shah, A.S.V., et al., 2016. High sensitivity cardiac troponin and the under-diagnosis of myocardial infarction in women: prospective cohort study. BMJ 354, i4840.

Hirsch, J.R., Waits, G., Li, Y., Soliman, E.Z., 2018. Racial differences in heart age and impact on mortality. Journal of the National Medical Association 110 (2), 169–175. Available at https://doi.org/10.1016/j.jnma.2017.08.003.

Huica, I., Botezatu, A., Iancu, I.V., Lupeanu, E., Anton, M., Rusanu, C.D.G., et al., 2011. Genetic and epigenetic aspects in cardio-vascular disease and ageing. Romanian Biotechnological Letters 16 (5), 6488–6496.

Hussain, M.S., Altamimi, A.S.A., Afzal, M., Almalki, W.H., Kazmi, I., Alzarea, S.I., et al., 2024. Kaempferol: paving the path for advanced treatments in aging-related diseases. Experimental Gerontology 188. Available at https://doi.org/10.1016/j.exger.2024.112389.

Ibrahim, N.E., Piña, I.L., Camacho, A., Bapat, D., Felker, G.M., Maisel, A.S., et al., 2020. Racial and ethnic differences in biomarkers, health status, and cardiac remodeling in patients with heart failure with reduced ejection fraction treated with sacubitril/valsartan. Circulation: Heart Failure 13 (11). Available at https://doi.org/10.1161/CIRCHEARTFAILURE.120.007829 E007829.

Jelenkovic, A., Sund, R., Hur, Y.M., Yokoyama, Y., Hjelmborg, J.V.B., Möller, S., et al., 2016. Genetic and environmental influences on height from infancy to early adulthood: an individual-based pooled analysis of 45 twin cohorts. Scientific Reports 6. Available at https://doi.org/10.1038/srep28496.

Ji, H., Kwan, A.C., Chen, M.T., Ouyang, D., Ebinger, J.E., Bell, S.P., et al., 2022. Sex differences in myocardial and vascular aging. Circulation Research 130 (4), 566–577. Available at https://doi.org/10.1161/CIRCRESAHA.121.319902.

Jiang, D., Zheng, D., Wang, L., Huang, Y., Liu, H., Xu, L., et al., 2013. Elevated PLA2G7 gene promoter methylation as a gender-specific marker of aging increases the risk of coronary heart disease in females. PLoS One 8 (3), e0179947. Available at https://doi.org/10.1371/journal.pone.0059752.

Jin, S.M., Youle, R.J., 2012. PINK1-and Parkin-mediated mitophagy at a glance. Journal of Cell Science 125 (4), 795–799. Available at https://doi.org/10.1242/jcs.093849.

Kalogeropoulos, A., Georgiopoulou, V., Psaty, B.M., Rodondi, N., Smith, A.L., Harrison, D.G., et al., 2010. Inflammatory markers and incident heart failure risk in older adults. Journal of the American College of Cardiology 55 (19), 2129–2137. Available at https://doi.org/10.1016/j.jacc.2009.12.045.

Kapuku, G., Treiber, F., Raouane, F., Halbert, J., Davis, H., Young-Mayes, S., et al., 2017. Race/ethnicity determines the relationships between oxidative stress markers and blood pressure in individuals with high cardiovascular disease risk. Journal of Human Hypertension 31 (1), 70–75. Available at https://doi.org/10.1038/jhh.2016.39.

Kessler, E.L., Rivaud, M.R., Vos, M.A., Van Veen, T.A.B., 2019. Sex-specific influence on cardiac structural remodeling and therapy in cardiovascular disease. Biology of Sex Differences 10 (1), 20426410. Available at https://doi.org/10.1186/s13293-019-0223-0.

Krittanawong, C., Khawaja, M., Rosenson, R.S., Amos, C.I., Nambi, V., Lavie, C.J., et al., 2022. Association of PCSK9 variants with the risk of atherosclerotic cardiovascular disease and variable responses to PCSK9 inhibitor therapy. Current Problems in Cardiology 47 (7). Available at https://doi.org/10.1016/j.cpcardiol.2021.101043.

Kypreos, K.E., Zafirovic, S., Petropoulou, P.I., Bjelogrlic, P., Resanovic, I., Traish, A., et al., 2014. Regulation of endothelial nitric oxide synthase and high-density lipoprotein quality by estradiol in cardiovascular pathology. Journal of Cardiovascular Pharmacology and Therapeutics 19 (3), 256–268. Available at https://doi.org/10.1177/1074248413513499.

Lagoumtzi, S.M., Chondrogianni, N., 2021. Senolytics and senomorphics: natural and synthetic therapeutics in the treatment of aging and chronic diseases. Free Radical Biology and Medicine 171, 169–190. Available at https://doi.org/10.1016/j.freeradbiomed.2021.05.003.

Lam, K.-K., Cheng, P.-Y., Hsiao, G., Chen, S.-Y., Shen, H.-H., Yen, M.-H., et al., 2009. Estrogen deficiency-induced alterations of vascular MMP-2, MT1-MMP, and TIMP-2 in ovariectomized rats. American Journal of Hypertension 22 (1), 27–34. Available at https://doi.org/10.1038/ajh.2008.306.

Lam, C.S.P., Cheng, S., Choong, K., Larson, M.G., Murabito, J.M., Newton-Cheh, C., et al., 2011. Influence of sex and hormone status on circulating natriuretic peptides. Journal of the American College of Cardiology 58 (6), 618–626. Available at https://doi.org/10.1016/j.jacc.2011.03.042.

Libby, P., Buring, J.E., Badimon, L., Hansson, G.K., Deanfield, J., Bittencourt, M.S., et al., 2019. Atherosclerosis. Nature Reviews Disease Primers 5 (1). Available at https://doi.org/10.1038/s41572-019-0106-z.

Lin, R., Kerkelä, R., 2020. Regulatory mechanisms of mitochondrial function and cardiac aging. International Journal of Molecular Sciences 21 (4), 1359. Available at https://doi.org/10.3390/ijms21041359.

Lin, C., Loke, W.H., Ng, B.H., Chin, Y.H., Chong, B., Goh, R.S.J., et al., 2023. Mortality, cardiovascular, and medication outcomes in patients with myocardial infarction and underweight in a meta-analysis of 6.3 million patients. American Journal of Cardiology 196, 1–10. Available at https://doi.org/10.1016/j.amjcard.2023.02.023.

Liu, C.-Y., Lai, S., Kawel-Boehm, N., Chahal, H., Ambale-Venkatesh, B., Lima, J.A.C., et al., 2017. Healthy aging of the left ventricle in relationship to cardiovascular risk factors: the multi-ethnic study of atherosclerosis (MESA). PLoS One 12 (6), e0179947. Available at https://doi.org/10.1371/journal.pone.0179947.

Liu, X., Sun, X., Zhang, Y., Jiang, W., Lai, M., Wiggins, K.L., et al., 2023. Association between whole blood–derived mitochondrial DNA copy number, low-density lipoprotein cholesterol, and cardiovascular disease risk. Journal of the American Heart Association 12 (20). Available at https://doi.org/10.1161/JAHA.122.029090.

Lobo, R., Jaffe, A.S., 2023. Sex-specific thresholds for cardiac biomarkers-we need to move forward. Reviews in Cardiovascular Medicine 24 (3). Available at https://doi.org/10.31083/j.rcm2403086.

Luo, T., Kim, J.K., 2016. The role of estrogen and estrogen receptors on cardiomyocytes: an overview. Canadian Journal of Cardiology 32 (8), 1017–1025. Available at https://doi.org/10.1016/j.cjca.2015.10.021.

Maldonado, E., Morales-Pison, S., Urbina, F., Solari, A., 2023. Aging hallmarks and the role of oxidative stress. Antioxidants 12 (3), 651. Available at https://doi.org/10.3390/antiox12030651.

Martin, T.G., Leinwand, L.A., 2024. Hearts apart: sex differences in cardiac remodeling in health and disease. Journal of Clinical Investigation 134 (13). Available at https://doi.org/10.1172/JCI180074.

Martinez, C.S., Zheng, A., Xiao, Q., 2024. Mitochondrial reactive oxygen species dysregulation in heart failure with preserved ejection fraction: a fraction of the whole. Antioxidants 13 (11). Available at https://doi.org/10.3390/antiox13111330.

Mijit, M., Caracciolo, V., Melillo, A., Amicarelli, F., Giordano, A., 2020. Role of p53 in the regulation of cellular senescence. Biomolecules 10 (3), 420. Available at https://doi.org/10.3390/biom10030420.

Misra, A., Shrivastava, U., 2013. Obesity and dyslipidemia in South Asians. Nutrients 5 (7), 2708–2733. Available at https://doi.org/10.3390/nu5072708.

Misra, A., Soares, M.J., Mohan, V., Anoop, S., Abhishek, V., Vaidya, R., et al., 2018. Body fat, metabolic syndrome and hyperglycemia in South Asians. Journal of Diabetes and its Complications 32 (11), 1068–1075. Available at https://doi.org/10.1016/j.jdiacomp.2018.08.001.

Miwa, S., Kashyap, S., Chini, E., von Zglinicki, T., 2022. Mitochondrial dysfunction in cell senescence and aging. Journal of Clinical Investigation 132 (13). Available at https://doi.org/10.1172/jci158447.

Moreau, K.L., Hildreth, K.L., Klawitter, J., Blatchford, P., Kohrt, W.M., 2020. Decline in endothelial function across the menopause transition in healthy women is related to decreased estradiol and increased oxidative stress. SGeroScience 42 (6), 1699–1714. Available at https://doi.org/10.1007/s11357-020-00236-7.

Morris, A.A., Zhao, L., Patel, R.S., Jones, D.P., Ahmed, Y., Stoyanova, N., et al., 2012. Differences in systemic oxidative stress based on race and the metabolic syndrome: the morehouse and emory team up to eliminate health disparities (META-Health) study. Metabolic Syndrome and Related Disorders 10 (4), 252–259. Available at https://doi.org/10.1089/met.2011.0117.

Mossavarali, S., Vaezi, A., Gholami, Z., Molaei, A., Yekaninejad, M.S., Asselbergs, F.W., et al., 2025. Determinants of artificial intelligence electrocardiogram-derived age and its association with cardiovascular events and mortality: a systematic review and meta-analysis. npj Digital Medicine 8 (1). Available at https://doi.org/10.1038/s41746-025-01727-7.

Mutchler, S.M., Kirabo, A., Kleyman, T.R., 2021. Epithelial sodium channel and salt-sensitive hypertension. Hypertension 77 (3), 759–767. Available at https://doi.org/10.1161/HYPERTENSIONAHA.120.14481.

Myerson, M., 2025. Statin use in women: unique aspects over the lifespan. Current Cardiovascular Risk Reports 19 (1). Available at https://doi.org/10.1007/s12170-024-00757-2.

Niu, L.L., Fan, H.L., Cao, J., Du, Q.X., Jin, Q.Q., Wang, Y.Y., et al., 2024. The impact of cardiovascular disease gene polymorphism and interaction with homocysteine on deep vein thrombosis. ACS Omega. Available at https://doi.org/10.1021/acsomega.4c05204.

North, B.J., Sinclair, D.A., 2012. The intersection between aging and cardiovascular disease. Circulation Research 110 (8), 1097–1108. Available at https://doi.org/10.1161/CIRCRESAHA.111.246876.

Pagan, L.U., Gomes, M.J., Gatto, M., Mota, G.A.F., Okoshi, K., Okoshi, M.P., 2022. The role of oxidative stress in the aging heart. Antioxidants 11 (2). Available at https://doi.org/10.3390/antiox11020336.

Pant, A., Chew, D.P., Mamas, M.A., Zaman, S., 2024. Cardiovascular disease and the mediterranean diet: insights into sex-specific responses. Nutrients 16 (4). Available at https://doi.org/10.3390/nu16040570.

Peck, R.N., Smart, L.R., Beier, R., Liwa, A.C., Grosskurth, H., Fitzgerald, D.W., et al., 2013. Difference in blood pressure response to ACE-Inhibitor monotherapy between black and white adults with arterial hypertension: a meta-analysis of 13 clinical trials. BMC Nephrology 14 (1). Available at https://doi.org/10.1186/1471-2369-14-201.

Pepin, M.E., Ha, C.M., Potter, L.A., Bakshi, S., Barchue, J.P., Asaad, A.H., et al., 2021. Racial and socioeconomic disparity associates with differences in cardiac DNA methylation among men with end-stage heart failure. American Journal of Physiology - Heart and Circulatory Physiology 320 (5). Available at https://doi.org/10.1152/AJPHEART.00036.2021.

Podolskiy, D.I., Lobanov, A.V., Kryukov, G.V., Gladyshev, V.N., 2016. Analysis of cancer genomes reveals basic features of human aging and its role in cancer development. Nature Communications 7. Available at https://doi.org/10.1038/ncomms12157.

Raghupathy, R., Szekeres-Bartho, J., 2022. Progesterone: a unique hormone with immunomodulatory roles in pregnancy. International Journal of Molecular Sciences 23 (3), 1333. Available at https://doi.org/10.3390/ijms23031333.

Raj, A., Chakole, S., Agrawal, S., Gupta, A., Khekade, H., Prasad, R., et al., 2023. The impact of menopause on cardiovascular aging: a comprehensive review of androgen influences. Cureus 15 (8). Available at https://doi.org/10.7759/cureus.43569.

Rajai, N., Medina-Inojosa, B.J., Hamidabad, N.M., Medina-Inojosa, J.R., Lewis, B.R., Sara, J.D., et al., 2025. Interplay of social determinants of health and traditional risk factors in predicting cardiac aging. Mayo Clinic Proceedings. Available at https://doi.org/10.1016/j.mayocp.2025.01.024.

Rame, J.E., Tam, S.W., McNamara, D., Worcel, M., Sabolinski, M.L., Wu, A.H., et al., 2009. Dysfunctional corin I555(P568) allele is associated with impaired brain natriuretic peptide processing and adverse outcomes in blacks with systolic heart failure: results from the genetic risk assessment in heart failure substudy. Circulation: Heart Failure 2 (6), 541–548. Available at https://doi.org/10.1161/CIRCHEARTFAILURE.109.866822.

Ramírez-Expósito, M.J., Sánchez-López, E., Cueto-Ureña, C., Dueñas, B., Carrera-González, P., Navarro-Cecilia, J., et al., 2014. Circulating oxidative stress parameters in pre- and post-menopausal healthy women and in women suffering from breast cancer treated or not with neoadjuvant chemotherapy. Experimental Gerontology 58, 34–42. Available at https://doi.org/10.1016/j.exger.2014.07.006.

Razgonova, M.P., Zakharenko, A.M., Golokhvast, K.S., Thanasoula, M., Sarandi, E., Nikolouzakis, K., et al., 2020. Telomerase and telomeres in aging theory and chronographic aging theory (review). Molecular Medicine Reports 22 (3), 1679–1694. Available at https://doi.org/10.3892/mmr.2020.11274.

Reeves, A.N., Odden, M.C., 2025. Racial disparities in hypertension subtype prevalence over the lifecourse: evidence of accelerated arterial ageing in a population representative cross-sectional study. BMJ Public Health 3 (1), e001993. Available at https://doi.org/10.1136/bmjph-2024-001993.

Ren, J., Sowers, J.R., Zhang, Y., 2018. Metabolic stress, autophagy, and cardiovascular aging: from pathophysiology to therapeutics. Trends in Endocrinology and Metabolism 29 (10), 699–711. Available at https://doi.org/10.1016/j.tem.2018.08.001.

Robinson, S., Ring, L., Oxborough, D., Harkness, A., Bennett, S., Rana, B., et al., 2024. The assessment of left ventricular diastolic function: guidance and recommendations from the British Society of Echocardiography. Echo Research and Practice 11 (1). Available at https://doi.org/10.1186/s44156-024-00051-2.

Roth, G.A., Mensah, G.A., Johnson, C.O., Addolorato, G., Ammirati, E., Baddour, L.M., et al., 2020. Global burden of cardiovascular diseases and risk factors, 1990-2019: update from the GBD 2019 study. Journal of the American College of Cardiology 76 (25), 2982–3021. Available at https://doi.org/10.1016/j.jacc.2020.11.010.

Salman, O., Zamani, P., Zhao, L., Dib, M.J., Gan, S., Azzo, J.D., et al., 2024. Prognostic significance and biologic associations of senescence-associated secretory phenotype biomarkers in heart failure. Journal of the American Heart Association 13 (17). Available at https://doi.org/10.1161/JAHA.123.033675.

Schafstedde, M., Nordmeyer, S., 2023. The role of androgens in pressure overload myocardial hypertrophy. Frontiers in Endocrinology 14. Available at https://doi.org/10.3389/fendo.2023.1112892.

Schlecht, N.F., Ben-Dayan, M., Anayannis, N., Lleras, R.A., Thomas, C., Wang, Y., et al., 2015. Epigenetic changes in the CDKN2A locus are associated with differential expression of P16INK4A and P14ARF in HPV-positive oropharyngeal squamous cell carcinoma. Cancer Medicine 4 (3), 342–353. Available at https://doi.org/10.1002/cam4.374.

Serio, S., Pagiatakis, C., Musolino, E., Felicetta, A., Carullo, P., Laura Frances, J., et al., 2023. Cardiac aging is promoted by pseudohypoxia increasing p300-induced glycolysis. Circulation Research 133 (8), 687–703. Available at https://doi.org/10.1161/circresaha.123.322676.

Shah, N.M., Lai, P.F., Imami, N., Johnson, M.R., 2019. Progesterone-related immune modulation of pregnancy and labor. Frontiers in Endocrinology 10. Available at https://doi.org/10.3389/fendo.2019.00198.

Shantsila, E., Shantsila, A., Gill, P.S., Lip, G.Y.H., 2016. Premature cardiac aging in South Asian compared to afro-caribbean subjects in a community-based screening Study. Journal of the American Heart Association 5 (11). Available at https://doi.org/10.1161/JAHA.116.004110.

Sucato, V., Coppola, G., Manno, G., Vadalà, G., Novo, G., Corrado, E., et al., 2023. Coronary artery disease in south asian patients: cardiovascular risk factors, pathogenesis and treatments. Current Problems in Cardiology 48 (8), 101228. Available at https://doi.org/10.1016/j.cpcardiol.2022.101228.

Sudre, C.H., Moriconi, S., Rehwald, R., Smith, L., Tillin, T., Barnes, J., et al., 2022. Accelerated vascular aging: Ethnic differences in basilar artery length and diameter, and its association with cardiovascular risk factors and cerebral small vessel disease. Frontiers in Cardiovascular Medicine 9. Available at https://doi.org/10.3389/fcvm.2022.939680.

Sun, Y., Zhang, J.N., Zhao, D., Wang, Q.S., Gu, Y.C., Ma, H.P., et al., 2011. Role of the epithelial sodium channel in salt-sensitive hypertension. Acta Pharmacologica Sinica 32 (6), 789–797. Available at https://doi.org/10.1038/aps.2011.72.

Tabaei, B.P., Chamany, S., Perlman, S., Thorpe, L., Bartley, K., Wu, W.Y., 2019. Heart age, cardiovascular disease risk, and disparities by sex and race/ethnicity among New York city adults. Public Health Reports 134 (4), 404–416. Available at https://doi.org/10.1177/0033354919849881.

Tamargo, J., Rosano, G., Walther, T., Duarte, J., Niessner, A., Kaski, J.C., et al., 2017. Gender differences in the effects of cardiovascular drugs. European Heart Journal - Cardiovascular Pharmacotherapy 3 (3), 163–182. Available at https://doi.org/10.1093/ehjcvp/pvw042.

Tan, S.C.W., Zheng, B.B., Tang, M.L., Chu, H., Zhao, Y.T., Weng, C., 2025. Global burden of cardiovascular diseases and its risk factors, 1990-2021: a systematic analysis for the global burden of disease study 2021. QJM: An International Journal of Medicine 118 (6), 411–422. Available at https://doi.org/10.1093/qjmed/hcaf022.

Teferi, H.M., Shelton, R.C., Conneely, K.N., De Vivo, I., Factor-Litvak, P., Kezios, K.L., et al., 2026. Allostatic load and biological aging among middle aged adults. Psychoneuroendocrinology 183. Available at https://doi.org/10.1016/j.psyneuen.2025.107661.

Topriceanu, C.C., Dev, E., Ahmad, M., Hughes, R., Shiwani, H., Webber, M., et al., 2023. Accelerated DNA methylation age plays a role in the impact of cardiovascular risk factors on the human heart. Clinical Epigenetics 15 (1). Available at https://doi.org/10.1186/s13148-023-01576-9.

Toussaint, L.L., Moriarity, D.P., Kamble, S., Williams, D.R., Slavich, G.M., 2022. Inflammation and depression symptoms are most strongly associated for black adults. Brain, Behavior, and Immunity - Health 26. Available at https://doi.org/10.1016/j.bbih.2022.100552.

Trexler, C.L., Odell, A.T., Jeong, M.Y., Dowell, R.D., Leinwand, L.A., 2017. Transcriptome and Functional profile of cardiac myocytes is influenced by biological sex. Circulation: Cardiovascular Genetics 10 (5). Available at https://doi.org/10.1161/CIRCGENETICS.117.001770.

Wakatsuki, A., Okatani, Y., Fukaya, T., 2002. Statin attenuates increase in C-reactive protein during estrogen replacement therapy in postmenopausal women. Circulation 106 (24). Available at https://doi.org/10.1161/01.cir.0000041808.84600.f6.

Waziry, R., Williams, O.A., Tiemeier, H., Miles, C., 2025. Vascular-related biological stress, DNA methylation, allostatic load and domain-specific cognition: an integrated machine learning and causal inference approach. BMC Neurology 25 (1). Available at https://doi.org/10.1186/s12883-025-04185-6.

Westin, E.R., Aykin-Burns, N., Buckingham, E.M., Spitz, D.R., Goldman, F.D., Klingelhutz, A.J., 2011. The p53/p21WAF/CIP pathway mediates oxidative stress and senescence in dyskeratosis congenita cells with telomerase insufficiency. Antioxidants and Redox Signaling 14 (6), 985–997. Available at https://doi.org/10.1089/ars.2010.3444.

Xu, P., Wang, M., Song, W.M., Wang, Q., Yuan, G.C., Sudmant, P.H., et al., 2022. The landscape of human tissue and cell type specific expression and co-regulation of senescence genes. Molecular Neurodegeneration 17 (1). Available at https://doi.org/10.1186/s13024-021-00507-7.

Xu, C., Qiu, Z., Guo, Q., Huang, Y., Zhao, Y., Zhao, R., 2025. The role of cellular senescence in cardiovascular disease. Cell Death Discovery 11 (1). Available at https://doi.org/10.1038/s41420-025-02720-5.

Yang, Q., Zhong, Y., Ritchey, M., Cobain, M., Gillespie, C., Merritt, R., et al., 2015. Vital signs: predicted heart age and racial disparities in heart age among U.S. Adults at the state level. MMWR. Morbidity and Mortality Weekly Report 64 (34), 950–958. Available at https://doi.org/10.15585/mmwr.mm6434a6.

Yin, L., Luo, M., Wang, R., Ye, J., Wang, X., 2021. Mitochondria in sex hormone-induced disorder of energy metabolism in males and females. Frontiers in Endocrinology 12. Available at https://doi.org/10.3389/fendo.2021.749451.

Ying, W., Post, W.S., Michos, E.D., Subramanya, V., Ndumele, C.E., Ouyang, P., et al., 2021. Associations between menopause, cardiac remodeling, and diastolic function: the CARDIA study. Menopause (New York, N.Y.) 28 (10), 1166–1175. Available at https://doi.org/10.1097/GME.0000000000001815.

Yusipov, I., Bacalini, M.G., Kalyakulina, A., Krivonosov, M., Pirazzini, C., Gensous, N., et al., 2020. Age-related DNA methylation changes are sex-specific: a comprehensive assessment. Aging 12 (23), 24057–24080. Available at https://doi.org/10.18632/aging.202251.

Zhang, L., Zhang, S., Yu, Y., Jiang, H., Ge, J., 2020. Efficacy and safety of rosuvastatin vs. atorvastatin in lowering LDL cholesterol: a meta-analysis of trials with East Asian populations. Herz 45 (6), 594–602. Available at https://doi.org/10.1007/s00059-018-4767-2.

Zitouni, K., Steyn, M., Lewis, J., Kelly, F.J., Cook, P., Earle, K.A., 2025. Variation in arterial stiffness and markers of oxidative stress in patients with type 2 Diabetes mellitus from different ethnic groups. Antioxidants 14 (7). Available at https://doi.org/10.3390/antiox14070858.

Chapter 17

Regenerative medicine and stem cell approaches in cardiac aging

Yumna Khan[1], Rabab Fatima[2], Amita Joshi Rana[3,4], Lina Eltaib[5], Mokhtar Rejili[6], Md. Sadique Hussain[7,8], Sumel Ashique[9], and Sabina Yasmin[10]

[1]*Saveetha Institute of Basic Medical Sciences (SIBMS), Saveetha Institute of Medical and Technical Sciences (SIMATS), Saveetha University, Thandalam, Chennai, Tamil Nadu, India,* [2]*School of Health Science & Technology, UPES, Dehradun, Uttarakhand, India,* [3]*College of Pharmacy, Graphic Era Hill University, Bhimtal, Uttarakhand, India,* [4]*Centre for Promotion of Research, Graphic Era (Deemed to be) University, Dehradun, Uttarakhand, India,* [5]*Department of Pharmaceutics, College of Pharmacy, Northern Border University, Rafha, Saudi Arabia,* [6]*Department of Biology, College of Sciences, Imam Mohammad Ibn Saud Islamic University (IMSIU), Riyadh, Saudi Arabia,* [7]*Uttaranchal Institute of Pharmaceutical Sciences, Uttaranchal University, Dehradun, Uttarakhand, India,* [8]*School of Pharmaceutical Sciences, Lovely Professional University, Phagwara, Punjab, India,* [9]*Department of Pharmaceutical Technology, School of Health and Medical Sciences, Adamas University, Kolkata, West Bengal, India,* [10]*Department of Pharmaceutical Chemistry, College of Pharmacy, King Khalid University, Abha, Saudi Arabia*

17.1 Introduction

Cardiac aging describes a gradual myocardial structural and functional deterioration, including rising ventricular stiffness, extracellular remodeling, calmodulin-choline impairments, mitochondrial impairment, and chronic, low-grade, inflammation (Pietri and Stefanadis, 2021). These changes lead to an increased vulnerability to heart failure, arrhythmias, and outcomes of ischemic injuries in the elderly (Paneni et al., 2017). Even though the turnover rates of cardiomyocytes remain low until death, multiple studies have shown that endogenous renewal is too small to compensate for age-related cell death, especially in the postmyocardial infarction period (Bergmann and Jovinge, 2014). Epigenetic aging, telomere erosion, and oxidative DNA damage deteriorate the proliferative capacity of resident cardiac cells and also constrain regenerative mechanisms (Grzeczka et al., 2023; Herman et al., 2021). The old myocardium, therefore, has limited ability to regenerate damaged tissue (Hashimoto et al., 2018), and hence the clinical significance of interventions that would move beyond standard pharmacological and surgical approaches is of therapeutic value.

Human cardiomyocytes have extremely little proliferative potential, which is further amplified with age. According to recent research, the rates of cardiomyocyte renewal are reduced to less than one percent per year in the elderly, underscoring the fact that the heart cannot regenerate damaged myocardium after injury (Bergmann et al., 2009). Mechanistic studies demonstrate that, in later life, mitochondrial dysfunction, metabolic rigidity, defective cytokinesis, and accruing genomic damage combine to inhibit the activity of the cardiomyocyte cell-cycle (Xu et al., 2025). These observations explain why exogenous regenerative therapies are necessary to have the potential to regenerate useful myocardial tissue.

Cellular senescence is a significant obstacle to the process of cardiac regeneration. Senescent endothelial cells, fibroblasts, and cardiomyocytes have altered secretome phenotypes, dysfunctional mitochondria, and defective reparative signatures (Lucas et al., 2023). Senescence-associated secretory phenotype (SASP) enhances chronic inflammation, extracellular matrix remodeling, and myocardial stiffening, thus establishing a regenerative-unfriendly microenvironment (Afzal et al., 2025). Research shows that a higher load of senescent cells in the aging heart is associated with impaired recruitment of stem cells, angiogenesis, and decreased cardiomyocyte turnover (Lewis-McDougall et al., 2019). These implications give a sound biological explanation for regenerative therapies that avoid or therapeutically control senescence-controlled deterioration.

The theoretical basis of cardiac regeneration was transformed dramatically during the last twenty years. Initial studies were on bone marrow-based cell transplant, but clinical results showed slight and irregular improvements when limited to

Cellular Senescence and Cardiac Aging. DOI: https://doi.org/10.1016/B978-0-443-45811-8.00014-3

cardiomyogenic differentiation and low retention of the cells (Zhu and Cheng, 2021). The introduction of induced pluripotent stem cells (iPSCs) transformed the landscape since it was possible to now create patient-specific cardiomyocytes with high fidelity and a platform to model disease and a prospective cell-based cure (Vo et al., 2025). Recent discoveries have turned to more direct cardiac reprogramming and paracrine-mediated regeneration, cellular cardiac tissue engineering, and gene-editing methods to boost endogenous repair (Haridhasapavalan et al., 2023; Tenreiro et al., 2021). Such advances are indicative of a growing body of knowledge of cardiac microenvironment biology, aging biology, and stem cell-based therapeutics.

The chapter gives a broad introduction to regenerative medicine and stem-cell-based interventions to cardiac aging. It starts by reviewing principal processes that inhibit regeneration in the aged myocardium: cellular senescence, metabolic dysfunction, and impaired stem cell niches. The following parts discuss the history of stem cell therapeutics, starting with the primitive trials based on bone marrow transplants and cell-based therapies, and the most powerful techniques based on iPSCs, synthetic heart tissues, and genome engineering scaffolds. Translational obstacles to aging, including loss of cell engraftment, immune stimulation, and microenvironmental limitations due to senescence, are also assessed in the chapter. Last, future therapeutic directions and future opportunities of reversing or limiting cardiac aging using regenerative interventions are also discussed.

17.2 Pathophysiology of cardiac aging relevant to regeneration

17.2.1 Cardiomyocyte loss, hypertrophy, and reduced turnover

These effects of cardiac aging include irreversible loss of cardiac muscle cells, mitochondrial dysfunction, and fibrosis. These cardiac processes are progressive; they have harmful effects on cardiac functions, and they provide fertile grounds for heart failure. Cardiac regeneration aims at reversing fibrosis, triggering vascularization, and replenishing the population of cardiac muscle cells, and can be utilized to reverse aging (Tracy et al., 2020). Even in the adult human heart, which is a regenerative tissue, the rate of cardiomyocyte renewal is about 0.5 per year; the death of myocytes per year is not compensated by corresponding increases in the formation of new myocytes, and compensatory hypertrophy must take the place of regenerative renewal (Tracy et al., 2020; Eschenhagen et al., 2017). Unlike skeletal muscle cells, adult cardiomyocytes cannot undergo robust proliferation to repair damaged myocardium, and the death of myocytes per year is not balanced by a comparable increase in the production of new myocytes (Castellan and Meloni, 2018). There was a peak of cardiomyocyte proliferation at embryonic days 10–12 and postnatal day 4, but a second burst of cardiomyocyte proliferation mainly occurs in the absence of cytokinesis to give rise to binucleate cardiomyocytes (Bishop et al., 2021). Cardiac size augmentation in cardiac muscles of the mammalian heart in old age is primarily brought about by cardiomyocyte hypertrophy and not by cardiomyocyte proliferation. The cardiomyocyte generation rate was 18–50 times lower in the end-stage heart failure patients than it is in the healthy heart, indicating the highly impaired regenerative ability of the diseased aging heart (Derks et al., 2025).

17.2.2 Changes in cardiac stem/progenitor cell niche

The sub-epicardium and the myocardial interstitium are found to consist of adult cardiac progenitor populations, such as c-Kit+ progenitor, Sca-1+ progenitor, cardiosphere, and cardiac colony-forming unit-fibroblasts (Valente et al., 2014). But there is a vast age-dependent effect on the proliferative response of cardiospheres and cardiac progenitor cells to myocardial infarction, where aging hearts exhibit practically no proliferative responses in comparison to young hearts. The aging effect tremendously increases the size of the hypoxic quiescent cardiac stem cells, and the smaller pool of cardiac stem cells available in the cycling mode keeps the cells in the cycling mode viable. Combinatorial regulation of multiple signaling pathways and low oxidative stress domains in the vasculature are related to the maintenance of Bmi1+ cardiac progenitor cells (Albericio et al., 2024). The young Sca-1+ cells can regenerate aging cardiac endothelial cells but not cardiomyocytes via the Cxcl12/Cxcr4 pathway, indicating the cardiac vascular niche is highly essential in maintaining the progenitor cells (Li et al., 2018).

17.2.3 Age-related extracellular matrix remodeling

An aged heart is defined by cellular and interstitial remodeling, apoptotic and necrotic processes leading to loss of myocytes, which causes hypertrophy of surviving myocytes and replacement fibrosis. The first is the aging-associated fibrosis, which is marked by the augmented deposition of collagen, which could be brought about by augmented collagen synthesis and diminished degradation because of inhibited matrix metalloproteinases. The increase in myocardial collagen in old age is best related to postsynthetic or degradative mechanisms rather than to an increase in collagen synthesis because both collagen type I and collagen type III mRNA levels tend to decline or do not change with age (Horn and Trafford, 2016). Aging of nascent collagen leads to cross-linking, and collagenase or another matrix metalloproteinase-resistant cross-linked collagen. Angiotensin II, ROS, TGF β1, and endothelin-1 signaling have significant roles in the process of fibrotic remodeling of the ageing heart. The constructive extracellular matrix (ECM) that becomes stiffer and has different composition forms an unfavorable microenvironment that hinders the processes of regeneration (Cheng et al., 2003).

17.2.4 Mitochondrial dysfunction and metabolic reprogramming

Symptoms of cardiac aging also include the dysfunction of mitochondria, such as defects in oxidative phosphorylation, decreased ATP generation, and increased oxidative stress (Martín-Fernández and Gredilla, 2016). Lipid uptake and oxidation imbalance led to cardiolipotoxicity and insulin resistance in aging, and more cardiac CD36 facilitates fatty acid uptake and maintains elevated cardiac lipid levels. Another rate-limiting enzyme of fatty acid oxidation is carnitine-palmitoyl transferase-1, which is drastically reduced in the heart of aged rats, and this could be the cause of the reduction of oxidative phosphorylation with age (Kliesow Remes et al., 2025). An increase in age-associated reduction of the levels of NAD+ inhibits the activity of SIRT1/3 and interferes with the PGC-1a-TFAM-mitochondrial biogenesis feedback loop (Imai and Guarente, 2014). Dysfunctional cardiac autophagy also leads to more advanced aging of the heart, cardiac hypertrophy, and the build-up of dysfunctional mitochondria. All these metabolic changes impaired the energy reserves required to promote regenerative activities.

17.2.5 Role of immune microenvironment in aging heart repair

The basal inflammatory response is mainly upregulated during aging, resulting in the process of inflammaging, among which the macrophage plays a prominent role, manifested by the increased number of cardiac macrophages and the increased levels of pro-inflammatory things such as tumor necrosis factor-a, interleukin-6, and matrix metalloproteinases. Heart resident cardiac macrophages eliminate senescent and dying cells and conduct electrical impulses, whereas in the aging heart, the macrophage phenotype changes to a pro-inflammatory subtype, resulting in inflammaging (Ma et al., 2018). Young adults' cardiac injury first triggered an inflammatory response of neutrophils and M1 macrophages at the site of injury, but in older hearts, the effect of the inflammatory response is dampened, causing a delay in the clearance of dead cells. The cardiac microenvironment following myocardial infarction is composed of ischemic and dead cardiomyocytes, which provide the release of damage-associated molecular patterns, such as cytokines that attract leukocytes to the heart (Anwar et al., 2024). Myocardial aging implies both intrinsic cardiac and extrinsic immunology, and lymphocytes play a role in creating a low-grade myocardial inflammation state to facilitate structural changes (Peterson et al., 2022).

17.2.6 Vascular aging and ischemic angiogenic responses

Vascular aging inhibits the ischemic angiogenic response such that the damaged or ischemic tissues are repaired more slowly and inefficiently. This is because of age-associated factors such as endothelial dysfunction, decreased concentration of signaling molecules such as VEGF, heightened oxidative stress, and changes in the extracellular matrix (Liu et al., 2024). The changes predispose older people to cardiovascular diseases and may slow down recovery from such events as a heart attack and stroke. The number of vascular progenitors declines dramatically with age, and aging inhibits vascular sprouting and branching to make fewer capillaries and poorly developed vascular networks. Senescent endothelial cells have less proliferation potential, which can restrain the ability to develop new vascular structures, and endothelial functional losses can also play a role in the development of atherosclerotic plaque (Jia et al., 2019). Old age is linked to endothelial dysfunction and a notable reduction of nitric oxide bioavailability that probably leads to defective angiogenesis and microvascular rarefaction. Isolated endothelial cells of aging subjects are down-regulated in the expression of the mRNA of the vascular endothelial growth factor and protein expression, and are deficient in their angiogenic potential. Older endothelial progenitor cells have less proliferation, survive, migrate/ mobilize, and senescence, and are vulnerable to oxidative stress because of low antioxidant enzyme levels (Lee et al., 2025). These vascular dysfunctions have a severe effect on the capacity to nourish recently regenerated cardiomyocytes.

17.3 Stem cell sources for cardiac regeneration

17.3.1 Embryonic stem cells

Embryonic stem cells (ESCs) are located within the inner cell mass of the human blastocyst, which is an initial formation of the developing embryo. The inner cell mass, however, ESCs that were obtained at the blastocyst stage can be cultured in the laboratory and will eventually grow indefinitely under the correct conditions (Andrews and Gokhale, 2024). The undifferentiated state of ESCs in culture can produce cells of all three layers of embryonic tissues. ESCs are considered early embryonic cells that possess the following two characteristics: self-renewal and the ability to differentiate to almost any cell type, which is known as pluripotency. Cell-based therapies and tissue engineering applications. The first use of human embryonic stem cells allowed the production of cardiomyocytes in high quantities to be used in cell-based therapies and tissue engineering applications. One of the key problems that ESC derivatives are faced with is that it is prone to host immune rejection, and most of the current studies involve immunosuppressive medication or immunodeficient animals to alleviate rejection effects (Huang et al., 2021). Cardiomyocytes derived from human pluripotent stem cells have demonstrated a strong ability to repair damaged heart, and procedures to induce human pluripotent stem cells to differentiate into cardiomyocytes have been established. ESC-derived cardiomyocytes have been shown to have therapeutic effects in small and large mammals by mechanisms of cardiac repair (Jiang and Lian, 2020). ESC-derived cardiomyocytes have been shown to exhibit therapeutic efficacy in small and large animal models by mechanisms of cardiac repair. In parallel, mechanistic studies show that the major objective of remuscularizing the heart with PSC-derived cardiac cells may be questioned by their alternative application as ex vivo sources of a biologically active extracellular vesicle-enriched secretome capable of equally improving heart performance by activating endogenous repair mechanisms. The sole utilization of this secretome would integrate the merits of a large-scale manufacturing that is more closely related to a biological drug, the probable absence of cell-associated immune and tumorigenicity risks, and the likelihood of intravenous injections that could be exposed to multiple doses (Menasché, 2020).

17.3.2 Induced pluripotent stem cells

Induced pluripotent stem cells (iPSCs) are stem cells cultured in labs, which are engineered embryonic-like cells created out of adult cells, such as skin or blood cells. Such cells are pluripotent cells or cells that could form any sort of cell in the body; hence, they are a valuable resource in modeling diseases, drug discovery, and possibly cell-based therapy without the same ethical issues as ESC (Ye et al., 2013). iPSCs are induced pluripotent cells that are based on embryonic stem cells. Autologous iPSC-derived cardiomyocytes can be a safer and more effective substitution to embryonic stem cell-derived cardiomyocytes, and allogenic transplantation of nonhuman primate hearts with iPSC-derived cardiomyocytes has been reported to be successful (Oikonomopoulos et al., 2018). The reprogramming of adult somatic cells into iPSCs, followed by an additional differentiation into cardiomyocytes, results in the production of patient-specific cardiomyocytes. Improved cardiac differentiation approaches have improved the quantity and quality of cardiac myocytes obtained with iPSCs by progressively adding growth factors to replicate the embryonic heart development (Azhdari et al., 2025). Cardiomyocytes developed using pluripotent stem cells are immature phenotypes of adult cardiomyocytes and may induce ventricular arrhythmias as an ectopic pacemaker (Garg et al., 2018). Human iPSC-derived cardiomyocytes and endothelial cells exhibited enhanced cardiac performance and a decreased arrhythmic outcome when treated jointly in the models of myocardial infarction (Cheng et al., 2023).

17.3.3 Adult stem cells

17.3.3.1 Mesenchymal stem/stromal cells

Mesenchymal stem/stromal cells (MSCs) can be found in nearly every tissue and are extremely mobile when reacting to the signals of tissue damage. They have the potential for immune regulation, pathogenesis, infection, mechanisms of action, and clinical uses. The MSCs can release numerous cytokines, chemokines, and growth factors that facilitate inflammation, angiogenesis, and fibrogenesis (Zhou and Shi, 2023). MSCs possess cardiac repair capability that seems greater than their differentiation into cardiac myocytes, and induce the proliferation and differentiation of native cardiac stem cells. By lowering the inflammatory levels, differentiating myocardial cells surrounding the areas of infarction and angiogenesis, enhancing resistance to apoptosis, and preventing fibrosis, mesenchymal stem cells are able to protect the myocardium (Hatzistergos et al., 2010). The MSCs of bone marrow and adipose are readily isolated and cloned, and immunologically tolerated, which makes possible an allogeneic off-the-shelf transplantation. MSCs decrease the size of an infarct by forming no more than about 8% of new myocardium that has been formed, most of the new tissue formed being by host processes, which necessitate interactions with the MSCs (Hoang et al., 2022).

17.3.3.2 Bone marrow-derived stem cells

The bone marrow-derived stem cells are a heterogeneous population of cells in bone marrow that are capable of differentiating to various types of cells, most famous being hematopoietic stem cells (which form blood cells) and mesenchymal stem cells (which form bone, cartilage, and fat cells) (Bhat et al., 2021). The cells find their application in the regenerative medicine field because they have the potential of repairing tissues, regulate the immune system, and treat diseases such as blood cancers, leukemia, and lymphoma by procedures such as bone marrow transplants. Allogeneic mesenchymal stem cells are safe and can be conveniently and effectively introduced into damaged myocardium, where they engraft and express proteins normally only found in cardiac myocytes, vascular endothelium, and smooth muscle (Amado et al., 2005). After the elevation of myocardial infarction, MSCs were transplanted and transformed into cardiomyocytes and vascularization, which induced host cardiomyocytes to proliferate in the initial two weeks after treatment. The cells derived from bone marrow have been reported to have a positive effect in direct differentiation as well as paracrine action, which stimulates intrinsic repair (Duran et al., 2013).

17.3.3.3 Adipose-derived stem cells

Markers of endothelial cells, vascular smooth muscle cells, and cardiomyocytes were the most frequent appearance rates of adipose-derived stem cells, and systemic transfusion had the highest functional recovery of cardiac activity following myocardial infarction (Nagata et al., 2016). In comparison to bone marrow-derived mesenchymal stem cells, adipose tissue stem cells have a higher yield, with 2 %of nucleated cells in processed lipoaspirate giving rise to a minimum of five times greater colony-forming units as compared to bone marrow MSCs (Prantl et al., 2021). Adipose-derived stem cells release substantial concentrations of vascular endothelial growth factor, hepatocyte growth factor, and transforming growth factor-b, and the conditioned media of hypoxic cells are of importance to enhance endothelial cell growth and decrease apoptosis considerably (Suchanecka et al., 2025).

17.3.4 Cardiac progenitor cells

The key participant in cardiac regeneration is the adult cardiac progenitor cells (CPC), which is a heterogeneous cluster of cells that are nonactivated in a physiological state and are activated in case of an injury, to assist in the renewal of cardiomyocytes. Their positive effect can be mediated by direct differentiation into heart cells and by exploiting resident stem cells, but primarily by paracrine discharge of factors. CPCs have the capability to release both cytokines, chemokines, and growth factors, and exosomes. CPCs are a type of cell in the heart that are progenitor cells. CPCs are mainly the various cardiac resident cellular pools such as c-Kit +, stem cell antigen-1 +, side population, cardiospheres, Islet 1 +, and stage-specific embryonic antigen-1 + CPCs (Quijada et al., 2020). They can have some overlapping surface markers and differentiation capacity, and embryonic CPCs can have some three cell lineage markers, such as cardiomyocytes, vascular smooth muscle cells, and endothelial cells. As a part of their regenerative repertoire, MSCs induce cardiac stem cell proliferation and differentiation, and MSC-conditioned medium alone has therapeutic effects (Hatzistergos et al., 2010). Nevertheless, the cardiomyogenic capacity of the cardiac progenitor cells is still debatable, and recent studies have cast doubt on their role in myocyte regeneration.

17.3.5 Endothelial progenitor cells

Endothelial progenitor cells (EPCs) have strong angiogenic and tissue repair capabilities. EVs are another effective component containing cellular proteins, microRNAs, messenger RNAs, and lncRNAs, and are secreted by various types of cellular organelles. The parent cells are the primary factors in the functions of EVs. Functional studies of EPC-derived EVs (EPC-EVs) have elucidated that they have therapeutic effects on most diseases, including cardiovascular disease, acute kidney injury, acute lung injury, and sepsis (Chen et al., 2022). EPCs are a type of cell that becomes specialized to perform a specific task, which could include tissue repair, creation of new blood vessels, and injury to sensorimotor neurons The endothelial progenitor cell population is formed by either the bone marrow cells or other tissues such as the spleen, vessel wall, adipose tissue and placenta and has been grouped according to whether they are hematopoietic or nonhematopoietic progenitor cells (Bianconi et al., 2018). EPCs have a high expression of angiogenic growth factors such as VEGF and stromal cell-derived factor-1 that have direct effects on mature endothelial cell migration and result in improved neovascularization following limb ischemia and myocardial infarction in animal models, but clinical studies have yielded mixed results (Huang and Huang, 2022).

17.3.6 Emerging cell types

17.3.6.1 Pericytes

Generally, pericytes have been implicated in the maintenance of vascular rheology and homeostasis, such as the regulation of blood flow, angiogenesis, structural stabilization of the vasculature, and vascular permeability. In addition, pericytes have also been found to be multipotent in their lineage, as is the case with other perivascular populations, such as MSCs and adventitial progenitor cells, to which they share common markers and properties (Ferland-McCollough et al., 2017). These diverse groups of people might be classified as PSCs. Such a property prefers the use of PSCs in regenerative medicine and tissue engineering. Nevertheless, the way PSCs are related to each other and the one that may be most suitable for cell therapies remains unclear to date. Angiogenic factors such as TGF, angiopoietin-1, and VEGF are also expressed by pericytes, which perform several activities, including angiogenesis, blood vessel support, flow regulation, maturation, vascular remodeling, and vessel patency (Avolio et al., 2017). Pericyte injection can reverse cardiac remodeling through inhibiting fibrosis and inflammation and stimulating angiogenesis, and pericytes are probably involved in controlling immune cell recruitment after myocardial infarction (Dabravolski et al., 2022). The multipotent progenitors in the embryonic epicardium are pericytes, fibroblasts, and vascular smooth muscle cells (Quijada et al., 2020). Pericytes are thus another cell type of a significant new type capable of therapeutic cardiac regeneration.

17.3.6.2 Epicardial-derived progenitors

Epicardial-derived progenitors (EPDCs), which begin as cells of the epicardium and are essential in heart development, give rise to several types of cells, such as smooth muscle cells, fibroblasts, and possibly endothelial cells. In embryonic development, the EPDCs migrate into the heart, developing the coronary vasculature and connective tissues (Matthes et al., 2009). Their contribution to cardiac repair may involve some of their work in adult heart regeneration, but this remains under research, and some studies have suggested that they may have a role in cardiac repair, albeit a minor role, compared to development.Wt1+ epicardial cells are new cardiomyocyte precursors that, in the development of the normal murine heart, become diverged to form cardiac fibroblasts, smooth muscle cells, and pericytes (Zhou et al., 2008). Temporal WT1 upregulation of the endothelial cells and epithelium in the infarcted region and border zone after cardiac injury indicates that WT1 could play an important role in cardiac repair and regeneration (Duim et al., no date). Cardiomyocyte aggregation, expansion, and organization in vitro can be directed by human IPS-derived preepicardial cells that express WT1, TBX18, SEMA3D, and SCX, which, upon coculture, results in cardiomyocyte aggregates that consolidate into connected beating syncytia with improved contractility (Tan et al., 2021). A highly promising future in heart regenerative medicine is epicardial-derived progenitors, which have both cellular and paracrine actions (Fig. 17.1).

17.4 Mechanisms of action in cardiac regeneration

Cardiac regeneration relies on coordinated mechanisms that restore cardiomyocyte viability, enhance angiogenesis, remodel the extracellular matrix, and fine-tune immune responses to create a repair-permissive environment. The mechanisms by which these regenerative processes are triggered by certain molecular and cellular pathways have been shown by several studies. For example, a study demonstrated that miR-210-enriched extracellular vesicles stimulated cardiomyocyte survival by targeting PTP1B and DAPK1, which also stimulated endothelial angiogenesis by inhibiting EFNA3 and reducing fibrosis, which enhanced functional recovery following ischemia/reperfusion injury (Song et al., 2021). Following the significance of immune regulation, another study discovered that neonatal cardiac B cells enhanced cardiomyocyte growth, activated angiogenesis, and inhibited inflammatory processes, and that their depletion slowed regeneration and boosted fibrosis, and age-dependent immune response cooperating with inherent cellular pathways (Tan et al., 2023). Moreover, cardiomyocyte cell-cycle reentry was also found to be inhibited by senescence-associated inhibitors like Rb1 and Meis2, which was also reported to be counteracted by immune responses to promote regeneration, demonstrating the synergistic effect of intrinsic antisenescence and immune control (Alam et al., 2019). Likewise, circMap4k2 was also reported to regulate cardiomyocyte proliferation by sponging miR-106a-3p and regulating Azin1, which is another molecular connection that activates inherent regenerative pathways after injury (Yan et al., 2024).

In addition to the cell cycle and immune modulation, it is important to create a repair-favorable microenvironment to regenerate. For instance, a study revealed that age-dependent Sirt6 loss caused cardiomyocyte senescence, telomere shortening, oxidative stress, and fibrosis, but Sirt6 loss inhibited these adverse effects and functional recovery, proving the significance of senescence regulation in cardiomyocyte regenerative competence (Pillai et al., 2021). Consistent with the interaction between immune cues and cardiomyocyte proliferation, Treg regulatory cells were discovered to secrete paracrine factors such as CCL24, GAS6, and AREG that promoted cardiomyocyte proliferation and inhibited fibrosis;

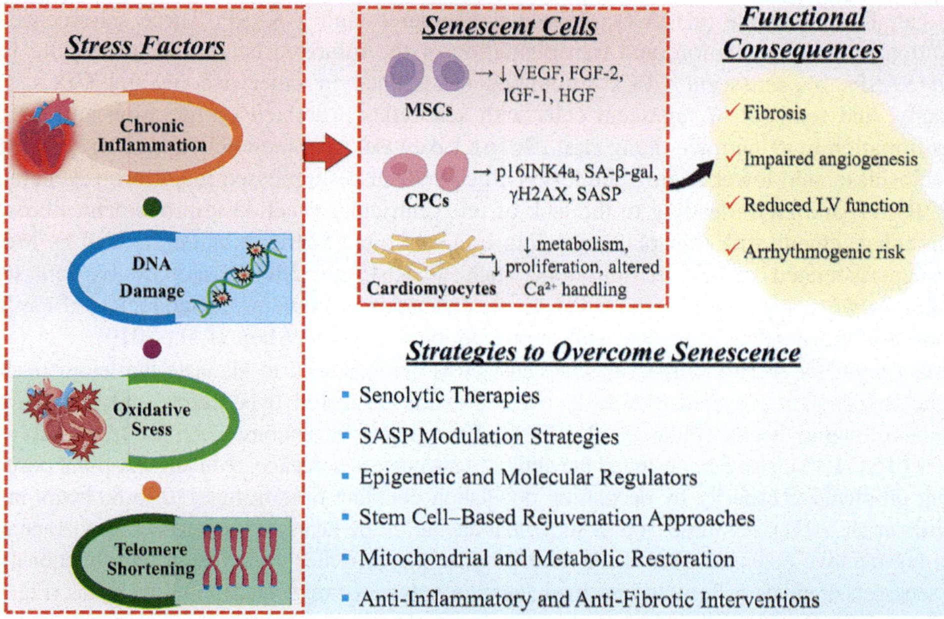

FIGURE 17.1 **Cellular senescence as a central driver of impaired cardiac regeneration during aging and therapeutic strategies to counteract its effects.** Conceptual diagram illustrating how cellular senescence acts as a central driver of impaired cardiac regeneration during aging. Stress factors such as chronic inflammation, DNA damage, oxidative stress, and telomere shortening induce senescence markers and the senescence-associated secretory phenotype (SASP), leading to fibrosis, reduced angiogenesis, impaired stem cell function, and diminished regenerative capacity; potential therapeutic strategies include senolytics, SASP modulation, epigenetic and metabolic interventions, stem cell–based therapies, and anti-inflammatory or antioxidant approaches.

Treg depletion impaired repair, and adoptive transfer restored regeneration, which is a synergistic effect between immune regulation and paracrine signaling (Li et al., 2019b). Complementing these findings, it was reported that ECM-glycopeptide hybrid hydrogel can induce M2 macrophage polarization, stimulate endothelial proliferation, and support cardiomyocyte survival, indicating that ECM remodeling as well as immune modulation can both induce an environment conducive to endogenous tissue repair (Kong et al., 2023).

Vascularization, progenitor cell activity, and structural remodeling further consolidate myocardial repair. It has been reported that human pluripotent stem cell-derived cardiovascular progenitor cells grow in vivo and differentiate into cardiomyocytes, endothelial cells, and smooth muscle cells, which mitigate fibrosis and inhibit adverse remodeling in response to myocardial infarction (Schwach et al., 2020). VEGF-B overexpression that coordinates with progenitor cell activity induced endothelial proliferation and coronary vessel development that increased perfusion and structural repair, proving the role of angiogenesis in supplementing cellular repair (Räsänen et al., 2021). Further, in zebrafish hearts, early ECM softening and dynamic remodeling have been found to offer a permissive biomechanical environment of cardiomyocyte proliferation and tissue reconstruction, offering a mechanistic setting of how structural signals interact with cellular and vascular responses to generate regeneration (Garcia-Puig et al., 2019).

17.5 Cellular senescence and its impact on regenerative therapies

The process of cellular aging and changes in the cardiac microenvironment are a key source of reduced efficacy of regenerative therapies. With age-associated alterations in the cardiomyocytes and the surrounding extracellular matrix, cardiomyocytes gradually lose the capacity to proliferate, respond to stress, and sustain functional homeostasis, which forms a less favorable microenvironment for repair (Ozcebe et al., 2021). Resident cardiomyocyte senescence also contributes to the functional impairment of metabolic activity, energy, calcium, and stress resilience, which affects cardiac dysfunction and arrhythmogenic risk (Lazzarini et al., 2022). In addition to these internal modifications, senescence influences the regenerative potential of therapeutic cells, which underscores several levels of obstacles to overcome before cardiac functionality can be restored in old hearts.

The impairment of regenerative outcomes by cellular senescence has been shown in various studies. More than half of the cells in people older than 70 were aged cardiac progenitor cells (CPCs) that had senescence markers, such as p16INK4a, SA-β-gal, DNA damage (γH2AX), shortened telomeres, and a SASP. These senescence CPCs exhibited defective proliferation and differentiation, and transplantation of the infarcted hearts did not restore functional ability. Notably, secreted SASPs by senescent CPCs stimulated senescence in otherwise healthy CPCs, further depleting regenerative capacity, and removal of senescent cells with senolytics promoted proliferation and activity in vitro and cardiomyocyte proliferation in aging mice (Fang et al., 2016). Likewise, aged mesenchymal stem cells (MSCs) had higher p16INK4a, β-galactosidase, and lower angiogenic levels. Such cells also suppressed major pro-regenerative factors, such as VEGF, FGF-2, IGF-1, and HGF, leading to the lack of left ventricular function improvement, fibrosis reduction, and neovascularization enhancement after transplantation into infarcted aged hearts (Khan et al., 2011). Mechanistic studies identified senescence-associated cell-cycle inhibitors, such as Rb1 and Meis2, as key barriers to cardiomyocyte proliferation. Inhibitors were suppressed, which led to the formation of smaller, mononucleated cardiomyocytes with greater regeneration ability, increased survival, and improved angiogenesis (Alam et al., 2019).

Sirt6, the important regulator of cardiomyocyte senescence, was demonstrated to alleviate the telomere-shortening process, oxidative stress, and mitochondrial dysfunction. Sirt6 overexpression decreased hypertrophy and fibrosis and helped in the functional restoration of aging hearts (Pillai et al., 2021). Emerging interventions, such as treatment with iPSC-derived extracellular vesicles (iPSC-EVs), also demonstrated the ability to attenuate senescence, reducing SA-β-gal positivity from 60% to 40% and improving proliferative capacity by decreasing population doubling time from >110 to 80 hours in senescent human fibroblasts (Rodrigues et al., 2021). All these studies show that ageing of the heart and senescence of therapeutic cells are major factors that restrict regenerative potential. One of the most promising approaches to the repair and functional delivery of aging hearts is to target senescence, either with molecular modulators, senolytic, or regenerative cell therapies (Fig. 17.1).

17.6 Preclinical studies in regenerative therapies for cardiac aging

Preclinical studies have explored a range of regenerative interventions aimed at mitigating cardiac aging and its functional decline. Mechanical stimulation, low-intensity ultrasound, stem cell-based therapies, exosomes, and senotherapeutics have shown the capacity to cause a reduction in apoptosis, a limitation in fibrosis, an increase in calcium handling, and reinforcement of mitochondrial and metabolic processes, which collectively lead to the improvement of cardiomyocyte survival and cardiac performance in aging models. These interventions included cardiac shock wave therapy (CSWT), which showed significant cardioprotective consequences. CSWT in a rat model of acute myocardial infarction (AMI) considerably reduced TUNEL-positive apoptotic cardiomyocytes, reduced cytochrome-c release, maintained mitochondrial integrity, and inhibited the JNK signaling pathway, enhancing cardiomyocyte survival (Zhang et al., 2018). Following this, in postinfarction heart failure, CSWT enhanced echocardiographic parameters, lowering serum BNP and NT-proBNP concentrations, reducing myocardial fibrosis, and activating the PI3K/Akt pathway, thereby suppressing adverse cardiac remodeling and emphasizing its antifibrotic and prosurvival properties (Wang et al., 2021). Similarly, low-intensity pulsed ultrasound (LIPUS) was also explored as a noninvasive modality to reverse myocardial injury. Optimal levels of LIPUS in myocardial models of ischemia/reperfusion (MI/R) decreased oxidative stress, inhibited inflammatory cytokines, maintained mitochondrial functionality, and inhibited apoptosis by down-regulating the ASK1/JNK pathway (Cao et al., 2023). In diabetic db/db mice with obesity, LIPUS also improved diastolic function, calcium handling, inhibited cardiomyocyte hypertrophy and interstitial fibrosis, and stimulated the eNOS-NO-cGMP-PKG signaling pathway, suggesting its ability to reverse age-related diastolic dysfunction (Monma et al., 2021).

The ability to target cellular senescence also showed a high level of regenerative potential. Local administration of ABT263 as biodegradable PLGA nanoparticles was demonstrated to eliminate stress-induced senescent cells in myocardial ischemia-reperfusion injury, suppressing the expression of pro-inflammatory molecules, promoting the reparative polarization of macrophages, and recovering the overall cardiac performance in the long term (Lee et al., 2021). Similarly, the senomorphic drug ruxolitinib suppressed senescence and SASP, improved mitochondrial dynamics, oxidative stress, and cardiac function in preclinical models of septic cardiomyopathy by inhibiting JAK2-STAT3, and low-dose therapeutics showed the same efficacy as senolytics (Yang et al., 2024). Furthermore, dasatinib + quercetin or sulforaphane treatment in a diet-induced cardiac aging model enhanced cardiac structure and function, reduced fibrosis, maintained mitochondria-endoplasmic reticulum contact sites, and alleviated the harmful effects of senescent cells (Silva-Palacios et al., 2025).

Table 17.1 summarizes additional preclinical studies investigating regenerative approaches for cardiac aging, including stem cell-, exosome-, and paracrine-based interventions across various models. These studies highlight the broad spectrum of strategies currently being explored to restore cardiac function, promote cardiomyocyte survival, and rejuvenate aged myocardium, reflecting the diverse therapeutic mechanisms demonstrated across multiple preclinical models.

TABLE 17.1 Experimental models of cardiac aging and regenerative interventions.

Model/Cell type	Intervention/Target	Key findings	Mechanism	Therapeutic effect	References
Human OMSCs, mouse MI model	UMSC-Exosomes	Reduced senescence; improved proliferation, migration, paracrine action; enhanced cardiac repair	miR-136 transfer suppressing Apaf1, reducing apoptosis and senescence	Improved cardiac function, reduced fibrosis, enhanced angiogenesis, reduced senescence and apoptosis	Zhang et al. (2020)
Aged cardiac MSCs	Aging-induced cMSC dysfunction	Increased SASP, reduced proliferation, monocyte recruitment, IL-1β production; impaired angiogenesis	CCR2-monocyte–IL-1β inflammatory loop sustaining senescence	Reduced SASP, improved proliferation and clonogenicity, restored angiogenic potential, normalized cMSC subpopulations	Martini et al. (2019)
Aged MSCs, rat MI model	Anti-miR-873-5p in AMSCs	Rejuvenated AMSCs; improved angiogenesis and cardiac function	Cab39/AMPK-mediated restoration of autophagy	Enhanced angiogenesis, improved cell survival, better cardiac function	Zhu et al. (2024)
Aged human CPCs	p16INK4A knockdown	Restored proliferation and antioxidant capacity; reduced ROS	Upregulation of BCL2, MCL1, CYGB, PRDX1; NFκB pathway involvement	Increased proliferation and survival, decreased ROS, improved antiapoptotic and antioxidant responses	Khatiwala et al. (2018)
HGPS iPSC-cardiomyocytes	Progeria-related cardiac aging model	Mitochondrial abnormalities; altered proteome; premature aging phenotype	Dysregulated ribosome, TCA cycle, and amino-acid biosynthesis pathways	Relevance for identifying targets for mitochondrial and metabolic rejuvenation	Monnerat et al. (2022)
Cynomolgus monkey ventricles (snRNA-seq)	FOXP1 deficiency	Cardiomyocyte loss, hypertrophy, senescence; dysfunction	Downregulation of FOXP1 and its target genes	Identification of FOXP1 as therapeutic target for preventing senescence and cardiac hypertrophy	Zhang et al. (2022)
H9c2 cells; D-gal rat model	MSC paracrine therapy	Reduced oxidative stress; decreased p53; increased SOD2; reduced fibrosis	IGF-1 mediated antioxidant and antisenescence signaling	Reduced oxidative stress and senescence, decreased fibrosis, improved cardiomyocyte survival	Hu et al. (2022)
Rat heart; hiPSC-CMs	6-Nitrodopamine	Enhanced catecholamine responses; improved inotropy	Modulation of adenylyl cyclase via CAP-1, CAP-2, STIM1	Improved contractility and responsiveness to catecholamines	Cipollone et al. (2025)
hiPSC-EC exosomes; murine MI model	EC-exosomes enriched in miR-100-5p	Improved cardiac function; reduced remodeling	Targeting PP-1β enhancing SERCA2a activity and Ca^{2+} homeostasis	Enhanced cardiac contractility, reduced left ventricular remodeling, improved cardiomyocyte survival	Li et al. (2023)
hiPSC-CMs (30–80 days)	Long-term culture maturation	Enhanced ICaL, INa, IK1; accelerated AP kinetics	IK1 identified as determinant of action potential maturation	Improved electrophysiological maturity and Ca^{2+} handling	Seibertz et al. (2023)
Adult bone marrow MSCs; aged mice	UC-EV rejuvenation	Increased self-renewal; telomere extension; improved regeneration	Transfer of PCNA and antiaging biomolecules	Enhanced regenerative potential, telomere length extension, improved angiogenesis, and tissue repair	Lei et al. (2021)

(Continued)

TABLE 17.1 (Continued)

Model/Cell type	Intervention/Target	Key findings	Mechanism	Therapeutic effect	References
Aged mice	Young Sca-1$^+$ BM cell transplantation	Increased angiogenesis, reduced senescence; endothelial rejuvenation	CXCL12–CXCR4 activation of Akt/FoxO3a signaling	Rejuvenation of cardiac endothelium, improved angiogenesis and tissue repair	Li et al. (2019a)
Aged rats, IR injury model	shFOXO4/SDF1 via UTMD	Senescent cell removal; improved BMSC homing; cardiac recovery	FOXO4 knockdown and SDF1-mediated recruitment	Reduced senescence, enhanced stem cell engraftment, improved cardiac function and neovascularization	Jiang et al. (2025)
DCM mouse model	UCB-MSCs	Improved cardiac structure and function; reduced apoptosis and fibrosis	Akt pathway activation; VEGF and IGF-1 upregulation	Enhanced cardiac function, reduced fibrosis and apoptosis, increased angiogenesis	Gong et al. (2016)
Aged hBM-MSCs, MI mouse model	miR-10a overexpression or KLF4 knockdown	Reduced apoptosis; improved angiogenesis; better cardiac function	KLF4–BAX/BCL2 pathway suppression	Improved survival of MSCs, enhanced angiogenesis, improved cardiac function	Dong et al. (2018)
ADMSCs; D-gal aging model	Alpinia oxyphylla Fructus preconditioning	Reduced senescence; improved cardiac performance	Enhanced mitochondrial protection, DNA repair, and paracrine activity	Reduced mitochondrial apoptosis, improved cardiac performance and paracrine signaling	Ruiz-Meana et al. (2020)
Aged hADSCs, MI mouse model	NDNF overexpression	Increased proliferation, migration, angiogenesis; preserved function	Restoration of NDNF-mediated survival and reparative signaling	Enhanced stem cell proliferation, migration, angiogenesis, and preserved cardiac function	Yang et al. (2019)
Aged C57BL/6 J mice	Young ADSC-sEVs	Reduced fibrosis, senescence, inflammation; improved diastolic function	Metabolic rejuvenation via sEV-mediated molecular reprogramming	Reduced fibrosis and inflammation, improved diastolic function, rejuvenated cardiac metabolism	Sanz-Ros et al. (2025)
Aged Wistar rats, MI	n-Butylidenephthalide preconditioning	Improved hADSC engraftment; reduced inflammation and fibrosis	Suppression of NLRP3 inflammasome and ROS regulation	Enhanced stem cell engraftment, reduced inflammation and fibrosis, improved cardiac repair	Lee et al. (2020)
Rap1-/- mice and human myocardium	Loss of Rap1; p53 inhibition	Telomere shortening, mitochondrial defects, hypertrophy; aging reversal via p53 inhibition	p53 suppression of PPARα impairs FA metabolism	Mitigation of telomere-driven aging, improved mitochondrial function, and reduced hypertrophy	Cai et al. (2021)
Patient-derived hiPSC-CMs	Telomere shortening model	Contractile dysfunction, senescence, mitochondrial defects	FOXC1 activation due to reduced TAD insulation in short-telomere cells	Improved contractility, reduced senescence, normalized mitochondrial function through FOXC1 modulation	Li et al. (2024)

TABLE 17.2 Evidence from clinical trials targeting cardiac aging in humans.

Study	Design and population	Intervention	Key findings	References
DREAM-HF	Phase III, randomized, double-blind, sham-controlled, patients with advanced chronic HF	Mesenchymal precursor cells	Reduced recurrent HF-related major adverse events; attenuated adverse ventricular remodeling	Borow et al. (2019)
IMPACT-CABG	Phase II, multicenter, randomized, 40 patients with chronic ischemic cardiomyopathy undergoing CABG	Autologous CD133^{+} stem cells intramyocardially	Safe and feasible; improvements in LV volumes and EF observed	Noiseux et al. (2016)
C-CURE	Prospective, multicenter, randomized, chronic ischemic HF patients	Cardiopoietic MSCs via endomyocardial injection	Improved LVEF and LV end-systolic volume; enhanced 6MWD and composite clinical score	Bartunek et al. (2013)
CRATUS	Randomized, double-blind, dose-finding, 30 elderly patients with aging frailty	IV allogeneic MSCs (100 or 200 million)	Improved physical performance and reduced systemic inflammation; safe	Tompkins et al. (2017)
CDR132L Phase 1b	Randomized, placebo-controlled, dose-escalation, 28 patients with chronic ischemic HF	CDR132L antisense oligonucleotide	Safe; dose-dependent reduction in miR-132 and NT-proBNP; trends toward improved cardiac function	Täubel et al. (2021)
L-arginine trial	Randomized, double-blind, placebo-controlled, 153 post-MI patients (77 ≥60 yrs)	L-arginine 3 TID for 6 months	No improvement in EF or vascular stiffness; higher mortality in the treated group	Schulman et al. (2006)
CUPID 1	Phase I/II, 39 patients with advanced HF	Intracoronary AAV1/SERCA2a gene therapy	The high-dose group showed reduced recurrent cardiovascular events; safe with a durable effect	Zsebo et al. (2014)

17.7 Clinical trials and human studies

Regenerative therapeutic studies in humans investigating cardiac aging have shown the possibility, safety, and functional utility of stem cell-based and gene-based intervention in patients with heart failure and ischemic cardiomyopathy. A TAC-HFT (*n*=19) and POSEIDON (*n*=30) clinical analysis was conducted to assess the effect of age on the efficacy of transendocardial mesenchymal stem cell (MSC) injection. Both younger and older patients were found to have improved in 6-minute walk distance and reduced Minnesota Living With Heart Failure Questionnaire (MLHFQ) levels with decreased scar size across age groups, demonstrating similar functional and structural benefits in both age groups (Golpanian et al., 2015). Likewise, a randomized, placebo-controlled trial (NCT01739777) was conducted to evaluate the intravenous allogeneic umbilical cord-derived MSCs (UC-MSCs) in 30 individuals with chronic stable heart failure and decreased ejection fraction. The administration of UC-MSC was well-tolerated, demonstrated significant positive changes in left ventricular ejection fraction (LVEF) at 3, 6, and 12 months, and improved quality of life but did not escalate adverse events (Bartolucci et al., 2017). The CONCERT-HF phase II trial (NCT02501811) also investigated the use of transendocardial autologous MSCs or c-kit$^+$ cardiac progenitor cells (CPCs) or a combination of both in chronic ischemic cardiomyopathy patients and found that all were feasible and possibly regenerative with respect to left ventricular remodeling, scar size, and functional capacity (Bolli et al., 2018). To date, there are very few human studies or clinical trials that specifically address cardiac aging, and clinical evidence is limited across the board. A summary of other clinical trials in regenerative therapies of cardiac aging, including study design, interventions, and clinical outcomes, is presented in Table 17.2

17.8 Future directions and emerging concepts

The current trend in regenerative medicine in cardiac aging is likely to go beyond repair to partial age-reversal of the old myocardium and its regenerative microclimates. There is some evidence emerging that hallmarks of aging can be counteracted to revive regenerative competence in aged cardiac cells. Other strategies, such as transient reprogramming using Yamanaka factors (Chen et al., 2021), senolytic and senomorphic therapies (Alum et al., 2025), NAD+ restoration (Katsyuba and Auwerx, 2017), and metabolic reprogramming, have provided evidence that they can elevate cardiac responses to stem cell-based therapies (Chen et al., 2021; Yamakawa and Ieda, 2021). Simultaneously, systems biology and artificial intelligence (AI)-based systems are finding more applications to combine multiomics data to discover important regulatory nodes in cardiac aging and regeneration (Lin et al., 2025). Predictive models that are made via machine learning are anticipated to hasten the process of identifying new regenerative targets, patient stratification optimization, and the application of personalized regenerative therapies to the aged population (Garmany and Terzic, 2024; Pandit et al., 2024). Last, newer frontiers in the same field are the transformative advances in whole-heart bioengineering and xenotransplantation, but they are ambitious and rapidly growing (Kawamura et al., 2015; Phillips et al., 2025). Recent advancements in decellularized cardiac scaffolds, 3D-bioprinting of vascularized myocardial tissue, and engineered porcine hearts with lower immunogenicity increased hopes of finding a solution to end-stage heart failure in older patients (Zheng et al., 2024; He et al., 2024; Barbulescu et al., 2022). Despite the notable technical, immunological, and ethical obstacles, the intersection of regenerative biology, computing sciences, and bioengineering is bound to remake therapeutic opportunities around cardiac aging in the next decades.

17.9 Conclusions

Multifactorial barriers to myocardial repair are caused by cardiac aging and altogether limit web-based regeneration and the effectiveness of regenerative therapy. The main point brought out in this chapter is that it will probably not be possible to restore cardiac structure and function in the old heart with a durable effect using single-modality interventions, but integrative approaches using techniques to rejuvenate the myocardial microenvironment and focus regenerative platforms will succeed in the future. New age-reversal strategies (transient or partial reprogramming, senolytic/senomorphic therapies, and metabolic or epigenetic modulation) can have positive possibilities to erase age-related molecular programs and make them more amenable to cell-and cell-free regenerative therapies. Simultaneously, application of multiomics, imaging, and clinical data integration due to systems biology and artificial intelligence will be able to reveal essential regulatory nodes of cardiac aging, allow for precise patient stratification, and streamline combinatoric regenerative regimens. On the translational borderline, whole-heart bioengineering (such as decellularized and bioprinted vascularized cardiac scaffolds) in combination with the highest levels of xenotransplantation (such as genetically engineered donor organ xenotransplants) can eventually find final solutions in end-stage heart failure in older patients.

Together, the convergence of rejuvenation biology, regenerative medicine, computational modeling, and bioengineering is likely to transform the paradigm of what therapeutic intervention should be in the case of cardiac aging and will help in designing a more effective, lasting, and personalized intervention in ten years.

References

Afzal, M., Rekha, M.M., Sahoo, S., Pandey, S.N., Maji, C., Goyal, K., et al., 2025. Targeting the senescence-associated secretory phenotype to modify osteoarthritis in aging. Inflammopharmacology 33 (11), 6555–6575. Available at https://doi.org/10.1007/s10787-025-02001-8.

Alam, P., Haile, B., Arif, M., Pandey, R., Rokvic, M., Nieman, M., et al., 2019. Inhibition of senescence-associated genes Rb1 and Meis2 in adult cardiomyocytes results in cell cycle reentry and cardiac repair post–myocardial infarction. Journal of the American Heart Association 8 (15). Available at https://doi.org/10.1161/jaha.119.012089.

Albericio, G., Higuera, M., Araque, P., Sánchez, C., Herrero, D., García-Brenes, M.A., et al., 2024. Development of a Bmi1+ cardiac mouse progenitor immortalized model to unravel the relationship with its protective vascular endothelial niche. International Journal of Molecular Sciences 25 (16). Available at https://doi.org/10.3390/ijms25168815.

Alum, E.U., Izah, S.C., Uti, D.E., Ugwu, O.P.C., Betiang, P.A., Basajja, M., et al., 2025. Targeting cellular senescence for healthy aging: advances in senolytics and senomorphics. Drug Design, Development and Therapy 19, 8489–8522. Available at https://doi.org/10.2147/DDDT.S543211.

Amado, L.C., Saliaris, A.P., Schuleri, K.H., St. John, M., Xie, J.S., Cattaneo, S., et al., 2005. Cardiac repair with intramyocardial injection of allogeneic mesenchymal stem cells after myocardial infarction. Proceedings of the National Academy of Sciences of the United States of America 102 (32), 11474–11479. Available at https://doi.org/10.1073/pnas.0504388102.

Andrews, P.W., Gokhale, P.J., 2024. A short history of pluripotent stem cells markers. Stem Cell Reports 19 (1), 1–10. Available at https://doi.org/10.1016/j.stemcr.2023.11.012.

Anwar, I., Wang, X., Pratt, R.E., Dzau, V.J., Hodgkinson, C.P., 2024. The impact of aging on cardiac repair and regeneration. Journal of Biological Chemistry 300 (9). Available at https://doi.org/10.1016/j.jbc.2024.107682.

Avolio, E., Alvino, V.V., Ghorbel, M.T., Campagnolo, P., 2017. Perivascular cells and tissue engineering: current applications and untapped potential. Pharmacology & Therapeutics 171, 83–92. Available at https://doi.org/10.1016/j.pharmthera.2016.11.002.

Azhdari, M., Rabbani, S., Tanideh, N., Boroumand, S., 2025. Cardiac regeneration revisited: enhanced cardiomyocyte differentiation and repair through composite microenvironments and small molecules. Biomedicine & Pharmacotherapy 192, 118642. Available at https://doi.org/10.1016/j.biopha.2025.118642.

Barbulescu, G.I., Bojin, F.M., Ordodi, V.L., Goje, I.D., Barbulescu, A.S., Paunescu, V., 2022. Decellularized extracellular matrix scaffolds for cardiovascular tissue engineering: current techniques and challenges. International Journal of Molecular Sciences 23 (21). Available at https://doi.org/10.3390/ijms232113040.

Bartolucci, J., Verdugo, F.J., González, P.L., Larrea, R.E., Abarzua, E., Goset, C., et al., 2017. Safety and efficacy of the intravenous infusion of umbilical cord mesenchymal stem cells in patients with heart failure. Circulation Research 121 (10), 1192–1204. Available at https://doi.org/10.1161/circresaha.117.310712.

Bartunek, J., Behfar, A., Dolatabadi, D., Vanderheyden, M., Ostojic, M., Dens, J., et al., 2013. Cardiopoietic stem cell therapy in heart failure: the C-CURE (cardiopoietic stem cell therapy in heart failURE) multicenter randomized trial with lineage-specified biologics. Journal of the American College of Cardiology 61 (23), 2329–2338. Available at https://doi.org/10.1016/j.jacc.2013.02.071.

Bergmann, O., Jovinge, S., 2014. Cardiac regeneration in vivo: mending the heart from within? Stem Cell Research 13 (3), 523–531. Available at https://doi.org/10.1016/j.scr.2014.07.002.

Bergmann, O., Bhardwaj, R.D., Bernard, S., Zdunek, S., Barnabé-Heide, F., Walsh, S., et al., 2009. Evidence for cardiomyocyte renewal in humans. Science 324 (5923), 98–102. Available at https://doi.org/10.1126/science.1164680.

Bhat, S., Viswanathan, P., Chandanala, S., Prasanna, S.J., Seetharam, R.N., 2021. Expansion and characterization of bone marrow derived human mesenchymal stromal cells in serum-free conditions. Scientific Reports 11 (1). Available at https://doi.org/10.1038/s41598-021-83088-1.

Bianconi, V., Sahebkar, A., Kovanen, P., Bagaglia, F., Ricciuti, B., Calabrò, P., et al., 2018. Endothelial and cardiac progenitor cells for cardiovascular repair: a controversial paradigm in cell therapy. Pharmacology & Therapeutics 181, 156–168. Available at https://doi.org/10.1016/j.pharmthera.2017.08.004.

Bishop, S.P., Zhou, Y., Nakada, Y., Zhang, J., 2021. Changes in cardiomyocyte cell cycle and hypertrophic growth during fetal to adult in mammals. Journal of the American Heart Association 10 (2). Available at https://doi.org/10.1161/jaha.120.017839.

Bolli, R., Hare, J.M., March, K.L., Pepine, C.J., Willerson, J.T., Perin, E.C., et al., 2018. Rationale and design of the CONCERT-HF trial (combination of mesenchymal and c-kit + cardiac stem cells as regenerative therapy for heart failure). Circulation Research 122 (12), 1703–1715. Available at https://doi.org/10.1161/CIRCRESAHA.118.312978.

Borow, K.M., Yaroshinsky, A., Greenberg, B., Perin, E.C., 2019. Phase 3 DREAM-HF trial of mesenchymal precursor cells in chronic heart failure. Circulation Research 125 (3), 265–281. Available at https://doi.org/10.1161/circresaha.119.314951.

Cai, Y., Liu, H., Song, E., Wang, L., Xu, J., He, Y., et al., 2021. Deficiency of telomere-associated repressor activator protein 1 precipitates cardiac aging in mice via p53/PPARα signaling. Theranostics 11 (10), 4710–4727. Available at https://doi.org/10.7150/thno.51739.

Cao, Q., Liu, L., Hu, Y., Cao, S., Tan, T., Huang, X., et al., 2023. Low-intensity pulsed ultrasound of different intensities differently affects myocardial ischemia/reperfusion injury by modulating cardiac oxidative stress and inflammatory reaction. Frontiers in Immunology 14. Available at https://doi.org/10.3389/fimmu.2023.1248056.

Castellan, R.F.P., Meloni, M., 2018. Mechanisms and therapeutic targets of cardiac regeneration: closing the age gap. Frontiers in Cardiovascular Medicine 5. Available at https://doi.org/10.3389/fcvm.2018.00007.

Chen, Y., Lüttmann, F.F., Schoger, E., Schöler, H.R., Zelarayán, L.C., Kim, K.P., et al., 2021. Reversible reprogramming of cardiomyocytes to a fetal state drives heart regeneration in mice. Science 373 (6562), 1537–1540. Available at https://doi.org/10.1126/science.abg5159.

Chen, K., Li, Y., Xu, L., Qian, Y., Liu, N., Zhou, C., et al., 2022. Comprehensive insight into endothelial progenitor cell-derived extracellular vesicles as a promising candidate for disease treatment. Stem Cell Research & Therapy 13 (1). Available at https://doi.org/10.1186/s13287-022-02921-0.

Cheng, T.H., Cheng, P.Y., Shih, N.L., Chen, I.B., Wang, D.L., Chen, J.J., 2003. Involvement of reactive oxygen species in angiotensin ii-induced endothelin-1 gene expression in rat cardiac fibroblasts. Journal of the American College of Cardiology 42 (10), 1845–1854. Available at https://doi.org/10.1016/j.jacc.2003.06.010.

Cheng, Y.C., Hsieh, M.L., Lin, C.J., Chang, C.M.C., Huang, C.Y., Puntney, R., et al., 2023. Combined treatment of human induced pluripotent stem cell-derived cardiomyocytes and endothelial cells regenerate the infarcted heart in mice and non-human primates. Circulation 148 (18), 1395–1409. Available at https://doi.org/10.1161/CIRCULATIONAHA.122.061736.

Cipollone, I., Monaco, V., Britto-Júnior, J., Lima, A.T., Antunes, E., Pupo, A.S., et al., 2025. The identification of adenylyl cyclase modulators as potential receptors for 6-nitrodopamine in human-induced pluripotent stem cell (hiPSC)-derived cardiomyocytes and their relevance in heart inotropism. Frontiers in Pharmacology 16. Available at https://doi.org/10.3389/fphar.2025.1597035.

Dabravolski, S.A., Markin, A.M., Andreeva, E.R., Eremin, I.I., Orekhov, A.N., Melnichenko, A.A., 2022. Emerging role of pericytes in therapy of cardiovascular diseases. Biomedicine & Pharmacotherapy 156, 113928. Available at https://doi.org/10.1016/j.biopha.2022.113928.

Derks, W., Rode, J., Collin, S., Rost, F., Heinke, P., Hariharan, A., et al., 2025. A latent cardiomyocyte regeneration potential in human heart disease. Circulation 151 (3), 245–256. Available at https://doi.org/10.1161/circulationaha.123.067156.

Dong, J., Zhang, Z., Huang, H., Mo, P., Cheng, C., Liu, J., et al., 2018. miR-10a rejuvenates aged human mesenchymal stem cells and improves heart function after myocardial infarction through KLF4. Stem Cell Research & Therapy 9 (1). Available at https://doi.org/10.1186/s13287-018-0895-0.

Duim, S.N., Goumans, M.J., Kruithof, B.P.T., 2016. WT1 in Cardiac Development and Disease. In: van den Heuvel-Eibrink, M.M. (Ed.), Wilms Tumor. Codon Publications.

Duran, J.M., Makarewich, C.A., Sharp, T.E., Starosta, T., Zhu, F., Hoffman, N.E., et al., 2013. Bone-derived stem cells repair the heart after myocardial infarction through transdifferentiation and paracrine signaling mechanisms. Circulation Research 113 (5), 539–552. Available at https://doi.org/10.1161/CIRCRESAHA.113.301202.

Eschenhagen, T., Bolli, R., Braun, T., Field, L.J., Fleischmann, B.K., Frisén, J., et al., 2017. Cardiomyocyte regeneration: a consensus statement. Circulation 136 (7), 680–686. Available at https://doi.org/10.1161/CIRCULATIONAHA.117.029343.

Fang, M., Wang, J., Li, S., Guo, Y., 2016. Advanced glycation end-products accelerate the cardiac aging process through the receptor for advanced glycation end-products/transforming growth factor-β-Smad signaling pathway in cardiac fibroblasts. Geriatrics & Gerontology International 16 (4), 522–527. Available at https://doi.org/10.1111/ggi.12499.

Ferland-McCollough, D., Slater, S., Richard, J., Reni, C., Mangialardi, G., 2017. Pericytes, an overlooked player in vascular pathobiology. Pharmacology & Therapeutics 171, 30–42. Available at https://doi.org/10.1016/j.pharmthera.2016.11.008.

Garcia-Puig, A., Mosquera, J.L., Jiménez-Delgado, S., García-Pastor, C., Jorba, I., Navajas, D., et al., 2019. Proteomics analysis of extracellular matrix remodeling during zebrafish heart regeneration. Molecular and Cellular Proteomics 18 (9), 1745–1755. Available at https://doi.org/10.1074/mcp.RA118.001193.

Garg, P., Garg, V., Shrestha, R., Sanguinetti, M.C., Kamp, T.J., Wu, J.C., 2018. Human induced pluripotent stem cell-derived cardiomyocytes as models for cardiac channelopathies: a primer for non-electrophysiologists. Circulation Research 123 (2), 224–243. Available at https://doi.org/10.1161/CIRCRESAHA.118.311209.

Garmany, A., Terzic, A., 2024. Artificial intelligence powers regenerative medicine into predictive realm. Regenerative Medicine 19 (12), 611–616. Available at https://doi.org/10.1080/17460751.2024.2437281.

Golpanian, S., El-Khorazaty, J., Mendizabal, A., DiFede, D.L., Suncion, V.Y., Karantalis, V., et al., 2015. Effect of aging on human mesenchymal stem cell therapy in ischemic cardiomyopathy patients. Journal of the American College of Cardiology 65 (2), 125–132. Available at https://doi.org/10.1016/j.jacc.2014.10.040.

Gong, X., Wang, P., Wu, Q., Wang, S., Yu, L., Wang, G., 2016. Human umbilical cord blood derived mesenchymal stem cells improve cardiac function in cTnTR141W transgenic mouse of dilated cardiomyopathy. European Journal of Cell Biology 95 (1), 57–67. Available at https://doi.org/10.1016/j.ejcb.2015.11.003.

Grzeczka, A., Graczyk, S., Kordowitzki, P., 2023. DNA methylation and telomeres-their impact on the occurrence of atrial fibrillation during cardiac aging. International Journal of Molecular Sciences 24 (21), 15699. Available at 10.3390/ijms242115699.

Haridhasapavalan, K.K., Borthakur, A., Thummer, R.P., 2023. Direct cardiac reprogramming: current status and future prospects. In: Turksen, K. (Ed.), Cell Biology and Translational Medicine, Volume 20. Advances in Experimental Medicine and Biology, vol 1436. Springer, Cham, pp. 1–18. Available at https://doi.org/10.1007/5584_2022_760.

Hashimoto, H., Olson, E.N., Bassel-Duby, R., 2018. Therapeutic approaches for cardiac regeneration and repair. Nature Reviews Cardiology 15 (10), 585–600. Available at https://doi.org/10.1038/s41569-018-0036-6.

Hatzistergos, K.E., Quevedo, H., Oskouei, B.N., Hu, Q., Feigenbaum, G.S., Margitich, I.S., et al., 2010. Bone marrow mesenchymal stem cells stimulate cardiac stem cell proliferation and differentiation. Circulation Research 107 (7), 913–922. Available at https://doi.org/10.1161/CIRCRESAHA.110.222703.

He, X., Good, A., Kalou, W., Ahmad, W., Dutta, S., Chen, S., et al., 2024. Current advances and future directions of pluripotent stem cells-derived engineered heart tissue for treatment of cardiovascular diseases. Cells 13 (24), 2098. Available at https://doi.org/10.3390/cells13242098.

Herman, A.B., Occean, J.R., Sen, P., 2021. Epigenetic dysregulation in cardiovascular aging and disease. Journal of Cardiovascular Aging 1 (1). Available at https://doi.org/10.20517/jca.2021.16.

Hoang, D.M., Pham, P.T., Bach, T.Q., Ngo, A.T.L., Nguyen, Q.T., Phan, T.T.K., et al., 2022. Stem cell-based therapy for human diseases. Signal Transduction and Targeted Therapy 7 (1). Available at https://doi.org/10.1038/s41392-022-01134-4.

Horn, M.A., Trafford, A.W., 2016. Aging and the cardiac collagen matrix: novel mediators of fibrotic remodelling. Journal of Molecular and Cellular Cardiology 93, 175–185. Available at https://doi.org/10.1016/j.yjmcc.2015.11.005.

Hu, W.S., Liao, W.Y., Chang, C.H., Chen, T.S., 2022. Paracrine IGF-1 activates SOD2 Expression and regulates ROS/p53 axis in the treatment of cardiac damage in d-galactose-induced aging rats after receiving mesenchymal stem cells. Journal of Clinical Medicine 11 (15). Available at https://doi.org/10.3390/jcm11154419.

Huang, H., Huang, W., 2022. Regulation of endothelial progenitor cell functions in ischemic heart disease: new therapeutic targets for cardiac remodeling and repair. Frontiers in Cardiovascular Medicine 9. Available at https://doi.org/10.3389/fcvm.2022.896782.

Huang, J., Feng, Q., Wang, L., Zhou, B., 2021. Human pluripotent stem cell-derived cardiac cells: application in disease modeling, cell therapy, and drug discovery. Frontiers in Cell and Developmental Biology 9. Available at https://doi.org/10.3389/fcell.2021.655161.

Imai, S.-ichiro, Guarente, L., 2014. NAD+ and sirtuins in aging and disease. Trends in Cell Biology 24 (8), 464–471. Available at https://doi.org/10.1016/j.tcb.2014.04.002.

Jia, G., Aroor, A.R., Jia, C., Sowers, J.R., 2019. Endothelial cell senescence in aging-related vascular dysfunction. Biochimica et Biophysica Acta - Molecular Basis of Disease 1865 (7), 1802–1809. Available at https://doi.org/10.1016/j.bbadis.2018.08.008.

Jiang, Y., Lian, X.L., 2020. Heart regeneration with human pluripotent stem cells: prospects and challenges. Bioactive Materials 5 (1), 74–81. Available at https://doi.org/10.1016/j.bioactmat.2020.01.003.

Jiang, X., Chao, L., Liu, K., Zhao, Y., Wu, J., Liu, C., et al., 2025. Cationic microbubbles loading shFOXO4/SDF1 rejuvenate the aged heart and alleviate myocardial ischemia-reperfusion injury in the elderly. Free Radical Biology and Medicine 237, 210–227. Available at https://doi.org/10.1016/j.freeradbiomed.2025.05.416.

Katsyuba, E., Auwerx, J., 2017. Modulating NAD + metabolism, from bench to bedside. The EMBO Journal 36 (18), 2670–2683. Available at https://doi.org/10.15252/embj.201797135.

Kawamura, M., Miyagawa, S., Fukushima, S., Saito, A., Toda, K., Daimon, T., et al., 2015. Xenotransplantation of bone marrow-derived human mesenchymal stem cell sheets attenuates left ventricular remodeling in a porcine ischemic cardiomyopathy model. Tissue Engineering Part A 21 (15-16), 2272–2280. Available at https://doi.org/10.1089/ten.tea.2014.0036.

Khan, M., Mohsin, S., Khan, S.N., Riazuddin, S., 2011. Repair of senescent myocardium by mesenchymal stem cells is dependent on the age of donor mice. Journal of Cellular and Molecular Medicine 15 (7), 1515–1527. Available at https://doi.org/10.1111/j.1582-4934.2009.00998.x.

Khatiwala, R.V., Zhang, S., Li, X., Devejian, N., Bennett, E., Cai, C., 2018. Inhibition of p16INK4A to rejuvenate aging human cardiac progenitor cells via the upregulation of anti-oxidant and NFκB signal pathways. Stem Cell Reviews and Reports 14 (4), 612–625. Available at https://doi.org/10.1007/s12015-018-9815-z.

Kliesow Remes, A., Ruf, T., Zurashvili, T., Ding, L., Meyer-Jens, M., Schwab, D.M., et al., 2025. AAV-mediated overexpression of CPT1B protects from cardiac hypertrophy and heart failure in a murine pressure overload model. Basic Research in Cardiology 120 (5), 975–989. Available at https://doi.org/10.1007/s00395-025-01123-y.

Kong, P., Dong, J., Li, W., Li, Z., Gao, R., Liu, X., et al., 2023. Extracellular matrix/glycopeptide hybrid hydrogel as an immunomodulatory niche for endogenous cardiac repair after myocardial infarction. Advanced Science 10 (23). Available at https://doi.org/10.1002/advs.202301244.

Lazzarini, E., Lodrini, A.M., Arici, M., Bolis, S., Vagni, S., Panella, S., et al., 2022. Stress-induced premature senescence is associated with a prolonged QT interval and recapitulates features of cardiac aging. Theranostics 12 (11), 5237–5257. Available at https://doi.org/10.7150/thno.70884.

Lee, T.M., Harn, H.J., Chiou, T.W., Chuang, M.H., Chen, C.H., Chuang, C.H., et al., 2020. Host pre-conditioning improves human adipose–derived stem cell transplantation in ageing rats after myocardial infarction: role of NLRP3 inflammasome. Journal of Cellular and Molecular Medicine 24 (21), 12272–12284. Available at https://doi.org/10.1111/jcmm.15403.

Lee, J.R., Park, B.W., Park, J.H., Lim, S., Kwon, S.P., Hwang, J.W., et al., 2021. Local delivery of a senolytic drug in ischemia and reperfusion-injured heart attenuates cardiac remodeling and restores impaired cardiac function. Acta Biomaterialia 135, 520–533. Available at https://doi.org/10.1016/j.actbio.2021.08.028.

Lee, C., Kim, M.-J., Kumar, A., Lee, H.-W., Yang, Y., Kim, Y., 2025. Vascular endothelial growth factor signaling in health and disease: from molecular mechanisms to therapeutic perspectives. Signal Transduction and Targeted Therapy 10 (1). Available at https://doi.org/10.1038/s41392-025-02249-0.

Lei, Q., Gao, F., Liu, T., Ren, W., Chen, L., Cao, Y., et al., 2021. Extracellular vesicles deposit to rejuvenate aged bone marrow–derived mesenchymal stem cells and slow age-related degeneration. Science Translational Medicine 13, eaaz8697.

Lewis-McDougall, F.C., Ruchaya, P.J., Domenjo-Vila, E., Shin Teoh, T., Prata, L., Cottle, B.J., et al., 2019. Aged-senescent cells contribute to impaired heart regeneration. Aging Cell 18 (3). Available at https://doi.org/10.1111/acel.12931.

Li, J., Li, S.H., Wu, J., Weisel, R.D., Yao, A., Stanford, W.L., et al., 2018. Young bone marrow Sca-1 cells rejuvenate the aged heart by promoting epithelial-to-mesenchymal transition. Theranostics 8 (7), 1766–1781. Available at https://doi.org/10.7150/thno.22788.

Li, J., Li, S.H., Dong, J., Alibhai, F.J., Zhang, C., Shao, Z.B., et al., 2019a. Long-term repopulation of aged bone marrow stem cells using young Sca-1 cells promotes aged heart rejuvenation. Aging Cell 18 (6). Available at https://doi.org/10.1111/acel.13026.

Li, J., Yang, K.Y., Tam, R.C.Y., Chan, V.W., Lan, H.Y., Hori, S., et al., 2019b. Regulatory T-cells regulate neonatal heart regeneration by potentiating cardiomyocyte proliferation in a paracrine manner. Theranostics 9 (15), 4324–4341. Available at https://doi.org/10.7150/thno.32734.

Li, H., Wang, L., Ma, T., Liu, Z., Gao, L., 2023. Exosomes secreted by endothelial cells derived from human induced pluripotent stem cells improve recovery from myocardial infarction in mice. Stem Cell Research & Therapy 14 (1). Available at https://doi.org/10.1186/s13287-023-03462-w.

Li, B., Xiong, W., Zuo, W., Shi, Y., Wang, T., Chang, L., et al., 2024. Proximal telomeric decompaction due to telomere shortening drives FOXC1-dependent myocardial senescence. Nucleic Acids Research 52 (11), 6269–6284. Available at https://doi.org/10.1093/nar/gkae274.

Lin, M., Guo, J., Gu, Z., Tang, W., Tao, H., You, S., et al., 2025. Machine learning and multi-omics integration: advancing cardiovascular translational research and clinical practice. Journal of Translational Medicine 23 (1). Available at https://doi.org/10.1186/s12967-025-06425-2.

Liu, L., Zhao, B., Yu, Y., Gao, W., Liu, W., Chen, L., et al., 2024. Vascular aging in ischemic stroke. Journal of the American Heart Association 13 (15). Available at https://doi.org/10.1161/jaha.123.033341.

Lucas, V., Cavadas, C., Aveleira, C.A., 2023. Cellular senescence: from mechanisms to current biomarkers and senotherapies. Pharmacological Reviews 75 (4), 675–713. Available at https://doi.org/10.1124/pharmrev.122.000622.

Ma, Y., Mouton, A.J., Lindsey, M.L., 2018. Cardiac macrophage biology in the steady-state heart, the aging heart, and following myocardial infarction. Translational Research 191, 15–28. Available at https://doi.org/10.1016/j.trsl.2017.10.001.

Martín-Fernández, B., Gredilla, R., 2016. Mitochondria and oxidative stress in heart aging. AGE 38 (4), 225–238. Available at https://doi.org/10.1007/s11357-016-9933-y.

Martini, H., Iacovoni, J.S., Maggiorani, D., Dutaur, M., Marsal, D.J., Roncalli, J., et al., 2019. Aging induces cardiac mesenchymal stromal cell senescence and promotes endothelial cell fate of the CD90 + subset. Aging Cell 18 (5). Available at https://doi.org/10.1111/acel.13015.

Matthes, S., Taffet, S., Delmar, M., 2009. Epicardium-derived cells as progenitors of cardiac fibroblasts: a possible role in arrhythmogenic right ventricular cardiomyopathy? Heart Rhythm 6 (11), 1692–1693. Available at https://doi.org/10.1016/j.hrthm.2009.09.052.

Menasché, P., 2020. Cell therapy with human ESC-derived cardiac cells: clinical perspectives. Frontiers in Bioengineering and Biotechnology 8. Available at https://doi.org/10.3389/fbioe.2020.601560.

Monma, Y., Shindo, T., Eguchi, K., Kurosawa, R., Kagaya, Y., Ikumi, Y., et al., 2021. Low-intensity pulsed ultrasound ameliorates cardiac diastolic dysfunction in mice: a possible novel therapy for heart failure with preserved left ventricular ejection fraction. Cardiovascular Research 117 (5), 1325–1338. Available at https://doi.org/10.1093/cvr/cvaa221.

Monnerat, G., Kasai-Brunswick, T.H., Asensi, K.D., Silva dos Santos, D., Barbosa, R.A.Q., Cristina Paccola Mesquita, F., et al., 2022. Modelling premature cardiac aging with induced pluripotent stem cells from a hutchinson-gilford Progeria Syndrome patient. Frontiers in Physiology 13. Available at https://doi.org/10.3389/fphys.2022.1007418.

Nagata, H., Ii, M., Kohbayashi, E., Hoshiga, M., Hanafusa, T., Asahi, M., 2016. Cardiac adipose-derived stem cells exhibit high differentiation potential to cardiovascular cells in C57BL/6 mice. Stem Cells Translational Medicine 5 (2), 141–151. Available at https://doi.org/10.5966/sctm.2015-0083.

Noiseux, N., Mansour, S., Weisel, R., Stevens, L.M., Der Sarkissian, S., Tsang, K., et al., 2016. The IMPACT-CABG trial: a multicenter, randomized clinical trial of CD133+ stem cell therapy during coronary artery bypass grafting for ischemic cardiomyopathy. Journal of Thoracic and Cardiovascular Surgery 152 (6), 1582–1588.e21582. Available at https://doi.org/10.1016/j.jtcvs.2016.07.067.

Oikonomopoulos, A., Kitani, T., Wu, J.C., 2018. Pluripotent stem cell-derived cardiomyocytes as a platform for cell therapy applications: progress and hurdles for clinical translation. Molecular Therapy 26 (7), 1624–1634. Available at https://doi.org/10.1016/j.ymthe.2018.02.026.

Ozcebe, S.G., Bahcecioglu, G., Yue, X.S., Zorlutuna, P., 2021. Effect of cellular and ECM aging on human iPSC-derived cardiomyocyte performance, maturity and senescence. Biomaterials 268, 120554. Available at https://doi.org/10.1016/j.biomaterials.2020.120554.

Pandit, S., Jamal, T., Ali, A., Parthasarathi, R., 2024. Multiscale computational and machine learning models for designing stem cell-based regenerative medicine therapies. Computational Biology for Stem Cell Research. Elsevier, India, pp. 433–442. Available at https://doi.org/10.1016/B978-0-443-13222-3.00027-7.

Paneni, F., Diaz Cañestro, C., Libby, P., Lüscher, T.F., Camici, G.G., 2017. The aging cardiovascular system: understanding it at the cellular and clinical levels. Elsevier USA, Switzerland. Journal of the American College of Cardiology 69 (15), 1952–1967. Available at https://doi.org/10.1016/j.jacc.2017.01.064.

Peterson, E.A., Sun, J., Wang, J., 2022. Leukocyte-mediated cardiac repair after myocardial infarction in non-regenerative vs. regenerative systems. Journal of Cardiovascular Development and Disease 9 (2). Available at https://doi.org/10.3390/jcdd9020063.

Phillips, K.G., Aljabban, I., Wolbrom, D.H., Griesemer, A., Leacche, M., McGregor, C., et al., 2025. Cardiac xenotransplantation: current state and future directions. Circulation 152 (1), 58–73. Available at https://doi.org/10.1161/CIRCULATIONAHA.124.070875.

Pietri, P., Stefanadis, C., 2021. Cardiovascular aging and longevity: JACC state-of-the-art review. Journal of the American College of Cardiology 77 (2), 189–204. Available at https://doi.org/10.1016/j.jacc.2020.11.023.

Pillai, V.B., Samant, S., Hund, S., Gupta, M., Gupta, M.P., 2021. The nuclear sirtuin SIRT6 protects the heart from developing aging-associated myocyte senescence and cardiac hypertrophy. Impact Journals LLC, United States Aging 13 (9), 12334–12342. Available at https://doi.org/10.18632/aging.203027.

Prantl, L., Eigenberger, A., Brix, E., Kempa, S., Baringer, M., Felthaus, O., 2021. Adipose tissue-derived stem cell yield depends on isolation protocol and cell counting method. Cells 10 (5), 1113. Available at https://doi.org/10.3390/cells10051113.

Quijada, P., Trembley, M.A., Small, E.M., 2020. The role of the epicardium during heart development and repair. Circulation Research 126 (3), 377–394. Available at https://doi.org/10.1161/CIRCRESAHA.119.315857.

Räsänen, M., Sultan, I., Paech, J., Hemanthakumar, K.A., Yu, W., He, L., et al., 2021. VEGF-B promotes endocardium-derived coronary vessel development and cardiac regeneration. Circulation 143 (1), 65–77. Available at https://doi.org/10.1161/CIRCULATIONAHA.120.050635.

Rodrigues, L., Pereira, R., Pogue, R., Carvalho, J., Rodrigues, V., 2021. Pluripotent stem cell exosomes: effective strategies for cellular senescence prevention and reversion. Cytotherapy 23 (4), 30–31. Available at https://doi.org/10.1016/j.jcyt.2021.02.089.

Ruiz-Meana, M., Bou-Teen, D., Ferdinandy, P., Gyongyosi, M., Pesce, M., Perrino, C., et al., 2020. Cardiomyocyte ageing and cardioprotection: consensus document from the ESC working groups cell biology of the heart and myocardial function. Cardiovascular Research 116 (11), 1835–1849. Available at https://doi.org/10.1093/cvr/cvaa132.

Sanz-Ros, J., Huete-Acevedo, J., Mas-Bargues, C., Romero-García, N., Dromant, M., van Weeghel, M., et al., 2025. Small extracellular vesicles from young adipose-derived stem cells ameliorate age-related changes in the heart of old mice. Stem Cell Research & Therapy 16, 138. Available at https://doi.org/10.1186/s13287-025-04255-z.

Schulman, S.P., Becker, L.C., Kass, D.A., Champion, H.C., Terrin, M.L., Forman, S., et al., 2006. L-arginine therapy in acute myocardial infarction: the vascular interaction with age in myocardial infarction (VINTAGE MI) randomized clinical trial. JAMA 295 (1), 58–64. Available at https://doi.org/10.1001/jama.295.1.58.

Schwach, V., Gomes Fernandes, M., Maas, S., Gerhardt, S., Tsonaka, R., van der Weerd, L., et al., 2020. Expandable human cardiovascular progenitors from stem cells for regenerating mouse heart after myocardial infarction. Cardiovascular Research 116 (3), 545–553. Available at https://doi.org/10.1093/cvr/cvz181.

Seibertz, F., Sutanto, H., Dülk, R., Pronto, J.R.D., Springer, R., Rapedius, M., et al., 2023. Electrophysiological and calcium-handling development during long-term culture of human-induced pluripotent stem cell-derived cardiomyocytes. Basic Research in Cardiology 118 (1). Available at https://doi.org/10.1007/s00395-022-00973-0.

Silva-Palacios, A., Zúñiga-Muñoz, A.M., Soria-Castro, E., Álvarez-León, E., Nieto, M., Navarrete-Anastasio, G., et al., 2025. Cardioprotective effect of senotherapy in chronically obese middle-aged female rats may be mediated by a MERCSs/Nrf2 interaction. The Journal of Nutritional Biochemistry 142, 109923. Available at https://doi.org/10.1016/j.jnutbio.2025.109923.

Song, B.W., Lee, C.Y., Kim, R., Kim, W.J., Lee, H.W., Lee, M.Y., et al., 2021. Multiplexed targeting of miRNA-210 in stem cell-derived extracellular vesicles promotes selective regeneration in ischemic hearts. Experimental and Molecular Medicine 53 (4), 695–708. Available at https://doi.org/10.1038/s12276-021-00584-0.

Suchanecka, M., Grzelak, J., Farzaneh, M., Azizidoost, S., Dari, M.A.G., Józkowiak, M., et al., 2025. Adipose derived stem cells – sources, differentiation capacity and a new target for reconstructive and regenerative medicine. Biomedicine & Pharmacotherapy 186, 118036. Available at https://doi.org/10.1016/j.biopha.2025.118036.

Tan, J.J., Guyette, J.P., Miki, K., Xiao, L., Kaur, G., Wu, T., et al., 2021. Human iPS-derived pre-epicardial cells direct cardiomyocyte aggregation expansion and organization in vitro. Communications 12 (1). Available at https://doi.org/10.1038/s41467-021-24921-z.

Tan, Y., Duan, X., Wang, B., Liu, X., Zhan, Z., 2023. Murine neonatal cardiac B cells promote cardiomyocyte proliferation and heart regeneration. npj Regenerative Medicine 8 (1). Available at https://doi.org/10.1038/s41536-023-00282-7.

Täubel, J., Hauke, W., Rump, S., Viereck, J., Batkai, S., Poetzsch, J., et al., 2021. Novel antisense therapy targeting microRNA-132 in patients with heart failure: results of a first-in-human Phase 1b randomized, double-blind, placebo-controlled study. European Heart Journal 42 (2), 178–188. Available at https://doi.org/10.1093/eurheartj/ehaa898.

Tenreiro, M.F., Louro, A.F., Alves, P.M., Serra, M., 2021. Next generation of heart regenerative therapies: progress and promise of cardiac tissue engineering. npj Regenerative Medicine 6 (1). Available at https://doi.org/10.1038/s41536-021-00140-4.

Tompkins, B.A., Difede, D.L., Khan, A., Landin, A.M., Schulman, I.H., Pujol, M.V., et al., 2017. Allogeneic mesenchymal stem cells ameliorate aging frailty: a phase II randomized, double-blind, placebo-controlled clinical trial. Journals of Gerontology - Series A Biological Sciences and Medical Sciences 72 (11), 1513–1521. Available at https://doi.org/10.1093/gerona/glx137.

Tracy, E., Rowe, G., LeBlanc, A.J., 2020. Cardiac tissue remodeling in healthy aging: the road to pathology. American Journal of Physiology - Cell Physiology 319 (1), C166. Available at https://doi.org/10.1152/ajpcell.00021.2020.

Valente, M., Nascimento, D.S., Cumano, A., Pinto-Do-Ó, P., 2014. Sca-1+cardiac progenitor cells and heart-making: a critical synopsis. Stem Cells and Development 23 (19), 2263–2273. Available at https://doi.org/10.1089/scd.2014.0197.

Vo, Q.D., Nakamura, K., Saito, Y., Akagi, S., Miyoshi, T., Yuasa, S., 2025. Induced pluripotent stem cells in cardiomyopathy: advancing disease modeling, therapeutic development, and regenerative therapy. International Journal of Molecular Sciences 26 (11). Available at https://doi.org/10.3390/ijms26114984.

Wang, L., Tian, X., Cao, Y., Ma, X., Shang, L., Li, H., et al., 2021. Cardiac shock wave therapy improves ventricular function by relieving fibrosis through PI3K/Akt signaling pathway: evidence from a rat model of post-infarction heart failure. Frontiers in Cardiovascular Medicine 8. Available at https://doi.org/10.3389/fcvm.2021.693875.

Xu, Q., Chen, X., Zhao, C., Liu, Y., Wang, J., Ao, X., et al., 2025. Cell cycle arrest of cardiomyocytes in the context of cardiac regeneration. Frontiers in Cardiovascular Medicine 12. Available at https://doi.org/10.3389/fcvm.2025.1538546.

Yamakawa, H., Ieda, M., 2021. Cardiac regeneration by direct reprogramming in this decade and beyond. Inflammation and Regeneration 41 (1). Available at https://doi.org/10.1186/s41232-021-00168-5.

Yan, J., Ai, C., Chen, Q., Wang, Q., Zhu, Y., Li, M., et al., 2024. CircMap4k2 reactivated by aneurysm plication alleviates residual cardiac remodeling after SVR by enhancing cardiomyocyte proliferation in post-MI mice. Journal of Advanced Research 65, 227–238. Available at https://doi.org/10.1016/j.jare.2023.11.034.

Yang, K., Song, H.F., He, S., Yin, W.J., Fan, X.M., Ru, F., et al., 2019. Effect of neuron-derived neurotrophic factor on rejuvenation of human adipose-derived stem cells for cardiac repair after myocardial infarction. Journal of Cellular and Molecular Medicine 23 (9), 5981–5993. Available at https://doi.org/10.1111/jcmm.14456.

Yang, B., Li, T., Wang, Z., Zhu, Y., Niu, K., Hu, S., et al., 2024. Ruxolitinib-based senomorphic therapy mitigates cardiomyocyte senescence in septic cardiomyopathy by inhibiting the JAK2/STAT3 signaling pathway. International Journal of Biological Sciences 20 (11), 4314–4340. Available at https://doi.org/10.7150/ijbs.96489.

Ye, L., Swingen, C., Zhang, J., 2013. Induced pluripotent stem cells and their potential for basic and clinical sciences. Current Cardiology Reviews 9 (1), 63–72. Available at https://doi.org/10.2174/157340313805076278.

Zhang, Y., Shen, T., Liu, B., Dai, D., Cai, J., Zhao, C., et al., 2018. Cardiac shock wave therapy attenuates cardiomyocyte apoptosis after acute myocardial infarction in rats. Cellular Physiology and Biochemistry 49 (5), 1734–1746. Available at https://doi.org/10.1159/000493616.

Zhang, N., Zhu, J., Ma, Q., Zhao, Y., Wang, Y., Hu, X., et al., 2020. Exosomes derived from human umbilical cord MSCs rejuvenate aged MSCs and enhance their functions for myocardial repair. Stem Cell Research & Therapy 11, 273.

Zhang, Y., Zheng, Y., Wang, S., Fan, Y., Ye, Y., Jing, Y., et al., 2022. Single-nucleus transcriptomics reveals a gatekeeper role for FOXP1 in primate cardiac aging. Protein & Cell 14 (4). Available at https://doi.org/10.1093/procel/pwac038.

Zheng, Z., Tang, W., Li, Y., Ai, Y., Tu, Z., Yang, J., et al., 2024. Advancing cardiac regeneration through 3D bioprinting: methods, applications, and future directions. Heart Failure Reviews 29 (3), 599–613. Available at https://doi.org/10.1007/s10741-023-10367-6.

Zhou, J., Shi, Y., 2023. Mesenchymal stem/stromal cells (MSCs): origin, immune regulation, and clinical applications. Cellular and Molecular Immunology 20 (6), 555–557. Available at https://doi.org/10.1038/s41423-023-01034-9.

Zhou, B., Ma, Q., Rajagopal, S., Wu, S.M., Domian, I., Rivera-Feliciano, J., et al., 2008. Epicardial progenitors contribute to the cardiomyocyte lineage in the developing heart. Nature 454 (7200), 109–113. Available at https://doi.org/10.1038/nature07060.

Zhu, D., Cheng, K., 2021. Cardiac cell therapy for heart repair: should the cells be left out? Cells 10 (3), 641. Available at https://doi.org/10.3390/cells10030641.

Zhu, W., Du, W., Duan, R., Liu, Y., Zong, B., Jin, X., et al., 2024. miR-873-5p suppression reinvigorates aging mesenchymal stem cells and improves cardiac repair after myocardial infarction. ACS Pharmacology & Translational Science 7 (3), 743–756. Available at https://doi.org/10.1021/acsptsci.3c00293.

Zsebo, K., Yaroshinsky, A., Rudy, J.J., Wagner, K., Greenberg, B., Jessup, M., et al., 2014. Long-term effects of AAV1/SERCA2a gene transfer in patients with severe heart failure. Circulation Research 114 (1), 101–108. Available at https://doi.org/10.1161/circresaha.113.302421.

Chapter 18

Future perspectives: personalized medicine and senescence modulation in cardiovascular health

Ishtiyaq Nissar[1], Jasreen Uppal[2], Mahak Gayaz[1], Apporva Chawla[2], Roohi Mohi ud din[3], Md. Sadique Hussain[4], Reyaz Hassan Mir[5], and Anas Ahmad[6]

[1]*Pharmacology Division, Department of Pharmaceutical Sciences, University of Kashmir, Hazratbal, Srinagar, Jammu and Kashmir, India,* [2]*School of Pharmacy, Khalsa University, Amritsar, Punjab, India,* [3]*Department of General Medicine, Sher-I-Kashmir Institute of Medical Sciences (SKIMS), Srinagar, Jammu and Kashmir, India,* [4]*Uttaranchal Institute of Pharmaceutical Sciences, Uttaranchal University, Dehradun, Uttarakhand, India,* [5]*Pharmaceutical Chemistry Division, Department of Pharmaceutical Sciences, University of Kashmir, Hazratbal, Srinagar, Jammu and Kashmir, India,* [6]*Department of Microbiology, Immunology and Infectious Diseases, Snyder Institute for Chronic Diseases, and Hotchkiss Brain Institute, Cumming School of Medicine, Foothills Medical Centre, University of Calgary, Calgary, AB, Canada*

18.1 Introduction

Cardiovascular diseases (CVDs) represent a serious health challenge globally, and the population may be susceptible to these diseases at different life stages. These are the leading causes of death across the globe, and the foundation of these diseases begins with atherosclerotic changes in childhood (Tong et al., 2024). These diseases remain the foremost cause of global mortality, significantly imposing a substantial source of health loss and major burden on population health and escalating healthcare costs. The primary risk factors leading to these diseases are preventable and may arise from unhealthy practices and harmful habits (Vaduganathan et al., 2022). Over the past three decades, CVDs prevalence showed up near two-fold rise, with prevalence climbing from 271 million to 523 million by 2019, with a corresponding increase in mortality (Brant et al., 2022). The prevalence of CVDs reached more than 500 million in 2019 globally, nearly twice the burden observed in 1990 (Tsampasian and Bloomfield, 2025). Recent estimates from 2019-22 indicate that tens of millions of people die from these diseases each year, with stroke and ischemic heart disease being the predominant contributors responsible for these deaths (Roth et al., 2020).

Globally, elevated systolic blood pressure remains the foremost preventable risk factor of premature cardiovascular mortality, responsible for about 10.8 million deaths from cardiovascular conditions and 11.3 million total deaths in 2021, with a particularly significant impact on cerebrovascular events, stroke, and ischemic heart disease (Razo et al., 2022). Dietary risks have been linked to about 6.58 million cardiovascular deaths in 2021, with dietary factors including high sodium, processed foods, low vegetables, and fruits (Vaduganathan et al., 2022).

CVD-related healthcare costs surpass 280 billion euros in Europe, and 400 billion dollars in the US, and projections indicate that US expenses on CVDs may rise to 1.34 trillion by 2050 (Luengo-Fernandez et al., 2023; Kazi et al., 2024) (Table 18.1).

As we age, many physiological systems undergo progressive decline, and the risk of developing CVDs gets higher. With advancing age, changes in vascular structure and functions take place, making the population more susceptible to CVDs (Cheng et al., 2023). The cardiovascular system undergoes cellular and molecular alterations like increased oxidative stress and inflammation, which play a core role in promoting the development of these diseases (Liberale et al., 2022).

Various studies have demonstrated that the possibility of development of these diseases escalates with age, highlighting advanced age as a major contributing factor. Changes in the cardiovascular system that take place with aging include

Cellular Senescence and Cardiac Aging. DOI: https://doi.org/10.1016/B978-0-443-45811-8.00008-8

TABLE 18.1 Biomarker-guided identification of age-related diseases associated with increased cellular senescence burden and potential responsiveness to senotherapeutics.

Disease	Description	Major causes	Risk factors	Common symptoms	Complications	Management/ Treatment	References
Cardiovascular diseases (e.g., Hypertension, Coronary Artery Disease)	Disorders of the heart and blood vessels common in older adults.	Arterial stiffening, high cholesterol, oxidative stress.	Age > 60, obesity, sedentary life, poor diet.	Chest pain, fatigue, breathlessness.	Stroke, myocardial infarction.	Lifestyle modification, antihypertensives, statins.	Farinha and Heaney (2025)
Type 2 diabetes mellitus	Chronic metabolic disorder with insulin resistance.	Genetic and lifestyle factors.	Obesity, inactivity, poor diet.	Polyuria, polydipsia, fatigue.	Neuropathy, nephropathy, retinopathy.	Diet control, oral hypoglycemics, insulin.	
Osteoarthritis	Degenerative joint disease-causing cartilage breakdown.	Aging, joint wear and tear.	Age > 50, obesity, joint stress.	Pain, stiffness, limited motion.	Disability, chronic pain.	Analgesics, physiotherapy, joint replacement.	Graham (2024)
Chronic obstructive pulmonary disease (COPD)	Progressive lung disease impairing airflow.	Long-term smoking, pollutants.	Smoking, air pollution, occupational exposure.	Cough, breathlessness, sputum.	Respiratory failure, heart strain.	Smoking cessation, bronchodilators, oxygen therapy.	Agusti and Vogelmeier (2023)
Alzheimer's disease/Dementia	Progressive brain disorder leading to cognitive decline.	Amyloid plaques, tau tangles, neuronal loss.	Age > 65, APOE ε4 gene, vascular disease.	Memory loss, confusion, behavioral changes.	Dependency, infections, malnutrition.	Cholinesterase inhibitors, cognitive therapy, caregiver support.	Jessen et al. (2024)
Chronic kidney disease (CKD)	Gradual loss of kidney function over years.	Diabetes, hypertension.	Aging, CVD, family history.	Swelling, fatigue, high BP.	Kidney failure, anemia, bone disease.	BP control, dialysis, transplant.	Iatridi et al. (2025)
Cancer (prostate, breast, colon, etc.)	Uncontrolled cell growth in tissues.	Genetic mutations, hormonal changes.	Aging, smoking, obesity, radiation.	Fatigue, weight loss, bleeding.	Metastasis, organ failure.	Surgery, chemo/ radiotherapy, immunotherapy.	Chhikara and Parang (2023)
Osteoporosis	Bone fragility due to loss of bone mass.	Estrogen loss, calcium deficiency.	Aging, inactivity, smoking.	Fractures, pain, posture change.	Disability, chronic pain.	Vitamin D, calcium, bisphosphonates.	Ye et al. (2024)
Parkinson's disease	Neurodegenerative disorder affecting motor control.	Dopamine neuron loss.	Age > 60, toxins, genetics.	Tremor, rigidity, bradykinesia.	Falls, dementia, depression.	Levodopa, physiotherapy, DBS.	McFarthing et al. (2024)
Atherosclerosis	Hardening and narrowing of arteries due to plaque.	Lipid buildup, inflammation.	Hypertension, smoking, diabetes.	Chest pain, fatigue, leg pain.	Stroke, heart attack.	Statins, exercise, dietary changes.	Saba and Libby (2025)

Source: No permission required.

stiffening of arteries and altered endothelial function, and these factors have a major role in driving CVDs in older adults (Cheng et al., 2023).

As the foremost cause of death globally, CVDs exert a significant influence on quality of life, patient morbidity, longevity, and daily functioning. Clinical data demonstrate that cellular senescence is closely linked to adverse cardiac outcomes in a wide range of CVDs, encompassing both ischemic and nonischemic forms of cardiomyopathy (Evangelou et al., 2023). As cardiac cells age and enter senescence, it contributes to the development of CVDs, including myocardial infarction, atherosclerosis, valvular heart disease, cardiac fibrosis, and others. While the loss of telomere length due to aging is a key driver of replicative senescence, other stressors, including disrupted metabolism, oxidative stress, and epigenetic alterations, can also promote a senescent state (Chen et al., 2022).

Cellular senescence is regarded as a condition that is identified as the permanent arrest of the cell cycle and is indicated to be induced by triggers like aging, damage to DNA, and elevated levels of reactive oxygen species (Davalli et al., 2016). Senescent cells exhibit defining characteristics of interconnected traits, including macromolecular damage, altered metabolism, resistance to apoptosis, noticeable alterations in cell morphology, cell cycle arrest, and expression of specific senescence-associated secretory phenotype (SASP) (Gorgoulis et al., 2019).

In the framework of cardiovascular conditions, senescence of cells enters the scene as a response to stress mechanism which is triggered by a plethora of stimuli, such as viruses, dysfunction of mitochondria, shortening of telomere, and impairment of autophagy (Gorgoulis et al., 2019). The onset and sustained presence of cellular senescence within the heart contribute significantly to exacerbating and advancing different cardiac-related conditions (Chen et al., 2022).

With the global shift from infectious to noncommunicable illnesses, cardiac disorders have become the primary contributors to the global health impact. The emergence of personalized medicine in cardiology has enabled new ways for integrated, personalized, and patient-oriented approaches in the prevention and therapeutic intervention of CVDs. Studies aimed to personalize the therapy for cardiac diseases have facilitated targeted strategies enabling early detection, timely clinical intervention, and minimal harmful outcomes (Sethi et al., 2023). By focusing on personalized interventions, designed for a particular patient or a particular group of individuals instead of the population as a whole, this approach provides the potential to improve cardiovascular health and reduce cardiovascular-related morbidity and mortality (Tingen et al., 2023). Personalized medicine is an emerging approach in healthcare practice that utilizes a patient's genetic information, environmental factors, and lifestyle data to design the diagnostic and therapeutic interventions for the particular patient (Krzyszczyk et al., 2018).

Current cardiology practice continues to focus on the diagnostic and therapeutic strategies that rely on population-derived information, with less priority given to the methodologies aimed at patient-specific treatment (Currie and Delles, 2018). But, with advances in computational technologies, the implementation of personalized medicine in cardiology has made a way to become practically applicable in the management of CVDs, as reflected by growing research in the past decade (Sethi et al., 2023). This approach addresses the reductionist perspectives in traditional medicine, which generalizes individuals with the same clinical symptoms as having a similar disease process and can be treated by the same therapeutic interventions rather than personalized strategies. The primary objective of personalized medicine lies in providing individualized treatment for the individual patient, considering a patient's unique genetic and physiological profile, rather than traditional approaches that rely on the profile of the average population (Leopold and Loscalzo, 2018).

The influence of both genetic and epigenetic determinants is gaining important recognition for their involvement in cardiovascular health, presenting opportunities for advancing the development of personalized cardiovascular therapies (Mensah et al., 2019). In cardiology, personalized medicine has the potential to influence every stage of disease progression, including assessment of risk, prevention strategies, and tailored therapeutic interventions (Currie and Delles, 2018).

18.2 Cellular senescence and cardiovascular pathophysiology

Cellular senescence represents a normal physiological phenomenon marked by a state of permanent arrest of the growth of cells (Wu et al., 2020). The accumulation of senescent cells in cardiovascular tissues results in endothelial damage, stiffening of arteries, and fibrosis, establishing a link between cellular aging and CVDs. Cardiac senescence is a normal process, potentially triggered by myocardial lesions, like cardiac insufficiency, myocardial ischemia, hypertension, hypertrophic cardiomyopathy, coronary artery disease, and pericardial disorders (Roh et al., 2016). Cardiomyocytes make up nearly 80% of cardiac cellular volume and 30%–40% of its total cells. Their senescence has a profound impact on the normal function of the heart. Senescent cardiomyocytes exhibit a state of abnormal shortening, increased pacing frequency, disrupted contractility, and SASP elaboration, which may influence nearby fibroblasts and other cardiac cells (Tang et al., 2020). Senescence is triggered by several stressful factors such as oxidative damage, DNA damage, telomere attrition, and mitochondrial dysfunction (Wu et al., 2020).

Alterations in DNA happen when structural changes occur in DNA that affect how it codes genetic information or interfere with processes like replication and transcription. Such damage can manifest in various forms, including single and double-strand DNA breaks, DNA–protein cross-linking, and insertion or deletion errors. Notably, double-strand breaks can activate the DNA damage signaling resulting in initiation of DNA repair pathways, programmed cell death, and cellular senescence (Wan et al., 2023). A feedback relationship exists between DNA damage and senescence. Senescent cells often fail to remove cytoplasmic DNA efficiently, which activates the cyclic GMP-AMP synthase (CGAS) pathway. This activation stimulates the interferon gene (STING) signaling cascade, which strengthens senescence and drives the production of senescence-associated secretory phenotype (SASP) factors (Yang et al., 2017).

Another cellular stressor known to trigger cellular senescence is telomere attrition. A telomere is a protective DNA protein complex located at the chromosome ends that maintains genomic integrity and is closely linked to aging (Lin and Epel, 2022). In normal somatic cells, during successive cell divisions, telomeres gradually shorten during successive rounds of cell division. Once the length of telomeres becomes critically short, they no longer sustain the protective loop structure at the ends of chromosomes, resulting in the fusion of chromosomes, and the cell undergoes apoptosis. This telomere uncapping triggers a DNA damage response, leading to stimulation of the p53-p21 tumor suppressor pathway and inducing cellular senescence (Zhu et al., 2019).

Mitochondrial dysfunction and senescence represent interdependent markers of aging. Functionally, mitochondrial dysfunction can be seen as declined respiratory efficiency, lower membrane potential, and enhanced oxidative stress. This dysfunction is a consequence as well as a triggering factor of senescence, forming feedback loops that reinforce the persistence of the senescent state (Miwa et al., 2022).

18.2.1 Senescence in atherosclerosis

Since the occurrence and frequency of atherosclerosis become more common with advancing age, it is characterized as an age-related condition and is demonstrated by numerous hallmark features of ageing within the atherosclerotic plaque. The ageing process is accompanied by elevated levels of senescent cells, DNA damage, telomere shortening, mitochondrial dysfunction, protein imbalance, inflammation, epigenetic alterations, impaired nutrient sensing, and depletion of progenitor cells (Uryga and Bennett, 2016).

Accumulating evidence demonstrates that the process of senescence has a central role in the initiation and pathogenesis of atherosclerosis. Different senescent cells that participate in atherosclerosis include senescent human vascular endothelial cells, senescent vascular smooth muscle cells, and senescent macrophages (Wu et al., 2020). The senescence of endothelial cells represents not only the adverse events but also as a triggering factor of endothelial dysfunction, which is recognized as an early predictor in the development of atherosclerosis (Bu et al., 2023). Endothelial dysfunction indicates the initiating stage in atherosclerosis, and it encompasses a broad spectrum of maladaptive changes in functional phenotype. These alterations significantly impact the maintenance of homeostasis and thrombosis, modulation of vascular tone, redox equilibrium, and inflammatory responses in the arterial wall (Gimbrone and García-Cardeña, 2016).

18.2.2 Endothelial dysfunction

The vascular endothelium forms a barrier between the blood and tissues and is vital for performing normal physiological functions. It is highly sensitive to pathogenic stimuli, and these stimuli can activate tumor suppressor signaling, contributing to cellular senescence. Senescent endothelial cells are usually active, secrete pro-inflammatory substances, and show a distinct morphological phenotype, and have been implicated in significantly contributing to the causation of a range of cardiovascular and metabolic diseases (Bloom et al., 2023). The tissues that contain abundant endothelial cells tend to have a significant presence of senescent cells (Yousefzadeh et al., 2020).

Experimental findings indicate that endothelial senescence disrupts vasodilation, angiogenic potential, and barrier integrity. Moreover, endothelial aging has been associated with several cardiovascular diseases, such as atherosclerosis, cardiac failure, pulmonary hypertension, and metabolic imbalance (Bloom et al., 2023). Accumulation of senescent endothelial cells could lead to inflammatory responses and oxidative stress, which leads to impairment in endothelial homeostasis and contributes to artery stiffening via SASP-mediated mechanisms (Lafargue et al., 2017).

18.2.3 Heart failure and myocardial fibrosis

According to the WHO, CVDs account for about 17.9 million deaths each year, corresponding to about 31% of all deaths (Makino et al., 2021). Among the different risk factors, cellular senescence is recognized as a key contributor to cardiovascular conditions like heart failure (Komalasari et al., 2019). With age, cardiac cell senescence contributes to a significant decline in cardiac function, which may result in the development of persistent heart failure, even in individuals with no prior heart disease. Cellular senescence in cardiomyocytes, endothelial cells, and fibroblasts collectively leads to impairment in cardiac performance and can precipitate heart failure (Luan et al., 2024). The build up of senescent cells within the myocardium is a key factor promoting inflammation, myocardial fibrosis, and endothelial dysfunction, which collectively impair cardiac contractility and diastolic function, resulting in worsening of heart failure. SASP, characterized by the elaboration of inflammatory cytokines and growth factors, contributes to heart failure pathogenesis by triggering systemic inflammatory responses within the myocardium (Xu et al., 2025).

18.2.4 Vascular calcification

The cardiovascular complications driven by rapid vascular stiffening and atherosclerosis represent the leading causes of illness and mortality in Western populations. These conditions are closely linked to vascular calcification, a pathological condition marked by the accumulation of calcium in blood vessels, which often occurs as an individual grows older. This process can occur in the arterial medial layer (medial calcification), within atherosclerotic plaques (intimal calcification), or in the heart valves (valvular calcification) (Villa-Bellosta, 2021).

Vascular calcification has been closely associated with atherosclerotic vascular disease. Moreover, stiffening of arteries, a functional impairment associated with vascular calcification, serves as an independent risk factor for cardiovascular-related mortality (Lee et al., 2020). As aging progresses, the prevalence of vascular calcification rises progressively and is strongly correlated with an elevated risk of the development of cardiovascular conditions and overall mortality, thereby significantly contributing to the global healthcare burden (Liu and Liu, 2024).

18.2.5 Role of senescent-associated secretory phenotype in chronic inflammation

Senescent cells release a diverse set of pro-inflammatory molecules, collectively referred to as SASP. If these persist for a long time, they may trigger chronic inflammation, disrupt immune function, and contribute to the development and progression of age-related health conditions (Sahu and Satapathy, 2025). The SASP, which gradually accumulates during normal aging and in age-related diseases, is recognized as a pivotal contributor to chronic inflammation and age-related phenotypic changes (Yue et al., 2022). Highly active senescent cells exhibit high metabolic and secretory activity and display SASP, the composition of which varies based on the type of cell and the specific triggers of senescence (Bloom et al., 2023).

Senescent cells secrete a set of canonical SASP factors, including chemokines, cytokines (IL-1β, IL-6, IL-8), growth factors, proteases, including matrix remodeling enzymes and plasminogen activators, bioactive lipids, extracellular vesicles, and reactive oxygen species in certain cases (Shosha et al., 2018; Calcinotto et al., 2019; Basisty et al., 2020).

SASP components can act in an autocrine signaling manner, reinforcing the senescent state of the cells that elaborate them. Furthermore, they can induce senescence in nearby cells through paracrine mechanisms, referred to as paracrine senescence. Chemokines released by senescent cells have been indicated in the recruitment and activation of immune cells, like natural killer cells and macrophages, which are responsible for facilitating the targeted elimination of senescent cells (Ohtani, 2022). The inflammatory mediators elaborated through the SASP may result in persistent inflammatory conditions that can accelerate aging of immune cells, diminishing immune efficiency, and reducing their capacity to remove inflammatory and senescent cells, thereby reinforcing a continuous cycle of inflammation and cellular aging (Li et al., 2023).

18.3 Personalized medicine in cardiovascular health

18.3.1 Genetic risk profiling and polygenic risk scores

Genes with uncommon mutations that cause heterozygous carriers to be several times more at risk for several prevalent diseases have been found. The existence of a familial hypercholesterolemia mutation in 0.4% of the population, which increases the risk of coronary artery disease (CAD) by up to three times, is a significant example (Abul-Husn et al., 2016). Risk can be considerably decreased by aggressively lowering circulating cholesterol levels in such carriers (Nordestgaard et al., 2013). The p.E508K missense mutation in HNF1A, which has a carrier frequency of 0.1% in the general population and 0.7% among Latinos, is another example (Lek et al., 2016). This mutation increases the risk of type 2 diabetes by up to five times (Estrada et al., 2014). Finding monogenic mutations can be very important for carriers and their families, most diseases are not caused by such variants.

Polygenic inheritance, which includes numerous common genetic variants with minor effects, is more important for most common diseases than uncommon monogenic mutations (Fisher, 1919; Gibson, 2012; Golan et al., 2014; Fuchsberger et al., 2016). The possibility of developing a genome-wide polygenic score (GPS) to identify people at clinically considerably elevated risk, for instance, comparable to levels imparted by uncommon monogenic mutations, has not been established yet (Chatterjee et al., 2013; Zhang et al., 2018).

Inadequate risk classification for clinical utility (e.g., identifying 20% of a group at 1.4-fold greater risk relative to the rest of the population) was a problem in earlier attempts to develop GPS (Ripatti et al., 2010). Three obstacles hindered these early efforts: (1) the small size of the first genome-wide association studies (GWAS), which impacted the accuracy of the estimated impact of individual variants on disease risk; (2) limited computational methods for creating GPS; and (3) the lack of large datasets required to validate and test GPS. In order to create a single quantitative and predictive measure of inherited risk, polygenic scores combine data from GWASs, which quantify the relationship between each of numerous common DNA variations and risk of disease. According to a polygenic score, several studies have found significant gradients in CAD risk, even among individuals with comparable clinical risk factor profiles (Inouye et al., 2018; Khera et al., 2018; Hindy et al., 2020; Sun et al., 2021). Polygenic scores are currently being used clinically in some biobanks and returned through direct-to-consumer testing platforms due to this potential (Hao et al., 2022; Maamari et al., 2022).

The predictive power of polygenic scores has advanced significantly over the last ten years, but score performance is still far below the theoretical maximum, which is the percentage of trait liability explained by common DNA variants, especially in people of non-European ancestry (Patel and Khera, 2023). According to simulation studies, even higher GWAS sample sizes may be able to enhance CAD scores by more precisely estimating the effect magnitude linked to each single nucleotide variant (Zhang et al., 2018). When compared to approaches that simply use GWAS data from a single ancestry source, polygenic scores that incorporate GWAS data from people of other ancestries in addition to the target population demonstrate a relative improvement in predicting accuracy (Ruan et al., 2022; Weissbrod et al., 2022). Additionally, using data from GWAS of related characteristics to improve polygenic prediction in the trait of interest is beneficial, according to the principles of genetic correlation (Bulik-Sullivan et al., 2015; Ning et al., 2020).

There is a great deal of enthusiasm, which is justified, for polygenic scores to usher in a new era of preventative clinical medicine, but there are also some important drawbacks. First, people of non-European origin exhibit lower prediction performance with polygenic scores. This is mostly due to various ancestries being somewhat under-represented in earlier GWAS discovery cohorts. Recent studies have focused on conducting GWAS in larger and more ancestrally diverse populations and creating methodologies that exploit ancestry-specific linkage disequilibrium patterns in order to improve score performance (Ruan et al., 2022; Weissbrod et al., 2022; The "All of Us" Research Program (2019); Fatumo et al., 2022). Second, current ratings are not as good at predicting incident disease, which would offer more therapeutic value in enabling targeted therapies, even though they have a strong association with prevalent disease (Elliott et al., 2020). Last but not least, most risk prediction models now in use rely on either clinical or genetic risk indicators; however, better integration of these modalities and the calculation of a clinically actionable risk estimate are needed (Riveros-Mckay et al., 2021; Weale et al., 2021).

18.3.2 Multiomics integration (epigenomics, proteomics, metabolomics)

The molecular causes of cardiovascular illnesses can be better understood thanks to omics data, such as proteomics, metabolomics, and genomes. A more thorough understanding of disease mechanisms, individualized treatment plans, and focused therapies based on each person's own genetic profile can all be made possible by incorporating this multiomics data into DT models.

For many years, "omics" technologies have simultaneously measured the activities of thousands of genes (transcriptomics), proteins (proteomics), or other molecular features from a mixed collection of cells. This results in high-dimensional complex data that is now referred to as "omics" data, which advances our understanding of the relationship between genotype and phenotype (Krassowski et al., 2020). The key idea is that people and their health are characterized by genetic, microbiological, proteomic, metabolic, clinical, and behavioral pathways (Collins and Varmus, 2015). This intrinsic individual variability may be eliminated by sophisticated computer methods for big datasets, allowing for more accurate clinical decision-making and intervention selection (Collins and Varmus, 2015). The idea of the "digital twin," which bases patient-specific therapy on employing a virtual replica (the "digital twin") to forecast treatment outcomes and customize a patient's prognosis (the "real-life twin"), is one way to realize the potential of precision medicine.

18.3.3 AI-driven prediction models and digital health technologies

The capacity for smooth learning and updating distinguishes a digital twin from a conventional modeling technique based on mechanical modeling. Using machine learning and artificial intelligence (AI) to handle continuous, complex, and heterogeneous input from a physical twin that is continuously changing and evolving is one way to update the state of a mechanistic model (Rathore et al., 2021). A digital twin built on the basis of physiology would provide better continuous risk assessments as a next-generation development of these mechanistic models by smoothly integrating data from sensors that get critical health measurements. Additionally, the mathematical framework acquired from current mechanistic modeling tools improves its interpretability (Lloyd-Jones et al., 2019).

Another crucial factor to take into account is the fact that mathematical modeling techniques frequently fail to produce precise risk assessments in circumstances where data are scarce, inaccurate, or the issue at hand is extremely complicated. AI may play a significant role in this situation since it can serve as a stand-in for mechanistic models when they are too complicated or unable to accurately depict specific aspects of a biological system (Mohamadou et al., 2020). AI may also act as a link between various data types and mechanistic models, allowing for the integration and interpretation of a broad range of information that would frequently be incompatible for mechanistic models to process, such as genetic and demographic data, comprehensive clinical measurements (such as 4-dimensional functional MRI), and real-time physiological data from ambulatory sensors. Furthermore, AI's dynamic learning capabilities allow it to change in step with fresh sensor data, effectively allowing a digital twin to be updated continuously to accurately reflect a person's changing health state.

The creation of the cardiovascular digital twin framework, which combines AI with cutting-edge mechanistic modeling techniques, is only relevant and useful for medicine when it accurately depicts the patient or individual based on data collected in a variety of formats, such as clinical data, genetic history, and patient characteristics (such as age and sex), and is continuously updated through sensor data. Startups and IT companies are likewise diligently addressing the same problems. For instance, Google is working with health delivery networks to develop big data prediction algorithms that alert medical professionals to high-risk diseases like heart failure and sepsis. AI-derived picture interpretation algorithms are being developed by Google, Enlitic, and several other businesses. Jvion provides a "clinical success machine" that determines which patients are most likely to react to treatment regimens and who are most at risk. Each of these could help doctors make decisions about the best diagnosis and course of therapy for their patients.

18.3.4 Clinical utility of personalized biomarkers and imaging tools

Senescence-associated beta-galactosidase (SA-β-gal) activity is the most prevalent marker used to indirectly identify senescent cells, while biological markers representing direct proof of cellular senescence have not yet been found. Lysosomal beta-galactosidase activity is typically measured at a low pH (often around pH 4), but in senescent cells, it becomes visible at a higher pH (pH 6) because the lysosomal compartment has significantly expanded (Dimri et al., 1995). High expression of p53, p16, p21, p38 mitogen-activated protein kinase (p38MAPK), and γH2AX, which indicates the activation of DNA damage responses, is an additional recognized indicator of cellular senescence (Iwasa et al., 2003; Muñoz-Espín and Serrano, 2014; Salama et al., 2014; Van Deursen, 2014).

Furthermore, cellular senescence is thought to be indicated by high mobility group A (HMGA) proteins or heterochromatin markers, such as HP1 and tri-methylated lysine 9 histone H3 (H3K9me3), which are identified as molecular markers of senescence-associated heterochromatin foci (Salama et al., 2014).

18.4 Senescence modulation: current and emerging therapeutics

18.4.1 Senolytics

Senolytic medications specifically cause senescent cells to undergo apoptosis. By targeting pathways like BCL-2 and p53 signaling that are particularly elevated in senescent cells, these medicines allow for selective removal while protecting normal cells. While navitoclax (ABT-263) has shown encouraging results in improving arterial stiffening and reducing plaque formation in mice (Karnewar et al., 2024; Mahoney et al., 2021), dasatinib and quercetin together have proven effective in lowering senescent cell populations in humans (Hickson et al., 2019; Nieto et al., 2024). Clinical trials for the treatment of solid tumors are examining navitoclax (Corcoran et al., 2024; Harrison et al., 2022). Clinical investigations are presently examining the potential of fisetin or dasatinib and quercetin combo to target cellular senescence and modify the SASP in cancer patients. Senolytics have also demonstrated potential in lowering SASP-driven systemic inflammation, which may benefit age-related diseases outside of the vasculature. These medications reduce pro-inflammatory cytokines like TNF-α and IL-6 by eliminating senescent cells, which lessens the chronic low-grade inflammation linked to ageing (Tchkonia et al., 2013). Senescent cells have beneficial effects on tissue repair, wound healing, and tumor suppression, even while they contribute to ageing and chronic inflammation through the SASP. For example, growth factors like VEGF and platelet-derived growth factor, which promote tissue regeneration and inhibit the growth of damaged cells, can be secreted by senescent cells.

By attracting immune cells to damage sites, these released substances support tissue remodeling and homeostasis (Demaria et al., 2014; Rhinn et al., 2019). In order to reduce off-target effects and preserve immune surveillance mechanisms crucial for tumor suppression and tissue regeneration, it is still difficult to optimize the specificity, dosage, and long-term safety of these treatments (Khosla, 2023; Kirkland and Tchkonia, 2020).

18.4.2 Senomorphics

Senomorphics, also known as senostatics, aim to reduce the inflammatory burden by modulating the SASP without eradicating senescent cells. These drugs, of which rapamycin is the gold standard, function by blocking important signaling pathways such as the NF-κB, MAPK, and mTOR pathways that are involved in the creation of SASP (Aziz et al., 2024; Ya and Bayraktutan, 2024). In order to reduce the pro-inflammatory milieu produced by senescent cells, other senomorphics, such as JAK inhibitors, target cytokine signaling cascades (Valenzuela, 2022; Xu et al., 2015, 2016). The Targeting Ageing with Metformin (TAME) experiment, one of the medications with senomorphic qualities, is presently investigating metformin's potential to postpone the onset of age-related illnesses in older persons (Abdelgawad et al., 2023; Barzilai et al., 2016).

Promising results from ongoing clinical trials investigating these medicines show enhanced physical function and decreased systemic inflammation in older persons. Long-term safety and possible side effects, like compromising immune responses or changing the milieu required for proper tissue homeostasis, are still up for debate. To optimize senotherapies and guarantee their safe and successful integration into clinical practice, more research is required.

18.4.3 Epigenetic and gene-editing approaches

Vascular senescence and its consequences are largely controlled by epigenetic changes. Aging-related changes in gene expression patterns include chromatin remodeling, histone alterations, and DNA methylation. For instance, hypomethylation of inflammatory genes might increase SASP-related signaling, whilst hypermethylation of promoters for DNA repair genes may worsen genomic instability (López-Otín et al., 2023; Sen et al., 2016). Another important mechanism that affects chromatin accessibility and transcriptional activity is histone acetylation, which is controlled by enzymes like histone acetyltransferases and deacetylases. This has an impact on pathways like p53/p21 and NF-κB activation (Sen et al., 2016; Hayakawa et al., 2015; Yao and Rahman, 2012).

Senescence is also influenced by metabolic and epigenetic pathways. The action of sirtuins, a family of NAD +-dependent deacetylases, is hampered by mitochondrial malfunction and decreased NAD+ levels. The control of metabolic pathways and histone modifications is also impacted by this disturbance. The signs of vascular ageing, such as oxidative stress, inflammation, and mitochondrial fragmentation, are made worse by this decrease in sirtuin function. Furthermore, the secretion of SASP factors is fueled by the reprogramming of glucose and lipid metabolism in senescent cells, which moves energy generation toward glycolysis and fatty acid synthesis. These metabolic and epigenetic alterations work together to provide a feedback loop that speeds up senescence and aids in the development of cardiovascular conditions, including hypertension and atherosclerosis (Evangelou et al., 2023; Banerjee et al., 2022; Katsuumi et al., 2018).

18.4.4 Current evidence from preclinical and early clinical trials

Endothelial cell (EC) senescence is a significant stage in the development of atherosclerosis and other CVDs and is acknowledged as a contributing factor to endothelial dysfunction. Research indicates that by reducing the milieu of SASP factors associated with senescence, genetic or pharmacological removal of senescence, particularly in ECs, can mitigate vascular dysfunction and illness in ageing. Our knowledge of how the endothelium reacts to ageing and how to treat endothelial dysfunction in this context has been improved as a result of these discoveries. In particular, focusing on endothelial cell senescence seems to be a viable approach to preserving endothelial functions and enhancing vascular health. Senolytic therapy is being investigated in the therapeutic context because preclinical research has demonstrated the potential of senolytics for the treatment and prevention of CVD. D&Q was administered three days a week for three weeks as part of a therapy regimen in an early clinical trial examining the efficacy of senolytics. The safety of this medication was confirmed by the high retention and completion rates in this research, which involved 14 patients with idiopathic pulmonary fibrosis (Justice et al., 2019). Treatment with D&Q resulted in noteworthy improvements in physical function, as seen by increases in 6-min walk distance and 4-min gait speed, despite no significant improvements in pulmonary function. Additionally, associations between improved physical function and a decrease in SASP components were found (Justice et al., 2019).

Additionally, senolytic therapy with D&Q decreased the burden of senescence in both adipose and epidermal tissue, according to preliminary data from a phase 1 clinical trial including patients with diabetic kidney disease. This decrease was linked to a drop in matrix metalloproteinases and circulating cytokines (Hickson et al., 2019). Neuroimaging endpoints and cognitive performance did not significantly vary in a 12-week pilot study of D&Q treatment for people with early-stage Alzheimer's disease (AD). However, the findings showed a decline in senescence-related cytokine and chemokine levels as well as a progressive rise in A42, a biomarker that is negatively correlated with the diagnosis of AD (Gonzales et al., 2023). Although more compounds with suitable safety, tolerability, and practicality need to be created and researched, these trials indicate promise for the therapeutic potential of senolytic therapy in removing senescence, alleviating SASP buildup, and lowering inflammation. Additionally, senolytic therapy should be used in clinical trials for cardiovascular disease and atherosclerosis. Research on the application and management of senolytic therapy in clinical settings for cardiovascular disorders is currently lacking. It's uncertain how senescent cell clearance—either systemically or cell-specifically—will affect the cardiovascular system, particularly regarding overall health span. Moreover, little is known about the systemic and tissue-specific long-term implications of senescent cell removal.

To address these issues, more research needs to be done. Although short-term removal of senescent endothelial and vascular smooth muscle cells improved atherosclerosis and cardiovascular function in preclinical models, more research is required to make sure that the removal of this cell population has no negative long-term or short-term effects on systemic function. However, senolytics' potential to extend health span and treat age-related cardiovascular illnesses is an interesting new area.

18.5 Convergence of personalized medicine and senescence biology

18.5.1 Patients suitable for senotherapeutics

The new family of medications known as senotherapeutics is intended to target and regulate cellular senescence, either by destroying senescent cells (senolytics) or by preventing their toxic secretions (senomorphics). Such treatments are most helpful for people whose illness development or cognitive dysfunction is significantly influenced by the growth of senescent cells, but not all patients are candidates for them (Zhang et al., 2023). Older people with functional decline. Senescent cells build up in areas including the immune system, muscle, and fat, people and long-term irritation. Senolytic treatment may be beneficial for elderly individuals exhibiting symptoms such as weariness, muscle loss, or poor wound healing in order to promote physical resilience and restore tissue health (Tchkonia et al., 2010).

18.5.1.1 Patients with age- related chronic diseases

Many chronic diseases are linked to senescent cell accumulation.

1. The buildup of senescent cells is connected to several chronic illnesses.
   - Osteoarthritis: Inflammatory chemicals released by senescent chondrocytes in joints cause cartilage destruction. Senescent vascular cells encourage the development and instability of plaque in atherosclerosis (Liu et al., 2022).
   - Type 2 Diabetes: Pancreatic cells and senescent fat interfere with glucose regulation.
   - Senescent glial cells have a role in neuroinflammation and neuronal damage in neurodegenerative illnesses, such as Alzheimer's. Senotherapeutics may slow the course of the illness and inflammation in these patients (Alshaebi et al., 2025).

1. After radiotherapy or chemotherapy, healthy cells frequently undergo senescence as a result of cancer therapy. These senescent cells brought on by therapy emit substances that can lead to tissue damage, exhaustion, and potentially the return of cancer. Senolytic treatment may improve healing and lessen long-term negative effects by removing these remaining senescent cells (Gewirtz et al., 2008).
2. Individuals with degenerative or fibrotic conditions, diseases like chronic kidney disease, liver cirrhosis, and idiopathic pulmonary fibrosis, are linked to senescent fibroblasts, which cause tissue scarring. By focusing on these cells, fibrosis can be decreased and organ function enhanced (Ren et al., 2023).
3. Treatment priority may be given to patients exhibiting molecular markers of a significant senescent cell load.
4. Elevated expression of p16^INK4a and p21^CIP1 is an important biomarker.
   - Increased activity of senescence-associated β-galactosidase (SA-β-gal).
   - Elevated inflammatory cytokine levels (TNF-α, IL-6, and IL-8).
   - DNA damage indicators (γ-H2AX foci) are present.

Clinicians can find patients who will react most favorably to senotherapeutics by using biomarker screening (Razo et al., 2022). Cellular senescence is accelerated by conditions such as metabolic syndrome, obesity, and chronic inflammatory disorders. The senescent-associated secretory phenotype (SASP) can be managed with senotherapeutic intervention, which also improves metabolic balance and reduces inflammation (Liu et al., 2020). Table 18.1

18.5.2 Biomarker-guided therapy selection

Using biological signs, or biomarkers, to determine which patients would benefit most from a certain medication is known as biomarker-guided therapy selection. This method is essential in senescence biology and personalized medicine to determine whether patients are a good fit for senotherapeutic therapies, which are medications that target senescent cells. Biomarkers are quantifiable chemicals or signals that show a disease condition, biological function, or reaction to therapy. These biomarkers, according to senescence biology, indicate the quantity, existence, or activity of senescent cells in tissues (Farinha and Heaney, 2025).

18.5.2.1 Key senescence biomarkers

Commonly used biomarkers to detect or quantify senescent cells include:

- p16^INK4a and p21^CIP1—cell cycle inhibitors that increase in senescent cells.
- SA-β-galactosidase (SA-β-gal)—an enzyme specifically active in senescent cells.
- SASP factors—inflammatory molecules such as IL-6, IL-8, TNF-α, and MMPs that are secreted by senescent cells.
- DNA damage markers like γ-H2AX or telomere shortening.
- Circulating microRNAs and metabolites linked with aging and inflammation (Chen et al., 2006).

18.5.2.2 How biomarkers guide therapy

Biomarker analysis helps clinicians stratify patients into groups based on their senescent cell burden and biological aging pattern:

- High senescent cell load → Senolytic therapy: To specifically eliminate senescent cells, patients with significant biomarker signs of senescence may be prescribed senolytic medications (such as Dasatinib + Quercetin or Navitoclax).
- Moderate senescent load → Senomorphic therapy: Senomorphics, such as Metformin or Rapamycin, can be administered to inhibit toxic secretions without destroying the cells if the biomarkers indicate mild senescence or persistent inflammation.
- Low or absent senescence indicators → No senotherapeutic intervention: Patients in this category may benefit from lifestyle or preventative interventions rather than senotherapy (Liberale et al., 2022).

18.5.2.3 Advantages of biomarker-guided selection

- **Precision:** Ensures the right therapy for the right patient.
- **Safety:** Avoids unnecessary drug exposure in patients.
- **Efficacy:** Improves treatment outcomes by targeting those most responsive.
- **Monitoring:** Biomarkers can also track treatment response over time—showing whether senescent cells are reduced or SASP levels are lowered (Boessen et al., 2014).

18.5.3 Personalized dosing and drug combinations

Personalized medication combinations and dosage are crucial in senotherapeutic treatment to guarantee that every patient receives the safest and most effective treatment possible, given their unique biology, age, disease condition, and senescent cell profile (Liu et al., 2016).

18.5.3.1 Personalized dosing

Senotherapeutics do not always have the same effect on people. Numerous factors, including immunological function, metabolism, senescent cell burden, and genetic background, might affect the dosage and frequency of a medication. Biomarker feedback, such as p16^INK4a or SASP levels, is used in adaptive dosing schemes to adjust the dose based on the patient's reaction. For instance, to eliminate senescent cell bursts without endangering healthy tissues, certain patients may require intermittent senolytic treatment, which consists of brief medication cycles. Tailored dosage reduces adverse effects, such inflammation or tissue damage from fast senescent cell clearance, while optimizing medication efficacy (Davalli et al., 2016).

18.5.3.2 Personalized drug combinations

Senotherapeutics often work better when combined with other drugs that target different pathways of aging or disease. Personalized medicine helps select the most suitable combinations based on molecular profiling. Examples include:

Senolytic combinations: Dasatinib + Quercetin: a well-known pair that removes different types of senescent cells. Navitoclax (BCL-2 inhibitor) combined with natural senolytics like Fisetin for broader activity.

Senolytics + Senomorphics: Combining senolytics (to clear senescent cells) with senomorphics (to suppress harmful secretions) for a balanced approach.

Senotherapeutics + antiinflammatory or metabolic drugs: Adding agents like Metformin or Rapamycin can reduce oxidative stress and enhance longevity effects (Vu et al., 2014).

18.5.3.3 Benefits of personalized approaches

- Optimized efficacy: Tailors therapy to each patient's biological needs.
- Reduced toxicity: Avoids overdosing or unnecessary combinations.
- Targeted aging control: Addresses multiple aging mechanisms simultaneously.
- Dynamic adjustment: Allows therapy to evolve as biomarkers or patient health change over time (Adams et al., 2020).
- In essence, the knowledge that cellular senescence and aging processes differ significantly across people gives birth to the idea of customized dosage and medication combinations. The way senotherapeutic medications operate in the body depends on several factors, including immunological function, metabolic rate, disease load, lifestyle, and heredity. A "one-size-fits-all" approach to treatment is therefore unproductive and sometimes dangerous. The goal of personalized methods is to adjust medication combinations, dosage, and treatment plans based on the molecular and clinical characteristics of each patient (Di Micco et al., 2021).

18.5.4 Digital twins and computational modeling for therapy

One of the most cutting-edge areas of personalized medicine and senotherapeutic research is the combination of digital twin technology with computational modeling. Without doing patient experiments directly, these technologies enable researchers and doctors to model biological processes, forecast treatment results, and enhance therapeutic approaches (Živanović and Filipović, 2024).

18.5.4.1 Concept of digital twins in medicine

An organ, tissue, or even a patient's complete body can be represented virtually by a digital twin, which is updated in real time utilizing clinical, molecular, and lifestyle data. It is constructed with data from sources such as:

Proteomic and metabolomic profiles; genomic and transcriptomic information (DNA, RNA sequencing); Clinical records (blood tests, medication history, etc.); wearable health devices (heart rate, glucose, activity); and medical imaging (MRI, CT, PET scans). Clinicians may test and simulate senotherapeutic therapies online before implementing them in real life, thanks to this digital model that replicates the patient's dynamic physiology (Currie and Delles, 2018).

18.5.4.2 Role of computational modeling

An organ, tissue, or even a patient's complete body can be represented virtually by a digital twin, which is updated in real time utilizing clinical, molecular, and lifestyle data. It is constructed with data from sources such as:

Proteomic and metabolomic profiles; genomic and transcriptomic information (DNA, RNA sequencing); Clinical records (blood tests, medication history, etc.); wearable health devices (heart rate, glucose, activity); and medical imaging (MRI, CT, PET scans) Clinicians may test and simulate senotherapeutic therapies online before implementing them in real life thanks to this digital model that replicates the patient's dynamic physiology (Leopold and Loscalzo, 2018).

18.5.4.3 Application in senotherapeutic optimization

Digital twins and computational models have several practical applications for optimizing senotherapeutic interventions:

1. Predicting patient response: The biology of each patient reacts differently to senolytic or senomorphic medications. Their molecular indicators, such as p16^INK4a, IL-6, or telomere length, are integrated into a digital twin, which forecasts how these levels will alter following medication exposure (Ren et al., 2025). For example, the model can simulate how Dasatinib + Quercetin clears senescent cells in the liver versus muscle tissue. It can predict whether the patient is at risk of excessive inflammation (a possible side effect of rapid senescent cell clearance) (Alharbi et al., 2022).

2. Personalized dosing optimization: Using computational pharmacokinetic and pharmacodynamic (PK/PD) models, digital twins can calculate: The ideal dosage that effectively removes senescent cells while minimizing toxicity. The best dosing frequency (e.g., once per month vs. weekly cycles). Inter-individual variability in drug absorption, metabolism, and clearance. Such models continuously adapt as new patient data (e.g., biomarker changes, organ function tests) are added, ensuring that therapy remains dynamically optimized (Huang et al., 2022).
3. Designing drug combinations: Digital twins can simulate the synergistic or antagonistic effects of multiple drugs on senescent cell pathways. For instance, estimating the impact of combining the BCL-2 inhibitor Navitoclax with the mTOR inhibitor Rapamycin on senescent fibroblast survival. Determining if taking metformin improves senomorphic activity or lowers inflammation caused by SASP. By reducing animal testing and saving time, these computer trials let clinicians make decisions on the safest and most efficient combinations (Barros et al., 2024).
4. Disease progression and aging simulation: Digital twins can forecast the course of diseases like osteoarthritis, fibrosis, or neurodegeneration by simulating the accumulation of senescence across tissues over time. This offers a new paradigm in preventative geriatric care by enabling researchers to act before symptoms manifest.

18.5.4.4 Tools and technologies used

Modern digital twin and computational modeling platforms integrate multiple technologies:

- Artificial intelligence (AI) and machine learning (ML): Analyze big biological data and predict therapy outcomes.
- Systems biology models: Use differential equations to represent complex cellular pathways (p53-p21, NF-κB, mTOR, etc.).
- Agent-based models (ABMs): Simulate interactions among individual cells, senescent cells, and immune cells in tissues.
- Multiscale models: Connect molecular events to organ-level responses (e.g., senescence in the liver influencing whole-body inflammation).
- High-Performance Computing (HPC): Enables real-time simulations for multiple therapy scenarios per patient (Mohanraj and Balaji, 2025).

18.5.4.5 Advantages of digital twin–based therapy design

- Precision: Enables therapy tailoring at the molecular level for each patient.
- Predictive power: Anticipates drug efficacy, toxicity, and aging trajectories before clinical use.
- Dynamic adjustment: Updates dosing and drug combinations as new clinical data emerge.
- Reduced risk: Decreases the likelihood of adverse events through prior in silico testing.
- Accelerated drug development: Reduces time and cost by optimizing senotherapeutics virtually before human trials (Vallée, 2024).

18.6 Future perspectives and challenges

18.6.1 Translational hurdles: efficacy, safety, and off-target effects

It is still difficult to translate senotherapeutics from successful laboratory trials to actual clinical practice, despite tremendous advancements in senescence biology and tailored medicine. Making sure these treatments are safe, effective, and precisely targeted without producing undesirable side effects is one of the main translational challenges (Lin et al., 2019).

18.6.1.1 Efficacy challenges

Althoug**h senotherapeutics (senolytics and senomorphics) have shown strong results in animal models, their effectiveness in humans is still u**nder investigation.

a. Biological complexity: Senescent cells are heterogeneous; they vary in different tissues, diseases, and even within the same organ. A drug that effectively removes senescent cells in the liver may not work the same way in the brain or heart. This heterogeneity makes it difficult to find a *universally effective* senotherapeutic (Kirkland and Tchkonia, 2015).
b. Translational gap between preclinical and human data: Although there is little data from human trials, some senolytic medicines, such as dasatinib + quercetin, have shown efficacy in animals. Because human and animal models mature differently, therapy timing, dosage, and duration may not necessarily correspond to actual circumstances (Jaba, 2022).
c. Biomarker limitations: There is ongoing work to find trustworthy biomarkers that appropriately indicate the load of senescent cells or the response to treatment. It's challenging to determine whether a medication actually helped or merely momentarily decreased inflammation in the absence of accurate indicators (Bu et al., 2023).

18.6.1.2 Safety concerns

Complete removal of senescent cells can occasionally be detrimental since they also have advantageous physiological functions, including tumor suppression, wound healing, and embryonic development (Calcinotto et al., 2019).

1. Loss of beneficial senescence: Senescent fibroblasts produce substances that aid in tissue repair during wound healing; indiscriminately removing all senescent cells may hinder regeneration or postpone recovery (Laphanuwat et al., 2023).
2. Toxicity and side effects: Because they affect healthy, nonsenescent cells, certain senolytic medications (such as Navitoclax) might result in thrombocytopenia and neutropenia. Dasatinib and other polypharmacological medications may disrupt regular kinase signaling, resulting in cardiovascular strain, nausea, or exhaustion. Prolonged usage may potentially alter the equilibrium of the immune system, leading to either excessive inflammation or immunological suppression (Yousefzadeh et al., 2020).
3. Dose-related toxicity: It is yet uncertain what dosage is best for long-term safety. While modest or occasional doses may not eliminate enough senescent cells, high or continuous dosing may cause organ damage (Lafargue et al., 2017).

18.6.1.3 Off-target effects

Senotherapeutics often target apoptosis resistance pathways or cell signaling cascades that are also active in healthy cells, leading to unintended consequences.

1. Nonselective targeting: BCL-2 and BCL-xL, which are necessary for the survival of healthy platelets and certain immune cells, are inhibited by medications such as Navitoclax. • Tissue damage, cytopenia, or weakened immunity can result from off-target harm (Skwarska and Konopleva, 2023).
2. Tissue-specific variations: Different microenvironments, drug permeability, or local enzyme activity might cause the same senolytic to function differently in different organs, which makes therapy prediction more difficult and patient variability higher (Kirkland and Tchkonia, 2020).
3. Drug interactions: Because of variations in local enzyme activity, drug permeability, or microenvironments, the same senolytic may function differently in different organs, making therapy prediction more difficult and patient variability higher (Luan et al., 2024).

18.6.1.4 Future perspectives

The research is quickly advancing toward precision senotherapy in order to overcome these obstacles, combining artificial intelligence, systems biology, and sophisticated modeling techniques like digital twins

1. Smart drug design: The creation of next-generation senolytics that target just senescent cells by identifying cell-specific markers or by using nanocarrier delivery mechanisms. Developing pro-drugs that only become active in the milieu of senescent cells (for example, in response to β-galactosidase activity) (Villa-Bellosta, 2021).
2. Improved safety through personalization: Customizing therapeutic intensity according to senescence load through biomarker-guided dosage. Using sporadic treatment plans to enable tissue repair in between medication cycles (Cho et al., 2024).
3. Advanced monitoring systems: Using wearable biosensors, AI-based pharmacovigilance, and digital twin models to continually evaluate treatment response and identify early toxicity (Ahire et al., 2024).

18.6.2 Ethical and regulatory considerations

Senotherapeutics and customized antiaging therapies present several ethical and legal issues that need to be properly resolved in order to guarantee safety, equity, and responsible innovation as they approach clinical use. These problems stem from the intricacy of customization, the novelty of addressing the biology of aging, and the long-term effects on people and society (Popescu et al., 2023).

18.6.2.1 Ethical considerations

1. Human experimentation and long-term safety: Senotherapeutics works at the level of cellular aging, which impacts almost every organ. Human research of such medications raises ethical questions because long-term consequences on longevity, cancer risk, and regeneration are not completely established. Eliminating senescent cells may have unexpected effects on immune regulation, wound healing, or tumor suppression; therefore rigorous ethical review and informed consent are crucial for early-stage trials, guaranteeing patients are aware of the potential risks and unknowns. Short-term clinical trials make it challenging to evaluate lifelong outcomes (Horowitz, 1977).

2. Enhancement versus therapy debate: A key ethical dilemma is distinguishing between using senotherapeutics for treating disease and for enhancing longevity in healthy individuals. Senotherapeutics affect nearly every organ at the level of cellular aging. Since the long-term effects on longevity, cancer risk, and regeneration are not fully known, human studies on such drugs present ethical concerns. Senescent cell removal may have unforeseen consequences for immunological control, wound healing, or tumor suppression; for this reason, thorough ethical evaluation and informed consent are essential for early-stage studies, ensuring that patients are aware of the unknowns and possible hazards. Lifetime outcomes are difficult to assess in short-term clinical studies (Raffaele and Vinciguerra, 2022).
3. Equity and accessibility: Personalized senotherapeutics rely heavily on genomic sequencing, biomarker profiling, and digital health data, which are costly and technologically demanding. There is a chance that these treatments will only be accessible in affluent environments, which would exacerbate health inequalities worldwide. To keep age-related medicines from becoming exclusive, it will be essential to provide fair access and cost control (Segun, 2024).
4. Privacy and data security: Large volumes of personal data, such as genetic, biometric, and lifestyle data, must be gathered for personalized treatment and digital twin models. One of the main ethical concerns is safeguarding this data from abuse or breaches. Before sharing data across AI or cloud-based systems, stringent measures for data ownership, confidentiality, and patient permission must be in place (Cellina et al., 2023).
5. Intergenerational and societal impact: Increases in human longevity and health might have a significant impact on resource management, healthcare expenditures, and population balance. The need for employment, pensions, and environmental resources may rise as people live longer. Ethical frameworks need to take into account how these treatments impact society as a whole in addition to individuals (Dowling and Lewandrowski, 2024).

18.6.2.2 Regulatory considerations

1. Classification of senotherapeutics: Regulatory agencies (including the FDA and EMA) currently classify pharmacological treatments under broad drug categories; nonetheless, there is no specific regulatory framework that applies just to antiaging or longevity drugs. It is still debatable whether senotherapeutics should be approved as remedies for illnesses or as antiaging measures. Determining the primary clinical outcomes, such as reduced senescent cell burden and improved functional health, is necessary for regulatory clearance (Ostios-Garcia et al., 2024).
2. Standardization of biomarkers and clinical trials: There is no universal standard for biomarkers of senescence or for measuring biological aging. Regulators require validated, repeatable biomarkers to evaluate the efficacy and safety of medications. In order to detect delayed side effects, clinical trials must be conducted for a sufficient amount of time, which poses administrative and budgetary challenges (Moqri et al., 2024).
3. Safety monitoring and risk management: Senotherapeutic drugs often target pathways involved in both normal and pathological processes. Regulators must ensure extensive toxicological testing and ongoing participant monitoring. The creation of transparent postmarket surveillance methods to identify unintended consequences following approval. The creation of dose recommendations to prevent abuse or overuse in the wellness sector (Kalimuthu et al., 2022).
4. Oversight of AI and digital twin integration: When digital twin models or AI systems guide therapy selection or dosing, regulatory oversight must verify their accuracy, transparency, and accountability. Therapy decision-making algorithms need to be proven and explicable. The law has to be clear about who is accountable in the event that a treatment choice made by AI results in damage (Strigari et al., 2025).
5. Ethical regulation of commercialization: The growing market for "antiaging" products risks premature commercialization of untested or exaggerated therapies. Strict advertising and labeling regulations are required to avoid deceptive promises, and regulatory bodies must establish clear distinctions between evidence-based senotherapeutics and untested longevity supplements (Perls, 2004).

18.6.2.3 Future directions

The creation of global ethical standards for studies on longevity (e.g., WHO, UNESCO partnership). Development of specialized regulatory processes for treatments aimed at aging, with precise standards for biomarker validation and trial length. Bioethics education and public involvement to make sure society is aware of the advantages and drawbacks of prolonging human life. Adopting data ethics guidelines for customized medicine powered by AI (Molnár-Gábor, 2014).

18.6.3 Cost-effectiveness and accessibility in precision cardiology

Though its broad adoption confronts significant obstacles linked to cost-effectiveness and accessibility, precision cardiology, a strategy that customizes diagnosis and therapy based on a patient's genetic, genomic, and lifestyle profile, has the potential to revolutionize cardiovascular care. Advanced imaging, high-throughput biomarker analysis, and genome sequencing are some of the costly and resource-intensive technologies that support precision cardiology. The initial investment in infrastructure, data analytics, and trained staff is significant, even though these techniques can enhance therapeutic results by facilitating early identification, precise risk classification, and customized drug selection. This mostly restricts access to high-income groups and sophisticated healthcare systems. Economically speaking, cost-effectiveness hinges on proving that tailored approaches lower long-term medical costs by averting consequences like myocardial infarction, harmful medication reactions, or cardiac failure. Many cost–benefit evaluations are still in the early stages, though, and extensive longitudinal research is required to verify long-term benefits. Since differences in patient education, digital infrastructure, and healthcare finance might result in a "precision gap," which excludes underprivileged groups from the advantages of improved care, accessibility also raises ethical questions. Healthcare organizations must make investments in open-source data sharing, scalable genomic platforms, and cost-cutting measures like AI-powered risk prediction tools and population-based screening panels in order to ensure fair adoption. Socioeconomic gaps can be closed by cooperative public–private projects, government funding, and international health regulations that support data democratization. In the end, whether this breakthrough becomes a cutting-edge technology or a universal norm for cardiovascular health will depend on providing affordable and easily available precision cardiology (Sethi et al., 2023).

18.6.4 Preventive cardiology through early senescence modulation

The goal of preventive cardiology is to lower the prevalence and progression of cardiovascular disease (CVD) prior to the onset of clinical symptoms. According to recent developments in senescence biology, the buildup of senescent cells in the cardiovascular system, including endothelial cells, vascular smooth muscle cells, and cardiomyocytes, speeds up the beginning of CVD by causing inflammation, arterial stiffness, and plaque formation. A unique preventative approach is provided by early control of cellular senescence with therapies like senolytics or senomorphics. These treatments can enhance vascular function, lower chronic inflammation, and restore tissue homeostasis by either selectively removing senescent cells or inhibiting the detrimental SASP. By identifying those with a high load of senescent cells, biomarker-guided screening makes it possible to focus therapies before irreversible harm is done. Combining lifestyle changes with senescence modulation, such as nutrition, exercise, and medication-assisted blood pressure or cholesterol management, can improve preventative results even further. By moving cardiovascular care away from reactive therapy and toward proactive maintenance, this strategy may postpone or prevent age-related heart conditions and lessen the financial load on healthcare systems. A promising area of precision preventive cardiology is early senescence-targeted interventions, which connect genetic discoveries to observable enhancements in cardiovascular health.

18.6.5 Interdisciplinary collaborations for clinical translation

Strong multidisciplinary cooperation between fundamental science, clinical medicine, computational biology, and regulatory science is necessary for the effective implementation of senescence-targeted treatments in clinical settings. Mechanistic insights into aging processes are provided by geoscientists and molecular biologists, and physicians use these discoveries in patient treatment plans. In order to enhance medication selection, dose, and timing, computational modelers and data scientists create digital twins, predictive models, and biomarker-based algorithms. Pharmacologists evaluate combinations of senolytics, senomorphics, and supportive agents as part of the medication development process. Health economists evaluate accessibility and cost-effectiveness, while regulatory specialists make sure experimental treatments adhere to safety, efficacy, and ethical requirements. Innovative delivery methods, including organ-specific senolytics or targeted nanoparticles, are made possible by cooperation with bioengineers and technology developers. Crucially, ethicists and patient advocacy organizations assist in directing fair implementation, guaranteeing that treatments are accessible, socially conscious, and safe. Interdisciplinary teams that bridge these fields expedite the translation of senescence-targeted therapeutics from the bench to the bedside, allowing for accurate, secure, and efficient cardiovascular treatments. Converting promising preclinical findings into clinically feasible preventive cardiology therapies requires such coordinated approaches.

18.7 Conclusion

By combining personalized medicine with senescence modulation, we can change the way we treat heart disease by personalizing treatments to each person's genetic, molecular, and aging profiles and moving from reactive to preventive techniques. Senolytics like dasatinib plus quercetin have been shown to work in elderly heart models by getting rid of old cells, increasing diastolic function, lowering inflammation, and lessening fibrosis. This opens the door for more targeted treatments. Senomorphics like rapamycin and metformin lower the levels of SASP, which helps with oxidative stress and mitochondrial dysfunction, which are both major causes of atherosclerosis, high blood pressure, and heart failure. Lab-on-a-chip technologies with patient-derived stem cells and AI-driven analysis enable personalized modeling of cardiovascular responses, optimizing senolytic dosing and predicting outcomes. Advanced biomarkers and clinical trials targeting high-risk groups must address barriers such as senescent cell heterogeneity, off-target effects, and regulatory hurdles. This convergence will eventually lead to less polypharmacy, better regeneration through stem cell therapies, and a longer healthy lifespan, changing the way we treat heart disease.

References

Abdelgawad, I.Y., Agostinucci, K., Sadaf, B., Grant, M.K.O., Zordoky, B.N., 2023. Metformin mitigates SASP secretion and LPS-triggered hyper-inflammation in Doxorubicin-induced senescent endothelial cells. Frontiers in Aging 4. Available at https://doi.org/10.3389/fragi.2023.1170434.

Abul-Husn, N.S., Manickam, K., Jones, L.K., Wright, E.A., Hartzel, D.N., Gonzaga-Jauregui, C., et al., 2016. Genetic identification of familial hypercholesterolemia within a single U.S. Health care system. Science 354 (6319). Available at https://doi.org/10.1126/science.aaf7000.

Adams, S.H., Anthony, J.C., Carvajal, R., Chae, L., Khoo, C.S.H., Latulippe, M.E., et al., 2020. Perspective: guiding principles for the implementation of personalized nutrition approaches that benefit health and function. Advances in Nutrition 11 (1), 25–34. Available at https://doi.org/10.1093/advances/nmz086.

Agusti, A., Vogelmeier, C.F., 2023. GOLD 2024: a brief overview of key changes. Jornal Brasileiro de Pneumologia 49 (6). Available at https://doi.org/10.36416/1806-3756/e20230369.

Ahire, Y.S., Patil, J.H., Chordiya, H.N., Deore, R.A., Bairagi, V.A., 2024. Advanced applications of artificial intelligence in pharmacovigilance: current trends and future perspectives. Journal of Pharmaceutical Research 23 (1), 23–33. Available at https://doi.org/10.18579/jopcr/v23.1.24.

Alharbi, K.S., Afzal, O., Altamimi, A.S.A., Almalki, W.H., Kazmi, I., Al-Abbasi, F.A., et al., 2022. A study of the molecular mechanism of quercetin and dasatinib combination as senolytic in alleviating age-related and kidney diseases. Journal of Food Biochemistry 46 (12). Available at https://doi.org/10.1111/jfbc.14471.

Alshaebi, F., Sciortino, A., Kayed, R., 2025. The role of glial cell senescence in Alzheimer's disease. Journal of Neurochemistry 169 (3). Available at https://doi.org/10.1111/jnc.70051.

Aziz, M.M., El-Sheikh, M.M., Mohamed, M.A., Abdelrahman, S.S., Mekkawy, M.H., 2024. The senomorphic impact of astaxanthin on irradiated rat spleen: STING, TLR4 and mTOR contributed pathway. International Journal of Immunopathology and Pharmacology 38. Available at https://doi.org/10.1177/03946320241297342.

Banerjee, P., Olmsted-Davis, E.A., Deswal, A., Nguyen, M.T.H., Koutroumpakis, E., Palaskas, N.L., et al., 2022. Cancer treatment-induced NAD+ depletion in premature senescence and late cardiovascular complications. The Journal of Cardiovascular Aging 3 (1), 28. Available at https://doi.org/10.20517/jca.2022.13.

Barros, M.T., Paci, M., Tervonen, A., Passini, E., Koivumaki, J.T., Hyttinen, J.A.K., et al., 2024. From multiscale biophysics to digital twins of tissues and organs: future opportunities for in-silico pharmacology. IEEE Transactions on Molecular, Biological, and Multi-Scale Communications 10 (4), 576–594. Available at https://doi.org/10.1109/TMBMC.2024.3442083.

Barzilai, N., Crandall, J.P., Kritchevsky, S.B., Espeland, M.A., 2016. Metformin as a tool to target aging. Cell Metabolism 23 (6), 1060–1065. Available at https://doi.org/10.1016/j.cmet.2016.05.011.

Basisty, N., Kale, A., Hee Jeon, O., Kuehnemann, C., Payne, T., Rao, C., et al., 2020. A proteomic atlas of senescence-associated secretomes for aging biomarker development. PLoS Biology 18 (1), e3000599. Available at https://doi.org/10.1371/journal.pbio.3000599.

Bloom, S.I., Islam, M.T., Lesniewski, L.A., Donato, A.J., 2023. Mechanisms and consequences of endothelial cell senescence. Nature Reviews Cardiology 20 (1), 38–51. Available at https://doi.org/10.1038/s41569-022-00739-0.

Boessen, R., Heerspink, H.J.L., De Zeeuw, D., Grobbee, D.E., Groenwold, R.H.H., Roes, K.C.B., 2014. Improving clinical trial efficiency by biomarker-guided patient selection. Trials 15 (1). Available at https://doi.org/10.1186/1745-6215-15-103.

Brant, L.C.C., Nascimento, B.R., Veloso, G.A., Gomes, C.S., Polanczyk, C., Oliveira, G.M.M.de, et al., 2022. Burden of cardiovascular diseases attributable to risk factors in Brazil: data from the "Global Burden of Disease 2019" study. Revista da Sociedade Brasileira de Medicina Tropical 55. Available at https://doi.org/10.1590/0037-8682-0263-2021.

Bulik-Sullivan, B., Finucane, H.K., Anttila, V., Gusev, A., Day, F.R., Loh, P.-R., et al., 2015. An atlas of genetic correlations across human diseases and traits. Nature Genetics 47 (11), 1236–1241. Available at https://doi.org/10.1038/ng.3406.

Bu, L.L., Yuan, H.H., Xie, L.L., Guo, M.H., Liao, D.F., Zheng, X.L., 2023. New dawn for atherosclerosis: vascular endothelial cell senescence and death. International Journal of Molecular Sciences 24 (20). Available at https://doi.org/10.3390/ijms242015160.

Calcinotto, A., Kohli, J., Zagato, E., Pellegrini, L., Demaria, M., Alimonti, A., 2019. Cellular senescence: aging, cancer, and injury. Physiological Reviews 99 (2), 1047–1078. Available at https://doi.org/10.1152/physrev.00020.2018.

Cellina, M., Cè, M., Alì, M., Irmici, G., Ibba, S., Caloro, E., et al., 2023. Digital twins: the new frontier for personalized medicine? Applied Sciences 13 (13), 7940. Available at https://doi.org/10.3390/app13137940.

Chatterjee, N., Wheeler, B., Sampson, J., Hartge, P., Chanock, S.J., Park, J.-H., 2013. Projecting the performance of risk prediction based on polygenic analyses of genome-wide association studies. Nature Genetics 45 (4), 400–405. Available at https://doi.org/10.1038/ng.2579.

Cheng, D.C.Y., Climie, R.E., Shu, M., Grieve, S.M., Kozor, R., Figtree, G.A., 2023. Vascular aging and cardiovascular disease: pathophysiology and measurement in the coronary arteries. Frontiers in Cardiovascular Medicine 10. Available at https://doi.org/10.3389/fcvm.2023.1206156.

Chen, J., Huang, X., Halicka, D., Brodsky, S., Avram, A., Eskander, J., et al., 2006. Contribution of p16 INK4a and p21 CIP1 pathways to induction of premature senescence of human endothelial cells: permissive role of p53. American Journal of Physiology-Heart and Circulatory Physiology 290 (4), H1575. Available at https://doi.org/10.1152/ajpheart.00364.2005.

Chen, M.S., Lee, R.T., Garbern, J.C., 2022. Senescence mechanisms and targets in the heart. Cardiovascular Research 118 (5), 1173–1187. Available at https://doi.org/10.1093/cvr/cvab161.

Chhikara, B.S., Parang, K., 2023. Global Cancer Statistics 2022: the trends projection analysis. Chemical Biology Letters 10 (1).

Cho, D., Lord, S.J., Ward, R., IJzerman, M., Mitchell, A., Thomas, D.M., et al., 2024. Criteria for assessing evidence for biomarker-targeted therapies in rare cancers—an extrapolation framework. Therapeutic Advances in Medical Oncology 16. Available at https://doi.org/10.1177/17588359241273062.

Collins, F.S., Varmus, H., 2015. A new initiative on precision medicine. New England Journal of Medicine 372 (9), 793–795. Available at https://doi.org/10.1056/NEJMp1500523.

Corcoran, R.B., Do, K.T., Kim, J.E., Cleary, J.M., Parikh, A.R., Yeku, O.O., et al., 2024. Phase I/II study of combined BCL-xL and MEK inhibition with navitoclax and trametinib in KRAS or NRAS mutant advanced solid tumors. Clinical Cancer Research 30 (9), 1739–1749. Available at https://doi.org/10.1158/1078-0432.CCR-23-3135.

Currie, G., Delles, C., 2018. Precision medicine and personalized medicine in cardiovascular disease. In: Kerkhof, P., Miller, V. (Eds.), Sex-Specific Analysis of Cardiovascular Function. Advances in Experimental Medicine and Biology, vol 1065. Springer, Cham, pp. 589–605. Available at https://doi.org/10.1007/978-3-319-77932-4_36.

Davalli, P., Mitic, T., Caporali, A., Lauriola, A., D'Arca, D., Courtney, M., 2016. ROS, cell senescence, and novel molecular mechanisms in aging and age-related diseases. Oxidative Medicine and Cellular Longevity 2016 (1). Available at https://doi.org/10.1155/2016/3565127.

Demaria, M., Ohtani, N., Youssef, S.A., Rodier, F., Toussaint, W., Mitchell, J.R., et al., 2014. An essential role for senescent cells in optimal wound healing through secretion of PDGF-AA. Developmental Cell 31 (6), 722–733. Available at https://doi.org/10.1016/j.devcel.2014.11.012.

Di Micco, R., Krizhanovsky, V., Baker, D., d'Adda di Fagagna, F., 2021. Cellular senescence in ageing: from mechanisms to therapeutic opportunities. Nature Reviews Molecular Cell Biology 22 (2), 75–95. Available at https://doi.org/10.1038/s41580-020-00314-w.

Dimri, G.P., Lee, X., Basile, G., Acosta, M., Scott, G., Roskelley, C., et al., 1995. A biomarker that identifies senescent human cells in culture and in aging skin in vivo. Proceedings of the National Academy of Sciences 92 (20), 9363–9367. Available at https://doi.org/10.1073/pnas.92.20.9363.

Dowling, Á., Lewandrowski, K.-U., 2024. The implications of low back pain on prolonged lifespan and future targeted care models to support the pursuit of healthy longevity. Regenerative Medicine & Peripheral Nerve Endoscopy. Bentham Science Publishers Ltd., pp. 1–15. Available at 10.2174/9789815274462124010003.

Elliott, J., Bodinier, B., Bond, T.A., Chadeau-Hyam, M., Evangelou, E., Moons, K.G.M., et al., 2020. Predictive accuracy of a polygenic risk score–enhanced prediction model vs a clinical risk score for coronary artery disease. The Journal of the American Medical Association 323 (7), 636. Available at https://doi.org/10.1001/jama.2019.22241.

Estrada, K., Aukrust, I., Bjørkhaug, L., Burtt, N.P., Mercader, J.M., García-Ortiz, H., et al., 2014. Association of a low-frequency variant in HNF1A with type 2 diabetes in a Latino population. The Journal of the American Medical Association 311 (22), 2305. Available at https://doi.org/10.1001/jama.2014.6511.

Evangelou, K., Vasileiou, P.V.S., Papaspyropoulos, A., Hazapis, O., Petty, R., Demaria, M., et al., 2023. Cellular senescence and cardiovascular diseases: moving to the "heart" of the problem. Physiological Reviews 103 (1), 609–647. Available at https://doi.org/10.1152/physrev.00007.2022.

Farinha, I., Heaney, L.G., 2025. Biomarker-guided versus symptom-guided therapy. ERS Monograph 2025 (108), 233–244. Available at https://doi.org/10.1183/2312508X.10014824.

Fatumo, S., Yakubu, A., Oyedele, O., Popoola, J., Attipoe, D.A., Eze-Echesi, G., et al., 2022. Promoting the genomic revolution in Africa through the Nigerian 100K Genome Project. Nature Genetics 54 (5), 531–536. Available at https://doi.org/10.1038/s41588-022-01071-6.

Fisher, R.A., 1919. XV.—The correlation between relatives on the supposition of mendelian inheritance. Transactions of the Royal Society of Edinburgh 52 (2), 399–433. Available at https://doi.org/10.1017/s0080456800012163.

Fuchsberger, C., Flannick, J., Teslovich, T.M., Mahajan, A., Agarwala, V., Gaulton, K.J., et al., 2016. The genetic architecture of type 2 diabetes. Nature 536 (7614), 41–47. Available at https://doi.org/10.1038/nature18642.

Gewirtz, D.A., Holt, S.E., Elmore, L.W., 2008. Accelerated senescence: an emerging role in tumor cell response to chemotherapy and radiation. Biochemical Pharmacology 76 (8), 947–957. Available at https://doi.org/10.1016/j.bcp.2008.06.024.

Gibson, G., 2012. Rare and common variants: twenty arguments. Nature Reviews Genetics 13 (2), 135–145. Available at https://doi.org/10.1038/nrg3118.

Gimbrone, M.A., García-Cardeña, G., 2016. Endothelial cell dysfunction and the pathobiology of atherosclerosis. Circulation Research 118 (4), 620–636. Available at https://doi.org/10.1161/CIRCRESAHA.115.306301.

Golan, D., Lander, E.S., Rosset, S., 2014. Measuring missing heritability: inferring the contribution of common variants. Proceedings of the National Academy of Sciences of the United States of America 111 (49), E5272. Available at https://doi.org/10.1073/pnas.1419064111.

Gonzales, M.M., Garbarino, V.R., Kautz, T.F., Palavicini, J.P., Lopez-Cruzan, M., Dehkordi, S.K., et al., 2023. Senolytic therapy in mild Alzheimer's disease: a phase 1 feasibility trial. Nature Medicine 29 (10), 2481–2488. Available at https://doi.org/10.1038/s41591-023-02543-w.

Gorgoulis, V., Adams, P.D., Alimonti, A., Bennett, D.C., Bischof, O., Bishop, C., et al., 2019. Cellular senescence: defining a path forward. Cell 179 (4), 813–827. Available at https://doi.org/10.1016/j.cell.2019.10.005.

Graham, P., 2024. Progressive hip osteoarthritis. Orthopaedic Nursing 43 (5), 300–302. Available at https://doi.org/10.1097/nor.0000000000001070.

Hao, L., Kraft, P., Berriz, G.F., Hynes, E.D., Koch, C., Korategere V Kumar, P., et al., 2022. Development of a clinical polygenic risk score assay and reporting workflow. Nature Medicine 28 (5), 1006–1013. Available at https://doi.org/10.1038/s41591-022-01767-6.

Harrison, C.N., Garcia, J.S., Somervaille, T.C.P., Foran, J.M., Verstovsek, S., Jamieson, C., et al., 2022. Addition of navitoclax to ongoing ruxolitinib therapy for patients with myelofibrosis with progression or suboptimal response: phase II safety and efficacy. Journal of Clinical Oncology 40 (15), 1671–1680. Available at https://doi.org/10.1200/JCO.21.02188.

Hayakawa, T., Iwai, M., Aoki, S., Takimoto, K., Maruyama, M., Maruyama, W., et al., 2015. SIRT1 suppresses the senescence-associated secretory phenotype through epigenetic gene regulation. PLoS One 10 (1), e0116480. Available at https://doi.org/10.1371/journal.pone.0116480.

Hickson, L.T.J., Langhi Prata, L.G.P., Bobart, S.A., Evans, T.K., Giorgadze, N., Hashmi, S.K., et al., 2019. Senolytics decrease senescent cells in humans: preliminary report from a clinical trial of Dasatinib plus Quercetin in individuals with diabetic kidney disease. EBioMedicine 47, 446–456. Available at https://doi.org/10.1016/j.ebiom.2019.08.069.

Hindy, G., Aragam, K.G., Ng, K., Chaffin, M., Lotta, L.A., Baras, A., et al., 2020. Genome-wide polygenic score, clinical risk factors, and long-term trajectories of coronary artery disease. Arteriosclerosis, Thrombosis, and Vascular Biology 40 (11), 2738–2746. Available at https://doi.org/10.1161/atvbaha.120.314856.

Horowitz, H.S., 1977. Ethical considerations in human experimentation. Journal of Dental Research 56 (3_suppl), 154–159. Available at https://doi.org/10.1177/00220345770560031001l.

Huang, W., Hickson, L.T.J., Eirin, A., Kirkland, J.L., Lerman, L.O., 2022. Cellular senescence: the good, the bad and the unknown. Nature Reviews Nephrology 18 (10), 611–627. Available at https://doi.org/10.1038/s41581-022-00601-z.

Iatridi, F., Carrero, J.J., Cornec-Le Gall, E., Kanbay, M., Luyckx, V., Shroff, R., et al., 2025. KDIGO 2024 clinical practice guideline for the evaluation and management of chronic kidney disease in children and adults: a commentary from the european renal best practice (ERBP). Nephrology Dialysis Transplantation 40 (2), 273–282. Available at https://doi.org/10.1093/ndt/gfae209.

Inouye, M., Abraham, G., Nelson, C.P., Wood, A.M., Sweeting, M.J., Dudbridge, F., et al., 2018. Genomic risk prediction of coronary artery disease in 480,000 adults: implications for primary prevention. Journal of the American College of Cardiology 72 (16), 1883–1893. Available at https://doi.org/10.1016/j.jacc.2018.07.079.

Iwasa, H., Han, J., Ishikawa, F., 2003. Mitogen-activated protein kinase p38 defines the common senescence-signalling pathway. Genes to Cells 8 (2), 131–144. Available at https://doi.org/10.1046/j.1365-2443.2003.00620.x.

Jaba, T., 2022. Dasatinib and Quercetin: Short-term Simultaneous Administration Yields Senolytic Effect in Humans. Issues and Developments in Medicine and Medical Research Vol. 2. Sciencedomain International, pp. 22–31. Available at 10.9734/bpi/idmmr/v2/15155D.

Jessen, F., Frölich, L., Hort, J., Winblad, B., Fortea, J., 2024. Amyloid-lowering treatment in Alzheimer's disease. The Lancet 404 (10468), 2161–2162. Available at https://doi.org/10.1016/s0140-6736(24)02510-8.

Justice, J.N., Nambiar, A.M., Tchkonia, T., LeBrasseur, N.K., Pascual, R., Hashmi, S.K., et al., 2019. Senolytics in idiopathic pulmonary fibrosis: results from a first-in-human, open-label, pilot study. EBioMedicine 40, 554–563. Available at https://doi.org/10.1016/j.ebiom.2018.12.052.

Kalimuthu, K., Keerthana, C.K., Mohan, M., Arivalagan, J., Christyraj, J.R.S.S., Firer, M.A., et al., 2022. The emerging role of selenium metabolic pathways in cancer: new therapeutic targets for cancer. Journal of Cellular Biochemistry 123 (3), 532–542. Available at https://doi.org/10.1002/jcb.30196.

Karnewar, S., Karnewar, V., Shankman, L.S., Owens, G.K., 2024. Treatment of advanced atherosclerotic mice with ABT-263 reduced indices of plaque stability and increased mortality. JCI Insight 9 (2). Available at https://doi.org/10.1172/jci.insight.173863.

Katsuumi, G., Shimizu, I., Yoshida, Y., Minamino, T., 2018. Vascular senescence in cardiovascular and metabolic diseases. Frontiers in Cardiovascular Medicine 5. Available at https://doi.org/10.3389/fcvm.2018.00018.

Kazi, D.S., Elkind, M.S.V., Deutsch, A., Dowd, W.N., Heidenreich, P., Khavjou, O., et al., 2024. Forecasting the economic burden of cardiovascular disease and stroke in the United States through 2050: a presidential advisory from the american heart association. Circulation 150 (4), e89. Available at https://doi.org/10.1161/CIR.0000000000001258.

Khera, A.V., Chaffin, M., Aragam, K.G., Haas, M.E., Roselli, C., Choi, S.H., et al., 2018. Genome-wide polygenic scores for common diseases identify individuals with risk equivalent to monogenic mutations. Nature Genetics 50 (9), 1219–1224. Available at https://doi.org/10.1038/s41588-018-0183-z.

Khosla, S., 2023. Senescent cells, senolytics and tissue repair: the devil may be in the dosing. Nature Aging 3 (2), 139–141. Available at https://doi.org/10.1038/s43587-023-00365-6.

Kirkland, J.L., Tchkonia, T., 2020. Senolytic drugs: from discovery to translation. Journal of Internal Medicine 288 (5), 518–536. Available at https://doi.org/10.1111/joim.13141.

Kirkland, J.L., Tchkonia, T., 2015. Clinical strategies and animal models for developing senolytic agents. Experimental Gerontology 68, 19–25. Available at https://doi.org/10.1016/j.exger.2014.10.012.

Komalasari, R., Nurjanah, Yoche, M.M., 2019. Quality of life of people with cardiovascular disease: a descriptive study. Asian Pacific Island Nursing Journal 4 (2), 92–96. Available at https://doi.org/10.31372/20190402.1045.

Krassowski, M., Das, V., Sahu, S.K., Misra, B.B., 2020. State of the field in multi-omics research: from computational needs to data mining and sharing. Frontiers in Genetics 11. Available at https://doi.org/10.3389/fgene.2020.610798.

Krzyszczyk, P., Acevedo, A., Davidoff, E.J., Timmins, L.M., Marrero-Berrios, I., Patel, M., et al., 2018. The growing role of precision and personalized medicine for cancer treatment. Technology 06 (03-04), 79–100. Available at https://doi.org/10.1142/s2339547818300020.

Lafargue, A., Degorre, C., Corre, I., Alves-Guerra, M.-C., Gaugler, M.-H., Vallette, F., et al., 2017. Ionizing radiation induces long-term senescence in endothelial cells through mitochondrial respiratory complex II dysfunction and superoxide generation. Free Radical Biology and Medicine 108, 750–759. Available at https://doi.org/10.1016/j.freeradbiomed.2017.04.019.

Laphanuwat, P., Gomes, D.C.O., Akbar, A.N., 2023. Senescent T cells: beneficial and detrimental roles. Immunological Reviews 316 (1), 160–175. Available at https://doi.org/10.1111/imr.13206.

Lee, Lee, Jeon, 2020. Vascular calcification-new insights into its mechanism. International Journal of Molecular Sciences 21 (8).

Lek, M., Karczewski, K.J., Minikel, E.V., Samocha, K.E., Banks, E., Fennell, T., et al., 2016. Analysis of protein-coding genetic variation in 60,706 humans. Nature 536 (7616), 285–291. Available at https://doi.org/10.1038/nature19057.

Leopold, J.A., Loscalzo, J., 2018. Emerging role of precision medicine in cardiovascular disease. Circulation Research 122 (9), 1302–1315. Available at https://doi.org/10.1161/CIRCRESAHA.117.310782.

Liberale, L., Badimon, L., Montecucco, F., Lüscher, T.F., Libby, P., Camici, G.G., 2022. Inflammation, aging, and cardiovascular disease: JACC review topic of the week. Journal of the American College of Cardiology 79 (8), 837–847. Available at https://doi.org/10.1016/j.jacc.2021.12.017.

Lin, J., Epel, E., 2022. Stress and telomere shortening: Insights from cellular mechanisms. Ageing Research Reviews 73, 101507. Available at https://doi.org/10.1016/j.arr.2021.101507.

Lin, A., Giuliano, C.J., Palladino, A., John, K.M., Abramowicz, C., Yuan, M.L., et al., 2019. Off-target toxicity is a common mechanism of action of cancer drugs undergoing clinical trials. Science Translational Medicine 11 (509). Available at https://doi.org/10.1126/scitranslmed.aaw8412.

Liu, W.W., Liu, M.L., 2024. Vascular calcification: where is the cure? Chinese Medical Sciences Journal 39 (3), 198–210. Available at https://doi.org/10.24920/004367.

Liu, S., Nikanjam, M., Kurzrock, R., 2016. Dosing de novo combinations of two targeted drugs: towards a customized precision medicine approach to advanced cancers. Oncotarget 7 (10), 11310–11320. Available at https://doi.org/10.18632/oncotarget.7023.

Liu, Z., Wu, K.K.L., Jiang, X., Xu, A., Cheng, K.K.Y., 2020. The role of adipose tissue senescence in obesity- and ageing-related metabolic disorders. Clinical Science 134 (2), 315–330. Available at https://doi.org/10.1042/cs20190966.

Liu, Y., Zhang, Z., Li, T., Xu, H., Zhang, H., 2022. Senescence in osteoarthritis: from mechanism to potential treatment. Arthritis Research & Therapy 24 (1). Available at https://doi.org/10.1186/s13075-022-02859-x.

Li, X., Li, C., Zhang, W., Wang, Y., Qian, P., Huang, H., 2023. Inflammation and aging: signaling pathways and intervention therapies. Signal Transduction and Targeted Therapy 8 (1). Available at https://doi.org/10.1038/s41392-023-01502-8.

Lloyd-Jones, D.M., Braun, L.T., Ndumele, C.E., Smith, S.C., Sperling, L.S., Virani, S.S., et al., 2019. Use of risk assessment tools to guide decision-making in the primary prevention of atherosclerotic cardiovascular disease: a special report from the American Heart Association and American College of Cardiology. Journal of the American College of Cardiology 73 (24), 3153–3167. Available at https://doi.org/10.1016/j.jacc.2018.11.005.

López-Otín, C., Blasco, M.A., Partridge, L., Serrano, M., Kroemer, G., 2023. Hallmarks of aging: an expanding universe. Cell 186 (2), 243–278. Available at https://doi.org/10.1016/j.cell.2022.11.001.

Luan, Y., Zhu, X., Jiao, Y., Liu, H., Huang, Z., Pei, J., et al., 2024. Cardiac cell senescence: molecular mechanisms, key proteins and therapeutic targets. Cell Death Discovery 10 (1). Available at https://doi.org/10.1038/s41420-023-01792-5.

Luengo-Fernandez, R., Walli-Attaei, M., Gray, A., Torbica, A., Maggioni, A.P., Huculeci, R., et al., 2023. Economic burden of cardiovascular diseases in the European Union: a population-based cost study. European Heart Journal 44 (45), 4752–4767. Available at https://doi.org/10.1093/eurheartj/ehad583.

Maamari, D.J., Brockman, D.G., Aragam, K., Pelletier, R.C., Folkerts, E., Neben, C.L., et al., 2022. Clinical implementation of combined monogenic and polygenic risk disclosure for coronary artery disease. JACC: Advances 1 (3), 100068. Available at https://doi.org/10.1016/j.jacadv.2022.100068.

Mahoney, S., Hutton, D., Rossman, M., Brunt, V., VanDongen, N., Casso, A., et al., 2021. Late-life treatment with the senolytic ABT-263 reverses aortic stiffening and improves endothelial function with aging. The FASEB Journal 35 (S1). Available at https://doi.org/10.1096/fasebj.2021.35.s1.02642.

Makino, K., Lee, S., Bae, S., Chiba, I., Harada, K., Katayama, O., et al., 2021. Absolute cardiovascular disease risk assessed in old age predicts disability and mortality: a retrospective cohort study of community-dwelling older adults. Journal of the American Heart Association 10 (24). Available at https://doi.org/10.1161/jaha.121.022004.

McFarthing, K., Buff, S., Rafaloff, G., Pitzer, K., Fiske, B., Navangul, A., et al., 2024. Parkinson's disease drug therapies in the clinical trial pipeline: 2024 update. Journal of Parkinson's Disease 14 (5), 899–912. Available at https://doi.org/10.3233/jpd-240272.

Mensah, G.A., Jaquish, C., Srinivas, P., Papanicolaou, G.J., Wei, G.S., Redmond, N., et al., 2019. Emerging concepts in precision medicine and cardiovascular diseases in racial and ethnic minority populations. Circulation Research 125 (1), 7–13. Available at https://doi.org/10.1161/CIRCRESAHA.119.314970.

Miwa, S., Kashyap, S., Chini, E., von Zglinicki, T., 2022. Mitochondrial dysfunction in cell senescence and aging. Journal of Clinical Investigation 132 (13). Available at https://doi.org/10.1172/jci158447.

Mohamadou, Y., Halidou, A., Kapen, P.T., 2020. A review of mathematical modeling, artificial intelligence and datasets used in the study, prediction and management of COVID-19. Applied Intelligence 50 (11), 3913–3925. Available at https://doi.org/10.1007/s10489-020-01770-9.

Mohanraj, R., Balaji, S.N., 2025. Digital twin technology: a comprehensive review of modeling, applications, challenges and future directions in complex system integration. Archives of Computational Methods in Engineering 33, 3291–3316. Available at https://doi.org/10.1007/s11831-025-10397-3.

Molnár-Gábor, F., 2014. The ethical mandate of UNESCO. Max Planck Yearbook of United Nations Law Online 18 (1), 332–367. Available at https://doi.org/10.1163/18757413-00180012.

Moqri, M., Herzog, C., Poganik, J.R., Ying, K., Justice, J.N., Belsky, D.W., et al., 2024. Validation of biomarkers of aging. Nature Medicine 30 (2), 360–372. Available at https://doi.org/10.1038/s41591-023-02784-9.

Muñoz-Espín, D., Serrano, M., 2014. Cellular senescence: from physiology to pathology. Nature Reviews Molecular Cell Biology 15 (7), 482–496. Available at https://doi.org/10.1038/nrm3823.

Nieto, M., Konigsberg, M., Silva-Palacios, A., 2024. Quercetin and dasatinib, two powerful senolytics in age-related cardiovascular disease. Biogerontology 25 (1), 71–82. Available at https://doi.org/10.1007/s10522-023-10068-5.

Ning, Z., Pawitan, Y., Shen, X., 2020. High-definition likelihood inference of genetic correlations across human complex traits. Nature Genetics 52 (8), 859–864. Available at https://doi.org/10.1038/s41588-020-0653-y.

Nordestgaard, B.G., Chapman, M.J., Humphries, S.E., Ginsberg, H.N., Masana, L., Descamps, O.S., et al., 2013. Familial hypercholesterolaemia is underdiagnosed and undertreated in the general population: guidance for clinicians to prevent coronary heart disease: consensus statement of the European Atherosclerosis Society. European Heart Journal 34 (45), 3478–3490. Available at https://doi.org/10.1093/eurheartj/eht273.

Ohtani, N., 2022. The roles and mechanisms of senescence-associated secretory phenotype (SASP): can it be controlled by senolysis? Inflammation and Regeneration 42 (1). Available at https://doi.org/10.1186/s41232-022-00197-8.

Ostios-Garcia, L., Pérez, D.M., Castelo, B., Herradón, N.H., Zamora, P., Feliu, J., et al., 2024. Classification of anticancer drugs: an update with FDA- and EMA-approved drugs. Cancer and Metastasis Reviews 43 (4), 1561–1571. Available at https://doi.org/10.1007/s10555-024-10188-5.

Patel, A.P., Khera, A.V., 2023. Advances and applications of polygenic scores for coronary artery disease. Annual Review of Medicine 74, 141–154. Available at https://doi.org/10.1146/annurev-med-042921-112629.

Perls, T.T., 2004. Anti-Aging medicine: the legal issues: anti-aging quackery: human growth hormone and tricks of the trade–more dangerous than ever. The Journals of Gerontology Series A: Biological Sciences and Medical Sciences 59 (7), B682. Available at https://doi.org/10.1093/gerona/59.7.B682.

Popescu, I., Deelen, J., Illario, M., Adams, J., 2023. Challenges in anti-aging medicine–trends in biomarker discovery and therapeutic interventions for a healthy lifespan. Journal of Cellular and Molecular Medicine 27 (18), 2643–2650. Available at https://doi.org/10.1111/jcmm.17912.

Raffaele, M., Vinciguerra, M., 2022. The costs and benefits of senotherapeutics for human health. The Lancet Healthy Longevity 3 (1), e67. Available at https://doi.org/10.1016/S2666-7568(21)00300-7.

Rathore, M.M., Shah, S.A., Shukla, D., Bentafat, E., Bakiras, S., 2021. The role of AI, machine learning, and big data in digital twinning: a systematic literature review, challenges, and opportunities. IEEE Access 9, 32030–32052. Available at https://doi.org/10.1109/ACCESS.2021.3060863.

Razo, C., Welgan, C.A., Johnson, C.O., McLaughlin, S.A., Iannucci, V., Rodgers, A., et al., 2022. Effects of elevated systolic blood pressure on ischemic heart disease: a burden of proof study. Nature Medicine 28 (10), 2056–2065. Available at https://doi.org/10.1038/s41591-022-01974-1.

Ren, L.-L., Miao, H., Wang, Y.-N., Liu, F., Li, P., Zhao, Y.-Y., 2023. TGF-β as a master regulator of aging-associated tissue fibrosis. Aging and Disease 14 (5), 1633. Available at https://doi.org/10.14336/ad.2023.0222.

Ren, Y., Pieper, A.A., Cheng, F., 2025. Utilization of precision medicine digital twins for drug discovery in Alzheimer's disease. Neurotherapeutics: The Journal of the American Society for Experimental NeuroTherapeutics 22 (3). Available at https://doi.org/10.1016/j.neurot.2025.e00553.

Rhinn, M., Ritschka, B., Keyes, W.M., 2019. Cellular senescence in development, regeneration and disease. Development 146 (20). Available at https://doi.org/10.1242/dev.151837.

Ripatti, S., Tikkanen, E., Orho-Melander, M., Havulinna, A.S., Silander, K., Sharma, A., et al., 2010. A multilocus genetic risk score for coronary heart disease: case-control and prospective cohort analyses. The Lancet 376 (9750), 1393–1400. Available at https://doi.org/10.1016/s0140-6736(10)61267-6.

Riveros-Mckay, F., Weale, M.E., Moore, R., Selzam, S., Krapohl, E., Sivley, R.M., et al., 2021. Integrated polygenic tool substantially enhances coronary artery disease prediction. Circulation: Genomic and Precision Medicine 14 (2). Available at https://doi.org/10.1161/CIRCGEN.120.003304 E003304.

Roh, J., Rhee, J., Chaudhari, V., Rosenzweig, A., 2016. The role of exercise in cardiac aging: from physiology to molecular mechanisms. Circulation Research 118 (2), 279–295. Available at https://doi.org/10.1161/CIRCRESAHA.115.305250.

Roth, G.A., Mensah, G.A., Johnson, C.O., Addolorato, G., Ammirati, E., Baddour, L.M., et al., 2020. Global burden of cardiovascular diseases and risk factors, 1990-2019: update from the GBD 2019 study. Journal of the American College of Cardiology 76 (25), 2982–3021. Available at https://doi.org/10.1016/j.jacc.2020.11.010.

Ruan, Y., Lin, Y.-F., Feng, Y.-C.A., Chen, C.-Y., Lam, M., Guo, Z., et al., 2022. Improving polygenic prediction in ancestrally diverse populations. Nature Genetics 54 (5), 573–580. Available at https://doi.org/10.1038/s41588-022-01054-7.

Saba, L., Libby, P., 2025. Myocardial infarction, stroke and arterial stenosis: time to reassess a major misunderstanding. Nature Reviews Cardiology. Available at https://doi.org/10.1038/s41569-025-01186-3.

Sahu, P., Satapathy, T., 2025. Immunopharmacology of senescence: targeting the senescence-associated secretory phenotype (SASP)—a mechanism-based review. Inflammopharmacology 33 (8), 4291–4310. Available at https://doi.org/10.1007/s10787-025-01867-y.

Salama, R., Sadaie, M., Hoare, M., Narita, M., 2014. Cellular senescence and its effector programs. Genes & Development 28 (2), 99–114. Available at https://doi.org/10.1101/gad.235184.113.

Segun, A.F., 2024. Advances in personalized medical therapeutics: leveraging genomics for targeted treatments. International Journal of Research Publication and Reviews 5 (10), 2921–2933. Available at https://doi.org/10.55248/gengpi.5.1024.2905.

Sen, P., Shah, P.P., Nativio, R., Berger, S.L., 2016. Epigenetic mechanisms of longevity and aging. Cell 166 (4), 822–839. Available at https://doi.org/10.1016/j.cell.2016.07.050.

Sethi, Y., Patel, N., Kaka, N., Kaiwan, O., Kar, J., Moinuddin, A., et al., 2023. Precision medicine and the future of cardiovascular diseases: a clinically oriented comprehensive review. Journal of Clinical Medicine 12 (5), 1799. Available at https://doi.org/10.3390/jcm12051799.

Shosha, E., Xu, Z., Narayanan, S.P., Lemtalsi, T., Fouda, A.Y., Rojas, M., et al., 2018. Mechanisms of diabetes-induced endothelial cell senescence: role of arginase 1. International Journal of Molecular Sciences 19 (4), 1215. Available at https://doi.org/10.3390/ijms19041215.

Skwarska, A., Konopleva, M., 2023. BCL-xL targeting to induce apoptosis and to eliminate chemotherapy-induced senescent tumor cells: from navitoclax to platelet-sparing BCL-xL PROTACs. Cancer Research 83 (21), 3501–3503. Available at https://doi.org/10.1158/0008-5472.CAN-23-2804.

Strigari, L., Schwarz, J., Bradshaw, T., Brosch-Lenz, J., Currie, G., El-Fakhri, G., et al., 2025. Computational nuclear oncology toward precision radiopharmaceutical therapies: ethical, regulatory, and socioeconomic dimensions of theranostic digital twins. Journal of Nuclear Medicine 66 (5), 748–756. Available at https://doi.org/10.2967/jnumed.124.268186.

Sun, L., Pennells, L., Kaptoge, S., Nelson, C.P., Ritchie, S.C., Abraham, G., et al., 2021. Polygenic risk scores in cardiovascular risk prediction: a cohort study and modelling analyses. PLoS Medicine 18 (1), e1003498. Available at https://doi.org/10.1371/journal.pmed.1003498.

Tang, X., Li, P.H., Chen, H.Z., 2020. Cardiomyocyte senescence and cellular communications within myocardial microenvironments. Frontiers in Endocrinology 11. Available at https://doi.org/10.3389/fendo.2020.00280.

Tchkonia, T., Morbeck, D.E., Von Zglinicki, T., Van Deursen, J., Lustgarten, J., Scrable, H., et al., 2010. Fat tissue, aging, and cellular senescence. Aging Cell 9 (5), 667–684. Available at https://doi.org/10.1111/j.1474-9726.2010.00608.x.

Tchkonia, T., Zhu, Y., van Deursen, J., Campisi, J., Kirkland, J.L., 2013. Cellular senescence and the senescent secretory phenotype: therapeutic opportunities. Journal of Clinical Investigation 123 (3), 966–972. Available at https://doi.org/10.1172/jci64098.

The All of Us Research Program Investigators, 2019. The "All of Us" Research Program. New England Journal of Medicine 381, 668–676. Available at 10.1056/NEJMsr1809937.

Tingen, H.S.A., VAN PRAAGH, G.D., Nienhuis, P.H., Tubben, A., Van Rijsewijk, N.D., Hove, D.T., et al., 2023. The clinical value of quantitative cardiovascular molecular imaging: a step towards precision medicine. British Journal of Radiology 96 (1152). Available at https://doi.org/10.1259/bjr.20230704.

Tong, Z., Xie, Y., Li, K., Yuan, R., Zhang, L., 2024. The global burden and risk factors of cardiovascular diseases in adolescent and young adults, 1990–2019. BMC Public Health 24 (1). Available at https://doi.org/10.1186/s12889-024-18445-6.

Tsampasian, V., Bloomfield, G.S., 2025. The evolving global burden of cardiovascular diseases: what lies ahead. European Journal of Preventive Cardiology 32 (11), 1016–1017. Available at https://doi.org/10.1093/eurjpc/zwae330.

Uryga, A.K., Bennett, M.R., 2016. Ageing induced vascular smooth muscle cell senescence in atherosclerosis. Journal of Physiology 594 (8), 2115–2124. Available at https://doi.org/10.1113/JP270923.

Vaduganathan, M., Mensah, G.A., Turco, J.V., Fuster, V., Roth, G.A., 2022. The global burden of cardiovascular diseases and risk: a compass for future health. Journal of the American College of Cardiology 80 (25), 2361–2371. Available at https://doi.org/10.1016/j.jacc.2022.11.005.

Valenzuela, N.M., 2022. Late phase endothelial cell inflammation is characterized by interferon response genes and driven by JAK/STAT, not NFκB. Vascular Pharmacology 146, 107090. Available at https://doi.org/10.1016/j.vph.2022.107090.

Vallée, A., 2024. Envisioning the future of personalized medicine: role and realities of digital twins. Journal of Medical Internet Research 26, e50204. Available at https://doi.org/10.2196/50204.

Van Deursen, J.M., 2014. The role of senescent cells in ageing. Nature 509 (7501), 439–446. Available at https://doi.org/10.1038/nature13193.

Villa-Bellosta, R., 2021. Vascular calcification: key roles of phosphate and pyrophosphate. International Journal of Molecular Sciences 22 (24), 13536. Available at https://doi.org/10.3390/ijms222413536.

Vu, T., Sliwkowski, M.X., Claret, F.X., 2014. Personalized drug combinations to overcome trastuzumab resistance in HER2-positive breast cancer. Biochimica et Biophysica Acta - Reviews on Cancer 1846 (2), 353–365. Available at https://doi.org/10.1016/j.bbcan.2014.07.007.

Wan, R., Srikaram, P., Guntupalli, V., Hu, C., Chen, Q., Gao, P., 2023. Cellular senescence in asthma: from pathogenesis to therapeutic challenges. EBioMedicine 94, 104717. Available at https://doi.org/10.1016/j.ebiom.2023.104717.

Weale, M.E., Riveros-Mckay, F., Selzam, S., Seth, P., Moore, R., Tarran, W.A., et al., 2021. Validation of an integrated risk tool, including polygenic risk score, for atherosclerotic cardiovascular disease in multiple ethnicities and ancestries. American Journal of Cardiology 148, 157–164. Available at https://doi.org/10.1016/j.amjcard.2021.02.032.

Weissbrod, O., Kanai, M., Shi, H., Gazal, S., Peyrot, W.J., Khera, A.V., et al., 2022. Leveraging fine-mapping and multipopulation training data to improve cross-population polygenic risk scores. Nature Genetics 54 (4), 450–458. Available at https://doi.org/10.1038/s41588-022-01036-9.

Wu, C.M., Zheng, L., Wang, Q., Hu, Y.W., 2020. The emerging role of cell senescence in atherosclerosis. Clinical Chemistry and Laboratory Medicine 59 (1), 27–38. Available at https://doi.org/10.1515/cclm-2020-0601.

Xu, C., Qiu, Z., Guo, Q., Huang, Y., Zhao, Y., Zhao, R., 2025. The role of cellular senescence in cardiovascular disease. Cell Death Discovery 11 (1). Available at https://doi.org/10.1038/s41420-025-02720-5.

Xu, M., Tchkonia, T., Ding, H., Ogrodnik, M., Lubbers, E.R., Pirtskhalava, T., et al., 2015. JAK inhibition alleviates the cellular senescence-associated secretory phenotype and frailty in old age. Proceedings of the National Academy of Sciences 112 (46). Available at https://doi.org/10.1073/pnas.1515386112.

Xu, M., Tchkonia, T., Kirkland, J.L., 2016. Perspective: targeting the JAK/STAT pathway to fight age-related dysfunction. Pharmacological Research 111, 152–154. Available at https://doi.org/10.1016/j.phrs.2016.05.015.

Yang, H., Wang, H., Ren, U., Chen, Q., Chena, Z.J., 2017. CGAS is essential for cellular senescence. Proceedings of the National Academy of Sciences of the United States of America 114 (23), E4612. Available at https://doi.org/10.1073/pnas.1705499114.

Yao, H., Rahman, I., 2012. Role of histone deacetylase 2 in epigenetics and cellular senescence: implications in lung inflammaging and COPD. American Journal of Physiology - Lung Cellular and Molecular Physiology 303 (7). Available at https://doi.org/10.1152/ajplung.00175.2012.

Ya, J., Bayraktutan, U., 2024. Senolytics and senomorphics targeting p38MAPK/NF-κB pathway protect endothelial cells from oxidative stress-mediated premature senescence. Cells 13 (15), 1292. Available at https://doi.org/10.3390/cells13151292.

Ye, C., Zhao, B., Leslie, W.D., Ruiz, J.I., Zhao, H., Abdel-Wahab, N., et al., 2024. Increase in major osteoporotic fractures after therapy with immune checkpoint inhibitors. BMJ Oncology 3 (1), e000398. Available at https://doi.org/10.1136/bmjonc-2024-000398.

Yousefzadeh, M.J., Zhao, J., Bukata, C., Wade, E.A., McGowan, S.J., Angelini, L.A., et al., 2020. Tissue specificity of senescent cell accumulation during physiologic and accelerated aging of mice. Aging Cell 19 (3). Available at https://doi.org/10.1111/acel.13094.

Yue, Z., Nie, L., Zhao, P., Ji, N., Liao, G., Wang, Q., 2022. Senescence-associated secretory phenotype and its impact on oral immune homeostasis. Frontiers in Immunology 13. Available at https://doi.org/10.3389/fimmu.2022.1019313.

Zhang, L., Pitcher, L.E., Prahalad, V., Niedernhofer, L.J., Robbins, P.D., 2023. Targeting cellular senescence with senotherapeutics: senolytics and senomorphics. The FEBS Journal 290 (5), 1362–1383. Available at https://doi.org/10.1111/febs.16350.

Zhang, Y., Qi, G., Park, J.-H., Chatterjee, N., 2018. Estimation of complex effect-size distributions using summary-level statistics from genome-wide association studies across 32 complex traits. Nature Genetics 50 (9), 1318–1326. Available at https://doi.org/10.1038/s41588-018-0193-x.

Zhu, Y., Liu, X., Ding, X., Wang, F., Geng, X., 2019. Telomere and its role in the aging pathways: telomere shortening, cell senescence and mitochondria dysfunction. Biogerontology 20 (1), 1–16. Available at https://doi.org/10.1007/s10522-018-9769-1.

Živanović, M.N., Filipović, N., 2024. System biology modeling for drug optimization. In: Filipović, N. (Ed.), Silico Clinical Trials for Cardiovascular Disease: A Finite Element and Machine Learning Approach. Springer Nature, Cham, pp. 105–137. Available at https://doi.org/10.1007/978-3-031-60044-9_5.

Index

Note: Page numbers followed by "*f*" and "*t*" refer to figures and tables, respectively.

D

E

F

G

H

I

L

M

N

O

P

R

T

Printed and bound by CPI Group (UK) Ltd, Croydon, CR0 4YY
14/09/2026
02216388-0003